Scientific Basis for Nuclear Waste Management IX

MATERIALS RESEARCH SOCIETY SYMPOSIA PROCEEDINGS

ISSN 0272 - 9172

Volume 1—Laser and Electron-Beam Solid Interactions and Materials Processing, J. F. Gibbons, L. D. Hess, T. W. Sigmon, 1981

Volume 2—Defects in Semiconductors, J. Narayan, T. Y. Tan, 1981

Volume 3—Nuclear and Electron Resonance Spectroscopies Applied to Materials Science, E. N. Kaufmann, G. K. Shenoy, 1981

Volume 4—Laser and Electron-Beam Interactions with Solids, B. R. Appleton, G. K. Celler, 1982

Volume 5—Grain Boundaries in Semiconductors, H. J. Leamy, G. E. Pike, C. H. Seager, 1982

Volume 6—Scientific Basis for Nuclear Waste Management, S. V. Topp, 1982

Volume 7—Metastable Materials Formation by Ion Implantation, S. T. Picraux, W. J. Choyke, 1982

Volume 8—Rapidly Solidified Amorphous and Crystalline Alloys, B. H. Kear, B. C. Giessen, M. Cohen, 1982

Volume 9—Materials Processing in the Reduced Gravity Environment of Space, G. E. Rindone, 1982

Volume 10—Thin Films and Interfaces, P. S. Ho, K.-N. Tu, 1982

Volume 11—Scientific Basis for Nuclear Waste Management V, W. Lutze, 1982

Volume 12—In Situ Composites IV, F. D. Lemkey, H. E. Cline, M. McLean, 1982

Volume 13—Laser Solid Interactions and Transient Thermal Processing of Materials, J. Narayan, W. L. Brown, R. A. Lemons, 1983

Volume 14—Defects in Semiconductors II, S. Mahajan, J. W. Corbett, 1983

Volume 15—Scientific Basis for Nuclear Waste Management VI, D. G. Brookins, 1983

Volume 16—Nuclear Radiation Detector Materials, E. E. Haller, H. W. Kraner, W. A. Higinbotham, 1983

Volume 17—Laser Diagnostics and Photochemical Processing for Semiconductor Devices, R. M. Osgood, S. R. J. Brueck, H. R. Schlossberg, 1983

Volume 18—Interfaces and Contacts, R. Ludeke, K. Rose, 1983

Volume 19—Alloy Phase Diagrams, L. H. Bennett, T. B. Massalski, B. C. Giessen, 1983

Volume 20—Intercalated Graphite, M. S. Dresselhaus, G. Dresselhaus, J. E. Fischer, M. J. Moran, 1983

Volume 21—Phase Transformations in Solids, T. Tsakalakos, 1984

Volume 22—High Pressure in Science and Technology, C. Homan, R. K. MacCrone, E. Whalley, 1984

Volume 23—Energy Beam-Solid Interactions and Transient Thermal Processing, J. C. C. Fan, N. M. Johnson, 1984

Volume 24—Defect Properties and Processing of High-Technology Nonmetallic Materials, J. H. Crawford, Jr., Y. Chen, W. A. Sibley, 1984

Volume 25—Thin Films and Interfaces II, J. E. E. Baglin, D. R. Campbell, W. K. Chu, 1984

Volume 26—Scientific Basis for Nuclear Waste Management VII, G. L. McVay, 1984

Volume 27—Ion Implantation and Ion Beam Processing of Materials, G. K. Hubler, O. W. Holland, C. R. Clayton, C. W. White, 1984

Volume 28—Rapidly Solidified Metastable Materials, B. H. Kear, B. C. Giessen, 1984

Volume 29—Laser-Controlled Chemical Processing of Surfaces, A. W. Johnson, D. J. Ehrlich, H. R. Schlossberg, 1984

Volume 30—Plasma Processing and Synthesis of Materials, J. Szekely, D. Apelian, 1984

Volume 31—Electron Microscopy of Materials, W. Krakow, D. Smith, L. W. Hobbs, 1984

Volume 32—Better Ceramics Through Chemistry, C. J. Brinker, D. E. Clark, D. R. Ulrich, 1984

Volume 33—Comparison of Thin Film Transistor and SOI Technologies, H. W. Lam, M. J. Thompson, 1984

Volume 34—Physical Metallurgy of Cast Iron, H. Fredriksson, M. Hillerts, 1985

Volume 35—Energy Beam-Solid Interactions and Transient Thermal Processing/1984, D. K. Biegelsen, G. Rozgonyi, C. Shank, 1985

Volume 36—Impurity Diffusion and Gettering in Silicon, R. B. Fair, C. W. Pearce, J. Washburn, 1985

Volume 37—Layered Structures, Epitaxy and Interfaces, J. M. Gibson, L. R. Dawson, 1985

Volume 38—Plasma Synthesis and Etching of Electronic Materials, R. P. H. Chang, B. Abeles, 1985

Volume 39—High-Temperature Ordered Intermetallic Alloys, C. C. Koch, C. T. Liu, N. S. Stoloff, 1985

Volume 40—Electronic Packaging Materials Science, E. A. Giess, K.-N. Tu, D. R. Uhlmann, 1985

Volume 41— Advanced Photon and Particle Techniques for the Characterization of Defects in Solids, J. B. Roberto, R. W. Carpenter, M. C. Wittels, 1985

Volume 42—Very High Strength Cement-Based Materials, J. F. Young, 1985

Volume 43—Coal Combustion and Conversion Wastes: Characterization, Utilization, and Disposal, G. J. McCarthy, R. J. Lauf, 1985

Volume 44—Scientific Basis for Nuclear Waste Management VIII, C. M. Jantzen, J. A. Stone, R. C. Ewing, 1985

Volume 45—Ion Beam Processes in Advanced Electronic Materials and Device Technology, F. H. Eisen, T. W. Sigmon, B. R. Appleton, 1985

Volume 46—Microscopic Identification of Electronic Defects in Semiconductors, N. M. Johnson, S. G. Bishop, G. D. Watkins, 1985

MATERIALS RESEARCH SOCIETY SYMPOSIA PROCEEDINGS

Volume 47—Thin Films: The Relationship of Structure to Properties, C. R. Aita, K. S. SreeHarsha, 1985

Volume 48—Applied Material Characterization, W. Katz, P. Williams, 1985

Volume 49—Materials Issues in Applications of Amorphous Silicon Technology, D. Adler, A. Madan, M. J. Thompson, 1985

Volume 50—Scientific Basis for Nuclear Waste Management IX, L. O. Werme, 1985

MATERIALS RESEARCH SOCIETY SYMPOSIA PROCEEDINGS VOLUME 50

Scientific Basis for Nuclear Waste Management IX

Symposium held September 9-11, 1985, Stockholm, Sweden

EDITOR:

Lars O. Werme
SKB, Stockholm, Sweden

MRS MATERIALS RESEARCH SOCIETY
Pittsburgh, Pennsylvania

A substantial portion of the funding for this symposium was provided by the United States Department of Energy and the United States Nuclear Regulatory Commission. However, the views and findings of the various papers are solely those of the authors, and do not necessarily represent the policy of either agency. The agencies encourage wide disseminaton of the technical information contained herein, with due respect for the Publisher's rights regarding the complete volume.

Published by:

Materials Research Society
9800 McKnight Road, Suite 327
Pittsburgh, Pennsylvania 15237
telephone (412) 367-3003

Library of Congress Cataloging in Publication Data

Main entry under title:

Scientific basis for nuclear waste management IX.

(Materials Research Society symposia proceedings, ISSN 0272-9172 ; v. 50)
"Proceedings of the Ninth International Symposium on the Scientific Basis for Nuclear Waste Management, arranged by the Materials Research Society"—Pref.
Includes bibliographies and indexes.
1. Radioactive waste disposal—Congresses.
I. Werme, Lars O. II. Materials Research Society. III. International Symposium on the Scientific Basis for Nuclear Waste Management (9th: 1985: Stockholm, Sweden) IV. Series.
TD898.S42 1985 621.48'38 85-32105
ISBN 0-931837-15-4

Manufactured in the United States of America

Manufactured by Publishers Choice Book Mfg. Co.
Mars, Pennsylvania 16046

Contents

* Invited speaker

PART VI: NEAR-FIELD CHEMISTRY

PART VII: A. THERMAL AND STRESS EFFECTS ON FLUIDS IN ROCKS

Preface

This volume contains the proceedings of the Ninth International Symposium on the Scientific Basis for Nuclear Waste Management arranged by the Materials Research Society. The Symposium was held September 9-11, in Stockholm, Sweden, as the second symposium in this series held in Europe and as the first one replacing the annual sumposia, since 1978 included in the Materials Research Society's Fall Meetings in Boston. On request of the Materials Research Society, the symposium was organized by the Swedish Nuclear Fuel and Waste Management Co., SKB. This series of symposia is widely recognized for providing a unique opportunity for an international gathering of scientists of different disciplines for discussions on the scientific basis for nuclear waste management.

The proceedings consist of 93 papers presented at the symposium. Each of the contributed papers was subjected to a peer review before being accepted for printing in this volume. The papers have been arranged into eleven secitons, including an overview section of invited papers. The sections of submitted papers are: fluids in crystalline rock; waste form performance-glass; waste form performance - ceramics and spent fuel; canisters - design, materials and performance, near-field chemistry; thermal and stress effects on fluids in rocks and diffusion and transport in clay buffers; transport in fractured rocks; radionuclide speciation and sorption; solute and colloid transport; geotechnical and transport modelling.

The organizers of the synposium and the editor of the proceedings wish to acknowledge with thanks the valuable help and unflagging support received from many individuals in making the ninth meeting a success.

The Program Committee contributed in preparing this symposium by selecting the papers, obtaing the peer reviews, assisting the organizers during the meeting and serving as Session Chairmen. The Program Committee members were:

P-E. Ahlstrom, SKB, Sweden (Chairman)
L.O. Werme, SKB, Sweden, (Scientific Secretary)
L.A. Boatner, Oak Ridge National Laboratory, USA
N.A. Chapman, British Geological Survey, UK
R.C. Ewing, University of New Mexico, USA
L.H. Johnson, Whiteshell Nuclear Research Establishment, AECL Canada
W. Lutze, Hahn-Meitner Institute, FRG
G. de Marsily, Ecole National Superieur des Mines de Paris, France
I. Neretnieks, Royal Institute of Technology, Sweden
D.M. Strachan, Battelle Pacific Northwest Laboratory, USA

Session Chairmen also included:

B. Allard, University of Linkoping, Sweden
F.T. Ewart, AERE, Harwell, UK
B. Grambow, Hahn-Meitner Institue, FRG

F. Karlsson, SKB, Sweden
I. McKinley, Federal Institute for Reactor Research, Switzerland
G. Ondracek, Kernforschungszentrum Karlsruhe, FRG
O. Stephansson, University of Luleå, Sweden

Besides the Materials Research Society and SKB, the symposium was sponsored by:

U.S. Department of Energy
U.S. Nuclear Regulatory Commission
The Swedish National Board for Spent Nuclear Fuel
The Swedish Nuclear Power Inspectorate
The National Institute for Radiation Protection

Thanks are due to all those who have contributed to the Symposium as authors and in discussions, who made it possible to run the symposium on a high scientific level. In particular, the outstanding work performed by Ms. Annica Nordmark during the meeting and the invaluable assistance in preparing this volume is gratefully acknowledged.

Lars O. Werme
November 1985

Opening Remarks

NUCLEAR ENERGY AND NUCLEAR WASTE MANAGEMENT

Opening remarks by Professor HANS G. FORSBERG
President of the Royal Swedish Academy of Engineering Sciences

I can honestly say that I am very proud to get this opportunity to open this symposium and provide a few opening remarks on nuclear energy and nuclear waste management. I was once a nuclear chemist myself. Moreover, the Royal Swedish Academy of Engineering Sciences, IVA, has since her start in 1919 been concerned with all aspects of energy questions and energy politics but particularly in recent decades with the nuclear energy issues.

Let us start by looking at nuclear energy and nuclear waste from an international point of view. At the end of 1984 there were 345 nuclear power units in operation in the world. Total operating experience at the same time was 3 467 years and 10 months. The exactness of these figures are, by the way, astounding. In which other energy area can you get such precise figures? Well, in ten years time we will almost have tripled the number of years of operating experience. Something like 10 000 years of operating experience seems to be a good basis for decisions on the future of nuclear energy. The figures of the amount of nuclear waste are not so precise, for obvious reasons.

According to the Swedish Nuclear Fuel and Waste Management Company the amount of high level waste in Sweden will with some 400 years of reactor operation amount to about 7 000 tons in the year 2010. That is equivalent of the space occupied by two ordinary single-family houses. For the whole world 300 000 tons would be a possible figure by the year 2010.

Before reverting to nuclear waste management I would like to give a few remarks to the energy situation in Sweden.

Sweden has one of the highest per capita levels of energy consumption in the world. This is quite reasonable given our severe climate and the size of our industrial establishment. The oil crises of the 1970s gave Sweden a nasty shock. We also felt the impact of the international debate on nuclear power, which reached us in the 1970s and persuaded a majority of the Swedish people to vote in a referendum to recommend decommissioning of the country's nuclear power plants. This was confirmed by a parliamentary resolution in the Swedish Riksdag putting the closing year in 2010.

The goals stipulated in the Riksdag resolution included the following:

* A successive development shall take place towards an energy system based largely on lasting, preferably renewable and indigenous of energy.

* Rational use of energy conservation shall be promoted and available sources of energy utilized in the best manner.

* The use of oil shall be reduced partly through energy conservation measures and partly by substitution with other kinds of energy.

* Nuclear power shall be utilized effectively until it is phased out, by the year 2010 at the latest. The start-up of the last reactors to be granted only if the utilities could prove one satisfactory way of managing the nuclear waste.

* The influence on the environment and on people shall be as low as possible.

* The cost to the national economy shall be as low as possible.

Unfortunately no clear lines were discernible as to how the first goals would be compatible with and balanced against the last two.

The 1970s were marked by new political interest in Sweden's energy supply; swarms of self-appointed experts put forward imaginative solutions, and every self-respecting organization had to have an energy policy programme. It was very difficult then to voice a calm opinion and have people to face the real situation.

After the energy policy decisions of 1981 there was a rapid decline in public discussion of energy matters. IVA then decided to undertake a comprehensive technical and economic study of the best way to meet Sweden's energy needs in the year 2011. Let me sum up the findings in this way:

* The energy system should be designed so that the country's resources are exploited in the optimum manner. Every deviation from this route means less total resources for Sweden's development. Limiting energy use at all costs should therefore not be an end in itself. Increased energy use may even be reasonable if it contributes to increasing the country's resources.

* An over-dimensioned energy system may mean increased costs, but an under-dimensioned system can be disastrous.

* The reduction of the use of oil needs no longer become an end in itself.

* Hydropower should be expanded with about 12 TWh; this can be done with less regulation of rivers and reservoirs than before. At least one river system should be left intact.

* Existing nuclear power plants should be exploited for as long as they are economic and meet health and environmental regulations.

* It is technically unjustified and indefensible in terms of the national economy to phase out nuclear power in 2010. The Riksdag will have to reconsider this decision in the 1990s.

So far the recommendation by IVA.

In the sessions to come during this symposium you will get some detailed information on the programmes for nuclear waste in various countries and a spectrum of research programmes will be presented.

The Swedish politicians have after the Referendum Parliament's Decisions given the last Swedish reactors the green lights to start up. This means Swedish authorities consider the way suggested by the utilities to take care of the waste a satisfactory one. Is there really any need for further research?

My answer is yes. In some ways, the nuclear waste problem is really a nice one to approach. The volume of the nuclear waste is small, in contrast to other kind of industrial waste. It is kept together solid, and current high level waste forms are comparatively well characterized. Furthermore, the nuclear power plants are new and well equipped and, last but not least, the final decisions to be made can still wait a few years.

In other ways, the question of what to do with the nuclear waste is not so simple. Even if authorities are satisfied in most countries one has to face a hostile public opinion. The Swedish Institute of Public Opinion Research has tested the following question on Swedes:

"Do you feel that we can safely store nuclear power waste in the bedrock in Sweden, or that we cannot?".

21 % thought that we could store it safely but 64 % thought that we could not. 15 % did not know. As a comparison, 49 % thought that it would be good for the country if we continued with for nuclear power and only 29 % thought it would be bad.

This means, at least here in Sweden, that people think nuclear power is not such a bad thing after all, but they still doubt that we will be able to take care of the nuclear waste in a safe way. A bit contradictory, I would say. Obviously, much more information to the public is necessary. That information must be based on even more scientific results than are available at present. A change in public opinion on nuclear waste is needed at least in this century if nuclear power should keep its place and expand in the next century.

The fact that the nuclear waste issue is a new issue gives us a headache when we try to set standards. Should we guarantee a safe storage for 10 000 years of for 100 000 years? And what do we mean by a "safe" storage? In discussions we often hear the ultimate goal described as "the safest storage possible". This is really a dangerous expression. It would lead to solutions too far from the reality of today's nuclear industry situation. Dangerous directly, because planning for these capital intensive nuclear power undertakings must be conservative and based on available knowledge, not on uncertain technical and trade systems, existant or not yet established. Dangerous also because they would cause a further erosion of public and political confidence if it were claimed that there is no other way to solve the controversial back end of the nuclear fuel cycle but a still more complicated development of the system. This does of course not preclude a future application of refined or new knowledge nor responses to changes in acceptance criteria.

Instead of "the safest storage possible" we must use phrases like "a realistically safe storage" and include economic considerations as well. Our job is thus to make a sober choice based on weighing the various factors, but without including any emotional attitudes.

The purpose of this international symposium is, in my opinion, to exchange information and experience in order to get a realisticly safe storage of nuclear waste. This does not mean that our goal is to find a single method, suitable for all countries. It is quite obvious that the conditions vary from country to country. Factors like the size of the nuclear programme, the availability of nuclear fuel cycle facilities e.g. geographic and geologic circumstances and not least the different legal requirements must be taken into account. Therefore, the methods for managing the nuclear waste could very well be quite different, even when based on the same broad criteria. Further research and exchange of research results is what is needed to solve the energy problems as a whole. I hope that this symposium will play its part by bringing us a little closer to safe solutions of the nuclear waste problem and that it will also bring you a little closer to each other and each other's problems.

PART I

Overviews

SWEDISH NUCLEAR HIGH-LEVEL WASTE PROGRAM—KEY RESEARCH AREAS

STEN BJURSTRÖM AND TÖNIS PAPP
Swedish Nuclear Fuel and Waste Management Co (SKB),
Stockholm, Sweden

INTRODUCTION

Nuclear power plays an important role in the Swedish power production. From this year on 45-50% of the Swedish electricity will be produced by nuclear power plants, the rest being hydro power.

This paper will summarize the background of the Swedish research in the field of radioactive waste and the strategy and structure of the planned future research activities.

The figure below shows the Swedish nuclear fuel cycle. Sweden has twelve reactor units at four sites with a total generating capacity of 9.500 MW

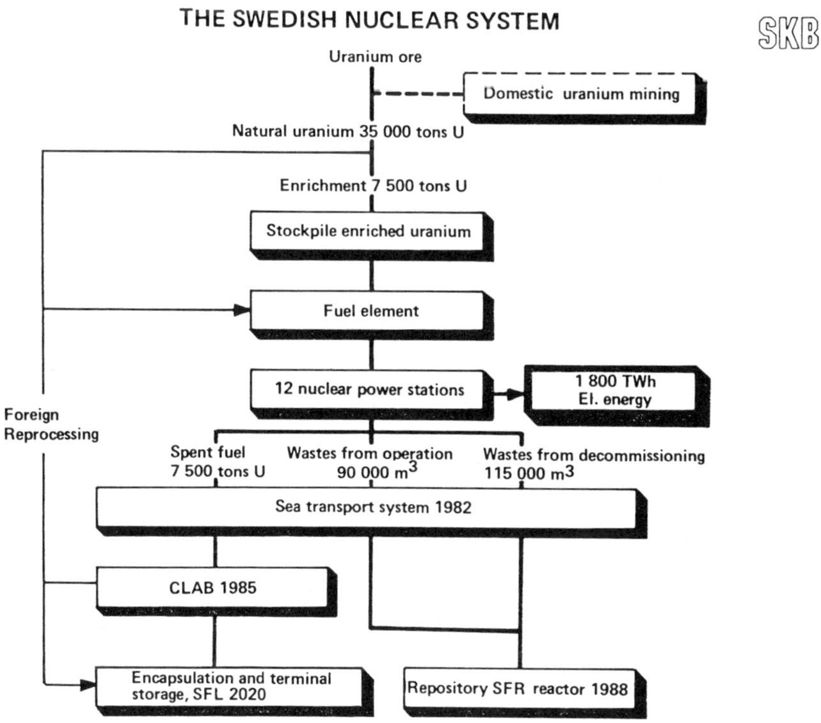

The decision by the Swedish Parliament to phase out nuclear power production in the year 2010 has defined the future need of uranium supply etc, and the timing and quantities of the different kinds of radioactive waste to be disposed of. This decision, of course, also has had great influence on the development and design of the Swedish scheme for transportation, storage and disposal of radioactive waste. SKB strategy is based on a fairly long period of interim storage prior to final and direct disposal of the spent fuel without reprocessing.

The radioactive residues generated by the nine Swedish BWR reactors and the three PWR reactors are spent fuel and various kinds of low and medium level waste from operation. In future decommissioning waste will be generated when the reactors are decontaminated and dismantled.

The back-end of the Swedish fuel cycle in the previous figure consists of two parts; one part is now in operation or under implementation with technology and systems already decided and designed in order to meet the requirements for uninterrupted power production till the year 2010.

In this category the seabound transportation system is in operation along the Swedish coast. There is also the storage facility for intermediate storage of used fuel, the CLAB, that was commissioned this year in July. And finally, there is the facility for final storage of low- and medium-level reactor waste, the SFR, now under construction.

The other part is the facilities needed for the conditioning and final disposal of the highly active and long-lived radioactive waste. For this, research and development work is going on for the optimization of the disposal method.

There are three principal motives for our research:

* the legislation on nuclear matters

* the question of public acceptance

* the economy

In the nuclear legislation of 1984, special consideration is given the research to be done on radioactive waste. The law stipulates that the <u>reactor owner</u> shall:

- before loading a new reactor with fuel, show the feasibility of a method for safe disposal of the spent fuel or high level waste,

- ensure all measures needed for a safe handling and disposal of radioactive waste generated

and finally

- carry out a comprhensive research programme in order to reach a safe solution acceptable to society.

Although according to opinion enquires only one of ten people in Sweden believes that Sweden will phase out nuclear power the year 2010 other polls and enquires show that more than two thirds of Swedish people do not believe that there is a solution available for the disposal of waste.

The fact that the Government in 1984 officially approved the KBS-3 as a safe method does not seem to have had any great influence on the public view.

Thus, the research that is carried out should also enhance the understanding by the general public.

Compared with the price of electricity, the methods proposed for the management of the back-end do not cost much. They amount, however, to a considerable sum of money, today estimated at 45 billion Swedish Crowns for the total back-end including the dismantling of the reactors.

Here, the nuclear industry takes the view that a comprehensive and high quality research programme is an effective way to reduce costs without reducing the safety margins.

THE SWEDISH RESEARCH PLAN

The starting point for the research planning is the research work that has been summerized in the so called KBS-3 concept, a disposal method with copper canisters deep down in granitic rock. This concept was approved as feasible for safe disposal by the Government in June 1984 in conncection with granting loading permits for our two last reactors.

From now on present tasks for SKB are:

a) to develop a practical and economical design for a repository,

b) to select a site, make a detailed site investigation and get the site licensed,

c) to build and license facilities needed and to operate them.

Through the intermediate storage possibilities offered by CLAB there is ample time for this. There is time for:

- public acceptance
- stepwise siting
- stepwise system development

Besides CLAB provides time for simplifying the solutions by reducing heat emission - and the development and use of the most economical solutions.

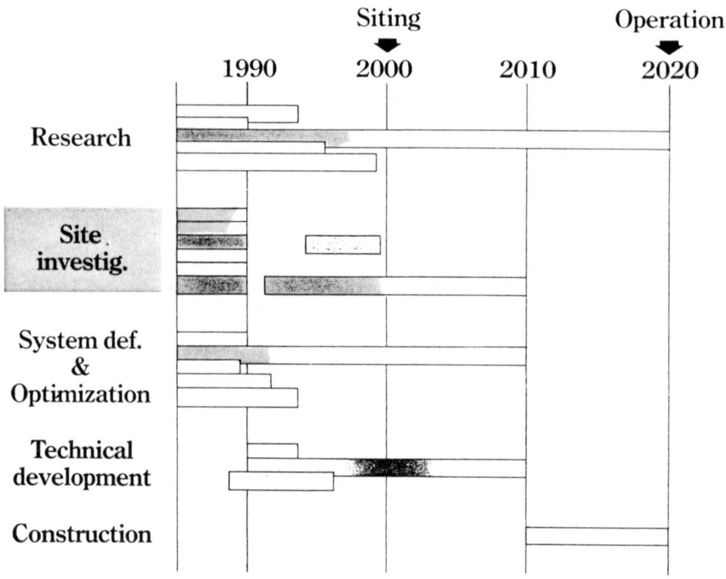

The figure above shows the timetable for future work. The research activities comprise the fundamental understanding of processes, models and databases.

The site investigations today aim at a broad survey of the geological possibilities in Sweden. Investigations have been initiated at more than ten places during this decade. At the end of the 80's a few sites will be selected for detailed investigations including shaft sinking and tunneling to verify the quality of the rock and relevant theories.

Parallel to this work the options for making man made barriers will be evaluated. A screening will be made in the middle of the 90's as a starting point for the design of an optimized site specific repository.

This work is supposed to result in a preliminary safety report and an application for siting in the years around 2000.

Design optimization will require technical development and adaption of procedures for conditioning and control of both waste and barriers. The main part of these efforts will be concentrated to the 1990-2000 period but may continue if the licence authorities so reguire.

When licenses have been acquired the construction work can start. During the period 2000-2020 it has been considered prudent to include substantial reserve - and spare time - to compensate for technical as well as political uncertainties.

KEY RESEARCH AREAS

As mentioned above we have the opinion that earlier research work has proven the feasibility of safe disposal. However, the need for compensation for uncertain factors and lack of knowledge make the proposed systems unnecessarily expensive and "over-safe". The research activities will therefore be focused on evaluation and intercomparison of possible alternative systems. Another main task is to reduce the uncertainty as well as to further develop the understanding of effects of beneficial phenomena that we have not today been able to take credit for, due to lack of data or models.

An example of a research area of importance for our planning is the ground water flow in jointed rock masses. For a safety analysis of a repository in crystalline rock, a detailed understanding of the flow of water in a jointed rock mass and associated transportation of nuclides is fundamental.

Up till now, most of the Swedish work in this area has been focused on the high quality, low permeability rock between fault zones which has been assumed almost open. There are, however, reasons to believe that also the more fissured and fractured zones can constitute high capacity barriers against the nuclide migration.

Beside a characterization and a deeper understanding of water flow in such zones, the main mechanism of interest is how the radio nuclides diffuse into fractured rock and its microfissures. In order to take advantage of this in the safety assessments, a research effort is planned to get some of these data in a several-year project on "flow and nuclide migration in fracture zones in crystalline rock".

Positive results would improve the safety margins for our concepts and give us increased flexibility in the selection of a site.

Another fundamental question that needs special consideration is the long time span involved. Here the geological system is one of the very few where the rate of change is so slow that realistic predictions can be made for time spans of serveral thousands of years.

This fact is reflected in many repository concepts. Efforts are often made not to disturb the natural environment and materials are often selected so the stability and interaction with the surroundings can be proven with geological evidence.

In this field, the investigations on naturally existing physical and chemical systems will be expanded to provide data for validation of models for processes that govern the radionuclide behaviour in the repository. Examples of processes considered most important for the validation are shown below.

KEY RESEARCH AREAS

- WATER FLOW IN JOINTED ROCK
- NUCLIDE MIGRATION IN FRACTURE ZONES
- NATURAL ANALOGUE SYSTEMS
  - REDOX FRONT
  - STABILITY OF CLAYS
  - SOLUBILITY OF UO_2
  - SEDIMENTATION PROCESSES
- INTERNATIONAL COOPERATION
  - MODELS FOR SAFETY ASSESSMENT

Of these processes the dissolution of the UO_2-fuel matrix is of primary interest and will require special efforts in fuel matrix characterization and identification of secondary solid phases that might limit the solubility.

A final example deals with measures for treatment of rock masses.

There will never be a possiblity to investigate a possible site in all details before it is fully excavated and we must be prepared to encounter zones of lower quality. Here various measures to treat the rock must be further developed.

Examples of this can be the control of groundwater flow in a repository by further developing existing grouting techniques or investigating the possibilities and potentials to introduce drain systems such as the hydraulic cage, used in another Swedish system under discussion, the WP-cave concept.

Prior to the introduction of such man made disturbances, however, there must be made extensive evaluations of the secondary effects these actions can have on the near-field and the interaction between the various components in the repository.

Many of the areas relevant for the R&D in this business demand international cooperation and consensus. This is of special importance for the society, when it has to decide on the acceptability of a suggested repository concept. It is essential that methods and praxis for safety asessments of repository systems are internationally accepted and that there is an international confidence in the models and the data bases used.

The international cooperation seems to be working very well in the field of radioactive waste management. A good example is the OECD Stripa Project, that has been going on for several years. Here the pooling of the skill and resources of many countries have produced know-how and data of unique value. for the disposal of radioactive wastes. The SKB will also in future allocate considerable resources into the international cooperation.

UNITED STATES CRYSTALLINE REPOSITORY PROJECT - KEY RESEARCH AREAS

EDWARD S. PATERA
United States Department of Energy, Crystalline Repository Project Office, 9800 S. Cass Ave., Argonne, IL 60439

ABSTRACT

The Crystalline Repository Project is responsible for siting the second high-level nuclear waste repository in crystalline rock for the United States Department of Energy. A methodology is being developed to define data and information needs and a way to evaluate that information. The areas of research the Crystalline Repository Project is involved in include fluid flow in a fractured network, coupled thermal, chemical and flow processes and cooperation in other nations and OECD research programs.

INTRODUCTION

The mission of the United States' Department of Energy's Crystalline Repository Project is to site candidates for the second high-level nuclear waste repository mandated by the Nuclear Waste Policy Act of 1982 [1]. It is the intent of the Department of Energy to construct a second repository, after receiving authorization from Congress, and that crystalline rocks be the primary target geology for this repository as specified in the United States Department of Energy Mission Plan [2].

Toward completing this mission the Crystalline Repository Project (CRP) can be divided into several phases: 1) the national survey, 2) the regional phase, 3) the area characterization phase, 4) the detailed site characterization phase, 5) construction, 6) operation, and 7) decomissioning. The national survey was completed and documented as OCRD-1 in April 1983 [3]. The CRP is concluding the regional phase which consists of screening approximately 240 rock bodies in three regions (northcentral, northeast, southeast) to approximately 15-20 areas for inclusion in the area phase. The regional phase is primarily a literature study of the three regions and screening is based on that data. The results of this screening will be documented in the Area Recommendation Report to be released in draft for Nov. 15, 1985.

The area phase of the project will consist of field characterization to provide data and information to select approximately three candidate sites to be nominated for detailed site characterization in the site phase. Area field work is slated to begin January 1987 and be completed in 36 months.

The purpose of this paper is to describe the research and development efforts being undertaken by CRP to support the siting program for the area phase. (The reader is refered to the Mission Plan for a summary of all research associated with the Civilian Radioactive Waste Management Program.) One activity undertaken by CRP related to identifying research needs was to convene the Geotechnical Assessment and Instrumentation Needs Symposium (GAIN 84). The proceedings of GAIN 84 [4] lists various research areas deemed necessary by the participants for a complete understanding of a High Level Waste (HLW) repository in crystalline rock.

SYSTEMS METHODOLOGY DEVELOPMENT

The geologic disposal of HLW is a profound and complex set of political, economic, social and technical problems. In order to address these complexities the CRP needs to develop a methodological way of collecting information and evaluating this information. Thus, decisions made are based on sound technical judgement and can be clearly followed by interested parties.

The CRP is currently developing a methodology that can be used during the area characterization phase. The objectives of the methodology are:

1) Provide a means of complying with regulatory requirements.
2) Identify data and information needs to complete the above objective.
3) Provide a mechanism for evaluating and comparing areas of interest in order to select candidate sites.
4) Track progress through area phase.
5) Allocate resources for siting and research needs.

The methodology is developed by technical discipline with respect to complying with the United States Department of Energy Siting Guidelines [5]. As a result sector models were developed for:

1) post-closure safety
2) pre-closure safety
3) environment impacts
4) socio-economic mitigation
5) transportation safety
6) overall costs

The use of sector models will allow the CRP to rate each candidate site on each subject of a sector model. Although not developed yet, the individual site ratings can be combined using some sort of preference model that will assign weights of importance to each of the outcome variables of each sector model. An overall rating based on all sector models can be determined and that information supplied to the decision maker for use in selecting sites for further development.

Because of the regulatory approach to the methodology significant weight will be attached to post-closure safety. Therefore, the collection of information and data necessary to conduct performance assessment activities become of paramount importance to the CRP.

RESEARCH ACTIVITIES

The Crystalline Repository Project is currently involved in a number of research projects both through international cooperation and those funded solely by CRP. The research areas include hydrogeology, geochemical transport, borehole geophysics and rock mechanics.

The CRP is involved in the OECD/NEA sponsored Stripa International Project as the United States representative. The Stripa Project is conducted in the Stripa iron ore mine in central Sweden. The areas of research include hydrogeologic characterization, tracer migration studies, cross-borehole characterization development, and backfill and buffer materials development.

The CRP is planning to cooperate extensively with the Canadians through the Atomic Energy of Canada, Ltd. (AECL) on four areas of mutual interest. These areas include: 1) Underground Research Laboratory (URL) shaft deepening and characterization; 2) URL experimental program; 3) Field characterization techniques development; and 4) Performance assessment activities.

The first area of cooperation is the extension of the URL shaft from its present level of 240 meters to approximately the 440 meter level. This would allow construction of the laboratory facilities at a depth more appropriate to a repository. In addition, characterization of the granite will be done down to this level in the shaft and a major sub-horizontal fracture zone will be characterized and stabilized.

The United States will also cooperate in the areas of testing within the URL. This includes the cooperative planning and performance of the experiments. Potential experiments include:

1) Excavation Response Experiment
2) Heated Block Experiment
3) Single Fracture Migration Experiment
4) Buffer/Container Experiment
5) Borehole/Shaft Sealing Investigation
6) Pressure Chamber Experiment

The third area of cooperation is the CRP's participation in AECL's hydrogeologic field characterization experiments. The CRP has two objectives for this task. First, is to assist in the field experiments to gain a better understanding of crystalline rocks through the collection of field data and subsequent performance analyses. The second objective is to test CRP's planned area characterization methods and equipment for use within the identified candidate areas in the United States.

The last area of cooperation is in performance assessment activities. These include doing assessments of the research areas as mentioned above for comparison with Canadian results, possible geochemical data generation and transfer of computer codes between countries.

The United States has bilateral agreements with other nations investigating crystalline rocks for use as a HLW repository. The CRP will be investigating cooperative efforts with Switzerland and France in the near future. The CRP is also interested in cooperating on an international scale in the area of natural analogues and their role in validating models of the behavior of geologic disposal systems.

The area of research that CRP conducts solely on its own are in hydrogeology and geochemistry. In these areas the research is devoted primarily to development of computer codes to model crystalline rocks to gain a better understanding of the physical and chemical processes that will be operating within a crystalline HLW disposal system.

It is felt that one of the most important aspects of demonstrating repository performance is determining the hydrogeologic processes. These include travel time to the acessible environment, most likely pathways, fluxes through the pathways and the repository itself, and the discharge points. All of these factors take place in fractured crystalline rock.

It is the nature of fluid flow within a fracture medium (versus a porous medium) that can represent a large uncertainty in being able to model the disposal system accurately. It is for these reasons that the CRP is devoting most of its research resources to develop an understanding of fluid flow in fractured systems.

The CRP approach to understanding fluid flow in fractures is three fold. The first is a theoretical treatment of fluid flow in a fracture network. The fracture network is generated randomly based on density, orientation, length, and apertures. The interconnections are calculated and then flow is determined through this network. The mathematical approach is designed to look at different parameters of the system and determine which are most important. This approach will also be used to determine when scale effects are important. For instance, it would be very desireable to know at what volume of fractured rock could one use porous medium equivalents to represent flow within crystalline rock to some specified degree of accuracy.

The second approach to modeling fractured rocks will be rooted in geologic field observations and the fracture or joint set data derived from surface and underground sources. Hopefully, this approach will enable one to predict fracture networks in a given geologic setting based on rock types, tectonic history, in-situ stress of the rock, and other pertinent data. This information together with the theoretical understanding of fracture networks will allow the CRP to construct conceptual models of sites with fractured rock and thus perform more confident analyses of the sites' performance.

The third approach is to use physical models of fracture networks in a laboratory. The objective would be to validate the theoretical understanding of fluid flow in fracture networks. This would be accomplished by setting up the requisite problems and constructuring physical models.

Research into the nature of hydrothermal reactions has been initiated by CRP. The objective of this research is to gain a better understanding of coupled thermal, chemical and flow processes. With information on how these processes interact, a source term of radionuclide release in the very near field can be constructed to accurately reflect what is happening in nature. This aspect of performance assessment is crucial to developing accurate and confident systems performance assessment in order to license a repository in crystalline rock.

The above hydrothermal experiments will be performed in a flow-through hydrothermal pressure vessel. This apparatus will allow the use of temperatures up to 500°C and 300 bars pressure. Fluid of a specified composition is flowed through the apparatus once and is collected for analyses at the end of the flow column at pressure and termperature. At this time granite and a granitic groundwater composition will be used. Initial runs will be at high temperatures to promote reactions and allow more experiments per unit time to be run. The information derived from the experiments will be 1) the change in groundwater composition as a function of both temperature and flow rate, 2) the change in alteration mineralogy with respect to temperature and flow rate and 3) the change in permeability of the system as a function of temperature, flow rate and total water flux through the system.

Although not planned at this time, this experimental apparatus could be used in conjunction with other materials (waste forms, cannister material, buffer/backfll material) to test the affects these materials would have on groundwater chemistry, host rock degradation, engineered barrier component degradation, and groundwater flow.

SUMMARY

The CRP will continue to use an integrated approach to identifying information needs to ensure the successful completion of its mission. The nature of crystalline rocks and how they will behave under repository conditions is a complex technical problem. The CRP will continue to support research and development in these areas both domestically and through international cooperation.

REFERENCES

1. Nuclear Waste Policy Act of 1982, Public Law 97-425, 1983.

2. United States Department of Energy, Mission Plan for the Civilian Radioactive Waste Management Program, Vol I, 1985.

3. Office of Crystalline Repository Development, A National Survey of Crystalline Rocks and Recommendations of Regions to be Explored for High-Level Radioactive Waste Repository Sites, OCRD-1, (Battelle Memorial Institute, Columbus, OH, 1983).

4. W.F. Ubbes and J.O. Duguid, Geotechnical Assessment and Instrumentation Needs for Isolation of Nuclear Waste in Crystalline Rocks, (Battelle Memorial Institute, Columbus, OH, in press).

5. United States Department of Energy, IOCFR Part 960 Nuclear Waste Policy Act of 1982; General Guidelines for the Recommendation of Sites for the Nuclear Waste Repositories, Federal Register, Vol 49, No. 236, p. 47714, 1984.

U.S. NUCLEAR DEFENSE WASTE PROGRAM
CURRENT DEVELOPMENT AND KEY RESEARCH AREAS

RAY D. WALTON, JR* AND KENNETH A. CHACEY**
* U.S. DOE, Defense Waste and Byproducts Management, Washington, DC 20545
** U.S. DOE, Savannah River Operations Office, P.O. Box A, Aiken, SC 29801

ABSTRACT

The United States has been engaged in defense nuclear activities for over 40 years. The primary goal of the Defense Waste and Byproducts Management Program is to develop technology which ensures the safe, permanent disposal of all defense nuclear waste. Programs are in place at each U.S. Department of Energy site which address long-term strategy for permanent disposal of waste generated as a result of defense operations. Technology is developed for assessing the hazards, environmental impact, and cost of each long-term disposition alternative for selection and implementation. This paper addresses the key research areas and major facilities associated with the long-term management of defense nuclear waste.

INTRODUCTION

The Defense Waste and Byproducts Management Office has the overall responsibility for managing defense nuclear wastes within the DOE. Defense nuclear wastes are generated from the reprocessing of spent nuclear fuel from defense production reactors, test reactors, naval propulsion reactors, and weapons fabrication [1]. Uranium, plutonium, and tritium are recovered and are used for new fuel fabrication and for weapons. Wastes, generated from the reprocessing and recovery process, are in interim storage pending final disposal. The current inventory of wastes stored over the past 40 years are approximately 450 thousand cubic meters (see Figures 1 and 2) [3].

High-level wastes (HLW) are stored at Savannah River Plant (SRP) in South Carolina, the Hanford Reservation (Hanford) in Washington State, and the Idaho National Engineering Laboratory (INEL) in Idaho (ID). HLW is the highly radioactive waste material requiring permanent disposal. Approximately 63 percent of the present HLW total, by volume, is stored at Hanford, 34 percent at SRP and the remaining 3 percent at INEL. This corresponds to approximately 62 percent of the HLW radioactivity stored at SRP, 33 percent stored at Hanford, and 5 percent at Idaho. By volume the defense waste is at present about 95 percent of the total inventory of U.S. nuclear waste, however, it represents only 9 percent of the radioactivity. The remaining U.S. HLW is contained in commercial spent nuclear fuel and a small amount of liquid at West Valley, NY. The low burnup of defense fuel contributes to the relatively large volume and low curie level. As required in the Nuclear Waste Policy Act an evaluation was made and the President determined that defense HLW should be disposed of in repositories provided for commercial HLW. Disposal studies are underway to determine appropriate charges for this service.

The transuranic (TRU) wastes generated from defense programs are in interim storage at the HLW sites and at Oak Ridge National Laboratory (ORNL) in Tennessee, Los Alamos National Laboratory (LANL) in New Mexico, and at the Nevada Test Site (NTS) in Nevada. The TRU wastes are generated from weapons fabrication, and reprocessing of spent fuel. They are generally in the form of used equipment, clothing, ion-exchange resins, empty containers, etc. The TRU waste is defined as containing alpha-emitting radionuclides with half-lives greater than 20 years and with concentrations greater than 100 nanocuries per gram. Approximately 68 percent, by volume, of TRU waste is stored at INEL, 17 percent at Hanford, 8 percent at LANL, 5 percent at SRP, and the remaining waste stored at ORNL and the NTS. This corresponds to approximately 53 percent of the radioactivity of TRU waste being stored at SRP, 22 percent at Idaho, 16 percent at LANL, 8 percent at Hanford, and the remainder at ORNL and NTS. A schematic showing the relative quantities and radioactivity of TRU and HLW are shown in Figures 1 and 2 [3].

Low-level waste (LLW) is located at all the defense facilities where HLW and TRU are produced, and results from the day-to-day handling of radioactive materials. LLW are radioactive materials not classified as HLW, TRU, spent fuel or byproduct material. There are approximately 90,000 cubic meters of LLW generated each year [2].

OBJECTIVE

Nuclear waste disposal objectives of the DOE are: (1) end interim storage, and (2) achieve permanent disposal. The Research and Development (R&D) efforts for long-term management of defense nuclear waste are ongoing. These technologies will be used for permanent disposal of TRU and HLW. Technology will also be developed for improved methods of LLW disposal. To accomplish this objective R&D programs have been established at DOE sites to develop waste forms suitable for permanent disposal of HLW, preparation of TRU waste for permanent isolation, and to obtain greater disposal techniques for LLW. Thus, major research and development program objectives are:

- End interim storage and achieve permanent disposal of high-level and transuranic wastes
- Develop technology for low-level waste disposal in compliance with applicable regulations

STRATEGY

The DOE R&D activities support three primary program elements: (1) interim storage operations, (2) long-term management, and (3) permanent disposal. The interim program involves safe processing and storage of defense TRU waste and HLW pending permanent disposal. The treatment and disposal of LLW is also part of this effort. The supporting research is associated with minimizing the quantity of waste generated and development of treatment methods to reduce the amount of waste requiring interim management and final disposal. The long-term management of defense waste addresses the disposal of HLW, TRU waste, and LLW and is being implemented sequentially, where possible, to maximize technology transfer, and to minimize capital expenditure peaks [1].

The HLW are stored as salts, liquids, and sludges which are highly radioactive. DOE has funded an extensive R&D program at each site to develop the technology for converting these wastes into forms which are acceptable for permanent disposal. Treatment processes at each site are being developed to reduce the volume of immobilized waste through separation of nonradioactive waste fractions and LLW components, and the selection of a waste form matrix capable of high radionuclide loadings while capable of meeting low repository release rates. The HLW technology program is supporting work at SRP for the Defense Waste Processing Facility (DWPF), and the selection of disposal strategies at Hanford and Idaho. At Hanford, design is underway on the Hanford Waste Vitrification Plant (HWVP). Treatment methods are being developed to reduce the volume

FIGURE 1: Volume and Radioactivity of High Level Waste Accumulated through 1984

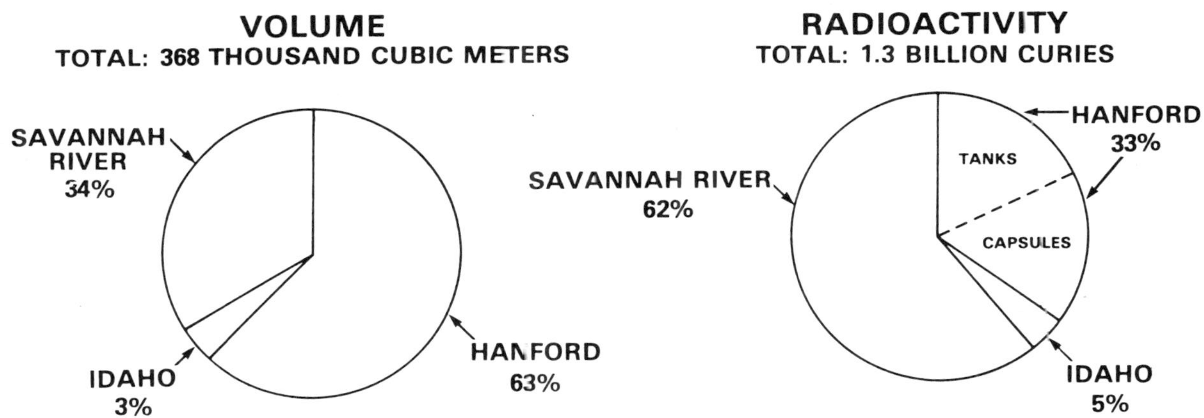

FIGURE 2: Volume and Radioactivity of Defense Transuranic Waste Accumulated through 1984

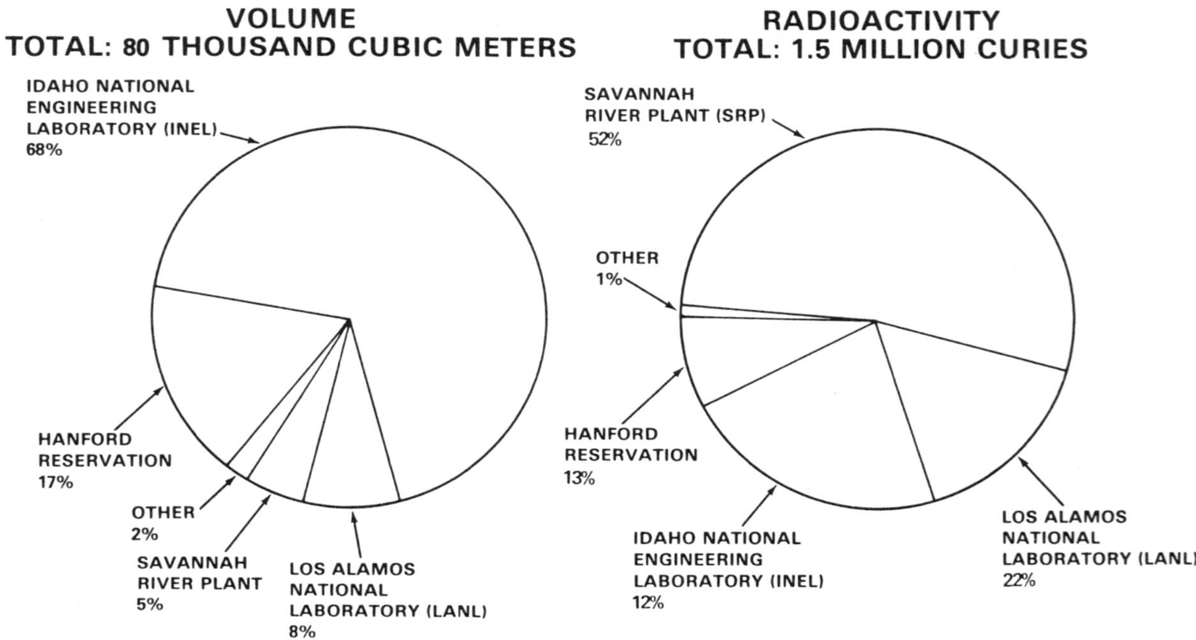

RW 0906

of TRU wastes and to assure that it meets repository waste acceptance requirements. Technology development started at Idaho because it has the largest inventory of stored TRU waste, and involves the development of a system to examine, process, and certify TRU waste for disposal in the Waste Isolation Pilot Plant (WIPP) [2]. Development for LLW has emphasized methods to reduce the amount of LLW generated, and methods for greater confinement and disposal. The strategy for long-term defense waste disposal is summarized below.

- Achieve orderly and cost-effective transition from interim storage to permanent disposal
- Technology transfer, immobilization, disposal in geologic repository, in-place disposal
- Support high-level waste processing facility at Savannah River
- Develop and select site-specific strategy for high-level waste disposal at Richland
- Develop and evaluate alternatives for high-level waste disposal at Idaho
- Establish waste acceptance specifications for disposal of high-level and transuranic wastes in a repository (criteria for CH-TRU established)
- Certify high-level and transuranic wastes for geologic disposal
- Design, construct, and operate waste processing and disposal facilities (DWPF, HWVP, WIPP, SWEPP, PREPP, WRAP, TWF, and WHPP)
- Reduce future waste generation

CURRENT RESEARCH AND DEVELOPMENT PROGRAM FOR HIGH-LEVEL WASTE

Savannah River Plant

The primary objective at SRP for long-term management of HLW is immobilization followed by off-site shipment to a repository. The approach developed through an extensive R&D program over the past several years is to remove the neutralized HLW from interim storage, separate out and decontaminate a salt stream, and immobilize the remaining sludge. The majority of the radionuclides and long-lived radioactivity is in the sludge which will be immobilized in borosilicate glass. The liquid and salt fraction will be decontaminated and handled as LLW. Key research areas have included the development and testing of the melter, glass withdrawal technique of partial pressure differential, the slurry receipt adjustment tank, formic acid pretreatment to prevent oxide formation in the melter, the off-gas system, canister cleaning and welding, glass formulation, and (laboratory and large scale) leaching studies to evaluate the release rate of radionuclides to the environment under repository conditions. Extensive research into the melter electrode material and placement has been required to control circulation and temperature patterns in the melt tank. The volatilization and production of gas in the melter has been minimized, and off-gas system research includes the development of a hydrosonic scrubber and sand filter which can reduce the overall contamination by a factor of 10^8. A 2' diameter by 10' length handling canister made of stainless steel has been designed, external decontamination methods developed, and canister closure by high current and high pressure welding developed.

Another key research program has developed test methods and supporting data to show HLW waste forms will meet the repository waste acceptance specifications. A 3 year radionuclide release mechanisms program directed by the Materials Characterization Center was completed in 1984 in which a number of mechanisms were proposed to model release rates under repository conditions. The experimental determination of long-term waste form performance is important to model radionuclide release rates 1,000 years after repository closure (based on an NRC regulatory requirement of $1x10^{-5}$ release per year of the radioactivity calculated to be present 1,000 years following closure).

At SRP simulated leach tests have been conducted on a laboratory scale, in a large scale leach test facility, and under simulated repository conditions (cup tests) in the hot cell. The large scale facility (shown in Figure 3) is used to determine static and repository specific leaching characteristics in a temperature controlled water bath. The first tests have evaluated the scaling factors used to relate small laboratory samples used in standardized test procedures to the full size waste form, and future tests will evaluate leaching under simulated repository conditions including the capability to simulate hydrostatic pressure. The interaction of the SRP waste form using the three candidate repository host rock cup tests (shown in Figure 4) has demonstrated the chemical stability of borosilicate glass under repository conditions and provided valuable information on radionuclide release rates to the environment. The major R&D accomplishments and future milestones of the SRP R&D program are summarized below.

- o Demonstrated canister closure techniques and frit blaster for canister decontamination
- o Started-up of full scale slurry receipt adjustment tank
- o Confirm design of large scale melter operation
- o Developed and demonstrated off-gas treatment with actual SRP HLW
- o Demonstrated pumping systems for transfer of HLW slurry
- o Define glass waste compositions which produce optimal product characteristics
- o Provide information on radionuclide release rate from the waste form under repository conditions

The DWPF has been designed and construction initiated. It will process the HLW generated from defense reprocessing operations at Savannah River. During the first 15 years of operation, the backlog of HLW will be processed resulting in about 500 canisters of glass produced each year. Following this phase the glass production rate will be based on normal operating levels. The major accomplishments and future milestones of the DWPF are summarized below.

- o DWPF construction started 1984
- o Complete 80 percent final design, October 1985
- o Start salt immobilization plant, August 1987
- o Complete construction, March 1989
- o Hot operation initiated, September 1989

Hanford Reservation (Hanford)

The conceptual designs for the Hanford Waste Vitrification Plant (HWVP) are underway for converting HLW to a solid waste form. HLW will be retrieved, a salt fraction removed, decontaminated, and incorporated in grout for near surface disposal as LLW [2]. The sludge fraction will be immobilized in borosilicate glass for repository disposal. Another alternative for long-term management of some previously designated HLW at Hanford is in-place stabilization and disposal where retrieval poses a major risk and large costs are not justified. The trade-offs of this alternative are being evaluated by Hanford and include an evaluation of on-site disposal versus retrieval, treatment, and immobilization.

FIGURE 3: Large Scale Leaching Facility

FIGURE 4: Simulated Repository Test

Hanford HLW research has focused on the technology development for the HWVP, in-place stabilization alternatives, and waste characterization. The key research areas have included the vitrification system, frit development, melter design, off-gas treatment, and canister sealing and decontamination. For in-place disposal major research efforts include: (1) barrier development (shown in Figure 5) and testing to reduce water recharge and wind erosion, (2) geologic life time marking systems, (3) radionuclide transport models, and (4) waste characterizations system. The major goals and objectives of the Hanford R&D program are summarized below.

- o Melter design and testing
- o Off-gas treatment development
- o Development of HWVP reference glass
- o One-third scale HWVP melter operation
- o Complete HDW-EIS
- o Technical exchange with other DOE sites
- o Develop information for repository release rates
- o Develop technology for in-place stabilization

The HWVP is being designed using technology for the DWPF to the maximum extent possible. The preconceptual design was initiated in 1983, research and development are underway, and definitive design will start in 1988. The HWVP construction is to be initiated in 1989 with operational capabilities by 1994.

Idaho National Engineering Laboratory

After careful review of development and all requirements, a reference strategy will be selected and development of a pilot-scale immobilization facility initiated. The data collected from this research will provide the information required for full scale implementation and process start-up by calendar year 2008.

As previously shown, a small portion of the total volume is currently contributed by the ICPP liquid and calcined waste. However, by year 2020 INEL will produce about 40 percent of the total volume of defense HLW [4]. Therefore, initial R&D at INEL is focused on volume reduction and processes for reducing ICPP HLW generation. Process options include incorporating calcine or high-level liquid waste into glass or glass/ceramic waste forms with high waste loadings, followed by geologic disposal. Key research areas will include laboratory scale tests to develop suitable waste forms (shown in Figure 6), vitrification studies for high fluoride waste, and the development of calcine retrieval methods from existing bin sets. The major accomplishments and goals of the INEL R&D program are summarized below.

- o The strategy planning document for long-term management of the ICPP HLW has been issued
- o Provide scoping studies and laboratory experiments by 1987 to evaluate alternative plant process options
- o Minimize HLW volumes through process control
- o Develop waste forms with high density loading to reduce the volume of waste requiring permanent disposal
- o Develop cold pilot-scale waste immobilization facility by 1990
- o Select a reference strategy by 1992
- o Initiate waste treatment facility construction by 2002

FIGURE 5: Barrier Development

FIGURE 6: ICPP Waste Forms

CURRENT RESEARCH AND DEVELOPMENT PROGRAM FOR TRANSURANIC WASTE

At each site technology is being developed to prepare newly generated TRU waste for permanent disposal in WIPP and to evaluate previously stored TRU waste for disposal requirements [2]. The WIPP facility is located in New Mexico and is a demonstration facility to obtain information on disposal of TRU waste in this bedded salt geology. The demonstration is scheduled for a 5-year period at which time a decision will be made regarding WIPP's suitability as a permanent geologic repository for defense TRU waste. Newly generated TRU waste will be certified (on a site-by-site basis) and shipped to WIPP starting in September 1988. The processing of stored TRU waste has begun at INEL due to the large inventory at that site. The Stored Waste Examination Pilot Plant (SWEPP) at INEL is constructed and awaiting the certification audit to enable future shipments of processed waste to WIPP. For waste which do not meet disposal requirements a Processing Experimental Pilot Plant (PREPP) facility at INEL is under construction for treatment and certification. At the present time NEPA documentation has been completed for only the retrievably stored TRU waste at INEL site and for newly generated TRU waste, however, the required documentation for retrievably stored TRU waste at the other sites is being prepared.

The TRU waste technology development has provided support to SWEPP and PREPP operations and the certification requirements for WIPP. Significant research efforts have also provided valuable information for TRU element separation from liquid streams, incineration, and final packaging techniques. Additionally, a mobile nondestructive examination facility is under design. The development of this project will allow small generators to test for curie level and physical contents without installing costly equipment. The major program objectives are listed below.

- Reduced waste generation
- Nondestructive assay and examination including Real-Time Radiography
- Size reduction
  - Plasma ARC Torch
  - Shredder
  - Incineration
  - Compaction
- Immobilization/grouting and In-situ Vitrification
- Greater confinement disposal
- Waste acceptance requirements for remote-handled transuranic waste
- Certification for WIPP acceptance
  - Contact-handled
  - Remote-handled

The construction of the Waste Isolation Pilot Plant (WIPP) is expected to be completed in 1986 and includes the above ground support and receiving facilities, the underground receipt facilities, and the emplacement tunnels. The total below ground disposal area for TRU waste covers about 100 acres, and includes 12 acres for testing and development studies [2]. The shafts are 10 to 19 feet in diameter and extend 2,150 feet below the surface.

The major R&D projects and future milestones for the WIPP test program are summarized below.

- o Construction completion December 1986
- o Begin routine contact-handled transuranic waste receiving October 1988
- o Initiate tests to determine the performance of remote-handled transuranic waste in bedded salt 1989
- o Determine the response of salt formation to excavation, heat, and radiation
- o Determine creep deformation on experimental room stability and long-term encapsulation of the waste
- o Determine methods for sealing shaft
- o Develop retrieval methods

Stored Waste Examination Pilot Plant; Process Experimental Pilot Plant

The Stored Waste Examination Pilot Plant (SWEPP) facility is an integral part of the program to provide permanent disposal of TRU waste material. During this phase nondestructive examination of TRU materials is conducted to determine if the radioactivity is greater than 100 nCi/g and to certify the waste to meet WIPP disposal requirements. The SWEPP facilities will start certification of TRU waste in 1986 and all waste that can be certified will be stored separately until WIPP start-up. Those wastes which are not certifiable will be sent to Process Experimental Pilot Plant (PREPP). The objective of the R&D at the PREPP facility is to demonstrate on a full scale basis methods for processing TRU waste for meeting the disposal criteria for WIPP. The initial experimental pilot processing in the PREPP facility will start up in 1987 and involve low speed shedding, a rotary kiln for waste incineration, and cement immobilization. During start-up, extensive tests using nonradioactive material will be conducted to increase process reliability and examine the quality of treated TRU material. The experimental tests in PREPP will begin in 1987 leading to full scale testing in 1988.

CURRENT RESEARCH AND DEVELOPMENT PROGRAM FOR LOW-LEVEL WASTE

The goal of the LLW technology program is to safely dispose of low-level nuclear waste generated as a result of defense operations. LLW are those wastes which are not HLW or TRU waste material, poses the least health hazard, and are normally disposed of as a dry solid in burial grounds at each DOE site. Technology development has improved monitoring methods, shallow land burial technology, and greater confinement disposal concepts. Additionally, treatment methods have focused on processes which will reduce volume and increase operational integrity of LLW for permanent disposal. Demonstrations to improve cover systems for LLW burial grounds have been conducted to improve hydraulic isolation.

The technology development for LLW has focused on obtaining improved treatment and disposal practices. This includes R&D for packaging, handling, treatment methods for Shallow Land Burial (SLB). The technology relating to SLB is applicable to about 90 percent of the defense LLW and greater confinement studies are associated with 10 percent of the potentially higher hazard LLW [2]. The waste treatment technology addresses methods for waste volume reduction, waste processing, and waste form testing. Included in waste form testing is leach mechanism development and identification of waste treatment processes. The specific areas of development for SLB include water management, monitoring, and predictive modeling. Through development of proper water management techniques the water infiltration and subsequent radionuclide transport mechanisms can be controlled. For LLW which is not suitable for SLB, Greater Confinement Disposal Tests (GCDT) are being performed on a full scale basis to demonstrate that these wastes can be efficiently disposed of at a depth of 100 feet. The basic design includes a 10 foot shaft, 120 feet deep. The LLW is stacked in the shaft and backfill is used to close the hole. Instruments are placed around the shaft and nonradioactive gaseous and liquid tracers injected into the waste package zone. The migration of the tracers is then monitored to determine the data for long range performance models. The major program objectives are listed below.

- o Minimization of subsidence (grout, compaction, etc.)
- o Stabilization of inactive disposal sites
- o Waste treatment (e.g., thermal destruction of nitrates)
- o Greater confinement disposal
- o Performance assessment (e.g., pathway analysis and dose to man evaluations)

SUMMARY

The U.S. Department of Energy has established a strategy for ending interim storage and achieving permanent disposal of defense waste. Significant progress is being made toward this goal. Research and development programs at each site are designed to establish the necessary technology required to ensure safe disposal of all nuclear wastes generated from defense programs. These research programs are currently focused on: 1) Developing design criteria for construction of facilities to treat high-level waste and transuranic waste; 2) Determine the performance of HLW forms under repository conditions; 3) Certification of transuranic waste for permanent isolation; 4) Demonstrating safe disposal of transuranic waste in a repository; and 5) Continued improvement of low-level waste disposal practices. By the year 2000 significant volumes of transuranic and high-level waste are to be processed for permanent disposal. Thus the next few years will require the Department of Energy's dedicated attention toward implementating and demonstration the technology which has developed in the long-term technology program.

REFERENCES

1. DOE/DP-0015, The Defense Waste Management Plan, U.S. Department of Energy, June 1983
2. DOE/DP-0016/2, FY 1986 Program Summary Document, Office of Defense Waste and Byproducts Management, U.S. Department of Energy, August 1985
3. DOE/RW-0006, Spent Fuel and Radioactive Waste Inventories, Projections, and Characteristics, U.S. Department of Energy, September 1984
4. WIN-138, Strategy Planning Document for the Long-Term Management of ICPP High-Level Radioactive Waste, Westinghouse Nuclear Company, Inc., March 1985

FRENCH HIGH LEVEL NUCLEAR WASTE PROGRAM — KEY RESEARCH AREAS

CLAUDE G. SOMBRET
CEA, Cen Valrho, BP 171, 30205 Bagnols sur Ceze Cedex, France

ABSTRACT

The most important aspects of this research program concern disposal safety: the long-term behavior and sensitivity of the materials to the variability inherent in industrial processes, and the characterization of the final product. This research required different investigations involving various scientific fields, and implements radioactive and non-radioactive glass samples as well as industrial scale glass blocks. Certain studies have now been completed; others are still in progress.

INTRODUCTION

Research on solidifying high level radioactive solutions began in France in the late 1950s, and has led to the development of various phosphate or other glass formulations. Because of their properties, borosilicate glasses were finally selected as the reference media for vitrifying the solutions produced in French reprocessing plants.

During this period, fabrication processes based on different techniques and with different capacities were also developed. Both continuous and batch processes using ceramic or metal furnaces were investigated to determine the optimum method for each type of waste disposal problem in France. Stages in this development program included the GULLIVER facility in 1963 (using a crucible batch process), the PIVER pilot facility in 1969 (metal pot batch process) and the AVM industrial facility at Marcoule (continuous metal furnace process) which has been used since 1978 to vitrify 918 m^3 of material with a total activity of 170 MCi, representing 414 metric tons of glass in 1709 containers. The R7 and T7 plants using the same continuous process are now under construction at La Hague (Figure 1), where R7 should begin operation early in 1987 (engineering work by SGN; plant operator: COGEMA).

Nevertheless, further specific research work is required in the area of radioactive waste containment, from production of the selected material to the final disposal conditions. Within the scope of this research, certain areas may be considered to be of vital importance justifying particular attention.

KEY RESEARCH AREAS

It is widely admitted, notably in France, that understanding the long term behavior of the material is essential: the ultimate repository must be designed on the basis of this knowledge. Other research areas are also highly important, however, as they affect long term disposal safety. These include assessing the degree to which the material and its manufacturing process are sensitive to the imponderables of industrial production, and characterizing the disposal materials.

Vitrification of fission product solutions generated by reprocessing LWR fuel is the most important problem in France because of the volume of these solutions compared with other high level wastes. The following discussion is thus primarily applicable to the glass selected for this purpose |1|, the formula for which is shown in Table I. Nevertheless, this research concept also applies to other glasses, particularly for graphite-gas and LMFBR wastes.

Figure 1. R7 and T7 Construction Site at La Hague

Intermediate and Long-Term Behavior

During intermediate storage (up to several decades) and subsequently in the geological repository, the radioactive glass will be submitted to highly specific conditions and phenomena. These include thermal effects, pressure, β and γ irradiation, α irradiation, and possible leaching.

Determining the effects of these factors requires very long investigation because of the complexity of certain phenomena involved in alteration of the vitreous medium and the multiple interactions between them. For example,

Table I. R7/T7 Glass Composition (ref. SON 68 18 17 L1 C2 A2 Z1)

OXIDES	wt %	OXIDES	wt %
SiO_2	45.5	Cr_2O_3	0.5
B_2O_3	14.0	P_2O_5	0.3
Al_2O_3	4.9	ZrO_2 (filings)	1.0
Na_2O	9.9	LiO_2	2.0
CaO	4.0	ZnO	2.5
Fe_2O_3	2.9	Actinide oxides	0.85
NiO	0.4	Fission product oxides	11.25

α irradiation may cause helium to be released, which may in turn modify the physical integrity of the material and therefore its mechanical properties, thus increasing the leach losses because of the increased exposed surface area, etc.

Investigations may be conducted on simulated glasses, but radioactive materials should be implemented whenever possible.

The effect of α emitters on glass is investigated using samples spiked with an actinide, principally ^{235}Np, ^{239}Pu, ^{238}Pu, ^{241}Am or ^{244}Cm. The test specimens are small rods or cylindrical blocks 80 mm high and 70 mm in diameter. The radionuclide content represents 0.85% of the oxide weight, which is exactly the theoretical sum of the actinide oxide contents. When a sample is prepared for standard laboratory examination the actinide oxides are replaced by UO_2 + ThO_2.

The actinide leach rates and irradiation effects on glass alterability are determined according to the following methodology (Figure 2):

· Simple experiments involving variations of a single parameter.

· Semi-integral experiments in which one or more parameters (e.g. pH, Eh or CO_2) are varied from their nominal values (Figure 3). Standard conditions are 10 MPa and 90°C.

· Integral experiments using a veritable geological mockup to simulate repository conditions: the glass is no longer in contact only with water, but also with all of the environmental materials.

Although no final decision has been made, granite is considered in France as the most favorable geological storage medium. A granite repository model was therefore developed to calculate the amounts of materials in the mockup in order to ensure accuracy in simulating the surface area proportions (Figure 4).

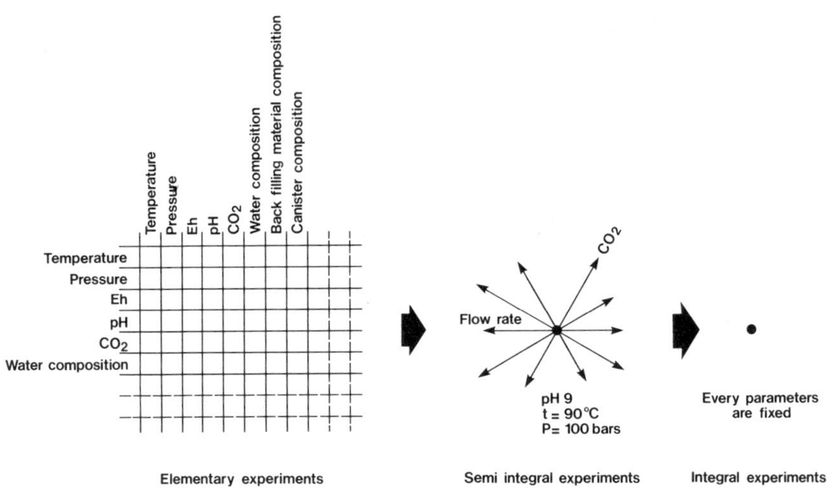

Figure 2. Test Methodology for Determining Glass Alterability

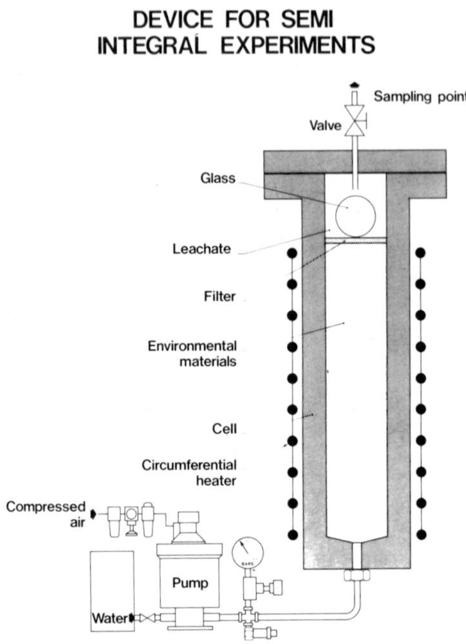

Figure 3. Semi-integral Experimental Leaching Device

On completion of these experiments, the solutions are analyzed, sample weight losses are measured and various surface analysis methods (SEM, EMP, STEM, X-ray) are implemented on the glass. Some findings are already available and have recently been published |2|. The actinide buildup in the hydrolyzed surface layer will soon be determined. Plans also call for resuming the determination of helium generated in situ |3| and for assessing possible energy accumulation by differential calorimetry.

The effects of β and Γ irradiation are investigated on samples irradiated using a 3 MeV electron accelerator: the samples are examined for structural alteration (infrared spectrometry), leaching behavior and physical properties (density, Young's modulus and fracture toughness). No effects have been detected to date at integrated doses of up to 2×10^{11} rads.

Effects of Industrial Conditions on the Material and the Manufacturing Process

Experience has shown that the characteristics of industrially produced solutions do not correspond exactly to those on which laboratory experiments are based, as determined by analysis of reprocessing plant flowsheets.

Moreover, even if equipped with reliable process controllers, industrial equipment is inevitably subject to a certain setpoint variation range.

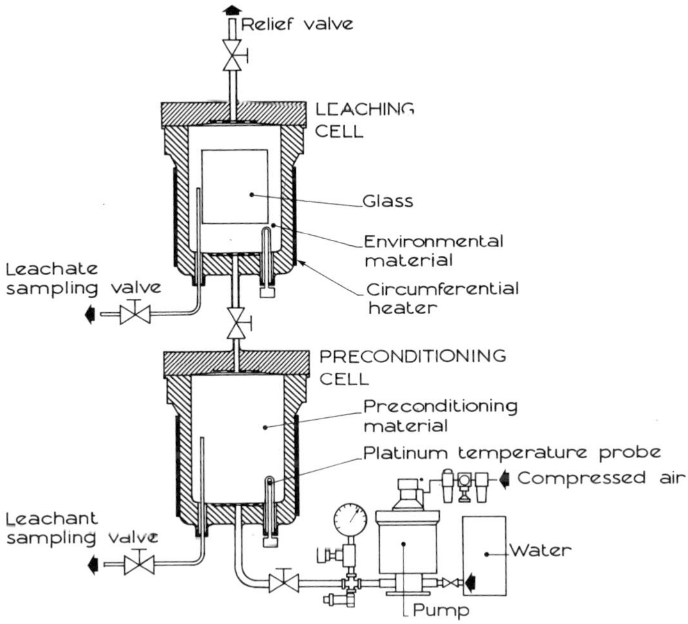

Figure 4. Integral Experimental Leaching Device

These two causes result in the following major effects:

- variations in the percentage amounts or in the element composition of the solutions in storage;
- composition variations in the raw materials required for vitrification;
- fluctuations in the calcinate or process solution feed rate to the vitrification facility;
- temperature variations at vital points in the heating apparatus.

It is thus indispensible to investigate the acceptable variation limits for these parameters. Laboratory tests and tests on industrial vitrification equipment are currently in progress at Marcoule.

The laboratory research program covers the following points. Gradual variations from the nominal chemical composition are tested element by element for various oxides: SiO_2, Al_2O_3, B_2O_3, Na_2O, Li_2O, CaO, ZrO_2, Fe_2O_3, P_2O_5 and ZnO, as well as for two additives (MgO and Gd_2O_3) not found in the nominal composition. Gradual variations in the raw material composition introduced industrially in the form of primary glass frit are also tested for all the elements in the frit, i.e. for SiO_2, Al_2O_3, B_2O_3, Na_2O, Li_2O, CaO and ZnO.

Subsequent tests cover variations of two elements, one in the frit and one in the solution: simultaneous increases or decreases are investigated as well as opposing variations of a fluxing agent and a refractory compound. Finally, since the vitrification process involves a calcining stage, samples are prepared with a frit + calcinate mixture in which the quantities of the two materials are allowed to vary.

The effects of a drop in the nominal process temperature will also be investigated.

Figure 5. Prototype R7/T7 Vitrification Equipment

The samples produced are non radioactive, and are submitted to a number of tests and measurements: viscosity between 950°C and 1200°C, alterability at 100°C (static and Soxhlet procedures), thermal stability, homogeneity.

When all of the results are available, certain tests will be conducted in hot cell conditions on a few radioactive glass blocks to confirm the effects of composition and temperature variations. Depending on the results obtained, tests will also be conducted on industrial size non radioactive blocks.

A complementary test program is also in progress on the full-scale industrial prototype of the R7/T7 vitrification process equipment at Marcoule (Figure 5) to assess the effects of operating conditions. All possible fault conditions are considered except for the offgas treatment, which has no effect on the glass composition.

A process failure analysis was conducted to evaluate the detection time and the effects of each type of fault on the glass composition. Safety features have been provided to ensure that the process alarm thresholds are quickly reached, so that the glass will sustain only slight composition variations -- no doubt within the limits that to be defined by the laboratory studies.

Characterization of the Glass Package

Glass characterization consists in a series of operations providing a comprehensive assessment of material properties, some of which must comply with basic safety regulations stipulated by the French authorities for the safety of nuclear facilities. Other properties must be determined prior to approval by the French agency responsible for long-term storage (ANDRA), and to provide basic data for repository design work by COGEMA clients.

Glass characterization covers the following areas: physical and thermal properties, homogeneity, thermal stability, leaching resistance, high temperature volatility and the effects of irradiation and insoluble metal inclusions due to fuel dissolution. Measurements are conducted on inactive laboratory glass specimens, industrial glass samples from the prototype facility, and radioactive laboratory glasses.

It should be noted that the thermal stability and leach resistance data do not necessarily reflect the actual long-term behavior, but only the glass properties at the beginning of intermediate storage. This is not the case, however, for the irradiation effects, which are determined by both research approaches.

Table II indicates a number of the measured physical and thermal properties.

SEM and EMP analysis shows satisfactory homogeneity for the R7/T7 glass: all the elements, including molybdenum, are uniformly distributed throughout the glass. The mean leach rate for the radioactive blocks is on the order of 10^{-7} g.cm^{-2}d^{-1} at equilibrium, as determined by tests described elsewhere |4|.

Leach tests on inactive samples are conducted according to the following procedure:

. Soxhlet and static modes at 100°C under atmospheric pressure for up to 91 days;

. static mode at 100°C and 100 bars for up to 28 days;

. static mode for up to 28 days at five different temperatures: 30°C, 50°C, 70°C and 90°C under atmospheric pressure, and 120°C at saturated steam pressure (approximately 2 bars).

The leach rate is expressed in terms of overall and element weight loss for Si, B, Al, Na, Mo, Ce, Zn, Cs, Zr, Sr, Li, Fe and U. For example, the overall weight loss after 91 days at atmospheric pressure is on the order of 18 x 10^{-6} g·cm^{-2}d^{-1} in static mode, and 150 x 10^{-6} g·cm^{-2}d^{-1} in Soxhlet mode.

The presence of 0.7 wt % of platinoid inclusions has no effect on the alterability or the mechanical strength of the glass |5|.

The volatility of powdered glass was unmeasurable using a thermogravity balance at 900°C. The mass loss was 0.3% after 8 hours at 1000°C.

Thermal stability is evaluated after heat treatment for 15 hours at temperatures between 590°C and 1180°C in 30°C steps, by analyzing the temperature ranges at which crystalline phases appear, the nature of the phases, the crystallization power and the degree of crystallization.

The presence of the container must also be taken into consideration. The characterization is therefore completed by a number of tests and examinations including full container cooling scenarios (corrosion by the glass, mechanical strains), package drop testing and compression strength measurements, and container metrology.

Table II. R7/T7 Glass Specification Data
(ref. SON 68 18 17 L1 C2 A2 Z1)

Specific gravity	2.754
Viscosity	8 Pa·sec at 1100°C
Transformation temperature	507°C
Mean expansion coefficient between 25 and 300°C	8.31 x 10^{-6}·°C^{-1}
Thermal conductivity at room temperature	1.09 W·m^{-1}·°C^{-1}
Young's modulus at room temperature	8.405 x 10^{10} N·m^{-2}
Fracture toughness	0.95 MPa·m$^{1/2}$

CONCLUSION

None of the key research areas may be considered predominant. All of these sectors are essential parts of a quality assurance program.

The three areas mentioned involve numerous scientific fields, and require constant coordination within a suitable organizational structure. This is the price that must be paid if the processing, conditioning and long-term storage of high level wastes is to become a credible option for routine implementation.

REFERENCES

1. C. Sombret, Nucl Energ J. British Nucl Energ Soc. 24, 2 (1985).

2. J.L. Noguès, E. Vernaz, N. Jacquet-Francillon and S. Pasquini, presented at the Int Seminar on Rad Waste Products, Julich (FRG), June 1985.

3. N. Jacquet-Francillon and E. Vernaz, presented at the MRS Int Symp on the Scientific Basis for Nuclear Waste Management, Boston, November 1981.

4. R. Bonniaud, F. Pacaud and C. Sombret, presented at Conf of Soc Française de Radioprotection, Versailles (France), 1974.

5. E. Vernaz, P. Deymier and N. Jacquet-Francillon, presented at 2nd Int Symp on Ceramics in Nuclear Waste Management, Chicago, April 1983.

GERMAN NUCLEAR HIGH-LEVEL WASTE PROGRAM - KEY RESEARCH AREAS

HELMUT GEIPEL

BMFT (Federal Ministry for Research and Technology),

Heinemannstraße 2, D-5300 Bonn 2, Federal Republic of Germany

ABSTRACT

Based on more than 10 years of research and development, vitrification and high level waste disposal have reached the status of demonstration projects in the Federal Republic of Germany: hot operation of the PAMELA vitrification plant is scheduled for october 1985, and a disposal test with 30 canisters of high active glass is being prepared in the ASSE salt mine. Safety studies for a model repository led to a good understanding of the relevant phenomena; they will continue using site-spezific parameters. Modelling and computer codes will be further developed in international cooperation. In addition to reprocessing, the technology for direct disposal of spent fuel will be developed and demonstrated in the next years.

INTRODUCTION

Since the beginning of commercial nuclear power exploitation in the Federal Republic of Germany, the management of spent fuel and of the arising wastes was of mayor concern and became a growing part in nuclear research and development. Important milestones were the German participation in the Eurochemic Reprocessing plant of the OECD, the construction of the domestic pilot reprocessing plant WAK in 1966-1969 and the commissioning of the test repository ASSE in an former salt mine for disposal of low and intermediate level waste in 1968. The R&D-institution operating the ASSE mine, the institute of deep geologic disposal of the GSF, is in these days looking back on 20 years of successful work.

In the same period of time, the nuclear research centers in Karlsruhe and Jülich together with the Hahn-Meitner-Institute in Berlin elaborated a broad basis for high level waste solidification including borsilicate and phosphor glass and glass ceramics.

The second half of the seventies saw growing political concerns about how and when the so-called "open questions" of the back end of the fuel cycle could be solved. This led to an amendment of the atomic energy act in 1976. In this act the responsibility for constructing and operating a nuclear waste respository was given to the federal government, while reprocessing and conditioning of waste remained within the responsibility ot the waste producers, which means primarily the utilities as the operators of nuclear power plants. For this purpose they founded the reprocessing company DWK.

Three years later, in 1979, another political milestone was set: A comprehensive concept for spent fuel and waste management was agreed upon by nearly all political leaders of the Federal Republic. As I shall come back to this concept later, I only want to

Mat. Res. Soc. Symp. Proc. Vol. 50. © 1985 Materials Research Society

state here that this concept foresees that all industrial spent fuel and waste management plants as well as the geologic repository are to go into operation before the end of this century.

As you surely know, the salt dome near Gorleben in Lower Saxony is actually investigated as a repository for all kinds of waste, especially heat-generating high level waste. The sinking of 2 exploratory shafts has begun last year and underground investitigation is expected to last until about 1992. And this is an important date since then all relevant parameters for repository design and waste packages must be fixed for the licensing procedure which has to follow immediately in order to meet the before-mentioned time schedule. This means spezifically for research on high level waste that final results are to be obtained by that year at the latest.

In order to give a complete view on the waste disposal activities in the Federal Republic of Germany, it is important to note that in addition to the Gorleben repository another repository project is well advanced: the licensing procedure for the former iron ore mine KONRAD started in August 1984, and commissioning is planned for 1989. This repository, however, will not accept high level waste but only low level and medium level waste without heat dissipation, especially decommissioning waste.

Coming back to high level waste research,I like to distinguish three mayor areas: conditioning and management of high level waste from reprocessing, conditioning of spent fuel for direct disposal and high level waste disposal.

HIGH LEVEL WASTE FROM REPROCESSING

In the Federal Republic of Germany, this kind of waste is to be looked at under 2 different aspects. On the one hand, liquid high level waste is actually collected in the WAK reprocessing plant and amounts to about 50 m^3. Larger quantities will be generated by the Wackersdorf reprocessing plant, which is expected to start hot operation in the middle of the 90s.

On the other hand, high level waste resulting from reprocessing abroad may be returned in vitrified form after 1990 and will be stored for about a decade before the Gorleben repository can go into operation. Research on interim storage of glass canisters, however, will not be significant as the dry storage cask technology also for glass blocks is already approaching the end of the licensing procedure.

The development of the vitrification technology in the Federal Republic of Germany had a milestone in 1978 when the reference concept was chosen with borosilicate glass as the most promising glass form, and the ceramic melter without separate calcination. The succeeding technology programm led to the design and construction of the Pamela vitrification plant that has meanwhile been constructed in Mol/Belgium under responsibility of DWK and in cooperation with Eurochemic. The German Federal Government has paid 80 % of the construction cost. The plant will start hot operation in October this year.

HIGH LEVEL WASTE DISPOSAL

High level waste disposal in salt is under investigation since more than 10 years, mainly in the ASSE salt mine. Tests included insertion of electric heaters in boreholes and - since the end of 1983 - the emplacement of a cobalt-60 radiation source. These tests showed among others the salt behavior under influence of heat and radiation.

Based on these results, a new test is being prepared in the ASSE mine: The emplacement of 30 canisters with 150 l of high active glass each. These canisters are already being fabricated at the Pacific Northwest Laboratories in the USA. They consist of borosilicate glass with certain contents of Caesium-137 und Strontium-90, so that after emplacement surface temperatures of 150, 200 and 250 °C and the adequate gamma radiation level can be reached.

The essential aims of this test are:

- the scientific investigation of the interactions between salt and waste packages, especially radiolysis in salt,
- the testing of the new handling equipment under mining conditions and under radiation,
- the demonstration of the whole system with respect to the Gorleben repository licensing procedure.

This test - eventually in combination with some canisters containing spent fuel - will be the last one in this field for the Gorleben licensing procedure. It is planned that with the expected results no further research activities in the later disposal mine will be necessary, although weigth limits in ASSE and the need for retrievability of the canisters make some compromises necessary.

Emplacement is scheduled for 1987, so that a five years' experience can be expected before the Gorleben licensing starts.

Another important aspect for high level waste disposal is the long term safety of the repository. It is evident that rock salt is a good host rock for heat producing waste due to its favourable thermal properties and its plasticity. Nevertheless, as a water inrush into the repository area cannot be excluded for all times, risk analyses must show that in this case the radiation exposure on the surface does not exceed allowable limits.

A comprehensive safety study was made in the frame of PSE (Projekt Sicherheitsstudien Entsorgung), based on a model repository and postulating a water path into the repository immediately after sealing at the end of the operation period. Calculations show that the closure of the backfilled galleries of the high level waste section is faster than the access of water so that it cannot reach the canisters and the high level waste packages retain their integrity.

This favorable result must, however, be checked using the site-specific design and geologic parameters when the are available. So, work will continue, including international cooperation under auspices of the European Communitiy and the Nuclear Energy Agency of the Organisation for Economic Cooperation and Development, OECD.

Pamela is a pilot plant with a capacity of 30 l/h of high active waste concentrate. For comparison, the planned reprocessing plant will generate about 50 l/h, so that the Pamela technology may easily be adapted. Cold testing of the bigger melter will take place at the Karlsruhe Nuclear Research Center.

In addition to this demonstration project of German vitrification technology there is some research of glass behavior and on canister materials. Details on these topics will be given in specific presentations of this conference. Moreover, further work will concentrate on the optimization of glass constituents with respect to specific high level waste compositions.

DIRECT DISPOSAL OF SPENT FUEL

Spent fuel from light water reactors was not designated as waste until the beginning of this year. And indeed, the atomic energy act prescribes generally the recovery and recycling of valuable material which means reprocessing of the fuel.

However, based on a four-years investigation, the Federal Government decided on Jan. 23 this year, that direct disposal technology should be further developed not only for specific fuel elements like high temperature reactor fuel but also for light water reactor fuel. The idea behind was that this technology may in the 90s become a complementary route in addition to the reprocessing plant under construction.

The first investigations from 1981 to 1984 included the detailed development of reference concepts for the disposal canister, the conditioning plant and a repository for spent fuel. These reference concepts formed the basis for safety analyses and assessments of e.g. feasability, economics and safeguards. Experimental work accompanied this program to a great extent.

The continuing programm is in the last phase of concurrence and will effectively start by the end of this year. For this program, the responsibility is strictly devided beween the Federal Government and Industry: The Federal Government will only promote work for disposal technology including a test program. Industry, primarily DWK will be responsible for canister and conditioning technology development.

Actual planning of DWK aims at the immediate start of a licensing procedure for a small pilot conditioning plant. This plant is designed to handle different fuel types and indifferent types of disposal canisters. Research of this field will concentrate on safety and economic optimization.

The disposal technology, too, needs some more development beyond the existing reference technology, and it concentrates on the demonstration of all relevant steps before the end of this decade. Although conditioning of spent fuel for disposal in salt has never taken place, it is expected that technology development and licensing will not imply unsuramountable obstacles because ist is based for the mayor part on existing and well-known technology.

There is to be demonstrated - in addition to the packaging itself - the safe handling of spent fuel canisters under disposal conditions, and the near-field-effects in the salt dome. For these nearfields effects, however, one can take credit of the knowledge already gained or expected with experiments for heat-producing waste.

SWISS HLW RESEARCH PROGRAMME: KEY RESEARCH AREAS

C. McCOMBIE
National Cooperative for the Storage of Radioactive Waste (NAGRA), Baden, Switzerland

INTRODUCTION

The increased emphasis put upon disposal of HLW in recent years led to the growth of large numbers of research programmes in many countries. Many results of this research have flowed into projects developing and analysing geologic repository concepts; the basic feasibility of implementing safe disposal facilities is now generally accepted by the technical community (though not always by the public!)

For identifying which areas could benefit from further research an extremely valuable procedure is to undertake a quantitative analysis of an entire repository system. This forces a synthesis of work from all of the diverse disciplines involved in repository development and allows better allocation of priorities to technical and scientific areas requiring further research. In Switzerland, legal requirements for the preparation and formal submission of such a synthesis task (Project Gewähr /1/) have led to intensive efforts being invested in field, laboratory and desk studios in recent years and has produced specific input for planning of future research work.

Before attempting to summarise research areas judged to be worthwhile, it is useful to recall that the aims of the research must be **project-oriented**, i.e. they must be tailored to the needs of a programme for development of geologic repositories which are **acceptable** with regard to safety and economics. Valid general aims are:

- Improving understanding of processes occuring in the disposal system in order to predict more accurately their long-term influence on system safety
- Improving system understanding, in order to increase confidence that no significant processes or interactions have been neglected
- Development of tools (experimental or analytical) for site selection and characterisation
- Refinement of methods and of data so that optimisation becomes more feasible.

If we take a project-oriented view of activities related to reposiory development (as opposed to focussing upon the different technical disciplines), we can identify three broad spheres of activity where research needs can arise. These closely coupled areas are site characterisation, system engineering, and system performance assessment. In the following sections it is apparent that the necessary breadth and depth of research in these areas is very different; advances are most required in the last since unusual scales of distance, complexity and time are involved in predicting repository performance.

RESEARCH NEEDS FOR SITE CHARACTERISATION

The host rock of a geological repository and its surrounding environment are important parts of the multi-barrier safety system proposed for radioactive waste disposal. It is important to adequately characterise the site at various levels of resolution and to quantify as far as possible its expected future evolution. This leads to development requirements in test methodology and in geologic interpretation. Because site investigations can, in general, be one of the most costly parts of repository development, it is also important to develop strategies for optimising programme design. These points are discussed briefly below with reference to the Swiss situation.

Test Methodology and Interpretation

Of high current importance for crystalline rock disposal are the methods for quantifying site hydrology and hydrochemistry since the release path via groundwater transport is of overriding importance. A good overview of the scope of the relevant methods which have been used in deep boreholes in Switzerland is given in reference [2]. In the deep borehole programme, much effort has been expended on hydraulic measurements and their interpretation in relatively extreme environments (e.g. temperatures, up to 115 °C) [3]. Further development here, in particular with respect to interpretation for fractured media and extrapolation to larger rock volumes is required. Improvements aimed at in hydrochemistry work include more reliable and representative downhole measurements (especially Eh and pH) as well as further development of some isotope measurement techniques which can contribute to understanding groundwater evolution (e.g. Kr-81). Advances in measurement technique such as those mentioned are needed to provide necessary raw data on groundwater properties. In some cases, there is a greater need for better understanding and interpretation of measured values in order to produce a more applicable database for the improved conceptual modelling of hydrogeochemical systems (which is discussed later).

With regard to techniques for geologic and mineralogic characterisation of crystalline host rock no pressing needs are recognised. The most important developments here are rather the collection, interpretation and synthesis of data on flow systems at different scales (micro fractures, weak or weathered zones, large disturbances). Not only the undisturbed host rock, but also the disturbed zones around excavations, must be characterised in this way.

It is possible that geophysical methods may contribute increasingly to the solution of these problems. At the small scale, advances continue to be made in borehole logging methods; at a larger scale tomographic techniques can contribute to locating inhomogeneities in geologic media; at the largest scale developments in surface seismic methods are needed to unambiguously map deep lying crystalline formations covered by extensive sediments.

Quantification of Future Scenarios

Fundamental to waste disposal planning is the issue of predicting future states of the repository system. Although the geologist in principle has one of the longest databases from which to extrapolate, he is scarcely accustomed to predictive geology and certainly not to quantitative, predictive geology. In recent Swiss project work, the concerted efforts made to specify possible future geologic developments [4] met with some success but advances are still possible. The areas to be developed are the collection of the appropriate data and its subsequent interpretation and extrapolation.

New data sources can be tapped by exploiting and encouraging advances in measurement methodology. For example, sensitive seismometer networks and accurate laser-based distance measurements can allow minute tectonic movements to be detected and enable a stable plinth of bedrock in which a repository can be placed to be identified.

An ultimate goal of improved tectonic understanding is the quantification of the probability of future geological events and processes. This information is needed to develop the consequence analyses most often performed today into full risk analyses as employed in other technical fields. It is doubtful, however, if sufficient, well founded data will be developed in the near future to allow a rigorous approach to be followed through here.

Development of Exploration Strategy

In fields where the targets and costs can be better quantified (e.g. resource prospection), exploration strategies can be optimised to decide borehole numbers, distribution etc. In waste disposal the complex process lying between geologic characterisation and repository performance prediction makes this problem very much more difficult. The prospect of being able to more objectively determine the intensity of data collection needed at different scales is, however, attractive enough to justify further studies in this area. The possible rewards to be reaped here are not only economic, in that fewer exploration holes may be needed at a given site, but also safety-related if an important aim is maximum characterisation with minimum disturbance of natural conditions.

DEVELOPMENTS IN ENGINEERING OF SYSTEM COMPONENTS AND REPOSITORY

Here no real **research needs** are perceived since projects today are based on existing, proven technology. On the other hand, there is considerable scope for **developments** which could contribute to improved cost/safety optimisation of the disposal system. Work on alternative HLW matrices (which is not being pursued in Switzerland) can be best justified in that improvements may lead to greater freedom in design of other barriers or of disposal concept (e.g. deep boreholes).

The future Nagra engineering programme will include container optimisation, development of waste package emplacement methods less dependent than the present on large amounts of compacted bentonite, refinement of repository design (better adaption to the local structure of the host rock), and study of excavation sealing methodology.

Over the whole field of engineering and technical development, increased attention is being paid to quality assurance procedures. This, however, involves research-type work in relatively few areas. Most lacking at present are system studies properly quantifying the detailed system quality requirements and some specific measurement techniques (e.g. for radionuclide assay in heavy matrices) which are not yet obtainable "off-the-shelf".

RESEARCH AREAS AIMED AT IMPROVING SYSTEM PERFORMANCE ASSESSMENT

As mentioned in the introduction, the most important research needs in the field of waste disposal can be identified through attempting to understand, quantify, and predict the behaviour of the total repository system. Requirements can be for methods to measure specific data, for mathematical models to simulate complex, interacting processes or for suitable approaches to validation of these models. In the following, we continue to follow our project-oriented discussion of research needs in that we first discuss in turn the subsystems normally analysed in quantitative performance assessments, before turning to issues involving overall system analyses.

The Near Field

The repository near field (including waste, overpack, backfill, and the immediately adjacent host rock) can provide a highly efficient radionuclide containment system for very long times. A convincing demonstration of this fact depends, however, in the Swiss case, on adequate understanding of the behaviours of borosilicate glass, steel overpack and bentonite backfill in deep crystalline groundwaters.

Our collaborative experimental research on glass leaching in low-flow conditions with inclusion of other relevant components will continue with increased emphasis on producing data for proper validation of models developed. Laboratory work on the iron container will concentrate on extentions to long times and more representative conditions, along with better assessment of the chemical form of corrosion products. Planned experimental work on bentonite includes further measurement of water uptake, diffusivities, gas permeability and rheological properties. A wide range of analytical measurements on aqueous solutions is needed as input for models of groundwater evolution in the near field. In particular, nuclide solubilities and speciation must be analysed in order to calibrate the chemical thermodynamic models increasingly being employed.

This type of modelling, which is of paramount importance in near-field performance assessment, requires not only data for validation of output predictions, but also an improved thermodynamics database as code input. Even more fundamental is the task of validating whether the basic concepts derived from equilibrium thermodynamics are sufficient for application in geochemical systems without explicit treatment of kinetic effects.

The rate and manner of water uptake in compacted bentonite may also be of importance in determining system behaviour. Models reliably accounting for all processes involved in describing resaturation in hygroscopic media like bentonite under a varying temperature field are not yet available. Finally a proper near-field transport model fully describing nuclide migration through fully wetted bentonite towards the disturbed host rock of the near field is yet to be fully developed.

<u>The Far Field</u>

In hard rock disposal concepts a major research area involves the prediction of deep groundwater movements [5]. The applicability of the commonly applied porous medium flow models, e.g. [6] has not been rigorously justified. Efforts are underway to identify methods which can adequately treat fractured media without requiring explicit fracture data more detailed than can be feasibly gathererd in the field. Most promising are models combining statistical representations of network flow paths with explicit treatment of major hydraulic weak zones which should be identifiable during thorough site characterisation.

The results from modelling of radionuclide transport within the flowing groundwater depend sensitively upon several concepts and geosphere parameters which must be further investigated. One example is hydrodynamic dispersion; at present, the characteristic length associated with its diffusion-type treatment cannot be properly chosen by any objective means.

Much more important in its influence upon safety analysis is the treatment of retardation processes. The common Kd and retention factor approach is a first approximation to quantifying a range of detailed processes (physical and chemical). With better understanding of the individual components, it may be possible to adopt more realistic and less grossly over-conservative approaches than currently used [7].

Even greater than the uncertainties due to chemical rock/water interactions are those determined by the fraction of the host rock available for sorption. If diffusion processes allow the nuclides to access all of the host rock within times comparable to the migration time, even a short geosphere path will provide an extremely effective barrier for almost all radionuclides [8]. Accordingly, it is important that more laboratory and field work be devoted to validation of matrix diffusion processes in the various types of crystalline rocks which can be encountered along the groundwater flow paths.

The final links in the modelling chain are biosphere transport and dose calculations. In the former area, the inherent uncertainties in predicting human behaviour over long times imply that only relatively crude modelling is justifiable. In spite of this, some parameters (e.g. specific transfer factors) are so poorly defined that research effort should be devoted to improvements. The important problem of understanding radiation effects especially at low doses will continue to be attacked by radiation biologists. For those planning and analysing disposal projects, however, the main task is to stay abreast of such developments and to be familiar with state-of-the-art radiation protection practice at all times.

Overall Disposal System

In the above we have attempted to pick out research objectives specific to the near- and far-field systems. A global analysis is, of course, ultimately required and some issues involved therein are also being actively further investigated. These fall into 2 categories: particular phenomena which could potentially have an influence throughout the system, and global questions affecting the total assessment methodology.

Specific potential problems not yet raised include the effects of colloids or complexes, possibly in conjunction with microbiological activity. The concern is that radionuclides in such forms may short-circuit retention barriers throughout the system, although little direct evidence of important effects is available. Pessimistic assumptions, necessitated by the lack of sufficient data, led in the Swiss case to consideration of the rôle of the bentonite buffer as a filter for colloids potentially formed at the waste. More work is, however, needed on the formation, stability and transport properties of potential radioactive colloidal matter in groundwater.

Other mechanisms which can affect all parts of the disposal system are included in the range of coupled processes [9]. Attention is being increasingly focussed upon thermal, mechanical, hydraulic, and chemical interactions. Experimental projects, especially in underground test facilities, are needed to provide data; modelling developments are also needed for some couplings, in particular those involving chemical interactions.

Global issues concerning our overall approach to assessing system performance include improvement of methodology and validation of predictions. In Switzerland, a largely deterministic approach has been used to date. This procedure was judged appropriate for the feasibility project because it reflected at the time both the state-of-the-art internationally and the official Swiss safety guidelines. But it is recognised that probabilistic methods are needed to quantify the uncertainties associated with calculated consequences of any release scenario and - more problematic - to combine scenarios of different probability.

The validation of our model concepts and their applicability over large distances and times is an important issue through all parts of the assessment. Complete rigorous validation will not be possible everywhere, but significant advances can be made by extended timescale experiments and by observation and analysis of natural systems. Although nature provides no exact analogue of an entire disposal system, many of the critical processes affecting repository performance have been occurring over long periods in natural systems so that important evidence is available [10]. Although adequate knowledge of initial and boundary conditions is often a stumbling block, much can be learnt. Processes which have already been studied to some extent, or which can be further investigated through analogue studies, include corrosion of various materials, chemical stability of bentonite, mobilisation of natural radionuclides in groundwaters and rocks, presence and stability of colloidal matter in groundwater etc.

FUTURE SWISS PROJECT WORK ON DEEP DISPOSAL OF HLW

In the preceding sections a long catalogue of research areas has been identified as valuable for progressing projects for developing and assessing deep geologic repository concepts. It must not, however, be concluded that no global assessments can be made before all of the listed points have been investigated. By substituting conservative assumptions and pessimistic empirical extrapolations for detailed understanding of certain processes, we can evaluate limiting consequences of repositories. This is valuable for judging basic feasibility of concepts. However, the results can be over-conservative and unsuited for any system optimization; accordingly research leading to improved understanding and better modelling is justified and the selection of the most appropriate projects is a challenging task.

The technical issues raised in the previous sections have been identified as possibly important in the course of Swiss work on Project Gewähr but the majority are applicable to all HLW programmes and many are being worked upon by various groups already. Most of the items mentioned will be tackled in future Swiss programmes - but with marked differences in weighting. The effort invested is determined not only by the importance attached to the particular area, but also by considering the work underway throughout the international community. It is a positive feature of waste disposal research today that information exchange and direct cooperation are widespread. This allows individual countries, and especially smaller countries, to try and concentrate their efforts into selected research activities and to round out their projects by complementary work from other sources.

For Nagra, the important milestones in HLW repository development following the submission of Project Gewähr are determined by the site-related activities. Table I gives an overview of the planning timescales. A continuous programme of laboratory work and studies will accompany the site investigations. In addition, work will proceed over the next years at our Grimsel underground research laboratory [11]. This has two objectives. One is to gather experience which must be available if we proceed to Phase III of the HLW programme which includes extensive underground testing at a chosen site; the second more immediate application is a test bed for methodology to be applied in the site investigations which should start soon at potential sites for low- and intermediate-leve waste repositories.

Table I. Milestones in Development of HLW Project

1978	Formulation of basic disposal concepts
1985	Submission of Project Gewähr 1985
1987/1988	Synthesis of regional studies (Phase I) in North Switzerland
1988-1995	Detailed site characterisation of one or more sites (Phase II)
1995-2010	Shaft sinking at one site, experiments in deep rock laboratory (Phase III)
2020	Repository operational

This latter programme is increasing in priority in Swiss disposal planning and, when Project Gewähr is brought to a conclusion, the intensity of effort on HLW repository planning will have to be reduced. This should be possible without technical quality suffering because of the relatively extended timescales given in Table I. In fact, the long-term nature of several of the proposed research areas of importance for HLW repository assessment would make it difficult to organise an accelerated programme.

REFERENCES

1. Nagra: Nagra Project Gewähr Reports, 8 Volumes NGB 85-01 to NGB 85-08, Baden, Switzerland, 1985

2. Nagra: "Sondierbohrung Böttstein - Untersuchungsbericht", Nagra Technical Report Series NTB 85-01, Baden, Switzerland, 1985

3. Grisak G.E., Pickens J.F., Belanger D.W., Avis J.D.: "Hydrogeologic Testing of Crystalline Rocks During the Nagra Deep Drilling Program", Nagra Technical Report Series NTB 85-08, Baden, Switzerland, 1985

4. Diebold P., Müller W.H.: "Szenarien der geologischen Langzeitsicherheit: Risikoanalyse für ein Endlager für hochaktive Abfälle in der Nordschweiz", Nagra Technical Report Series NTB 84-26, Baden, Switzerland, 1984

5. Herbert A.W., Hodgkinson D.P., Jackson C.P., Lever D.A., Rae J, Robinson P.C., Sharland S.M, Tasker P.W.: "Some current problems in groundwater flow and radionuclide transport modelling", AERE TP 1131, Harwell, UK, 1985

6. Kiraly L.: "FEM 301 - A Three Dimensional Model for Groundwater Flow Simulation", Nagra Technical Report Series NTB 84-49, Baden, Switzerland, 1984

7. McKinley I.G., Hadermann J.: "Radionuclide Sorption Database for Swiss Safety Assessment", Nagra Technical Report Series NTB 84-40, Baden, Switzerland, 1984

8. Hadermann J., Rösel F.: "Radionuclide chain transport in inhomogeneous crystalline rocks: Limited matrix diffusion and effective surface sorption", Nagra Technical Report Series NTB 85-40, Baden, Switzerland, 1985

9. LBL: "Coupled thermo-mechanical-hydrochemical processes associated with a nuclear waste repository", Panel Report LBL-18250, Berkeley, USA, 1984

10. Chapman N.A., McKinley I.G., Smellie J.A.T.: "The potential of natural analogues in assessing systems for deep disposal of high-level radioactive waste", Nagra Technical Report Series NTB 84-41, Baden, Switzerland, 1984

11. Nagra: "Felslabor Grimsel - Rahmenprogramm und Statusbericht", Nagra Technical Report Series NTB 85-34, Baden, Switzerland, 1985

A REVIEW OF PROGRESS IN THE CANADIAN NUCLEAR FUEL WASTE MANAGEMENT PROGRAM

R.B. LYON and L.H. JOHNSON
Atomic Energy of Canada Limited, Whiteshell Nuclear Research Establishment, Pinawa, Manitoba ROE 1LO, Canada

ABSTRACT

The Canadian Nuclear Fuel Waste Management Program is reviewed, illustrating the progress that has been made in assessing the concept of disposal of nuclear fuel waste in plutonic rock of the Canadian Shield. Research is being conducted into used fuel storage and transportation, fuel waste immobilization, site characterization and selection methods, and performance assessment modelling. Details of achievements in these areas are outlined, and results of the most recent interim assessment are discussed.

INTRODUCTION

In 1981, the Canadian government approved a ten-year program for research and technology development pertaining to the immobilization and disposal of nuclear fuel wastes in a deep vault in plutonic rock in the Canadian Shield. The basic components of this program include development of

(1) technology for the storage and transportation of used fuel,

(2) immobilization techniques for used fuel and fuel recycle wastes prior to disposal, and methods for sealing a disposal vault,

(3) techniques necessary to characterize and select a suitable disposal site, and

(4) a methodology and supporting research base for assessment of the environmental impact of the operation of immobilization and disposal facilities and of the long-term impact of the waste vault contents on man and the environment.

The last three components, as well as overall program coordination, are the responsibility of Atomic Energy of Canada Limited, while Ontario Hydro, the principal nuclear utility in Canada, oversees the used fuel storage and transportation aspects of the program.

Other participants in the program include the Department of Energy, Mines and Resources, Environment Canada, and a number of Canadian universities and private consultants. Ongoing scientific review of the program is provided by a Technical Advisory Committee. This group, consisting of members nominated by the various professional societies throughout Canada, publishes annual reports reviewing the progress of the research and development activities [1].

The ultimate focal point of the program is concept assessment, which involves the completion of an environmental assessment of the disposal concept, without reference to a specific site for disposal, and submission of the assessment documentation to the Canadian government for review in 1988. Evaluation of the proposed concept will consist of government and regulatory review, followed by a public hearing.

At the halfway point in the program, all essential major research facilities are in place, and the programs are now focusing on providing the necessary research base for concept assessment. In this paper, the full range of research activities providing supporting data for the assessment is reviewed, with emphasis on key recent developments. The methodology used in the performance assessment is then described along with some results from the most recent interim assessment.

STORAGE AND TRANSPORTATION OF USED FUEL

Used fuel storage and transportation studies are aimed at optimizing fuel storage and developing a commercial used fuel transportation cask. Used fuel is currently stored in water-filled bays at the generating stations, and experience indicates that this approach should be suitable for at least 50 years [2].

A decision on building additional storage bays or implementing dry storage will be made in 1987. Various designs for dry storage are being examined, including dry vault, concrete canister and cast-iron cask. In addition, a concrete integrated cask is being designed that is suitable for storage, transportation and disposal [3]. Half-scale models of concrete integrated casks have been constructed, and impact and fire tests are in progress. The dry vault and cast-iron cask concepts have been tested and licensed in other countries and appear to be cost-effective options for storage. The concrete canister approach has been studied in Canada for several years, and fuel stored in such canisters for over three years at 150°C has shown no significant signs of deterioration [3]. Concrete storage of used fuel at the Gentilly-1 reactor in Quebec has been approved, and the canisters are under construction.

The objective of the transportation studies is to develop a cask for large-scale transportation of used fuel. Design, construction and licensing of the cask are expected to be complete by 1988. Detailed design and testing of the reference cask, which would carry 192 fuel bundles (4560 kg), are well underway [3].

FUEL WASTE IMMOBILIZATION

Research programs in fuel waste immobilization are concerned with the development and characterization of all man-made components of the nuclear fuel waste disposal system. Included are the characterization of waste forms (both used fuel and immobilized fuel recycle wastes), development and testing of waste containers, and evaluation of barrier materials for backfilling and sealing the disposal vault.

Container Development and Materials Evaluation

A number of different designs and materials are being considered for containment of the waste form. Most of the research over the past few years has concentrated on development of a container for used fuel. Options studied include stressed-shell containers, designed to withstand the hydrostatic pressures in the disposal vault, and several supported-shell designs. For the latter approach, cast metal (e.g., lead), packed particulate, or structural supports (e.g., steel tubes surrounding the fuel bundles) provide the necessary internal support [4]. More recently, design of a container for waste glass has been undertaken as part of a comprehensive engineering study for concept assessment. Several candidate container metals are being examined to determine their weldability and corrosion performance under disposal conditions.

The performance of some of the container designs under temperature and pressure conditions expected in the disposal vault has been evaluated in the Hydrostatic Test Facility [5]. Half-scale lead-matrixed, and full-scale particulate-packed and structurally supported, containers have been tested. All three performed well and only minor deformation occurred. When deliberate voids were introduced, local deformation resulted, but the integrity of the containers was not compromised [5].

Research into methods for remote closure welding of containers has progressed well. In particular, tests of resistance-diffusion bonding of titanium and electron-beam welding of copper suggest that these are promising techniques [6].

The corrosion behaviour of ASTM grades 2 and 12 titanium, copper, Hastelloy C-276, and Inconel 625 is being studied over a range of salinity, temperature and irradiation conditions. Capabilities for performing corrosion studies in γ-radiation fields have been enhanced significantly by completion of the Immobilized Fuel Test Facility (IFTF), which permits large numbers of long-term corrosion experiments to be performed in γ-radiation fields comparable to those occurring at the surface of a disposal container [7]. Studies indicate that, although grade-2 titanium is susceptible to crevice corrosion at temperatures higher than $\sim 100^{\circ}C$, this may be inhibited by γ-radiation [8]. Localized corrosion of nickel-based Hastelloy C-276 and Inconel 625 has been studied as a function of temperature and chloride concentration. The results suggest that, below $100^{\circ}C$, localized corrosion should not occur even in 20 wt.% NaCl solution [9].

Some work on the corrosion behaviour of materials that could be used for very long-term containment is also being performed. Based on corrosion tests, alumina, zirconia, titania, graphite, cermet, concrete and porcelain have been selected for further study [10].

Used Fuel Characterization

The properties of used fuel relevant to its performance as a waste form are being examined in detail, in particular the leaching and dissolution behaviour. The amounts of Cs-137 and I-129 in the fuel-sheath gap and at the UO_2 grain boundaries have been determined as a function of fuel irradiation history. Leaching studies indicate that short-term releases in dissolution tests (termed "instant release") result from complete removal of the gap inventory and partial release of the grain boundary fraction [11]. Used fuel dissolution experiments at $150^{\circ}C$ have shown that fission product and uranium releases to solution are strongly redox dependent, with lower releases occurring under reducing conditions [12]. The products of α-radiolysis of water have been shown to cause some oxidation of UO_2 [13,14], although it is not yet clear if this will significantly influence the performance of used fuel as a waste form.

Waste Immobilization

Waste immobilization research is focused on developing processes and products for immobilizing the wastes arising from recycling of used CANDU fuel. Process development studies are examining evaporation, calcination, vitrification, ion-exchange and off-gas treatment methods. Waste forms suitable for disposal are being developed for high-level wastes, I-129, C-14 and Zircaloy fuel bundle components.

Horizontal and vertical wiped-film evaporators have been used successfully to concentrate liquid wastes. The Waste Immobilization Process Experiment, consisting of a rotospray calciner, an electromelter and an off-gas system, and capable of producing 10 kg/h of non-radioactive sodium borosilicate glass, has completed several months of successful operation [15].

Methods have been developed for removal of I-129, C-14 and Kr-85 gases from process streams. Both corona-discharge and photochemical scrubbing techniques have been applied to remove iodine from air [16]. Molecular sieve adsorbents have proven satisfactory for the removal of Kr-85 from air [17]. Carbon dioxide (containing C-14) has been separated from air using calcium hydroxide. A number of potential hosts for I-129 have been examined, and it appears that the phase assemblage Bi_2O_3 + Bi_5O_7I may be a satisfactory waste form for I-129 [18]. Lead and bismuth hydroxy-carbonates may be suitable hosts for C-14.

Glass, glass-ceramic and ceramic waste forms are being examined in terms of their fabrication parameters and leaching performance. Temperature-viscosity relationships in the sodium borosilicate system have been studied [19], and miscibility limits have been determined for various multicomponent borosilicate glasses. The durability of aluminosilicate glasses as a function of composition has been examined, and the results indicate that these glasses have low leach rates over a wide range of composition [20].

Sphene-based glass-ceramics have been prepared by melting at 1250°C, followed by controlled crystallization, and by the use of sintering techniques at 950°C. The latter approach appears promising and may permit production of this waste form at temperatures that minimize radionuclide volatilization and eliminate refractory corrosion. Leaching studies show that sphene-based glass-ceramics are highly durable in saline groundwaters. No significant enhancement in leaching due to γ-irradiation (4 Gy/h) was observed in a 485-day test at 100°C. Leach tests of metamict sphene, produced by heavy-ion bombardment, simulating α-decay, indicate that the durability is lower than that of undamaged sphene by a factor of 5. Sphene-based glass-ceramics, however, show no such leach rate enhancement for similar irradiation conditions [21].

Vault Chemistry

The vault chemistry program is examining chemical processes that could occur in the near-field of a disposal vault. Such processes include the sorption of radionuclides by buffer materials, the hydrothermal alteration of bentonite, and the interactions of waste forms with container materials, buffer materials, groundwaters and host rock under disposal temperature and pressure conditions.

Because early assessment results suggested that I-129 could be one of the principal contributors to radiation dose from the disposal of used fuel, research on sorption of radionuclides in the near-field has focused on developing buffer additives for removal of iodide from solution. Although several metal oxides and sulfides show an affinity for iodide, their ability to sorb iodide is greatly reduced in the presence of bentonite. This effect is due to the destabilization of metal oxy-iodides by carbonate released from the clay [22].

The diffusion of strontium, plutonium, and americium in bentonite/sand buffer materials has been studied over a range of buffer density and temperature conditions. The results show that reasonable agreement is obtained between measured diffusion profiles in clays and those calculated using a simple diffusion model [23].

Interactive tests involving waste forms (both used fuel and fuel recycle waste glasses), container materials, buffer materials, groundwaters and host rock are being performed in a series of experiments in the Immobilized Fuel Test Facility [7]. The first test, which contained 18 separate experiments, was recently terminated after 6 months of operation. These experiments are designed to aid in the development of source-term models for SYVAC, (Systems Variability Analysis Code - a code for predicting radionuclide doses to man from nuclear waste disposal) and to provide information on possible synergistic interactions among near-field components that may influence radionuclide release from waste forms.

The reactions of bentonite with various synthetic groundwaters have been studied over a broad range of temperature and pH conditions [24]. Recent studies have examined the interaction of bentonite and groundwaters with various cements that might be used in a disposal system, and results suggest that the presence of bentonite accelerates the rate of dissolution of the cements [25].

Disposal Vault Sealing

Research is being carried out to develop methods and materials suitable for backfilling and sealing a disposal vault. This includes development of a clay-based buffer material to surround the waste container, backfill material to fill rooms, drifts and shafts, and borehole and shaft seals designed to ensure that, after vault closure, the pattern of water flow through the rock is controlled by the natural rock features.

Various buffer and backfill materials have been evaluated, including mixtures of crushed granite or quartzite with sodium bentonite, calcium bentonite, crushed illitic shale or glacial lake clay. The mineralogical and chemical composition of the clays has been determined. Extensive studies of the physical properties of buffer materials have been performed, including measurements of swelling potential, hydraulic conductivity, compaction, shrinkage and long-term creep. The swelling pressure exerted by a confined buffer material depends on the effective clay density (mass of clay/(volume of clay + the volume of voids)). Based on an evaluation of the physical and engineering properties, a reference buffer material has been selected, composed of bentonite and sand in a 1:1 dry weight ratio. This material has an effective clay density of 1.3 Mg/m^3 and exhibits a swelling pressure of greater than 800 kPa in saline groundwater [26]. Similar testing has resulted in a 75/25 crushed granite/clay mixture being selected as a reference backfill material.

A number of conceptual designs for shaft seals have been prepared [27], and the performance of a variety of grouting materials has been evaluated [28].

GEOSCIENCE RESEARCH

The objectives of the geoscience program are to develop techniques suitable for site screening and evaluation, and to contribute to the general assessment of the concept of disposal in plutonic rock of the Canadian Shield. The field research program is evaluating relevant physical, chemical and mechanical properties of the rock, and chemical and hydrogeological properties of the rock-groundwater system. A variety of laboratory geochemistry programs are examining radionuclide-rock-groundwater interactions and the thermomechanical stresses that would occur in rocks near a disposal vault.

Field Research

Deep exploratory drilling and surface mapping are being carried out to obtain the data necessary to develop site selection criteria. Various new and existing techniques are being evaluated at five field research areas. The Chalk River, Atikokan and Whiteshell research areas are in granitic rock, whereas those at East Bull Lake and Overflow Bay are in gabbroic rock.

Methods of correlating subsurface fracture characteristics to surface fractures are required in order to select a suitable disposal site. Airborne and surface geophysical monitoring techniques and air photo lineament analysis are being used to predict subsurface fracture conditions, and drilling is being carried out to test the predictions [29].

The mineralogy, geochemistry and relative chronology of fracture fillings in plutons are being studied. Uranium disequilibrium and O-18 analyses of calcites, clays and goethites found in open fractures in granite indicate that they have reacted with waters during the last 250 000 years, suggesting that these minerals are the products of recent rock-water interactions [30].

Brackish and saline waters have been found at depths of greater than 300 m at all research areas, with salinities ranging from 3 g/L total dissolved solids in gabbro to 40 g/L in granite. If correction is made for dilution due to surface waters, concentrations of over 140 g/L are estimated for a depth of 1 km in granite [31].

Underground Research Laboratory

A hydrogeological monitoring network has been established at the Underground Research Laboratory (URL), located north of the WNRE site, to determine the effects of the URL construction on the groundwater system. Groundwater inflow rates, water chemistry variations and piezometric levels were monitored continuously as the URL shaft excavation proceeded. The shaft has been completed to the target depth of 240 m. Models were developed, using data from the drilling program done before shaft sinking, to predict the response of the groundwater system to the excavation of the URL. Comparison with the data obtained during excavation shows good agreement on water inflow rates and excellent agreement on the extent and degree of water-level drawdown [32].

An extensive program of experiments is planned for the URL. This includes determination of the rock response to thermal and mechanical loading, testing of buffer and backfill emplacement techniques, and borehole, drift and shaft sealing experiments.

The URL is the first subsurface facility to be excavated below the water table in undisturbed plutonic rock. Although the URL was designed to meet the needs of the Canadian Nuclear Fuel Waste Management Program, the facilities will be made available to other nations, groups or organizations. This involvement may be on the basis of shared costs for experiments of mutual interest, or on a commercial basis.

Vault Engineering Studies

A variety of alternative designs for a disposal vault have been examined, involving single-level, multi-level and long-hole emplacement concepts [3]. In addition, a comprehensive engineering study has been completed that examined acquisition, transportation, emplacement methods, and costs of buffer and backfill materials for a disposal vault [33].

For a used fuel disposal vault, the results show that construction costs would be significantly higher for a multi-level vault than for a single-level vault, due to the need for lower loading density in the former to meet thermal constraints. However, a multi-level vault for immobilized fuel recycle waste is feasible, since the cost per container would be only 11% higher than for a single-level vault. The option of constructing a multi-level vault may be valuable, if rock quality or pluton size limits the area available for a vault. Potential problems with short-term container retrieval and buffer emplacement quality control would appear to rule out a long-hole disposal method.

Geochemistry

The processes that will influence migration of radionuclides in fractured plutonic rock are being studied to provide the basis for detailed radionuclide transport models. Short-term laboratory and field experiments provide information on the parameters affecting radionuclide sorption and transport. Field geochemistry studies have examined the end results of natural geological processes, to obtain data on rock alteration and radionuclide migration.

Significant progress has been made in the development of suitable chemical equilibrium codes and compilation of associated thermodynamic data for calculating groundwater composition in equilibrium with mineral assemblages. Work on an internally consistent, thermodynamic data base for minerals in the system Na_2O-K_2O-CaO-MgO-FeO-Fe_2O_3-Al_2O_3-SiO_2-TiO_2-H_2O-CO_2 is almost complete [34]. A thermodynamic data base for neptunium has been completed [35], and heat capacity data for a number of actinides, fission products and groundwater species have been compiled [36].

The Tc(IV)/Tc(III) redox couple has been studied in aqueous bicarbonate solutions [37]. It appears that Tc(IV) carbonate complexes will be the most important species in reducing groundwaters. Sorption studies, however, indicate that these complexes are strongly sorbed by iron oxides [38].

Chemical stripping techniques have been used to determine the minerals responsible for radionuclide sorption. Iron oxide and oxyhydroxide surface coatings on minerals appear to play a major role in radionuclide sorption [38].

A technique has been developed to measure pore space in rocks, and this has been applied to highly altered rock adjacent to a water-bearing fracture. The results indicate an increase in porosity of the altered rock by a factor of two [39].

Matrix diffusion of non-reactive species in granite has been studied, and this retardation mechanism is being evaluated for possible inclusion in the geosphere submodel used in SYVAC.

Natural analog studies provide information that can be used in predicting the behaviour of radionuclides in rock surrounding the disposal vault. Uranium deposits in Northern Saskatchewan are being examined as analogs of a nuclear fuel waste disposal vault. Although water-rock interactions have occurred in the ore bodies a number of times since the deposits formed almost a billion years ago, reducing conditions limited uranium migration in the clay surrounding the deposits to less than 5 metres [40].

ENVIRONMENTAL RESEARCH

The environmental research program is studying pathways by which radionuclides may migrate in the biosphere, and the ultimate radiation dose to man due to these various exposure routes. Potential exposure routes being examined include ingestion of well water, terrestrial plants and animals, surface water and aquatic plants and animals, inhalation of air, and external exposure from soil, air and water.

The locations of groundwater discharge zones relative to lakes and rivers, and the discharge rates, will determine the environmental concentrations of the radionuclides and the predominant exposure pathway. Airborne thermal imagery appears to be a promising method for locating such discharge zones [41]. Identification of submerged discharge zones is being examined using a lakebed drag with temperature and conductance probes.

Studies of dispersion in contaminant plumes are providing data on the importance of this phenomenon in controlling radionuclide concentrations in sediments. Results show that aquifer stratification is the dominant control on large-scale dispersion, and that transverse dispersion within strata is a factor of 5 to 10 lower than longitudinal spreading [42]. Ground-probing radar has proven valuable in defining stratigraphic features that may influence dispersion in unconsolidated sediments [41].

The relative contributions of groundwater flow and surface runoff to streamflow could influence radionuclide concentrations in surface waters. Studies using deuterium as a tracer indicate that, even in the spring, streamflow is still predominantly composed of groundwater [41]. Sedimentation-resuspension processes in surface waters are also being examined, as these influence radionuclide concentrations in lakes and rivers.

Organic soils occur in low-lying areas in the Canadian Shield where groundwater discharge is expected to occur, and they may have a dominant effect on radionuclide migration in the biosphere. Migration of a variety of radionuclides in moss and peat cores is being studied in field tests. Sorption of a number of radionuclides by soils is being examined for both aerobic and anaerobic conditions. The long-term dynamics of chemical elements in Shield soils are being evaluated through studies of weathering and podzolization [41].

The uptake of a number of radionuclides by plants has been studied over a soil concentration range of three orders of magnitude. Results show good confirmation of a linear concentration ratio model [41].

Potential atmospheric exposure pathways are being studied, including soil suspension by wind, and gaseous and particulate suspension due to forest fires and burning wood for fuel.

A food chain model, LIMCAL, has been incorporated into SYVAC to calculate dose equivalents for man. LIMCAL includes all the major ingestion pathways leading to man, including terrestrial, fresh-water and salt-water food types, as well as man's and animal's drinking water [43]. The importance of inhalation pathways for both animals and man has been evaluated, and inhalation by animals was shown to be a potentially important pathway, particularly for some actinides [44]. A review of the importance of the soil ingestion pathway for animals indicated that, for cattle, this exposure pathway is much more important than root uptake by plants for many radionuclides [45].

PERFORMANCE ASSESSMENT MODELLING

The evaluation of risks associated with nuclear fuel waste disposal requires consideration of both the pre-closure and post-closure periods. Pre-closure activities include all aspects of operation of the disposal facilities, up to and including closure of the vault. The pre-closure assessment, which is conventional in nature, involves examination of the socio-economic and environmental impacts associated with fuel waste transportation, immobilization and emplacement, as well as vault sealing and decommissioning of surface facilities. Post-closure assessment involves determination of the long-term risks associated with migration of radionuclides from the vault to the biosphere.

The approach used in the post-closure assessment has been widely discussed. It involves use of the SYVAC code, which links together submodels representing processes affecting radionuclide mobility in the vault, geosphere and biosphere (see Figure 1) [46]. Rather than adopting conservative single values for parameters in the submodels, distributions are defined, based on field and laboratory research, that take into account uncertainty and variability in the data. Sampling from each parameter defines one scenario for the system. The models are then executed in sequence, resulting in a dose consequence. This is repeated many times to give a histogram of frequency vs. consequence.

The research programs, in addition to their role in developing appropriate technologies for fuel waste transportation, immobilization and disposal, provide the conceptual basis and supporting data for the detailed submodels used to describe radionuclide dissolution and transport processes within the vault, geosphere and biosphere. Substantial progress has been made in incorporating research results into the detailed assessment models. Since SYVAC1, an early version of the code, was completed, the number of sampled parameters has increased by an order of magnitude to 540 in the recently completed SYVAC2. The research programs are thus moving rapidly to provide data and updated models for the assessment. A summary of results from SYVAC2, the most recent assessment, is now presented.

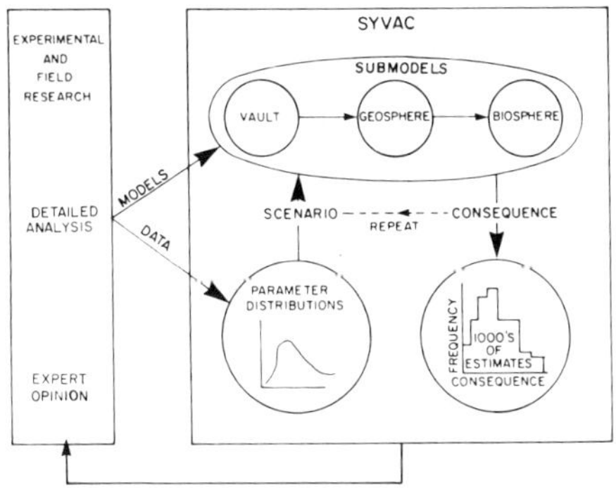

Fig. 1 Schematic representation of the SYVAC computer code and the performance assessment procedure

Pre-Closure Assessment

The potential radiation dose to man has been estimated for normal operation of a disposal facility and for worst-case scenarios. Relative to natural background radiation, both airborne and waterborne emissions would generally result in small dose contributions. In the case of dilution in small lakes, however, the doses could reach 75% of the natural background radiation dose. Used fuel transportation was estimated to result in a dose of less than 2% of that due to natural background radiation. For "worst-case" accidents possible during fuel transportation and operation of a disposal facility, individual doses to members of the public were estimated to be well below the emergency dose limit recommended by the ICRP [47].

Post-Closure Assessment

The revised version of SYVAC, called SYVAC2, has been used to assess the disposal of both used fuel and fuel recycle wastes [48]. For the latter case, the analysis presented here assumed that the high-level wastes were immobilized in calcium aluminosilicate glass, the I-129 in bismuth oxyiodide, and the C-14 in bismuth carbonate, and that zirconium fuel bundle components were compacted.

The assessment results are presented in terms of the maximum annual effective dose equivalent to a member of the most exposed group. A dose of 10^{-10} mSv, about ten orders of magnitude lower than the dose from natural background radiation, is arbitrarily defined as a "zero" consequence.

Maximum doses calculated for the disposal of used fuel for periods of 10^5, 10^6 and 10^7 years are shown in Figure 2. No doses exceeding the zero value were calculated until 24 000 years. In Figure 3, distributions of doses are shown for individual radionuclides and for total dose from all radionuclides. Iodine-129 is by far the most significant contributor to dose, while Tc-99 contributed to maximum doses in 6% of the cases.

The principal factors responsible for the generally low doses are the low dissolution rate of the used fuel and the long water transit times in the geosphere. The predominance of I-129 as the principal contributor to radiation dose is largely due to the "instant release" of a portion of the inventory of this radionuclide from used fuel. All doses greater than 0.3 mSv resulted from simulations involving a well-water pathway.

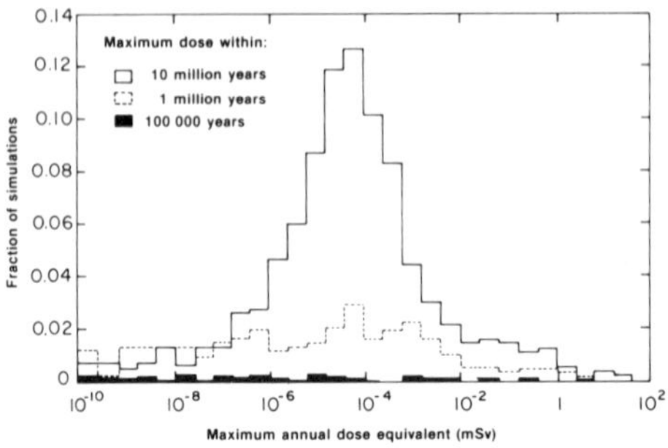

Fig. 2 Distribution of maximum annual dose equivalent for used fuel disposal

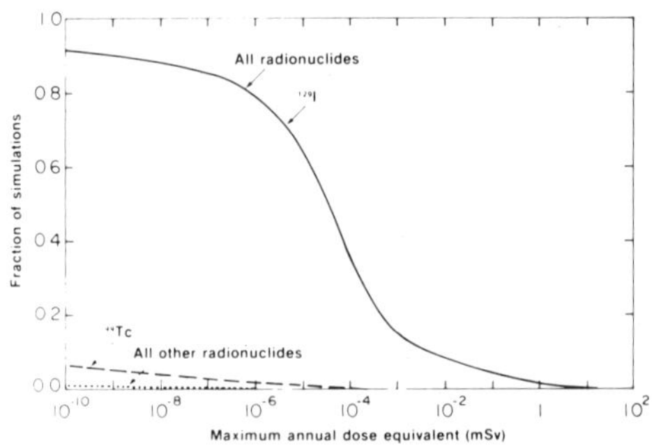

Fig. 3 Downward cumulative distributions of maximum total annual dose equivalent and maximum doses from individual radionuclides within 10^7 years, for used fuel disposal

For fuel recycle wastes, the maximum annual dose equivalents for periods of 10^5, 10^6 and 10^7 years are shown in Figure 4. Technetium-99 was the most significant contributor to dose, followed by I-129. Most radionuclides were released more rapidly by the glass; the release of I-129 from the bismuth oxyiodide waste form, however, was much lower than from used fuel.

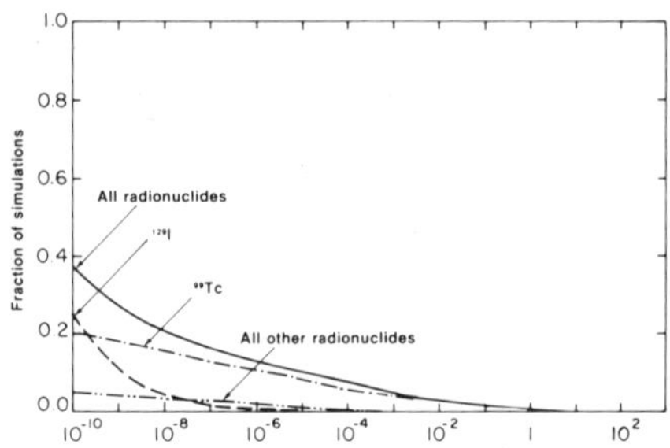

Fig. 4 Downward cumulative distributions of maximum total annual dose equivalent and maximum doses from individual radionuclides within 10^7 years, for fuel recycle waste disposal

The data presented, although preliminary, indicate that radiation doses resulting from disposal of nuclear fuel waste are likely to be far lower than those resulting from natural background radiation ($\sim$ 1mSv/y). They therefore support the view that the basic disposal concept is sound.

Although the basic approach used in the SYVAC code has not changed significantly, the submodels in SYVAC2 are more detailed and generally better supported by research data than those used in SYVAC1. All aspects of the detailed models are being examined to ensure that they are conceptually sound, and the research programs are being critically reviewed to ensure that they are focused on providing the data and understanding necessary for the development of SYVAC3, an improved version of the code, in preparation for the final concept assessment.

CONCLUSIONS

The basic components of the Nuclear Fuel Waste Management Program, including research and development related to used fuel storage and transportation, fuel waste immobilization, site characterization and performance assessment modelling have been described. Preliminary performance assessment results support the view that disposal in plutonic rock in the Canadian Shield appears to be a promising approach for long-term management of nuclear fuel wastes. The program remains on schedule, and research is directed at the completion of documentation on concept assessment for submission to the Government of Canada in 1988.

REFERENCES

1. L.W. Shemilt (Chairman), Fifth Annual Report of the Technical Advisory Committee on the Canadian Nuclear Fuel Waste Management Program, Report TAC-6, 1985. Available from Dr. L.W. Shemilt, McMaster University, Hamilton, Ontario L8S 4L7.

2. J.A. Remington, R.C. Oberth and C.E.L. Hunt in Proceedings of the IAEA International Conference on Radioactive Waste Management, Seattle, Washington, 1983 May, 3, p. 107.

3. R.B. Lyon and D.B. McConnell, editors. The Canadian Nuclear Fuel Waste Management Program - 1984 Annual Report, Atomic Energy of Canada Limited Report, AECL-8393, 1985 (in preparation).

4. K. Nuttall, J.L. Crosthwaite, P. McKay, P.M. Mathew, B. Teper, P.Y.Y. Maak and M.D.C. Moles, in Scientific Basis for Nuclear Waste Management VI, edited by D.G. Brookins, (Elsevier, 1983), p. 677.

5. L.J. Hosaluk, J.L. Crosthwaite and B. Teper in Proceedings of The Nineteenth Information Meeting of the Nuclear Fuel Waste Management Program, Atomic Energy of Canada Limited Technical Record TR-350 (in press).

6. P.Y.Y. Maak, see Ref. 5.

7. J.L. Crosthwaite, in Proceedings of the Sixteenth Information Meeting of the Nuclear Fuel Waste Management Meeting, Atomic Energy of Canada Limited Technical Record, TR-218*, 1984.

8. B.M. Ikeda, see Ref. 5.

9. J.L. Postlethwaite, see Ref. 5.

10. M. Onofrei, D.K. Raine, L. Brown and F. Stanchell in Scientific Basis for Nuclear Waste Management VIII, edited by C.M. Jantzen, J.A. Stone, and R.C. Ewing (Materials Research Society, 1985), p. 395.

11. L.H. Johnson, N.C. Garisto, and S. Stroes-Gascoyne, in Waste Management '85, 1, edited by R.G. Post 1985, p. 479, and references therein.

12. S. Stroes-Gascoyne, L.H. Johnson, P.A. Beeley and D.M. Sellinger, this volume.

13. M.G. Bailey, L.H. Johnson and D.W. Shoesmith, Corros. Sci. 25, 233 (1985).

14. D.W. Shoesmith, S. Sunder, L.H. Johnson and M.G. Bailey, this volume.

15. C.B. Saunders, see Ref. 5.

16. A.C. Vikis, Atomic Energy of Canada Limited Report, AECL-7819, 1984.

17. D.M. Ruthven, J.S. Devgun, F.H. Tezel and T.S. Sridhar, Can. J. Chem. Eng., 62, 526 (1984).

18. P. Taylor, see Ref. 5.

19. J.C. Tait, D.L. Mandolesi and H.E.C. Rummens, Atomic Energy of Canada Limited Report, AECL-8163, 1984.

20. J.C. Tait and D.L. Mandolesi, Atomic Energy of Canada Limited Report, AECL-7803, 1984.

21. P.J. Hayward, this volume.

22. D.W. Oscarson, R. Taylor, H. Miller, S.C.H. Cheung, Appl. Clay Sci. 1, (1985).

23. R.W. Gillham, H.M. Johnston, M.J.L. Robin and H.D. Sharma, see Ref. 5.

24. R.M. Johnston and H.G. Miller, Atomic Energy of Canada Limited Report, AECL-8366, 1984.

25. N.C. Burnett, R.D. Hooton, R.B. Heimann and M. Onofrei, this volume.

26. D.A. Dixon and M.N. Gray, see Ref. 5.

27. P.H. Seymour and M.N. Gray, see Ref. 5.

28. A.M. Crawford, H.T. Chan and R.D. Hooton, see Ref. 5.

29. P.A. Brown and N.A.C. Rey, in Proceedings of the Seventeenth Information Meeting of the Nuclear Fuel Waste Management Program, Atomic Energy of Canada Limited Technical Record, TR-299*, 1985.

30. J.J. Cramer and M. Gascoyne Geological Association of Canada - Mineralogical Association of Canada Annual Meeting, program with abstracts, 1984, p. 55.

31. M. Gascoyne, C.C. Davison, J.D. Ross and R. Pearson, in Geol. Assoc. of Canada Special Publication, Origin and Evolution of Saline Groundwaters, edited by P. Fritz and S.K. Frape (in preparation).

32. C.C. Davison (private communication).

33. W.L. Wardrop and Associates Ltd., Canadian Mine Services Ltd., and Hardy Associates Ltd., Atomic Energy of Canada Limited Technical Record, TR-341*, 1985.

34. R.G. Berman and T.H. Brown, Atomic Energy of Canada Limited Technical Record (in preparation).

35. R.J. Lemire, Atomic Energy of Canada Limited Report, AECL-7817, 1984.

36. P.P.S. Saluja, Atomic Energy of Canada Limited Report, AECL-8447, 1984.

37. J. Paquette and W.E. Lawrence, Can. J. Chem. (in press).

38. F.B. Walton, J. Paquette, J.P.M. Ross and W.E. Lawrence, submitted to Nucl. and Chem. Waste Management.

39. T.W. Melnyk (private communication).

40. J.J. Cramer, see Ref. 29.

41. S.L. Iverson in Proceedings of the Eighteenth Information Meeting of the Nuclear Fuel Waste Management Program, Atomic Energy of Canada Limited Technical Record, TR-320*, 1985.

42. R.W. D. Killey and G.L. Moltyaner, EOS, 65, 206 (1984).

43. R. Zach and G.R. Sherman, Atomic Energy of Canada Limited Technical Record, TR-232*, 1983.

44. R. Zach, Heath Physics (in press).

45. R. Zach and K.R. Mayoh, Health Physics 46, 426 (1983).

46. K.W. Dormuth and G.R. Sherman, Atomic Energy of Canada Limited Report, AECL-6814, 1981.

47. K. Johansen, W.E. Dunford, K.J. Donnelly, J.H. Gee, B.J. Green, J.S. Nathwani, A.M. Quinn, B.G. Rogers, M.A. Stevenson and J.A. Tamm, Atomic Energy of Canada Limited Report, AECL-8373-3, 1985.

48. E.L.J. Rosinger, B.W. Goodwin and D.M. Wuschke, in Waste Management '85, 1, edited by R.G. Post, 1985, p. 47.

---

* Unrestricted, unpublished report available from SDDO, Atomic Energy of Canada Limited Research Company, Chalk River, Ontario KOJ 1JO.

PART II

Fluids in Crystalline Rocks

DIFFUSIVITIES IN CRYSTALLINE ROCK MATERIALS

KRISTINA SKAGIUS AND IVARS NERETNIEKS
Department of Chemical Engineering, Royal Institute of Technology, S-100 44 Stockholm, Sweden

ABSTRACT

The diffusion of non-sorbing species in different crystalline rocks and fissure coating materials has been studied. The results show that the effective diffusivity of iodide, Uranine and Cr-EDTA in rock materials with fissure coating material is of the same magnitude or higher than in granites and gneisses. The results also show that it is not possible to assign one value to the diffusivity of a rock from a given area. The variations in properties are too large. The estimated effective diffusivity of iodide in rocks without fissure coating material was found to be in the range $1 \cdot 10^{-14}$ m^2/s to $70 \cdot 10^{-14}$ m^2/s.

To simulate the stress that exists in the bedrock at large depths, diffusion experiments with iodide and electrical resistivity measurements in rock materials under mechanical stress have been performed. It was found that the diffusivity in rock samples at 300 - 350 bars stress was reduced to 20 - 70% of the value in samples under atmospheric pressure.

INTRODUCTION

In crystalline rocks which are fissured, radionuclides released from an underground repository will be transported with the moving groundwater along fissures in the rock. A significant retardation of the nuclides can be expected if the nuclides migrate into the micropores of the rock by molecular diffusion [1]. If the nuclides are also sorbed on the rock material or react chemically with it, the diffusion into the micropores will lead to a reaction or sorption not only at the fissure surfaces but at the micropore surfaces within the rock matrix as well.

To determine the diffusivity in the micropores, diffusion experiments with "non-sorbing" species (iodide, Uranine and Cr-EDTA) in different rocks and fissure coating materials have been performed. The diffusivity of iodide and the electrical resistivity in rock materials under mechanical stress, corresponding to repository depths, have also been studied. This was to investigate if the high stress causes changes in the properties of the rock material which would influence the diffusivity. Electrical resistivity measurements may be an indirect and much faster method to determine effective diffusivities.

This paper includes some results that have been presented earlier [6].

EXPERIMENTAL

The rock materials used in the experiments were granites and gneisses from different areas of Sweden (Finnsjön, Stripa, Karlshamn, Gideå, Svartboberget, Fjällveden, Studsvik) and the fissure coating materials were from Finnsjön and Stripa.

The diffusion experiments with iodide, Uranine and Cr-EDTA were performed in the following way: The rock piece was fixed between two chambers. One chamber was filled with a solution containing either iodide (1 mol/l or 0.1 mol/l), Uranine (~10 g/l) or Cr-EDTA (~8 g/l). The other chamber was filled with distilled water or, in the experiments with 0.1 mol/l iodide, 0.1 mol/l sodium nitrate. The sodium nitrate solution was used to obtain equal ionic strength on either side of the rock piece. The concentration increase in the chamber, that at the outset was free from the diffusing component, was determined.

Before the diffusion experiments were started the porosity of the rock samples were determined by measuring the weight difference between the rock samples when saturated with water and when oven-dried (water-saturation method). After the diffusion experiments had been performed the rock samples were saturated with a solution of iodide, Uranine or Cr-EDTA of known concentration. The amount of component each sample contained in the saturated condition was determined by leaching out the component from the sample with distilled water. From this information the pore volume and the porosity of the sample was determined (leaching method).

In the diffusion experiments with iodide in rock samples under mechanical stress, a plate of stainless steel with circular channels was placed on each side of the sample. The channels in each plate were connected to a circulation system by tubes (Fig. 1). The solution in each circulation system flowed through a tube into the centre of the plate, then circulated in the channels which were in contact with the rock sample and then out to a storage bottle via the tube connected to the periphery of the plate. The rock sample with the plates were confined in a layer of polyurethane and placed in a pressure cell. The pressure in the cell was raised by pressurizing the cell with hydraulic oil. The polyurethane is deformable enough to transmit the pressure from the oil to the rock sample. During the diffusion experiment the pressure cell was placed in a water bath at a temperature of 25°C.

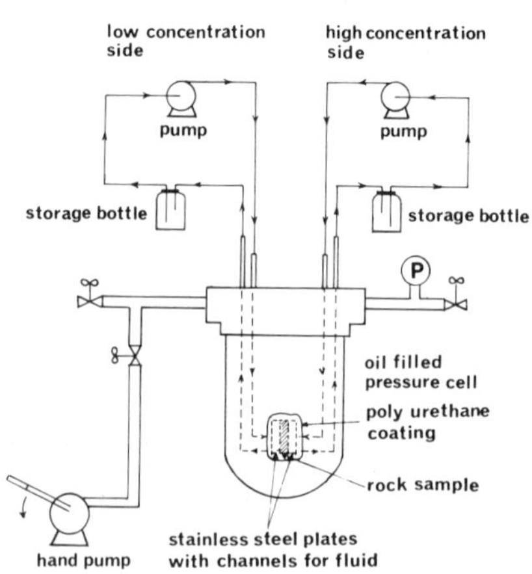

Fig. 1. The apparatus used for the diffusion experiments under stress.

The apparatus used in the electrical resistivity measurements was similar to the apparatus that was used in the diffusion experiments under stress. Rock cores were saturated with 1 mol/l NaCl solution. The length of the cores (l = 30 mm) and the concentration of the NaCl solution were chosen on the basis of earlier experiments [2].

On either side of the saturated core a stainless steel plate was fastened. The plates acted as electrodes, and they could also, in small channels, take up the pore solution which would be pressed out from the core during the experiment. Between the core and the electrodes a thin compressed slab of porous foam rubber, saturated with the same NaCl solution as the core, was placed. This ensured good electrical contact between the rock core and the electrodes. The whole package with the rock core and the electrodes was then cast in polyurethane after ensuring that the core was surface dry.

The rock core with the electrodes was placed in a pressure cell and the electrodes were connected to a conductivity meter. The pressure in the cell was raised by pumping in hydraulic oil. At various pressures the resistance in the saturated rock core was measured at a frequency of 50 Hz.

DETERMINATION OF THE DIFFUSIVITY

The total amount of diffusing substance Q which has passed through the rock sample in time t can be calculated by solving the diffusion equation for the appropriate initial and boundary condition. The result is [3]

$$\frac{Q}{\ell\cdot C_1} = \frac{D_e\cdot t}{\ell^2} - \frac{\alpha}{6} - \frac{2\cdot\alpha}{\pi^2}\sum_{n=1}^{\infty}\frac{(-1)^n}{n^2}\exp\left(-\frac{D_e\cdot n^2\cdot\pi^2\cdot t}{\ell^2\cdot\alpha}\right) \qquad (1)$$

ℓ is the thickness of the sample, C_1, is the inlet concentration, $D_e = D_p\cdot\varepsilon_p$ is the effective diffusivity, D_p the pore diffusivity, ε_p the porosity, $\alpha = \varepsilon_p + K_d\cdot\rho$ is the rock capacity factor, K_d is the sorption coefficient and ρ is the density of the sample. Eq. (1) is only valid if the outlet concentration is negligible compared to the inlet concentration.

As $t \to \infty$ eq. (1) becomes

$$Q = \frac{C_1\cdot D_e}{\ell}\cdot t - \frac{C_1\cdot\ell\cdot\alpha}{6} \qquad (2)$$

This is the equation for a straight line with the slope $C_1\cdot D_e/\ell$ and an intercept on the time axis $t = \ell^2\cdot\alpha/6\cdot D_e$. From the slope of the concentration versus time curve at longer times the effective diffusivity D_e can be determined, and the intercept of the line on the time axis gives α. If the diffusing component does not sorb on the rock material then $\alpha = \varepsilon_p$.

There are some indications that electrical conductivity and molecular diffusion may depend in the same way on the formation factor $\varepsilon_p\cdot\delta_D/\tau^2$ [4]. This means that

$$\frac{D_p\cdot\varepsilon_p}{D_v} = \frac{\varepsilon_p\cdot\delta_D}{\tau^2} = \frac{R_o}{R_s} \qquad (3)$$

where R_s is the resistivity of the salt-water-saturated rock core and R_o is that of the salt-water. δ_D is the constrictivity and τ the tortuosity of the pores in the rock sample. D_v is the bulk phase diffusivity of the component. From electrical resistivity measurements the formation factor for a rock sample can be calculated and an estimate of the effective diffusivity of a component in the rock sample can be made.

RESULTS AND DISCUSSION

In Table I the range of the determined effective diffusivities of iodide, Uranine and Cr-EDTA in fissure coating materials, granites and gneisses are given. The number of samples studied are presented within brackets.

Table I. Determined effective diffusivities of iodide, Uranine and Cr-EDTA.

	Effective diffusivity, $D_e \cdot 10^{-14}$ m^2/s		
	Iodide	Uranine	Cr-EDTA
Fissure coating material	0.2 - 6000 (14)	0.7 - 4000 (10)	0.6 - 50 (3)
Granites	4 - 66 (38)	0.2 - 7 (7)	0.6 - 5 (6)
Gneisses	1 - 36 (9)	0.3 - 8 (7)	0.6 - 3 (4)

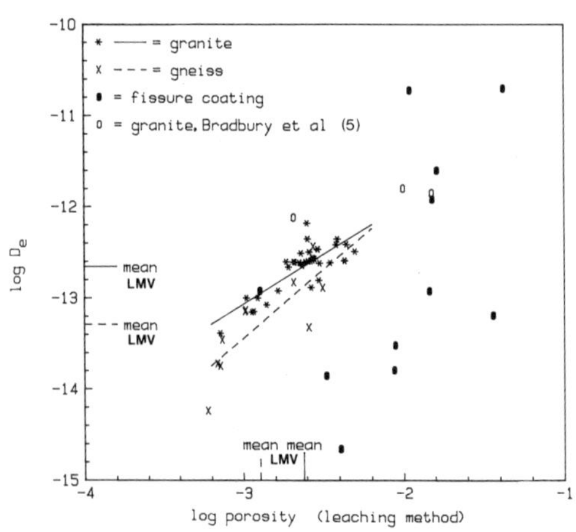

Fig. 2. Effective diffusivities versus porosities determined by the leaching method.

The Table shows that the effective diffusivity in the fissure coating materials is of the same order of magnitude or higher as in the granites and the gneisses.

In Figure 2 the logarithmic value of the diffusivities of iodide in the granites, gneisses and fissure coating materials are plotted versus the logarithmic value of the experimental porosities determined by the leaching method. Results obtained by Bradbury et al. [5] for iodide diffusion in different granites from

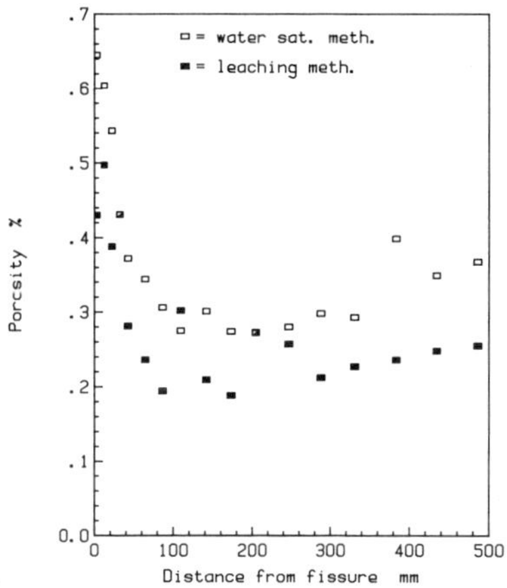

Fig.3. Porosity versus distance from fissure surface.

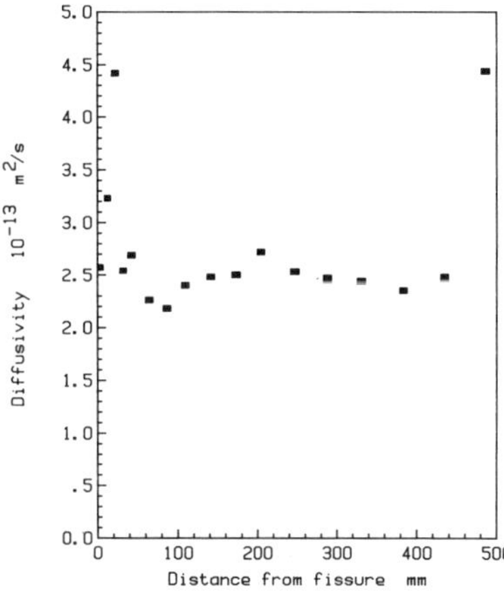

Fig. 4. Effective diffusivity versus distance from fissure surface.

the United Kingdom are also presented for comparison. For the granites and the gneisses a linear regression has been made and the mean values, both logarithmic (LMV) and arithmetic (AMV), of the effective diffusivities and porosities have been calculated. The effective diffusivity in the granites, LMV = $22.0 \cdot 10^{-14}$ m^2/s and AMV = $25.2 \cdot 10^{-14}$ m^2/s, is higher than in the gneisses, LMV = $5.1 \cdot 10^{-14}$ m^2/s and AMV = $9.2 \cdot 10^{-14}$ m^2/s.

The porosities are also higher for the granites, LMV = 0.24% and AMV = 0.26%, than for the gneisses, LMV = 0.13% and AMV = 0.15%.

The Figure also show that the fissure coating materials that have an effective diffusivity that is of the same order of magnitude as the granites and the gneisses have much higher porosity values. This could be due to a higher "storage" porosity in the fissure coating materials, or to a lower pore diffusivity in the fissure coating materials compared with the granites and the gneisses. The measured porosity is a total porosity including both "transport" pores and "storage" pores. The effective diffusivity is only dependent on the "transport" porosity.

Some of the samples in the iodide diffusion experiments were taken at different distances from a fissure surface. In one case neither the porosity nor the effective diffusivity showed any obvious variation with the distance from the fissure surface. In another case the porosity decreased with distance up to about 80 mm from the fissure surface and then remained rather constant or increased slightly (Fig. 3). The effective diffusivity, however, showed no obvious variation with the distance from the fissure surface (Fig. 4). A higher total porosity thus does not have to give a higher diffusivity. The "transport" porosity could have about the same value even if the total porosity increases.

In Figure 5 the outlet concentration of iodide versus time for diffusion through a gneiss sample from Svartboberget is shown. The experiment

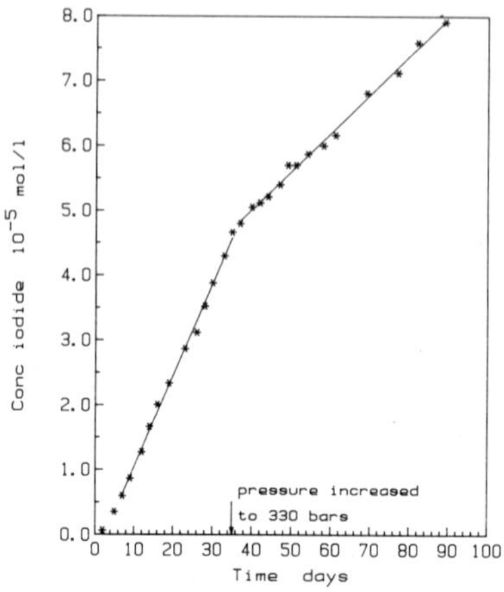

Fig. 5. Outlet concentration of iodide versus time.

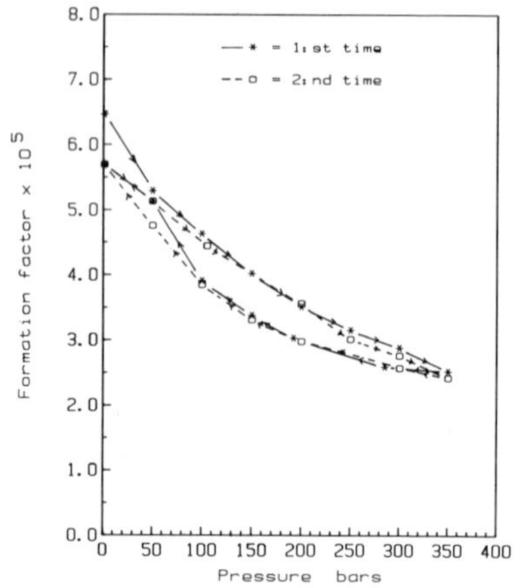

Fig. 6. Formation factor versus pressure for a granite sample.

was performed in a pressure cell. After 35 days the pressure was increased from 1 to 330 bar. When the pressure was increased the slope of the concentration versus time curve decreased. According to eq. (2) this means that the effective diffusivity decreased.

In the resistivity measurements the pressure was raised in steps and the resistance was measured at each level. From the resistance the resistivity was calculated, and then the formation factor was determined by eq. (3). The resistance was also measured as the pressure was lowered from the maximum value down to atmospheric pressure. Figure 6 shows the formation factor versus pressure for a granite sample from Finnsjön, where the procedure with increasing and decreasing the pressure was made two times on the same rock core. The formation factor decreased with increasing pressure, and then increased again when the pressure was lowered down to atmospheric pressure, however, not to the same values. The second time the procedure with increasing and decreasing the pressure was performed, about the same values were obtained for pressures higher than 100 bars. The decrease in the formation factor with pressure, which also means a decrease in the effective diffusivity according to eq. (3), indicates that the crossectional area of the pores in the sample is decreased when the sample is under mechanical stress.

The effective diffusivities in samples at a stress of 300 - 350 bars relative to the effective diffusivities in the same samples at atmospheric pressure obtained in the diffusion experiments are in fair agreement with those determined from the electrical resistivity measurements. The diffusivity in samples at a stress of 300 - 350 bars was found to be 20 - 70% of the value in unstressed samples.

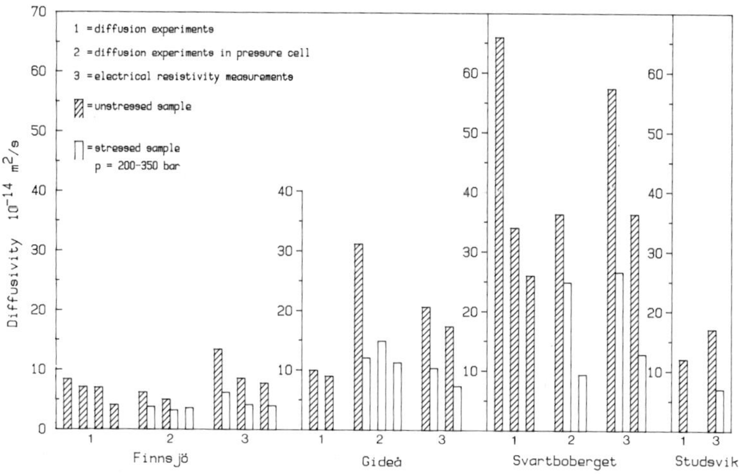

Fig. 7. Effective diffusivities in granites.

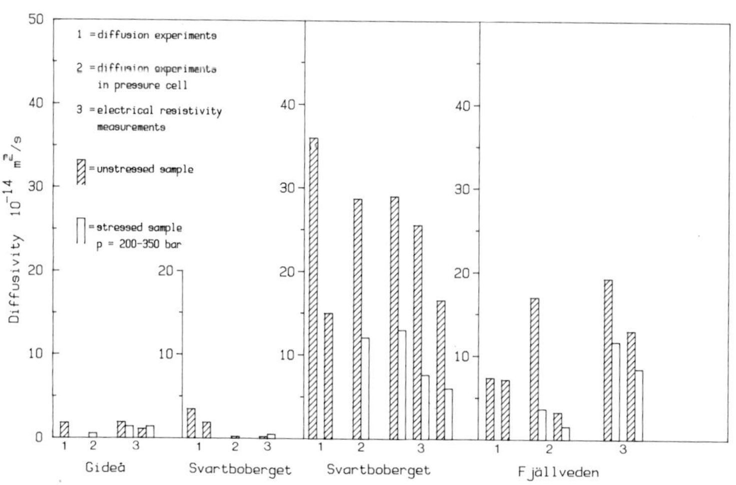

Fig. 8. Effective diffusivities in gneisses.

In Figures 7 and 8 the results from the two types of iodide diffusion experiments and the electrical resistivity measurements are summarized in the same diagram. Figure 7 shows the diffusivities determined in granites and Figure 8 the diffusivities determined in gneisses. The open bars represents the diffusivities in samples under stressed conditions and the filled bars the diffusivities in samples under atmospheric pressure. Adjacent filled and open bars represent the same sample under unstressed and stressed conditions. Otherwise each bar represent one sample.

The Figures shows that there is a large variation in the diffusivities between granites from different areas and also between gneisses from different areas. The granite and the gneiss from Gideå were taken from the same drill core with a difference in depth of about 5 meters. Also the granite and the two gneisses from Svartboberget were taken from the same drill core. The diffusivity in the granites from Gideå and Svartboberget is markedly higher than in the gneisses from the same areas. This shows the difficulty in giving just one value of the diffusivity for a rock from a certain area.

CONCLUSIONS

The results show that the "non-sorbing" species iodide, Uranine and Cr-EDTA may be transported through different rock materials by diffusion in the micropores. The effective diffusivity of iodide was found to be in the range $1 \cdot 10^{-14}$ m^2/s to about $70 \cdot 10^{-14}$ m^2/s. The estimated diffusivities showed large differences between granite and gneiss taken from the same drill core. Therefore it is not possible to give one precise value of the diffusivity in a rock material from a certain area.

The results also show that iodide, Uranine and Cr-EDTA are able to diffuse through rock materials with fissure coating material. The diffusivity in the rock with fissure coating material is of the same order of magnitude or higher than in the rock without fissure coating material. This means that diffusion can transport species from the moving groundwater in fissures in the rock, through the fissure coating material and into the rock matrix.

The diffusion experiments and the electrical resistivity measurements show that the diffusivity and thus the formation factor decreases when the samples are put under mechanical stress. This is probably a combined effect of a decrease in the crossectional area of the pores in the material and increases in tortuosity and constrictivity.

REFERENCES

1. Neretnieks I., Diffusion in the Rock Matrix: An Important Factor in Radionuclide Retardation ?, J. Geophys. Res., 85, 1980.
2. Skagius K., Neretnieks I., Diffusion in Crystalline Rocks of some Sorbing and Non-Sorbing Species, KBS Technical Report 82-12, 1983-03-09.
3. Crank J., The Mathematics of Diffusion, 2nd ed, Oxford University Press, 1975, p 50-51.
4. Klinkenberg L.J., Analogy between Diffusion and Electrical Conductivity in Porous Rocks, Geol. Soc. Am. Bull., 12, 1951, p 559.
5. Bradbury M.H., Lever D., Kinsey D., Aqueous Phase Diffusion in Crystalline Rock, Scientific Basis for Nuclear Waste Management V, 11, 1982, p 569-578.
6. Skagius K., Neretnieks I., Porosities of and Diffusivities in Crystalline Rock and Fissure Coating Materials, Scientific Basis for Nuclear Waste Management VII, 26, 1983, p. 835-842.

DIFFUSION AND PERMEABILITY BASED SORPTION MEASUREMENTS IN INTACT ROCK SAMPLES

MICHAEL H. BRADBURY AND I.G. STEPHEN
Chemistry Division, Harwell Laboratories Oxon, OX11 ORA, England

ABSTRACT

Distribution ratios have been determined for Sr, Cs and Tc (as pertechnetate) over a range of concentrations for a Darley Dale sandstone by three different methods: crushed rock batch tests, with W/R ratio and particle size range as independent variables, diffusion-sorption and permeability-sorption experiments on intact rock samples. Comparisons of R_d - values obtained from batch and diffusion-sorption experiments indicate that crushed rock tests can overestimate sorption by as much as one or two orders of magnitude.

INTRODUCTION

Measurements of the sorption of radionuclides on geological materials is important in the study of the migration rates of active species away from a repository through the host rock. Sorption data, usually presented as simple distribution ratios, R_d, or Fruendlich isotherms, have largely been generated by batch or column tests employing crushed rock. Even where the preparation of the rock has been the same, agreement between the two methods has been variable (e.g. [1] and [2]). The use of crushed rock leads to a simplified experimental procedure and a relatively easy and rapid means of producing large quantities of data. Often the variables inherent in the technique (solution volume/rock mass ratio (W/R) and particle size range (PSR)) are chosen more or less arbitrarily and mainly for convenience. This, amongst other factors, has lead to wide variations in the reported sorption data, even in laboratory intercomparison studies [3].

A possibly more fundamental criticism is that crushing leads to the creation of fresh fracture surfaces and the implantation of strain energy within the near surface regions which could seriously modify sorption behaviour. Moreover, fractured surfaces may undergo relatively rapid "weathering" effects once in contact with water [4] which could lead to time dependent changes in sorption. This, combined with the well documented variation of sorption with W/R ratio and PSR [5,6,7] must lead to some doubts regarding the usefulness of such data to quantify sorption in intact rock in the field.

With the above in mind, a program of work was started whose purpose was twofold; (i) to develop experimental techniques for the measurement of sorption through intact rock samples and to identify the difficulties and limitations of the methods, and (ii) to generate sorption values from crushed rock batch tests, where the W/R ratio and the PSR were independent variables, and to compare these with sorption data obtained from intact rock tests.

One of the main advantages of using intact rock is that its geometrical and geochemical characteristics, i.e. its internal pore structure, is preserved, and thus approximates to the in situ condition better than crushed rock. Laboratory methods based on diffusion-sorption (D-S) and permeability-sorption (P-S) have been tried.

EXPERIMENTAL PROCEDURE

Batch Sorption Tests

Batch sorption tests on Darley Dale sandstone were carried out in contact with air for two W/R ratios, three PSR's at four or more concentrations for each combination. Other similar measurements were made on anhydrite and upper magnesian limestone, though these results are not reported here. The aim was to choose rocks which were thought to have very different sorption, diffusion and permeation properties so as to make eventual final comparisons as wide as possible.

In the absence of the appropriate groundwaters, all experiments were conducted with AnalaR grade distilled water "equilibrated" with crushed rock over periods of approximately 5 months. Prior to use the simulated groundwaters were sequentially filtered through 0.2 and 0.05 μm Amicon filters.

Mixtures of spiked solutions containing I-125, Sr-85 and Cs-137, or I-125 and Tc-95m, were contacted with crushed rock contained in polypropylene test tubes which were gently agitated. Experiments were carried out in triplicate for each combination of concentration, W/R ratio and PSR. Concentrations in one of the three tests were monitored at regular intervals for times of up to 2000 hours. Once "equilibrium" had been established the other two test tubes were centrifuged at 9000 rpm for 30 minutes, sampled and then counted on a Philips PW4800 γ-counter interfaced with a Commodore 8032 computer.

Diffusion-Sorption on Intact Rock

Discs of sandstone 25mm in diameter and ~5mm thick were sealed within the experimental arrangement shown schematically in Figure 1.

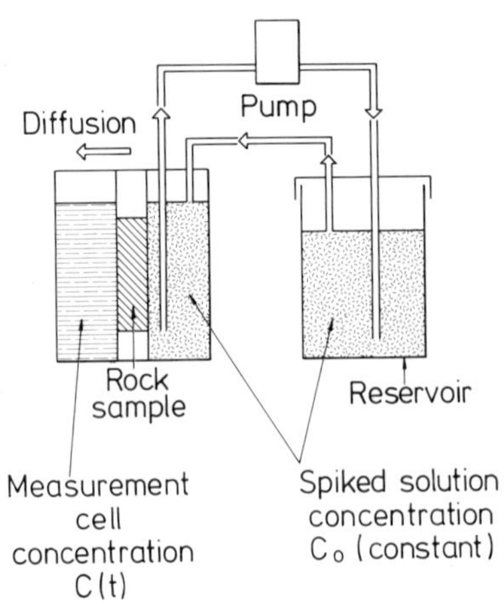

Figure 1 : Experimental arrangement of diffusion-sorption experiments.

Mixtures of spiked solutions containing Cs-137, I-125 and Sr-85 (or I-125 and Tc-95m) were continuously circulated from the reservoir (vol ~ 400 ml) past one face of the sample. Because sorption and transport through the rock would tend to reduce the source concentration in the limited cell volume (~ 25 ml) re-circulation through a large external reservoir was carried out in order to minimise any change. Samples were taken at frequent intervals from the measurement cell and counted as already described. Aliquots used for counting were replaced by an equal volume of simulated groundwater. The theory and method of analysis for D-S experiments have been given previously [8,9] and as shown in the above references a plot of C_t against t yields the intrinsic diffusion coefficient, D_i, from the slope, and the rock capacity factor α, from the intercept on the time-axis of the extrapolated linear region, e.g. see figure 3.

$$\alpha = \varepsilon + \rho R_d \tag{1}$$

where ε is the porosity, ρ is the rock density and R_d is the mean distribution ratio between the source concentration C_o and the final measured concentration, C_f. Thus for a non-sorbing species such as iodide, $\alpha = \varepsilon$ ($R_d = 0$), and for a sorbing species R_d may be calculated from the value of α.

Permeability-Sorption on Intact Rock

The arrangement for P-S experiments on sandstone discs 70mm in diameter and ~ 20mm thick using mixtures of inactive iodine and caesium solutions (Cs conc - 5×10^{-5} M) is illustrated in Figure 2.

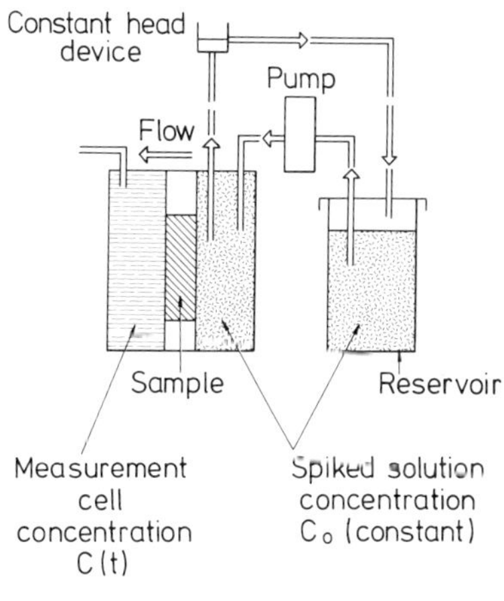

Figure 2 : Experimental arrangement of permeability-sorption experiments.

A re-circulation system via an external renewable reservoir (vol ~ 5ℓ) was used for the same reason as given for the D-S experiments and also to maintain a constant head of ~ 1m of the solution across the sample.

Bulk water flow through the sandstone was monitored continuously (flow rate ~ 1.3ml hr^{-1}) and Cs concentration increases were measured by emission spectroscopy on 10ml aliquots taken from the measurement cell (each sample volume removed was replaced by an equal volume of simulated groundwater). Iodide concentrations were measured in situ by ion specific electrodes.

Assuming to a first approximation that "plug" flow is occurring and that sorption is linear and reversible, then the breakthrough time is given simply by:

$$t_B = \frac{\alpha\,\ell^2}{K\,\Delta\,P} \tag{2}$$

where ℓ is the thickness, K is the hydraulic conductivity and ΔP is the static pressure head; and

$$C(t) = \frac{\dot{F}\,t\,C_o}{V} \quad \text{for } C_t \ll C_o \text{ and } t > t_B \tag{3}$$

where $\dot{F}$ is the volume flow rate, and V is the measurement cell solution volume.

A plot of C(t) against t is thus linear (see for example the iodide results in Figure 4) and the intercept on the time-axis of the extrapolated linear region gives the breakthrough time, t_B. Sorption data can be calculated from equation (2) or by comparing the breakthrough times for a non-sorbing and sorbing species provided that the flow rate remains constant throughout, i.e.

$$\frac{t_B \text{ (Iodide)}}{t_B \text{ (Caesium)}} = \frac{\varepsilon}{\alpha_{cs}} \qquad (4)$$

It is clear from (3) that $\dot{F}$ can be obtained from the slope and thus the permeability of the sample can be calculated [10]. The intercept, t_B, will depend on the sorption properties of the radionuclide whereas the gradient should be the same for all radionuclides. The latter provides a useful internal check on the data and assumptions in the analysis since the permeability calculated from the iodide and caesium data should be the same and correspond to the value obtained from the flow measurements.

RESULTS

Batch Sorption

All batch results were plotted as Fruendlich isotherms with W/R ratios and PSR's as independent variables. The values of the constants, a and b, in the equation

$$C_R = a\, C_W^{\,b} \qquad (5)$$

where C_R is the "equilibrium" concentration on the rock and C_W is the "equilibrium" concentration in the solution, were obtained from a least square analysis of the log/log transformed data and are given in table I.

The general trends in the results were that R_d values increased as the W/R ratio increased and the PSR decreased. Similar observations have been reported previously [5,6,7].

Table I : Values of the Fruendlich isotherm constants 'a' and 'b' (equation 5)) for sandstone batch sorption results (pH = 6.8 to 7.1; C_W (mg/ml), C_R (mg/g)).

CONC RANGE (M)	W/R RATIO	PARTICLE SIZE RANGE (mm)					
		< 0.1		1 - 2		COUPONS	
		a	b	a	b	a	b
Cs 7.0 E-14	20:1	100	0.75	79.4	0.71	52.5	0.76
to 7.2 E- 6	3:1	316	0.94	50.1	0.77	39.8	0.75
Sr 7.7 E-14	20:1	186	0.97	79.8	0.95	101	0.99
to 1.1 E- 6	3:1	358	1.07	59.1	0.99	48.4	1.0
Tc 4.2 E-14	20:1	8.0 E-3	0.88	2.4 E-4	0.70	3.8 E-3	0.70
to e.0 E-11	3:1	2.9 E-3	0.94	5.7 E-4	0.78	6.3 E-4	0.66

Diffusion-Sorption Results

The D_i, α and R_d values, summarised in Table II, were calculated from plots such as those shown in Figure 3. A porosity of 0.1 was used in the calculations for sandstone which was obtained from previous diffusion studies on this rock using 0.1 and 1.0M inactive KI solutions [11]. In some cases, iodide, at these low concentrations, seems to be exhibiting slight sorption, though not in all cases (see Table II).

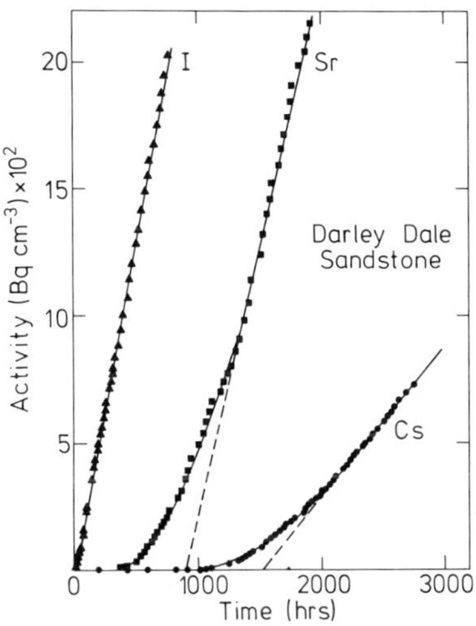

Figure 3 : Diffusion curves for I, Sr and Cs through sandstone.

The R_d values listed are not associated with a unique concentration but rather are interpreted as mean distribution ratios valid between the source concentration (C_o) and the final concentration in the measurement cell (C_f). The two concentrations represent the end points of the final linear concentration profile existing across the samples within the pore water under steady state conditions. Hence these two concentrations are given in Table II.

Table II : Summary of Diffusion-Sorption Results for sandstone.

SAMPLE	NUCLIDE	CONC RANGE (M) C_o	C_f	D_i ($m^2 s^{-1}$)	α	R_d ($ml\ g^{-1}$)
DDS 9	I	1.1 E-10	6 E-12	3.3 E-11	0.6	0.2
	Sr	1.7 E-11	3.3 E-12	1.4 E-10	51	21
	Cs	8.1 E-10	2.3 E-12	NO BREAKTHROUGH AFTER 3000 HRS		
DDS 19	I	1.3 E-10	4.4 E-12	2.4 E-11	0.3	0.006
	Sr	1.4 E- 6	2.9 E- 7	1.2 E-10	59	25
	Cs	1.9 E- 5	8.7 E- 7	2.4 E-11	21	9
DDS 4	I	8.9 E-11	3.8 E-12	2.5 E-11	0.06	-
	Tc	7.4 E-11	2 E-12	1.9 E-11	0.3	0.07
DDS 1	I	9.1 E-11	1.8 E-12	1.6 E-11	0.8	-
	Tc	8 E-11	2.6 E-12	1.7 E-11	0.19	0.04

Allowing for some sample to sample variation, the values of D_i measured for I-125, Tc-95m and Cs-137 are remarkably consistent and compare well with the results from previous inactive KI diffusion experiments [11]. For the case of Sr-85 the D_i value is higher by a factor of between 4 and 5 compared with iodide. Though the data base is limited, this is considered to be a genuine result and not due to experimental error. In diffusion related work on granites and compacted Bentonite [12,13] other workers have observed similar apparently more rapid diffusion rates for Sr and also for Cs. (Increases by an order of magnitude and more compared with iodide were reported). As yet no satisfactory explanation for this behaviour has been given and, unfotunately, none can be offered here.

Permeability-Sorption Results

A comparison between the I and Cs breakthrough curves is given in Figures 4 and 5. Analysis of the iodide curve, assuming "plug" flow, yielded a porosity of ~ 0.1 which is similar to values found from diffusion measurements, and a permeability of 1.9×10^{-16} m^2. This compares well with a permeability of 2.5×10^{-16} m^2 calculated from bulk water flow rates and previously measured values of 2.4×10^{-16} m^2 [10]. Thus for iodide, the assumption of "plug" flow and the simple analysis given appears to produce consistent results (some evidence for dispersion effects in the early part of the curve are apparent, but the effect is small). However, for the Cs breakthrough curve the situation is more complex. The form of the curve is flattened, and calculations of permeability from the apparently linear portion of the curve yielded values at least one order or magnitude lower than given previously. One or both of the initial assumptions seems to be inappropriate. In an attempt to gain futher insight into the problem, the experiment was stopped and the sample sub-cored. The sub-cores were then sectioned perpendicular to the transport axis, dissolved and analysed for Cs. The concentration profiles obtained are shown in Figure 6. These profiles, which appear to be linear across the sample, represent predominantly the quantities of Cs sorbed onto the rock since in each section the Cs in the pore water (porosity ~ 0.1, maximum concentration 5×10^{-5}M) constitutes only a small fraction of the measured value.

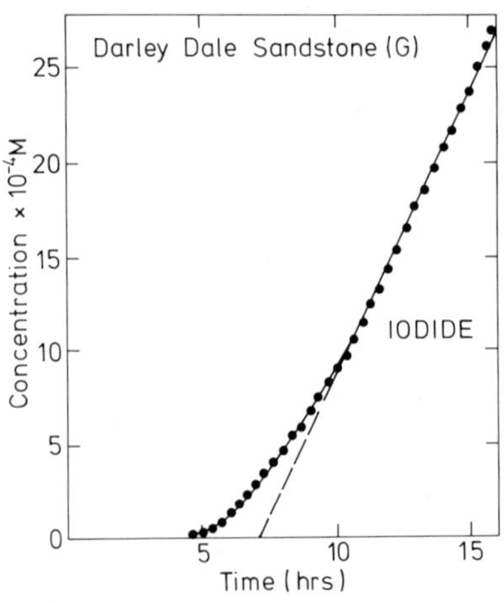

Figure 4 : Iodide breakthrough curve for permeability-sorption experiments on sandstone.

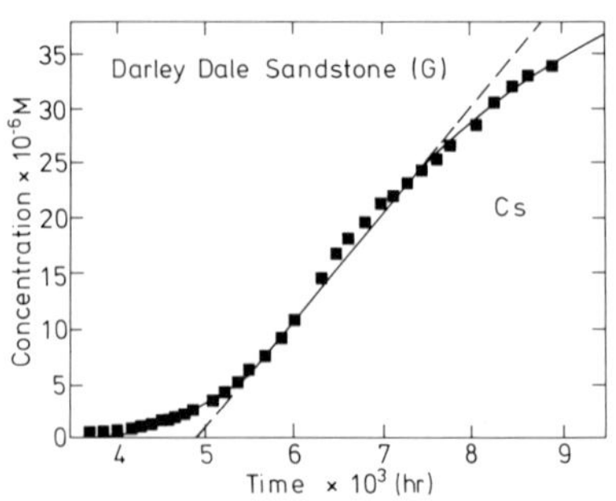

Figure 5 : Caesium breakthrough curve for permeability-sorption experiments on sandstone.

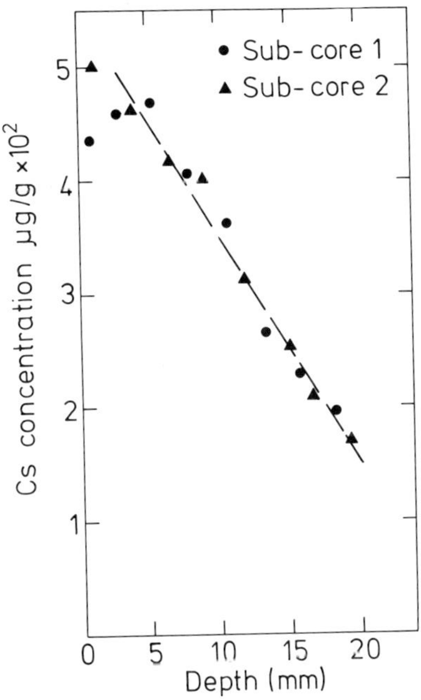

Figure 6 : Caesium concentration profile through sandstone after completion of a permeability-sorption experiment.

DISCUSSION

Comparison of Batch and D-S Sorption Data

The basis selected for comparison between R_d values calculated from batch and D-S experiments (Table III) was the following:

Concentration: R_d values were calculated from Fruendlich isotherms at concentrations corresponding to the source concentration, C_o, and the final measured concentration, C_f, in the diffusion experiments since the R_d's derived from this method were interpreted as mean values valid between these concentration limits.

W/R Ratio and PSR: In the batch tests the extremes in behaviour were represented by a W/R ratio of 20:1 combined with a PSR of $<$ 0.1mm (highest R_d's) and a W/R ratio of 3:1 with coupons (lowest R_d's). Other combinations of W/R ratios and PSR's fell mainly within the bounds of the above two Fruendlich isotherms. These two isotherms were used to calculate the bounding batch R_d values at C_o and C_f given in Table III.

Implicit in the comparison is how well separate determinations of R_d values (from crushed rock) and diffusion parameters (from intact rock) can be combined to predict diffusion controlled mass transfer through small scale intact rock samples.

The following may be noted from Table III:

(i) Sorption measurements for Cs show the largest discrepancy between the two methods (factors of 20-100).
(ii) The discrepancies measured for Sr are smaller (between 2 and 20).
(iii) Tc shows only very weak sorption on sandstone though the discrepancies between D-S and batch tests can still be high (factors of up to 20).
(iv) R_d's measured for Sr by the D-S method seem to be virtually constant within the concentration range measured.
(v) The magnitude of the differences between the two techniques increases as the concentration decreases. This is a reflection of the generally strong dependence of R_d with concentration found in batch tests (particularly for Cs) whereas R_d's from D-S experiments appear to vary much less with concentration.

Table III : Comparison of batch and diffusion-sorption R_d values for sandstone.

NUCLIDE	DIFFUSION			BATCH			
	C_o (M)	C_f (M)	R_d (ml g^{-1})	W/R RATIO	PSR (mm)	R_d's at C_o	R_d's at C_f
Cs	8.1 E-10	2.3 E-12	NO BREAK-THROUGH	20:1	< 0.1	5500	23000
				3:1	COUPONS	2200	9800
	1.9 E- 5	8.7 E- 7	9	20:1	< 0.1	450	970
				3:1	COUPONS	170	380
Sr	1.7 E-11	3.3 E-12	21	20:1	< 0.1	370	390
				3:1	COUPONS	46	46
	1.4 E- 6	2.9 E- 7	25	20:1	< 0.1	250	270
				3:1	COUPONS	48	47
Tc	7.4 E-11	2 E-12	0.007	20:1	< 0.1	0.08	0.13
				3:1	COUPONS	0.37	1.25

(vi) In the majority of cases R_d's from coupon experiments at low W/R ratios were nearest to those found from D-S experiments and also yielded the most conservative sorption data from batch experiments.

The reason for these discrepancies may lie in full or in part with the factors discussed with respect to batch tests in the introduction, or on other factors only indirectly studied here, e.g. surface areas.

For the radionuclides and rock types studied here it is apparent from Table III that batch measurements can over-estimate sorption by as much as one or two orders of magnitude. The magnitude of the difference between the results found from the two methods depends on the radionuclide, its concentration, the rock, and the W/R ratio and PSR parameters chosen for the batch tests.

Permeability-Sorption

The P-S technique was evolved to provide a second method for determining R_d values in intact rock which was more appropriate for strongly sorbing species than the D-S tests. The flow rate through a sample, proportional to $\Delta P/\ell$, together with sorption determines the breakthrough time. It might seem, therefore, that by the suitable choice of ℓ and ΔP, experiments with sorbing species could be carried out on almost any time-scale considered experimentally convenient. However, such an approach may give rise to misleading results because implicity in it is the assumption of rapid or instantaneous equilibrium during migration across the sample. If this is not so, then sorption kinetics may play an important role. For example, in a first series of P-S experiments, Cs breakthrough occurred after only ~80 hours (corresponding to an apparent R_d of ~ 3 ml/g) whereas parallel crushed rock batch tests were indicating equilibration times of many hundreds of hours and R_d - values in the range 140-350 ml/g. Consequently the non-attainment of equilibrium was considered, and in a second series of experiments conditions were altered to increase the transit time for a non-sorbing species by approximately one order of magnitude. Under the new conditions the apparent breakthrough time for Cs increased to ~5000 hours (equivalent to an R_d of ~ 25) and the time scale was much greater than required for "equilibrium" in the batch tests. In the first series of P-S experiments sorption kinetics seem to have played a decisive role whereas in the second series this was much less likely though kinetic effects cannot be completely ruled out without further work. Thus the conditions under which such experiments are conducted are extremely important. The above considerations would also apply to crushed rock column tests as well.

One additional point is that the application of high pressure gradients across intact rock samples may lead to time dependent changes in the pore structure (and hence sorption and permeability) by material re-arrangement within pores and pore flushing [10]. This may be an additional major fault with intact rock column experiments designed to give rapid breakthrough curves for sorbing species by applying high pressure gradients.

Finally, if the simple assumptions of plug flow and linear equilibrium sorption are applied, then the breakthrough times predicted from calculations using crushed rock R_d values are between 5-15 times longer than found experimentally i.e. approximately the same order of difference found between R_d-values in crushed rock and D-S experiments. However, because of the comments made in the "Peremeability-Sorption Results" section such a direct comparison for P-S experiments may not be valid. Nevertheless, what is clear is that a simple combination of flow rate and crushed rock sorption data cannot predict the migration of a sorbing species across a relatively small sample (70 mm diameter, 20 mm thick) under well defined and well controlled conditions.

In addition, estimates of dispersion parameters calculated from the iodide results (Figure 4) could not alone account for the form of the Cs breakthrough curve (Figure 5) i.e. dispersion was too small, and no explanation considered so far can account for the apparently linear sorption profile given in Figure 6. Analyses are continuing on these results but in order to fully understand them more work is clearly required.

CONCLUSIONS

1. Crushed rock batch tests may overestimate sorption distribution ratios by as much as one to two orders of magnitude.

2. The diffusion-sorption technique is a feasible means of measuring sorption on intact rock samples but is probably not practical for strongly sorbing species because of the long experimental times required.

3. The intrinsic diffusion coefficients measured for low source concentrations of I, Cs and Tc showed good internal consistancy and compare well with results from previous measurements using inactive KI.

4. The diffusion coefficients measured for Sr were 4-5 times higher than those of iodide. This is not what would be anticipated for aqueous phase diffusion and a different mechanism or process may be occurring.

5. The most conservative batch data was obtained on coupons at low solution volume to rock mass ratios.

6. For the permeability-sorption experiments the simple assumptions of plug flow and linear equilibrium sorptions could not account for the results obtained for Cs.

7. Application of high pressure gradients to promote rapid breakthrough in permeability-sorption experiments on intact rock samples may cause misleading results because of sorption kinetic effects and changes in the pore structure.

ACKNOWLEDGEMENTS

This work has been commissioned by the Department of the Environment, as part of its radioactive waste management research programme. The results will be used in the formulation of Government policy, but at this stage they do not necessarily represent Government policy.

The authors would also like to thank K. Daveridge for performing some of the batch tests and Dr. D. Lever for his helpful comments.

REFERENCES

1. J.F. Relyea, R.J. Serne and D. Rai., Battelle, Pacific Northwest, PNL Report No. 3349 UC-70 (1980).

2. In: Sorption - Modelling and Measurement for Nuclear Waste Disposal Studies. NEA Workshop, p.70, Paris (1983).

3. J.F. Relyea and R.J. Serne., Battelle, Pacific Northwest, PNL Report No. 2872 UC-70 (1979).

4. F.B. Walton, T.W. Melnyk, R.G. Abry and K. Fleury. Chem. Geol. 36, 115, (1982).

5. B. Allard, H. Kipatsi and J. Rydberg., Chalmers University of Technology, Göteborg. KBS 55 (1977).

6. R.J. Serne and J.F. Relyea., Battelle, Pacific Northwest, PNL Report No. 3997 (1982).

7. W.R. Daniels., Los Alamos Labs, LA-8586-PR (1980).

8. M.H. Bradbury, D. Lever and D. Kinsey in: Scientific Basis for Nuclear Waste Management V, edited by W. Lutze, (North Holland, N.Y., 1982), pp. 569-578.

9. M.H. Bradbury in: Sorption - Modelling and Measurement for Nuclear Waste Disposal Studies. NEA Workshops, p.62, Paris (1983).

10. M.H. Bradbury. Harwell Report AERE R 10836 (1983).

11. M.H. Bradbury unpublished results.

12. K. Skagius and I. Neretnieks., Royal Institute of Technology, Stockholm. KBS Technical Report 82-12 (1982).

13. T. Eriksen and A. Jacobsson., Royal Institute of Technology, Stockholm. KBS Technical Report 81-06 (1981).

AN ANALOGUE VALIDATION STUDY OF NATURAL RADIONUCLIDE MIGRATION IN CRYSTALLINE ROCKS USING URANIUM-SERIES DISEQUILIBRIUM STUDIES

J.A.T. SMELLIE*, A.B. MacKENZIE** and R.D. SCOTT**.
* Swedish Geological Co., Box 1424, 751 44 Uppsala, Sweden.
** Scottish Universities Research Reactor Centre, East Kilbride, Glasgow G75 0QU, Scotland.

ABSTRACT

Concentrations and isotope ratios of natural decay series radionuclides have been studied in three contrasting crystalline rock drillcore sections intersecting water-conducting fractures deep in the bedrock. Radioactive disequilibria resulting from rock-water interactions were observed in two of the cores. These indicated uranium migration along distances of 40 cm or more on a timescale of 10^6 years in conjunction with thorium immobility under the same conditions. Fracture surface minerals showed a high affinity for radionuclide retardation and a limit of about 3 cm is suggested for the migration of radionuclides from fracture fluids into the saturated rock. This limit may correspond to enhanced matrix porosities resulting from earlier hydrothermal activity along the same channels.

INTRODUCTION

Assuming a "worst case" situation in which progressive corrosion of the near-field engineered barriers would result in the slow release of radioactive waste into meteoric waters circulating within the host bedrock, far-field retardation processes would qualitatively become the most important factors in controlling the release of radionuclides to the biosphere. During transport along hydraulically connected fractures in the host bedrock, radionuclides in solution will be subject to retardation processes such as sorption and migration into the water saturated rock matrix.

To date, much of the work and model derivation relating to radionuclide migration has been based upon short-term laboratory experiments. An obvious means of validating such work is to study the migration of natural radionuclides during geological timescales within crystalline bedrock, and the nuclides of the natural uranium decay series present a useful range of physical and chemical properties for this purpose. In particular, the various parent/daughter equilibria can be used to provide information on rock-water interactions. Since the initial work on this topic [1] such methods have become well established and have been applied in studies of a wide range of geological environments [2, 3, 4].

As part of a jointly-funded programme of research between the Swedish Nuclear Fuel and Waste Management Co. (SKB) and the Swiss Nationale Genossenschaft für die Lagerung radioaktiver Abfälle (NAGRA), drillcore lengths intersecting suitable water-conducting fracture zones were chosen from three contrasting hydrogeological environments; two in Switzerland (Böttstein and Grimsel) and one in Sweden (Kråkemåla), all located in granite bedrock. The samples analysed represented profiles from the intersected fracture face into the bedrock.

SAMPLING AND ANALYSIS

Drillcore lengths of 40 - 55 cm were chosen from the respective areas; these varied in diameter from 56 mm (Swedish sample) to 76 mm (Swiss sample). A thin rock slice was removed along the drillcore length for mineralogical and related studies. A portion of the core was retained for reference and the remainder was sectioned into 1 cm slices for analysis. The mineral coatings scraped from the inside surface of the fracture zone constituted another sample. The individual slices were powdered and subdivided for natural decay series analysis, INAA and Fe^{2+} determinations. Natural decay series measurements and INAA analysis were carried out using proven techniques [5, 6].

GEOLOGICAL SETTING AND MINERALOGY OF THE DRILLCORES

Kråkemåla Drillcore (Kl: 317.85-318.40 metres)

The Kråkemåla specimen was selected at a depth of approximately 318 m from one of three test holes drilled in the Götemar granite, located along the south-eastern coast of Sweden [7, 8]. The chosen core is traversed by a single, open fracture zone at 45° to the vertical drillcore length; from geophysical logging investigations the fracture zone is considered water-bearing. This is partly supported by fracture plane coatings of hematite and FeOOH-oxides. In addition, the feldspathic content marginal to the fracture zone shows evidence of alteration extending to 2-3 cm from the fracture plane.

The drillcore is representative of the Götemar coarse-grained granite and is macroscopically homogeneous. At greater resolution the rock is mineralogically variable consisting primarily of quartz (15-45 vol. %), plagioclase (30-42 vol. %) and potassium feldspar (30-50 vol. %) with subordinate biotite, muscovite and chlorite. Accessory minerals include interstitial amounts of magnetite, apatite, zircon, fluorite, sphene and monazite. Throughout the length of the drillcore sample the granite is weakly but uniformly hydrothermally altered; this is generally typical to the depths of maximum drilling in the massif [8].

The fractured area occupies a width of about 1 cm and is composed of a series of parallel to sub-parallel micro-fissures easily distinguished by FeOOH-oxide coatings. The central and largest fracture (max. 1.5 mm wide) is also partly infilled with hematite and subordinate chlorite. Macroscopically the influence of the fracture zone extends up to 2 cm into the host granite. This is apparent as FeOOH-oxide dustings present both interstitially within the rock matrix and included within altered to partly altered potassium feldspar phases; biotite and magnetite are also more markedly altered to chlorite and hematite respectively.

Böttstein Drillcore (BOE: 618.34-618.70 metres)

The Böttstein area is located in central northern Switzerland about 30 km north-west of Zurich and 5 km from the German border. Drilling in the area [9] encountered crystalline basement at a depth of approximately 300 metres. This consists of the Böttstein granite which extends down to at least 1500 m, the maximum depth of the drillhole. This granite is typically coarse-grained, rich in biotite, and porphyritic in texture accompanied by large phenocrysts of potassium feldspar. The chosen drillcore section was selected at a depth of approximately 618 m from a fracture intersection which was suspected as being water-conducting. The fracture was located near the contact between the granite and thin intruded horizons of pegmatite (up to 40 cm wide) and aplite (up to 1.5 m wide); the fracture occurs in the pegmatite only a few centimetres from the granite contact.

The selected drillcore length investigated consists of about 7 cm of pegmatite in contact with 13 cm of biotite-rich granite which is representative of the Böttstein porphyritic granite variety. In terms of average modal content (vol. %), the granite comprises 27% quartz, 26% plagioclase, 38% potassium feldspar and 8% biotite. Apatite and zircon are the most frequent accessories present.

The pegmatite containing the fracture zone is fresh, coarse-grained, and consists of quartz, potassium feldspar and albite, together with muscovite and tourmaline. Regarding the fracture faces, the dominant mineral phases are calcite, fine-grained quartz and clays (mostly interlayerings of illite and smectite) and sporadic apatite. Hydrothermal alteration of the granite extends for some 8 cm from the pegmatite contact. This is characterized by strong argillisation with illite/smectite as the dominant alteration phases; some calcite is contained in potassium feldspar and finely dispersed FeOOH-oxides give the rock a reddish tinge. A sealed fissure (0.5 - 2.0 mm wide) containing some calcite and clay minerals occurs at 8 cm and thereafter the granite is less altered (although the feldspars are still pigmented with iron-oxide) and the biotite content is higher.

Grimsel Drillcore (FLG: 94.12-94.52 metres)

The Grimsel area is located near the Grimsel Pass situated in central south Switzerland some 15-20 km from the Italian border. The site, which now constitutes the NAGRA underground test laboratory [10], was initially developed as an underground hydroelectric power scheme. The drillcore sample investigated was selected from a near-horizontal exploration borehole (SB 80.001:94.12-94.52 m) drilled into the bedrock from the access tunnel to the power station: the altitude of the borehole level is 1730 metres.

The granite is typically grey-coloured, fresh, mineralogically homogeneous in hand specimen, and medium- to coarse-grained in texture; the granite is also characteristically gneissose. The chosen fracture is fresh and intersects the core axis at 40°; no alteration on a macro scale was observed and no evidence of any fracture filling minerals. Its water-conducting properties are suspected on the basis of a large release of groundwater encountered while penetrating the fracture zone during drilling. The age of these fractures is uncertain, at least late Alpine to recent; the area is still considered tectonically unstable and the continuous formation of new fractures/joints cannot be excluded.

The average modal content of the granite core (expressed in vol. %) is 33% quartz, 34% potassium feldspar, 21% plagioclase, 7% biotite/chlorite, 1-2% muscovite, 1% hornblende and 2% epidote. The associated accessory phases are sphene, orthite, apatite, zircon, calcite and opaques. The feldspars have undergone some alteration to sericite, and the biotite to chlorite; magnetite is partly oxidized.

RESULTS

The following section summarizes the main observations relating to the natural decay series measurements. These results, together with the INAA data, particle track analysis and more detailed mineralogy, will be presented in full upon completion of this study.

Kråkemåla Drillcore

The drillcore contents and distributions of thorium and uranium and its daughter decay products are shown in Figure 1. The average uranium contents of uranium (15.67 ppm) and thorium (65.3 ppm) are anomalously high when compared with the average contents for world-wide granites. The average Th/U ratio of 4.4 indicates a general depletion of uranium which is consistent with previous work on the Götemar granite [8]. Most of the uranium is associated with zircon and monazite with subordinate amounts contained in sphene, magnetite, hematite, and ilmenorutile-type phases. Small amounts are also associated with sericite, epidote and FeOOH-oxide dispersions.

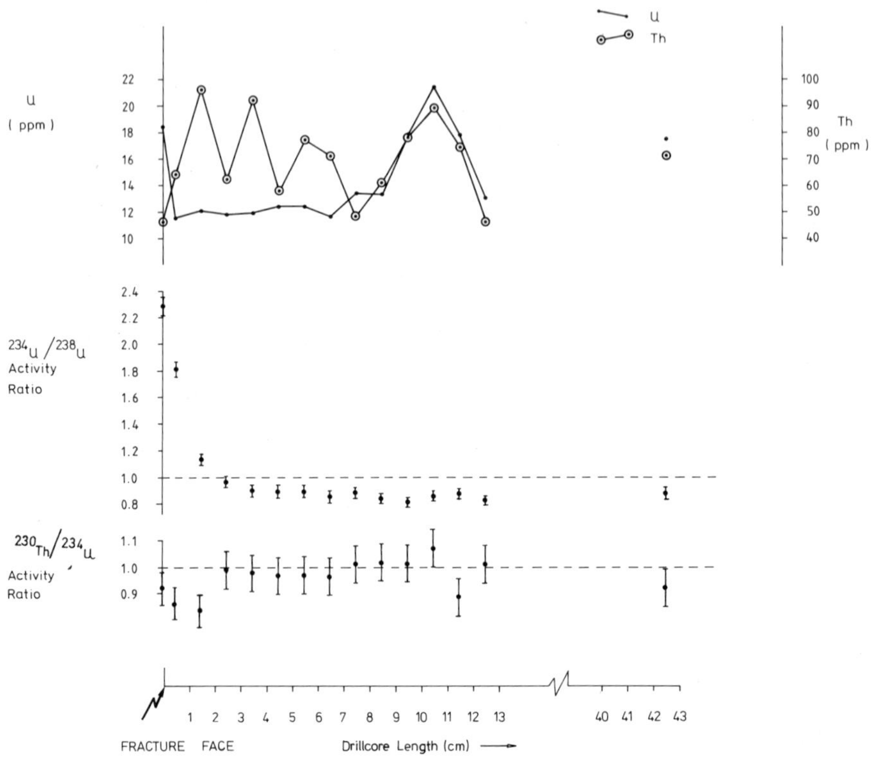

Figure 1. Contents and distributions of thorium and uranium and its daughter decay products (Kråkemåla Drillcore).

The distribution of uranium and thorium along the drillcore from the fracture zone shows that apart from the fracture face, which indicates an increase, the uranium content is fairly uniform for a distance of some 8 cm into the granite until a broad peak occurs over a length of 4 cm. In comparison, thorium indicates an antipathetic decrease at the fracture face, and then shows an extremely irregular fluctuation until it too forms a broad peak sympathetic to that observed for uranium. The similarity of the Th profile with those of the rare-earth elements (unpublished data) suggests that thorium has remained relatively immobile throughout the history of the granite, and its irregular distribution thus reflects the irregular distribution of the dominant Th-bearing mineral phase, monazite, which is of primary origin. With regards to uranium, which in comparison is easily mobilized and leached from granite provided the system has been open to fluid movement, Figure 1 shows some interesting features. There is a sympathetic increase in uranium and thorium from approximately 8 to 13 cm along the drillcore which shows that this part of the rock has probably retained its original uranium and thorium signature, although the intergranular labile uranium has almost certainly been removed. In contrast, the absence of any sympathetic uranium/thorium relationship within the initial 8 cm of the drillcore indicates that uranium has at some stage been leached from even the more resistate radioactive mineral phases possibly during some hydrothermal event, a process supported mineralogically by the partial alteration of monazite along this part of the core. However, uranium removal is not indicated from the fracture surface sample which instead shows a marked increase. This suggests that the groundwater/fracture uranium accumulation rather than the leaching is taking place.

Isotopically the $^{234}U/^{238}U$ ratio along the major length of the drillcore is less than unity indicating uranium leaching. In strong contrast, the granite marginal to the fracture zone is characterized by a systematic increase in the $^{234}U/^{238}U$ ratio to a value of 2.29 at the fracture face which indicates an accumulation of uranium. The $^{230}Th/^{234}U$ ratio is unity throughout most the core with a possible decrease being apparent close to the fracture.

The source of uranium giving rise to the enhanced concentrations and $^{234}U/^{238}U$ ratio values close to the fracture could be either uranium leached from the granite (as indicated by the $^{234}U/^{238}U$ ratios for the rest of the core), or from uranium transported in solution within the fracture migrating into the water-saturated rock. If the former process is responsible, then uranium migration over at least 40 cm is occurring whereas a constraint of about 3 cm can be applied if the latter process is taking place. The two processes are not mutually exclusive.

The $^{230}Th/^{234}U$ equilibrium and $^{234}U/^{238}U$ disequilibrium suggest that the process involved is either continuous or else indicates a time-scale long enough to allow re-equilibration of ^{230}Th with ^{234}U but short enough to maintain ^{234}U disequilibrium with ^{238}U, i.e. a time-scale of about 10^6 years.

The mechanism giving rise to uranium deposition at and near the fracture could be due to any or all of: a) precipitation at a boundary between differing physico-chemical conditions, b)sorption on the fracture filling minerals, and c) solution to solid recoil transfer.

Böttstein Drillcore

The drillcore contents and distributions of thorium and uranium and its daughter decay products are shown in Figure 2. Although not presented, higher amounts of Na and lower amounts of K and Rb support the mineralogical observations and help to distinguish the pematite from the granite. The pegmatite uranium contents range from 2.42 - 4.43 ppm and the thorium from 0.95 - 7.12 ppm (mean Th/U ratio of 1.06). In the adjacent granite the corresponding ranges are 3.37 - 5.73 ppmU and 21.69 - 33.49 ppmTh respectively (mean Th/U ratio of 6.35). The sampled core section indicates a significant depletion of uranium; thorium appears to have been immobile. Most of the uranium and thorium is believed to be associated with the accessory mineral contents which include zircon and apatite. Minor amounts of uranium and thorium may be accomodated by the mafic constituents such as biotite/chlorite, or as matrix sorptions associated with finely dispersed iron-oxides.

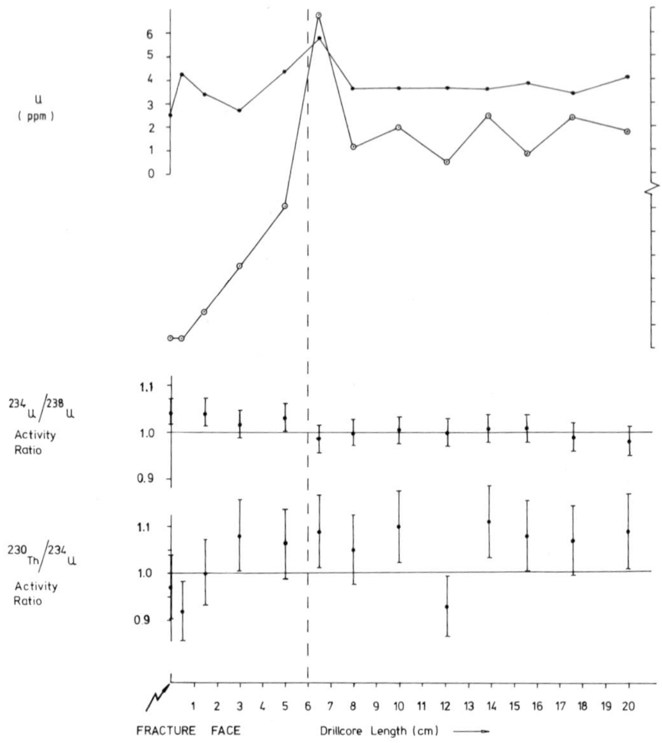

Figure 2. Contents and distributions of thorium and uranium and its daughter decay products (Böttstein Drillcore).

Within the granite portion of the drillcore the uranium content is uniformly distributed until the pegmatite is approached whereupon a small increase is observed. In the pegmatite the uranium decreases only to increase again some 3 cm from the fracture surface: a small decrease occurs at the fracture surface. The thorium is essentially sympathetic to that of uranium except that in the pegmatite a continuous decrease is observed.

As inferred by the mineralogy, the granite represented by this part of the drillcore has been subjected to hydrothermal temperatures in association with the emplacement of the adjacent pegmatite and aplite horizons. This appears to have remobilized and removed part of the available uranium in the rock resulting in low but uniform levels within both the granite and pegmatite. The observed increase of uranium and thorium at the granite/pegmatite contact may be due to limited remobilisation and concentration of both elements marginal to the pegmatite.

The $^{234}U/^{238}U$ ratio is unity for most of the core length although a slight increase is observed close to the fracture face. The above discussion of the Kråkemåla $^{234}U/^{238}U$ ratio distribution can again be applied in this case but the degree of uranium mobilisation involved is obviously less. The $^{230}Th/^{234}U$ results for Böttstein exhibit a value sligthly greater than unity except close to the fracture. This presents an obvious contrast to the Kråkemåla core and suggests solution to rock transfer of ^{230}Th produced by the decay of ^{234}U in solution. The difference between the Kråkemåla and Böttstein $^{230}Th/^{234}U$ results could be partly explained by a much more rapid fluid transport through the Kråkemåla core than in the Böttstein core.

Grimsel Drillcore

The uranium distribution along the drillcore length varies irregularly between about 5 and 9 ppm and this is even more accentuated by the thorium which ranges from 18 to 26 ppm. This variation is not affected by the fracture and can be explained by mineralogical inhomogeneities. There is no obvious relationship between the uranium and thorium, which suggests that the uranium and thorium contents are controlled by radioactive minerals which are distinctly uranium or thorium-bearing. Alternatively, it may reflect a redistribution of uranium within the rock; small additions of uranium are suggested by a mean Th/U ratio of 2.96. In contrast thorium has probably been immobile and has retained its original signature.

The consistent isotope values around unity indicate little or no isotopic mobilisation and thus an absence of rock/water interaction within the interval of the last 2 Ma. In all probability this fracture is young, having occurred at any stage from Alpine to recent times. A fairly recent joint may not have had time to react sufficiently with the groundwater which is now characterized by a chemical and isotopic signature obtained from other parts of the granite bedrock.

CONCLUSIONS

Although the data are incomplete, the most important conclusions resulting from this study are:

1. Two out of three of the investigated samples indicate that naturally-occurring uranium radionuclides have been remobilised by rock/water interaction processes within normal compact granite marginal to a water-conducting fracture zone at repository depths. Thorium does not exhibit mobilisation under the same conditions.

2. Rock/water interaction has taken place under predominately reducing conditions within measureable geological time, i.e. within the last 0.5 Ma as determined by the ^{238}U - ^{234}U - ^{230}Th decay series and may still be active at the present time.

3. Isotopic movement has been mostly from the granite to the fracture zone through distances of at least 40 cm.

4. Some radionuclide movement has also taken place from the fracture zone (via the groundwater) into the granite; the extent of penetration has been from 1-3 cm and this measure could therefore act as an upper limit in migration models. However, it may be an important prerequisite for penetration that the fracture system has been previously subject to hydrothermal processes resulting in a zone of enhanced porosity around the fracture. Furthermore, the mineralogical alteration products would make the rock more conducive to radionuclide sorption/retardation mechanisms.

5. Sorption and isotopic fractionation has occurred at the fracture/ groundwater interface in association with fracture zone minerals such as hematite, chlorite, clay minerals and FeOOH-oxides. The major source of these radionuclides is believed to be the groundwaters.

6. It is not known at the moment the mechanism of the radionuclide migration, i.e. by fracture-flow or by matrix diffusion processes. Some indication may be obtained when the data are complete.

ACKNOWLEDGEMENTS

This project was jointly funded by the Swedish Nuclear Fuel Supply Company (SKB) and the Swiss Nationale Genossenschaft für die Lagerung Radioaktiver Abfälle (NAGRA), whose support at all stages of the study is gratefully acknowledged. With regards to the Swiss samples and for constructive comments on the manuscript, the considerable help of Dr. Urs Frick (NAGRA) is also acknowledged.

REFERENCES

1. V.V. Cherdynstev, in Absolute Age Determinations of Geological Formations. Proc. Third Session Comm. 1955, p. 175.

2. H.P. Schwarz, M. Gascoyne, D.C. Ford, Chem. Geol. 36, 86-102 (1982).

3. J.N. Rosholt, J. Geophys. Res. 88, 7315-7330 (1983).

4. J.A.T. Smellie, J.N. Rosholt, Lithos, 17, 215-225 (1984).

5. M.P. Bacon, J.N. Rosholt, Geochim. Cosmochim. Acta, 46, 751-666 (1982).

6. A.B. MacKenzie, R.D. Scott, I.G. McKinley, J.M. West, Inst. Geol. Sc. Rep. No. FLPU 83-6 (1983).

7. P. Kresten, J. Chyssler, Geol. Fören. Stockh. Förh. 98, 155-161 (1976).

8. J.A.T. Smellie, J.S. Stuckless, Chem. Geol. (In press).

9. NAGRA Technische Berichte, NTB 85-01, (1985).

10. NAGRA Technische Berichte, NTB 85-34, (1985).

ESTIMATIONS OF CHANGES IN MICROCRACK-POPULATION, ELASTIC MODULUS, AND PERMEABILITY, DUE TO DIFFERENTIAL-THERMAL EXPANSION IN PLUTONIC ROCK SURROUNDING A NUCLEAR FUEL WASTE DISPOSAL VAULT

B.J. WILKINS*, Y. LINER* AND G.L. RIGBY*
*Atomic Energy of Canada Limited, Pinawa, Manitoba ROE 1L0

ABSTRACT

Atomic Energy of Canada Limited is studying the disposal of nuclear-fuel-waste in a vault in plutonic rock. An important part of this study is to develop models to assess the rate of escape of waste components to the biosphere. In these models time-dependent changes in the system are important. One such change is the increase in temperature of the rock mass immediately surrounding the vault, and in the far-field. Differential-thermal expansion between rock constituents (crystals), and the expansion of trapped water, will produce microcracking and, hence, possibly changes in intact-rock elastic modulus and permeability. This paper describes the model devised to estimate the extent of microcracking in rock surrounding a vault, and the anticipated changes in rock elastic modulus and permeability. The model is applied to a possible vault, located at various depths in plutonic rock typical of the Canadian Shield. It estimates the dimensions of an envelope outside of which the influence of the vault on microcracking will be negligible.

INTRODUCTION

Atomic Energy of Canada Limited is studying the disposal of nuclear fuel waste in a vault in plutonic rock, and is developing models to assess the rate of escape of waste to the biosphere. Part of the work is to estimate how heating of the rock mass may produce microcracking and, hence, changes the intact rock elastic modulus (E) and permeability (B). A model and a computer code MCROC have been developed for these purposes [1,2]. Important features of the model are 1) that the influence of water is accounted for and 2) that the extent of microcracking and the resulting changes in E and B are overestimated. This paper describes the application of the model to a rock mass containing a disposal vault. The vault is designed to accept Canadian nuclear fuel waste.

THE MODEL

The model considers that a crack exists in each crystal facet, throughout the rock mass. These cracks are thin and penny-shaped, with an inital diameter of 2Lo. The ratio of crack diameter to facet diameter is 2Lo/W. This ratio and the facet diameter are different at each facet.

As the rock heats up, elastic strains are developed due to differential- thermal expansion (DTE), which occurs at each crystal facet. The strain magnitude, at a given facet, depends on the local temperature increase (ΔT) and the difference in thermal expansivity ($\Delta\alpha$) between the crystals sharing the facet.

The local elastic strain energy, due to DTE, is transformed into new crack surface (energy) at each facet. The kinetics of the process i.e., the velocity (V) of crack extension, at each facet, is given by

$$V = dL/dt = AK_I^n \tag{1}$$

where A and n are constants for a given rock type and environment, K_I is the stress intensity factor in tensile-opening mode, L is the crack radius and t is time [3]. K_I depends on Lo, W and the local stress (σ), due to DTE. Crack extension reduces the local values of stiffness (S) and σ; hence, K_I and V are altered. In the code MCROC the following equations are solved:

$$dU/dt = dT/dt\ (\Delta\alpha W)/2 \tag{2}$$

$$\text{and } dL/dt = AY^n\ (US/W)^n\ L^{n/2} \tag{3}$$

where U is displacement, σ = US/W, and Y is a local geometry constant related to L and W only, to give K_I, σ, V and L as functions of time and of thermal history experienced by the facet.

The rock mass is represented by a statistically chosen (Monte Carlo method) facet-microcrack array. The parameters specified are Lo/W, W and $\Delta\alpha$. MCROC predicts how the microcrack array develops, with time, at various points (different thermal histories) in the rock mass. Specifically, it predicts the distributions of L from which $\bar{L}^2$ and $\bar{L}^3$ are estimated. The changes in intact rock elastic modulus (E/Eo) and permeability (B/Bo) are then estimated from:

$$E/Eo = 1/(1+16\ N\ \bar{L}^3/9) \tag{4}$$

and,

$$B/Bo = \bar{L}^2/\bar{L}o^2 \tag{5}$$

where Eo is the value of E for uncracked rock and Bo is the value of B at time zero [2].

Water enhances stress-dependent corrosion and, therefore, the rate of microcracking. Its influence is accounted for by the constants A and n in equation (1). In addition, the expansion of water, trapped in closed cracks, can also produce microcracking. Hence, in the model, every microcrack is assumed to be closed and full of water, at t = 0. As the temperature increases, each crack is extended to accommodate the water expansion. For this calculation the water temperature is taken to be ΔT plus 1.5 x 10^{-2} °C•m^{-1} depth. Water expansion increases $\bar{L}^3$ by a factor of $\sim(1+\alpha^*\ \Delta T/2)^3$, where α^* is the average volume thermal expansivity for water, for the temperature range concerned (0 to 100°C). The factor is $\sim$1.09, using the α^* value for 85°C, and assuming ΔT = 85°C. Similarly, a factor of $\sim$1.06 is found for the increase in $\bar{L}^2$.

MODEL APPLICATION

The model is applied to a disposal vault designed to accept Canadian nuclear fuel waste. Using the code HOTROK [4], the rock mass temperature is calculated as a function of time and position. The following assumptions are made: the vault is 1400 x 1400 x 10 m (height) in size and is located at a depth of 500 to 1000 m; the rock mass has isotropic thermal properties with thermal diffusivity and conductivity of 43.2 m^2•a^{-1} and 3.0 W•m^{-1}•°C^{-1}, respectively; the vault contains used CANDU† fuel (cooled for 10 years), producing an initial thermal loading of 18 W•m^{-2} [5]; the

† Canada Deuterium Uranium

surface of the earth remains at 0°C; the rock is wet and the intact rock elastic modulus (E) is 64.3 GPa. Hence, the isotherms generated, at t > 0, are symmetrical about a vertical line through the vault centre (two-planes of symmetry), and are almost symmetrical about the horizontal plane.

The HOTROK results are shown in Figures 1 and 2. In Figure 1 are shown two envelopes (not isotherms) that surround the vault at the 750-m depth. Each envelope is for a different maximum temperature reached during the heating and cooling cycle. The time at which the maximum temperature occurs is different at all points on each envelope, except for mirror-image points in the horizontal plane. The thermal histories for the highest elevations on the two envelopes correspond to the heating curves H and I, shown in Figure 2. The maximum temperatures seen by the inner and outer envelopes are 62.8 and 43.3 °C, respectively. The results for vault depths from 500 to 1000 m are similar. At 500 m, the influence of the surface of the earth is greatest, and the tops of the envelopes are nearer to the horizontal plane of the vault. The tops of the inner and outer envelopes are ∼61 and ∼180 m, respectively, above the vault plane at 500 m, in contrast to ∼65 and ∼215 m for the 750-m deep vault. Also, shown in Figure 2 is the thermal history (J) of the vault centre. The maximum temperature increase at this point is 85°C. This is to keep the temperatures of the nuclear fuel waste containers at $\leq$ 100°C.

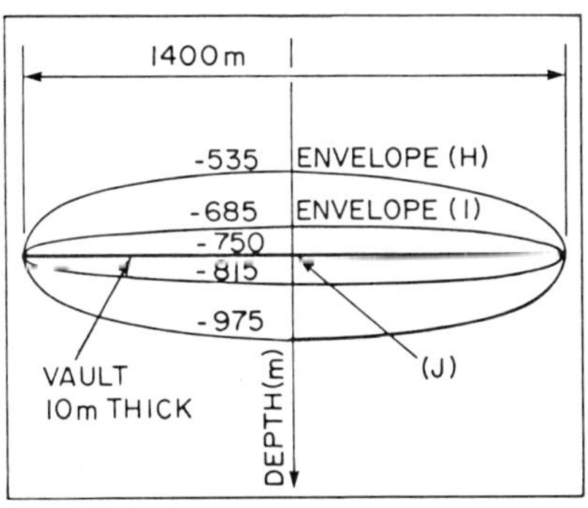

Figure 1. Two envelopes (not isotherms), surrounding a vault, at which maximum temperatures are reached regardless of time. The shapes of the envelopes are approximately ellipsoidal

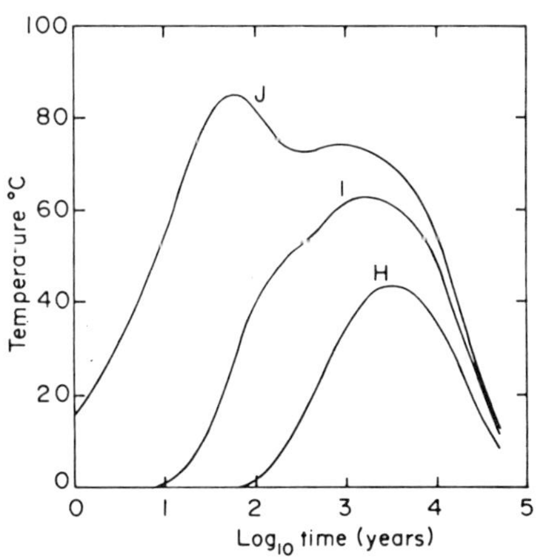

Figure 2. Thermal history for the vault centre (J), and the two envelopes (I and H) shown in Figure 1

The parameters Lo/W, W and $\Delta\alpha$ are assumed to have triangular-shaped frequency distributions. This is for convenience and because the distribution form is a reasonable approximation. The distribution is specified in terms of minimum, mode and maximum values of the variables. Table I shows how the rock parameters are assumed to be typically distributed in plutons in the Canadian Shield. This set of parameters is referred to, in this paper, as the standard case. The maximum value of $\Delta\alpha$ is set at 1.5×10^{-5} °C^{-1}, because it is the largest value expected in plutonic rock i.e., the largest thermal expansivity difference possible between quartz and feldspar at < 100°C [6].

Table I. DISTRIBUTION DETAILS FOR THE ROCK PARAMETERS THE STANDARD CASE FOR TYPICAL PLUTONIC ROCK

PARAMETER	MINIMUM	MODE	MAXIMUM
Lo/W	0	0.05	0.1
$\Delta\alpha$(°C^{-1})	0.30E-5	0.10E-4	0.15E-4
W(m)	0.25E-3	0.10E-2	0.25E-2

RESULTS AND DISCUSSION

E/Eo and B/Bo have been estimated, for t = 0, 1000 and 50 000 years, for the ten cases shown in Table II. The letters STD mean the rock has the parameters shown in Table I. Otherwise, new distribution details for the parameter(s) are shown. The results are shown in Tables III and IV, and in Figures 3 and 4. These ten cases were chosen to include the three temperature histories shown in Figure 2, and to show the influence of different Lo/W, W and $\Delta\alpha$ distributions. The two A and n pairs shown are for granite, taken from the Lac du Bonnet batholith (n=29, $A=10^{-178}$), and a gabbro (n=19, $A=10^{-124}$), both in water at 80°C [3].

Figures 3 and 4 show how L/W and σ vary with time at the vault centre (case 4/Table II). In each estimation the sample array consists of one-hundred facet-microcracks. For clarity, in these figures, the behaviour of only twenty facet-microcracks are shown.

The results in Tables III and IV show that the extent of microcracking increases with the parameters Lo/W, W and $\Delta\alpha$.

Cases 1, 2 and 3 in Table II are for the vault centre, and for the two maximum-temperature envelopes shown in Figure 1 (see also Figure 2). These cases assume the rock parameters are distributed as shown in Table I. They represent the best estimates of the extent of microcracking, and its consequences, in the rock mass. The results (Tables III and IV) show that, even at the vault centre after 50 000 years, the estimated changes in E and B are negligible ($\sim$0.02 and $\sim$3 percent, respectively).

Cases 4, 5 and 6 in Table II are as cases 1, 2 and 3, except that the W distribution has been changed. Now the minimum, mode and maximum values for crystal diameter are 0, 10 and 30 mm, respectively. The average crystal diameter is (40/3) mm. Although crystals can be this large (or even larger) in plutonic rock, they are not typical. These cases were chosen to illustrate the influence of crystal size, and to produce an overestimation of the extent of microcracking. Even so, the changes in E and B at the 65-m (62.8°C) envelope are not excessive, and at the 215-m (43.3°C) envelope are unimportant.

Table II. DETAILS OF CASES EXAMINED

CASE IDENTITY	DISTRIBUTION DETAILS	n	$-\log_{10}A$	TEMPERATURE HISTORY (CODE* AND MAXIMUM TEMPERATURE)
Gabbro				
1	STD (see Table I)	19	124	J (85°C)
2	STD	19	124	I (62.8°C)
3	STD	19	124	H (43.3°C)
4	W(0,.5E-2,.15E-1)	19	124	J
5	W(0,.5E-2,.15E-1)	19	124	I
6	W(0,.5E-2,.15E-1)	19	124	H
7	W(0,.5E-2,.15E-1) Δα(.3E-5,.15E-4,.15E-4)	19	124	H
8	Lo/W(0, .1, .5)	19	124	H
Granite				
9	STD	29	178	I
10	W(0,.5E-2,.15E-1)	29	178	I

* see Figure 2

Cases 9 and 10 in Table II show the behaviour of granite taken from the Lac du Bonnet batholith. These cases may be compared directly to cases 2 and 5 for the gabbro. The results show that the granite is more prone to develop microcracks than the gabbro.

Table III. E/Eo AT VARIOUS TIMES FOR CASES SHOWN IN TABLE II

Case Identity	E/Eo Time in Years 0	1000	50 000
Gabbro			
1	0.99926	0.99923	0.99922
2	0.99926	0.99924	0.99924
3	0.99926	0.99925	0.99925
4	0.99925	0.82503	0.79634
5	0.99925	0.94727	0.88496
6	0.99925	0.99924	0.99060
7	0.99925	0.99924	0.94732
8	0.93614	0.93333	0.83405
Granite			
9	0.99926	0.99924	0.99923
10	0.99925	0.79756	0.78746

Table IV. B/Bo AT VARIOUS TIMES FOR CASES SHOWN IN TABLE II

Case Identity	B/Bo Time in Years 0	1000	50 000
Gabbro			
1	1	1.033	1.033
2	1	1.023	1.024
3	1	1.009	1.014
4	1	37.78	44.05
5	1	9.140	23.70
6	1	1.009	2.436
7	1	1.009	9.095
8	1	1.025	20.46
Granite			
9	1	1.023	1.024
10	1	43.52	46.54

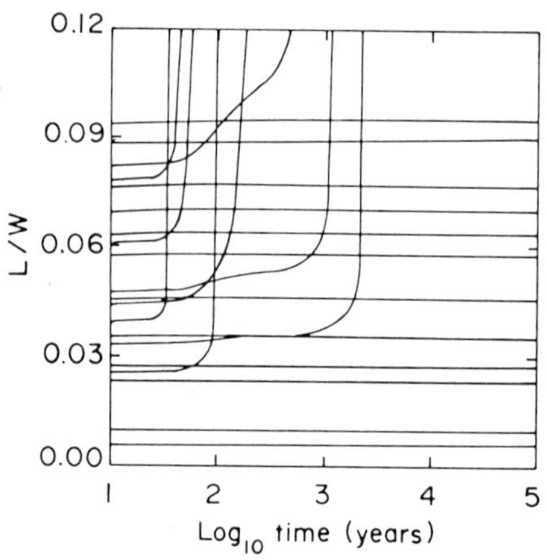

Figure 3. The ratio of half-crack length to facet diameter (L/W) versus time for case 4 in Table II. When L = W/2 a facet is completely cracked

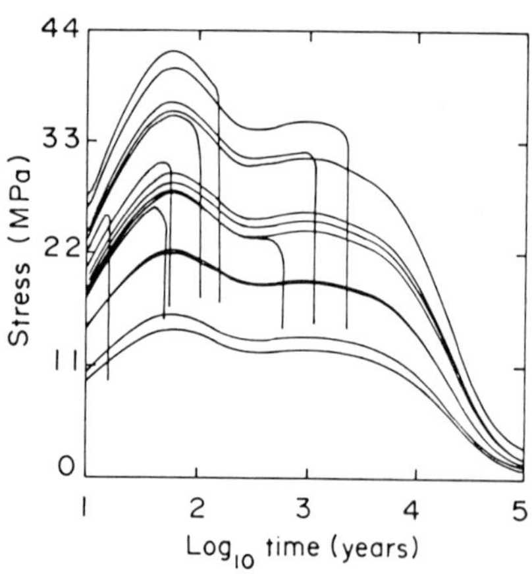

Figure 4. The development of stresses, due to differential-thermal expansion, the individual cracks shown in Figure 3

CONCLUSIONS

A typical pluton, in the Canadian Shield, containing a disposal vault designed for Canadian nuclear fuel waste, will experience a thermal perturbation for $\sim$50 000 years, after waste emplacement.

In the centre of the vault and beyond, assuming typical plutonic rock (either wet gabbro or wet granite), changes in the intact rock elastic modulus and permeability, due to microcracking arising from the thermal perturbation, will be negligible.

Increasing the crystal sizes of rock constituents increases the extent of microcracking, due to differential-thermal expansion. However, even an entire rock mass composed of large crystals (40/3 mm in diameter on average) could be tolerated. It would produce an intact rock elastic modulus decrease of $<$ 22 percent, in 50 000 years, 65 m beyond the vault. The corresponding increase in intact rock permeability would be less than a factor of 47.

REFERENCES

1. B.J.S. Wilkins, G.L. Rigby and J.R. Dryden, ISRM Symposium Aachen, pp. 1013 to 1019, 1982.

2. B.J.S. Wilkins, Y. Liner and G.L. Rigby, 26th U.S. Symposium on Rock Mechanics, Rapid City, South Dakota, 1985.

3. B.J.S. Wilkins, A.R. Reich and W.R. Wallace, Atomic Energy of Canada Limited Technical Record TR-264*, 1984.

4. W.G. Mathers, Atomic Energy of Canada Limited Technical Record, TR-336*, 1985.

5. L.J. Clegg and J.R. Coady, Atomic Energy of Canada Limited Report, AECL-4436/1, 1977.

6. R.W. Davidge, J.R. McLaren and I. Titchell, in Fracture Mechanics of Ceramics, edited by R.C. Bradt, D.P.H. Hasselman, and F.F. Lange, (Plenum Publishing Corporation, New York, N.Y.) Vol. 5, 6 (1982).

---

* Unrestricted, unpublished report, available from SDDO, Atomic Energy of Canada Limited Research Company, Chalk River, Ontario, K0J 1J0.

VALIDATION OF HYDROGEOLOGIC MODELS TO DESCRIBE GROUND-WATER FLOW IN THE CRYSTALLINE BASEMENT OF NORTHERN SWITZERLAND

ROBERT W. ANDREWS*, FRANCESCO KIMMEIER**, PIERRE PERROCHET** AND LASZLO KIRALY**
*INTERA Technologies, Inc., 6850 Austin Center Blvd., Austin, TX 78731
**Centre d' Hydrogeologie, Universite de Neuchatel, CH 2000 Neuchatel, Switzerland

ABSTRACT

Two hydrogeologic conceptual models have been developed to describe the ground-water flow regime in the crystalline rocks of northern Switzerland. The purpose of these models has been to characterize the regional and local ground-water flow systems and to define the ground-water travel paths and velocities from potential waste disposal areas to the biosphere. Several parameters have been used as validation tools in the modeling process. These include potentiometric data, volumetric discharge rates, infiltration rates, bulk and isotopic geochemistry, and temperature distributions. The results of utilizing these validation tools are described.

INTRODUCTION

The terms verification and validation have been applied by various researchers to indicate how well a model approximates reality. The term <u>verification</u> is used to describe the ability of a numerical model to approximate a mathematical model (e.g., comparison of analytical and numerical solutions of a mathematical model to ensure that the equations are solved as intended). The term <u>validation</u> is used to describe the ability of a mathematical model to describe a physical system (e.g., the appropriateness of Darcy's law to describe ground-water flow in a porous or fractured media) and the ability of a conceptual model to reproduce the observed conditions ("reality").

Validation of hydrogeologic conceptual models is difficult. This is due to the large number of degrees of freedom in the model (e.g., permeability distribution in space and applied boundary conditions) and the sparsity of available information with which to compare the simulated results. As a result, several models can be formulated which adequately explain the available data. In addition, it is difficult to rule out the possibility that a single measurement may only be representative of very local hydrogeologic conditions which are impossible to incorporate in large scale models. Given these difficulties, the present paper will describe the results of initial attempts to validate the conceptual models of ground-water flow in northern Switzerland.

HYDROGEOLOGIC SETTING

The area of investigation in the present study is north-central Switzerland. This area has been identified by NAGRA (Swiss National Cooperative for the Storage of Radioactive Wastes) as being potentially suitable for radioactive waste disposal [1]. The area of interest is illustrated on Figure 1. The crystalline bedrock in this area is being investigated as the potential host rock for waste storage. While the crystalline is exposed in the southern Black Forest (Figure 1), the only means of evaluating the geologic, geotechnical, and hydrogeologic characteristics of this rock mass in northern Switzerland is by exploratory drilling. To date, six boreholes have been drilled (five have encountered the crystalline) by NAGRA and these boreholes represent the bulk of the information available on the crystalline, although numerous other oil/gas exploration boreholes and water wells have been drilled through the sedimentary sequence overlying the crystalline (Figure 1).

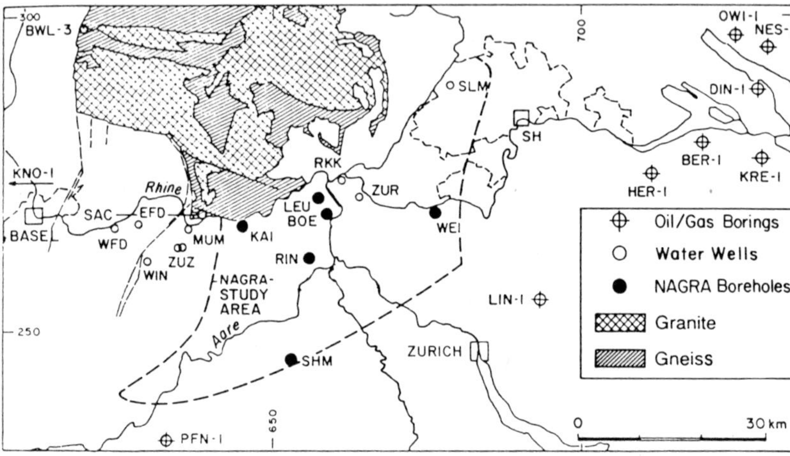

Fig. 1 NAGRA Study Area and Deep Exploratory Borings

The geology of north-central Switzerland is characterized by gently dipping sedimentary rocks overlying a crystalline basement. In places, sediments have been deposited in deep, structural troughs within the crystalline. Several faults displace the sedimentary sequence, although in general these features are believed to die out in the plastic salt/anhydrite of the Middle Muschelkalk which overlies the crystalline basement in the study region [2]. The sedimentary rocks in the study area may be broadly segregated into aquifers and aquitards based on their regional hydrologic characteristics. The available data to define the permeability of each hydrostratigraphic layer are presented in [3]. As these data are uncertain, a range of permeability values has been used in [3] to address the sensitivity of the model results to this parameter. The permeability of the crystalline generally decreases with depth, although some transmissive zones do exist at relatively great depths [4].

HYDROGEOLOGIC MODELS

The ground-water flow regime of north-central Switzerland has been modeled using the three-dimensional finite element program FEM301, which solves for steady-state ground-water flow in an equivalent porous media [5]. Two hydrogeologic models have been constructed to evaluate the ground-water flow regime in the area of interest. A regional model has been constructed to account for all recharge to and discharge from the crystalline within the boundaries of the model. This model extends from the Aare Massif in the central Alps on the south to the Black Forest on the north and from Lake Constance on the east to the Aare River on the west. The regional model encompasses an area of 23,000 km^2 and is comprised of about 15,000 nodes and 5,600 three-, two-, and one-dimensional quadrilateral elements. In order to evaluate the effects of greater structural, stratigraphic and topographic complexity in the area of principal interest a more refined, local model has also be constructed. The local model encompasses an area of about 900 km^2 and consists of 20,000 nodes and about 6,000 quadrilateral elements. Both of these models are described in detail in NAGRA Technical Report 84-50 [3].

The hydrogeologic models indicate that the principal recharge to the crystalline occurs at the crystalline outcrop in the Aare Massif and Black Forest. If the crystalline is not hydrologically continuous between these two outcrops, significant vertical leakage through the Tertiary Molasse in central Switzerland can recharge the crystalline [3]. The principal discharge of the crystalline is to the Rhine River along the border of Switzerland and Germany. The regional flow system in the crystalline is controlled by the regional confinement provided by the Middle and Lower Muschelkalk aquitard. If the permeability of this confining layer is increased (from 10^{-11} to 10^{-9} m/sec), then more local flow systems extend into the crystalline and crystalline ground waters may discharge to the Aare, Reuss, and Limmat Rivers. The simulated equipotentials and general ground-water flow directions in the crystalline are illustrated in Figure 2 for the regional model.

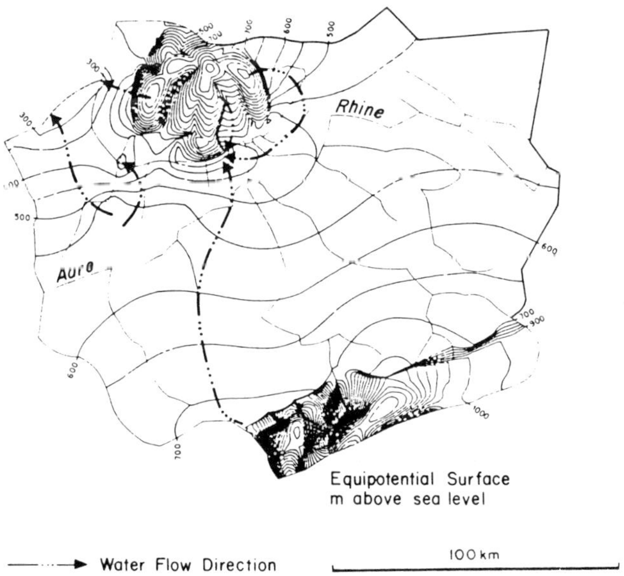

Fig. 2 Simulated Equipotentials and Ground-Water Flow Directions in the Crystalline

MODEL VALIDATION

The hydrogeologic conceptual model should be validated to the extent feasible prior to using the results from such a model for predictions of repository performance. This validation may take several forms. At a scale of kilometers or tens of kilometers, the model should be able to reproduce the general ground-water flow directions and overall volumetric flow rates within and between the various transmissive hydrogeologic units. At a scale of meters or tens of meters one would like the model to predict the travel path(s) and velocities of dissolved radionuclides which may escape the engineered containment facility. Validation of the travel path(s) and velocities is virtually impossible without extensive in-situ testing and in fact, there will always be some residual uncertainty in defining the precise flow path. However, validation of the general flow regime is possible using the following observations:

- potentiometric data
- volumetric discharge rates
- infiltration rates
- bulk geochemistry
- isotopic (stable and radioactive) geochemistry
- temperature

While each of these observations has an associated uncertainty, combining them should yield a consistent "picture" of the ground-water flow system. A brief summary of the status of the efforts to validate the regional groundwater flow moedl in northern Switzerland is described below.

Potentiometric Data

The observed potentiometric data of the deep aquifers in the study region is presented in [3]. Numerous additional data exist for the shallow alluvial aquifers, but as the heads in these aquifers essentially mimic the surface topography they are not considered here. Of the 80 wells in the study area, 14 are completed in the crystalline. In some cases, the wells are open over several formations, hence a mixed head is measured. In other cases, numerous tests have been performed in a single formation within an individual well. This is particularly true of the testing performed by NAGRA in the crystalline at Boettstein, Weiach, Schafisheim, Kaisten and Leuggern.

Measured heads are subject to many uncertainties. The potential error associated with heads measured by short term pressure tests is discussed by Grisak and Pickens [4]. Additional errors exist in the data from the oil industry (generally drill stem tests) and from thermal and mineral water wells where the exact test conditions are often not known or well construction, test procedures, or borehole pressure history effects may impact the measured head. For these reasons the reported heads should only be considered accurate to within plus or minus 10 m, with this error being up to several tens of meters for many of the wells in the study region.

Comparison of observed and simulated heads in the crystalline for three regional model runs is presented in Table 1. Run E06V01 and E11V01 differ principally by the permeability of the uppermost aquifers in the Alps. Run E19V02 utilized prescribed infiltration rates as opposed to a prescribed head boundary condition. With the exception of Schafisheim the agreement between observed and simulated heads is generally quite favorable. It has been suggested [3] that the low head measured in the crystalline at Schafisheim is due to a "short circuit" through the confining sedimentary strata.

Volumetric Discharge Rates

The volumetric discharge rates calculated along all modeled rivers are compared to observed base level stream flows at five gauging stations (Table 2). We assume that the lowest observed stream flow is totally supplied by groundwater discharge as opposed to surface runoff. In both

Table I

Comparison of Observed and Simulated Heads within the Crystalline

Well	Code	Observed Head	Simulated Head (m) E06V01+	E11V01	E19V02
Wintersingen	WIN	<385	389	397	481
Weierfeld	WFD	<281	311	363	430
Engerfeld	EFD	<380	311	363	430
Sackingen a	SAC	292	317	316	324
Sackingen b	SAC	289	317	316	324
Kaisten - 310*	KAI	323			
- 482		326	319	320	338
- 798		354			
- 999		351			
Zurzach Z1	ZUR	>344	395	398	453
Bottstein - 394	BOE	370			
- 399		368			
- 456		371			
- 621		368	358	373	442
- 791		363			
- 1326		414			
- 1498		396			
Zuzgen I	ZUZ	<412	360	355	402
Zuzgen II	ZUZ	<406	360	355	402
Weiach	WEI	>415	433	455	480
Badenweiler 3	BWL	>415	417	388	414
Schafisheim	SHM	365	464	526	535
Leuggern	LEU	362	351	330	418

*-310: Indicated the depth (m)
+ see text

Table II Comparison of Observed and Simulated Stream Discharge

Stream Gauging Station	Observed Base Flow (m^3/sec)	% of Catchment Within Model Boundaries	Base Flow Within Model Boundaries (m^3/sec)	Simulated Discharge (m^3/sec) *E06V01	E11V01	E11V02	E19V02
Basel	202.	54.	109.	9670.	791.	479.	104.
Rekingen	120.	47.	56.	2230.	263.	116.	38.
Baden	26.	100.	26.	2960.	154.	89.	18.
Mellingen	29.	100.	29.	3780.	220.	130.	18.
Brugg	93.	37.	34.	661.	102.	91.	26.

* See text

Run E06V01 and E11V01 the water table surface is prescribed. Therefore decreasing the permeability of the uppermost aquifer in the Alps (run E11V01) decreases the recharge and discharge. Run E11V02 differs from E11V01 in that the prescribed heads in the Alpine recharge areas are decreased by 500 m, causing a reduction in recharge and discharge. Run E19V02 utilizes prescribed infiltration rates which have been adjusted to better approximate the observed discharges [3]. It is interesting to note that, while applying infiltration rates improves the fit between observed and simulated stream discharges (Table 2), the fit worsens between the observed and simulated heads in the area of interest (Table 1). This reinforces the decoupled nature of the shallow and deep ground-water flow regimes.

The ability to directly compare observed and simulated groundwater discharges as a means of model validation must be treated with caution. First, the observed base flow may be controlled by the alluvial aquifers not the deeper sedimentary or crystalline aquifers. Second, the simulated discharge is almost totally a function of the permeability and gradients in the outcropping aquifers, not the relatively small fluxes in the deeper confined aquifers. Finally, the degree of averaging incorproated in the model implies that small drainage systems which do contribute to the observed discharges are not explicitly modeled. In summary, it is difficult to place a great deal of confidence in the comparison of observed and simulated discharge rates as a validation tool. It does indicate if the model is order of magnitude correct but nothing more precise. It should be noted that detailed water balance assessments over small areas also aid little due to the averaging required in the models.

Infiltration Rates/Water Table Surface

The boundary condition on the upper surface of a ground-water flow model may be either a prescribed infiltration rate or a specified head. If the infiltration rate is assigned, the simulated water table may be compared to the observed water table to ensure some degree of compatibility. If the water table surface is specified, the calculated infiltration rates may be compared to observed recharge rates to again check the validity of a particular simulation. Because in both cases the results are strongly controlled by the permeability of the uppermost element, any differences between observed and simulated near-surface heads or infiltration rates is just as likely due to permeability uncertainty as it is to the prescribed boundary condition.

Only a qualitative significance can be placed on the comparison of observed and simulated near-surface hydrogeology. First, there are no direct observations of infiltration rates in Switzerland. Even if there were, they would only represent very local conditions which would be impossible to extrapolate given the variable geology, topography, meteorology and vegetation. Secondly, there are no direct observation of heads in the recharge areas. We may assume these heads mimic the surface topography, but to what extent is difficult to define (especially in the Alps).

In summary, great care must be utilized in attempting to validate a large scale hydrogeologic model by the use of near-surface observations. Even if a good "match" is obtained, this implies very little with regards to the deep groundwater flow regime which is of greatest interest. The approach taken in [3] has been to vary the upper boundary conditions to evaluate their impact on the deep flow system of interest.

Bulk Geochemistry

The spatial variability of the dissolved constituents in groundwater can be used to provide qualitative statements concerning the general groundwater flow directions. The following conclusions have been made by Schmassmann and others [6]:

1) If one assumes the salinities observed in the crystalline at Schafisheim (about 8 g/l) are representative of crystalline

groundwaters south of the Permo-Carboniferous trough, then the low salinities (1 to 2 g/l) observed at Kaisten, Boettstein (upper 500 m) and Leuggern imply recharge from the north (Figure 3).

2) Gypsum saturation variations within groundwaters of the Upper Muschelkalk indicate a generally south to north component of flow in the region between the Laegern and the Rhine.

These observations have been used to insure that the general flow regime simulated by the model agrees with the geochemistry (compare Figures 2 and 3). However, as observed geochemical differences can often be the result of local hydrogeologic conditions (for example mixing of different formation waters) some care must be utilized in interpreted regional flow regimes from only geochemical evidence.

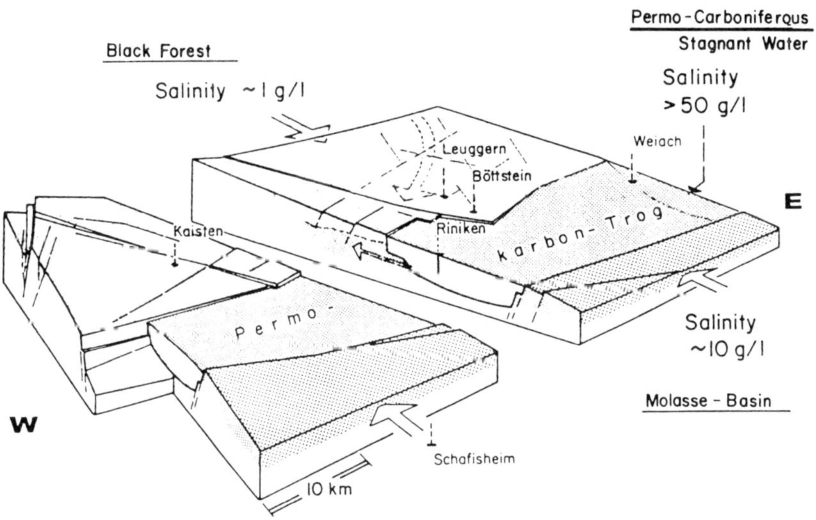

Fig. 3 Flow Regimes in the Crystalline Based on Salinity

Isotopic Geochemistry

Pearson [7] has concluded from his evaluation of the isotopic data from Boettstein that (1) the residence time of the Muschelkalk groundwater is 17'000 years (± 6'000 years) based on corrected C-14 and (2) the residence times of C-14 in the Buntsandstein/Upper crystalline groundwater range between 6'000 and 12'000 years.

Great care must be utilized in the interpretation of "residence times" generated from isotopic measurements due to the effects of possible geochemical reactions, matrix diffusion, or channeling on measured groundwater "ages". In addition, given the wide variability in the permeability and flow porosity along a likely travel path from the recharge area to the measurement point, it is impossible to utilize groundwater "ages" to unambiguously validate the results of a hydrodynamic model. Given these precautions, the above ages are at least consistent with the results of the regional and local hydrogeologic models [3].

Temperature Distribution

The groundwater temperature may also provide some indication of the flow regime. A severe limitation with temperature (as is also the case with reactive tracers), however, is the non-conservative nature of its transport. The measured temperature at any point, whether a spring or a well, is strongly controlled by the thermal properties of the rock mass. Except in extremely rapidly flowing groundwater systems, the observed temperature is more a function of the conductive transfer of heat through the rock than the convective flow of heat with the water.

Anomalous temperatures do exist in the study region. In particular, the thermal springs at Baden (in the Muschelkalk) have a temperature of about 47°C. This indicates the discharging water at Baden must come from great depths. The probable source of this water is from the south where the Muschelkalk is much deeper (and hence warmer). This south to north flow in the Muschelkalk has been reproduced in the regional model [3].

CONCLUSION

While the initial effort to validate the regional hydrogeologic model is encouraging, the work is not complete. As additional data are being collected and interpreted (esp. hydrochemical, isotopic, thermal, and hydrogeologic) this study must be treated as preliminary.

At the present stage in the characterization of the groundwater flow regime, sensitivity analysis is an important component in safety analysis. The impact of uncertain parameters on the ability to validate the conceptual model and on the predicted ground-water flow paths and flow rates must be considered. These sensitivity analyses are reported on in [3].

REFERENCES

1. NAGRA, 1985. Endlager fuer Hochaktive Abfaelle: Das System der Sicherheitsbarrieren. NGB 85-04, NAGRA, Baden, Switzerland.

2. W. H. Muller and A. Isler, 1985. Erlauterungen zur Geologischen Karte der Nordschweiz. NTB 84-25, NAGRA, Baden, Switzerland.

3. F. Kimmeier, P. Perrochet, L. Kiraly, R. Andrews, 1985. Simulation par Modele Mathematique des Ecoulements Souterrains entre les Alpes et la Foret Noire. NTB 84-50, NAGRA, Baden, Switzerland.

4. G. Grisak and J.F. Pickens, 1985. Hydrogeological Testing of Crystalline Rocks during the NAGRA Deep Drilling Program. NTB 85-08, NAGRA, Baden, Switzerland.

5. L. Kiraly, 1985. FEM 301 - A Three-Dimensional Model for Groundwater Flow Simulation, NTB 84-49, NAGRA, Baden, Switzerland.

6. H. Schmassmann, W. Balderer, W. Kanz, A. Pekdeger, 1984. Beschaffenheit der Tiefengrundwasser in der Zentralen Nordschweiz und den Angrenzenden Gebieten, NTB 84-21, NAGRA, Baden, Switzerland.

7. F. J. Pearson, Jr., 1985. Sondierbohrung Bottstein-Results of Hydrochemical Investigations: Analysis and Interpretation, NTB 85-05, NAGRA, Baden, Switzerland.

CONFIDENCE IN SOLUTIONS OF FLOW THROUGH STOCHASTICALLY GENERATED HARD ROCK FORMATIONS

CAROL BRAESTER* AND ROGER THUNVIK**
*Israel Institute of Technology, Haifa 32000, Israel
**Royal Institute of Technology, S-100 44 Stockholm, Sweden

ABSTRACT

Confidence in solutions of flow through stochastically generated hard rock formations was studied with the aid of a simplified synthetic model. The formation is conceptualized as a fracture network intersecting an impervious rock mass. The geometrical properties of the fracture network were assumed to be known while fracture transmissivities were considered a stochastic process.

First, network fracture transmissivities were generated using a given probability distribution function. This a priori model was considered the "true formation". In a second step, only a limited amount of information, similar to that obtained in reality from boreholes, was used to construct a conditioned-by-measurement model.

Identical flow tests were performed on a formation constructed with limited data and on the "true formation". The ratio of the rates of flow resulting from these tests was considered a measure of confidence in the stochastically generated formation. The results, with this model and a particular data set, show uncertainty values between 47% to 63%, corresponding to fracture sample sizes of 11% and 2% respectively, from the total number in the network.

INTRODUCTION

Burial of radioactive waste in fractured hard rock formations is being studied in Sweden at the Swedish Nuclear Fuel Supply Company. Safety analysis of radioactive waste repositories includes, among others computations, prediction of flow patterns and calculation of travel times of water particles from the repository, possibly contaminated, to the upper part of the aquifer. Such calculations, conducted either by deterministic or stochastic partial differential equations, require that rock properties, such as transmissivity, be described deterministically or stochastically over the entire flow domain. Practically, such information is limited and interpolation methods have to be used. This is especially true for radioactive waste repository sites where excessive drilling should be avoided.

Field investigations for transmissivity determination were carried out by the Swedish Geological Co. Among other methods, the double packer test within an interval of two metres was extensively used. The analysis of the results indicates that transmissivities are lognormally distributed (Carlsson et.al. [1]). Possible spatial correlation of transmissivity was not checked.

The true properties of a formation, at the scale of discretization required in flow solutions, are inaccessible, and predicted flow patterns using expected property values are subject to uncertainty.

To gain an insight into the confidence in flow solutions through stochastically generated formations -even a rough one- we carried out a study on a synthetically generated formation. The geometry of the fracture network was assumed to be known while fracture transmissivities were considered a stochastic process.

In the last decade, stochastic methods were extensively applied to the study of conventional groundwater aquifers and, recently, also to fractured rocks. Long et.al. [2] studied continuum equivalents for stochastically generated fracture networks, Andersson et.al. [3] studied stochastically generated fracture network geometry with constant fracture transmissivity, Schwartz et.al. [4] and Smith and Schwartz [5] used stochastic models to investigate mass transport through a fracture network.

THE MODEL

In the model, the vertical cross-section of a fractured rock formation is represented as a two-dimensional fracture network intersecting the impervious rock mass. For convenience, we considered an orthogonal network with a distance of two metres separating each pair of fractures. However, the same results obtained with this pattern hold for steady-state flow through an irregular quadrilateral fracture network with identical values of the individual fracture coefficients T/L (fracture transmissivity over fracture length).

The considered flow domain of 100m x100m (Fig.1) was discretized into an interconnected network of 51x51 one-dimensional fracture elements, resulting in a total of 5100 fractures.

Following Carlsson et.al. [1], fracture transmissivities were considered to be lognormally distributed. For lack of other evidence, it was assumed that transmissivities are not spatially correlated, so that fracture transmissivity realizations were generated by a Monte Carlo simulation technique.

The classical lognormal distribution function ranging between zero and infinity was transformed to bound the transmissivity variation between T_{min} and T_{max}, the lower and upper transmissivity values, respectively.

The probability density function f(T) with the cumulative distribution function F(T) is written

$$f(T) = \frac{1}{2\pi\sigma} \; \frac{T_{max} - T_{min}}{(T_{max}-T)(T - T_{min})} \exp \left(- \frac{(\ln(T-T_{min}(/(T_{max}-T) - \mu)^2}{2\sigma^2}\right)$$

$$F(T) = \int_{T_{min}}^{T} f(u)\, du$$

where μ is the mean and σ the standard deviation.

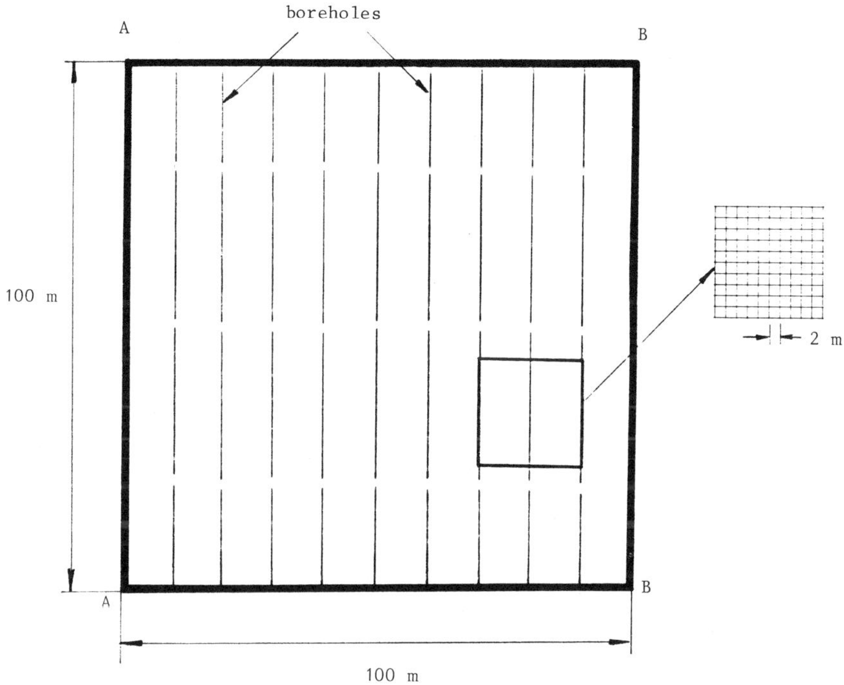

Fig. 1. The investigated flow domain

The four parameters which characterize the lognormal distribution, i.e., mean, variance, and minimum and maximum transmissivity values, calculated from one of the reported transmissivity histograms (Magnusson et.al.[6]), are:

- minimum transmissivity (T_{min}) = 1.3x10-8 m^2/s
- maximum transmissivity (T_{max}) = 1.3x10-4 m^2/s
- lognormal mean (μ) =-4.987
- lognormal variance (σ) = 2.584

A formation generated with these data was considered the "true formation" (a priori model). Other formations were generated using a limited amount of information at selected fictitious borehole locations (Fig.1). These were varied between 2 and 11 locations.

Boreholes are located along vertical fractures, so that only horizontal fractures are intercepted. Thus, except for the boundaries where only 51 fractures are intercepted, each borehole intercepts 102 fractures (51 at the left hand side and 51 at the right hand side). Owing to the random transmissivity generation using the same function for both fracture directions, and considering each formation the result of an average of a number of realizations, interception of only horizontal fractures should not represent a limitation of the model.

Assuming that the samples obtained at the selected boreholes also obey a lognormal distribution, the four parameters of the probability density function were calculated using these sample transmissivities. The obtained function was used to generate conditioned-by-measurement fracture transmissivities, i.e., in all realizations, the known transmissivities at the selected "borehole" locations were preserved.

Identical flow tests under steady-state conditions were performed for every realization of the "true" formation and for those generated with limited information. In the present case, a number of eight realizations was found to lead to satisfactory statistical results.

The governing equation of flow through an individual fracture is

$$q_i = - T_i d\varphi/dl$$

where q is rate of flow, T is fracture transmissivity, l is coordinate along the fracture, and φ is piezometric head (= $p/\rho g+z$). Writing the equation of conservation of mass at each node and assembling the equations of flow through the individual fractures, one obtains the matrix form equation

$$[A][\varphi]=[0]$$

This equation was solved under constant head boundary conditions along A-A and B-B (Fig.1). The calculations were performed with the aid of a computer program.

RESULTS

The ratios of the rates of flow resulting from flow tests with formations constructed with sample data and with the "true formation" are presented in Table I. The results with different numbers of tested sections, varying from two to eleven, located at a distance of 100 metres from each other in the first case, and 10 metres in the last case, show errors in the rate of flow varying from 61% to 46%. An increase in the number of tested sections from 5 to 11 does not significantly decrease this error.

Table I: Results with the different runs.

Run No.	No. of bore-holes	Borehole location x (m): 0	10	20	30	40	50	60	70	80	90	100	No. of tested fractures	% tested fractures	% flow rate error
1	2	x										x	102	2	61.0
2	3	x				x						x	153	3	54.9
3	4	x				x				x		x	204	4	50.9
4	5	x		x		x				x		x	255	5	47.3
5	6	x		x		x		x		x		x	306	6	47.2
6	11	x	x	x	x	x	x	x	x	x	x	x	561	11	45.9

As already mentioned, the same results hold for steady-state flow through regular or irregular quadrilateral networks with identical fracture coefficients, T/L, which means that, except for the requirement of a quadrilateral network, the geometry of the fracture network is irrelevant.

The results obtained should obviously be restricted to the assumed fractured rock model, the assumed probability function, and the size of the sample.

CONCLUSIONS

To study the confidence in solutions of flow through stochastically generated hard rock formations we considered a simple problem of flow through a simplified rock model. Fracture transimissivity was assumed to be a stochastic variable while the geometry of the fracture network was assumed to be known.

Formations conditioned-by-measurement of fracture transmissivities at pseudo-boreholes showed confidence values in the range from 40 to 54 per cent corresponding to distances between adjacent boreholes of 100 and 10 metres, respectively. A distance of 10 m between the boreholes may be considered rather short, implying that from a practical point of view the upper confidence level was reached. Furthermore, the confidence level will become lower if also other parameters such as the length, orientation and spacing between the fractures will be considered.

Calculation of groundwater and heat flow have so far been carried out using permeabilities based on an envelope through the highest values, implying that higher rates of flow are allowed through the formation. Consequently, the results of the calculations should be conservative. The results of the present study suggest that stochastically generated formation properties may lead to non-conservative results. Further investigations should be carried out to compare the two different approaches.

NOMENCLATURE

g - acceleration of gravity
l - coordinate along fracture
L - fracture length
p - pressure
q - rate of flow through the individual fracture
T - fracture transmissivity
z - vertical cartesian coordinate, positive upwards
φ - piezometric head $(z+p/\rho g)$

Parameters of the fracture transmissivity distributions
μ - mean
ρ - fluid density
σ - standard deviation
σ^2 - variance

REFERENCES

1. Carlsson, L., Winberg, A., Grundfelt, B., Model calculations for the groundwater flow at Finnsjön, Fjällveden, Gideå and Kamlunge, "SKBF-KBS-TR-83-45, Swedish Nuclear Fuel Supply Company, KBS Division, (1983).
2. Long, J.,C.,S., Remer, J.S., Wilson, C.R., Witherspoon, P. A., Porous media equivalents for networks of discontinuous fractures, Water Resources Research, (18) 645-658, (1982).
3. Andersson, J., Shapiro, A.,M., Bear, J., A stochastic model of a fractured rock conditioned by measured information, Water Resources Research, (20), 79-88 (1984).
4. Schwartz, W., F., Smith, Crowe, A., A stochastic analysis of macroscopic dispersion in fractured media, Water Resources Research, (19) 1253-1265 (1983).
5. Smith, L., Schwartz, W., F., An analysis of influence of fracture geometry on mass transport in fractured media, Water Resources Research, (20), 1241-1252 (1984).
6. Magnusson, K-Å, Duran, O., Comparative study of geological, hydrological and geophysical borehole investigations, SKB-KBS-TR-84-09, Swedish Nuclear Fuel Supply Company, KBS Division, (1984).

AKNOWLEDGEMENT

The present investigation was sponsored by the Swedish Natural Science Council and the Swedish Nuclear Fuel and Waste Management Company (SKB).

THE ORIGIN OF SALINE GROUNDWATERS IN GRANITIC ROCKS: EVIDENCE FROM HYDROTHERMAL EXPERIMENTS

DAVID SAVAGE, MARK R. CAVE AND ANTONI E MILODOWSKI
Fluid Processes Research Group, British Geological Survey, Keyworth Nottinghamshire, NG12 5GG, United Kingdom.

ABSTRACT

Hydrothermal rock-water experiments at 80°, 150° and 250° C have been carried out to investigate the origin of saline groundwaters in the Carnmenellis Granite, Cornwall, UK which have previously been attributed to be derived via dissolution of biotite and plagioclase feldspar by dilute meteoric groundwaters. The nature of product fluid and solid phases in the experiments was dominated by the dissolution of plagioclase and potassium feldspars, coupled with the precipitation of laumontite, calcite and smectite. However, the release of chloride from the rock was minimal, suggesting minor dissolution of the major chloride bearing mineral, biotite. The relative inertness and stability of biotite during the experiments was borne out by direct physical observation of reacted biotites and by consideration of the fluid phase composition in terms of thermodynamic mineral stability diagrams. It is concluded that the experimental and natural systems are buffered under different chemical conditions, which may be due to acidity generated via oxidation of localised sulphide zones within the granite and absent in the experiments. Dissolution of biotite may be an important mechanism for enhancing the salinity of groundwaters in granitic rocks if groundwaters are buffered to be outside the stability of biotite and convective circulation is significant.

INTRODUCTION

Recent interest in the groundwater chemistry of granitic and crystalline rocks because of researches into radioactive waste disposal or geothermal energy has revealed the widespread occurrence of saline fluids within such rocks. For example, salinities of 325,000mg/l, 19,000 mg/l and 1400 mg/l have been recorded in crystalline rocks of the Canadian Shield [1], the Carnmenellis Granite, UK, [2], and the Stripa Granite, Sweden [3], respectively. Since granitic rocks are prime candidates as hosts for high level radioactive waste repositories in many countries' disposal options, an understanding of the origin of such fluids is necessary, if the post disposal geochemical evolution of the groundwater is to be modelled. A number of hypotheses have been put forward for their origin:-

1. Invasion of the crystalline rock by a saline fluid, such as seawater or a brine from sedimentary rocks [4]

2. Breakdown of fluid inclusions within the rock [3]

3. Mixing of intergranular pore fluids [5]

4. The interaction of meteoric groundwater with the rock, releasing salts via silicate dissolution reactions [2, 18]

5. Radiolytic decomposition of water via uranium decay, thus increasing the salinity of the residual fluid [6].

This communication discusses the results of hydrothermal experimental work designed to investigate the origin of saline fluids associated with the Carnmenellis Granite, Cornwall, UK which have previously been ascribed to an origin by rock-water interaction [2, 18].

GEOLOGICAL SETTING

The Carnmenellis Granite is one of a number of cupola structures with an approximately circular outcrop in Cornwall, UK which was intruded into Devonian sediments some 290 m.y. ago. Although demonstrably connected at depth to form part of a 200 Km batholith, recent geophysical evidence [17] would appear to indicate inhomogeneity within the granite bodies. The granite is rich in boron, lithium, fluorine and chlorine, and hydrothermal mineral veins containing Sn, Cu, Pb and Zn sulphides are abundant. The average geothermal gradient in the granite is 30° C/km which is above average for the UK as a whole. Consequently the granite is the site of the first European Hot Dry Rock geothermal energy experiment [7], which currently involves injection and production wells at depths of 2 - 2.5 Km, where ambient temperatures are of the order of 80° C. Summaries of the geochemistry of the artificial circulation system have been given by McCartney [8] and Edmunds et al [9]. The granite also contains the only thermal groundwaters (up to 52° C) known to occur in granites in the UK. The groundwaters are often saline and are unusually enriched in Li which may range up to 125 mg/l. Edmunds et al [2] have ascribed the origin of this salinity to the weathering of biotite and plagioclase feldspar to kaolinite. From volumetric considerations they noted that fluid inclusions could not contribute significantly to the salinity and that oxygen isotope ratios precluded any contribution from seawater.

EXPERIMENTAL AND ANALYTICAL METHODS

Three rock-water experiments at 80°, 150° and 250° C were carried out in direct sampling autoclaves [10,11,12] for run durations of 200, 150 and 100 days respectively. A constant pressure of 50MPa was maintained throughout each experiment. Starting materials were a sample of drill chippings (up to 4mm in size) from a depth of 1800 metres from the Geothermal Energy Project's production well (RH11) at Rosemanowes Quarry in the Carnmenellis Granite (Table I) and a sample of fluid from a local quarry which is the usual make-up fluid for the geothermal circulation system, (Table II).

Table I Chemical and modal analysis of Carnmenellis drill chippings

	Drill Chippings (%)		
SiO_2	73.57	Quartz	33.6
TiO_2	0.13	K-feldspar	25.9
Al_2O_3	14.88	Plagioclase	30.4
Fe_2O_3(T)	1.28	Biotite	2.2
MnO	<0.1	Muscovite	6.8
MgO	0.24	Tourmaline	1.1
CaO	0.63	Andalusite)	
Na_2O	3.45	Apatite)	
K_2O	4.84	Monazite)	- traces
Cl	0.01	Zircon)	
F	0.13	Opaques)	
		Topaz)	
		Fluorite	
	99.16		

The rock/water ratio used was 10:1 (mass) and 325 cm^{-1} (BET area/volume). Reaction progress was monitored by chemical analysis of 8 solution samples extracted at constant temperature and pressure together with an additional sample extracted after the termination of the experiment. Solution samples were filtered through a 0.2 micron filter and analysed for pH, ΣCO_2, Cl, F, and Br, by electrode, titration, colorimetry and ion chromatography techniques. The remaining solution was acidified and analysed for Na, K, Ca, Mg, B, Li, SO_4, SiO_2, Sr, Ba, Fe and Al by inductively coupled plasma emission spectrometry (ICP). Calculations of in-situ pH and distribution of species calculations were made using the geochemical software package EQ3/6 [13]. Solids were characterised before and after reaction using a scanning electron microscope fitted with an energy dispersive X-ray analyser.

Table II: Chemical analyses in ppm of final solution samples from the hydrothermal experiments, Trolvis Quarry Water(TQW) and HDR fluid [18], and an example of Carnmenellis groundwater [18].

Sample No.	TQW	91/8	95/8	92/8	HDR	SC1
Temperature (°C)	25	80	150	250	37*	40
Time (days)		202	151	98		
pH	7.3	8.2	8.7	7.8	8.3	6.7
Na	19.4	30.9	52.3	51.6	72.0	3520
K	3.4	11.6	6.9	17.9	2.8	153
Ca	9.6	16.2	15.0	2.4	17.6	1840
Mg	3.2	0.7	<0.05	<0.01	0.48	55
Al	0.1	0.4	0.8	4.1	0.015	<0.1
SiO_2	0.2	64.8	177.4	567	28.0	34.2
Cl	31.7	37.0	33.0	33.9	70.0	9280
SO_4	19.7	21.6	23.2	23.4	8.2	129
ΣCO_2	15	98.0	146.0	82.0	88.0	60
B	0.03	0.05	0.11	0.39	0.22	13.9
F	0.8	5.14	9.65	13.46	6.0	3.0
Br	0.155	0.17	0.18	0.23	0.27	35
Fe	0.013	<0.008	<0.008	0.028	0.064	4.2
Sr	<0.001	0.037	0.019	0.005	0.13	30
Ba	<0.011	0.034	0.022	0.054	0.006	0.94
Li	0.01	0.13	0.28	1.39	0.209	107
TDS	103	286	464	797	294	15250

*Surface discharge temperature.

RESULTS

Limitations of space preclude a full presentation of all data acquired from the experiments, so emphasis will be placed upon those chemical components most directly related to the origin of salinity in the Carnmenellis porefluids (i.e. halogens, Li). A full description of the fluid and mineral chemistry from the experiments will be presented elsewhere [14,15], but a brief summary of the major conclusions is relevant to the present discussion.

A summary of the final fluid compositions from the experiments is presented in Table II. The behaviour of the majority of chemical components in the fluid phase was dominated by the development of steady-state concentration levels at all temperatures. These components included SiO_2, CO_2, K, Ca, Mg, Al, Sr and Fe. In addition, certain chemical components reached steady-state at only one or two temperatures e.g. B (at 80° and 150° C), SO_4 (at 250°C) Na (at 80°C) and Li (at 80°C). Other elements were dominated by dissolution kinetics in that concentrations increased with time, e.g. Li (at 250°, 150°C) Na (at 250°,150°C) F,B,(at 250°C) and Ba (at all temperatures). The development of steady-states may be interpreted as being governed by dissolution of primary mineral phases coupled with the precipitation of secondary phases. The secondary phases were directly observable at 250°C, but were only inferred to have precipitated at 80° and 150° C through consideration of the chemistry of the fluid phase together with mineral saturation states calculated by EQ3/6 from available thermodynamic data. Secondary solid phases which have been identified or inferred to have precipitated during the experiments, together with their steady-state chemical component in parenthesis, were:- amorphous silica (SiO_2); calcite (Ca, ΣCO_2, Sr); laumontite (Ca,Al,SiO_2); and smectite (Na,K,Mg,Al,Fe,SiO_2). Anhydrite also precipitated at 250° C which may have acted as a buffer for sulphate concentration. The primary phases which appeared, by SEM, to have been most affected by dissolution were plagioclase and potassium feldspars, tourmaline and quartz. Quartz grains showed decrepitation of fluid inclusion trails which apparently had little effect upon fluid phase salinity. Micas appeared relatively unaffected.

Consequently, important reactions at all temperatures appear to have been the dissolution of feldspars and quartz coupled with the precipitation of laumontite, calcite, smectite and amorphous silica. The differences between the experimental temperatures manifested themselves in the times to achieve steady states and the magnitude of those steady states. Steady states were achieved in 10 days at 250°C, 50 days at 150°C and 100-200 days at 80°C. The fluid chemistry observed is very similar to that revealed in the Rosemanowes circulation system at 80°C (Table II), and secondary solid phases are also similar to those observed along natural fractures within the Rosenamowes boreholes.

In order to consider the nature and origin of salinity in the experimental fluid phases, full data for the halogen elements Cl, F and Br are presented in Fig.1. Chloride shows very little difference between the experimental and starting fluid compositions, with perhaps a 1-4 ppm increase due to reaction during the experiment. There are no discernible differences between reaction temperatures, and concentrations are essentially time invariant. Bromide behaves in a similar manner to chloride, although data at 250°C suggest a possible increase then decrease in concentrations at approximately 50 days reaction time. This irregularity is associated with a change in analytical technique from a colorimetric to a chromatographic method and may be spurious. Fluoride on the other hand, behaves quite differently. Fluoride shows progressive concentration increases with time, although the increases are parabolic rather than linear which may reflect an approach to fluorite solubility, as indicated by saturation index calculations. The marked differences in behaviour of the halogens may be reconciled by consideration of their sources within the rock. The main source of chloride in the granite is biotite, there being 2 modal percent biotite present with a chloride content of 0.1 wt percent [15].

Although fluoride is also present in biotite (up to 0.9 wt percent), the main sources of fluoride are trace grains of topaz and fluorite. This conclusion is reached from mass balance considerations concerning both lithium and fluoride concentrations in the fluid [15]. As yet there are no data on mineral contents of bromide available. It may therefore be inferred that biotite has shown minimal reaction during the experiments, whereas topaz and fluorite have reacted readily.

Since the breakdown of biotite was identified by Edmunds et al [2,5,18] being the most significant reaction governing the salinity of both the Carnmenellis groundwaters and the Rosemanowes geothermal circulation system, it is relevant to compare those data with the data presented here from hydrothermal experiments. Fig 2 is a bivariate plot of Li versus Cl and is in part after Edmunds et al [5]. Analytical data for the principal lithium and chloride bearing minerals in the drill chippings used in the experiments have also been added. It may be seen from Fig. 2 that the hydrothermal experiments display quite a different trend from that observed for the groundwaters in the Carnmenellis Granite. The groundwaters show increasing Li with increasing Cl so that the Li/Cl ratio is approximately 0.01. The rock-water experiments show progressive enrichment in Li with minimal enrichment in Cl. It may be inferred that the Li content observed in the experimental fluid at 250°C should coexist with a Cl content of approximately 100ppm if the same reaction mechanism operates in the

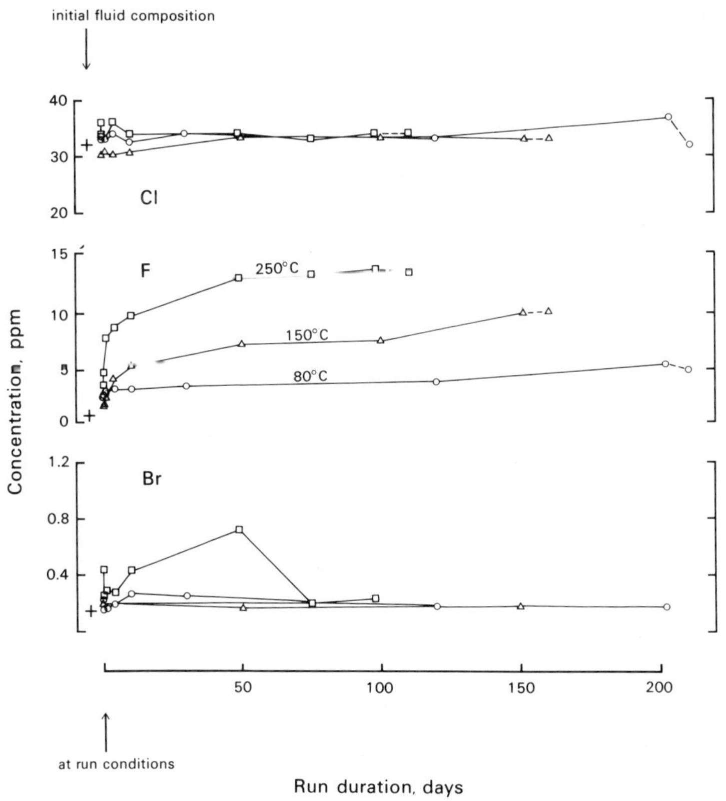

Fig. 1. Cl, F and Br fluid phase data versus time. The dashed tie lines link data points from the solution extracted after cooling of the autoclave.

experiments and groundwaters. So although there exists a good similarity between the fluids in the experiments and the HDR circulation system fluid, it is clear that there is a marked divergence between the experimental fluids and the compositions of the groundwaters. The trend observed for the experimental fluids may be accounted for by dissolution of muscovite, rather than biotite.

This discrepancy between the experimental fluid and groundwaters in terms of biotite reactivity may be explained by consideration of the stability diagram for the system K_2O-MgO-Al_2O_3-SiO_2-HCl-H_2O at 25°, 80°, 150° and 250° C presented in Fig.3. The arrowed trends for the experimental fluids all plot within the stability field of the magnesian end-member of the biotite solid solution, phlogopite. Only the final fluid compositions at 150° and 250° C approach the muscovite-phlogopite reaction boundary. By comparison, groundwater data at 25° C plot across the kaolinite-muscovite fields indicating that biotite would be unstable in such fluids. Although groundwaters plotted on Figure 3 were sampled at temperatures vaying from 15° to 52° C [2, 18], a mean temperature of 25° C was chosen for the thermodynamic data in Figure 3. Construction of the same diagram but using thermodynamic data at 50° C would result in a net

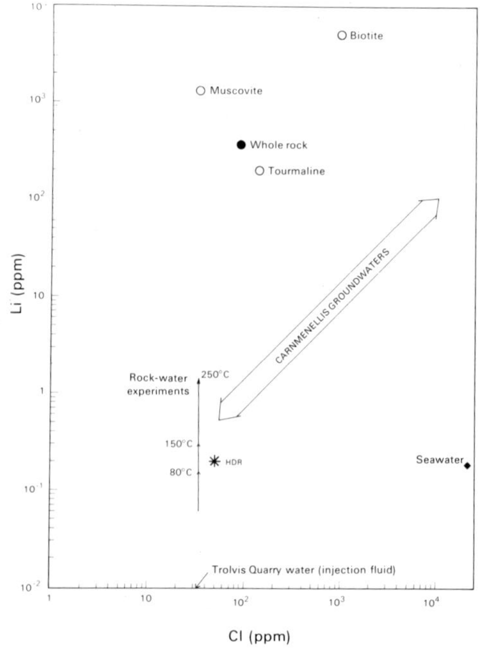

Fig. 2. Li/Cl plot of the experimental solutions, Carnmenellis groundwaters, Rosemanowes geothermal circulation system (HDR) and the analysed mineral phases. Data for the groundwaters and HDR system are from ref. [18]

movement of reaction boundaries by approximately -0.3 log units, which would still imply biotite instability in the Carnmenellis groundwaters. It is also noteworthy that the composition of the fluids, circulating at the Rosemanowes site plot well within the stability field of phlogopite at the appropriate temperature. Figure 3 offers a partial explanation for the discrepancy between the hydrothermal experiments and the groundwaters in terms of biotite stability, and hence chloride release to the fluid phase. Quite clearly, from the evidence produced in Figure 3, one would not expect biotite to be unstable in the experimental solutions, whereas the diagram bears out Edmunds et al's supposition that biotite is unstable in the Carnmenellis groundwaters. In qualitative terms, the experimental solutions are too alkaline (high $Mg^{2+}/(H^+)^2$ and K^+/H^+) for effective biotite dissolution.

Therefore it may be concluded that the natural and experimental systems are chemically buffered under quite different conditions. This may be due to a number of factors, but possibilities are :-

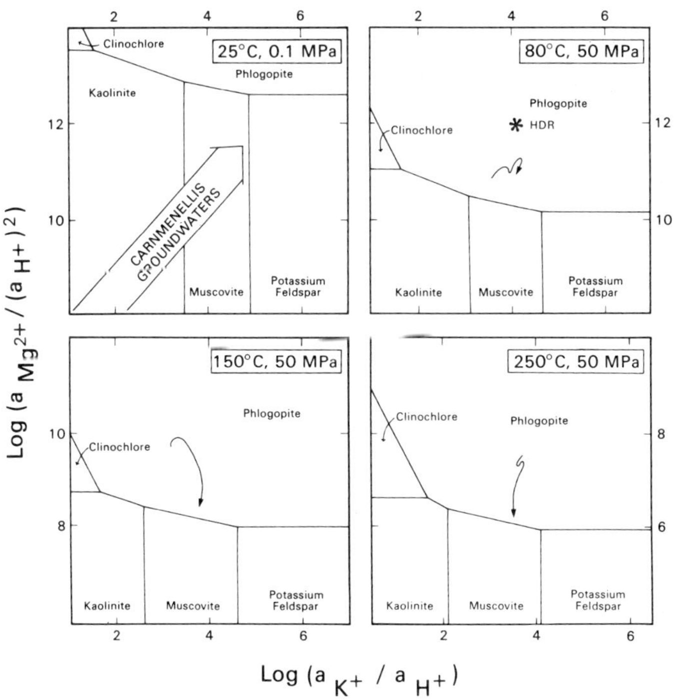

Fig. 3. Activity diagrams for the system K_2O-MgO-Al_2O_3-SiO_2-HCl-H_2O constructed from the data of Helgeson et al [16] at 25°,1MPa and 80°, 150° and 250° C,50MPa. Experimental data are shown as arrowed curved, indicating increasing time. Data for the Carnmenellis groundwaters and the Rosemanowes geothermal circulation system are from refs. [2] and [18] respectively.

1. Extensive sulphide mineralised zones exist within the Carnmenellis Granite. The circulation of oxidised groundwaters through these zones may provide an effective means of generating and buffering acidity within the natural groundwaters and hence, promoting biotite dissolution.

2. High groundwater flow rates and convective recirculation would enhance the capacity of an acidic fluid to generate salinity via biotite dissolution. Low-flow or static hydrogeological regimes would rapidly 'titrate' any fluid acidity against feldspars and micas within the granite and thus inhibit biotite dissolution.

CONCLUSIONS

Hydrothermal experiments involving Carnmenellis Granite and a meteoric groundwater at 80°, 150° and 250°C have not reproduced fluid salinities observed in deep groundwaters within the granite, and it is concluded that this is due to different chemical buffering mechanisms between the natural and synthetic systems.
However, biotite dissolution may be a viable mechanism for enhancing the salinity in granitic groundwaters particularly if-

1. The biotite is relatively rich in chloride ($\geqq$0.1 wt. percent)
2. The chemical composition of the groundwater is externally buffered to be outside the stability field of biotite.
3. Fluid flow rates and convective recirculation are pronounced.

REFERENCES

1. S.K. Frape, P.Fritz and R.H. McNutt, Geochim. Cosmochim. Acta, 48, 1617-1628 (1984).
2. W M Edmunds, J.N. Andrews, W.G. Burgess, R.L.F. Kay and D.J. Lee, Min. Mag..,48, 407-424 (1984).
3. L. Carlsson, T. Olsson, J.N. Andrews, J-C Fontes, J.L. Michelot and K. Nordstrom, KBS-Technical Report 1, 1983.
4. P Fritz and S.K. Frape, Chem. Geol., 36, 179-190 (1982).
5. C.O. Grigsby, J.W. Tester, P.E. Trujillo, D.A. Counce, J.Abbott, C.E. Holley and L.A. Blatz, J. Volcanol. Geotherm. Res., 15, 101-136(1983).
6. I.F. Vovk, Saline Brines in Crystalline Basement Rocks, (Ukrainian Academy of Science, Kiev, 1982), pp 163.
7. A.S. Batchelor and C.M. Pearson, Trans. Inst. Mining Metall., 88, 51-56 (1979).
8. R.A.McCartney, PhD thesis, Camborne School of Mines, 1984.
9. W.M. Edmunds, R.L.F. Kay and R.A.McCartney, Chem. Geol. (in press).
10. W.E. Seyfried, P.C. Gordon and F.W. Dickson, Amer. Mineral., 64, 646-649 (1979)
11. D. Savage, British Geological Survey Report, FLPU 84-2, 1984.
12. D. Savage, British Geological Survey Report, FLPU 84-3, 1984.
13. T J Wolery, Lawrence Livermore Laboratory Report, UCRL-52658, 1979.
14. D. Savage, M R Cave and A. E. Milodowski, Trans. Inst. Mining, Metall. (in press).
15. M R. Cave, A. Forshaw, I.George, A.E. Milodowski, D. Savage and R. Wilmot, manuscript in preparation.
16. H. C. Helgeson, J.M. Delany, H.W. Nesbitt and D. K. Bird, Amer. J.Sci, 278A, 1-229 (1978).
17. M. Brooks, J.J. Doody and F.R.J. Al-Rawi, J. Geol. Soc. London, 141, 97-103 (1984)
18. W.M. Edmunds, R.L.F. Kay, D.L. Miles & J.M.Cook. Spec. Pub. Geol. Assoc. Canada (in press).

EFFECTS ON FISSURE CALCITE BY RECHARGE WATER - SOME PROBLEMS RELATED TO ^{14}C-DATING OF GROUNDWATER

EVA-LENA TULLBORG
Sveriges Geologiska AB, Kungsgatan 4, S-411 19 GÖTEBORG

ABSTRACT

The modelling of water flow in the bedrock is of especial importance for safety assessment of radioactive waste disposal in crystalline rocks. Age dating of the water has been tried for the validation of ground water transport models. Thus, several tritium and ^{14}C-analyses have been carried out on water samples from deep boreholes drilled within SKB's (Swedish Nuclear Fuel and Waste Management Company) study sites. However, several processes will influence the final content of ^{14}C in the groundwater measured and correction for enrichment and dilution are hard to carry out. Another contribution to the understanding of recharge water transport by using frequency and stable isotope analyses of calcite fissure fillings is suggested.

AGE DATING OF GROUNDWATER

Water analyses of tritium and ^{14}C are available from eight study sites (Fig. 1). Both recharge and discharge areas were sampled at various depths (100 to 650 metres). Figure 2 shows a tritium/^{14}C-age plot. It can be seen that ^{14}C-ages in the range of 1000 to 2000 years have even been recorded for waters with high tritium contents. This contradiction can be explained as the result of mixing of waters of different origin and disturbances of the ^{14}C-system.

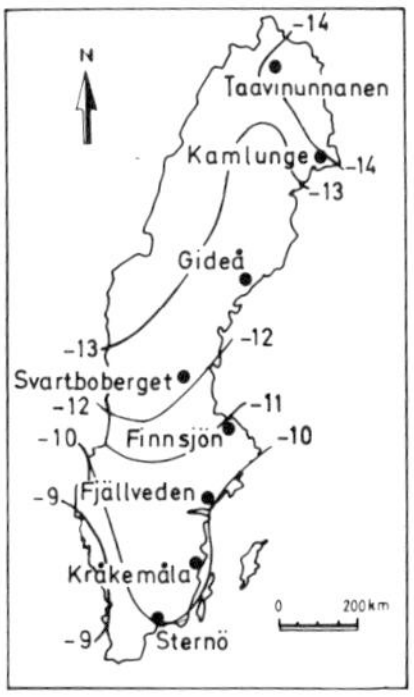

Fig. 1 Location of study sites which have been sampled for water analyses of tritium and ^{14}C. Isolines shown represent $\delta^{18}O_{SMOW}$-values of the precipitation [1].

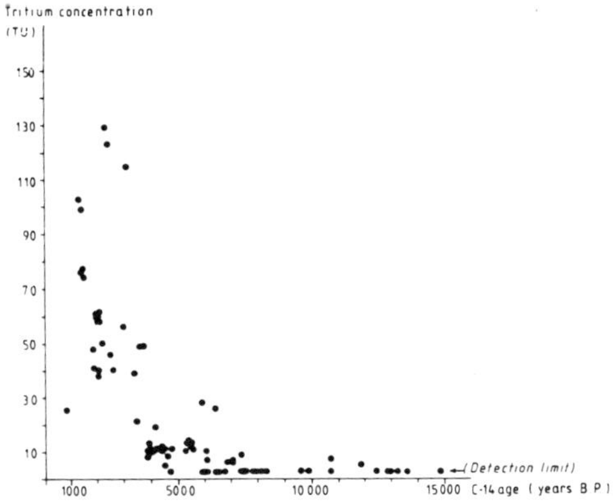

Fig. 2 ^{14}C- age versus tritium content for water samples.

These disturbances are e.g. due to dissolution of old calcite, lacking ^{14}C, within the unsaturated zone in the recharge area. Several methods have been used to correct for dissolution of old calcites; e.g. [2]. These corrections are based on the $\delta^{13}C$ values of the water, of the soil-CO_2 and of the dissolved old carbonate [3]:

$$\delta_T = \frac{C_0}{C_T} (\delta_0 - \delta_L) + \delta_L \qquad (1)$$

C_0 = initial carbonate content in the groundwater
C_T = total content of dissolved carbonate
δ_0, δ_T, δ_L = $\delta^{13}C$ of initial carbonate, of total dissolved carbonate and carbonate added to the groundwater by dissolution respectively

$$\delta^{13}C = \left(\frac{(^{13}C/^{12}C)_{sample} - (^{13}C/^{12}C)_{standard}}{(^{13}C/^{12}C)\ standard} \right) \cdot 10^3$$

Groundwater samples from SKB's study sites were measured for HCO_3- and pH. Generally the pH was within the range of 7 to 9 units. If it is assumed that $HCO_3^- = C_T$, a plot of δ_T as a function of $1/HCO_3^-$ gives δ_L as an intercept [3]. Figure 3 shows this plot for recharge areas. A slight correlation can be seen for two of these areas (Finnsjön and Fjällveden). From the other study sites too few analyses are available to get significant results.

Most carbonate present within the crystalline rocks at the study sites is calcite coating on fracture and fissure surfaces. It has been found that $\delta^{13}C$ for these calcites vary within a wide range [4, 5]. It is therefore evident that calculations using equation (1) only give a rough estimate of $\delta^{13}C$ in the dissolved carbonate from these areas.

It has been pointed out [6] that methane producing bacteria will also affect the $\delta^{13}C$ content of the dissolved carbonate. Further complications are introduced if the assumption of a soil CO_2 charge of the infiltration is not valid. The latter has probably been the case during the last glaciation of Sweden (c 10 000 - 20 000 B.P.).

No correlation can be seen when plotting $1/HCO_3^-$ versus $\delta^{13}C$ for discharge areas (Fig. 4). The probable explanation for this is that ^{14}C has been removed from the water, which is supersaturated in respect of calcite. Enrichment of ^{14}C in precipated calcite has been observed by e.g. Evans [7]. Other processes affecting ^{14}C are sorption on calcite surfaces [8] and diffusive exchange with the pore water of the rock matrix as suggested by Neretnieks [9].

Thus, the combined effect of processes controlling ^{14}C distribution in a complex water-rock system, like those investigated, could be almost impossible to calculate.

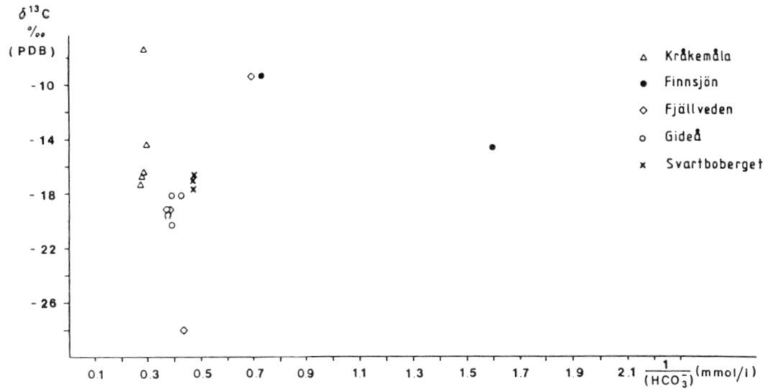

Fig. 3 $\delta^{13}{}_{HCO_3^-}$ versus $\frac{1}{(HCO_3^-)}$ for water samples from recharge areas.

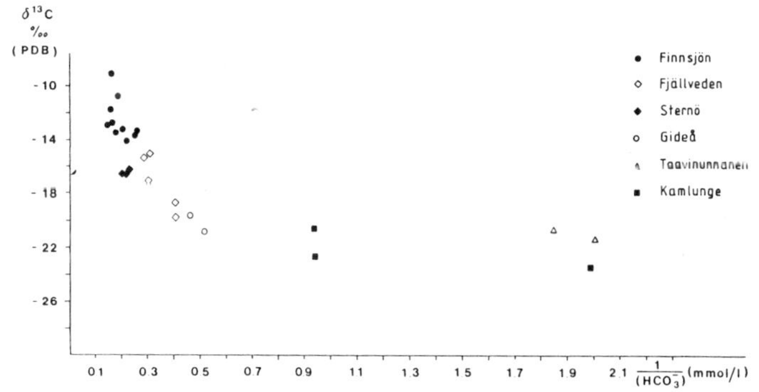

Fig. 4 $\delta^{13}C_{HCO_3^-}$ versus $\frac{1}{(HCO_3^-)}$ for water samples from discharge areas.

FISSURE FILLING CALCITE

To be able to discuss the relevance of ^{14}C dating of groundwater it is important to know whether dissolution or precipitation of calcite is prevailing at sampled levels. As groundwater analyses from recharge areas usually do not give a clear-cut answer whether calcite saturation has been reched or not at sampled levels (due to e.g. inflow of water from other levels during the pumping procedure),fissure filling calcites were studied. $\delta^{18}O$ as well as $\delta^{13}C$ have been analysed for calcite samples from boreholes from several of SKB's study sites. The samples refered to in this paper have been taken from "open" fissures in more or less typical inflow areas. As can be seen infigures 5 and 6, the $\delta^{18}O$ contents can roughly be correlated to the geographical latitude (i.e. to the $\delta^{18}O$ in the precipitation as shown in figure 1). This indicates that the isotopic character of oxygen of the calcite sampled, refers to present climatological and hydrological conditions.

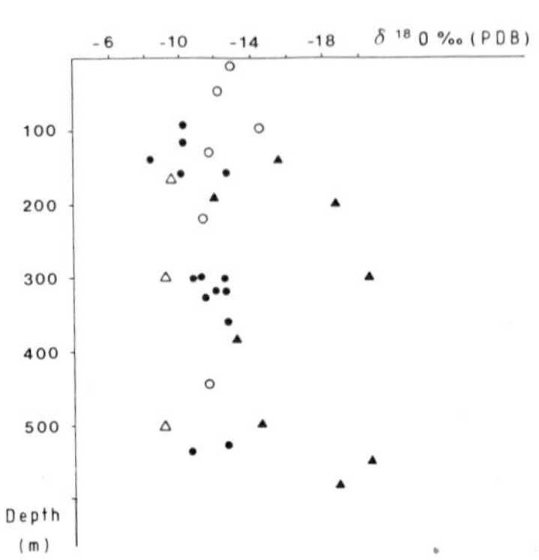

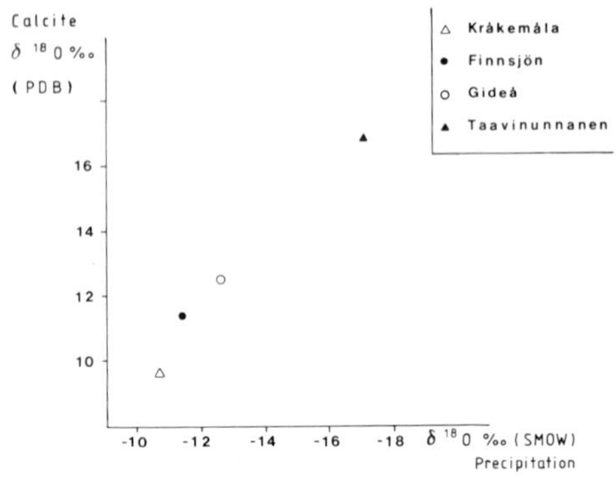

Fig. 5. $\delta^{18}O$ versus depth for fissure filling calcites within open fissures (cf. also Fig. 1).

Fig. 6. Average $\delta^{18}O$ in calcites from open fissures versus $\delta^{18}O$ values in precipitation [1, 11].

As this climatological condition (including small variation, <2^{O}C) probably has prevailed for the last 5,000 years [10], the modern calcites probably have precipitated within this time interval.

Figure 7 shows hydraulic conductivity and frequency of calcite coated fissures versus depth for three inflow areas (Gideå, Taavinunnanen and Finnsjön) as well as $\delta^{13}C$ in fissure filling calcite from open fissures in these areas. In Gideå as well as Taavinunnanen the hydraulic conductivities are highest at a shallow depth (<100 m).

Different patterns can be seen concerning frequency of calcite coated fissures. In Gideå the maximum is recorded at a depth of c. 100 m, which corresponds to the depth showing the highest hydraulic conductivity. In Taavinunnanen a maximum frequency is recorded at a depth of c. 430 m. The lack of calcite fissures in the upper part of the borehole in Taavinunnanen is interpreted as due to a rapid recharge of water down to considerable depth before calcite saturation is reached. However, in Gideå calcite saturation was reached at a shallow depth due to a slow recharge of the water. These differences are *inter alia* due to a dominance of horizontal fractures in Gideå in contrast to Taavinunnanen, where steep fractures are dominating (Fig. 8).

The $\delta^{13}C$ versus depth plot for Taavinunnanen indicates a minimum for $\delta^{13}C$ (close to equilibrium with the groundwater present) at a depth of c. 400 metres. A recent deposition of low $\delta^{13}C$ is further indicated by the ^{18}O-values as shown in figure 9.

For Gideå a similar correlation can be seen. Thus, the lowest $\delta^{13}C$ values correspond to the fissure calcite frequency peak at c. 100 to 150 metres depth.

Concerning the Finnsjön test site a high hydraulic conductivity corresponds to low $\delta^{13}C$ calcites at a depth of c. 300 m, which corresponds to a distinct fracture zone considered as a major pathway for the recharge water. The frequency pattern of calcite coated fissures is further complicated by the large amount of hydrothermal calcite precipitates found in fissures at Finnsjön [4].

To find out the age of calcites of the $\delta^{13}C$ peak, ^{14}C-dating will be carried out. This work, in progress, will probably verify a recent origin of these calcites.

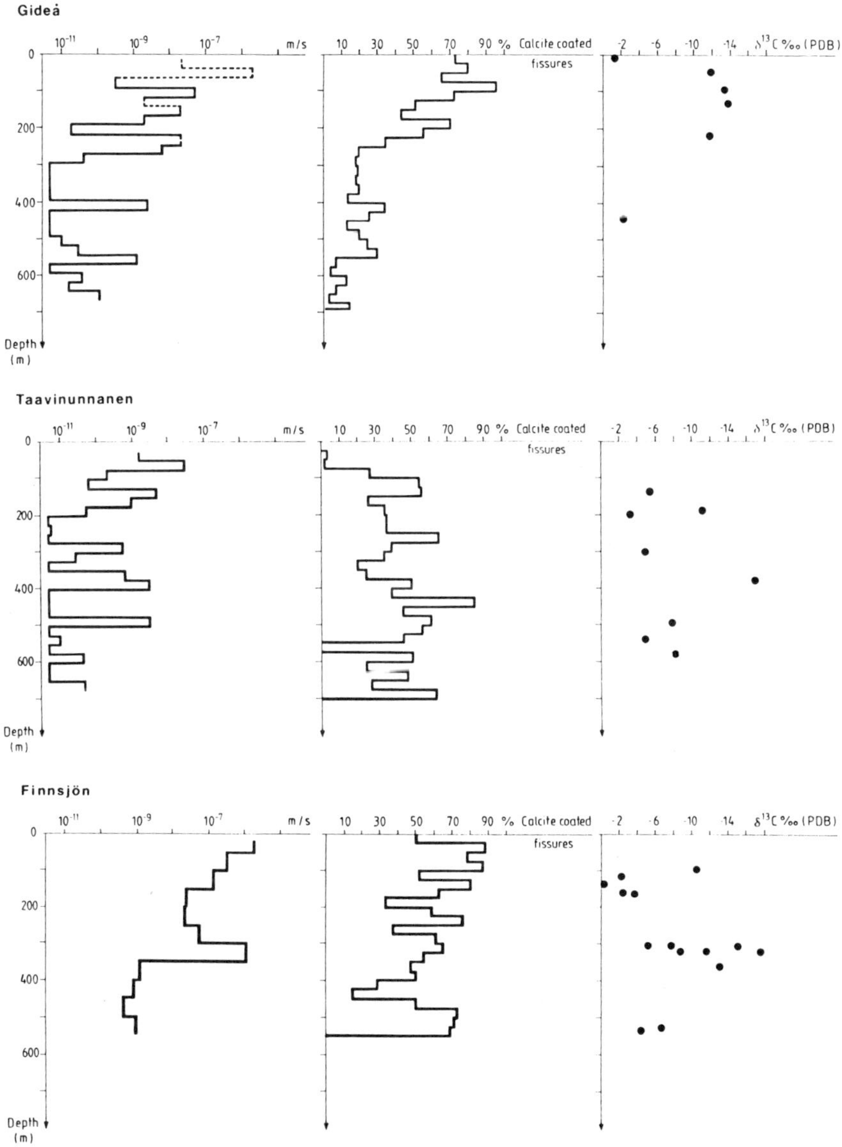

Fig. 7 Hydraulic conductivity [12, 13, 14], frequency of calcite coated fissures and $\delta^{13}C$ for calcite in open fissures versus depth from three inflow areas. Hydraulic conductivity and calcite frequency is shown for 25 m sections except for the hydraulic conductivity measured at Finnsjön where 50 m sections are shown.

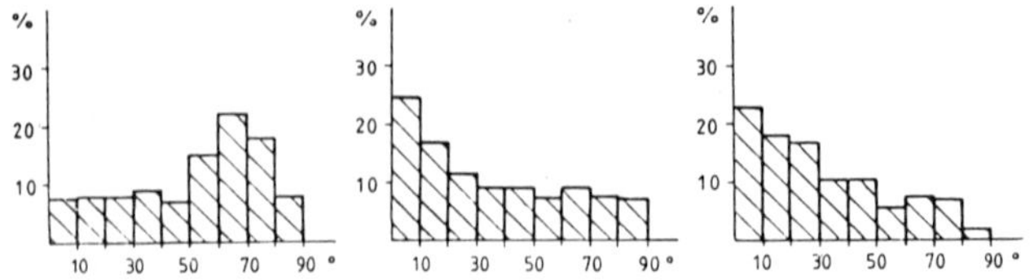

Fig. 8 Deviation from vertical shown in degrees for fissures from Gideå (left), Taavinunnanen (centre) and Finnsjön (right).

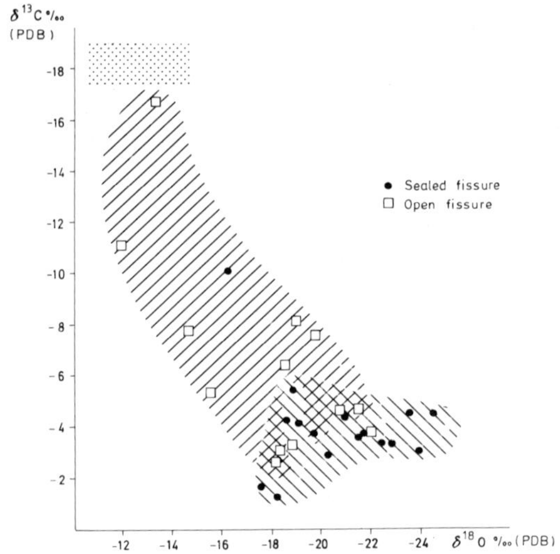

Fig. 9 $\delta^{13}C$ versus $\delta^{18}O$ for calcite coatings from open and sealed fissures at Taavinunnanen. Rectangle represents composition of calcites which can be precipitated by water present within the borehole. Ruled area at the bottom includes calcites sampled from sections with hydraulic conductivities lower than $1.9 \cdot 10^{-11}$ m/s (the detection limit). The other ruled area includes samples from sections with hydraulic conductivities higher than $1.9 \cdot 10^{-11}$ m/s.

CONCLUSIONS

Two major difficulties make ^{14}C-dating of groundwater hazardous:

1) Old calcite (lacking ^{14}C) with various $\delta^{13}C$ values will be dissolved by the recharge water making relevant corrections difficult.
2) ^{14}C will be enriched in precipitated calcites at super saturation of the groundwater. Sorption of ^{14}C on present calcites will also take place.

Influence from precipitation of new calcite and dissolution of old calcite will lower the ^{14}C content of the groundwater, thus resulting in too high "ages".

As has been shown precipitation of calcite will start at various depths, in some cases already at very shallow depths. This means that good ^{14}C-dating of groundwater at moderate to deep levels is very hard to carry out within a complex system of soils and old metamorphic bedrock. However, ^{14}C-analyses combined with water chemistry will characterise the groundwater.

On the other hand results from Gideå and Tavinunnanen show that investigations of frequency and origin of fissure filling calcite (e.g. by using stable isotopes) is an useful complement for tracing water pathways and water flow in crystalline rocks. Studies of fissure fillings can also be used to verify whether a water sample is representative for the sampled level or not.

REFERENCES

1. J.O. Burgman, E. Eriksson, F. Westman. Report from Division of Hydrology, Department of Physical Geography, University of Uppsala, Sweden (1983).

2. B. Johansson, SKB Technical Report, 80-08 (1980).

3. A.S. Talma, I.C. Vogel, T.H.K. Meatas, Isotope Hydrology IAEA. 1983,481.

4. E-L. Tullborg, S.Å. Larson, SKB Technical Report 82-20 (1982); 83-74 (1983).

5. S.Å. Larson, E-L. Tullborg, SKB Technical Report. 84-08 (1984).

6. J.F. Barber, P. Fritz, Isotope Hydrology IAEA 1978, 661.

7. G.V. Ewans, R.L. Otlet, L.L. Wesell, A.M. Bath, Isotope Hydrology 1983, IAEA, Vienna. 577 (1984).

8. B. Allard, B. Torstenfelt, K. Andersson, Third International Symposium on the Scientific basis for Nuclear Waste Management, Boston, Nov., Proceedings plenum 465, (1981).

9. I. Neretnieks, Water. res. research 17 (2), 421 (1981).

10. P. H. Lundegårdh, J. Lundqvist, M. Lindström, Berg och Jord i Sverige, Almqvist & Wiksell, 1983).

11. E-L. Tullborg, A. Winberg, International Symposium on Hydrochemical Balances of Freshwater Systems. Uppsala, Sweden, (1984).

12. K. Ahlbom, B. Albino, L. Carlsson, G. Nilsson, O. Olsson, L. Stenberg, H. Timje, SKB Technical Report 83-52, (1983).

13. B. Gentzschein, S.Å. Larson, E-L. Tullborg, SKB Technical Report 85-02, (1985).

14. L. Carlsson, B. Gentzschein, G. Gidlund, K. Hansson, T. Svenson, U. Thoregren, SKB Technical Report 80-10, (1980).

THE ASSESSMENT OF REDUCING CONDITIONS AT DEPTH IN GRANITIC ROCK

PETER WIKBERG
Department of Inorganic Chemistry,
The Royal Institute of Technology,KTH,
100 44 Stockholm, Sweden

ABSTRACT

Deep groundwaters from sites potentially suitable for the disposal of spent nuclear fuel have been chemically characterized. The waters are classified according to their contents of main constituents and pH as shallow or deep groundwaters. New equipment consisting of a mobile field laboratory and downhole pH and Eh probes has been successfully used in the investigations.

The deep groundwaters contain appreciable amounts of iron and sulphide and Eh values monitored downhole are more negative than those obtained from calculations based on the ferrous iron ferric hydroxide system. Laboratory simulations of the water rock interaction indicate that the dissolved oxygen is quickly removed when the water is in contact with the rock.

INTRODUCTION

The disposal of high level radioactive waste in deep geologic formations has been suggested. The Swedish concept presented in the KBS-3 study [1] is to isolate the waste in copper canisters embedded in bentonite in excavated granitic rock repositories at 500 m depth. Dissolution and transport by the groundwater is then by far the most important dispersion mechanisms for the radionuclides eventually released from the waste. The groundwater will also have an influence on the corrosion of the waste canister material. Consequently the chemistry and flow of the groundwater play an important role in the safety evaluation.

The release and transport of many radionuclides are strongly dependent on their oxidation states which are determined by a chemical system including the waste matrix, the groundwater and the surrounding rock. Uranium, neptunium and technetium are more soluble and less sorbed on the rock in the oxidation states VI, V and VII respectively than in more reduced states. In the safety evaluation presented in the KBS-3 study a comparison was made between two cases with otherwise equal conditions except for different redox conditions. In the case where the actinides assumed to be released from the waste were reduced in the geosphere the calculated radiation dose to man above a proposed repository was two orders of magnitude lower than if the actinides were supposed to be transported through the geosphere in an oxidized form. The dose levels are tolerable in both cases but if it is possible to demonstrate that the actinides are present in reduced form in the geosphere there is indeed a very broad safety margin for this concept.

THE CHEMICAL CHARACTER OF THE GROUNDWATERS IN SWEDISH CRYSTALLINE ROCK

Groundwater can be divided into two main classes: deep and shallow groundwater. The chemical character of shallow groundwater, in contrast to deep groundwater, is more affected by continuous mixing than by the groundwater-rock interaction. A condensed overview of groundwater-rock reactions is given by Fritz and Frape [2].

Shallow Groundwater

Water percolating through the soil takes up carbon dioxide from the root zone which has an enhanced carbon dioxide partial pressure due to the biological activity. The carbon dioxide which is dissolved in the water is transformed into carbonic acid making the water acidic and agressive [3]. When this water enters the bedrock the easily dissolved calcite minerals give the water a $Ca-HCO_3$ character. The pH of the water is determined by the bicarbonate in the water, the input of carbon dioxide giving the water a pH around neutral.

The degree of the calcite dissolution and the input of CO_2 is of course dependant on the local hydrological conditions, the biological activity and the soil cover [4].

Water entering the soil is saturated by oxygen. Some of the oxygen will be reduced by the organic material in the soil. Depending on the rate of percolation and the thickness of the soil cover water entering the rock will be slightly reducing or oxidizing. The redox conditions of the shallow groundwaters should be interpreted as controlled by reaction sequences involving oxygen, nitrate, iron, manganese and sulphide as described by Champ et.al.[5] and Edmunds et.al.[6].

Deep Groundwater

Water of the $Ca-HCO_3$ type will slowly change its character in the low conductive fractures of the rock. The weathering of feldspars will affect the chemistry of the water raising the pH and the sodium content of the water, resulting in saturation and precipitation of calcium carbonate. This system is no longer open to input of CO_2 and the pH is determined by the reactions between the bicarbonate content of the water and the solid calcite. These waters have a pH around 8.5 and can be described as $Na-Ca-HCO_3$ type of water [7].

Considerable increase in salinity and pH with depth are frequently observed. The origin of this salinity is debated, however, the mere existence of the highly saline waters, of Na-Ca-Cl type are indications of a very long residence time for the deep groundwaters.

The $Na-Ca-HCO_3$ dominated waters tend to contain a fair amount of ferrous iron whereas the Na-Ca-Cl waters have a much lower iron content but instead a fairly high sulphide concentration. Both ferrous iron and sulphide will contribute to the reducing character of the water.

EQUIPMENT FOR THE INVESTIGATION OF GROUNDWATER CHEMISTRY

During the past five years extensive investigations of potential nuclear waste disposal sites have been performed in Sweden. The experience obtained from these investigations has been used to develope new equipment for the groundwater chemical characterization, see figure 1. This equipment consists of a mobile field laboratory for main and redox sensitive constituent analysis, downhole pH and Eh measuring device and a supporting umbilicle hose for data transmission, groundwater transport and pump and packer operation.

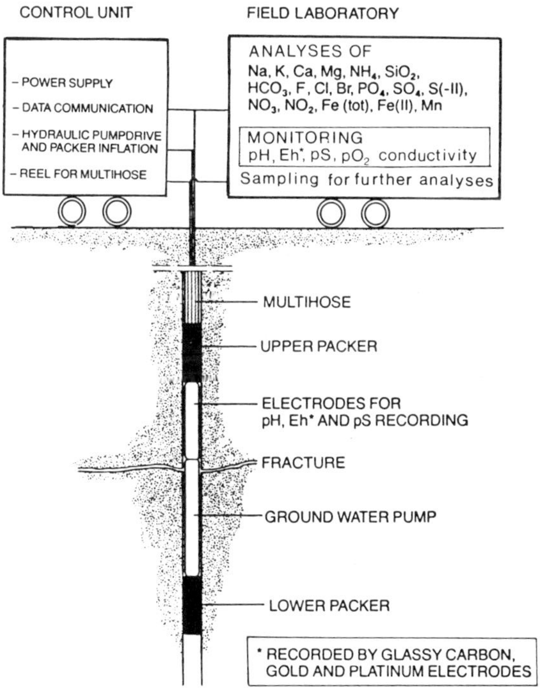

Figure 1.

The equipment for the investigation of deep groundwater chemistry, comprising a mobile field laboratory, a downhole pH, Eh and pS measuring probe, a groundwater pump and packers, all connected by an umbilical hose.

The Mobile Field Laboratory

From previous work we found that some of the chemical analyses should be made in the field. The main reasons are:

- o High permeability parts of the rock are frequently highly contaminated by drilling water. Extensive sampling of these disturbed waters can be avoided if the amount of drilling water residue can be analyzed already in the field, see table 1.
- o The borehole penetrating horizontal fracture zones with different piezometric pressures may cause a flow of shallow groundwater into deeper parts of the rock. This is especially true in recharge areas where the piezometric pressures decrease by depth. The concentrations of the main constituents Na, K, Ca, Mg, Cl, HCO_3 and SO_4 can be used to detect such failures, as the shallow groundwaters are rich in Ca and HCO_3.
- o The ratios of ferrous/ferric iron and sulphide/sulphate might change during the time elapsed between sampling and analysis due to oxidation by atmospheric oxygen even if the samples are appropriately collected and preserved.

The analytical procedures with detection limits are listed in table 1. In addition to the field analyses, samples are also collected and sent to external laboratories for special analysis of trace elements, filter residues, isotopes, organic substances and gases.

Table 1. Analyses of main costituents and trace and redox sensitive elements performed in the mobile field laboratory. The detection limits are given in mg/l.

constituent	detection limit	constituent	detection limit
ION CHROMATOGRAPHIC DETERMINATIONS			
Sodium	0.1	Fluoride	0.2
Potassium	0.1	Chloride	0.1
Ammonium	0.1	Bromide	0.05
Nitrate	0.05	Sulphate	0.05
Nitrite	0.05	Phosphate	0.1
TITRATIONS			
Calcium	2	Alkalinity	0.6
Magnesium	0.4	as HCO_3	
SPECTROPHOTOMETRIC DETERMINATIONS			
Iron (II)	0.005	Silicon (SiO_2)	1
Iron (total)	0.005	Sulphide (tot)	0.01
Manganese	0.01	Phosphate	0.005
		Nitrite	0.003
SPECTROFLUORIMETRIC DETERMINATION			
Uranine	<0.1 % drilling water contamination		

Downhole pH and Eh Measurements

Very long times were needed for the Eh electrodes to stabilize when the measurements were carried out in a flow through cell at the surface. It seemed as if residual oxygen in the cells affected the electrodes for days after onset of the groundwater flow [8]. Also occasional pump failures resulted in an increase in the measured Eh values, likely due to reactions with oxygen diffusing into the flow cell. In order to avoid these difficulties we realized that the Eh measurements had to be made in the guaranteed absence of oxygen, i.e. in the test section down in the borehole. A previously developed open hole probe was modified to fit in between the packers.

RESULTS OF THE CHEMICAL INVESTIGATIONS

The redox conditions of the groundwaters

The chemical interaction between the groundwater and the rock determine the character of the groundwater. This can be seen as a gradual change in the composition according to the previous outline.

Along with the change in the composition of main constituents there is also a change in the redox conditions, i.e. the recorded Eh value and the concentrations of ferrous iron and sulphide. The water in the highly permeable parts of the rock has a fairly high ferrous iron content whereas the water in the less permeable parts of the rock has much lower iron but a rather high sulphide content. An example of the difference between these two types of water can be seen in table 2.

Table 2. The chemical composition of groundwater from boreholes Fj 2, 468m and high hydraulic conductivity and Fj 7, 722 m and low hydraulic conductivity. All concentrations are given in mg/l.

Element	Fj 2	Fj 7	Element	Fj 2	Fj7
pH	6.9	9.0	Eh(mV)	-220	-350
HCO_3	182	16	Na	36	300
Cl	3.7	470	K	1.4	1.3
F	0.6	6.2	Ca	27	40
SO_4	2.0	0.5	Mg	3.0	0.2
S(-II)	0.2	0.6	Mn	0.2	.007
NO_3	<.001	.009	Fe(II)	1.0	<.005
NO_2	<.001	<.001	NH_4	.05	.04

The Eh values of the waters in table 2 are very low. This is a general feature found as a rule since the downhole Eh measurements were introduced. In figure 2 the Eh measured downhole are plotted verus pH for a number of waters. In the figure is also included the theoretical Eh versus pH dependence for some mineral equilibria.

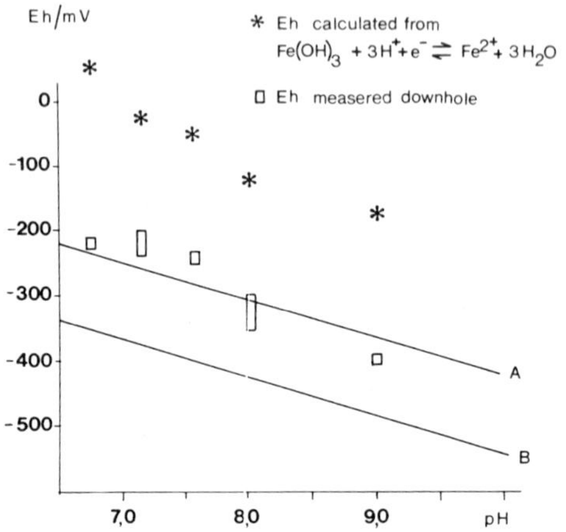

Figure 2.

Eh versus pH plot of the downhole measured potentials. The bars indicate the maximum variation of the Eh values of the platinum, gold and glassy carbon electrodes.

Line A: $2Fe_3O_4 + 2H_2O = 3Fe_2O_3 + 2H^+ + 2e^-$

Line B: $Fe_2O_3 + 2FeS + H_2O = Fe_3O_4 + FeS_2 + 2H^+ + 2e^-$

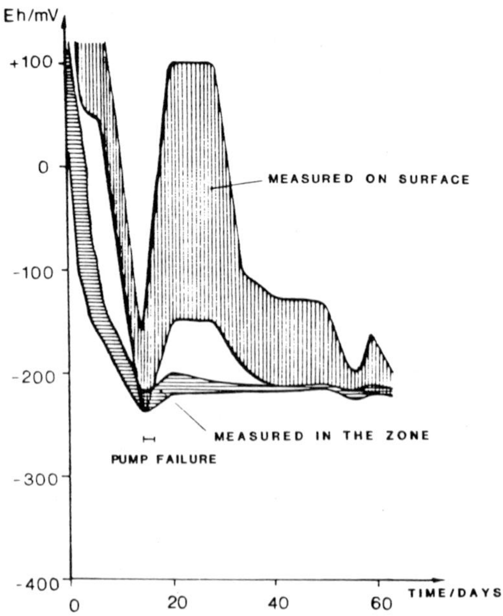

Figure 3.

Eh measured in the packed off section and in a flow through cell at the surface during pumping in Fj2, 468 m. The shadowed areas include Eh measured with platinum, gold and glassy carbon electrodes.

The downhole measurements are as expected less sensitive to disturbances in the function of the equipment. This supports our idea that the electrodes on the surface are affected by traces of oxygen in the measuring cell. In figure 3 both the surface and downhole registered Eh values are illustrated as a function of time. It is obvious that a pump failure has a very small effect on the downhole Eh, whereas the surface Eh changes considerably due to the diffusion of oxygen into the measuring cell.

Laboratory investigation of the groundwater-rock interaction

During the field investigations we have in many cases sampled water with ten per cent or more drilling water contamination. Despite this the water has mostly been reducing with an appreciable content of ferrous iron. This simply indicates that the kinetics for the reduction of oxygen initially dissolved in the drilling water is fast enough to deplete the water of oxygen within a timespan of a few months.

In order to test these observations on a laboratory scale we have run a number of experiments where we have the groundwater in contact with drillcores of the rock in well sealed glass vessels. The water phase is circulated continuously by a magnetic piston pump inside the vessel and the pH and Eh values are recorded. The vessel is filled with approximately 5 l of water and the surface area of the rock is about 30 dm^2. The results of some of these experiments are illustrated in figure 4 as Eh versus time curves. The pH curves are not included as there was no significant change during the duration of the experiments. A comparison between the experiments where the water initially was saturated by air and the one where the water was deoxygenated clearly indicate that there is a definitive delay in Eh drop in the case where air saturated water was used.

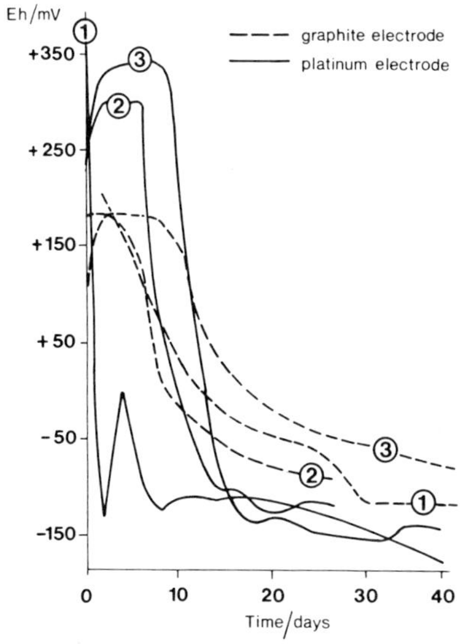

Figure 4.

The Eh versus time curve for three groundwater-rock interaction experiments:

1. Deaerated natural groundwater
2 and 3. Air saturated deionized water + HCO_3.

Dissolved oxygen is quickly reduced when water is in contact with the rock. The preliminary results of our laboratory experiments on redox kinetics in the groundwater-rock system indicate that the rate of reduction is not dependent on the amount of iron in the water. This may indicate that the dominating mechanism is the diffusion of oxygen into the rock where the ferrous iron minerals are oxidized.

ACKNOWLEDGEMENT

I am very grateful to the people involved in the field work and to my colleagues for the good work they have performed. The financial support from SKB as well as the encouragement of the SKB staff is acknowledged.

REFERENCES

1. Final Storage of Spent Nuclear Fuel - KBS-3, (SKBF/KBS Part I-IV, Stockholm 1983)

2. P. Fritz and S.K. Frape, "Saline Groundwaters in the Canadian Shield", Chem. Geol., 36, 179 (1982)

3. Gunnar Jacks, "Ground Water Chemistry at depth in granites and gneisses", KBS TR 88, Stockholm 1978

4. Gunnar Jacks, "Chemistry of some groundwaters in igneous rocks", Nord. Hydrol., 4, 207 (1973)

5. D.R. Champ, J. Gulens and R.E. Jackson, "Oxidation-Reduction Sequences in Groundwater flow Systems", Can. J. Earth Sci., 16, 12 (1979)

6. W.M. Edmunds, D.L. Miles and J.M. Cook, "A comparative Study of sequential Redox Processes in Three British Aquiferes from the United Kingdom", in Proceedings of the Uppsala Symposium on Hydrochemical Balances of Freshwater Systems 1984., Ed. Erik Eriksen. IAHS-AISH Publication No.150.

7. W. Stumm and J. Morgan, Aquatic Chemistry, 2nd ed,(John Wiley & Sons, New York 1981)

8. P. Wikberg, I. Grenthe and K. Axelsen, "Redox conditions in groundwaters from Svartboberget, Gideå, Fjällveden and Kamlunge", KBS TR 83-40, (SKBF/KBS Stockholm 1983)

9. Carol M. Jantzen, "Methods of Simulating low Redox Potential (Eh) for a Basalt Repository", Mat. Res. Symp. Proc., 26, 613 (1984)

10. Ivars Neretnieks, "Some Uses for Natural Analogues in Assessing the Function of a HLW Repository" in Natural Analogues to the Conditions around a final Repository for high-level radioactive Waste. Proceedings of a workshop held at lake Geneva, Visconcin, USA October 1-3, 1984. Ed. John Smellie. KBS TR 84-18 (SKB, Stockholm 1985)

11. R. Lindberg and D. Runnels, "Ground Water Redox Reactions", Science, 225, 925 (1984)

INVESTIGATIONS OF FRACTURE ZONES IN CRYSTALLINE ROCK BY BOREHOLE RADAR

OLLE OLSSON*, LARS FALK*, OLOF FORSLUND**, LARS LUNDMARK** AND ERIC SANDBERG*
* Swedish Geological Company, Box 1424, S-751 44 Uppsala, Sweden
** Swedish Geological Company, Box 118, S-930 70 Malå, Sweden

ABSTRACT

A borehole pulse radar system has been developed as part of the International Stripa Project with the objective to identify and characterize fracture zones at a considerable distance from boreholes. The radar uses very short pulses, which are transmitted and received by dipole antennas inserted into the boreholes. The pulses are extremely broadband with center frequencies of 25-60 MHz corresponding to wavelengths of a few meters in the rock. At 25 MHz the attenuation in the Stripa granite is 28 dB/100 m and the pulse velocity is approximately 128 000 km/s. Reflection measurements have been used to identify fracture zones and determine their position and orientation. The zones often cause strong and well-defined reflections. Improvements in the pulse form and numerical filtering of the data have consequently made the radar a very efficient instrument for locating fracture zones. During measurements in Stripa reflections from fracture zones have been observed more than 100 m from the borehole.

INTRODUCTION

A large part of the groundwater flow in crystalline rock takes place in fracture zones. The Crosshole program of the International Stripa Project has the objective to develop methods which can give information about the location, extent, and physical properties of fracture zones. The Crosshole program includes the development of radar as one of the methods with potential to give at least part of the required information. The radar system is designed for investigation of volumes where the distance from the boreholes is of the order of 50 m. The aim is to be able to detect relatively small fracture zones with a thickness down to a few decimeters.

THE RADAR SYSTEM

The new radar system consists of four different parts [1];

- a microcomputer with two 5 inch floppy disc units for control of measurement, data storage, data presentation and signal analysis.
- a control unit for timing control, storage and stacking of a single radar measurement.
- a borehole transmitter for sending out short radar pulses.
- a borehole receiver for detection and digitization of radar pulses.

Mat. Res. Soc. Symp. Proc. Vol. 50. © 1985 Materials Research Society

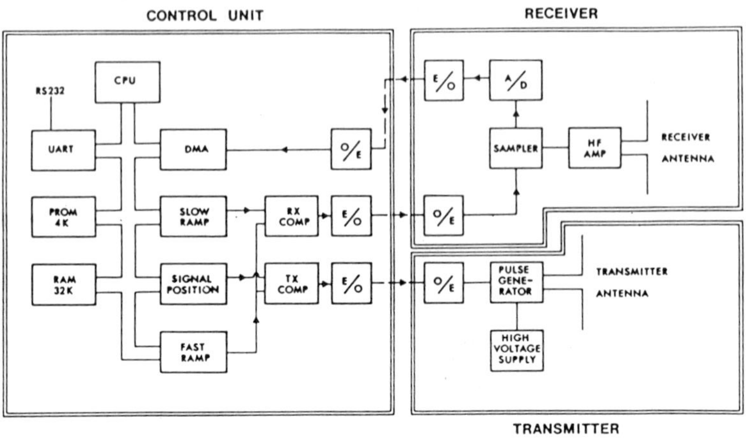

Figure 1. Block diagram of the borehole radar system.

The block diagram of the control unit, transmitter and receiver is shown in Figure 1. The system works in principle like a sampling oscilloscope, i.e. only one sample of the received signal is measured for each pulse produced by the transmitter. When the next pulse is received a new sample is taken at a slightly displaced time. Thus after a number of samples a replica of the entire signal is recorded as a function of time.

The borehole probes communicate with the control unit by optical fibers. This will eliminate any disturbing reflections due to wave propagation along electrically conducting cables. Two fiber links are used to transmit timing information to the transmitter and the receiver for the sampling process and one fiber link is used to transmit measured data in digital form from the receiver to the control unit.

As there is no direct connection between the transmitter and the receiver the radar may be used for both single-hole and crosshole measurements. The system also provides absolute timing of the transmitted pulses which makes it possible to measure the travel time of the radar pulses in a crosshole measurement.

Care has been taken in the design of the system to make it sufficiently flexible to allow variation of the centre frequency, the sampling frequency, and the length and position of the sampled time interval. The measurement procedure is relatively simple. After initialization of a data disc and selection of the variable parameters the probes are inserted into the borehole and the measurement is started. When the measurement of a trace is completed the computer unit gives an audio signal to indicate that it is time for the operator to move the probes to the next measuring position. In the horizontal holes at Stripa the probes are pushed into the boreholes by glass fiber rods. Measurements have normally been carried out with a separation of 1 m. One measurement takes about 20-40 s depending on the number of samples and stacks. This corresponds to about 100 m of measured borehole per hour in one meter steps.

The system has now been in operation since summer 1984 and during that time single and cross-hole measurements have been performed at Stripa corresponding to a total borehole length of almost 20 km.

SINGLE HOLE RADAR MEASUREMENTS

The principle of a radar reflection measurement is depicted in Figure 2. The distance to a reflecting object is determined by measuring the difference in arrival time between the direct and the reflected pulse. The basic assumption that the speed of propagation is the same everywhere in the rock mass has been verified in several crosshole measurements. As the radar is pushed step by step into the borehole the time difference will vary in a characteristic manner typical of the reflector.

The two basic patterns are point reflectors and plane reflectors as shown in Figure 2. As an example the point reflector pattern corresponds rather closely to the reflection from the drift in Figure 3, because the drift extends in a direction almost perpendicular to the boreholes. The planar pattern characterizes the fracture zones and is found in all radar maps. It should be remembered that a planar reflector works like a mirror and is therefore the most efficient type of reflector and the only one likely to be observed at large ranges.

The demand on range has necessitated the use of a pulse which is extremely broad band in frequency in order to have both a short pulse and frequencies low enough to allow propagation in rock. Any reduction in frequency would lead to a broader pulse with poorer resolution and an increase in frequency would give a decrease in range. At present the pulse width is about 5 m with a centre frequency of 25 MHz.

The borehole geometry has forced us to use dipole antennas, so that the radar in Figure 2 is cylindrically symmetric. Consequently one can not obtain the complete orientation of a fracture plane from measurements in a single borehole. The orientation can however be determined by combining results from several boreholes. A future possibility is the developement of directional antennas that would be sufficiently efficient to overcome the limitations introduced by small borehole diameters.

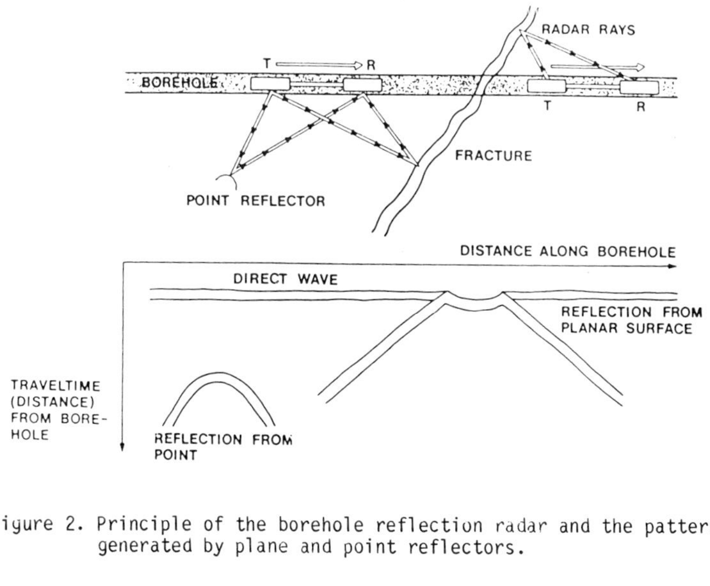

Figure 2. Principle of the borehole reflection radar and the patterns generated by plane and point reflectors.

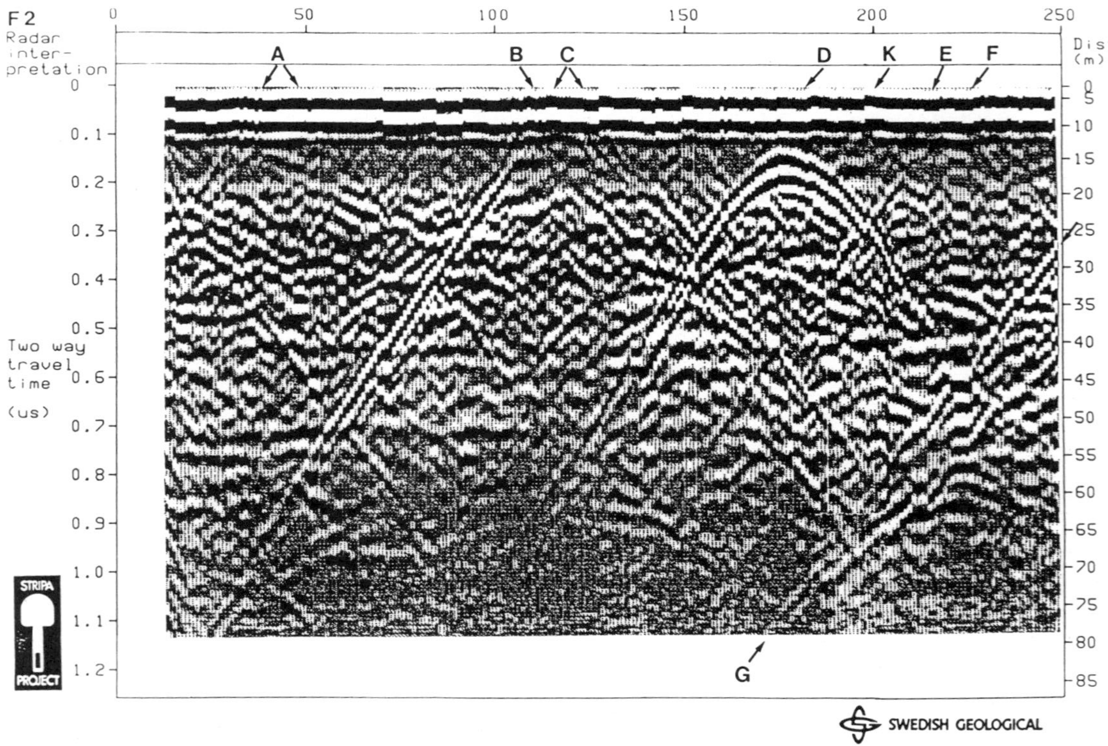

Figure 3. Radar picture from borehole F2 with the fracture zones indicated. The drift passing near the borehole is clearly seen at a depth of 175 m.

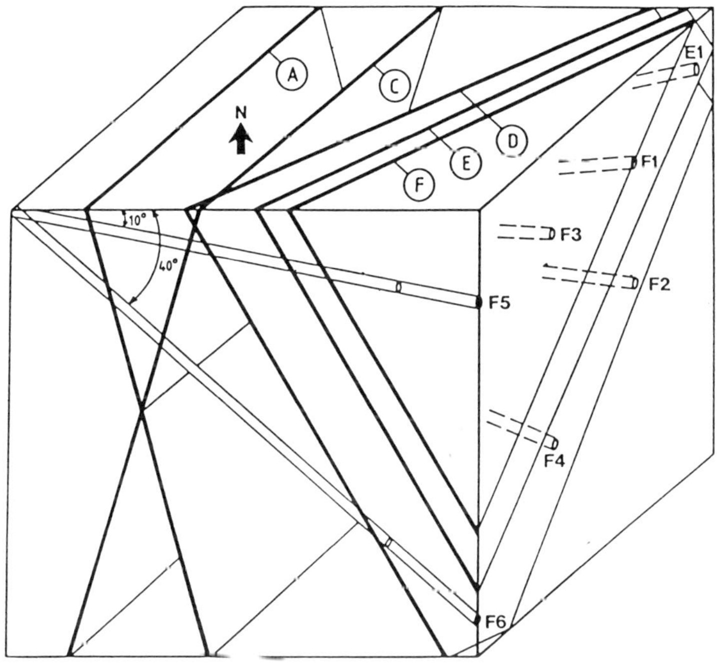

Figure 4. The configuration of the boreholes and some of the fracture zones.

At the Cross-hole Site at the 360 m level in Stripa there are seven boreholes extending in a fan-like fashion from the site (Figure 4). Logging data and core maps have previously been used to construct a picture of the position and orientation of fracture zones by identifying their position in several boreholes [2]. A similar procedure was adopted for the radar measurements in order to compare the two methods in an independent way.

The radar maps from the different boreholes (see Figure 3) show that fracture zones produce significant reflections of radar pulses. The identification is considerably simplified by the fact that the fracture zones appear to be plane and therefore produce a characteristic pattern as the radar is pushed into the borehole. This pattern, which is the basis of our analysis, may be used to determine the angle of intersection with the borehole as well as the point of intersection.

It is sufficient that the fracture plane intersect the extension of the borehole since the visible part of a pattern can often be extended far beyond the ends of the borehole. There are several examples of fracture zones that have been identified in this way. The intersection angles can be determined with varying precision: fracture planes that intersect the borehole almost orthogonally are difficult to separate while oblique zones have clearly defined angles. On the other hand oblique zones are often more difficult to observe in the clutter generated by small reflectors in the rock.

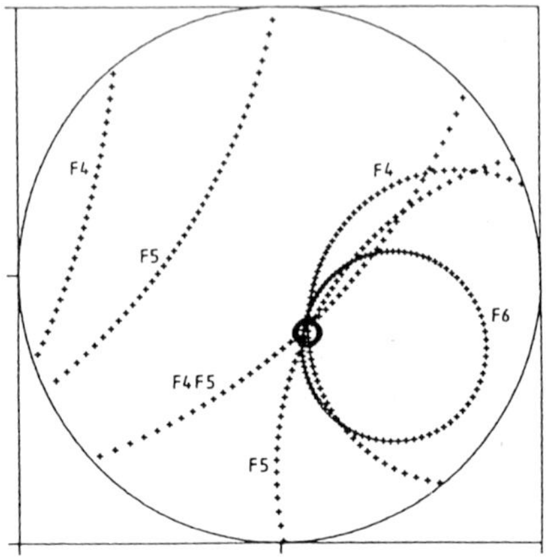

Figure 5. The Wulff diagram used to determine the orientation of zone K from intersection points and intersection angles. The ring marks directions 5 degrees off the determined direction.

With the use of a nomogram where the reflection pattern is depicted for different angles of intersections it is possible to obtain a set of intersection points and intersection angles corresponding to all fracture zone observed in the boreholes. This set is quite large and complicated for the Crosshole Site where many fracture zones are visible in the maps. It was found that the analysis is best performed step by step in order to check that less probable candidates are not introduced too early in the analysis which usually causes confusion.

The analysis of the fracture zones is performed by a computer program which can calculate the possible orientations and positions derived from an intersection angle or two points of intersection in different boreholes. The orientation of the plane may be described by plotting its normal in a stereographic Wulff diagram. Using this diagram one can step by step find out which reflections correspond to a certain fracture zone.

An example of such an identification is given in the diagram of Figure 5 corresponding to zone K, a zone not identified in the logging data [2]. One reason for this is that the zone only intersects three boreholes but can be seen in the radar data in the extension of at least one more borehole. It is prominent only in borehole F4, but with the radar it can of course be followed far from the boreholes. A very good determination of the orientation of this plane is obtained by combining the intersection angles from boreholes F4, F5 and F6. This result is confirmed by plotting the possible orientations obtained from the intersection points in different boreholes. Figure 5 shows that in this way we obtain an almost perfect determination of the orientation of zone K, since the curve corresponding to intersections in F4 and F5 gives exactly the same result as the intersection angles.

Results from a crosshole measurement between E1 and N1

The objective of the crosshole measurement between N1 and E1 was to obtain data on the attenuation of radar waves in the Stripa granite and to obtain the velocity of propagation. The transmitter and the receiver were positioned at the same borehole depth in each of the boreholes. The boreholes are perpendicular and approximately straight. In this configuration the relative position of the antennas is always the same so no correction has to be made for the directional dependence of the antenna radiation lobes.

The resulting time-distance plot is shown in Figure 6. The amplitude of the received signal is plotted in terms of different grey scales, where black corresponds to positive values, grey to zero, and white to negative values. The travel time of the direct wave increases as the borehole probes are moved into the boreholes as shown by the increasing time delay.

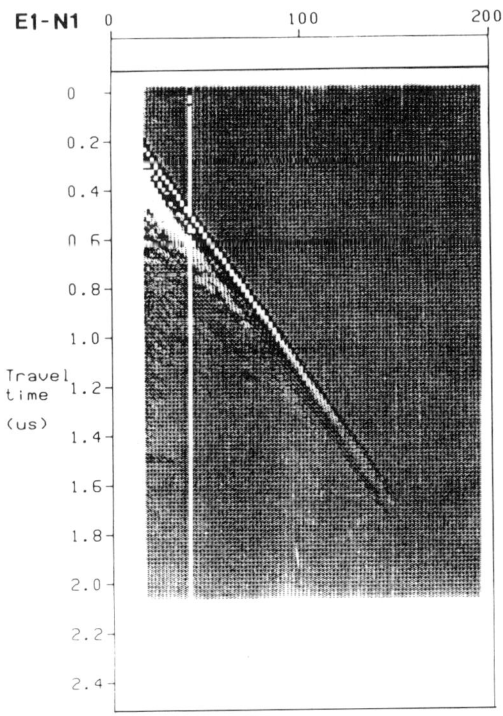

Figure 6. Time-distance diagram from a crosshole measurement between the orthogonal boreholes E1 and N1.

From this measurement the velocity of radar waves in the Stripa granite was estimated to be 128 m/us. This corresponds to a relative dielectric constant of 5.5. Determinations of the dielectric constant on core samples gave an average value of 5.4 at a frequency of 20 MHz for samples classified as unaltered granite. The agreement between the core sample data and the field data can thus be considered excellent. The variation of the peak to peak amplitude of the direct wave as a function of distance gives an attenuation for the radar waves of about 28 dB/100 m.

A number of experiments have been performed with various antenna designs and changes have been made in the impedance matching between antennas and electronics in order to reduce the ringing. The best result obtained so far is shown in Figure 7, where the pulse is shown after propagation through 50 m of rock. The centre frequency of the pulse is 23 MHz corresponding to a wavelength of 5.3 m in Stripa granite.

Results from a crosshole measurement between E1 and F2

A crosshole experiment with dense ray coverage has been performed between the boreholes E1 and F2. These boreholes originate from essentially the same point and the angle between them is 15°. The transmitter was positioned in F2 and moved in 10 m steps through the borehole interval 60-200 m. For each transmitter position the receiver was moved in two meter steps in E1 through the borehole interval 60-200 m. This gives a total of 1065 rays covering the plane between the boreholes.

Tomographic inversion has been performed on the travel time and attenuation data for the direct wave. The inversion was made with a computer program developed by the Swedish National Defence Research Institute (FOA) [3]. The differences in travel time proved to be too small to give any geologic information. The variations in attenuation were significant and the result of the inversion is shown in Figure 8. The dark color corresponds to areas with increased attenuation. One of these corresponds to a major fracture zone through the experimental site. A drift passing between the two boreholes is observed as an area of high appearent attenuation due to scattering of the radiation by the drift.

CONCLUSIONS

A new borehole radar system has been constructed and operated within the framework of the Stripa Project. During the field tests the equipment has performed well and has also proved durable and suitable for work in the mine environment. The measurement procedure is fairly simple and measurements are performed rapidly.

The results obtained so far have demonstrated the great potential of the borehole radar system for the investigation of fracture zones. It is particularly encouraging that most zones seem to reflect radar waves very well. All zones previously derived from core maps and logging data were discovered during the analysis of the radar measurements and several new zones were also found. All major reflections have been taken into account, which proves that the method works consistently.

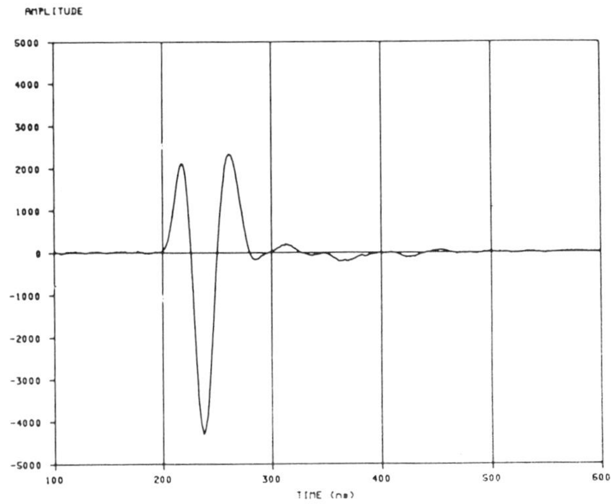

Figure 7. Radar pulse after propagation through 50 m of granite between E1 and N1.

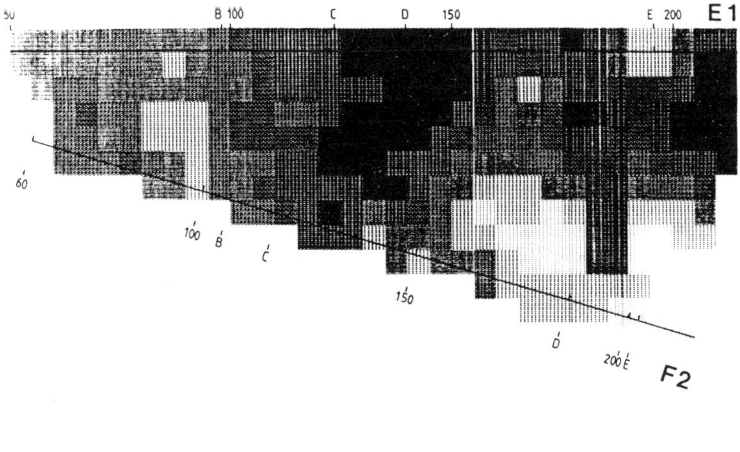

Figure 8. Result from tomographic inversion of attenuation data obtained in the crosshole measurement between E1 and F2.

The reflection patterns indicate that the fracture zones are plane and often occur in systems with approximately the same orientation. The reflections give remarkably well-defined lines, even when the zones are known to be wide. The natural explanation for this is that the reflections are caused by the sharp change in resistivity at the edge of a zone and this is confirmed by results of geophysical logging.

The orientation of a zone can be determined by correlating data from several boreholes in a step by step procedure supported by a computer program. This procedure has been shown to work consistently for all zones seen from the Crosshole Site. The interpretation technique used in studying the radar pictures often turns out to be just as important as the quality of the measurements. Digital filtering improved the identification of zones considerably, especially since they could be followed out to larger range.

The attenuation of radar waves in the Stripa granite has been found to be 28 dB/100m and the velocity to be 128 m/us. The obtained velocity is in excellent agreement with the dielectric constant determined on core samples. The high attenuation limits the range in reflection measurements to about 100 m and in crosshole measurements to about 250 m. A significant increase in the range requires an increase in the dynamic range by orders of magnitude.

In summary the radar system has proved very efficient in detecting and describing fracture zones. Much work remains on the physical description of fractured regions but it is clear that the purely geometrical problem of describing their orientation can already be reliably handled.

ACKNOWLEDGEMENTS

This project has been funded through the International Stripa Project. The support and encouragement from the Project Management and the members of the Technical Subgroup has been of great value. The pioneering work on borehole radar in Sweden was performed by Bruno Nilsson, Boliden Mineral AB, and we greatly appreciate that he has shared his profound experience with us.

REFERENCES

1. Olsson, O. and Sandberg, E., Crosshole investigations - Preliminary design of a new borehole radar system. Stripa Project, IR-84-04, SKB, Stockholm, 1984.

2. Carlsten, S., Magnusson, K.-Å., Olsson, O., Crosshole investigations - Description of the small scale site. Stripa Project, IR-85-05, SKB, Stockholm, 1985.

3. Ivansson, S., Cross-hole investigations - Tomography and its application to crosshole seismic measurements. Stripa Project, IR-84-08, SKB, Stockholm, 1984.

EVALUATION OF SINGLE-HOLE HYDRAULIC TESTS IN FRACTURED CRYSTALLINE ROCK BY STEADY-STATE AND TRANSIENT METHODS: A COMPARISON

J-E. ANDERSSON AND O. PERSSON
Swedish Geological Co., Box 1424, 751 44 Uppsala, Sweden

ABSTRACT

The results from a large number of single-hole packer tests in crystalline rock from three test sites in Sweden have been analysed statistically. Average hydraulic conductivity values for 25 m long test intervals along boreholes with a maximal length of about 700 m are used in this study. A comparison between steady state and transient analysis of the same test data has been performed.

The mean vaule of the hydraulic conductivity determined from steady state analysis was found to be about two to three times higher compared to transient analysis. However, in some cases the steady state analysis resulted in 10 to 20 times higher values compared to the transient analysis. Such divergence between the two analysis methods may be caused by deviations from the assumed flow pattern, borehole skin effects and influence of hydraulic boundaries.

INTRODUCTION

Data on hydraulic conductivity from single-hole double-packer measurements from three test sites in Sweden in crystalline rock have been used in the current study. The different test sites are Fjällveden, Gideå and Kamlunge. At the Fjällveden test site the main type of rock is a veined gneiss with a northeast-southwest strike of foliation. The fracture frequency is 4 fractures per meter within the upper 100 m and 1.8 fractures per meter below 300 m /1/. The Gideå test site consist of a veined gneiss of northeast-southwest strike of foliation. The fracture frequency is more than 4.0 fracture per meter to a depth of 400 m. Below the 500 m level the frequency is 2.0 fractures per meter /1/. At the Kamlunge test site the most common rock types are gneisses and red granite. The fracture frequency is more than 4.0 fractures per meter down to 200 m level. Below 300 m level the fracture frequency is approximately 2.0 /1/.

HYDRAULIC TESTS

In the Swedish standard program for site investigation for disposal of nuclear waste, single-hole packer tests are performed along each of the deep boreholes. The borehole length normally ranges from 600 m to 800 m and the diameter is 56 mm. The standard test intervals (packer spacing) are 20 or 25 m long. These tests are followed by detailed tests of intervals with high hydraulic conductivity using a packer spacing of 5 or 10 m.

Water injection tests with constant head followed by a fall-off phase are performed. The injection and fall-off phase normally have a duration of 2 hours each. During the injection phase the decline in flow rate is monitored continously. In the fall-off phase the pressure decline after shut-in is monitored.

ANALYSIS METHODS

Steady-state analysis

The hydraulic conductivity can be calculated from steady-state tests using an equation of the form /2/.

$$K = \frac{Q}{L \cdot H_0} \times C \qquad (1)$$

where K = hydraulic conductivity (m/s)
Q = flow rate (m^3/s)
L = length of test interval (m)
H_0 = injection head (m)

C is a dimensionless constant which varies depending on the assumed flow pattern about test interval and the estimated radius of influence of the test. The flow pattern can either be elliptical, radial or alternatively radial near the borehole becoming spherical at a certain distance from the active borehole. The relevance of the factor C in testing of fractured non-porous rock is questionable /3/.

In this study approximative values of the hydraulic conductivity have been calculated according to equation (1) assuming a steady flow rate by the end of the injection tests and assuming C is equal to one /4/. Two different conductivity values are calculated from each test by using the actual flow rate at two different times on the transient flow rate curve, see Fig 1.

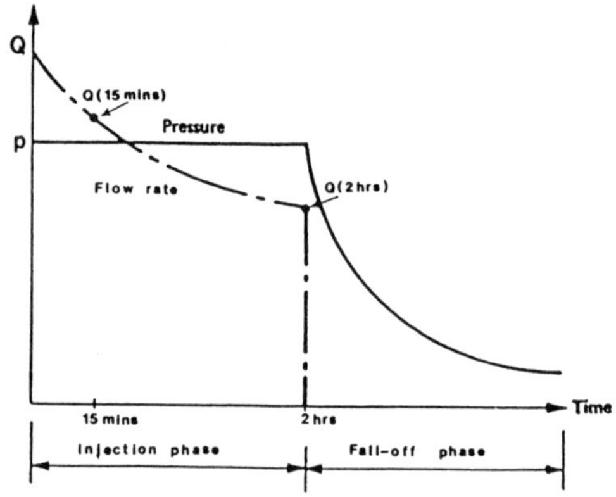

Figure 1 Different phases in single-hole hydraulic testing.

The hydraulic conductivity (KPR) corresponding to the flow rate at 2 hours of injection is defined as

$$KPR = \frac{Q(2\ hrs)}{L \cdot H_0} \qquad (2)$$

To compare the results with those obtained from shorter injection times the hydraulic conductivity (K15) corresponding to 15 minutes of injection is defined:

$$K15 = \frac{Q(15\ \text{mins})}{L \cdot H_0} \tag{3}$$

Transient analysis

The theoretical decline of flow rate during a constant head test is shown in Fig. 2. If the tested interval has an infinite extent in the radial direction the response will follow the curve marked $r_{eD} = \infty$ (infinite dimensionless outer radius). This curve corresponds to the Jacob and Lohman solution /5/. If the interval is of finite extent the flow rate will decline rapidly. If leakage occurs into the fracture system or rock matrix the flow rate will tend to stabilize.

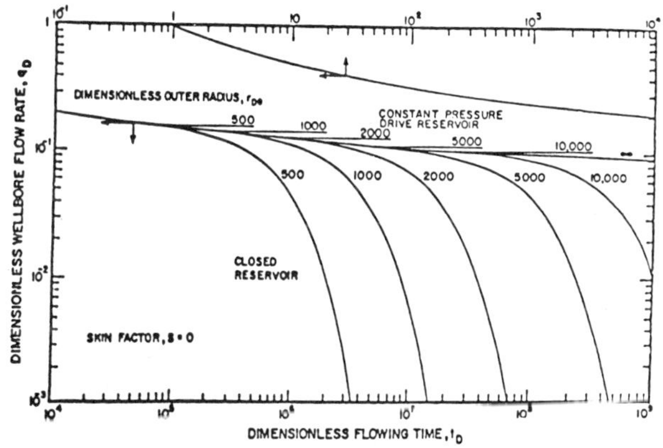

Figure 2 Type curves for radial flow in aquifers with infinite, open and closed boundaries. Constant-head tests.

The test data are plotted on various graphs to identify different flow regimes, i.e. linear, radial and spherical flow /6/. The problems of identifying the actual flow regime during a test are discussed by Ershaghi and Woodbury /7/. The reciprocal flow rate 1/Q(t) and flow rate Q (t) during injection is plotted versus time t in semilogarithmic and logarithmic graphs. Also 1/Q(t) is plotted versus the root of time in linear graphs to detect linear flow periods. During the fall off phase, the residual pressure is plotted in a Hornerdiagram. The pressure change related to the actual pressure at injection stop is plotted versus time in a logarithmic diagram.

The transient analysis of the single-hole tests is based on the continuum approach using porous media assumption. The hydraulic conductivity of the rock mass is generally determined from the slope of the straight line in the semilogarithmic diagrams assuming radial or pseudoradial flow. If the radial flow regime is not reached during the test, the hydraulic conductivity is estimated by type curve matching in the logarithmic diagram. From the injection phase, the conductivity is calculated from the following equation assuming radial flow /8/:

$$KI = \frac{0.183}{L \cdot H_0 \Delta(1/Q)} \tag{4}$$

where KI = hydraulic conductivity from injection phase (m/s)
L = length of test interval (m)
H_o = injection head (m)
$\Delta(1/Q)$ = change of reciprocal flow rate per log cycle (s/m^3)

From the fall-off phase the hydraulic conductivity is generally calculated from the Horner-diagram assuming radial flow by the following equation /9/:

$$KU = \frac{0.182\ Q_p}{L \cdot \Delta H} \tag{5}$$

where KU = hydraulic conductivity from fall-off phase (m/s)
Q_p = flow rate at injection stop (m^3/s)
ΔH = change in head per log cycle (m)

In addition, the effective borehole radius or alternatively the specific storage coefficient is estimated for each test and also the static head from the Horner-diagram. Due considerations are taken to wellbore storage effects in low-permeability test intervals during the fall-off phase. It should be pointed out that the transient analysis is more based on simple rules of thumb than a rigorous analysis in each case due to the large number of tests.

Comparison of analysis methods

The simplified steady-state analysis used in this study is based on the assumptions of steady-state flow conditions and the factor C = 1. For infinite aquifers and fractures, steady conditions are never reached in practice. In Moye´s equation for steady-state flow the factor C is defined /10/:

$$C = \frac{1 + \ln(L/2r_w)}{2\pi} \tag{6}$$

For a test interval of 25 m length and a borehole diameter of 56 mm the factor C is close to 1. Moye´s equation assumes radial flow near the well becoming spherical at a distance of L/2.

The transient analysis of hydraulic conductivity is mainly based on the slope of a straight line through many data points assuming radial or pseudoradial flow. Also tests with linear or spherical flow can be analysed by transient methods. In addition, transient analysis allows identification of different flow regimes and outer boundary conditions from the test response. From this information the fracture geometry may be deduced.

Doe and Remer /3/ made a theoretical comparison of steady-state and transient analysis. They found that the error in the steady-state analysis generally increases with lower hydraulic conductivity. The calculated hydraulic conductivity will in general be a factor of two or three higher by steadystate analysis compared to transient and occasionally an order of magnitude higher. A similar study was made by Almén et al /11/. These studies did not consider the borehole skin effect /12/.

CROSSPLOTS OF HYDRAULIC CONDUCTIVITY

To compare transient analysis results with steady-state analysis results different crossplots of hydraulic conductivity were prepared using values from each test site and also for all sites together.

In addition, the ratios of KPR/KI, K15/KI, KPR/KU, K15/KU, KI/KU and K15/KPR are calculated and plotted versus KI and KU respectively in a semi-logarithmic diagram.

The lower measurement limit for hydraulic conductivity with the actual equipment configuration can for practical purposes be considered to 5.10^{-12} m/s. Hydraulic conductivities below this limit are omitted from this study. All conductivity values used are calculated from tests with 25 m packer spacing.

RESULTS

Selected crossplots are shown in Fig. 3. The first group of crossplots show a very good correlation between different hydraulic conductivity values. Statistical data for the different conductivity ratios described above are shown in Table 1. Comparing mean values from steady-state analysis with transient analysis of injection test data (KPR/KI) it is concluded that the conductivity KPR, calculated from steadystate analysis after 2 hours of injection, generally is about a factor of 2 higher than the corresponding transient value, see Table 1 (total). This is to be expected considering the assumption of a true steady-state involved in the steady-state analysis which is generally not achieved during tests. The difference increases as the injection time decreases as shown from the mean values of the ratio K15/KI. The same conclusions can be drawn from the comparison of the steady-state analysis results with the results from transient analysis of the fall-off tests (the ratios KPR/KU and K15/KU) but the differences in mean values are somewhat larger in this case.

However, in a few cases the conductivity value from the steady-state analysis is lower than that from the transient analysis, c.f. Fig. 3 (K15/-KI vs log KI). This may be due to a positive skin in the tested section which will result in a too low conductivity value from the steady-state analysis because of a reduced flow rate due to skin damage. On the other hand, a negative skin factor will improve the flow rate and the hydraulic conductivity will thus be further overestimated compared to transient analysis. The hydraulic conductivity value determined from the latter analysis should be unaffected by the skin effect unless the fractures are completely plugged.

A comparison of the standard deviation and the 95 % confidence interval for steady state and transient analysis methods shows that the scatter in the results increase significantly with decreasing test time, which also could be expected. The crossplots of the ratio KPR/KI versus log KI and the ratio KPR/KU versus KU show that the preliminary conductivity KPR may sometimes be up to an order of magnitude higher than the values KI and KU from the transient analysis. The K15-values may occasionally be 10-20 times higher compared to the transient analysis. However, data points showing such great deviation between steady-state and transient analysis are relatively few, c.f. Fig. 3. No firm conclusions regarding the conductivity intervals in which large deviations occur most frequently can be drawn from the present material although there is a tendency towards larger deviations in lower conductivity intervals. In such intervals the deviation may also be due to inaccuracies in either test data or analysis, regardless of the analysis method used.

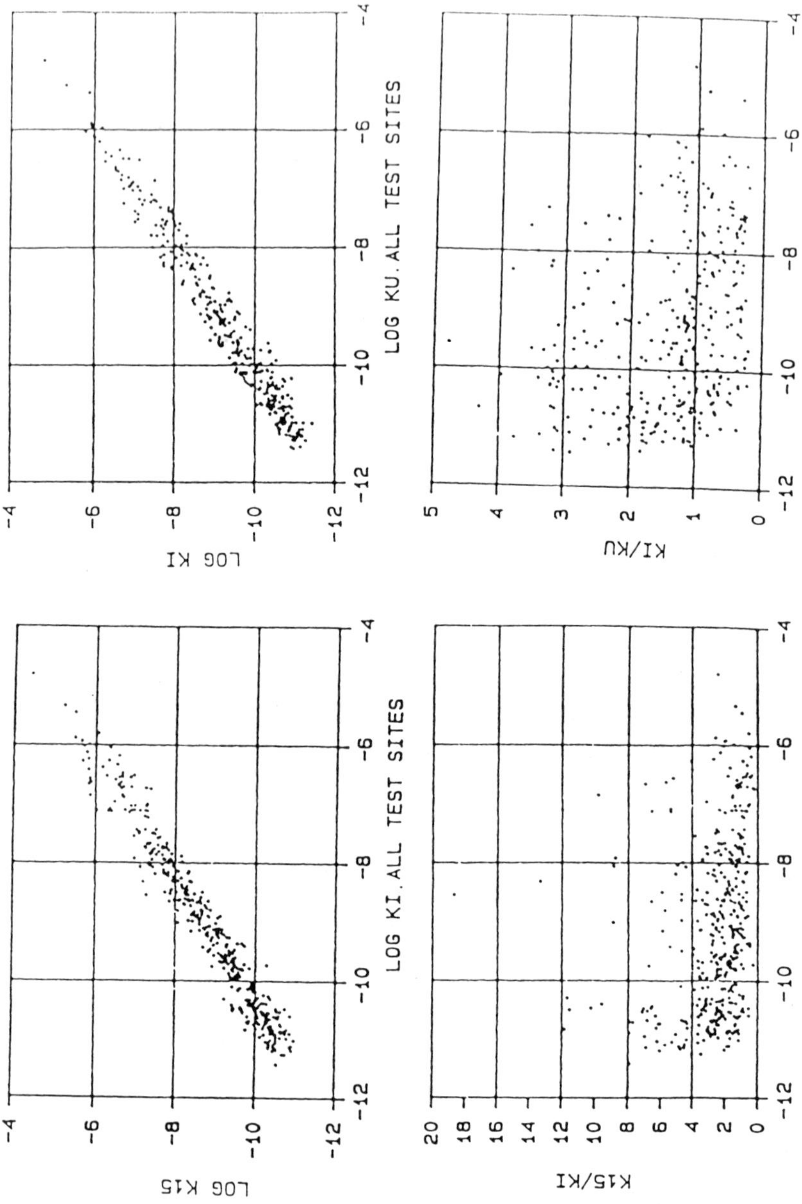

Figure 3 Correlation between hydraulic conductivity from steady state and transient analysis and distribution of different conductivity ratios as a function of actual hydraulic conductivity.

One reason for large errors in KPR and K15, particularly in higher-conductivity intervals, can be the occurence of linear flow during a substantial part of a test. This causes a large change in flow rate (or head) with time contrary to the assumption of steady-state. A linear flow response may be caused by flow in high-conductive channels in the rock, c.f. channeling. Another reason can be effcts of (barrier) boundaries during the test. Such effects are generally not considered in the analysis of steady-state tests. With transient analysis methods such effects can be identified and taken into account by the interpretation. On the other hand, if spherical flow occurs during a test, the change in head is much smaller and the steady-state analysis should be more accurate.

Short time constant head injection tests (c. 15 mins) are generally analysed by steady-state methods whereas longer tests (2 hours or more) are analysed by transient methods. The most relevant comparisons of the two test methods should therefore be the ratios K15/KI and K15/KU which are shown in Fig. 3.

Table 1 Statistical parameters for the hydraulic conductivity from different tests sites.

	N	MEAN	STDEV	SE MEAN	95.0 PERCENT	C.I.
Fjällveden						
KPR/KI	154	2.03	1.36	0.11	(1.81,	2.24)
KPR/KU	126	2.67	1.73	0.15	(2.36,	2.97)
K15/KPR	159	1.35	0.63	0.05	(1.25,	1.45)
KI/KU	126	1.51	0.98	0.09	(1.34,	1.68)
K15/KI	150	2.83	2.57	0.21	(2.41,	3.24)
K15/KU	125	3.68	2.71	0.24	(3.20,	4.16)
Gideå						
KPR/KI	186	1.83	1.12	0.082	(1.664,	1.988)
KPR/KU	147	2.36	1.54	0.13	(2.11,	2.61)
K15/KPR	191	1.43	0.85	0.06	(1.31,	1.55)
KI/KU	131	1.32	0.82	0.07	(1.17,	1.46)
K15/KI	161	2.54	2.06	0.16	(2.22,	2,86)
K15/KU	130	3.26	2.85	0.25	(2.77,	3.76)
Kamlunge						
KPR/KI	124	1.647	0.791	0.071	(1.506,	1.787)
KPR/KU	91	2.34	1.43	0.15	(2.05,	2.64)
K15/KPR	118	1.61	1.17	0.11	(1.40,	1.82)
KI/KU	81	1.41	0.90	0.10	(1.22,	1.61)
K15/KI	112	2.62	1.74	0.16	(2.30,	2.95)
K15/KU	78	3.70	2.70	0.31	(3.09,	4.31)
All sites						
KPR/KI	463	1.85	1.14	0.053	(1.745,	1.953)
KPR/KU	364	2.46	1.59	0.083	(2.300,	2.627)
K15/KPR	468	1.45	0.88	0.04	(1.37,	1.53)
KI/KU	338	1.42	0.91	0.05	(1.32,	1.51)
K15/KI	423	2.67	2.18	0.11	(2.46,	2.87)
K15/KU	333	3.52	2.76	0.15	(3.22,	3.82)

A comparison of mean values for transient analysis of the injection and fall-off tests shows that the mean value of the previous tests is about a factor of 1.5-2 higher than the latter (ratio KI/KU). The reason for this deviation is not completely known at present. One possible reason is that the fall-off test data sometimes may be less reliable due to insufficient duration of the previous injection period, particularly in test intervals with low conductivity /9/. According to the authors experience from the actual tests, fall-off data are more difficult to interprete than injection as more reliable. The ratio KI/KU from transient analysis also shows some scatter particularly towards low-conductive intervals.

Comparing mean values for different steady-state analysis methods (ratio K15/KPR), this ratio is relatively constant with the K15-values generally about 50 % higher than the KPRvalues. However, a few peak values may occur. As an example, a value of 4.5, caused by linear flow, was determined.

A comparison between the test sites shows that the conductivity ratios between steady-state and transient analysis are greatest for the Fjällveden site and smallest for the Kamlunge site. Also the deviations of the ratios show the same pattern, c.f. Table 1.

CONCLUSIONS

A comparison between steady-state analysis and transient analysis of the injection phase show that the steadystate analysis in general gives K-values which are about 2-3 times higher, depending on the test duration, and occasionally 10-20 times higher than the transient analysis. This is in good agreement with the conclusions by Doe and Remer /3/. Corresponding maximal deviation for steady-state versus transient falloff data is a factor of 15-25.

Comparing steady-state analysis using different injection times (K15 and KPR) shows that the shorter test time analysis (K15) results in a higher K-value by about 50 %. The difference between the transient and steady-state analysis will generally increase when linear flow and flow barriers occur. On the other hand, spherical flow and/or effects of recharge boundaries will decrease the difference.

The results of this study may assist in the design of the hydraulic test program for site investigations in crystalline rock.

REFERENCES

1. Ahlbom, K., Carlsson, L. and Olsson, O., : Final disposal of spent nuclear fuel - geological, hydrogeological and geophysical methods for site characterization. SKBS Technical Report TR 83-43. 1983.

2. Ziegler, T.W., : Determination of Rock Mass Permeability. U.S. Army Eng. Waterways Exp. Station, Tech. Report S-76-2. 1976.

3. Doe, T. and Remer, J., : Analysis of Constant-head Well Tests in Nonporous Fractured Rock. Proceedings of 3rd International Well-testing Symposium, Berkeley, California. 1980.

4. Banks, D.C., : In situ measurements of permeability in basalt. Proceedings. Symposium on Percolation through fissured rock. ISRM, IAEG, Stuttgart. 1972

5. Jacob, C.E. and Lohman, S., Nonsteady flow to a well of constant drawdown in an extensive aquifer. Trans. Am. Geophys. Union (Aug. 1952) 559-569. 1952.

6. Gringarten, A.C., Flow Test Evaluation of Fractured Reservoirs. Paper presented at the Symposium on Recent Trends in Hydrology, Geological Soc. of America, Berkeley, CA, Feb. 8-9 1979.

7. Ershagi, I. and Woodbury, J.J., Examples of Pitfalls in Well Test Analysis. J. Pet. Tech. (Feb. 1985) 335-341. 1985.

8. Uraiet, A.A. and Raghavan, R., Unsteady Flow to a Well Producing at constant Pressure. J. Pet. Tech. (Oct. 1980) 1803-1812. 1980.

9. Uraiet, A.A. and Raghavan, R., Pressure Buildup Analysis for a Well Produced at constant Bottomhole pressure. J. Pet. Tech (Oct. 1980) 1813-1824. 1980.

10. Moye, D.G., Drilling for Foundation Exploration. Civil Eng. Trans., Inst. Eng. Australia (Apr. 1967) 95-100. 1967.

11. Almén, K-E., Andersson, J.-E., Carlsson, L. and Hansson, K., Hydraulic tests. Part 5: Steady state injection tests (In Swedish). KBS Progress report AR 84-30, Sept. 1984.

12. Earlougher, R.C., Advances in Well Test Analysis, Soc. Pet. Engr. Monograph Series, Vol. 5 (1977) SPE, Dallas. 1977.

CHEMICAL THERMODYNAMIC DATA SET FOR MINERALS ASSOCIATED WITH GRANITE

A.B. MULLER*, B. FRITZ**, A. WYMAN*** and M. SNELLMAN****
*OECD Nuclear Energy Agency, Paris, France
**Institut de Géologie, Université Louis Pasteur, Strasbourg, France
***Department of Geoscience, University of Nevada, Las Vegas, USA
****Technical Research Center of Finland, Helsinki, Finland

ABSTRACT

The OECD Nuclear Energy Agency is coordinating the development of a critically reviewed CODATA compatible chemical thermodynamic data base for ten radioelements important to various areas of nuclear technologies. Such a data base in itself is insufficient for modeling the speciation and solubility behaviour of radioelements in natural rock/water systems. Data must also be available for describing the rock phases present, which determine to a large extent the physico-chemical environment. Since a number of NEA Member countries are contemplating nuclear waste disposal in granitic environments, a data base for seventy-three minerals associated with granite is presented here, as a complement to the data base work on radioelements.

INTRODUCTION

Over the last two years the OECD Nuclear Energy Agency has been involved in coordinating the development of a CODATA compatible chemical thermodynamic data base for aqueous species and solid compounds of ten radioelements important to various nuclear technologies [1]. These data are used in thermodynamically-based models of speciation and solubility. It is recognised that such a data base is in itself insufficient for modelling the behaviour of these materials in natural water systems, such as those associated with the geological disposal of nuclear wastes. The physico-chemical environment of such systems, particularly in the far field, is determined primarily by the geologic materials in contact with the water. Therefore, the chemical thermodynamic characteristics of the minerals in the system as well as those of the major solutes must be known.

As a complement to the NEA data base being developed for the ten radioelements at the NEA Data Bank in Saclay (France), the current data set was compiled and critically reviewed for seventy-three minerals (85 compounds) which occur in granite, weathered granite and granite fracture fillings. Granite and related minerals were chosen because a number of NEA Member countries are considering the disposal of nuclear wastes in granitic formations or in formations of similar mineralogy. The objectives of this paper are 1) to determine which minerals should be included in a "granite" chemical thermodynamic data set and 2) to present selected values for the parameters $\Delta_f G°$, $\Delta_f H°$, $S°$ and Cp at 298 K. Their temperature functions have not been presented due to space restrictions, but appear in a supplementary publication [2].

SELECTION OF MINERALS

There is no unanimous agreement among geologists regarding a precise definition of a granite. The three criteria used most commonly are mineral composition, texture and geologic mode of occurrence. The discussion herein emphasizes mineral composition, about which there is the greatest disagreement among geologists and mineralogists.

Recent attempts at defining granite in terms of mineral composition were made by the International Union of Geological Sciences Commission on Petrology, Subcommission on the Systematics of Igneous Rocks [3,4,5]. This paper is consistent with their internationally accepted comprehensive system for igneous rock classification, based on actual mineral content of the rocks in volume percent.

There are three primary mineral groups in granites. These are quartz (Q), alkali feldspar including albite (A) (0-5% An) and plagioclase feldspar (P) (anorthite content > than An_5). Recalculations are made so that the light-colored (felsic) minerals will equal 100 percent and can be plotted on the triangular diagram in Figure 1. To be included in the

Figure 1. The Classification Of Plutonic Igneous Rocks. Rocks with compositions from the shaded area are considered in the present data base. (After Streckeisen, 1973, 1976)

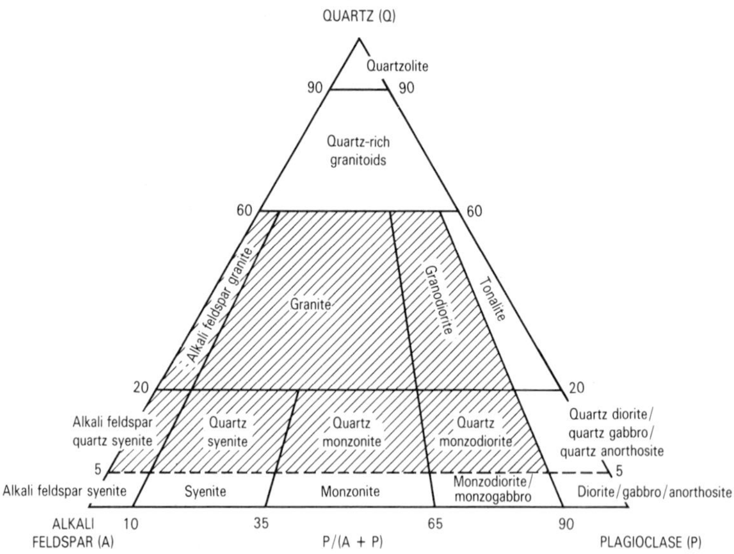

triangle the igneous rocks must also be plutonic and phaneritic. A rock with as little as 10 percent Q, A and P could be plotted, but none of these rocks would fall into the granite-type categories.

Primary minerals are those which have crystallized from a magma. They can be classified into two categories: essential and accessory. In granites it is the relative percentages of light-colored constituents which are important and which are used in specific naming. The accessory minerals are used only to indicate a variety (e.g., hornblende-biotite granite).

According to the I.U.G.S. classification system, by definition, a granite must contain between 20 and 60 percent quartz, and P/(A+P) between 10 and 65 percent [6] as the primary essential minerals. Rocks in which Q = 5 to 20 percent are called quartz-bearing syenites, monzonites and diorites. What was formerly called quartz monzonite is now included in granite by most North American and Western European authors. However, many geologists (particularly those in the USSR) still use the term in this way, so that it has become confusing and ambiguous [6]. In addition to the portion of granite which can be called quartz monzonite and in which there is a higher percentage of quartz (Q > 20%), the term has also been used for rocks in which Q = 5 to 20 percent. In both cases, A and P are in approximately equal propor- tions.

The term "granite" as defined here is expanded to include alkali feldspar granite, granite and granodiorite (all with Q < 20%) and alkali feldspar quartz syenite, quartz syenite, quartz monzonite and quartz monzodiorite (all with Q = 5 to 20 percent). Thus, rocks with Q = 5 to 60 percent and A and P varying between 10 and 90 percent are all included. The actual mineral species do not differ among these rocks, only the relative percentages of Q, A and P, as shown in Figure 1. Typical chemical compositions (reported as oxides) appears in Table I.

Primary accessory minerals, like primary essential minerals, have also crystallized from magma. They range from "major", usually ferromagnesian (mafic) minerals such as biotite and hornblende, to "very rare", those that would not generally be found in granite and are included only because of their place in solid solution series. The following general volume percentages are used in this paper to quantify the terms describing accessory mineral occurrences:

Major accessory	.2-25%
Minor accessory	1-2%
Rare accessory	0.1-1%
Very rare accessory	< 0.1%

The primary essential and various primary accessory minerals for which chemical thermodynamic data appears in this paper are listed in Table II.

A second criterion for defining granite is texture. Here only phaneritic rocks will be considered as granite; that is, rocks in which the mineral grains are large enough to be distinguished individually by the naked eye or by the use of a ten-power lens. Phaneritic rocks, while being holocrystalline and equigranular, may also be porphyritic (i.e., containing phenocrysts larger than the groundmass). These are included in granites and are called granite porphyry if the phenocrysts are greater than 50 percent of the rock and porphyritic granite if the phenocrysts are less than 50 percent.

Table I. Chemical Compositions of 202 granite-type rocks from worldwide localties. Rock names as defined by [3]. (After [6], p.307)

	Granite and quartz monzonite (97 analyses)		Granodiorite and quartz monzodiorite (105 analyses)	
	Average	Range	Average	Range
SiO_2	71.93	62.0 - 77.48	66.37	55.81 - 73.00
TiO_2	0.26	0.04 - 0.73	0.53	0.12 - 0.98
Al_2O_3	14.58	12.44 - 17.95	15.90	13.36 - 20.49
Fe_2O_3	0.78	0.05 - 4.0	1.07	0.02 - 3.46
FeO	1.42	0.34 - 3.78	2.87	0.76 - 4.71
MnO	0.05	0.01 - 0.15	0.08	0.02 - 0.13
MgO	0.66	0.02 - 2.70	1.66	0.14 - 3.97
CaO	1.86	0.34 - 4.7	3.42	1.37 - 7.42
Na_2O	3.57	2.62 - 4.73	3.51	2.05 - 5.4
K_2O	4.02	2.7 - 5.41	2.91	1.14 - 5.10
P_2O_5	0.08	0.005 - 0.4	0.15	0.04 - 0.28
CO_2	0.05	0.0 - 0.2	0.07	0.03 - 0.3
H_2O	0.545	0.12 - 1.47	0.68	0.2 - 1.2
Rb, ppm	225	71-532	123	40-202
Sr, ppm	109	0-359	282	118-690

The third criterion is geologic occurrence. Granite rocks are plutonic, thus their geologic mode of occurrence is in batholiths, large intrusive plutons with steeply dipping walls and typically lacking any known floor. They are commonly composed of felsic rocks and in outcrops they range in size from about one hundred to several thousand square kilometers. They are also not characteristically made up of a single rock type, but of smaller intrusions of granite, quartz monzonites, granodiorites and other felsic rocks [8].

In selecting the minerals included in the chemical thermodynamic data set presented in this paper, some assumptions and special considerations have been made:

1. The candidate granite contains no ore mineralization. Most ore minerals, including those found in hydrothermal veins and pegmatites, have not been considered.

2. Since this data base is intended to support primarily far field calculations, weathering is considered to be the principle source of secondary minerals. Nevertheless, some hydrothermal alteration minerals that may form due to the heat generated by waste emplacement have been considered.

3. Similarly, primarily weathering product minerals are considered as fracture filling materials, although some minerals expected due to the long exposure of fracture surfaces to warm water from near the waste emplacement are considered.

4. Minerals not found in the geological environment of unmineralized granite batholiths are excluded. These include most of the metamorphic minerals, minerals which are found only in sedimentary rocks and rocks in which the silica percent is less than 66 percent (intermediate, mafic and ultramafic rocks).

Table II Minerals found in granites and non-granite auxiliary minerals (in italics) considered in this review.

Categories are numbered as follows:

1 Primary essential mineral
2 Major accessory mineral
3 Minor accessory mineral
4 Rare accessory mineral
5 Very rare accessory mineral
6 Secondary (weathering or hydrothermal) mineral
7 Fracture filling mineral
8 Non-granite auxiliary mineral

In generic formulae, W, X, Y and Z represent discrete chemical elements; A and B represent a list of elements that may be found in ionic substitution. Valence information is given under Generic Formula heading.

Mineral Groups/Species	Category	Generic Formula and notes	Specific Formula used in data base
Sulfides		X_mZ_n; X=S	
galena	8		PbS(c)
würtzite	8		ZnS(wurt)
sphalerite	8		ZnS(spha)
covellite	8		CuS(c)
pyrite	3,7		FeS_2(pyri)
pyrrhotite	3,7	$Fe_{1-x}S$ (x=0-0.17)	
		x = 0	FeS(pyrr)
		x = 0.12	Fe_7S_8(c)
alabandite	8		MnS(c)
bornite	8		Cu_5FeS_4(c)
Oxides			
tenorite	8		CuO(c)
hematite group		X_2O_3; X=Al,Fe(III),Ti	
corundum	3	α	Al_2O_3(coru)
hematite	3	α	Fe_2O_3(c)
ilmenite	3	Fe(II)	$FeTiO_3$(c)
rutile group		XO_2; X=Mn,Ti	
rutile	3		TiO_2(ruti)
pyrolusite	6		MnO_2(c)
spinel group		XY_2O_4; X=Fe(II),Mg; Y=Fe(III)	
magnetite	3		Fe_3O_4(c)
magnesioferrite	8		$MgFe_2O_4$(c)
chrysoberyl	4		$BeAl_2O_4$(c)
Hydroxides			
goethite group		XO(OH); X=Fe(III),Al	
goethite	6	α	FeOOH(c)
diaspore	6	α	AlOOH(dias)
boehmite	6	α	AlOOH(boeh)
gibbsite	6		$Al(OH)_3$(gibb)
limonite	6,7	$FeOOH{\bullet}nH_2O$	
Halides			
fluorite	3,6		CaF_2(c)
Carbonates			
calcite group		ACO_3; A=Ca,Mn(II),Zm	
calcite	7		$CaCO_3$(calc)
rhodocrosite	8		$MnCO_3$(c)
smithsonite	8		$ZnCO_3$(c)
aragonite group		ACO_3; A=Ca,Sr,Pb	
aragonite	8		$CaCO_3$(arag)
strontianite	8		$SrCO_3$(c)
cerussite	8		$PbCO_3$(c)

Table II (cont'd)

Mineral Groups/Species	Category	Generic Formula and notes	Specific Formula used in data base
Sulphates			
alunite	6		$KAl_3(SO_4)_2(OH)_6(c)$
jarosite	6	$KFe_3(SO_4)_2(OH)_6$	
anhydrite	6		$CaSO_4(c)$
gypsum	6,7		$CaSO_4 \cdot 2H_2O(c)$
Phosphates			
anhydrous normal group		AXO_4; A=Y,Ce,La,Nd,Th; X=P	
xenotime	5		$YPO_4(c)$
monazite	4	$(Ce,La,Nd,Th)PO_4$	$CePO_4(c)$
			$LaPO_4(c)$
			$NdPO_4(c)$
anhydrous halogenated group		$A_5(XO_4)B$; A=Ca; X=P; B=OH,F	
hydroxylapatite	5		$Ca_5(PO_4)_3OH(c)$
fluorapatite	3		$Ca_5(PO_4)_3F(c)$
Nesosilicates		Si:O=1:4 $(SiO_4)^{4-}$	
olivine group		A_2SiO_4; A=Fe(II)	
fayalite	5		$Fe_2SiO_4(c)$
garnet group		$X_3Y_2(SiO_4)_3$; X=Fe(II),Ca; Y=Al,Fe(III)	
almandine	4		$Fe_3Al_2(SiO_4)_3(c)$
andradite	4	Fe(III)	$Ca_3Fe_2(SiO_4)_3(c)$
zircon group		$XSiO_4$; X=Zr,Hf,Th	
zircon	3		$ZrSiO_4(c)$
thorite	5		$ThSiO_4(c)$
topaz	5	$Al_2SiO_4(F,OH)_2$	$Al_2SiO_4(OH)_2(c)$
			$Al_2SiO_4F_2(c)$
titanite (sphene)	3		$CaTiSiO_5(c)$
Sorosilicates		Si:O=2:7 $(Si_2O_7)^{6-}$	
epidote group		$A_2B_3(SiO_4)_3OH$; A=Ca,Ce,Y; B=Al,Fe,Mn	
zoisite	7		$Ca_2Al_3(SiO_4)_3OH$(zois)
clinozoisite	7		$Ca_2Al_3(SiO_4)_3OH$(clin)
epidote	6,7		$Ca_2FeAl_2(SiO_4)_3OH(c)$
piemontite	7	$Ca_2(Al,Mn,Fe)_3(SiO_4)_3OH$	
Cyclosilicates		Si:O=1:3 $(Si_6O_{18})^{12-}$	
beryl	5	$Be_3Al_2(Si_6O_{18})$	
tourmaline group		$WX_3Y_6(BO_3)_3Si_6O_{18}(OH,F)_4$; W=Na; X=Fe; Y=Al	
schorl	4	$NaFe_3Al_6(BO_3)_3Si_6O_{18}(OH)_4$	
Inosilicates (single chain)		Si:O=1:3 $(SiO_3)^{2-}$	
pyroxene group		ABZ_2O_6; A=Ca,Fe(II),Mg,Na; B=Al,Fe,Mg,Ti; Z=Si,Al	
hypersthene	3	$(Mg,Fe)_2Si_2O_6$	$Mg_2Si_2O_6(c)$
			$Fe_2Si_2O_6(c)$
augite	3	$(Ca,Na)(Mg,Fe,Al,Ti)(Si,Al)_2O_6$	$CaAl_2SiO_6(c)$
acmite	3		$NaFeSi_2O_6(c)$
Inosilicates (double chain)		Si:O=4:11 $(Si_4O_{11})^{6-}$	
amphibole group		$W_{0-1}X_2Y_5Si_8O_{22}(OH,F,Cl)_2$; W=Ca,Na; X=Mg,Fe(II); Y=Al,Fe(III)	
calcic amphiboles			
hornblende	2	$Ca_2(Fe,Mg)_4Al_2Si_7O_{22}(OH)_2$	
tremolite			$Ca_2Mg_5Si_8O_{22}(OH)_2(c)$
alkali amphiboles			
riebeckite	3	$Na_2Fe_5Si_8O_{22}(OH)_2$	
arfvedsonite	3	$Na_3Fe_5Si_8O_{22}(OH)_2$	

Table II (cont'd)

Mineral Groups/Species	Category	Generic Formula and notes	Specific Formula used in data base
Phyllosilicates		Si:O=2:5 $(Si_2O_5)^{2-}$	
serpentine group			
antigorite	8		$Mg_3Si_2O_5(OH)_4$(anti)
chrysotile	8		$Mg_3Si_2O_5(OH)_4$(chry)
sepiolite	8		$Mg_4Si_6O_{15}(OH)_2.6H_2O$(c)
kaolinite group			
kaolinite	6,7		$Al_2Si_2O_5(OH)_4$(kaol)
dickyte	8		$Al_2Si_2O_5(OH)_4$(dick)
halloysite	8		$Al_2Si_2O_5(OH)_4$(hall)
talc	8		$Mg_3Si_4O_{10}(OH)_2$(c)
montmorillinite (smectite)	6	$(Na,Ca,K,Mg)_{0.33}(Al,Mg)_2Si_4O_{10}(OH)_2$	nH_2O
		Ca	$CaAl_{14}Si_{22}O_{60}(OH)_{12}$(c)
		K	$K_2Al_{14}Si_{22}O_{60}(OH)_{12}$(c)
		Mg	$MgAl_{14}Si_{22}O_{60}(OH)_{12}$(c)
		Na	$Na_2Al_{14}Si_{22}O_{60}(OH)_{12}$(c)
illite	6	$(K,H)(Al,Mg,Fe)_2(Si,Al)_4O_{10}[(OH)_2,H_2O]$	$K_2Al_{10}Si_{14}O_{40}(OH)_8$(c)
			$K_3MgAl_9Si_{14}O_{40}(OH)_8$(c)
chlorite group		$A_{5-6}Z_4O_{10}(OH)_8$; A=Mg,Fe,Ca; Z=Al,Si	
clinochlore	6,7	$(Mg,Fe)_5Al(Si_3Al)O_{10}(OH)_8$	$Mg_{40}Al_{16}Si_{24}O_{80}(OH)_{64}$(c)
			$Mg_{26}Fe_8Al_{20}Si_{24}O_{80}(OH)_{64}$(c)
			$Mg_8Fe_{26}Al_{25}Si_{20}O_{80}(OH)_{64}$(c)
prehnite	8		$Ca_{16}Al_{16}Si_{24}O_{80}(OH)_{16}$(c)
mica group		$XA_{2-3}B_4O_{10}(OH,F)_2$; X=K,Na; A=Al,Mg,Fe(II),Fe(III),Li,V; B=Al,Si	
(muscovite series)			
muscovite	2	$KAl_2(AlSi_3O_{10})(OH,F)_2$	$KAl_3Si_3O_{10}(OH)_2$(c)
paragonite	5	$NaAl_2(AlSi_3O_{10})(OH)_2$	$NaAl_3Si_3O_{10}(OH)_2$(c)
phengite	7	$K[Al_{1.5}(Mg,Fe(II))_{0.5}](Si_{3.5}Al_{0.5})O_{10}(OH,F)_2$	
sericite	6	$KAl_2(AlSi_3O_{10})(OH,F)_2$	
(biotite series)			
biotite	2	$K(Mg,Fe(II))_3(Al,Fe(III)(Si_3O_{10})(OH,F)2$	$K_4Mg_4Fe_7Al_6Si_{11}O_{40}(OH)_8$(c)
phlogopite	3	$KMg(AlSi_3O_{10})(OH,F)_2$	$KMg_3AlSi_3O_{10}(OH)_2$(phlo)
			$KMg_3AlSi_3O_{10}F_2$(c)
annite	3	$K(Fe(II)_3AlSi_3O_{10}(OH,F)_2$	$KFe_3AlSi_3O_{10}(OH)_2$(c)
Tectosilicates		Si:O=1:2 $(SiO_2)^0$	
SiO_2 group			
quartz	1		SiO_2(quar)
chalcedony	7		SiO_2(chal)
opal	7	SiO_2 nH_2O	$SiO_2.0.5H_2O$(am)
feldspar group		AB_4O_8; A=Ca,K,Na; B=Al,Si	
(alkali feldspars)			
orthoclase	1		$KAlSi_3O_8$(orth)
microcline	1		$KAlSi_3O_8$(micr)
(plagioclase feldspars)			
albite (low)	1		$NaAlSi_3O_8$(l-al)
anorthite	8	hexagonal	$CaAl_2Si_3O_8$(hexa)
		triclinic	$CaAl_2Si_3O_8$(tric)

5. A number of minerals not expected in a granitic environment as defined above are included in Table II as auxiliary data. Their names appear in italics in this paper.

SELECTED DATA SET

A number of minerals are identified in Table II as being important to granite but for which no "specific formula" appears in Table III. This "specific formula" is the unique (usually stoichiometrically integral) formula under which data is stored in the NEA data base. The absence of a formula in this column in Table II indicates that although the mineral was found to be important to granite, no reliable chemical thermodynamic data was found for that phase. This lack of selected data occurs primarily in the secondary (weathering or hydrothermal) mineral and fracture fill mineral categories. Examples of such minerals are limonite, jarosite and some of the phyllosilicates. Further data collection is also needed for several pyroxenes and amphiboles.

An important conclusion of this study is that the US Geological Survey chemical thermodynamic data (as represented in [9, 10, 11] and related publications) are compatible with CODATA key values [12], with some well defined exceptions. The most important of these exceptions are:

1) Although the data for most solid phases in the USGS data base are CODATA compatible, those for aqueous species need not be. This has no direct impact on the data set presented here.
2) The USGS data set uses P(red) as the phosporus reference phase rather than P(white) used by CODATA. All data presented here have been converted to the CODATA reference phases when necessary.
3) Fundamental differences exist between USGS and CODATA Gibbs energy and enthalpy data for Sr. The data reported here for *strontianite* is consistent with [13] and CODATA [12].

A selected, internally consistent data set is presented in Table III. Here, data which may not be compatible with the USGS data set are presented in parentheses. The selected data are classified into recommended and preferred categories. Recommended data is thought to be within the stated uncertainty of an eventual CODATA value. Preferred data is considered by the reviewers to be the best available, but not recommendable. For further explanation of these categories, the rounding conventions used and the rationale for the 95% confidence interval as an uncertainty, see [1]. More digits than are significant have been presented throughout Table III. The explicit uncertainty intervals are thought to prevent this from being misleading.

In a number of cases, the minerals have no unique, well defined stoicheometric formula. In such cases, data for mixing or solid solution endpoints have been given when available even though the end member does not necessarily occur in granite. For example, data for FeS(c) and $Fe_{0.875}S$(c) are given for pyrrhotite. The compilation of thermochemical data for phyllosilicates is particularly difficult because of the behaviour of these solid phases as non-stoichiometric minerals, which resembles solid solution phases with a wide range of possible compositions within each family. The question is still under discussion as to whether they are really solid solutions or only complex mixtures of microphases.

In order to study the solution-phyllosilicate interactions, one may either define a very large number of defined compounds in relation to field observations and measurements or laboratory experiments or one may try to

Table III. Selected chemical data for minerals found in granite and non-granitic minerals (see Table 2)

Compounds for which the data are <u>recommended</u> are preceded by hyphens. The remaining compounds are <u>preferred</u>. Data in parentheses are not necessarily consistent with USGS data set. Asterisks indicate selected temperature function which appears in [2]

Specific formula used in data base	$\Delta_f G°$(298.15 K) (kJ.mol^{-1})	$\Delta_f H°$(298.15 K) (kJ.mol^{-1})	S°(298.15 K) (J.K^{-1}.mol^{-1})	Cp(298.15 K) (J.K.$^{-1}$.mol^{-1})
- PbS(c)	-96.08 ± 0.84*	-97.71 ± 0.96*	91.38 ± 1.25*	49.25 ± (1.00)*
ZnS(wurt)	-190.22 ± 1.52*	-194.57 ± 1.50*	58.84 ± 0.12*	45.88 ± (1.00)*
- ZnS(spha)	-202.50 ± 1.78*	-206.90 ± 1.70*	58.66 ± 0.13*	45.76 ± (1.00)*
CuS(c)	-49.08 ± 4.27	-48.58 ± 4.18*	66.65 ± 1.67*	(47.64 ± 2.00)*
- FeS_2(pyri)	-160.23 ± 1.72*	-171.54 ± 1.67*	52.93 ± 0.13*	62.17 ± (1.00)*
FeS(pyrr)	(-101.24 ± 2.00)	(-100.96 ± 2.00)*	(60.33 ± 1.00)*	(54.65 ± 2.00)*
Fe_7S_8(c)	(-748.50 ± 1.50)	(-736.40 ± 1.50)	(485.80 ± 1.00)*	(398.57 ± 1.00)*
MnS(c)	-218.16 ± 1.00*	-213.86 ± 0.84*	78.20 ± 1.67*	49.96 ± (1.00)*
Cu_5FeS_4(c)	(-372.6 ± 10.0)	(-354.9 ± 10.0)	(380.7 ± 12.0)	(238.5 ± 12.0) *
- CuO(c)	-129.56 ± 1.30*	-157.32 ± 1.26*	42.63 ± 0.21*	42.30 ± (1.00)*
- Al_2O_3(coru)	-1582.24 ± 1.10*	-1675.71 ± 1.00*	50.92 ± 0.20 *	79.34 ± (0.04)*
- Fe_2O_3(c)	-745.27 ± 1.30*	-827.15 ± 1.30*	87.48 ± 0.32*	103.80 ± 0.11 *
- $FeTiO_3$(c)	-1159.17 ± 1.63*	-1236.62 ± 1.59*	105.86 ± 1.25*	99.50 ± (1.00)*
- TiO_2(ruti)	-889.45 ± 1.03*	-944.75 ± 1.26*	50.29 ± 0.17*	55.10 ± (1.00)*
- MnO_2(c)	-465.14 ± 0.88*	-520.03 ± 0.84*	53.05 ± 0.42*	54.02 ± (1.00)*
- Fe_3O_4(c)	-1014.14 ± 0.82*	-1117.26 ± 0.89*	146.01 ± 0.50*	151.08 ± 0.17 *
- $MgFe_2O_4$(c)	-1317.00 ± 1.88*	-1428.42 ± 1.84*	123.85 ± 0.84*	143.72 ± (1.00)*
$BeAl_2O_4$(c)	-2178.46 ± 2.80	-2300.78 ± 2.80	66.28 ± 0.13	
FeOOH(c)	-488.55 ± 0.75	-599.33 ± 0.73	60.38 ± 0.63	
- AlOOH(dias)	-920.90 ± 0.36*	-999.46 ± 0.37*	35.34 ± 0.09*	53.10 ± 0.09 *
- AlOOH(boeh)	-915.77 ± 0.68*	-990.42 ± 0.72*	48.43 ± 0.51*	65.49 ± 0.22 *
- $Al(OH)_3$(gibb)	-1155.10 ± 0.64*	-1293.33 ± 0.63*	68.44 ± 0.34*	91.73 ± 0.16 *
CaF_2(c)	-1176.92 ± 0.42*	-1229.26 ± 0.42*	68.87 ± 0.34*	68.59 ± (0.50)*
- $CaCO_3$(calc)	-1130.61 ± 1.50*	-1209.06 ± 1.50*	91.78 ± 0.44*	82.99 ± 0.16 *
- $MnCO_3$(c)	-816.05 ± 1.38*	-889.27 ± 1.21*	100.00 ± 2.10*	75.52 ± (1.00)*
$ZnCO_3$(c)	-731.48 ± 2.97	-812.78 ± 2.93	82.42 ± 1.25	(79.90 ± 2.00)*
- $CaCO_3$(arag)	-1130.07 ± 1.50*	-1209.66 ± 1.50*	87.95 ± 0.43*	81.92 ± 0.32 *
- $SrCO_3$(c)	-1144.73 ± 1.00	-1225.79 ± 1.10	97.20 ± (1.70)*	81.42 ± (1.50)*
$PbCO_3$(c)	-625.34 ± 1.55	-699.02 ± 1.85	130.96 ± 3.40	(87.40 ± 6.00)*
$KAl_3(SO_4)_2(OH)_6$(c)	(-4659.0 ± 10.0)	(-5169.7 ± 10.0)	328.02 ± 1.80*	372.60 ± (2.00)*
- $CaSO_4$(c)	-1321.70 ± 4.18*	-1434.11 ± 4.23*	106.69 ± 1.67*	99.66 ± (1.50)*
$CaSO_4.2H_2O$(c)	-1797.20 ± 4.60	-2022.63 ± 4.64	194.14 ± 1.25	
YPO_4(c)	(-1830.0 ± 10.0)	(-1958.00 ± 5.00)	(66. ± 37.)	
$CePO_4$(c)	(-1817.70 ± 2.00)	(-1942.60 ± 2.00)	(101.71 ± 9.0)	
$LaPO_4$(c)	(-1829.00 ± 2.00)	(-1956.50 ± 2.00)	(80.5 ± 10.0)	
$NdPO_4$(c)	(-1817.70 ± 2.00)	(-1950.0 ± 15.0)	(80.0 ± 20.0)	
$Ca_5(PO_4)_3OH$(c)	(-6298.19 ± 5.90)	(-6686.86 ± 6.00)	(425.69 ± 3.00)	385.10 ± (3.00)*
$Ca_5(PO_4)_3F$(c)	(-6467.88 ± 5.90)	(-6837.46 ± 6.00)	(423.24 ± 3.00)	375.93 ± (3.00)*
- Fe_2SiO_4(c)	-1380.22 ± 1.10*	-1479.17 ± 1.38*	151.54 ± 1.07*	131.99 ± 0.17 *
$Fe_3Al_2(SiO_4)_3$(c)	(-4955.73 ± 10.0)	(-5271.03 ± 8.00)	(340.41 ± 5.00)	(313.96 ± 5.00)*
$Ca_3Fe_2(SiO_4)_3$(c)	(-5428.65 ± 10.0)	(-5778.12 ± 10.0)	(293.42 ± 5.00)	(270.0 ± 20.0)
- $ZrSiO_4$(c)	-1918.89 ± 1.04*	-2033.40 ± 1.00*	84.03 ± 1.25*	98.60 ± (1.30)*
$ThSiO_4$(c)	(-2048.8 ± 10.0)	(-2160.3 ± 10.0)	(106.70 ± 2.00)	
$Al_2SiO_4(OH)_2$(c)	(-2693.25 ± 5.00)	(-2904.66 ± 5.01)	(112.04 ± 1.00)	(153.72 ± 2.15)
$Al_2SiO_4F_2$(c)	(-2910.66 ± 5.00)	(-3084.42 ± 5.01)	(105.40 ± 1.00)	(143.66 ± 2.02)
$CaTiSiO_5$(c)	-2459.86 ± 2.43	-2601.40 ± 2.38	129.20 ± 0.84	139.00 ± (1.00)*
- $Ca_2Al_3(SiO_4)_3OH$(zois)	-6495.30 ± 2.02*	-6891.12 ± 2.08*	295.88 ± 0.72*	650.86 ± 0.48 *
$Ca_2Al_3(SiO_4)_3OH$(clin)	(-6495.60 ± 3.00)*	(-6891.50 ± 3.00)*	(295.60 ± 0.70)*	(650.86 ± 0.50)*
$Ca_2FeAl_2(SiO_4)_3OH$(c)	(-6072.43 ± 10.0)	(-6461.90 ± 10.0)	(314.97 ± 4.00)	
$Mg_2Si_2O_6$(c)	(-2921.77 ± 4.90)	(-3095.50 ± 4.86)	(135.72 ± 0.84)	(164.18 ± 3.00)
$Fe_2Si_2O_6$(c)	(-2240.84 ± 5.51)	(-2395.11 ± 5.50)	(189.97 ± 1.20)	

Specific formula used in data base	$\Delta_f G°$(298.15 K) (kJ.mol^{-1})	$\Delta_f H°$(298.15 K) (kJ.mol^{-1})	S°(298.15 K) (J.K^{-1}.mol^{-1})	Cp(298.115 K) (J.K.$^{-1}$.mol^{-1})
$CaAl_2SiO_6$(c)	(-3105.73 ± 5.50)	(-3280.31 ± 5.50)	(146.44 ± 2.50)	
$NaFeSi_2O_6$(c)		(-2575.90 ± 8.10)		
$CaMg_5Si_8O_{22}(OH)_2$(c)	-116279.00 ± 17.4	(-12355.10 ± 17.3)	(548.9 ± 1.25)	
$Mg_3Si_2O_5(OH)_4$(anti)	-4039.00 ± 2.00*	-4361.00 ± 2.00*	220.30 ± 1.10*	655.63 ± 2.00 *
- $Mg_3Si_2O_5(OH)_4$(chry)	-4032.36 ± 2.00*	-4482.04 ± 2.00*	221.34 ± 1.13*	273.86 ± 0.76 *
$Mg_4Si_6O_{15}(OH)_2.6H_2O$(c)	(-9251.60 ± 10.0)	(-10086.30 ± 10.0)	(613.40 ± 3.00)	
- $Al_2Si_2O_5(OH)_4$(kaol)	-3799.61 ± 0.98*	-4119.78 ± 1.06*	204.97 ± 1.02*	246.14 ± 0.77 *
- $Al_2Si_2O_5(OH)_4$(dick)	-3795.95 ± 1.54*	-4118.47 ± 1.24*	197.06 ± 3.07*	239.79 ± 0.74 *
- $Al_2Si_2O_5(OH)_4$(hall)	-3780.37 ± 1.51*	-4101.03 ± 1.20*	203.33 ± 3.07*	245.25 ± 0.76 *
- $Mg_3Si_4O_{10}(OH)_2$(c)	-5536.05 ± 4.35*	-5915.90 ± 4.33*	260.83 ± 0.63*	321.70 ± (1.00)*
$CaAl_{14}Si_{22}O_{60}(OH)_{12}$(c)	(-32200. ± 150.)	(-34430. ± 150.)	(1536.0 ± 30.)	
$K_2Al_{14}Si_{22}O_{60}(OH)_{12}$(c)	(-32123. ± 150.)	(-34365. ± 150.)	(1591.6 ± 30.)	
$MgAl_{14}Si_{22}O_{60}(OH)_{12}$(c)	(-32016. ± 150.)	(-34245. ± 150.)	(1536.4 ± 30.)	
$Na_2Al_{14}Si_{22}O_{60}(OH)_{12}$(c)	(-32076. ± 150.)	(-34313. ± 150.)	(1576.6 ± 30.)	
$K_2Al_{10}Si_{14}O_{40}(OH)_8$(c)	(-21747.2 ± 18.0)			
$K_3MgAl_9Si_{14}O_{40}(OH)_8$(c)	(-21773.2 ± 20.0)	(-23281.9 ± 20.0)	(1127.80 ± 5.00)*	(1291.20 ± 5.00)*
$Mg_{40}Al_{16}Si_{24}O_{80}(OH)_{64}$(c)	(-65760. ± 250.)	(-70993. ± 253.)	(3600. ± 60.)	(4138. ± 60.)*
$Mg_{26}Fe_8Al_{20}Si_{24}O_{80}(OH)_{64}$(c)	(-62547. ± 28.)			
$Mg_8Fe_{26}Al_{25}Si_{20}O_{80}(OH)_{64}$(c)	(-58320. ± 28.)			
- $Ca_{16}Al_{16}Si_{24}O_{80}(OH)_{16}$(c)	-46531.4 ± 13.1 *	-49549.0 ± 13.6 *	2389.92 ± 5.76*	2648.88 ± 3.52 *
- $KAl_3Si_3O_{10}(OH)_2$(c)	-5600.67 ± 3.29	-5976.74 ± 3.24	306.40 ± 0.61*	326.10 ± (0.80)*
$NaAl_3Si_3O_{10}(OH)_2$(c)	(-5555.00 ± 7.00)	(-5935.28 ± 7.03)	(278.00 ± 2.00)	(320.42 ± 3.00)*
$K_4Mg_4Fe_7Al_6Si_{11}O_{40}(OH)_8$(c)				(1396.1 ± 15.0)*
$KMg_3AlSi_3O_{10}(OH)_2$(phlo)	(-5860.0 ± 20.0)	(-6244.2 ± 20.)	319.70 ± (4.50)	(356.60 ± 5.00)*
- $KMg_3AlSi_3O_{10}F_2$(c)	-6053.07 ± 3.80	-6392.88 ± 3.66	336.30 ± 2.10*	342.40 ± (2.50)*
$KFe_3AlSi_3O_{10}(OH)_2$(c)	(-4810.00 ± 12.0)	(-5173.50 ± 12.1)	(373.00 ± 5.00)	(348.20 ± 5.00)*
- SiO_2(quar)	-856.29 ± 0.90*	-910.70 ± 0.90*	41.46 ± 0.40*	44.75 ± 0.04 *
SiO_2(chal)	(-854.70 ± 1.00)	(-909.12 ± 1.01)	(41.34 ± 0.50)	(44.75 ± 0.50)*
$SiO_2.0.5H_2O$(am)	(-848.90 ± 2.00)	(-897.75 ± 2.00)	(60.00 ± 5.00)	
$KAlSi_3O_8$(orth)	(-3747.70 ± 5.50)	(-3973.11 ± 5.50)	(214.05 ± 3.00)	(202.83 ± 3.50)*
- $KAlSi_3O_8$(micr)	-3742.33 ± 3.40	-3967.69 ± 3.37	214.20 ± 0.41*	202.40 ± (1.00)*
- $NaAlSi_3O_8$(l-al)	-3711.72 ± 3.44	-3935.12 ± 3.42	207.40 ± 0.40*	(205.10 ± 1.00)*
$CaAl_2Si_2O_8$(hexa)	-4001.42 ± 3.28	-4222.58 ± 3.12	214.80 ± 1.30	
$CaAl_2Si_2O_8$(tric)	-4003.33 ± 3.14	-4229.10 ± 3.12	199.30 ± 0.30	(211.40 ± 2.00)*

investigate the solid solution models for each family of phyllosilicates(kaolinite-type, smectites, illites, chlorites, etc.). In the latter approach, the characteristics of end-members must be defined. Some of these are existing defined compounds (such as kaolinite, talc, pyrophyllite and muscovite), but others are only theoretical end-members or occur very rarely as pure mineral (such as greenalite, Fe^{3+}-pyrophyllite, etc.), generally corresponding to one very insoluble aqueous species (like Fe^{3+}). This approach requires first a theoretical basis for describing the multi-component solid solution phase, and then a data base of measured or estimated thermochemical data for end-members (even if they are numerous). Different models have been proposed using different approaches [14 to 20]. They differ not only in the thermochemical data used but also in the models themselves. For that reason, it was considered premature to introduce these complex mineral phases in the selected data base presented here. The choice of model type and the choice of the data for the end-member are not independent and must be done simultaneously. Data given in Table III are therefore limited to physically existing end-members.

Due to space restrictions, the detailed rationale behind the selection of each datum could not be presented here. The compiled data set upon which the selected data in Table III is based and the origin of the selections are presented in [2]. Although each source of data cannot be cited in the present paper, since in the compiled data set several thousand references were used either directly or via other critical reviews, the papers of Helgeson [21, 22] must be cited as a significant complement to the USGS data. No heat capacity data were used directly from Helgeson's work because of potential inconsistencies with the USGS data. Thermal functions for data marked with an asterisk in Table III also appear in the report [2].

ACKNOLEDGEMENTS

The advice of Niel Plummer of the USGS is greatly appreciated. This compilation and critical review would not have been possible without the dedication of Dr. Pierre Nagel of the NEA Data Bank at Saclay, France and of Nathalie Cramer.

REFERENCES

1 A.B. Muller, Radioactive Waste Management and the Fuel Cycle, 6(2), 131 (1985).
2 A.B. Muller, NEA Compilation of chemical thermodynamic data for minerals associated with granite, RWM-5, OECD Nuclear Energy Agency, Paris.
3 A.L. Streckeisen, Earth Science Review 12, 1 (1976).
4 A.L. Streckeisen, Geology 7, 331 (1979).
5 A.L. Streckeisen, Geotimes 18, 26 (1973).
6 R.L. Bates and J.A. Jackson, Glossary of Geology 2nd ed., American Geological Institute, Church Falls, VA. (1962).
7 D.W. Hyndman, Petrology of Igneous and Metamorphic Rocks, 2nd ed., McGraw-Hill Book Co., New-York (1985).
8 E.G. Ehlers and H. Blatt, Petrology: Igneous, Sedimentary and Metamorphic, W.H. Freeman and Co., San Francisco (1982).
9 R.A. Robie, B.S. Hemingway and J.R. Fisher, U.S. Geological Survey Bulletin 1452 (1968).
10 B.S. Hemingway, J.L. Haas, J.R. and G.R. Robinson, U.S. Geological Survey Bulletin 1544 (1982).
11 G.R. Robinson et al., U.S. Geological Survey Open File Report 83-79 (1982).
12 CODADA recommended key values for thermodynamics 1977, CODATA Bulletin 28, Paris and pre-publication version of extended key value data set soon to be published by CODATA, J.D. Cox and D.D. Wagman, personal communication (1985).
13 E. Busenberg, L.N. Plummer and V.B. Parker, Geochim. Cosmochim. Acta, 48, 2021 (1984).
14 P. Aaraard and H.C. Helgeson, Clays and Clay Minerals 31, 203 (1983).
15 S.V. Mattigod and G. Sposito, Geochim. Cosmochim. Acta 42, 1753 (1978).
16 E. Merino and B. Ranson, Clays and Clay Minerals 30, 29 (1982).
17 J. Nriagu, Amer. Min. 60, 834 (1975).
18 R.K. Stossell, Geochim. Cosmochim. Acta 43, 1151 (1975) and R.K. Stossell, Geochim. Cosmochim. Acta 45, 1733 (1982).
19 Y. Tardy and R.M. Garrels, Geochim. Cosmochim. Acta 38, 1101 (1974).
20 Y. Tardy and B. Fritz, Clay Minerals 16, 361 (1981).
21 H.C. Helgeson, Amer. J. Sci. 207, 729 (1969).
22 H.C. Helgeson et al., Amer. J. Sci. 278-A (978).

PART III

Waste Form Performance—Glass

STATIC LEACHING OF FULLY RADIOACTIVE WASTE GLASS AT 90°C IN THE PRESENCE OF BENTONITE, GRANITE, AND STAINLESS STEEL CORROSION PRODUCTS

HANS-PETER HERMANSSON*, INGA-KARI BJÖRNER*, HILBERT CHRISTENSEN*, TOSHIAKI OHE** AND LARS WERME***
* Studsvik Energiteknik AB, S-611 82 Nyköping, Sweden
** Central Research Institute of Electric Power Industry, 2-11-1, Iwato-kita, Komae-shi, Tokyo 201, Japan
*** Swedish Nuclear Fuel and Waste Management Co, P O Box 5864 S-102 48 Stockholm, Sweden

ABSTRACT

The results from the JSS phase II static glass leaching experiments are presented. The experiments have been performed in the presence of bentonite, granite and stainless steel corrosion products at 90°C for 28 days. The leaching behaviour of Cs has been analysed using gamma-spectrometry. The surface of the leached glass samples has been analysed by infrared reflection spectrometry.

The presence of uncompacted bentonite seems to have a very small, increasing, effect on the leach rate of the glass compared to when no bentonite is present. In the presence of compacted bentonite, a lower leach rate has been found than in the presence of uncompacted bentonite. (For Cs 0.3 g x m^{-2} x d^{-1} and 0.7 g x m^{-2} x d^{-1} respectively.) The addition of stainless steel corrosion products seems to have a slightly increasing effect on the leach rate in the presence of compacted bentonite. The beneficial effect of granite on the leach rate seems to be overshadowed by the presence of uncompacted bentonite.

INTRODUCTION

The JSS-project [1, 2] is a joint research programme initiated by Japan, Switzerland and Sweden. It is sponsored through the Central Research Institute of Electric Power Industry (CRIEPI), Tokyo, Japan, Nationale Genossenschaft für die Lagerung Radioaktiver Abfälle (NAGRA), Baden, Switzerland, the Swedish Nuclear Fuel and Waste Management Company (SKB), Stockholm, Sweden, and Studsvik Energiteknik AB, Nyköping, Sweden. The work is being performed at Studsvik Energiteknik AB.

The major effort concerning HLW Glass within the next few years will be to study the interactions between the glass/canister/overpack/granite components of the repository. These studies will generate information that will be useful in making reliable predictions of glass corrosion throughout the service life of the repository. In order to accomplish this objective, it is necessary to obtain a thorough understanding of the influence on leaching of the different species that exist in the ground water, as well as species that originate from the interaction of ground water with bentonite, quartz sands, and granite. The JSS-programme will be concentrated along these lines. The results of phase I of this project have been reported [1, 2]. The present article gives results from phase II of the project, which concerns bentonite, granite and stainless steel corrosion product interactions.

Mat. Res. Soc. Symp. Proc. Vol. 50. © 1985 Materials Research Society

EXPERIMENTAL

Glass

The radioactive glass used in this study was manufactured by CEA/-Marcoule and is the same fully radioactive glass as was used in phase I of the JSS-project. The glass composition and preparation of the specimens are described in [1] and [2]. The final dimensions of the glass pieces were 3x9x9 mm and the surface finish was 600 grit. Each specimen weighed approximately 0.6 g and contained approximately 4 GBq of radioactivity. The specimens used for the phase I and phase II experiments originated from the same glass block, and each specimen was identified according to its original position in the glass block.

Leaching procedure

All the tests have been performed at $90^{o}C$ with a leaching time of 28 days and with doubly distilled water as the leachant. The ratio SA/V was 0.1 cm^{-1}. Bentonite, (Volclay MX-80), compacted or uncompacted, was present in all the experiments, and granite and corrosion products (metal oxides from stainless steel SS 2333, produced in our laboratory [3]) in some of them. Granite from the Stripa mine was crushed to give a surface area of 0.2 m^2/g as determined by the BET method.

The complete experimental matrix is shown in Table I. The ratios SA glass/volume bentonite and SA glass/volume granite were 1 cm^{-1}.

Table I. Test matrices

Two experiments have been carried out for each matrix position.

Matrix 1

1.1 Compacted bentonite, granite, water.

1.2 Uncompacted bentonite, granite, water.

1.3 As for 1.1, but with stainless steel corrosion products added.

1.4 As for 1.2, but with stainless steel corrosion products added.

Matrix 2

2.1 Compacted bentonite, granite, water; space filled with uncompacted bentonite. (See Figure 1).

2.2 Uncompacted bentonite, water. Arrangements to separate bentonite and glass. (See Figure 2).

2.3 As for 1.2, but carried out in gilded stainless steel containers.

2.4 Bentonite saturated water, gilded stainless steel containers (+ 2 blanks).

Due to the expected high swelling pressure from compacted bentonite special pressure resistant cells of stainless steel have been used in all the experiments, except where otherwise stated. Before closing the pressure cell the specimen and bentonite were added as shown in Figures 1 and 2.

After evacuating the air, the leachant was sucked into the cell, the volume recorded and the cell was then closed and placed in the heating cabinet. The gold plated stainless steel containers from phase I [1, 2] have been used in two of the tests (2:3 and 2:4, see Table I). The purpose of these tests has been to compare the two experimental systems.

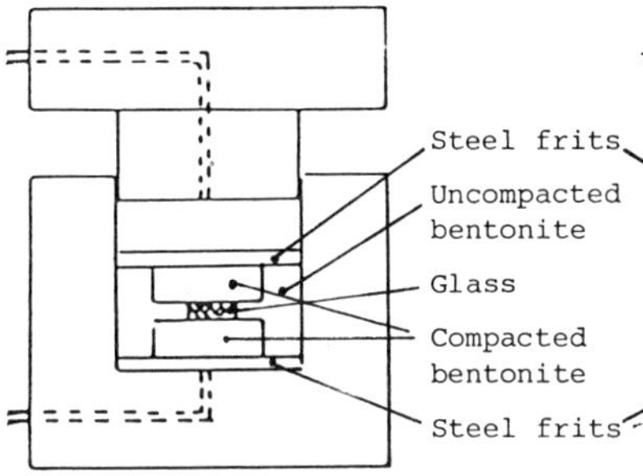

Figure 1. Stainless steel pressure resistant leaching container: arrangement for experiment 2:1.

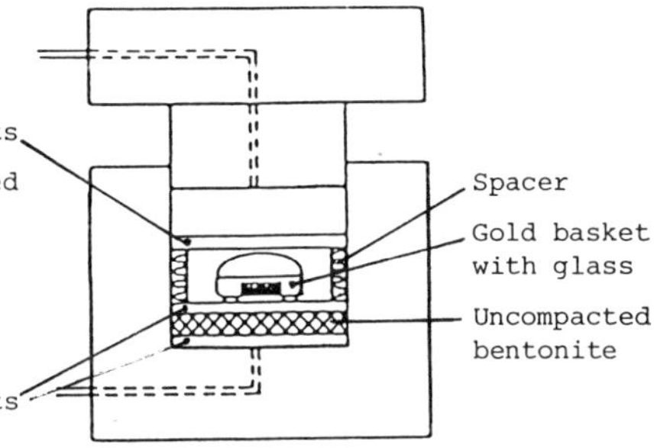

Figure 2. Pressure resistent leaching container with special arrangements for experiment 2:2.

After leaching the glass specimen was removed from the bentonite, cleaned with water and acetone and dried in the air. Infrared reflection spectra (IRRS) of the glass were recorded and the glass was weighed. Gamma spectrometric measurements were performed on the bentonite cakes and cleaning solutions.

In experiments 2:3 and 2:4 the container and the remaining bentonite were treated with dilute nitric acid and heated to 90°C for 4 hours. The bentonite slurry was filtered through filter paper (Munktell 00H), pore size approximately 3µm. Gamma spectrometric measurements were made on the filtrate, on the cleaning solution, and on the remaining bentonite. Inductively Coupled Plasma (ICP) analysis of B and Mo was performed on the filtrate only.

RESULTS AND DISCUSSION

Weight loss

Total mass losses for experiments 1:1 through 1:3 were determined after drying the glass specimens at 110°C for 1 h. For the remaining experiments the weight after leaching was recorded after only drying in air, in order to ensure better conditions for SIMS measurements which were carried out by Swiss Federal Institute for Reactor Research (EIR), Würenlingen, Switzerland. The reported total mass losses may be too low in some cases because bentonite still adhered to the sample surface. No weight increases were observed, as was the case in a few experiments during phase I when only granite was present. (See Table II.)

Table II. Summary table of results from phase II, results from phase I are included. All experiments 90°C, 28 days, SA/V=0.1 cm^{-1}, doubly distilled water. (For experimental code, see Table I)

Experi-ment	Toal mass loss(g/m^2)	Norm. mass loss(g/m^2) Cs-134	Cs-137	B	Mo	IRRS-data Peak pos. (cm^{-1})	Intensity (%)
1.1	1.2 ± 0.6	7.2 ± 0.1	7.6 ± 0.3	-	-	1050	23
1.2	8 ± 4	18 ± 1	20 ± 1	-	-	(1062)	8
1.3	14 ± 1	24 ± 2	28 ± 1	-	-	(1053)	9
1.4	14 ± 10	20 ± 1	22 ± 0	-	-	(1065)	11
2.1	9 ± 3	16 ± 5	16 ± 5	-	-	1048	18
2.2 a)	9.7	19	18	-	-	*	7
2.3	7 ± 6	25 ± 3	25 ± 3	10 b)	2 b)	(1058)	13
2.4	9.1 ± 0.9	21 ± 1	22 ± 1	22 b)	5 b)	1055	40
Phase I							
IA1	6.9	17	17	9.6	11	(990)	9
IB	9.5	16	16	10	12	(1025)	22 c)
IA2	-1.3	0.3	0.2	0.2	0.2	1010	14

a) One sample only
b) Measured on filtered leachate, no analysis of bentonite
c) Large variations between the two sides (± 10).
( ) Peak position values within brackets are inprecise due to low intensity or peak broadening.
* Value of peak position cannot be given due to the low intensity
IA1 = experiment without granite, Ar- atmosphere
IB = " " " , Air
IA2 = " with " , Ar-atmosphere

Gamma spectrometric data

Gamma spectrometric measurements were performed on the moist bentonite cakes, and on the solutions used to clean the glass sample after leaching. In experiments 2:3 and 2:4 the filter cake, the filtrate and the cleaning solutions were measured. In most cases it was only possible to measure Cs-134 and Cs-137. In Table II the $(NL)_i$ values from the experiments in phase II are compared to the corresponding experiments in phase I.

Infrared reflection data (IRRS)

Infrared Reflection Spectrometry (IRRS) has been demonstrated by Clark et al [4, 5] to be a useful technique for distinguishing glasses with different compositions, and for semiquantative analysis of glass corrosion. During leaching, peak position and intensity (% reflection) change, depending on the degree of corrosion [4].

In the current experiments a deposit of bentonite on the surface may have influenced the intensity. Large differences between the spectra obtained from the two sides of a specimen indicate variations in the roughness which could be an effect of bentonite deposits. However, in the non-radioactive experiments the SEM-EDS analysis indicated that the increased roughness was not caused by the deposition of bentonite, but rather by inhomogeneous attack of the surface [3]. Figure 3 shows spectra of the JSS-A glass before leaching and after leaching in bentonite saturated water. The IR spectra from this experiment show a very high reflectance and a sharp peak similar to that of fused silica, see Figure 4. This phenomenon could be due to repolymerization of silica. Infrared data are also reported in Table II.

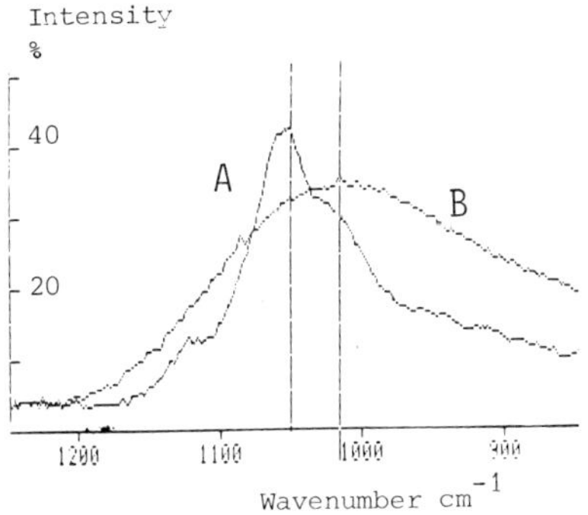

Figure 3. IRRS spectrum of glass leached in bentonite saturated water (A), and of glass before leaching (B).

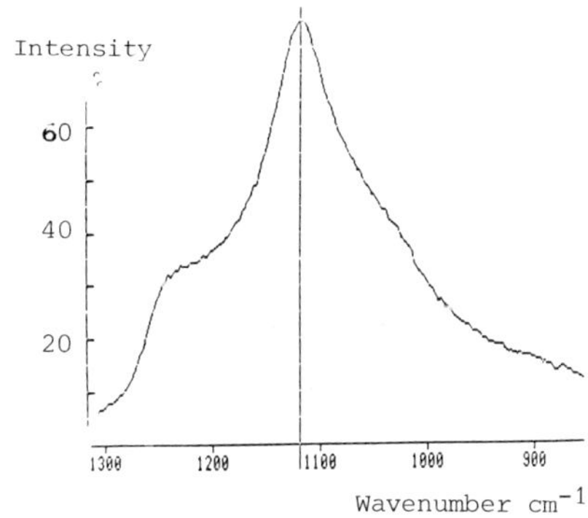

Figure 4. IRRS spectrum of vitreous silica.

Leachate analyses

In experiments 2:3 and 2:4 the bentonite or bentonite saturated water was treated with dilute nitric acid (0.5 %) and then filtered. The filtrates were analysed for B and Mo using the ICP. The results are reported in Table II. In the bentonite or bentonite saturated water B and Mo are originally present only in low concentrations, see Tables III and IV. No analysis was made for B and Mo in the bentonite phase of experiments 2:3 and 2:4.

Table III. Contents of B, Cs and Mo in bentonite and granite

Solid phase	Contents, ppm B	Cs	Mo
Bentonite	20 ± 10	< 1	1 ± 1
Granite	5 ± 2	4 ± 1	2 ± 1

Table IV. Concentrations of elements (in ppm) in the bentonite water used in method 2:4 (equilibrated at 45°C).

Na	Ca	Si	B	Mo	Cs	HCO_3	pH
540	90	65	0.2	0.2	<0.001	190	9.0

Effects of repository components

In the non-radioactive tests it was obvious that the glasses most severely attacked were those leached in bentonite saturated water, or those leached in the presence of corrosion products [3]. This was confirmed both by total mass loss and elemental mass loss data, by IRRS and by SEM-EDS analyses. The difference is not as evident in the case of the JSS-A glass, but the tendency seems to be the same.

The gamma-spectrometric measurements of Cs provide the most reliable basis for comparison between the results of the current experiments and those from phase I [1, 2], where either leachant only or leachant and granite were used. Both total mass loss and IR data could be affected by deposits of bentonite on the glass surface causing the values of total mass loss and reflection intensity to be too low. The only exception is experiment 2:2 where bentonite and glass were separated from each other by a special arrangement.

From the data in Table II it is obvious that the results from the phase I and phase II experiments are of the same order of magnitude. The presence of compacted bentonite apparently decreases the leach rate, compared to the experiments of phase I when leachant only was present. This is confirmed by total mass loss, Cs and IR data and is probably due to the fact that the system is more or less dry. In all other cases the weight loss and the mass loss of Cs are somewhat higher in the phase II experiments than in those experiments of phase I when leachant only was present. Granite, which was added in some experiments of phase I [2], apparently had a beneficial effect probably caused by buffering effects and/or silica saturation, see experiment no IA2, Table II.

The addition of corrosion products seems to have an increasing effect on the leach rate especially in the case of compacted bentonite, see experiments 1:3 and 1:4 in Table II. This increase may be due to removal of silicon by adsorption on iron colloids [6].

The effect of bentonite-saturated water seems to be similar to that of uncompacted bentonite. The slight increase in leach rates in both cases could be caused by the higher pH and/or the higher HCO_3^- concentration in the two cases, see Table IV.

The normalized mass loss of B is higher than the total mass loss value for those experiments where leachate analyses have been performed. The Mo values are lower, which could be due to adsorption on the bentonite. Adsorption of Mo, and to some extent also of B, on the bentonite was found in the non-radioactive experiments [3]. Small amounts of B dissolved from the bentonite were also found in those experiments.

In the case of uncompacted bentonite the B value is of the same order as that found in distilled and silicate water in phase I. In the experiment in bentonite water the B value is significantly higher compared with the phase I case.

Comparison with other glasses

Table V shows some of the results from the current tests, together with the corresponding results from the experiments performed with the Swedish simulated waste glasses ABS-39, ABS-41 and ABS-118, the last of which has a composition very close to JSS-A. The composition of these three glasses is given in [3].

The similarity in corrosion behaviour between the JSS-A glass and ABS-118 that was found in phase I can also be observed in phase II (Table V). It is also obvious from this table that both JSS-A and ABS-118 exhibited better leaching resistance than either ABS-41 or ABS-39.

Table V. Comparison between JSS-A and other glasses

Uncompacted bentonite plus granite	JSS-A pressure cell	JSS-A gilded cont.	ABS-118	ABS-41	ABS-39
Total mass loss (g/m^2)	8 ± 4	7 ± 6	6.4 ± 1.2	28 ± 1	40 ± 2
Normalized elemental mass loss (g/m^2):					
Cs in filtrate	-	12	10 ± 4	26 ± 2	62 ± 4
Cs in bentonite	-	13	5.5 ± 3.6	14 ± 2	26 ± 2
Cs total	19	25	16	40	88
B in filtrate	-	9.6	7.2 ± 0.5	30 ± 2	65 ± 1
B in bentonite	-	-	1.1 ± 1.1	1.4 ± 0.1	8.1 ± 2.6
Mo in filtrate	-	2.0 ± 0.1	2.6 ± 3.2	19 ± 6	46 ± 8
Mo in bentonite	-	-	25 ± 25	23 ± 3	40 ± 7
IRRS max intensity	8	13	7	7	7
peak position	(1062)	(1058)	-	(1035)	(1021)
Uncompacted bentonite + corrosion products	**JSS-A pressure cell (including granite)**		**ABS-118 (excluding granite)**	**ABS-41**	**ABS-39**
Total mass loss (g/m^2)	14 ± 10		26 ± 5	42 ± 1	74 ± 8
Normalized mass loss of Cs(g/m^2)	21 ± 1		33 ± 16	36 ± 1	77 ± 15
IRRS max intensity	11		10	27	6
peak position	(1065)		1028	1042	(1035)
Compacted+uncompact bentonite	**JSS-A (including granite)**		**ABS-118 (excluding granite)**	**ABS-41**	**ABS-39**
Total mass loss (g/m^2)	9 ± 3		1.7 ± 0.3	36 ± 15	31 ± 2
IRRS max intensity	18		24	16	17
peak position	1048		1055	1026	1040
Bentonite saturated water	**JSS-A gilded container**		**ABS-118**	**ABS-41**	**ABS-39**
Total mass loss (g/m^2)	9.1 ± 0.9		8.9 ± 1.1	23 ± 2	32 ± 2
Normalized mass loss (g/m^2) of B	22 ± 6		14 ± 4	31 ± 5	26 ± 4
of Mo	5.0 ± 0.6		12 ± 4	27 ± 6	70 ± 6
IRRS max intensity	40		9	6	9
peak position	1055		1025	1020	1035

CONCLUSIONS

Some conclusions can be drawn from the data presented in this paper. All conclusions are based on data from 90°C and 28 d of leaching.

The gamma spectrometric measurements of Cs have produced highly reproducible data constituting a good basis for drawing conclusions. Weight loss and IRRS measurements may be subject to analytical errors due to bentonite deposits on the glass surface. This could cause the values of weight loss and IRRS reflection intensity to be too low.

The leach rate seems to be slightly higher in the presence of uncompacted bentonite than in the absence of bentonite, although the difference is too small to be statistically significant without further experimental verification.

Based on weight loss and Cs data a lower corrosion rate is found in the presence of compacted bentonite than in the presence of uncompacted bentonite, which is also confirmed by IRRS data.

The addition of stainless steel corrosion products in the presence of compacted bentonite seems to increase the leach rate. This is indicated by Cs data, by weight loss data, and by IRRS data.

The beneficial effect of granite, found in the phase I experiments is completely overshadowed by the influence of bentonite.

No effect of the difference in experimental procedure using pressurized or gilded containers has been observed.

No difference can be observed between the results obtained in bentonite water and in the presence of uncompacted bentonite.

An IRRS peak displacement from 1020 cm^{-1} to approximately 1050 cm^{-1} is generally observed, which indicates the formation of a silica rich layer. This is very clear in the case of bentonite saturated water.

Comparison of radioactive JSS-A glass and non-radioactive ABS-118 generally shows equivalent levels of leaching.

ACKNOWLEDGEMENTS

The radioactive JSS-A glass was provided by CEA/Marcoule, France. Samples for leaching studies were cut and polished by the Studsvik Hot Cell Laboratory, applying the technique developed at the Hahn-Meitner Institute, Berlin. Technical assistance in the leaching and analyses was provided by Karin Åsberg, Rune Lundqwist, Mårten Hillver from STUDSVIK and Hayaichi Yokoyama from CRIEPI. Karen Gott has corrected the English and participated in the general administration of the work. The manuscript was typed by Evelyn Gäfvert. Funding was provided by the JSS participants.

REFERENCES

1. H-P Hermansson, H. Christensen, I-K. Björner, D.E. Clark, H. Yokoyama and L. Werme in Scientific Basis for Nuclear Waste Management VII, edited by G.L. McVay (Elsevier Science Publishers, New York, 1984), pp. 671-679.

2. H-P. Hermansson et al, JSS-Project Phase I Report. Studsvik Report NW-84/653.

3. H. Christensen, H-P. Hermansson and I-K. Björner, Studsvik Report NW-84/638.

4. D.E. Clark, C.G. Pantano Jr and L.L. Hench, Corrosion of Glass (Books for Industry, 777 Third Ave., New York, N. Y., 1979), p. 11.

5. D.E. Clark, E.C. Ethridge and M.F. Dilmore, Glass Technology, 18 4, 121-124 (1977).

6. G.L. McVay and C.Q. Buckwalter, J. Am. Ceram. Soc., 66, 170 (1983).

GLASS/WATER REACTION WITH AND WITHOUT BENTONITE PRESENT - EXPERIMENT AND MODEL

B.GRAMBOW*, H.P.HERMANSSON**, I.K.BJÖRNER**, L.WERME***
* Hahn-Meitner-Institut für Kernforschung GmbH, 1 Berlin-39, Glienickerstr. 100, FRG
** Studsvik Energiteknik AB, S-61182 Nyköping, Sweden
*** Swedish Nuclear Fuel and Waste Management CO. SKB, Box 5864 S-10248 Stockholm, Sweden

INTRODUCTION

In nuclear waste repository design bentonite has been included as part of the backfill for its sorbtive capacity and low water permeability. Nevertheless, it cannot keep the waste form dry once intrusion of groundwater has occured [1]. Leach experiments [2], [3] with the radioactive nuclear waste form borosilicate glass JSS-A have been performed with and without bentonite present to provide a database which allows the long term stability of the glass in aqueous environment to be forecasted and the chemical effects of bentonite to be studied.

Quality and completeness of the data base can only be judged by a consistent and quantitative understanding the overall glass leaching mechanism as it responds to changes in environmental conditions and to the presence or absense of repository system components, such as bentonite. Even the "simple" reaction of glasses with deionized water can only be described sufficiently if a network of various reversible and irreversible reactions is considered. Recently, computer assisted geochemical reaction path models were applied successfully to this reaction [4],[5],[6].

However, understanding of the glass/water reaction in presence of bentonite is much more complicated, because the data base is limited and the modeling of bentonite/water reactions has to face some fundamental unsolved scientific questions [7]. The effect of bentonite on glass leaching is judged either as detrimental [8] or as benificial [9]. In this paper, we try to overcome these difficulties by linking experiment and model as close as possible in order to evaluate the effect of bentonite on glass leaching.

RATE EQUATION AND THERMODYNAMIC MODEL

The model for the glass/bentonite/water system is developed from two subsystems: glass/water and bentonite/water. Communication between the submodels for the overall system occurs through contact of the solution with the solids, because solution chemistry plays a key role in both models.

Glass/water model

The glass/water model, described in detail elsewhere [6], combines kinetic and thermodynamic constraints. The following processes are considered: (1) dissolution of the glass matrix and subsequent release of soluble species - rate limiting step is the desorbtion of silanol groups from the glass matrix (reaction $SiO_2 + 2H_2O \rightarrow H_4SiO_4$), (2) hydrolysis and complexation of species in the aqueous phase, (3) solubility controlled formation of stable and metastable phases within a growing surface layer, (4) saturation of silicon as H_4SiO_4 in solution with respect to silanol groups on the glass surface causing a decrease in the rate of reaction to a finite value, and (5) alteration of metastable phases constituing the driving force of the long term reaction. The glass dissolution reaction is described by a rate equation similar to first order dissolution (Eq.1)

$$r_m(\xi) = k_+ \left(1 - IAP^*(\xi)/K^*\right) \qquad (1)$$

where k_+ is the rate constant for the forward rate, K^* is the stability constant for the rate limiting reaction, and $IAP^*(\xi)$ is the ion activity product for this reaction ($IAP^* = a_{H4SiO4}/a^2_{H2O} \approx a_{H4SiO4}$).

The influence of environmental constraints and additional reactions on the kinetics of the glass/water reaction is mainly described by their influence on k+ and IAP*. Within a limited pH regime and at constant temperature even k+ may be treated as a constant [6] in various experiments. The dependence of the activity of H_4SiO_4 on the reaction progress is calculated using the geochemical code PHREEQE [10],[3], considering all reactions consuming or producing silicic acid. From these data a time/concentration relationship is derived by numerical integration of Eq.1.

Bentonite/water model

The bentonite/water model describes the effect of bentonite on solution chemistry using equilibrium thermodynamics. The active component of the bentonite is montmorillonite, the formula of which for this study is [11]

$$Na_{0.3}\,(Si_{3.96}Al_{0.04})(Al_{1.55}Fe(III)_{0.20}Fe(II)_{0.01}Mg_{0.24})O_{10}(OH)_2 \quad .$$

It is assumed that the solution is in equilibrium with montmorillonite, although the equilibration takes some time (after 4 days 1/2 saturated, saturated after 100 days at 25°C, [12]). Dissolution of the minor components of bentonite: quartz, felspar, calcite, and siderite is neglected. In contrast, the model has to consider the dissolution of sulfate and chloride containing minerals, since high concentrations of these anions are found in solutions contacting bentonite (see below, table II). It was assumed that sulfate and chloride come from highly soluble salts, which dissolve instantaneously according to the weight fraction in the bentonite. This weight fraction is determined from solution analysis (see below). For the computer simulations of the bentonite/water reaction, the respective amount of sulfate and chloride was added to the initial solution composition.

The assumption of an equilibrium for montmorillonite has two aspects: the solubility equilibrium and the ion exchange equilibrium of Na and H. It is assumed, that both equilibria are coupled, i.e. the solubility depends explicitly on the concentration of Na in solution.

For the solubility equilibrium of wyoming montmorillonite at 25°C, thermodynamic data are available in the literature [13]. For calculating the temperature dependence of solubility a respective ΔH value is used from the Ca-montmorillonite (PHREEQE data base). For the ion exchange equilibria, the exchange constant is derived from the fit to experimental data (pH, Na concentration) of the bentonite/water system. No temperature dependence of ion exchange is considered yet.

In addition to the reactions mentioned above, precipitation of ferrihydrite and gibbsite is considered. Equilibrium with CO_2 from the air is established.

The glass/bentonite/water model

The overall system is composed of the two subsystems. As a function of the amount of glass dissolved, the computer model uses solubility as a criterion to select phases out of a list of potential alteration products including gibbsite and ferrihydrite. Montmorillonite and CO2 from the air are considered to be saturated throughout the reaction.

Fig.1 shows schematically the overall system. There are three sources for the ions in solution: the glass, the bentonite and the gas phase, and there are three sinks for these ions: the bentonite, the alteration products and the solution species. As in the glass/water subsystem, the resulting dependence of the silicic acid concentration on reaction progress is used to compute the time dependence of reaction progress and hence for all concentrations of interest (including pH).

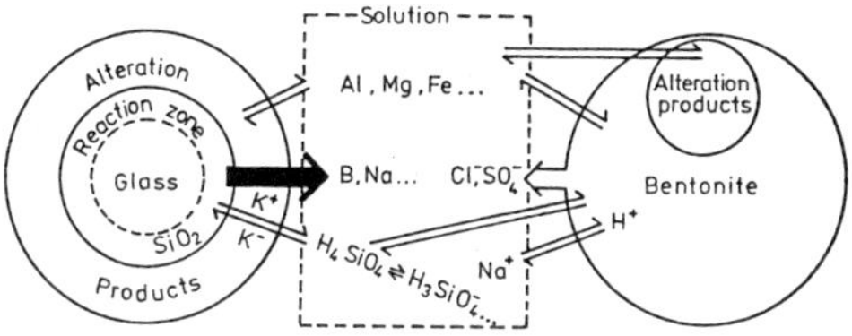

Fig.1. Scheme of the glass/bentonite/water thermodynamic and kinetic model

APPLICATION TO THE EXPERIMENTAL DATA

The model is first applied to experiments with only glass and aqueous solution present in order to derive the forward rate constant, the saturation concentration and the list of potential secondary alteration products. Thereafter, the bentonite/water system is analysed in terms of the needs of the model, and finally, the modelling is focussed on the overall system. The experimental details are reported in a paper given by Hermansson et al. in this volume [3].

Time dependence of the reaction of JSS-A glass in aqueous solutions

The data used to derive the required parameter for the model comprise experiments in deionized water with and without air present, and experiments in silicate water and in granitic ground water ("Allard water"). The application of the model to the glass/water reaction is performed in three steps: 1. Chemical modelling of leachates for a list of potential secondary alteration products. 2. Determination of k+ , K*, and the longterm rate. 3. Use the reaction path/time model to calculate the time dependence of leaching. Due to the large volume of data analysed, only one example is presented here, the complete data evaluation will be published as part of the JSS-project.

Step 1: For up to 300 minerals contained in an extended PHREEQE database [10], the degree of saturation was calculated for the various solutions resulting from glass leaching. These calculations are very important to create a list of potential secondary alteration products, which can be used for the simulation of precipitation reactions. The general finding from the chemical modelling of the leach data is, that some minerals are always oversaturated (i.e. leonhardite, chlorite, or goethite), whereas others are subsaturated and may or may not become saturated with time. Only subsaturated and saturated minerals are used within the list of potentially precipitating phases, in order to maintain the observed supersaturation also in the computer simulations. The list of phases shown in table I is based on the evaluation of various leachates obtained in JSS-phase I.

Step 2: The forward rate constant k_+ and the stability constant K* for the rate limiting step ("saturation concentration") can be obtained by plotting (Fig.2) the measured dissolution rates of the glass matrix in various tests against the IAP*, i.e. against the respective concentration

Table I. List of potential secondary alteration products for computer simulation of precipitation in case solubility limits are exceeded

Phases for Si:	$SiO2$ (log K(90C) = -3.160)
Al:	gibbsite (Log K lower by 0.6), analcime, wyoming montmoril.
Fe:	amorphous iron(III) hydroxide (ferrihydrite)
Ba,Mo,Nd:	$BaMoO_4$, $CaMoO_4$, Barite, $Nd(OH)_3$, $Nd_2(CO_3)_3$
Ca,Sr:	calcite, Ca-montmorillonite, strontianite, celestite
Mn:	rhodochrosite, pyrolusite
P:	hydroxyapatite, strengite, vivianite
Mg:	sepiolite, artenite, brucite

of orthosilicic acid, which is calculated from total silica in solution by the chemical model in step 1. The data contained in Fig.2 are derived from leach tests in deionized water with and without contact of air, silicate water (marked by "△"), "Allard water", and solutions with granite present.

According to Eq.1 the data should follow a straight line. The data for silicate water deviate from this line. No satisfactory explanation is available yet, but uncertain carbonate and chloride content of the actual silicate water made the calculations of IAP* ambiguous. At the intersection of the line with the y-axis (IAP* = 0) for k_+ a value of 1.5 g/m2d-1

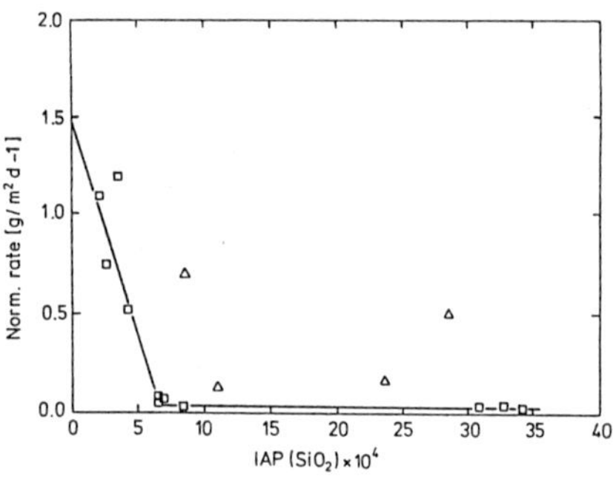

Fig.2. Effect of saturation on the leach rates in experiments of JSS-phase I

is obtained, and at the intersection with the x-axis (r_m = 0, IAP*=K*) for K* a value of 7×10^{-4} is obtained.

It is an interesting result of this study, that even in supersaturated solutions (IAP* $> 7 \times 10^{-4}$) the reaction of the glass with the aqueous solution does not cease (Fig.2), as would be expected from Eq.1. In contrast, in these cases glass leaching continues with a low rate. The rate under supersaturated conditions may be identified as the long term rate of reaction, which may be driven by the slow alteration of silica rich metastable reaction products or by the alteration of the reaction zone (see [6]). A "long term rate" of 0.04 g/m2d-1 is obtained from this plot. Applying the general rate equation for glass dissolution [6] this corresponds to a chemical affinity for the long term driving force of 25 cal/mol.

Step 3: Using the list of potential secondary alteration products (table I) and k_+, K*, and the long term rate as input, the reaction path/time model is used to simulate the various experiments. The starting solution may be deionized water, CO2 saturated water, or any other solution, provided the ionic strength does not exceed 0.1. For the present communication only results for deionized water are presented. Figure 3 shows the comparison between the reaction path/time model calculations and the experimental results. Good aggreement is obtained, although, average parameters were calculated from all the experiments. Better aggreement can be achieved if the parameters were specific only to deionized water. However, a consistent data base for all experiments was considered to be better than individual fits to experimenal data.

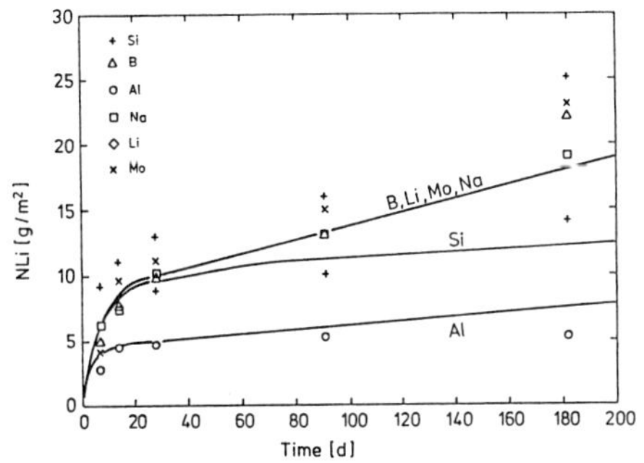

Fig.3 Comparison of leach results of JSS-A glass in deionized water (MCC-1 test, 90 C, 10 m-1) with the computer simulation

Reaction of bentonite and water

Before the reaction model can be applied to the bentonite/water experiments, the analytical data have to be evaluated to determine the amount of sulfate and chloride released per given quantity of bentonite and to determine the controls for carbonate. Several solutions from the bentonite water experiments were investigated (Table II). Solutions I or Ia were used in various experiments of JSS-phase II [3]. They were prepared by leaching bentonite at 45°C for ca. 1 day (solution I) or for 8 h (solution Ia). In addition, the other solution analyses (all 25°C experiments) were used to evaluate the quality of the data of solution I and Ia and to study the effect of the clay/water ratio and of time on solution composition. The solutions IVa-d were taken from ref.[14].

Table II. Bentonite solution analyses (mg/l) and modelling results (*initial solution III/IV in mg/l: 70/59 Cl, 9.6/8.6 SO_4, 65/55 Na, 123/115 HCO3, 18/19 Ca, 5.6/3.7 Si; ** Fe modified for control by $Fe(OH)_3$)

Solution	I	Ia	II	III*	IVa* 7d	IVb* 30d	IVc* 90d	IVd* 180d
Clay/water wt.ratio	0.07	0.07	0.23	0.11	0.03	0.03	0.03	0.03
pH(25°C)	9.2	9.0	8.8	9.0	9.5	9.6	9.1	9.1
Na	540	297	670	383	130	220	260	290
K	n.d.	n.d.	10.5	n.d.	3.1	2.6	3.0	3.9
Ca	90	3.5	6.6	4.8	11	1.4	2.4	7.0
Mg	n.d.	6.3	n.d.	n.d.	3.3	0.6	0.5	6.0
Si	65	106	n.d.	n.d.	5.6	4.7	7.1	9.9
Al	n.d.	39.0	n.d.	n.d.	0.14	0.10	0.15	0.19
Fe	n.d.	9.8	n.d.	n.d.	1.4	0.07	0.20	2.0
CO_3	190	280	n.d.	417	200	300	513	613
SO_4	n.d.	320	872	307	60	50	18	7
Cl	n.d.	1.5	132	107	56	65	63	72
charge balance (meq/l)	24	0.4	21.9	-0.02	-0.4	0.6	0.16	0.05
log pCO_2	-3.54	-3.47	--	-3.44	-4.30	-4.24	-3.45	-3.39
log IAP/K Calcite	1.65	0.41	--	0.65	1.15	0.42	0.51	1.04
" " Montmorillonite	--	--	--	--	-1.15	-2.06	0.09	0.52**
log Na^+/H^+	7.57	7.04	7.26	7.22	7.20	7.52	7.10	7.16

In addition to solution analyses, Table II contains some results of chemical modelling of these solutions, that may help in the evaluation of the data.

The calculated charge imbalance (Table II) may be used as a measure of the completeness of the analyses. The comparatively high imbalance of solutions I and II results from the missing data for sulfate and chloride or carbonate respectively. The concentrations of Ca and Si in solution I and of Si, Al and Fe in solution Ia appear to be too high in comparison with the other analyses. This may indicate the presence of colloids (solution I filtered by .45 um and Ia centrifuged at 17000g). The molar ratio of Si, Mg, Al and Fe in solution Ia just matches the composition of hypothetical wyoming-montmorillonite colloids.

The solution concentrations of Na and SO_4 depend critically on the clay/water ratio, while its influence on chloride release is less pronounced. An average release value of 0.4 g SO_4 and 0.03 g Cl per 100 g of bentonite can be obtained from the data in Table II. Since the concentration of Na in solution is likely to be controlled by the Na^+/H^+ exchange ratio with an average value of log Na/H = 7.2, it is the response of pH to the changing clay/water ratio, which influences Na release.

The carbonate content of all solutions is likely to be controlled by air, since except for the solutions IVa,b carbonate is close to its equilibrium value with respect to the partial pressure of CO_2 (log pCO_2 = -3.52). In contrast, carbonate control by calcite is unklikely because all solutions are oversaturated with respect to calcite (see Table II: log IAP/K >0 indicates supersaturation).

The hypothesis of a solubility equilibrium for montmorillonite is tested for the solutions IV. As can be seen from Table II, initially the solutions are subsaturated while at 90 or 180 days of reaction the solutions become saturated.

In the next step, the bentonite/water submodel is used to describe quantitatively the formation of solutions Ia, II and IV based on the montmorillonite solubility and ion exchange equilibrium (log Na/H = 7.2). For adjusting the model to the specific clay/water ratio, the SO_4 and Cl content of the solution analyses (Table II) was used as initial solution composition. Carbonate is calculated by assuming equilibrium with air.

Table III. Modelled bentonite equilibrium solution in mg/l (experimental SO_4 + Cl used for adjustment to the different solutions)

Solution	Clay/water ratio	Na	CO_3	Si	Al	Fe	Mg	Ca	pH(25C)
Ia	0.07	338	439	15.6	0.04	0.002	0.7	0.6	9.1
II	0.23	608	259	16.1	0.03	0.001	0.7	0.6	8.9
IV	0.03	280	514	15.4	0.05	0.002	0.7	0.6	9.2

The resulting solution compositions are given in Table III. For example solution Ia is formed by dissolving 0.15 mmol/kg H_2O of montmorillonite and 14 mmol/kg H_2O of ion exchange. With an exchange capacity of 0.64 meq of Na/g dry bentonite [11] and a clay/water ratio of 0.07 this result signifies, that the bentonite has lost 25% of its Na content. Additionally, 0.026 mmol of $Fe(OH)_3$ and 0.2 mmol of gibbsite were precipitated.

The clay/water ratio has a significant effect on the release of Na. At a high clay/water ratio the release of sulfate and chloride is high and consequently the pH is lower. Due to ion exchange a high Na-release results. A comparison of Tables II and III shows that the model describes the Na release well. The calculated silica equilibrium concentration is slightly higher than that measured for solution IV (Table II) but is much lower than that in solution I, which may include colloids. Similarly, colloids are likely the cause for the deviation of Fe and Al solution data from the calculations. Low calulated Ca and Mg concentrations indicate that these ions may participate in ion exchange. Respective exchange constants have to be derived.

The results of modelling give some insight into the time dependence of the bentonite water reaction observed 14 with the solutions IV. The rate limiting step may be the carbonate uptake from the air (see the approach of CO_2 with time towards its equilibrium value of log CO_2 = -3.52, Table II). The dependence of the Na release on time may only be a response of the ion exchange equilibrium on the decrease of pH, which is caused by the uptake of carbonate from the air. The decrease of sulfate with time indicates the precipitation of sulfate phases not considered in the model.

Results for modelling the glass/bentonite/water system

Modelling of the time dependent reaction path of glass dissolution in presence of bentonite was performed with the help of the results of the previous sections. Analcime was excluded from the list of potential alteration products because analcime cannot coexist in equilibrium with montmorillonite under the conditions of the experiment. For the leaching of JSS-A glass in presence of bentonite at 90°C, Fig.4 shows the calculated normalized elemental mass loss for B and Si. For comparison, the experimental B leach data are included for JSS-A glass and its nonradioactive simulant ABS-118. The experiments for JSS-A glass include reactions with compacted and uncompacted bentonite, whereas for ABS-118 only uncompacted bentonite was considered. For experimental details see ref. [3]. To show the effect of bentonite on glass leaching, Fig.4 includes results calculated for B leached from the glass by deionized water alone (see also Fig.3).

The sequence of alteration product formation is ferrihydrite, Nd(OH)3 pyrolusite, calcite, hydroxyappatite, SiO_2, strontianite and gibbsite. The solution is a factor of 10 oversaturated with respect to analcime. The pH is buffered at 9.0 throughout the reaction as a result of the uptake of Na by bentonite. The lower rate of the release of Si as compared to the release of B is attributed to the reprecipitation

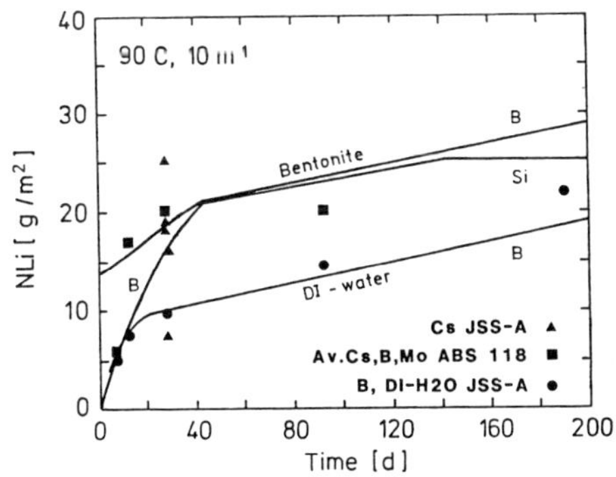

Fig.4. The effect of Bentonite on glass leaching: model and experiment

of some initially dissolved montmorillonite. Formally, this process may be considered as sorption of silica on bentonite [15]. Nevertheless, as a result of glass leaching the silica concentration raises and finally saturation will be reached, which leads to a decrease of the overall rate of glass dissolution. However, saturation will be achieved later than in deionized water alone due to the higher pH (=9.0 throughout the reaction) in the presence of bentonite and to the consumption of silica by montmorillonite. This effect is probably the main cause for considering the presence bentonite as detrimental [8] to glass leaching performance, but the calculations show, that it is only a short term effect.

SUMMARY AND CONCLUSIONS

A model is developed for leaching of a highly radioactive glass in the presence of bentonite. It is based on two submodels: a thermodynamic-kinetic model for glass corrosion in ground water and an equilibrium model (ionexchange, solubility) for the bentonite/water system. A few rate and stability constants and potential secondary alteration products are derived from a thermodynamic and kinetic analyses of experimental results. Using these data, both submodels describe adequately the experimental reality for a variety of repository relevant conditions.

Interesting results for the submodels are obtained: Leaching of glass continues with a low rate even if the solution concentration of silica is supersaturated with respect to the glass surface. Total dissolution of soluble salts in the bentonite/water system causes equilibrium solution composition to change with the clay/water ratio.

Modelling of the effect of bentonite on the corrosion of the glass matrix shows good agreement with measured release of boron. The detrimental effect of bentonite on glass leaching performance reported in literature [8] is expected to be a short term effect, which results from the later achievement of silica saturation decreasing the overall reaction rate. Modelling of the release of individual glass constituents in the presence of bentonite requires further research.

REFERENCES

1. T. H. Pigford, et al.; Waste Isolation Systems Panel, National Research Council, National Academy Press, Washington, D.C. (1983)
2. H. P. Hermansson, I. K. Björner, H. Christensen, and H. Yokoyama; JSS-Project - Technical Report 84-01, SKB, Stockholm, Sweden (1985)
3. H. P. Hermansson et al. (1985), this volume
4. B. Grambow, in Advances in Ceramics 8, edited by G. Wicks, W. Ross (Am. Ceram. Soc. Columbus, Ohio, 1984) pp. 474-481
5. D.M. Strachan, K.M. Krupka, B. Grambow, Nucl. Chem. Waste 5, 87-99 (1984)
6. B. Grambow, in Scientific Basis for Nuclear Waste Management VIII edited by C. M. Jantzen, J. A. Stone, R. C. Ewing, (Elsevier Science Publishers, New York, 1985) (in press)
7. J. A. Kittrick, in ACS Symposium Series 93, edited by E.A.Jenne, Am. Chem. Soc. (1979)
8. H. Christensen, H. P. Hermansson, D. E. Clark, L. Werme, in Advances in Ceramics 8, edited by G. Wicks and W. Ross, (Am. Ceram. Soc.), pp. 346-357 (1984)
9. J.L. Nogues, E.Y. Vernaz, N. Jacquet-Francillon, S. Pasquini, in Scientific Basis for Nuclear Waste Management VIII, eds. C. M. Jantzen, J. A. Stone, R. C. Ewing, (Elsevier Science Publishers, New York, 1985) (in press)
10. D. L. Parkhurst, D. C. Thorstensen, L. N. Plummer, (original PHREEQE) Water Resources Investigations 80-96 (U.S. Geol. Surv. Reston Virginia, 1980) extension to run precipitation controlled reaction by P. Offermann,HMI; updated with MINTEQ - database by W. Howden, K. Kruppka, PNL (1984)
11. M. Müller Vonmoos, G. Kahr, Technischer Bericht 83-12, Nagra, Baden (1983)
12. A. Lerman, F. T. Mackenzie, O. P. Bricker, Earth and Planetary Letters 25, 82-88, (1975)
13. Y. Tardy, B. Fritz, Clay Minerals 16, 361 (1981)
14. M. Smellman, personal communication, VTT, Finland (1985)
15. R. Siever, N. Woodford, Geo. Cosmochim. Acta, 37, 1851-1880 (1973)

RESULTS FROM THE LONG-TERM INTERACTION AND MODELING OF SRL-131 GLASS WITH AQUEOUS SOLUTIONS

DENIS M. STRACHAN, L.R. PEDERSON, and R.O. LOKKEN
Pacific Northwest Laboratory, Richland, WA 99352, USA

INTRODUCTION

During the various processing activities in the production of plutonium for the U.S. national defense programs, waste by-products were generated. These wastes, which have been generated over about the past 40 years at sites such as Savannah River Plant in South Carolina and Hanford in Washington, have been stored in underground tanks. Plans for the final disposal of the Savannah River Plan waste have now been documented [1] and call for the decontaminated soluble fraction of the waste to be mixed with cementitious materials and disposed of in near-surface trenches [2]. The radionuclide fraction of the soluble waste and the insoluble fraction of the waste will be converted to glass and sent to a geologic repository. In order for the repository to accept the glass, certain performance criteria must be met. Although not fully defined, it is certain that these criteria will include some specification on the rate of release of nuclides from the glass waste form when contacted by water.

To defensively predict whether glass or any other waste form can meet anticipated radionuclide release criteria, leaching tests are needed which are carried out for times as long as possible and are coupled with modeling calculations. Modeling calculations must be based upon a physical and chemical understanding of glass-water interactions, and not merely a mathematical fit to early-time leaching data. Geochemical models can be used to predict what solution concentrations ought to be, based upon alteration product formation, their related solubilities, and groundwater complexities. These calculations can then be tested against results obtained from long-term experiments. Unfortunately, the number of reported leaching experiments decreases rapidly with increased test length. Very few tests extend beyond one year. In this study, leaching experiments on a Savannah River Laboratory glass (SRL-131) were carried out to two years in four different leachants and at two temperatures. The results were then compared to calculations using the PHREEQE geochemical code [3].

Since the test began, the glass formulation for SRP waste has been changed from SRL-131 to SRL-165 [4]. Results from the long-term testing of SRL-131 glass will, however, still be useful in documenting the long-term performance of SRL-165. The SRL-165 glass formulation is also subject to change as adjustments are made to the overall process that will be used to treat the waste and to produce the glass.

EXPERIMENTAL METHODS

Materials. Samples of SRL-131 glass were produced at Savannah River Laboratory by melting the nonradioactive constituent oxides, carbonates, and nitrates in a laboratory-scale ceramic melter. Depleted uranium was included in the composition. After melting at 1150°C, the molten glass was poured into 0.25L stainless steel beakers. These were subsequently annealed at an unspecified temperature and time before cooling to room temperature. The as-received samples were cut into specimens of approximately 400mm^2 surface area (approximately 13mm x 11mm x 2mm), washed in an ultrasonic bath, dried, and weighed. The washes included one in acetone, one in ethanol, and one in deionized water.

Randomly selected specimens were analyzed by crushing the specimen into a fine powder and fusing a portion with Na_2O_2 and another portion with KOH.

The fusates were then dissolved in 6M HCl and diluted to a known volume. The results from the analyses of these specimens are shown in Table I.

Table I. Composition of SRL-131 Glass

Al_2O_3	3.05	Cs_2O	0.30	MgO	1.24	SiO_2	38.6
B_2O_3	9.84	CuO	0.02	MnO_2	4.17	SrO	0.14
BaO	0.01	Fe_2O_3	14.28	Na_2O	14.8	TiO_2	0.75
CaO	1.04	K_2O	0.12	Nd_2O_3	0.45	UO_2	1.62
CeO_2	0.24	La_2O_3	0.33	NiO	1.54	ZrO_2	0.31
Cr_2O_3	0.18	Li_2O	3.89	P_2O_5	0.10		
						TOTAL	97.00

Leach Test. Four synthetic leachants were used in this experiment: deionized water, a silicic acid/sodium bicarbonate water, a nearly saturated salt brine, and simulated groundwater. The first three leachants were those recommended by the Materials Characterization Center [5]. The simulated groundwater composition was that specified by Jones [6]. The composition of these leachants, except for the deionized water, are shown in Table II.

Table II. Leachant Composition (mg/L)(a)

	Silicate Water	Basalt Water	Salt Brine	
B (0.01)	0.04	-	17	(0.1)
Ca (0.01)	-	2.4	302	(0.1)
Fe (0.005)	-	-	0.3	(0.05)
K (0.3)	-	3.58	26800	(300)
Li (0.004)	-	-	35.2	(0.04)
Mg (0.06)	-	-	30800	(60)
Na (0.01)	47.9	364	34400	(10)
Si (0.02)	28.9	30.4		(0.2)
Sr (0.002)	-	0.004	10	(0.02)

(a) The numbers in parentheses are the detection limits in low ionic strength solutions (left) and the salt brine (right). All concentrations are as analyzed.

Although the MCC-1 Static Leach Test Method is described in detail elsewhere [5], it is briefly described here. The specimens described above were placed in Teflon PFA jars containing a Teflon PFA specimen support and volume of leachant equal to 10.0 times the specimen surface area. These jar/specimen assemblies were weighed and placed in mechanical convection oven at either 40°C or 90°C for times up to two years. The test matrix that was used is shown in Table III.

Periodically, assemblies were removed from the oven and weighed. Only one assembly was found to have lost more than 10 mass% water; this assembly was discarded. At the specified time a jar/leachant/specimen assembly was removed from the oven and weighed. The specimen was removed from the jar, washed, air dried, and weighed. The solution, now referred to as the leachate, was allowed to cool to room temperature in the resealed jar. After cooling, the pH of an aliquot of the leachate was measured using a calomel combination electrode. The solution analyses were made using an Inductively Coupled Plasma Spectrometer. A dilution of the brine samples of 1:10 was made prior to these analyses. Cesium analysis on the undiluted leachates were performed using an Atomic Absorption Spectrometer.

Table III. Test Matrix for the 2-yr Leach Test at 40°C and 90°C

Leachant	Number of Specimens per Leachant										Total No. of Tests
Test Duration (d)	3	7	14	28	56	91	182	364	546	728	
Deonized Water	1	1	1	3	2	3	2	3	3	3	22
Silicate Water				3	2	3	2	3			13
Salt Brine				3	2	3	2	3	3(a)	3(b)	19
Simulated Groundwater				3	2	3	2	3	3(b)	3(b)	19
											73

(a) Leached only at 40°C.
(b) Leached only at 90°C.

During the first six months of the test and after removal of the specimens, the leachates were acidified using concentrated acids to 1 vol% HNO_3 for the non-brine leachates and 1 vol% HCl for the brine leachates, returned to a 90°C oven for about 16 hours, and then analyzed. For test durations longer than six months, the leachates were filtered through 2 nm filters, acidified, and analyzed. The jars were filled with a 1 vol% acid solution, returned to the 90°C oven for about 16 hours, cooled, and the resulting solution analyzed for sorbed material. No significant quantities of sorbed material were found.

RESULTS AND DISCUSSION

Because of the large volume of data that were obtained in this test only the overall results will be discussed and the data presented in summary plots. A document with a more complete data analysis is available [7].

Results from the long-term leaching of SRL-131 at 40°C and 90°C are shown in Figures 1 and 2 for B and Si and for selected elements in Table IV. The plotted data are the normalized elemental mass losses. Comparison of the B leach data at 40°C in the four leachants would indicate that the leachability of SRL-131 glass is greatest in the deionized water leachant, followed in turn by silicate water, brine, and the simulated groundwater. The silica concentrations for both the silicate water and the simulated groundwater are dominated by those initially present in these two solutions. Both the silicate water and the simulated groundwater had about the same concentrations of silica at the end of the one year of leaching and about a factor of two and ten more than in the deionized water and the brine solution, respectively. Because of the initially high concentrations of Si and Na, the normalized mass losses are subject to a higher degree of error. The high Na value for the simulated groundwater at 40°C relative to the B value is probably a result of this type of error.

Solid state analyses revealed gel layers that were thin, generally less than 5μm after leaching for 728d, and were amorphous in character. No evidence for crystalline material was observed either under SEM examination or from x-ray analyses. Modeling results for the major elements using the PHREEQE geochemical code indicated that, with the exception of elements such as Al, Fe, and Mn, the glass leached congruently in deionized water and silicate water. The phases used for modelling were ferrihydrite, montmorillonite, rhodochrosite, sepiolite, schoepite, and chalcedony. Calcium behavior was modelled assuming Ca^{++} sorbtion.

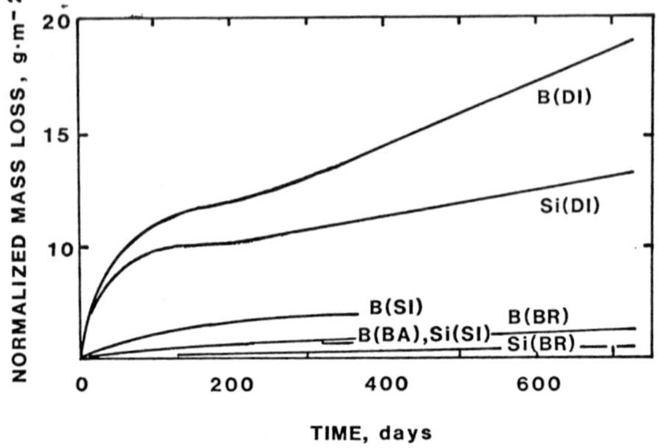

Fig. 1 Normalized Si and B mass losses for SRL-131 Glass from the 2-yr leach test at 40°C (DI=deionized water, SI=silicate water, BR=brine, and BA=simulated groundwater; Si(BA) not shown see Table IV).

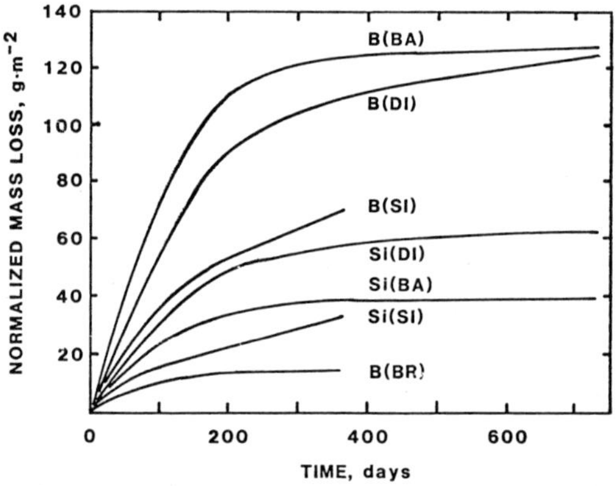

Fig. 2 Normalized Si and B mass losses for SRL-131 Glass from the 2-yr leach test at 90°C (DI=deionized water, SI=silicate water, BR=brine, and BA=simulated groundwater; Si(BR) not shown see Table IV).

Table IV Normalized Elemental Mass Losses from 364 and 728 Days of Leaching

Leachant	Deionized Water				Silicate Water		Brine			Simulated Groundwater		
Element	40°C		90°C		40°C	90°C	40°C		90°C	40°C	90°C	
Time (d)	364	728	364	728	364	364	364	728	364	364	364	72
B	9.4	14	131	125	1.7	67	1.1	2.2	14	0.6	124	133
Na	9.6	14	114	110	2.4	60	--	--	--	4.6	91	101
Si	6.7	8.2	70	63	0.69	35	0.8	2.8	2	0.3	39	39
Al	2.2	0.65	14	14	DL	6.7	DL	4	DL	0.4	1.3	2.9
Ca	1.5	2.5	--	2.6	0.49	DL	DL	DL	DL	DL	DL	DL
U	1.4	--	3.4	--	0.95	8	DL	DL	DL	DL	1.5	1.4
pH	8.7	8.7	10.0	9.9	8.6	9.8	7.8	7.1	7.2	9.0	10.0	10.0

DL = Detection Limits

At 90°C, the glass leached most in deionized water then, in decreasing order, in silicate water, simulated groundwater and brine. As indicated by the solution results, much thicker alteration layers were observed under the SEM for these samples relative to those tested at 40°C. Evidence for the formation of crystalline material was also seen under microscopic examination, but no x-ray diffraction was found for any of the specimens. The "crystals" that were observed under the SEM were generally enriched in Mn with significant amounts of Fe and Si. Of note, however, were the crystallites that formed in the gel layer of the specimen which was leached for 728d in deionized water. In this specimen a thin band of crystallites, approximately 1μm in diameter and containing only U and, presumably, oxygen, were found about 10μm from the unleached glass. When the specimen that had been leached for 364d was examined, only a diffuse region enriched in U was found, also at about the same distance from the unleached glass. Since the glass had leached about 10μm more between 364d and 728d, the U-rich region had migrated with the leaching front and began to crystallize. Recent MCC [8] results indicate that there is a diffuse area of lower Eh in the alteration layer near the unleached glass than at the layer/solution interface. Therefore, it would appear that because the leaching of the glass had slowed sufficiently after 1 to 1.5 years, that diffusion of U to an area of minimum solubility was possible. It did not appear that the mass of this phase that precipitated was sufficient to control the solution concentration of U. It is, however, likely that similar recrystallization of other elements will occur with time since most of the alteration layer is amorphous.

It is apparent from the solution data and from the SEM examination (not shown) that there was precipitation of significant quantities of material on the surface of the glass specimens leached in brine at 90°C. The examination of the surface indicated that a Mg-silicate had precipitated in isolated clumps on the surface of specimens leached for up to about 56d. By 364d these isolated clumps disappeared and had been replaced by an apparently amorphous layer, also enriched in Mg. The fact that the initial alteration product(s) did not form a continuous layer on the surface of the glass and yet the leach rate was dramatically depressed relative to the deionized water results clearly indicates that there was no protection of the glass by a thick layer in which transport of species was important. A large portion of this reduction may simply be due to a decrease in the saturation concentration of the glass in brine, similar to that observed for silica [9], and in the lower pH of the brine solution; the pH was 10 for the deionized water and about 7 for the brine. Based on the comparison of the deionized water and brine results at 40°C one might argue that a decrease in the saturation concentration is only part of the explanation. Alternatively, the observed decrease may be a kinetic effect caused by dependence of either or both the forward or reverse reaction rates on ionic strength. If only the forward reaction rate is affected then the rate would have to decrease with increasing ionic strength. Since the glass would dissolve at a lower rate, both B and Si, and hence the concentration of H_4SiO_4, would go into solution at a lower rate. However, the time to reach saturation would not change and is dependent only on the reverse reaction rate. If, however, only the reverse reaction, i.e. condensation of silanol groups [10], were affected by the increasing ionic strength, this rate would have to increase with increasing ionic strength. This would be consistent with the work of Weres, Yee, and Tsao [11] who showed that the condensation of silanol groups increases with increasing ionic strength for silica. The leaching results in brine may also be indicative of slow transport of silica through an extremely thin layer (<1μm). The effect of salinity on the reaction kinetics is expected to be complex since, in addition to the dependence of the kinetics of the activated complex on ionic strength, the overall driving force, i.e. the Gibbs free energy for the reaction glass + water = reaction products + leachate, depends also on the ionic strength.

Geochemical calculations showed that in all experiments at 90°C there was sufficient reaction that a number of elements were expected to exhibit incongruent dissolution due to the formation of mineral phases. A typical result of these calculations is shown in Figure 3. This plot shows the results of the calculation as lines and data points for the actual solution

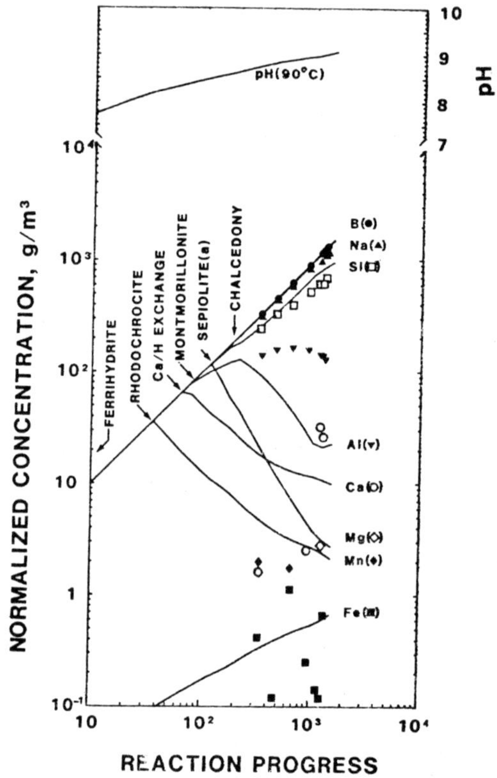

Fig. 3 Calculated and Observed Data From SRL-131 Glass in Deionized Water at 90°C.

data. The phases that were used in this calculation are typical of the phases that were used for all of the solutions. As indicated above, these calculations were not carried out for the brine because of the high ionic strength. In the example shown in Fig. 3, rhodocrosite was used as the control for Mn, and yielded a reasonable fit to the observed data. Other Mn phases could have been used but they are either much more soluble by one or more orders of magnitude or much less soluble by the same amount. It is also possible that the Mn was sorbed in the alteration layer. Calcium sorption [12] was used for the Ca data. Aluminum solution concentrations are initially controlled by gibbsite, and then by montmorillonite. The fit to the experimental data is quite good. The Si concentrations are controlled by montmorillonite, sepiolite and chalcedony and by slow dissolution from the alteration layer. Also shown in Fig. 3 is the calculated pH as a function of reaction progress at 90°C. These pH values were not recalculated to 25°C where the actual measurements were made.

CONCLUSIONS

Leaching studies of SRL-131 simulated defense nuclear waste glass have been carried out to two years duration, in leachants that simulate groundwaters of different ionic strengths. The leachability of SRL-131 glass followed the trend: deionized water > silicate water > salt brine = simulated groundwater at 40°C and deionized water = simulated groundwater > silicate water > salt brine at 90°C. The results are in general agreement with calculations using the PHREEQE geochemical code and indicate a sometimes complex sequence temperature dependent of mineral precipitation and redissolution.

The rather low leachability in the salt brine leachant both at 40°C and 90°C suggest that there is a significant effect of ionic strength on the mechanism of leaching proposed by Grambow [10]. Weres, Tsao, and Yee [11] suggest that the rate of silanol group condensation increases rapidly with increasing ionic strength. This would be consistent with Grambow's mechanism and may explain the low leach rates observed when leaching glasses in salt brine solutions. However, specific experiments need to be done to investigate the effect of ionic strength on the kinetics of the glass/solution interaction. It should be noted that alteration product formation also plays an important role in the kinetics of glass leaching [13], since when these phases precipitate they may control the pH and/or the Si concentration, and should not be overlooked when assessing the leachability of a glass [4].

The calculations done in this study have illustrated the complexity of the precipitation sequence of mineral phases as a function of reaction progress. Complex, temperature dependent, precipitation/dissolution sequences have also been observed in experiments using natural materials, such as sea water and basaltic glass [15,16].

The results of this two-year study point to the need for further work in establishing the thermodynamics of the observed phases and the relationship of these phases to the thermodynamically favored suite of phases for any given waste form/host rock/groundwater system. In order to study these phases, very long-term experiments may be necessary so that the phases grow to sizes which can studied. In addition, geochemical codes should be used to better understand the experiments, to design leach test matrices, and to help predict the long-term results of experiments. The combination of long-term experiments and geochemical codes should lead to an improved method for forecasting the long-term behavior of a nuclear waste repository.

ACKNOWLEDGEMENTS

The hard work of Mr. Wayne Richey and Mr. Paul Martin is gratefully acknowledged. This work was performed for the US Department of Energy under contract DE-AC06-76RLO-1830 with funding from the Savannah River Laboratory.

REFERENCES

1. Department of Energy, DOE Report No. DOE/EIS-0082, 1982.

2. C.A. Langton, M.D. Dukes, and R.V. Simmons, in Scientific Basis for Nuclear Waste Management VII, edited by G.L. McVay (Elsevier Science Publishers, New York, 1984).

3. D.L. Parkhurst, D.C. Thorstenson, and N.L. Plummer, US Geological Survey Publication No. USGS/WRI-80-96, 1980.

4. P.D. Soper, et al., Bull. Amer. Ceram. Soc. 62, 1013-1018 (1983).

5. Materials Characterization Center, DOE Report No. DOE/TIC 11400, 1981.

6. T.E. Jones, Rockwell Hanford Operations Report No. RHO-BW-ST-37P, 1982.

7. D.M. Strachan, L.R. Pederson, and R.O. Lokken, Pacific Northwest Laboratory (PNL) Report No. PNL-SA-12879. 1985.

8. Materials Characterization Center, PNL Report No. PNL-5157, 1985.

9. W.L. Marshall and J.M. Warakomski, Geochim. Cosmochim. Acta 44, 915-924, 1980.

10. B.E. Grambow, Glastech. Berich. 56, 566-571, 1983.

11. O. Weres, A. Yee, and L. Tsao, J. Colloid Interf. Sci. 84, 379-402, 1981.

12. D.M. Strachan, K.M. Krupka, and B.E. Grambow, Nuc. Chem. Waste Manag. 5, 87-99, 1984.

13. B.E. Grambow and D.M. Strachan, in Scientific basis for Nuclear Waste Management VII, edited by G.L. McVay (Elsevier Science Publishers, New York, 1984).

14. R.P. Haaker, G. Malow, and P. Offermann, idib.

15. J. Crovisier et al, Geochim. Cosmochim. Acta 47, 377-387, 1983.

16. J. Crovisier and B.E. Gambow, this volume.

INVESTIGATIONS ON THE ROLE OF SURFACE LAYERS IN HLW GLASS LEACHING

R. CONRADT, H. ROGGENDORF, AND H. SCHOLZE
Fraunhofer-Institut für Silicatforschung, Neunerplatz 2, D-8700 Würzburg, Federal Repuplic of Germany

ABSTRACT

A corrosion test series was performed to clarify the role of reaction product layers on the corrosion of a simulated HLW borosilicate glass in a salt brine under hydrothermal conditions. The layers were unprotective at 200°C. At 120°C, slight protective effects ocurred when the leachant contained dissolved reaction products. The consequence for the long term behaviour between 120 and 200°C is a constant glass dissolution rate.

INTRODUCTION

When the long term behaviour of HLW glass is discussed, the question arises whether the diverse layers known to form on the glass surface during leaching have any protective function or not. The effects of surface layers were discussed in several papers. A growing reaction product layer [1] controlling the glass dissolution would yield the well known Tammann law, i.e. a $\sqrt{t}$ law controlling the normalized weight loss (boron release) for long times unless saturation effects become dominant [2]. A variation of the surface to solution volume ratio "s" in otherwise identical experiments would yield a dependance on $s\cdot\sqrt{t}$, not on $s\cdot t$, as can be concluded from [3]. Chick and Pederson [4] demonstrated for four different glasses in water (90°C, $s = 0.1\ cm^{-1}$) that reaction product layers are not or only slightly protective. Grambow and Strachan [5] reported similar results for corrosion in water (same conditions), but found protective layers in Mg containing solutions. Recently, Haaker et al. [6] presented some results suggesting a protective function of analcime crystals in such a layer for the glass underneath (NaCl brine, 200°C, $s = 0.014\ cm^{-1}$). Further, an evaluation of a broad set of data suggests that reaction layers are rate controlling in systems where extremely high initial rates occur [7]. Beside the reaction product layers, the thin hydration or ion exchange zone [8] underneath should be taken into account. Although this hydration zone is very thin, it is known to be depleted of network modifiers (and as a consequence enriched in silica) and it could have protective properties for this reason. This paper presents results on the German glass SM 58 LW 11 in Q brine, a Na-K-Mg-chloride-sulfate brine (respective compositions see "glass 2" and "Q-solution" in [9]), performed to clarify the role of the different layers during the corrosion process.

CORROSION BEHAVIOUR OF THE GLASS SM 58 LW 11 IN Q BRINE

The understanding of the experimental concept of this paper requires previous informations on the main corrosion mechanisms of the investigated glass. These are:

- the dissolution of the glass network,
- the formation of a layer of reaction products on top of the surface of the undissolved (residual glass),
- the alteration of a comparably thin surface zone of the residual glass by an ion exchange process (hydration zone).

The material dissolved from the glass is partly present in the reaction layer, partly in the leachant. During the inavoidable cooling procedure after the corrosion tests, some material is expected to precipitate from the solution, thus forming a residue. The situation after a corrosion experiment can be characterized by the following sketch:

leachant + residues ⁝ reaction layer // hydration zone / bulk glass (hydration zone / bulk glass = residual glass)

The dotted stroke indicates that the reaction layer is porous and contains some leachant; the double stroke marks a sharp transition and the single stroke a gradual one. In general, the reaction layers do not flake off during the tests; they can, however, be removed afterwards without problems (e.g. by a rubber policeman) and in a reproducible way (proved by the reproducibility of XPS measurements on the residual glass). Thus the corrosion progress in terms of the normalized weight loss q in $mg \cdot cm^{-2}$ can be determined by two independant measurements: firstly by directly weighing the residual glass and secondly by determining the concentration of highly soluble elements (boron, lithium) in the leachant. Note that in the considered system, the pH remains constant during corrosion (approx. 5.8).

THE ROLE OF REACTION PRODUCT LAYERS

Experiments in Q brine

In order to clarify the role of reaction product layers for the corrosion in Q brine (both, fresh brine and old brine containing corrosion products), a test program of periodically interrupted tests was performed. The whole program consisted of three test series at 120 and 200°C, respectively, all at 13 MPa, with five samples per series. The samples (cut to a size of 1.2 cm x 0.9 cm x 0.2 cm, polished to optical quality, etched in $HF\text{-}HNO_3$ solution) were corroded in 90 ml of Q brine at $s = 0.034\ cm^{-1}$. Approximately every 42 d, the tests were interrupted for the interim manipulations and examinations described below; the total corrosion time was 297 d.

- type A: the reaction layer is wiped off and placed back into the brine; the residual glass is dried under vacuum at room temperature, weighed, and placed back into the brine;
- type B: the reaction layer is wiped off, dried at 110°C, weighed, and placed back into the brine; residual glass sample: see type A;
- type C: the sample remains covered with the reaction layer and is exposed to a fresh brine; the old brine is analysed;
- type D: the reaction layer is wiped off, dried at 110°C, and weighed; the residual glass sample is weighed and then exposed to a fresh brine; the old brine and the reaction layer are analysed;
- type E: no manipulations, i.e. the sample remains covered with the reaction layer and remains exposed to the old brine (quasistatic tests).

After 297 d, a complete analysis (by atomic emission and inductively coupled plasma emission spectroscopy) of the layers and brines was performed for every type. The systems had to be cooled down to room temperature, before the above manipulations could be carried out, and heated up again afterwards. In order to check the effect of this inavoidable procedure, in every autoclave a complete set of type A to E tests was performed with the type E test as an internal reference. Results from undisturbed static tests (type F) were used as an external reference. The results at 120 and 200°C are shown in Figure 1 and 2, respectively.

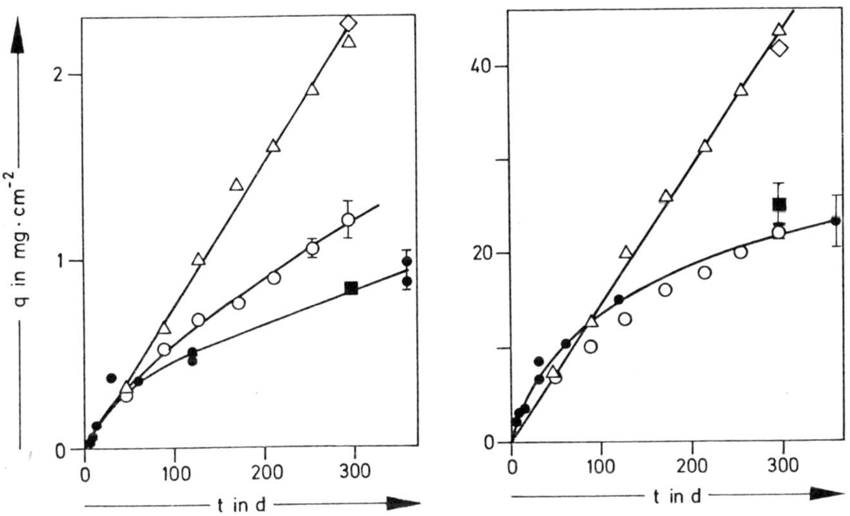

Fig. 1-2. Normalized weight loss of SM 58 LW 11 corroded at 120°C (Fig. 1) and 200°C (Fig. 2), 13 MPa, and s = 0.034 cm^{-1} in Q brine, as a function of time;
○ : stripped samples in old brine (means of types A and B);
◇ : covered samples in fresh brine (type C);
△ : stripped samples in fresh brine (type D);
■ : covered samples in old brine, quasistatic tests (type E);
● : undisturbed static tests (type F);
the data points refer to five samples for the undisturbed static tests and to three samples for the rest

The first result refers to the reliability of the test procedure itself. The type E results agree well with the results from the undisturbed static tests. The ratio E:F is approx. 1.0 at 120°C and 1.1 at 200°C. This means that the interruptions themselves do not (or only slightly) affect the corrosion process in the type E tests. The same is expected to hold for the other tests, too. Further, a comparison of the normalized weight loss data from type A and B tests leads to A:B = 0.95±0.06 at 120°C and 0.99±0.02 at 200°C. The interim drying of the reaction layer in case B has no effect. This is another indication for the quick equilibration of disturbances in the system.

A comparison between the undisturbed static (type F) and the type A,B tests shows no significant difference at 200°C (see Fig. 2). This means firstly that the reaction layers are not protective against the attack of an old brine and secondly that the deviations from a direct proportionality between weight loss q and time t are not due to these reaction layers. The idea of a diffusion barrier can be dismissed in this case. At 120°C (see Fig. 1), the situation is somewhat more complex. The reaction layers have a partially protective function against the attack of an old brine; the ratio A,B:F is 1.45±0.15 after 297 d. As to the deviations from a direct proportionality between q and t, there is a minor contribution by the reaction layers and a major one by some other reason, which can be seen from a comparison between the three curves in Figure 1.

The normalized weight loss data from the type C and D tests are very different from the rest; however, they agree among each other: 6.65±0.2 and 129±4 mg after 297 d at 120 and 200°C, respectively. Lithium and boron analysis data (not available at all interruptions) show that this is valid for the interim intervals, too:

C:D = 0.99±0.10 (Li, 200°C, based on data from 5 intervals)
1.07±0.14 (B, 200°C, based on data from 3 intervals)
1.13±0.20 (Li, 120°C, based on data from 5 intervals, 1 outlier)

(The boron analysis for the 120°C results failed.)

The conclusion is that the reaction layers are not protective against the attack of a fresh brine, neither at 120 nor at 200°C. A further conclusion can be drawn from a comparison of the type A,B to the type C,D data. While the former follow a non-linear time law equal (at 200°C) or similar (at 120°C) to that of the static tests, a linear time law is valid for the latter. Thus, the deviations from linearity in the A,B as well as in the static case have to be attributed to the accumulation of reaction products in the brine.

Experiments in NaOH solution

Although non-protective reaction layers on HLW glasses seem to be a normal case (see above, [4], and [5]), there are cases where the contrary is valid without doubts. This has been concluded indirectly, e.g. for the system SM 58 LW 11 + NaOH solution, by an evaluation of the initial rates, time exponents, and activation energies of the weight loss [7] and can now be demonstrated directly in Figure 3. Two test series were performed with only one interruption after 10 d, and a total duration of 18 and 26 d, respectively. The agreement between the quasistatic (type E) and the undisturbed static (type F) tests is sufficiently good. The reaction layers are strongly protective: The normalized weight losses in the type A tests (stripped samples, old solution) do not only exceed the values of the static tests but also reach those of stripped samples in fresh solution

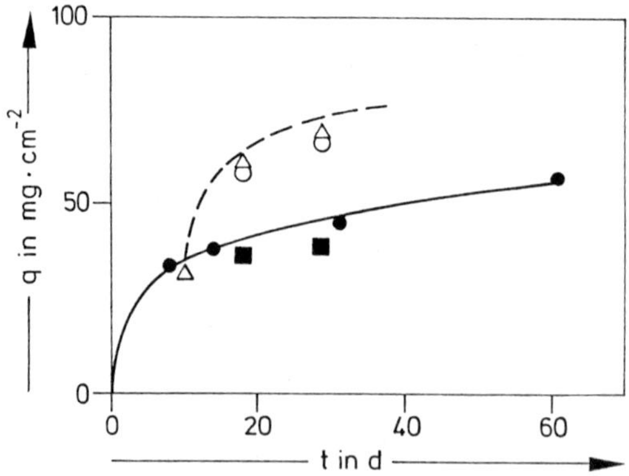

Fig. 3. Normalized weight loss of SM 58 LW 11 corroded at 200°C, 13 MPa, and s = 0.034 cm^{-1} in NaOH solution (0.01 mol/l), as a function of time; the symbols have the same meaning as in Figure 1; the dashed curve is a duplicate of the solid one shifted to the value at 10 d

(type D). This means that the accumulation of reaction products in the solution does not significantly affect the corrosion progress during the observation time. Thus, in this particular case, the deviations from a linear time law must not be attributed to the accumulation of reaction products in the solution but to the presence and/or formation of the reaction layers on the glass surface.

Characterization of reaction layers formed in Q brine

The investigation of the reaction layers involved the following steps: The layers were removed from the samples, carefully rinsed with water to wash off highly soluble salt residues, dried at 110°C, weighed, and chemically analysed. Due to the small amounts of matter, only a relatively low precision is achieved. Figures 4 and 5 show the increase or reproduction of the layers in terms of a normalized mass gain per surface area q_l (integrated over the corrosion time) for type B,C, and E tests.

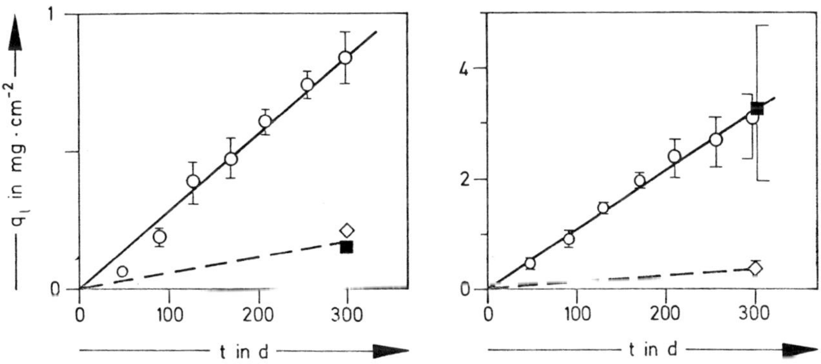

Fig. 4-5. Normalized mass gain of reaction layers formed on SM 58 LW 11 during the corrosion at 120°C (Fig. 4) and 200°C (Fig. 5), 13 MPa, and s = 0.034 cm^{-1} in Q brine, integrated over the corrosion time; the symbols have the same meaning as in Figure 1

Firstly, a comparison between layer growth and glass dissolution: The type B data can be approximated by straight lines with an average reproduction rate in old brine of 0.12±0.08 and 0.45±0.18 mg·cm^{-2} per 42 d (average time for a single test interval) at 120 and 200°C, respectively. This is in contrast to the more pronounced non-linear behaviour of the normalized weight loss of type B in Figures 1 and 2, suggesting that glass dissolution and layer growth are not directly correlated. (For comparison: The reproduction rates in fresh brine (type D) are 0.23±0.08 and 0.56±0.30 mg·cm^{-2} per 42 d). Secondly, as to the layer growth in old brine: In the quasistatic case (type E) at 120°C, the amounts of layer produced during 297 d are as low as those reproduced within 42 d in a type B test. In other words, no significant layer growth takes place beyond the first 42 d. At 200°C, however, a continuous layer growth must have taken place throughout the whole 297 d, which can be concluded from the congruence between the type B and E data in Figure 5. Thirdly, as to the layer growth in fresh brine, i.e., when the old brine is periodically replaced by a fresh one (type C): In this case, the layers "cease to grow" at 120 as well as at 200°C; the amounts after 297 d do not significantly exceed those reproduced during a single 42 d interval in a type D test. Finally, as to the results of the chemical analysis: The Mg content of the layers was much higher than the corresponding amounts dissolved from the glass. This may be due to a rest of precipitated salt from the brine (most likely $MgCl_2 \cdot 2H_2O$ after drying at 110°C), which must not be regarded as a component of the layers, or to the formation of magnesium silicates. The true layer masses may be systematically lower than those in Figures 4 and 5. If only the first case applies, the average bias is 40 % of the latter data. Further results of the chemical analysis of the reaction layers and the leachants are shown in Tables I and II. Except for the SiO_2 values in Table II, the balance between analysed and calculated amounts is fair; and so is the agreement between the sum of the components in the layers and the total masses of the layers (Fig. 4-5). Between the reproduced layers (Table I) and those generated under quasistatic conditions (Table II), there are only differences as to the absolute amounts, but not to the mass ratios of the components. In both cases, the ratio of the major components SiO_2 and TiO_2 is roughly 1:1. Only 5% of the dissolved SiO_2 takes part in the build-up of the layer, whereas 60 to 100% of the TiO_2 and 50 to 80% of the ZrO_2 are found there. The components Nd_2O_3 and SrO cannot be detected in the layers (respective detection limits 0.04 and 0.001 mg). These results suggest that the chemical nature of the layers does not change during the accumulation of reaction products in the brine. The data do not suggest a specific mineralogy.

Table I. Chemical analysis of the reacted matter in type D tests at 200°C; average values for a single 42 d interval

component	amount in mg		
	in the layer	in the brine incl. residues	calc. by weight loss
SiO_2	0.47	9.4	10.1
TiO_2	0.44	0.34	0.74
ZrO_2	0.06	0.03	0.13
SrO	b.d.	0.01	0.01
Nd_2O_3	b.d.	0.08	0.11
sum	0.97	9.9	11.2

b.d. = below detection limit

Table II. Chemical analysis of the reacted matter in type E tests at 200°C; values for the whole period of 297 d

component	amount in mg		
	in the layer	in the brine incl. residues	calc. by weight loss
SiO_2	2.1	*	40
TiO_2	2.7	0.7	2.9
ZrO_2	0.39	0.04	0.51
SrO	b.d.	0.03	0.04
Nd_2O_3	b.d.	0.30	0.42
sum	5.2	-	44

b.d. = below detection limit; * = analysis failed

DISCUSSION

The results on the protective properties of reaction layers presented above agree with the indirect conclusions drawn in a previous paper [7] on the basis of activation energies. In NaOH solution, energies as low as 20 to 30 kJ/mol were found reflecting the transport of dissociated matter across the aqueous phase in the reaction layer as rate controlling mechanism. In Q brine, approx. 70 kJ/mol were found reflecting the glass dissolution reaction. The partially protective properties of the layers at 120°C are compatible with the finding that below approx. 120°C a shift towards lower activation energies is observed. In the temperature range between 120 and 200°C, the glass dissolution is the dominating process. According to Grambow [10], a rate r_0 for conditions far from saturation and a rate r_∞ for saturation conditions are expected in this case. Figure 6 shows an evaluation of recent and previous [7] data at 200°C, equilibrium pressure (some 30 d tests at 13 MPa), and s = 0.003 or s ≈ 10 cm^{-1}, respectively. In both cases, a linear time dependence i.e. a constant rate is found according to the theory, although in the intermediate stages (see Figure 2), a non-linear behaviour is observed. Due to the presented results, the latter behaviour has to be considered as a transition stage between r_0 and r_∞, attributed to the increasing concentration of reaction products in the brine, provided that 1. the composition of the rest glass surface remains constant after an initial period and 2. the hydration zone is not a rate controlling diffusion barrier. The first point is still unsolved, which is often overlooked although it may be important for long term predictions [11]. The second one appears to be a matter of fact when one compares the network dissolution solution depths (1 to 100 µm range; normalized weight loss divided by glass density, 2.6 $g \cdot cm^{-3}$) to the hydration depths (sub-µm range [8]) for the undisturbed static tests in Figures 1 and 2. However, under conditions close to saturation, the situation is not that simple. This is due to the considerable reduction of the rates during the corrosion process. From Figure 6, r_0 and r_∞ can be estimated as 0.4 and 0.005 $mg \cdot cm^{-2} \cdot d^{-1}$, respectively, corresponding to network dissolution velocities $v_0 \approx 1.5$ $µm \cdot d^{-1}$ and $v_\infty \approx 20$ $nm \cdot d^{-1}$. The dissolution depths after 120 d are 175 and 2.4 µm, respectively. According to the classic theory of glass corrosion, e.g. [12],

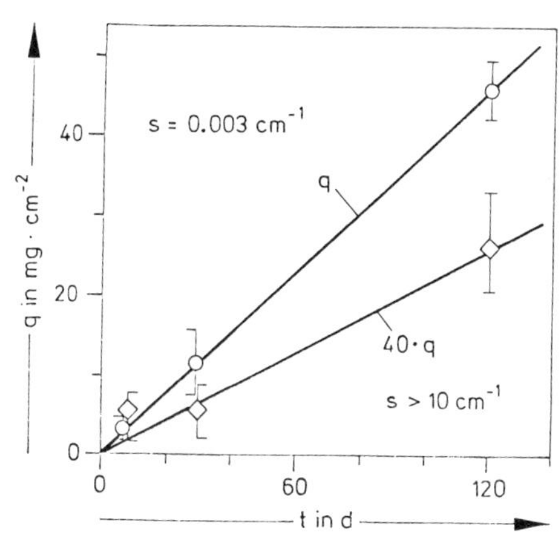

Fig.6. Normalized weight loss of SM 58 LW 11 in Q brine at 200°C;
○ : s = 0.003 cm^{-1} ; ◇ : s ≈ 10 cm^{-1} (values multiplied by 40)

the resulting hydration depths x_0 and x_∞ are inversely proportional to the respective dissolution velocities. Thus, $x_0 \approx 200$ nm observed under conditions far from saturation suggests a value of $x_\infty \approx 14$ µm for conditions close to saturation, which is appreciably larger than the corresponding dissolution depth after 120 d. Therefore, the role of the hydration zone may be more significant in the latter case than in the former one.

CONCLUSIONS

Corrosion tests, periodically interrupted for interim manipulations and examinations, helped to clarify the questions for the protective properties of the diverse layers formed during the corrosion of the glass SM 58 in Q brine. At temperatures sufficiently above 120°C and up to 200°C, none of these layers affect the corrosion process. The rate controlling factors are the concentration of reaction products in the brine and probably the concentrations of the glass components at the rest glass surface.

ACKNOWLEDGEMENTS

This publication presents results obtained under contracts with the Bundesminister für Forschung und Technologie, Federal Republic of Germany, and with the Commission of the European Communities under respective contract nos. KWA 830313 and WAS-323-83-53 D(B).

LITERATURE

1. W.L. Kuhn and R.D. Peters, in Scientific Basis for Nuclear Waste Management VI, edited by D.G. Brookins (Elsevier Science Publishers, New York, 1983), p. 167
2. R.M. Wallace and G.G. Wicks, ibid., p. 23
3. A.J. Machiels and C. Pescatore, ibid., p. 209
4. L.A. Chick and L.R. Pederson, ibid. VII, edited by G.L. McVay (Elsevier Science Publishers, New York, 1984), p. 635
5. B. Grambow and D.M. Strachan, ibid., p. 623
6. R. Haaker, G. Malow, and P. Offermann, ibid. VIII, edited by C.M. Jantzen, J.A. Stone, and R.C. Ewing (Elsevier Science Publishers, New York, 1985), p. 121
7. R. Conradt, H. Roggendorf, and H. Scholze, ibid., p. 155
8. R. Conradt, H. Roggendorf, and H. Scholze, ibid. VII, edited by G.L. McVay (Elsevier Science Publishers, New York, 1984), p. 9
9. H. Scholze, R. Conradt, H. Engelke, and H. Roggendorf, ibid. V, edited by W. Lutze (Elsevier Science Publishers, New York, 1982), p. 173
10. B. Grambow, ibid. VIII, edited by C.M. Jantzen, J.A. Stone, and R.C. Ewing (Elsevier Science Publishers, New York, 1985), p. 15
11. W. Lutze, private communication
12. Z. Boksay, G. Bouquet, and S. Dobos, Phys. Chem. Glasses 9, 69(1968)

LEACHING OF ACTINIDES FROM THE FRENCH LWR REFERENCE GLASS

S. FILLET, JL. NOGUES, E. VERNAZ and N. JACQUET-FRANCILLON
CEA - CEN/Valrho, BP 171, 30205 Bagnols sur Ceze, France

ABSTRACT

The French LWR reference glass was spiked successively with ^{237}Np, ^{239}Pu, ^{238}Pu and ^{241}Am to study the specific leaching behavior of each actinide, as part of a larger program investigating the effects of α decay in this glass. This paper describes the preparation of the α doped glasses and presents the initial leaching results for neptunium, plutonium and americium.

Glass samples were leached for up to six months at 90°C under static conditions. Weight loss measurements were performed, and the leachate was analyzed by ICP for the inactive elements, and by α spectrometry for the actinides. The results provide a basis for comparing the retention capability of the glass for each actinide, and the alterability of the bulk glass.

Finally, a comparison between glass specimens doped with ^{239}Pu and with ^{238}Pu, having the same chemical composition but very different specific activities, allows the effects of radiolysis on leaching to be estimated.

INTRODUCTION

Although they are present only in minute quantities in fission product confinement glasses, the long half-life actinides constitute the major longterm hazard if they are released into the biosphere as a result of leaching by underground water. They also represent the potentially most serious source of irradiation damage for the material [1]. Investigations have therefore been undertaken for several years in order to study the actinide leaching mechanisms in nuclear glasses and the effects of α decay. A fabrication and test program was therefore undertaken with samples of the SON 68 18 17 L1C2A2Z1 glass selected in France for vitrification of LWR solutions at La Hague. The samples were spiked successively with 0.85 wt % of various actinide oxides : this is the specified concentration for actinides, including uranium, in the industrial glass. It is therefore reasonable to assume that the distribution of the α emitters in the samples is representative of the actual glass.

The program objectives are to investigate the individual behavior of each actinide, the dose rate (and particularly α radiolysis) effects by comparing glass samples of very different specific activities, and the cumulative dose effect. The preparation of glass samples doped with U + Th, ^{237}Np, ^{239}Pu, ^{238}Pu and ^{241}Am is described here and the first leaching results are given.

EXPERIMENTAL CONDITIONS

Glass Preparation

The composition of the SON 68 18 17 L1C2A2Z1 glass is shown in Table I. Five doped glass samples were prepared. The total actinide oxide content was the same (0.85 wt %) in each case :

Sample N°1 : 0.52 wt % UO_2 + 0.33 wt % ThO_2
Sample N°2 : 0.85 wt % NpO_2 (^{237}Np)
Sample N°3 : 0.85 wt % PuO_2 (^{239}Pu)
Sample N°4 : 0.85 wt % PuO_2 (Pu enriched in ^{238}Pu)
Sample N°5 : 0.85 wt % AmO_2 (^{241}Am)

These glass formulations cover the bulk of the alpha hazard in real glass, since the curium hazard becomes negligible after a relatively short time. The actinides were tested separately to determine the specific leaching properties of each oxide

A nitric acid solution of the actinide was first added to the simulated inactive fission product solution, then mixed with the glass frit prepared in the form of sludge. The mixture was then calcined in consecutive batches in the Inconel melting pot heated in an electrical resistance furnace, then cast into graphite crucibles in the form of test bars measuring 25 x 25 x 100 mm.

The glass was next annealed for one hour at 525°C and cooled at a rate of less than 50°C.h^{-1}. The samples were cut into specimens measuring 25 x 25 x 3 mm using a diamond-impregnated saw in a glove box.

Table I. SON 68 18 17 L1C2A2Z1 calculated glass composition

Oxide	wt%	Oxide	wt %	Oxide	wt %
SiO_2	45.48	NiO	0.74	SnO_2	0.02
Al_2O_3	4.91	Cr_2O_3	0.51	Sb_2O_3	0.01
B_2O_3	14.02	Cs_2O	1.42	TeO_2	0.23
Na_2O	9.86	SrO	0.33	BaO	0.60
CaO	4.04	Y_2O_3	0.20	La_2O_3	0.90
ZnO	2.50	MoO_3	1.70	Ce_2O_3	0.93
Li_2O	1.98	MnO_2	0.72	Pr_2O_3	0.44
Fe_2O_3	2.91	CoO	0.12	Nd_2O_3	1.59
P_2O_5	0.28	Ag_2O	0.03	transuranic	
ZrO_2	2.65	CdO	0.03	+ oxides	0.85
				Total	100.00

A separate glove box was used to fabricate and cut up the ^{237}Np-doped samples to prevent any contamination.

The specific activity of the glass samples is indicated in Table II. Since absolutely pure actinides could not be obtained, the total specific activity of the glass is listed together with the specific activity due to the principal isotope. These values are to be compared with real glasses activity which is about 1.6 10^{-3} Ci.g^{-1} after 300 years.

Table II. Glass specific activity ($Ci.g^{-1}$ of glass)

Actinide	Total specific Activity	Specific Activity Due to Actinide
^{237}Np	6.11×10^{-6}	5.98×10^{-6}
^{239}Pu	1.04×10^{-3}	7.29×10^{-4}
^{238}Pu	3.57×10^{-2}	3.43×10^{-2}
^{241}Am	2.97×10^{-2}	2.76×10^{-2}

Leaching conditions

The experiments were conducted under the following conditions :

Leaching mode	: static in teflon container
Leachant	: double distilled water
Temperature	: 50°C and 90°C
SA/V	: $0.5\ cm^{-1}$

The solution pH was measured after cooling to room temperature. The leachants were then acidified in the leach vessel, with 14 N nitric acid, and analyzed by plasma torch emission spectrometry for the following elements : Si, Al, B, Na, Li, Ca, Zn, Mn, Ce, Nd and Fe. The actinides present in the leachate were analyzed by and spectrometry. After drying, the specimens were weighed to determine the mass loss.

RESULTS

Behavior of the glass matrix

The normalized elemental mass losses for the inactive elements were similar for all five glass samples at each of the two leaching temperatures investigated, as shown in Tables III through VI.

Table III. Normalized elemental mass losses (x 10^{-6} $g.cm^{-2}$) for 5 doped glasses after 7 days at 50°C

Samples	Mass Loss	Si	Na	B	Al	Sr	Nd	Zr	pH
1	<66	43	91	61	37	47	0.8	0.9	8.43
2	<66	31	91	39	25	32	1.6	0.9	8.40
3	<66	40	47	48	27	36	1.1	ND	7.99
4	<66	40	36	37	21	26	1.5	0.7	7.5
5	<66	50	48	50	32	34	ND	1.4	6.61

Table IV. Normalized elemental mass losses (x 10^{-6} $g.cm^{-2}$) for 5 Doped glasses after 7 days at 90°C

Samples	Mass Loss	Si	Na	B	Al	Sr	Nd	Zr	pH
1	145	197	298	301	110	126	1.41	0.77	9.4
2	165	212	337	306	91	124	0.80	0.66	9.1
3	185	189	294	272	98	124	ND	1.83	9.4
4	119	183	252	253	79	99	ND	ND	9.0
5	172	197	257	274	109	103	ND	0.7	8.8

Table V. Normalized elemental mass losses (x 10^{-6} $g.cm^{-2}$) for 5 doped glasses after 28 days at 50°C

Samples	Mass Loss	Si	Na	B	Al	Sr	Nd	Zr	pH
1	<66	64	105	92	49	65	1.28	0.66	8.6
2	<66	82	110	111	37	64	1.44	0.77	8.6
3	<66	54	130	65	37	53	9.44	ND	7.7
4	<66	69	74	75	41	52	3.5	1.0	7.2
5	<66	58	69	67	18	42	ND	0.8	5.1

Table VI. Normalized elemental mass losses (x 10^{-6} $g.cm^{-2}$) for 5 doped glasses after 28 days at 90°C

Samples	Mass Loss	Si	Na	B	Al	Sr	Nd	Zr	pH
1	178	219	344	343	117	128	1.62	1.8	9.3
2	185	250	465	404	78	114	1.24	0.64	9.1
3	192	214	347	324	101	128	ND	6.47	9.3
4	175	237	331	345	88	109	ND	ND	8.6
5	152	211	321	330	42	66	ND	0.8	7.90

The slight mass loss variations observed are not significant, considering the very low measured values.

On the basis of previous work with inactive samples [3-6], the elements constituting nuclear glass were classified into three groups :

. "mobile" elements, found in significant quantities in solution (e.g. B, Na, Li)

. "intermediate" elements, only a fraction of which enters solution (e.g. Si, Al, Sr) ; the remainder is found in the surface layer.

. "slightly mobile" elements found in minute amounts in the leachate (e.g Nd, Zr, Ce) and which constitute the bulk of the interface layer, probably in the form of hydroxides and/or hydrates.

This classification is valid for the five glass samples tested in this study. The presence of the actinides investigated therefore has no effect on glass alterability.

Actinide behaviour

Table VII shows the evolution of the actinide concentration in the leachate with temperature and time.

Table VII. Actinide concentration in the leachate (ppb)

Time	U		^{237}Np		^{239}Pu		^{238}Pu		^{241}Am	
(days)	50°C	90°C	50°C	90°C	50°C	90°C	50°C	90°C	50°C	90°C
1	<20	<192	29	146	10	3.6	18	11.5	3.7	0.76
3	<20	61	81	150	14	5.4	67	11.7	3.3	0.84
7	<20	63	113	180	12	7.9	26	18.9	3 5	0.76
14	<20	75	189	210	11	6.5	37	16 5	-	-
28	24	66	238	236	25	9.1	48	29 6	-	-
56	53	95	266	240	27	9.1	-	-	-	-
91	66	126	421	320	-	10	-	-	-	-

Table VIII shows the evolution of the actinide normalized mass losses with temperature and time

Table VIII. Normalized elemental mass losses for actinides ($\times 10^{-6}$ g.cm^{-2})

Time	U		^{237}Np		^{239}Pu		^{238}Pu		^{241}Am	
(days)	50°C	90°C	50°C	90°C	50°C	90°C	50°C	90°C	50°C	90°C
1	<10	<90	8	38	4	1.5	4	3.5	1	0.2
3	<10	29	21	40	6	2.2	21	3.5	0.9	0.2
7	<10	30	29	47	5	3.3	8	5.7	1	0.2
14	<10	35	48	53	4	2.8	11	5	0.8	0.4
28	12	31	62	60	10	3.8	14	9	1.5	0.3
56	36	45	73	63	11	4.2	12	12	-	-
91	32	58	109	81	6	9.0	-	-	-	
182	48	46	119	41	-	-	-	-	-	-

It may be observed that at both temperatures the actinide classification by order of decreasing mobility is the same, ie

$$^{237}Np > U > ^{238}Pu > ^{239}Pu > ^{241}Am$$

For Uranium and Neptunium in no case are solubility limits reached. This is in good agreement with the high solubility limit expected from the literature [7] in oxidizing condition.

For Plutonium at 90°C a maximun concentration of about 10 ppb (4 10^{-8} M) is about one order of magnitude higher that the one reported by Bruno [8] for spent fuel. However a higher solubility limit for amorphous material than for cristalline one is not surprising. It will be noticed that this result may come from non soluble plutonium since the leachate is acidified

For Americium at 90°C a maximum concentration of about 1 ppm (4 10^{-9} M) is in good agreement with Rai results [9] for a pH of 8.8.

When the actinide leach values (Table VII) are compared with the leach values for certain glass component elements (Tables III through VI), a number of observations may be made.

Neptunium is one of the "intermediate" elements. It is leached more slowly than silicon, and resembles strontium or aluminum from this standpoint. It is thus likely to be found partly in the hydrolyzed layer.

Plutonium is one of the low-mobility elements, but still dissolves to a greater extent than cerium or neodynium.

Americium is characterized by very low release even less than the least mobile of the inactive elements such as Zr or Nd. It thus probably remains entirely trapped in the hydrolyzed layer.

Temperature effects

From Table VIII the following features can be seen :

- for uranium a faster release is obtained for the higher temperature
- for neptunium and plutonium no temperature effect is observed on the leach value
- for americium a lower release is obtained for the higher temperature.

This reverse behavior with regard to the matrix elements has already been observed for these actinides [10]

Radiolysis effects

Alpha radiolysis effects can be assessed by comparing the glass samples doped with ^{239}Pu and ^{238}Pu. The actinide quantity was the same (0.85 wt %) in both cases ; only the specific activity differed.

It may be observed from Table VIII that plutonium is found in solution in larger quantities when the specific activity rises. The ratio of the cumulative leach rates for ^{238}Pu and ^{239}Pu was 2.5:1 after 14 days at 50°C compared with 1.8:1 after 14 days at 90°C.

The pH values were measured after the experiments. The pH was lower in both cases for ^{238}Pu than for ^{239}Pu.

It may be postulated that radiolysis slightly lowers the pH, increasing plutonium solubility and thus the normalized plutonium mass loss. Such radiolysis effect has been argued before [7].

Conversely, radiolysis has no effect on leaching of the major glass components (Tables III through VI).

CONCLUSION

The initial results obtained for SON 68 18 17 L1C2A2Z1 glass spiked with various actinides are in good agreement with previous findings on inactive simulated glass [3]. After a few months at 90°C the solution concentrations of most of the elements show no further significant variations, and the corrosion rate appears to be very low.

The specific activity range of the glass samples investigated allowed the α radiolysis effect to be evaluated. After 28 days, no effect was observable on the glass alteration rate (Si, B, Na etc...). However, a slight difference was noted between the leach rates for ^{239}Pu and ^{238}Pu ; this difference diminishes as corrosion progresses.

It was clearly shown in these experiments that the actinides investigated behave differently than the matrix elements. After 28 days of static leaching in double distilled water at 90°C, the following bulk leach rates were obtained ($\times 10^{-6}$ $g.cm^{-2}$ day^{-1}) :

Si, B, Na :	10
Neptunium :	2
Plutonium :	0.2
Americium :	0.01 (extrapolation)

These results show that all of the actinides are leached at much lower rates than the mobile elements in the glass, and that they probably contribute to the formation of the interface layer. The actual retention factor may vary by several orders of magnitude, however, depending on the specific actinide considered. In distilled water, they may be classified as follows, by order of increasing mobility :

$$Am < Pu < Np$$

This classification, which is no doubt related to the solubility of the actinide hydroxides in the leachate, is certainly highly dependent on any variations in the electrochemical conditions of the environment. These initial results may be considerably modified in the presence of environmental materials.

REFERENCES

1 E. Vernaz, N. Jacquet-Francillon, R. Bonniaud Sciences et Recherche , Echo du CEA 1, (1982)

2 WJ. Weber, FP. Roberts Nuclear Technology, October 1981

3 JL. Nogues, E. Vernaz, N. Jacquet-Francillon Scientific Basis for Nuclear Waste Management VIII Edited by CM. JANTZEN - JA. STONE - RC. EWING (Elsevier Science Publishers, New York 1985 p 89-98

4 Takashi Murakami, Tsunetaka Banba Nuclear Technology 67 Dec. 1984, N°3 p 419-428

5 Tsunetaka Banba, Takashi Murakami Submitted to Nuclear Technology

6 R.M. Wallace, GG.Wicks Scientific Basis for Nuclear Waste Management VI edited by D.G Brookins (Elsevier Science Publishers, New York, 1983) p 23-28

7 M. Krupka, D. Rai, R.W. Fulton, RG. Strickert Scientific Basis for Nuclear Waste Management VIII edited by CM JANTZEN, JA STONE, RC EWING (Elsevier Science Publishers, New York, 1985) p 753-760

8 J. Bruno, RS. Forsythe, L. Werme Scientific Basis for Nuclear Waste Management VIII edited by CM JANTZEN, JA STONE, RC EWING (Elsevier Science Publishers, New York, 1985) p 413-420

9 D. Rai, JL. Ryan Scientific Basis for Nuclear Waste Management VII edited by GL. Mc VAY (Elsevier Science Publishers, New York 1984) p 805-815.

10 R.A. Bonniaud, N.R Jacquet-Francillon, C.G. Sombret Scientific Basis for Nuclear Waste Management II edited by Clyde J.M. Northrup Jr p 117 - 125

THE ROLE OF GROUNDWATER OXIDATION POTENTIAL AND RADIOLYSIS ON WASTE GLASS PERFORMANCE IN CRYSTALLINE REPOSITORY ENVIRONMENTS

CAROL M. JANTZEN AND NED E. BIBLER
E. I. du Pont de Nemours and Company, Savannah River Laboratory
Aiken, South Carolina 29808

ABSTRACT

Laboratory experiments have shown that groundwater conditions in a granite repository will be as reducing as those in a basalt repository. Chemical analysis of the reduced groundwaters confirmed that the Fe^{2+}/Fe^{3+} couple controls the oxidation potential (Eh). The reducing groundwater conditions were found to decrease the time-dependent release of soluble elements (Li and B) from the waste glass. However, due to the lower solubility of multivalent elements released from the glass when the groundwaters are reducing, these elements have significantly lower concentrations in the leachates.

Gamma radiolysis reduced the oxidation potential of both granitic and basaltic groundwater in the absence of both waste glass and oxygen. This occurred in tests at atmospheric pressure where H_2 could have escaped from the solution. The mechanism for this decrease in Eh is under investigation but appears related to the reactive amorphous precipitate in both groundwaters. The results of these tests suggest that radiolysis may not cause the groundwaters to become oxidizing in a crystalline repository when abundant Fe^{2+} species are present.

INTRODUCTION

The Savannah River Laboratory is investigating the performance of nuclear waste glass under conditions representing potential repository environments. For crystalline rock repositories, such as granite and basalt, the groundwaters are expected to be reducing, e.g., have negative oxidation potentials (Eh). The object of the current study was to assess the effect of groundwater oxidation potential on waste glass durability and to evaluate the effect of radiolysis on the oxidation potential.

Anoxic Leaching

The oxidation potential of the groundwater has a large effect on multivalent elements in the waste glass. The adsorption and migration of these elements in geologic media are often highly dependent upon their oxidation states in the groundwater. Because long lived nuclides such as Tc, Pu, Np, and U are less soluble in the reduced state, reducing groundwaters can be beneficial for repository performance. The rate of release of soluble elements from the glass under anoxic conditions is a better measure in determining whether the actual leaching kinetics are slower in reduced groundwaters.

Previous studies have demonstrated that oxidizing leachants are more aggressive than reducing solutions and increase the release of multivalent elements from nuclear waste glasses. White et al. [1] showed that in anoxic solutions Tc concentrations (from dissolution of simulated defense waste reference glass) were lower than in oxic solutions. However, these experiments were carried out in deionized water with a palladium scrubber and at pH extremes of 3 and 11. Peters and Diamond [2] reported that Np is leached less by a factor of 10, while Pu and U are leached less by a factor of 2 to 5 from simulated commercial high-level waste glass (PNL 76-68) at oxidation potentials of 0 to +0.1 V compared to leaching rates in air-saturated (+0.3 V) solutions. Similar results were achieved in both deionized water and in simulated brines. Apted and Myers [3] and Coles and Apted [4] examined the behavior of simulated commercial high-level waste glass (PNL 76-68) in basaltic groundwaters at between 200-300°C. The elements Na, Mo, Ca, B, and Sr were leached at one-half the level observed in oxic groundwaters [3]. Technetium was released from the anoxic solutions a factor of 4 less than in the oxic solutions [4]. Although the oxidation potential was not measured in these experiments, the Eh was inferred to be reducing since the system was closed and contained crushed glass, crushed basalt, and solution. Jantzen and Wicks [5] examined the durability of defense waste glass (SRL 165) in basaltic groundwaters after 28 days at oxidation potentials between +0.23 and -0.30 V. The redox sensitive elements Fe, Ni, and Mn appeared to be released by a factor of 4 to 12 less under anoxic conditions while the U concentrations were barely above the analytic detection limit for all tests. These trends can be attributed to the lower solubility of the reduced species of these elements since there was little change in the leachability of the glass matrix elements B and Si. However, all of these elements, except U, appear in the rock-equilibrated water and trends of silicon and boron release are difficult to discern. In basaltic groundwater, the only reliable indicator element that is released from the waste glass is lithium. Lithium appeared to have concentrations in the reducing groundwaters less by a factor of 3 than in the oxidizing groundwaters after 28 days [5].

Effects of Radiolysis

Since groundwater is mostly water, its radiolysis should dominate the effects of radiation. Radiolysis of water forms equivalent amounts of oxidizing and reducing species. Oxidizing species are H_2O_2 molecules, and $\cdot OH$ and possibly $\cdot HO_2$ radicals depending on the type of radiation. Reducing species are H_2 molecules, thermalized hydrated electrons, and H atoms [6]. If these species all remain in the radiation field, a steady state is quickly achieved and there is no net change in the redox properties of the water. It has been postulated [7] that in a repository H_2 will escape, leaving an excess of oxidizing species. As a result of this, the multivalent radionuclides could be oxidized and become more soluble. Recently, Neretnieks [8,9] has modeled the effect of an oxidizing front caused by radiolysis in the flowing groundwater of a repository in granite. The movement of such an oxidizing front is postulated to decrease the retention of redox active radionuclides in the repository.

EXPERIMENTAL

The release of soluble components from defense waste glass (SRL 165) was determined in basaltic groundwater (GR-4) and in granitic groundwater (Stripa). Groundwater samples were equilibrated with rock and reduced before the leaching tests began. An x-ray amorphous precipitate, possibly a colloid, enriched in Fe^{2+} was identified in both groundwaters. The Fe^{2+} precipitate has a high surface area and is possibly the reactive species controlling the solution redox. The precipitate reactivity and the role of Fe^{2+} during radiolysis were studied. The change in oxidation potential due to groundwater radiolysis in the absence of the waste form and in the absence of air was determined. The Eh of the reduced groundwaters was measured after irradiation by Co-60 gamma rays. Various gas volumes were used in the irradiation vessels in order to simulate H_2 volatilization from the solution, as had previously been predicted for a repository environment.

The experimental cell for laboratory equilibration of simulated silicate groundwater with crushed rock consists of a 2-liter glass reaction vessel on a hot plate. The apparatus is enclosed in a glove box purged with 99.999% argon gas. The reaction vessel is filled with simulated groundwater solution and sparged with the ultrapure argon while heating to about 70°C. For determination of the relative reactivity of the various solids as a function of reaction time and solids surface area, continuous Eh and pH measurements were made at 70°C. The details of this procedure are described elsewhere [5,10].

Solid-solution residence times necessary to establish reducing groundwater conditions are about 90-120 hours depending on solids reactivity and particle size [11]. When the exact (SA/V)(t) parameter has been established for a particular solid and simulated groundwater, new solutions are equilibrated without continuous Eh and pH monitoring. Continuous monitoring is undesirable because of the considerable in-leakage of KCl from the measuring probes at 70°C. Oxygen concentrations are too low to be detected with a dissolved oxygen sensor.

The Eh and pH of rock-equilibrated leachants are measured in the reaction vessel after the specified solid-solution residence time. Aliquots are measured at the initiation and completion of the leaching or irradiation experiments. Aliquot Eh measurements are stable but introduce errors of +0.1 V. This may be attributed to the inhomogeneous distribution of the reactive precipitate. All measurements are made in the ultrapure argon atmosphere. The amorphous precipitate was not filtered from the rock equilibrated solutions during the leaching experiments. The radiolysis experiments were carried out with unfiltered samples.

Teflon® (Du Pont) leach vessels containing simulated SRP waste glass were filled with rock-equilibrated groundwaters to evaluate the interactions of glass and anoxic groundwaters under static MCC-1 conditions in the absence of rock. The solution was kept at 70°C during filling to minimize any temperature effects on silica saturation. All the vessels were filled in the environmental chamber. Duplicate 7-, 14-, 21-, and 28-day samples were left in an oven in the environmental chamber while a duplicate set of samples were removed to an oven in an oxidizing (air) environment for the same length of time. Although the vapor space in these vessels was filled with argon gas, the samples removed to the oxidizing environment exchanged air for argon in less than 1 hour.

The leachants, leachates, and irradiated solutions were analyzed for both cations and anions. Cations were determined by inductively coupled plasma induced spectroscopy and anions were determined by ion chromatography. The concentration of Fe^{2+} was determined spectrophotometrically using 1,10-phenanthroline as the indicator [12]. Analysis of standard Fe^{2+} solutions gave a molar absorptivity of 12×10^3 at 508 nm, in good agreement with published values [12]. Preparation of the solutions for spectrophotometric analysis was performed in an argon atmosphere with deaerated reagents to prevent air oxidation of the Fe^{2+}. Total Fe was measured by plasma-induced spectroscopy. All solutions were acidified with either H_2SO_4 or HNO_3 to dissolve any colloid present prior to analysis.

Irradiations were performed with a Co-60 gamma ray source. The dose rate to the solutions was 10^5 rad/hr as determined by the Fricke dosimeter. Temperature during the irradiation was 30°C. The rock-equilibrated solutions for irradiation were placed in 316 SS vessels while in the argon glovebox. The Parr vessels containing Teflon® gaskets were sealed and irradiated in a Co-60 source. Oxidation potential measurements on identical solutions and vessels stored in the absence of radiation confirmed that the vessels did not leak during the irradiation. After being irradiated, each vessel was returned to the glovebox where it was opened and the solutions were prepared for analysis.

RESULTS AND DISCUSSION

Equilibrated Stripa and Grande Ronde Groundwaters

Simulated Stripa granitic groundwater (Table I) was equilibrated with crushed Stripa granite from borehole No. 13. The Eh/pH changes as a function of time (Figure 1) are very similar to those reported for Grande Ronde groundwater in the presence of crushed Umtanum basalt [5]. The final at temperature Eh value for the granitic groundwater was -0.45 V at a pH of 8.8. The granite was more reactive than basalt achieving a lower Eh in about 70 hours rather than 90-120 hours for basalt (Table I). The oxidation potential of the granitic groundwater equilibrated in the laboratory was lower than either field measurements or calculated values reported by Wickberg et al. [13] for the Stripa repository (Figure 1).

Reproducible Eh and pH values for Grande Ronde-4 groundwater with Umtanum basalt have been in the -0.34 to -0.40 V range at pH values between 9.1 and 10.1 [5,11]. The rock-equilibrated GR-4 groundwater used in this study had an Eh value of -0.40 V at a pH of 9.1 at 65°C (Table I) which compares exactly to Eh values of -0.40 V at pH of 9.1 ± 0.5 calculated for a basalt repository at 70°C [14].

Both the granite and basalt equilibrated groundwaters contained a precipitate. Table I summarizes the differences between the starting groundwater compositions and the rock-equilibrated solutions. Acidified and unacidified rock-equilibrated solutions were analyzed to determine the chemical contribution of the precipitate phases. This is particularly important because previous work [11] had demonstrated that the precipitate phases in rock-equilibrated basaltic solutions could take as long as three months to dissolve after the solutions were acidified according to MCC-1P procedures [15]. In the unfiltered and acidified solutions, Fe^{2+} was definitely detected by spectrophotometric analysis. Ferrous iron appears to amount to 20-70 percent of the total Fe present. The presence of Fe^{2+} confirms that it is the Fe^{2+}/Fe^{3+} couple that is being measured, and that the Fe^{2+} is controlling the groundwater Eh.

TABLE I. Final Compositions of Reduced Groundwaters Equilibrated with Rock and Unfiltered

		Stripa Granitic Groundwater			Grande Ronde Basaltic Groundwater		
		Recipe*	Not Acidified	Acidified***	Recipe**	Not Acidified	Acidified***
Cations (ppm)	Na	65	112	102	334	408	382
	K	3.9	N.A.	N.A.	13	N.A.	N.A.
	Si	12	17	15	45	72	68
	Ca	18	5.8	8.4	2.2	2.8	4.5
	Mg	4.3	0.082	0.25	-	0.26	0.6
	Li	-	0.047	0.044	-	0.05	0.04
	B	-	1	0.98	-	2	2
	Al	-	0.56	0.62	-	0.77	1.7
	Fe(total)	-	0.09	1.0	-	1.0	6.9
	Fe^{2+}	-	N.D.	0.69	-	0.32	1.2
Anions (ppm)	F	-	0.4	N.A.	20	24	21
	Cl	70	156	N.A.	405	475	558
	SO_4	9.6	13	N.A.	4	7	<12
	HCO_3^-	123	N.A.	N.A.	120	N.A.	N.A.
pH (T°C)		8.2 (25°C)	8.8 (75°C)		9.4 (25°C)	9.1 (65°C)	
Eh (T°C)		+0.30 (45°C)†	-0.45 (75°C)††		+0.30 (45°C)†	-0.40 (65°C)¶	

* Lars Werme - personal communication
** GR-4 composition
N.A. = not analyzed
N.D. = not detected
*** Acidified with either HNO_3 or H_2SO_4
† Reading after sparging with Ar for 2 hours
†† After 70 hours equilibration
¶ After 90 hours equilibration

Role of Amorphous Precipitates in Reducing Groundwaters

To determine the role of the precipitate phases in the rock-equilibrated solutions, samples were analyzed after being filtered through 0.45 micron filters. Since the precipitate is inhomogeneously distributed in the solution, large volumes (about 500 mL) were filtered. After filtration, both the starting solution and the filtrates were acidified and analyzed. The filtrates were lower in Al, Ca, total Fe, Fe^{2+}, Mg, Si, and sometimes Na or K.

The filter papers contained several grams of material. It was dried at 90°C for 1 hour and then analyzed by x-ray diffraction. For basalt, a crystalline feldspar of the $NaAlSi_3O_8$-$CaAl_2Si_2O_8$ type was identified. For granite, a K-Ca rich feldspar, a Ca rich iron pyroxene and a chlorite phase were identified. A mass balance subtracting out the alkali, alkaline earth, aluminum, and silicon associated with the crystalline phases suggested that a residual x-ray amorphous precipitate rich in Fe, Mg, and Si with some Ca was present for granite as well as basalt [10].

Interestingly, an amorphous precipitate of the same composition had been found for GR-4 groundwater or deionized water that had been equilibrated with an SRL 165 waste glass. The glass had an Fe^{2+}/Fe^{3+} ratio of 0.5. The SRL waste glass had achieved Eh and pH values similar to those of basalt rock [5]. These final Eh and pH values are given in Table II. The residual (Fe + Mg)/Si ratio of the amorphous precipitate for GR-4 groundwater equilibrated with basalt and SRL 165 waste glass is 2.6, and for Stripa groundwater equilibrated with granite is about 3.6.

The significance of the Fe^{2+} silicate colloid is apparent on the Eh-pH diagram (Figure 1). All the final Eh-pH values for basalt, granite, and SRL waste glass fall in the thermodynamically calculated stability field for $FeSiO_3$. Jacobs and Apted [14] and others [16] had suggested that dissolved Fe^{2+} glass in basalt mesostasis was in equilibrium with amorphous silica and secondary magnetite in the repository environment, implying that the quartz-pyroxene magnetite (QPM) equilibrium controls the groundwater Eh in the stability field of $FeSiO_3$.

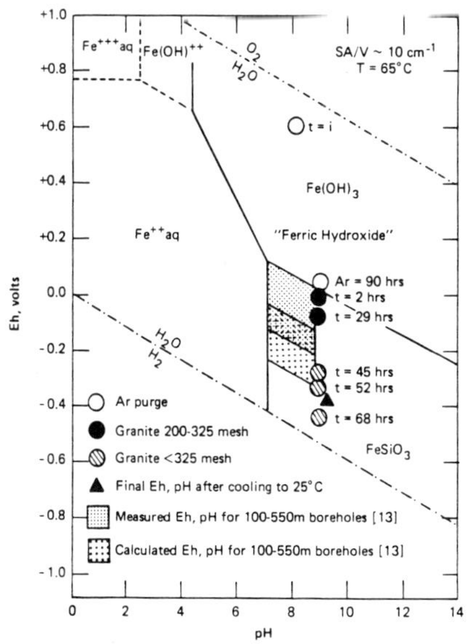

FIGURE 1. Stripa Granite/Granitic Groundwater Interactions

Although borosilicate waste glass is an Fe^{2+} glass, granite does not contain an Fe^{2+} intergranular phase. Recent studies by White et al. [17,18] have demonstrated that the driving force behind cation-electron mass transfer and aqueous reduction reactions is the oxidation of in situ Fe^{2+} iron on or beneath crystalline solid surfaces, e.g., reduced iron minerals act as electron donors. Moreover, White [18] demonstrated that O_2 uptake rates for basalt and magnetite in aqueous solution are comparable. Magnetite is a major Fe^{2+} phase in basalt and granite but not in waste glass. However, the redox active species, Fe^{2+}, is common to all the solids and thus they all achieve the same Eh and pH values in either simulated groundwater or deionized water (Table II).

The observed Eh and pH values of Table II lie between the QPM equilibrium boundary suggested by Jacobs and Apted [14] at 65°C and the Fe_3O_4-Fe_2O_3 boundary [19]. Although the magnetite in basalt and granite may be the redox active phase dissolving, magnetite is metastable relative to $FeSiO_3$ in a system saturated with respect to amorphous silica. Thus the equilibrium diagram is expected to be controlled by the Fe_2O_3-$FeSiO_3$ couple [19]. However, Drever [20] suggests that initial precipitation of species in an aqueous system would be amorphous and have a higher free energy than the anhydrous species. Hence the $Fe(OH)_3$-$FeSiO_3$ boundary plotted in Figure 1 represents the equilibrium of $Fe(OH)_3$ in the presence of SiO_2 glass.

The measured Eh and pH values from Table II for basalt, granite, and SRL glass-equilibrated solutions are lower than either the $Fe(OH)_3$-$FeSiO_3$ boundary or the Fe_2O_3-$FeSiO_3$ couple. However, the half reaction for $Fe(OH)_3$ in the presence of silicic acid

$$Fe(OH)_3 + H_2SiO_3 + H^+ \longrightarrow FeSiO_3 + H_2O \quad (1)$$

has a negative free energy of reaction in agreement with the data of White et al. [17] for quantitative reduction of aqueous Fe^{3+} and H^+ concurrent with loss of oxygen in solution. It is assumed that $FeSiO_3$ and $Fe(OH)_3$ species are amorphous and highly reactive precipitates and that an aqueous activity for these species can be estimated from the known Fe^{2+} and Fe^{3+} concentrations (Table I). This yields a $Fe(OH)_3$-$FeSiO_3$ metastable couple parallel to the couple drawn on Figure 1, but passing through the measured Eh and pH values. Calculation of the pO_2 for the complete metastable reaction including the oxygen contribution is

$$2\ FeSiO_3 + 1/2\ O_2 + 5\ H_2O \longrightarrow 2\ Fe(OH)_3 + 2\ H_2SiO_3 \quad (2)$$

The K_{sp} and the analytic concentrations suggest a calculated pO_2 of 10^{-66} atm in agreement with the 10^{-70} atm value calculated from the measured Eh and pH. This agreement thus supports the hypothesis that an amorphous reactive ferrous silicate precipitate formed by reacting Fe^{2+} species in saturated silica solutions establishes and maintains the reduced groundwater oxidation potentials of crystalline repository groundwaters.

TABLE II. Eh and pH for Solutions Equilibrated with Redox Solids*

Initial Solution	Redox Solid: Basalt Rock Eh (V)	Basalt Rock pH	Granite Rock Eh (V)	Granite Rock pH	SRL 165 Glass Eh (V)	SRL 165 Glass pH
GR-4 water	-0.4	9.1	-	-	-0.39†	9.5†
Deionized water	-0.42**	8.4**	-	-	-0.38	9.0
Stripa water	-	-	-0.45	8.8	-	-

* Measurements taken with mixtures were in the range 60° - 80°C
** Reference 10
† Reference 5

Effect of Eh on Glass Leaching

The time-dependent leaching of the soluble elements Li and B from SRL 165 waste glass under oxic and anoxic conditions is shown in Figure 2. Lithium releases could be accurately determined for both the Stripa and Grande Ronde groundwater but B releases could only be determined for the Stripa groundwater. The B concentration in the basalt equilibrated groundwater solution was too high and no changes were detected due to glass dissolution. This was also true for Na, Si, and other cations in the solutions. Results in Figure 2 are the averages of duplicate experiments. The error bars indicate the range of the results. The pH of the oxic and anoxic solutions remained the same during the experiments. This occurs because the leachants are dominated by the groundwater chemistry.

Leaching was consistently lower in the low Eh solutions by a factor of about 2. Thus reducing groundwaters have only a small effect on the durability of the glass per se. However, other studies have shown that low Eh significantly lowers the concentration of many redox active elements. This is apparently due to solubility rather than increased durability of the glass.

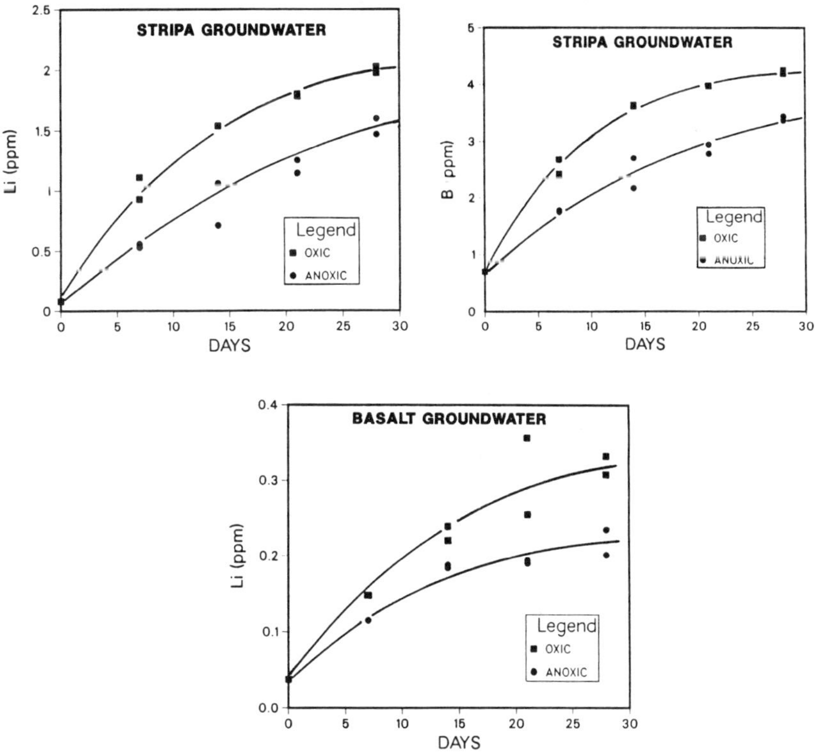

FIGURE 2. Time Dependent Release of Li and B from SRL Waste Glass in Reduced and Oxidized Groundwaters

Effect of Radiation of Low Eh Groundwater Solutions

Results of irradiating samples of the reducing groundwaters in the absence of air or glass in sealed vessels with Co-60 gamma rays for 20 hours are shown in Table III. In contrast to predictions of a recent model of groundwater radiolysis [7], the solutions did not become oxidizing as a result of accumulation of H_2O_2. As indicated in Table III, radiolysis caused the solutions to become more reducing, i.e., the Eh decreased for both solutions. Ten other samples of the groundwaters were irradiated for various times ranging from 5 to 20 hours. In all cases, the Eh decreased to approximately -0.5 V. As shown in Table III, the concentrations of soluble anions and cations in the groundwaters did not change significantly as a result of radiolysis. Apparently, radiolysis of the Teflon® gasket at these total doses was not producing sufficient F^- ions to change their concentration in the groundwater. The total Fe concentrations of the solutions were different because different amounts of Fe containing particulates were removed from the stock solutions for each test.

To determine if Fe was being leached from the vessels during radiolysis and whether this may cause the solutions to become more reducing, samples of Ar saturated deionized water were irradiated in the stainless steel vessels and the total amount of Fe was analyzed. Increases in Fe concentration due to radiolysis were less than 0.02 ppm; thus, it appears that only small amounts of Fe were leached from the steel vessels. In these irradiation tests with deionized water, radiolysis did cause the Eh of the water to decrease from ~0.0 V to -0.1 V although the solutions were not well poised. This decrease is less than that observed with the groundwaters (Table III).

The effect of gas volume over the solution was investigated to see if a larger free volume, which would allow more H_2 to escape from the solution, would cause the solution to become oxidizing. In all the above tests, the gas volume was only one-tenth the solution volume; thus H_2 could not readily escape the solution. In two other tests, the gas volume relative to the solution volume was increased by a factor of 10. Again, radiolysis caused the Eh to decrease for both the granitic and basaltic groundwater to -0.5 V after only 5 hours irradiation. The reason for the decrease in Eh is not immediately apparent, but it may be due to a change in the Fe^{2+}/Fe^{3+} ratio. In these initial experiments, the change in this ratio due to radiolysis could not be determined because the ratio in the solution before radiolysis was not accurately known. The stock solutions were not stirred while sampling and results for total Fe in the stock solutions were not reproducible due to inhomogeneous distribution of the amorphous precipitate.

TABLE III. Effect of Radiation on Reduced Stripa and Basalt Groundwaters*

		Stripa		Basalt	
		Unirradiated	Irradiated	Unirradiated	Irradiated
Eh (V)		-0.28	-0.48	-0.23	-0.48
pH		9.5	8.5	10.1	10.1
Cations (ppm)	Na	106	103	441	429
	Li	0.030	0.034	0.043	0.052
	B	1.0	1.0	2.2	2.1
	Si	17	16	73	71
	Ca	6.2	6.1	3	3.3
	Fe (total)	0.29	1.34	1.2	1.4
	Fe ($^{2+}$)	N.D.	0.84	0.57	0.83
Anions (ppm)	F	0.4	1.0		
	Cl	181	217		
	SO_4	18	13		

* Irradiated 20 hr at 10^4 rads/hr. Solution to gas volume = 0.1 T = 31°C
Final samples acidified with H_2SO_4 before cation analysis.

Two tests were performed to confirm the observed decreases in Eh with irradiation. In the first test, a sample of an irradiated solution which had a measured Eh of -0.45 V was exposed to air and the Eh was measured again. In a matter of minutes, the Eh increased to +0.37 V, the value for oxygenated groundwaters. In the second test, a deaerated sample of the standard Zobell solution was irradiated for 65 hours. The Eh decreased from +0.43 V to -0.46 V. This solution is an equimolar solution of ferro- and ferri-cyanides at neutral pH. For the test the pH was adjusted to 9.5 to correspond to that for a basalt groundwater. Radiolysis also caused the pH to decrease to 8.8, probably because of the hydrolysis of Fe ions released from the CN^- complexes. The irradiated solution was very blue, and a precipitate of Prussian Blue was observed. These tests confirmed that the Eh decreases which were observed were definitely due to radiolysis.

Changes in oxidation potential due to radiolysis have been measured in another system. This was an acidic nuclear process solution containing large concentrations of fission products [21]. The dose rate to the solution (primarily from fission product beta particles) was 1.5 x 10^5 rad/hr. In this solution, which initially had all the Fe in the +2 state, the Eh increased from +0.55 V to +1.0 V in only 10 minutes. Prior to this 10 minute period, however, the Fe^{2+} concentration decreased linearly from 0.018 M to 0.0005 M in 8 hours while there was little change in Eh. Thus, it appears that with appropriate changes in the Fe^{2+}/Fe^{3+} ratio the Eh can change rapidly. In the experiments reported in this study, the total Fe concentration was very low; thus, the change in the Fe^{2+}/Fe^{3+} ratio even though small, may have caused a significant change in the Eh at low radiation doses.

The stability of Fe^{2+} in neutral solutions during irradiation has been demonstrated by Matthews [22]. In his tests, the solution was a deaerated solution of $FeSO_4$ at pH ~6. Radiolysis caused the Fe^{3+} concentration to increase until a steady state was reached at a Fe^{2+}/Fe^{3+} ratio of 0.42. The calculated H_2 concentration in solution was 3.4 x 10^{-5} M corresponding to a H_2 partial pressure of only 40 torr above the solution. In these solutions, it was postulated that the Fe^{2+} and Fe^{3+} ions were associated with a colloid and that this colloid did not hinder the reactivity of the ions. The Eh of the final solution after steady state was reached was not measured. Based on the Fe^{2+}/Fe^{3+} ratio, the Eh was probably negative. In the experiments of Matthews [22], H_2O_2 was being produced as a molecular product by the radiolysis of water. Apparently, this production did not cause complete oxidation of Fe^{2+} to Fe^{3+}.

To measure the pressure changes during irradiation, two aliquots of basalt groundwater were irradiated in stainless steel vessels attached to pressure transducers. Both the pressure and temperature were monitored. As shown in Figure 3, the pressure above each increased linearly after an initial rapid rise due to the temperature change caused by gamma ray heating. The total pressures in the two tests were different due to different initial Ar pressures. As with the other tests, the Eh for both solutions decreased from -0.3 to -0.5 V. The observed pressure increases due to radiolysis were ~30 torr. Careful analysis of the composition of the gas after the tests indicated that the pressure increases were due solely to the production of H_2. No other gases were formed by the radiolysis. The G value for the production of H_2 for both tests was 0.64 ±0.03 molecules/100 eV. This value is somewhat larger than 0.45, the value for the molecular yield of H_2 from gamma ray radiolysis of water [5]. Perhaps some of the aqueous electrons produced in the bulk of the solution are recombining to form H_2 but this appears unlikely based on the relative rate constants for the various reactions involved [6]. It does appear that H_2 escapes from the solution during radiolysis, but even with this occurring, the Eh of the solution decreases rather than increases. More experiments need to be performed before these observations can be fully explained.

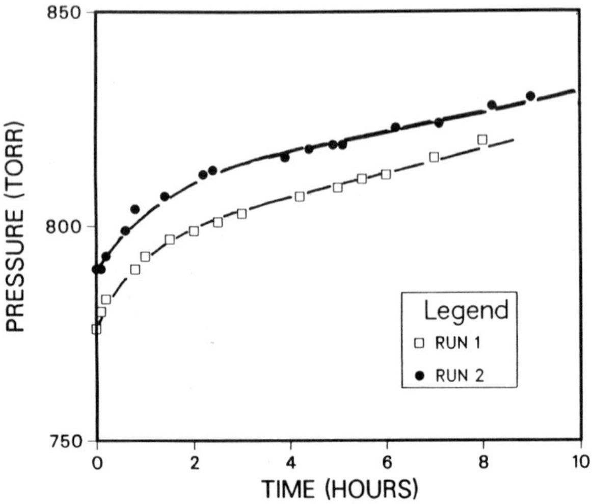

FIGURE 3. **Pressure Buildup Inside a Stainless Steel Reaction Vessel from Gamma Irradiation of Reduced Basalt Groundwater**

IMPLICATION FOR THE RESULTS TO CRYSTALLINE REPOSITORY PERFORMANCE

The model that has been proposed [7] suggests that the reducing groundwaters of crystalline rock repositories will become oxidizing as a result of radiolysis from the waste package. The basic assumption of the model is that H_2 will be transported away from the solution by diffusion and convection leaving an excess of oxidizing species in the solution. The results presented in this study contradict this. These results show that radiolysis causes the solution to be less oxidizing even under conditions where H_2 would be expected to escape from the solution. In a repository, the escape of H_2 would be expected to be hindered by the high lithostatic and hydrostatic pressure. However, it needs to be shown experimentally with solutions actually containing U, Np, Tc, or other redox active radionuclides that oxidation of these species does not take place. For example, if the Fe^{2+} concentration is low and the radionuclide concentration is high, the OH radicals may oxidize the radionuclides rather than Fe^{2+}.

These experiments have to be extended to alpha radiolysis where the yields of H_2 and H_2O_2 are larger than with gamma radiation because of the higher linear energy transfer of alpha radiation. The effects of alpha radiation will be investigated in future experiments. The results of those experiments will be significant to the application of the existing models to repository performance. If the repository groundwaters remain reducing, the most soluble waste glass elements will leach less by a factor of 2 than in oxidizing groundwaters, and the concentrations of redox-sensitive elements will be low due to reduced solubility.

CONCLUSIONS

The data presented in this paper lead to the following three conclusions:

1. Solutions simulating the predicted granite and basalt repository groundwater conditions, including low Eh, can be prepared in controlled laboratory conditions.

2. The major role of Eh in glass leaching is to reduce the solubility of redox-active elements and only slightly improve the durability of the glass. This conclusion is based on the leachability of Li and B from the glass. The solubility of these elements is not affected by the Eh.

3. Radiolysis resulting from gamma and beta radiation from the waste glass may not cause reducing groundwaters to become oxidizing as previously postulated. Alpha radiolysis remains to be investigated.

Acknowledgment

The information contained in this article was developed during the course of work under Contract No. DE-AC09-76SR00001 with the U.S. Department of Energy.

REFERENCES

1. W. B. White in Final Report of the Defense High-Level Waste Leaching Mechanisms Program, PNL-5157, Battelle Pacific Northwest Laboratory, Richland, WA, 4.1-4.49, August 1984.
2. R. D. Peters and H. Diamond, PNL-3971, Battelle Pacific Northwest Laboratory, Richland, WA, 21 pp, October 1981.
3. M. J. Apted and J. Myers, RHO-BW-ST-38P, Rockwell Hanford Operations, Richland, WA, 78 pp, July 1982.
4. D. G. Coles and M. J. Apted, Sci. Basis for Nuclear Waste Management, VII, G. L. McVay (ed.), Elsevier Publ. Co., New York, 129-136 (1984).
5. C. M. Jantzen and G. G. Wicks, Sci. Basis for Nuclear Waste Mgt. VIII, C. M. Jantzen et al. (eds), Materials Research Society, Pittsburgh, PA, 29-35 (1985).
6. I. G. Graganic and Z. D. Draganic, The Radiation Chemistry of Water, Academic Press, New York (1971).
7. H. Christensen and E. Bjergbakke, Sci. Basis for Nuclear Waste Management VI, D. G. Bookins (ed), Elsevier Publ. Co., New York, 429-436 (1983).
8. I. Neretnieks, Sci. Basis for Nuclear Waste Management VII, G. L. McVay (ed), Elsevier Publ. Co., New York, 1009-1022 (1984).
9. I. Neretnieks, Nuclear Technology, 62, 110-115 (1983).
10. C. M. Jantzen, Sci. Basis for Nuclear Waste Mangement VII, G. L. McVay (ed), Elsevier Publ. Co., New York, 613-621 (1984).
11. G. L. Pine and C. M. Jantzen, "Implications of One-Year Basalt Weathering/Reactivity Study for a Basalt Repository Environment." DP (in draft), E. I. du Pont de Nemours and Company, Savannah River Laboratory, Aiken, SC.
12. E. B. Sandell, Colorimetric Determination of Traces of Metals, Interscience Publishers, Inc., New York (1959).
13. P. Wikberg, I. Grenthe, K. Axelsen, SKBF/KBS 83-40, Royal Institute of Technology, Stockholm, Sweden (1983).
14. G. K. Jacobs and M. J. Apted, EOS Transactions, 62, 1965 (1981).
15. Nuclear Waste Materials Handbook Waste Form Test Methods, MCC-1P Static Test, Prepared by Materials Characterization Center, J. E. Mendel, Manager, USDOE Report DOE/TIC-11400, Pacific Northwest Laboratory, Richland, WA (1981).
16. D. L. Lane, M. J. Apted, C. C. Allen, J. Myers. Sci. Basis for Nuclear Waste Management VII, G. L. McVay (ed), Elsevier Publ. Co., New York, 95-104 (1984).
17. A. F. White and A. Yee, Geochim. Cosmochim. Acta, 49, 1263-1275 (1985).
18. A. F. White, A. Yee, and S. Flexser, Chem. Geology, 49, 73-86 (1985).
19. R. M. Garrels and C. L. Christ. Solutions, Minerals, and Equilibria, Harper Row Publ., New York (1965).
20. J. I. Drever. The Geochemistry of Natural Waters, Printice Hall, Inc., Englewood Cliffs, NJ (1982).
21. N. E. Bibler, Nuclear Technology, 34, 412-415 (1977).
22. R. W. Mathews, Aust. J. Chem., 36, 1305 (1983).

CHARACTERIZATION OF LEAD-IRON PHOSPHATE NUCLEAR WASTE GLASSES

EDWIN SCHIEWER*, WERNER LUTZE*, LYNN A. BOATNER**, BRIAN C. SALES**
*Hahn-Meitner-Institut für Kernforschung Berlin GmbH, Glienicker Straße 100
** Solid State Division, Oak Ridge National Laboratory,+ Oak Ridge, Tennessee 37831

INTRODUCTION

Lead-iron phosphate glasses have recently been proposed as a new primary disposal medium for both commercial nuclear reactor wastes and some types of high-level radioactive U.S. defense wastes [1-4]. The initial work on the characterization of lead-iron phosphate nuclear waste glasses, hereafter named phosphate glasses, concentrated on the preparation, thermal properties, and devitrification behavior (both during and after preparation) of these glasses. Aqueous corrosion rates were measured at 90°C in distilled water, acidic and basic solutions, low eH distilled water, and in the "reference" natural ground water appropriate to disposal in a tuffaceous formation (i.e., J-13 well water [5]). The results of these initial investigations were extremely encouraging and indicated that homogeneous, highly leach resistant (at 90°C) phosphate glasses loaded with either commercial or high-level U.S. defense wastes could be prepared at relatively low temperatures (process temp.: ≤ 1050°C). The chemical stability of the phosphate glasses was found to be primarily due to the stabilizing effect of iron on the structure of lead phosphate glass. This effect is illustrated in Fig. 1 where the leachate conductivity at 90°C is plotted versus time for phosphate glasses with varying iron concentrations. The addition of 9 wt.% iron oxide to lead metaphosphate glass [$Pb(PO_3)_2$] increases the chemical durability of the glass in water by a factor of about 10^4. The tendency of pure lead phosphate glasses to crystallize is also greatly suppressed by the addition of iron oxide [3,4]. Lead metaphosphate glass can be completely crystallized by heating the glass in air at 300°C for several hours. Lead-iron phosphate glasses, however, can be heated in air for several hundred hours at a temperature as high as 500°C without exhibiting any evidence of crystallization.

The effects of iron oxide on the structural properties of lead-iron phosphate glasses have been investigated using the techniques of Mössbauer spectroscopy, electron paramagnetic resonance, Raman and infrared spectroscopy [6], EXAFS, low angle x-ray scattering, liquid chromatography [7], and transmission electron microscopy. From these measurements it is concluded that the improved chemical durability of lead-iron phosphate glasses is related to the ability of iron to strengthen the cross bonding between the polyphosphate chains and/or to strengthen the bonding of the ends of these chains to the surrounding glass structure.

The present article concentrates on the properties of a lead-iron phosphate glass loaded with simulated commercial waste (Table I). Corrosion experiments have been continued at 90°C and higher temperatures (150 and 190°C) in both deionized water and in a saturated NaCl solution with the goal of obtaining a better understanding of the corrosion mechanisms of these glasses. Quantitative solution analysis, along with optical and scanning electron microscopy and energy dispersive x-ray and microprobe analysis of the glass surface, were used to investigate the corrosion process. Measurements of the electrical resistivity of the glass melt, differential thermal analysis results, and devitrification data are also reported.

+Operated by Martin Marietta Energy Systems, Inc. for the U.S. Department of Energy under contract DE-AC05-84OR21400.

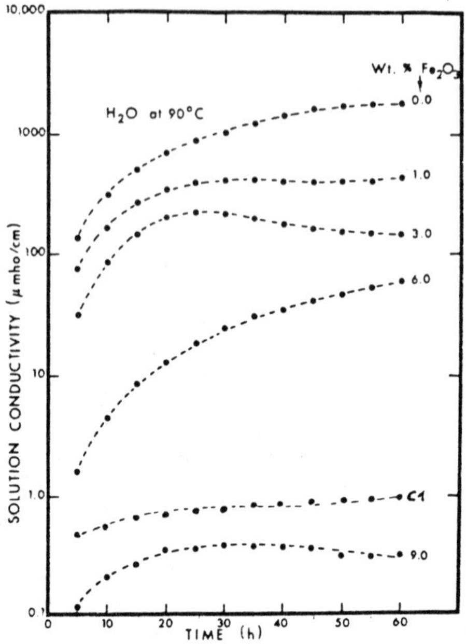

Fig. 1 Leachate conductivity versus time for a lead metaphosphate glass $[Pb(PO_3)_2]$ doped with 0.0, 1.0, 3.0, 6.0, 9.0 percent Fe_2O_3 and glass with 6.4 wt.% waste oxides. The leachate conductivity is roughly proportional to the total ionic concentration released by the glass into solution.

EXPERIMENTAL

The composition of the phosphate glass investigated here is given in Table I. This glass contains 6.4 wt.% commercial waste and has a density of 4.9±0.1 g/cm^3. Some experiments were performed with slightly different compositions and these will be specifically addressed in the text.

Table I Composition of Lead-Iron Phosphate Commercial Waste Glass C1

Compound	wt.%	Simulated Waste Constituents wt.%					
P_2O_5	32.2	Fe_2O_3	4.1	P_2O_5	1.8	La_2O_3	4.0
PbO	50.6	U_3O_8	6.2	Rb_2O	0.5	CeO_2	9.0
Fe_2O_3	10.8	NiO	0.8	Y_2O_3	1.6	PrO_2	4.0
Simulated waste	6.4	Cs_2O	7.8	RuO_2	5.0	Nd_2O_3	12.3
		SrO	2.9	PdO	1.8	Sm_2O_3	2.5
		ZrO_2	13.4	Ag_2O	0.2	Eu_2O_3	0.5
		MoO_3	14.1	TeO_2	2.0	Gd_2O_3	0.4
		CrO_3	0.6	BaO	4.3		
	100			99.8			

Phosphate glasses were prepared by first thoroughly mixing the appropriate amounts of lead oxide (PbO), ammonium dihydrogen phosphate ($NH_4H_2PO_4$), and iron oxide (Fe_2O_3) with a powdered form of the simulated nuclear waste. Ammonium dihydrogen phosphate, which decomposes to P_2O_5 at about 450°C, was used in preference to P_2O_5. The well-mixed powders were melted in a platinium crucible at temperatures between 900 and 1050°C for 3-5 h. The molten glass, which could be poured at temperatures as low as 800°C, was cast in a preheated mold of either boron nitride or graphite. The glass was then annealed for 1-2 h and cooled to room temperature over a few hours. The electrical resistivity of the glass melts was measured at temperatures between 800 and 1120°C. The phosphate glass was analyzed using differential thermal analysis (DTA) to measure the transformation temperature T_g and the temperature of the maximum crystallization rate T_c. The glass melt was slowly cooled through T_c and later analyzed at room temperature for crystallite formation in order to evaluate its devitrification properties.

Chemical corrosion experiments were performed in deionized water and in saturated NaCl solution at 90, 150, 190°C for 30 days. At 90°C the experiments were performed according to the MCC-1P procedure [8], at 150°C and 190°C the EC Static High-Temperature Leach Test [9] was used. The surface area to solution volume ratio was 10 m^{-1}. Sample surfaces were investigated by optical microscopy, energy dispersive x-ray and microprobe analysis. The leachates were analyzed quantitatively using inductively coupled plasma emission spectroscopy. Several corrosion experiments were performed using glass powder. x-ray diffraction patterns were taken from the corroded samples for phase identification.

RESULTS AND DISCUSSION

Fig. 2 shows the electrical resistivity for phosphate glass with 6.4% HLW (C1) and with an additional 0.6 wt.% of Na_2O(C1+Na). The curve for the German PAMELA borosilicate glass [10] is included. A comparison shows that the phosphate glass has a sufficiently high conductivity to be melted in a direct-heated electric melter.

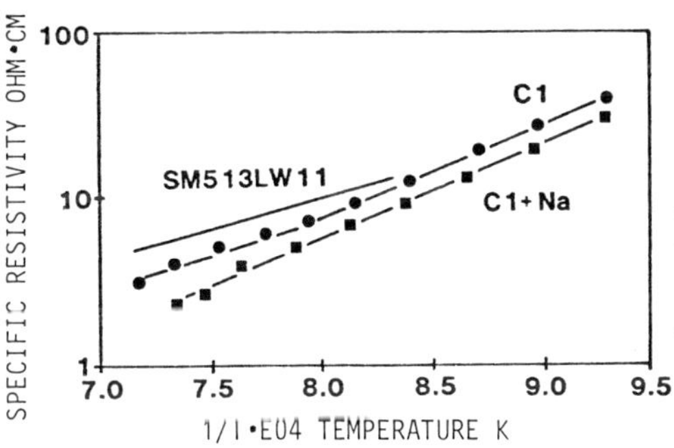

Fig. 2: Specific resistivity of the HLW phosphate glasses (C1 and C1+Na) and one borosilicate glass (SM513LW11) as a function of temperature in the range of 1120-800°C.

From a DTA measurement, the transformation temperature for glass C1 was found to be 494°C, and the temperature of the maximum crystallization rate was 640°C when the glass powder was heated at a rate of 10°/min. The glass was substantially crystallized after heating in air at 600°C for one day but did not crystallize after heating for several hundred hours in air at 500°C. When glass C1 was cooled at a rate of 1°/min. through the temperature range 680-600°C, over half of the sample crystallized. The iron concentration, however, was found to have a marked influence on the crystallization behavior. A glass was prepared with essentially the same composition as C1 but with the iron oxide concentration reduced from 10.8 wt.% to 7 wt.%. This new glass was then cooled at a rate of 1°/min. through the temperature range where the crystallization rate was high and the sample showed no evidence of bulk crystallization. From conductivity cell measurements [11], the corrosion resistance of the glass is about the same as that of glass C1 in deionized water at 90°C.

The results of the corrosion experiments performed with glass C1 are summarized in Table II. All experiments were run in quadruplicate and the reported data are the averaged values. Table III lists the phases found following corrosion of the phosphate glass. Some of these phases are shown in Fig. 3.

In previous work [1] the corrosion rate of the phosphate glass was reported to be extremely low in deionized water, 10^{-2} to $10^{-3}gm^{-2}d^{-1}$ at 90°C, compared with initial rates on the order of ≤ 1 $gm^{-2}d^{-1}$ for borosilicate glass [12]. With increased sensitivity it is possible to detect the ele-

Table II Results of Corrosion Tests with Glass C1 in Deionized Water and in Saturated NaCl (* = data below limit of detection)

	H_2O / 30 days			NaCl / 30 days		
T(C)	90°	150°	190°	90°	150°	190°
pH	6.8 ±0.05	6.8 ±0.05	2.0± 0.1	2.8±0.1	1.8± 0.1	1.4± 0.1
NLG (gm^{-2})	0.11±0.01	3.2 ±0.4	164. ±25.	44. ±1.	390. ±80.	980. ±40.
NLi (gm^{-2})						
P	0.1 ±0.02	2.2 ±0.3	65. ± 8.	12. ±1.	92. ±20.	185. ± 8.
Pb	*	0.03±0.01	0.4± 0.1	10. ±1.	38. ± 8.	104. ±20.
Fe	*	0.1 ±0.01	7.6± 0.1	0.4±0.1	1.8± 0.8	3. ± 1.
Cs	0.4 ±0.05	1.5 ±0.4	14. ± 2.	-	-	-
Mo	0.4 ±0.05	6.3 ±0.8	3.3± 0.5	*	4. ± 1.5	18. ± 7.
Sr	0.2 ±0.04	0.4 ±0.05	79. ± 4.	54. ±1.	100. ±10.	126. ±16.

NLG = normalized mass loss of the sample after MCC-1P
NLi = normalized elemental mass loss after MCC-1P

Table III Phases Formed During Corrosion of Glass C1 (++ almost certain, + some evidence, - possible)

30 days	H_2O	NaCl
Temp.(C)	Phases	
90°	++ $Pb_3(PO_4)_2$ - $Fe(OH)_3$	++ $Pb_5(PO_4)_3Cl$ (Pyromorphite) - $FePO_4 \cdot 2H_2O$ (Strengite) - $Fe(OH)_3$
150°	++ $Pb_3(PO_4)_2$ ++ $FePO_4 \cdot 2H_2O$ - $Fe(OH)_3$	++ $Pb_5(PO_4)_3Cl$ + $NaFeP_2O_7$ + PdTe-compound + $Fe_3(PO_4)_2(OH)_2$ (Lipscombite) + $Fe_3(PO_4)_2(OH)_3$ ++ $FePO_4 \cdot 2H_2O$ ++ $Pb_3(PO_4)_2$
190°	++ $Pb_3(PO_4)_2$ ++ $FePO_4 \cdot 2H_2O$ + $REPO_4 \cdot H_2O$ + $Pb_5(PO_4)_3OH$	++ $Pb_5(PO_4)_3Cl$ + $NaFeP_2O_7$ ++ $FePO_4 \cdot 2H_2O$ ++ $Pb_3(PO_4)_2$ + $REPO_4 \cdot H_2O$ - $(UO)_3(PO_4)_2$ or $U_3(PO_4)_4$

ments, P, Cs, Mo and Sr in the leachate after 30 days (Table II). A small amount of crystalline reaction product, $Pb_3(PO_4)_2$, is found on the sample surface after 30 days in deionized water at 90°C. The morphology of these crystals can be seen in Fig. 3.1. The material on the glass surface was physically removed, dissolved in dilute HNO_3 and analyzed. As expected, major elements of the glass matrix not detected in solution are found in the layer. The material contained in the surface layer should not be considered in calculating the extent of glass corrosion. It is important to remember that the extent of corrosion can be derived from solution analysis data only when the detected elements do not participate in phase formation, as e.g. molybdenum. This element is known to form $BaMoO_4$ but not at NLi values as low as reported in Table II for 90°C. Barium molybdate can form at NL(Mo)=400g/m^2 as was calculated with the help of the computer code

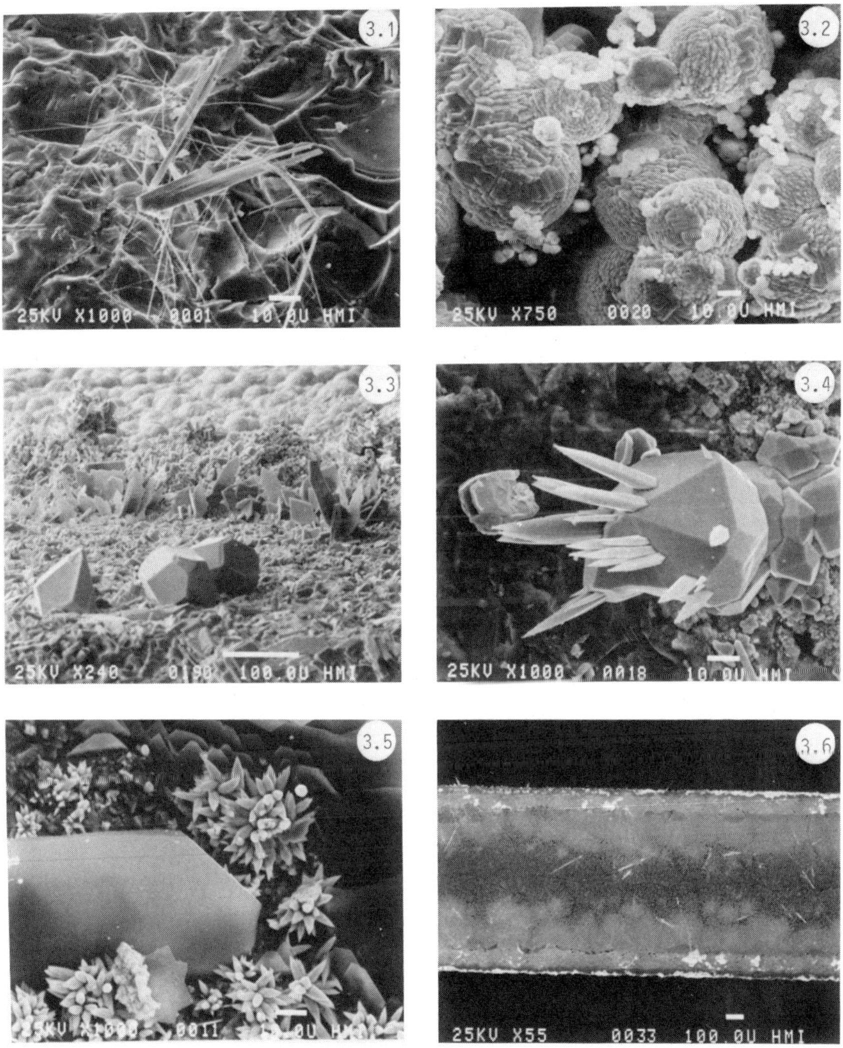

Figs. 3.1-3.6: Corroded lead-iron phosphate glass, 30 days.

Fig. 3.1: Needles of $Pb_3(PO_4)_2$ on the glass surface (90°C,H_2O).

Fig. 3.2: Large agglomerates are likely to be $Pb_5(PO_4)_3OH$. Small white crystals are $REPO_4 \cdot H_2O$ (190°C,H_2O).

Fig. 3.3: $Pb_5(PO_4)_3OH$ in the back. Large crystals in the front are a lead compound of unknown stoichiometry (190°C,H_2O).

Fig. 3.4: Pyromorphite, $Pb_5(PO_4)_3Cl$, large crystal. White spot is a Pd-Te-compound. Needles are likely to be $Fe_3(PO_4)_2(OH)_2$ (Lipscombite) or $Fe_3(PO_4)_2(OH)_3$ (150°C,NaCl).

Fig. 3.5: Large crystal is $NaFeP_2O_7$. Rosettes are pyromorphite (190°C,NaCl).

Fig. 3.6: Cross section of a glass specimen. Completely corroded (190°C,NaCl).

PHREEQE [13]. With NL(Mo)=0.4g/m^2 one obtains a corrosion rate of $1.3 \cdot 10^{-2}$ $gm^{-2}d^{-1}$ and a "concentration" of about 4 mg glass/L leachate. The rate of $1.3 \cdot 10^{-2} gm^{-2}d^{-1}$ is in good agreement with a value of $0.6 \cdot 10^{-2} gm^{-2}d^{-1}$ from a 270 day experiment (Tab. 5 in ref. [3]) considering that the latter value was obtained with phosphate glass, 1D, using Na and Cs solution analysis data. These results show that the corrosion rate is constant over longer periods of time. Glass 1D exhibits similar stability as C1. In a closed or slow flow system the corrosion rate of borosilicate glass decreases in the course of a few months by one to two orders of magnitude from the initial rate to the long-term rate [12]. Accordingly, under closed system conditions where saturation effects control or reduce the borosilicate corrosion rate, the difference between the corrosion rates for phosphate and borosilicate glasses is lowered or vanishes.

As expected, corrosion rates increase with temperature. At 150°C for phosphate glass in water the same phases form as at 90°C. In addition strengite is found (Table III). The reaction progress is NL(Mo)=6.3g/m^2, i.e. about 16 times that at 90°C. The surface layer has a brown color, probably from iron-hydroxide. Under the SEM the surface looks very much the same as at 90°C, Fig. 3.1. The chemical analysis of the layer gives evidence for the presence of P, Pb, Fe, and Nd but very little Mo.

At 190°C the corrosion process is relatively fast. When comparing the NLG values in Table II about 50 times as much material has reacted as at 150°C but the sample is still intact. The maximum possible value of NLG was calculated to be about 2000g/m^2. The pH value dropped from about neutral to 2.0 indicating the formation of phosphoric acid. Additional phases form. Their chemical composition is given in Table III. The morphology of the sample surface and some reaction products are shown in Figs. 3.2 and 3.3. The entire glass beneath the surface layer shows evidence of alteration similar to that illustrated in Fig. 3.6.

The corrosion of phosphate glass in saturated NaCl solutions is much different from that in deionized water. The pH values in the leachates are lower at all temperatures. At 90°C, the glass surface is covered with a layer of poorly crystallized or amorphous material together with some pyromorphite crystals, $Pb_5(PO_4)_3Cl$ (Tab. III). Phosphorus, lead and strontium are found in solution and their concentrations give evidence for a stronger corrosion of the glass in NaCl compared with water (Table II). This became quite clear when the surface layer was dissolved and analyzed. Iron with NLi=106 and neodymium with NLi=121g/m^2 show that the reaction progress is at least 300 times that in water at the same temperature.

After 30 days at 150°C in NaCl solution much more material has been dissolved than even at 190°C in water. To see this, one can compare the respective NLG values in Table II. As per the MCC procedure [8] these values were determined from the mass loss before and after the experiment with the layer remaining on the sample. These data give a minimum value for how much glass has dissolved. Some 80 wt.% of the sample is still pristine glass. Table III shows that the reaction products constitute a great variety of solid phases. Fig. 3.4 shows a large crystal of pyromorphite. The needles on this crystal are an iron phosphate, probably strengite.

Finally, at 190°C, over half of the sample has dissolved and the rest has, as in water, undergone substantial reaction with the solution. The list of solid reaction products is given in Table III. There is one phase which has not been detected in the previous experiments, a uranium phosphate, the composition of which is only tentative. Fig. 3.6 shows a cross section of the sample at a low magnification. Large white crystals at the rim and needles throughout the bulk can be seen. The sample surface can be seen in Fig.3.5. The large crystal is $NaFeP_2O_7$. Pyromorphite is present in a different morphology (compare with Fig. 3.4).

Phase formation as a function of the reaction progress, at 90°C in deionized water was calculated with the help of the computer code PHREEQE

[13]. The result is that the formation of $Fe(OH)_3$ is possible in addition to lead phosphate. The quantity was, however, too small to be identified as $Fe(OH)_3$ in the experiment. After a reaction progress of 8 mg glass dissolved per liter, a third phase, strengite, would, according to the calculations, form and $Fe(OH)_3$ redissolve.

With an average chain length of about 2.6 [3] and a metal to phosphorus ratio of almost 2 the structure of the phosphate glass should contain a lot of "terminal" groups $(O{=}PO_3)^{2-}$, some "middles" $(O{=}PO_3)^-$ and very few "branches" $(O{=}PO_3)$ [14]. This glass has a structure similar to that of an iron-aluminium phosphate nuclear waste glass the corrosion of which was investigated in an earlier study [15]. In the lead-iron phosphate glass the links between the chains and their ends are made by Pb-O-P, Fe-O-P bonds and with the metal ions from the waste which are present in comparatively low concentrations. Since every second octahedral position throughout the glass is occupied by iron [3], P:Fe=4:1, larger structural units are tied together by iron. These units are connected by Pb-O-P bonds, P:Pb=2:1.

In glasses of this structure, there are several types of dissolution mechanisms possible: hydrolysis, hydration and acid/base reactions. Hydrolysis leads to the cleavage of P-O-P bonds and hydration to disentanglement and transport of polymeric phosphate chains into solution. Reactions with P-O-Me groups (e.g. Me=Pb,Fe) along the chains and at their ends are acid/base reactions [14]. They affect the ionic interactions between the chains.

Upon dissolution of the phosphate glass in deionized water at 90°C and higher temperatures, orthophosphate (Tab. III) is the only phosphorus bearing anion found in the solid reaction products. In NaCl, in two cases, pyrophosphate was found. These findings are the same as in [15] and indicate, that cleavage of the phosphate chains is one of the chemical reactions involved in the process of glass dissolution. It can however, not yet be decided whether this reaction occurs predominantly in the glass or in solution. In any case the release of phosphate species into solution supposes that bonds are opened in the glass between the chains or at their ends. The stabilizing effect of iron on the glass structure and thus on the corrosion, in this picture, is that the acid/base reaction, i.e. the exchange of cations for protons is slow in the case of iron. Water may then predominantly interact with Pb-O-P bonds.

The data base for the phosphate glass is by far not as complete as that for some borosilicate nuclear waste glasses. Hence, an attempt at discussing the leaching mechanism for this particular phosphate glass is not free of speculation. If an analogy can be made with silicate glasses where the formation of $Si(OH)_4$ is the rate-determining step [16], a lead species could be assumed instead for the phosphate glass. The lead species would be Pb^{++} and $PbOH^+$ in solution with a ratio of about 5 at pH=7. On the glass surface P-O-H groups would remain (zeroth order dissolution reaction) and the back reaction (first order) could then be condensation between the $PbOH^+$ species and a P-O-H group. The net reaction rate would be expected to become constant very soon, since the Pb^{++} concentration and thus the concentration of $PbOH^+$ is controlled by the solubility of lead orthophosphate which is the least soluble lead bearing compound at pH=6.8 in low ionic strength solutions. An increase of the net reaction rate with decreasing pH, as observed experimentally [1], is in agreement with this model. The lower the pH the less $PbOH^+$ species exist in solution and the slower is the back reaction. Additionally, the rate of the hydrolysis reaction increases with decreasing pH [17].

One of the obvious differences between the glass corrosion in water and in NaCl-solution is, that pyromorphite, $Pb_5(PO_4)_3Cl$, is the reaction product for lead, instead of lead orthophosphate (Table III). Pyromorphite controls the concentration of Pb^{++} in solution but this concentration is lower by at least a factor of 10^3 [18] than in the case of $Pb_3(PO_4)_2$.

The back reaction should, therefore, be slower than in water. In a saturated NaCl-solution the dominating lead species are lead chloride complexes rather than Pb^{++} [18]. The greater the reaction progress the more the solution pH decreases as shown in Table II, which in turn accelerates the pH-dependent acid/base and the hydrolysis reactions. Hence, with the higher initial reaction rate of the glass and the induced pH changes the enhanced corrosion of phosphate glass in an NaCl solution can be accounted for.

CONCLUSIONS

1. The electrical conductivity of lead-iron phosphate glass melts can be made sufficiently high to allow these glasses to be melted in a direct-heated electric melter.
2. The glass composition can be adjusted to avoid bulk crystallization at a cooling rate as low as 1°C/min.
3. Phosphate glass corroded in deionized water at temperatures below 100°C exhibits very low leach rates. In contrast with borosilicate glasses saturated NaCl solution jeopardizes the high corrosion resistance of the phosphate glass, since lead chloride complexes form and the concentration of Pb in solution is controlled by pyromorphite.

REFERENCES

1. B.C. Sales and L.A. Boatner, Science 226, 45 (1984).
2. B.C. Sales and L.A. Boatner, Material Lett. 2, 301 (1983).
3. B.C. Sales and L.A. Boatner, Oak Ridge National Laboratory Report , ORNL-6168, UC-70, 1985.
4. B.C. Sales and L.A. Boatner, J. Non-Cryst. Solids, in press.
5. V.M. Oversby, Lawrence Livermore National Laboratory Report UCRL-53326, 1982.
6. B.C. Sales, M.M. Abraham, J.B. Bates, and L.A. Boatner, J. Non-Cryst. Solids 71, 103 (1985).
7. B.C. Sales, R.S. Brazell, J.B. Bates, and L.A. Boatner, J. Non-Cryst. Solids - submitted.
8. MCC-1P Static Leach Test Method, Report DOE/TIC-11400, 1981.
9. Report EUR-9772, 1985.
10. E. Schiewer, Radioactive Waste Management, to be published.
11. B.C. Sales, M. Petek, and L.A. Boatner, in Scientific Basis for Nuclear Waste Management VI, edited by D.G. Brookins (Elsevier Science Publishers, New York, 1983) pp. 251-258.
12. E. Freude, B. Grambow, W. Lutze, H. Rabe, and R.C. Ewing, in Scientific Basis for Nuclear Waste Management VIII, edited by C.M. Jantzen, J.A. Stone, and R.C. Ewing (Materials Research Society, Pittsburgh, 1985) pp. 99-106.
13. D.L. Parkhurst, D.C. Thorstensen, and L.N. Plummer, Water-Resources Investigations 1980, 80.
14. J.R. Van Wazer, J. Am. Chem. Soc. 72, 644 (1950)
15. B. Grambow and W. Lutze, in Scientific Basis for Nuclear Waste Management II, edited by C.J.M. Northrup (Plenum Press, New York, 1980) pp. 109-116.
16. B. Grambow, in Scientific Basis for Nuclear Waste Management VIII, edited by C.M. Jantzen, J.A. Stone, and R.C. Ewing (Materials Research Society, Pittsburgh, 1985) pp. 15-27.
17. R.K. Osterheld, in Topics in Phosphorus Chemistry VII, edited by E.J. Griffith (John Wiley & Sons, New York, 1972), pp. 103-254.
18. W.L. Lindsay, Chemical Equilibria in Soils (John Wiley & Sons, New York, 1979), pp. 329-341.

AN ASSESSMENT OF THE LONG-TERM DURABILITY OF CONCRETE IN RADIOACTIVE WASTE REPOSITORIES

A. ATKINSON, D.J. GOULT AND J.A. HEARNE
Materials Development Division, Building 552, AERE Harwell, Didcot, Oxfordshire, OX11 ORA, United Kingdom.

ABSTRACT

A preliminary assessment of the long-term durability of concrete in a repository sited in clay is presented. The assessment is based on recorded experience of concrete structures and both field and laboratory studies. It is also supported by results of the examination of a concrete sample which had been buried in clay for 43 years.

The engineering lifetime of a 1 m thick reinforced concrete slab, with one face in contact with clay, and the way in which pH in the repository as a whole is likely to vary with time have both been estimated from available data. The estimates indicate that engineering lifetimes of about 10^3 years are expected (providing that sulphate resisting cement is used) and that pH is likely to remain above 10.5 for about 10^6 years.

INTRODUCTION

Many conceptual designs for repositories for disposal of radioactive waste propose extensive use of hydraulic cements and concretes. These materials may be used for the waste form matrix itself, as a constructional material for walls and floor, as a grout to back-fill gaps between waste packages and as a sealant of shafts and tunnels. A simplified design [1] for a repository being considered in the U.K. is shown in Fig. 1. This particular example is a shallow structure, sited in clay, for the disposal of β,γ wastes having very low levels of α-emitting isotopes. Concrete may be used to provide casks for the drums of waste and the structure itself is essentially a box with walls of reinforced concrete. The outer surface of a wall is in contact with the clay.

The cement and concrete repository components serve many useful functions, such as controlling water flow (by having low permeability) and radionuclide migration (by having low diffusion coefficients and some sorption capacity for radionuclides) in the near field. The principal functions of the concrete are, however, probably mechanical and chemical. Mechanically, the structural components must support the required loads so that the repository structure does not collapse. The chemical function is to maintain the pH in the near field at a high value for as long as possible. In the relatively short term this provides an environment which minimises the corrosion of steel containers and, in the long term, ensures low aqueous solubility limits for any traces of long-lived radionuclides.

In this paper we consider how long the cement and concrete in the repository will be able to fulfil these mechanical and chemical functions. For the mechanical function, we consider the behaviour of a 1 m thick reinforced concrete wall whose outer surface is in contact with clay at the edge of the repository. The estimated lifetime is based on recorded experience with real structures, accelerated laboratory tests and field experiments. The predicted behaviour is supported by examination of concrete which has been exhumed after being buried for 43 years in a clay formation.

For the chemical function we estimate how the pH of a repository as a whole is likely to evolve over very long timescales. The estimates are based on simple groundwater transport models and the chemistry of cement systems.

ASSESSMENT OF ENGINEERING LIFETIME OF CONCRETE

Studies of ancient structures (e.g. Roman), in which calcareous cements were used, have indicated the potential longevity of cement and concrete [2]. The factors which have controlled the behaviour of these materials are, however, likely to be different from those which will determine the performance of modern materials in very different environments. For example, ancient cements show extensive carbonation by reaction with CO_2 from the atmosphere, whereas in underground repositories the environmental reactions will be with groundwater constituents. It is therefore necessary to consider the processes which may lead to changes in the concrete in the environment in question.

Space only permits a brief description of the assessment, the details of which are reported elsewhere [3]. The following causes of concrete alteration were considered;

(a) internal changes (e.g. crystallisation, alkali-aggregate reactions),
(b) reactions with the environment (clay) through groundwater,
(c) physical changes from wet/dry or freeze/thaw cycles,
(d) corrosion of reinforcement,
(e) action of micro-organisms,
(f) radiation effects (radiolysis, radiation damage).

Of these it was concluded that (b) and (d) were the most likely sources of deterioration of the concrete structure.

For the purpose of the assessment we define the 'engineering lifetime' of the concrete section as the time by which its load-bearing capacity has been reduced by a factor of two.

Reactions with groundwater

Potential reactions between the constituents of a clay groundwater and concrete are summarised in Table I, together with their deleterious consequences. Reported experience of concrete structures in similar environments [4] indicates that reactions with Mg^{++} and SO_4^{--} ions are likely to be the most severe. The SO_4^{--} ions, together with Ca^{++} from the cement environment, attack the hydrated calcium aluminates to form sulpho-aluminates:

$$3CaO.Al_2O_3.6H_2O + CaSO_4 + 6H_2O = 3CaO.Al_2O_3(CaSO_4).12H_2O \quad \text{(monosulphate)} \qquad (1)$$

$$\text{Monosulphate} + 2CaSO_4 + 20H_2O = 3CaO.Al_2O_3.3(CaSO_4).32H_2O \quad \text{(ettringite).} \qquad (2)$$

These reactions involve considerable increases in the volume of solid, which leads to disruption of the concrete.

The Mg^{++} ions attack all the Ca-containing material to produce solid $Mg(OH)_2$, which is far less soluble than $Ca(OH)_2$, and hydrated magnesium silicates. Since the Ca compounds are responsible for binding the concrete, the reactions result in loss of strength.

In 1970 the Building Research Establishment (Watford, U.K.) initiated a long-term programme to assess the performance of various concretes under exposure to sulphate-bearing groundwater. The programme employed both laboratory and field tests and the results of exposure for up to 5 years have been reported [5]. In the laboratory, the attack was accelerated by increasing the sulphate concentration in the water to 0.19 mole ℓ^{-1}, which is an order of magnitude greater than expected in a clay environment. It should be noted that such acceleration may change the dominant mechanisms of alteration. The accelerated laboratory tests resulted in a visible deterioration zone whose depth, X_s, is given approximately by the empirical expression [3].

$$X_S(cm) = 0.55.C_A(\%).([Mg] + [SO_4]).t(y). \tag{3}$$

In this equation C_A is the percentage (by weight) of $3CaO.Al_2O_3$ in the unhydrated cement and it appears because its hydration products are those which are principally under attack. The ion concentrations are in mole ℓ^{-1} and t is the time in years. The variability observed in the laboratory tests was typically $\pm$ 30% [5] (spanning maximum and minimum depths of attack).

Table I. Summary of potentially deleterious interactions between clay groundwater and concrete

Groundwater Constituent		Concrete Constituent	Possible Consequence
Conc in clay water			
2×10^{-2} M	SO_4^{--}	$3\ CaO.Al_2O_3\ 6H_2O$	Expansion and cracking.
1×10^{-3} M	CO_2	$CaO - SiO_2 - H_2O$ gel	Loss of pH.
1×10^{-2} M	Mg^{++}	$Ca(OH)_2$	Loss of strength and pH.
-	H_2O	Soluble Ca^{++}	Loss of strength and pH.
3×10^{-4} M	O_2	Steel reinforcement	Loss of strength. Expansion and cracking.
3×10^{-3} M	Na^+, K^+	SiO_2 aggregate	Expansion and cracking.

The attack in the field experiments was assessed by measuring sulphate penetration profiles into the concrete [4]. The sulphate penetration depths for OPC concretes were in the range 0.8 to 2 cm. Equation (3) satisfactorily accounts for the observations [3]; predicting approximately 1 cm of attack on OPC concrete after 5 years in an environment where [Mg] + $[SO_4] = 5 \times 10^{-2}$ M.

Equation (3) was therefore used to calculate a range of engineering lifetimes (i.e. when X_S = 50 cm) for concrete in groundwater of composition given in Table I and the results are summarised in Table II. Ordinary Portland Cement (OPC) contains between 5 and 12% $3CaO.Al_2O_3$ (8% in the experiments referred to) whereas Sulphate Resisting Portland Cement (SRPC) contains less than 3% (1.2% in the experiments). This variability in composition, when compounded with the observed variability in behaviour, leads to uncertainties in estimating concrete behaviour. Furthermore, the estimates also involve an extrapolation based only on empiricism. A sound mechanistic understanding of the mechanism of sulphate attack will be

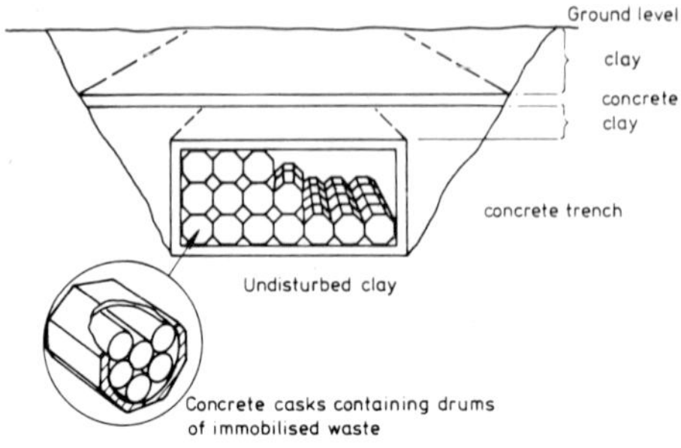

Fig. 1 Conceptual design of a shallow repository in a clay bed for the disposal of intermediate level waste having low α-ctivity (ref. 1).

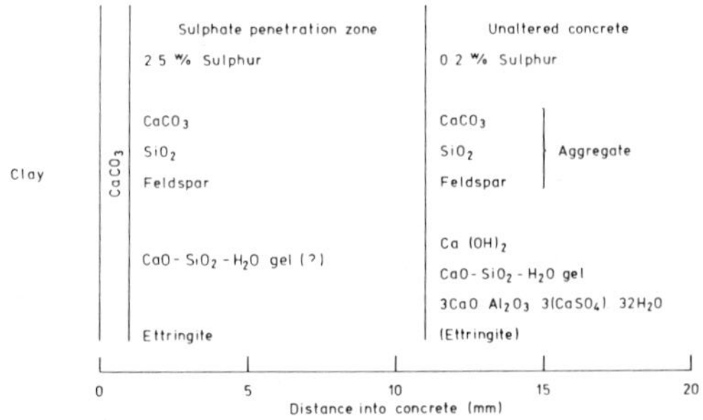

Fig. 2 Schematic diagram showing the composition and depth of the alteration zone on concrete which had been buried in Oxford clay for 43 years. (The sulphur concentrations are in percentage of cement by weight.)

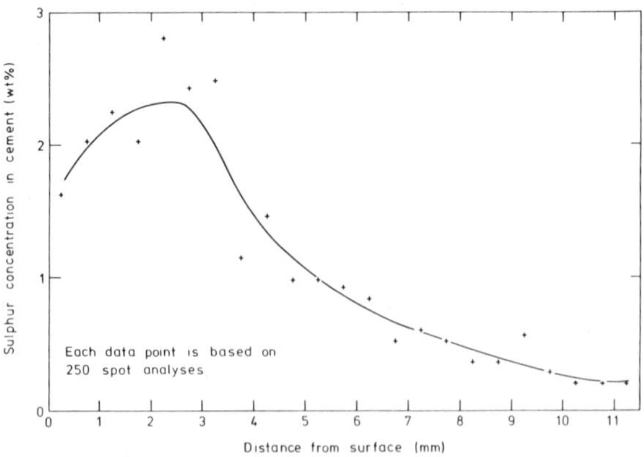

Fig. 3 Concentration of sulphur in cement phase (determined semi-quantitatively by electron probe microanalysis) as a function of depth into concrete after burial in clay for 43 years.

Table II. Estimated engineering lifetime* of an average reinforced concrete in clay

Degradation Process	Lifetime (years)			Comments
	Min.	Probable	Max.	
Sulphate attack	180	400	800	Ordinary Portland Cement
	700	2500	-	Sulphate Resisting Cement
Corrosion of reinforcement		900		Cracking
			3000	No cracking

* Time for the loss of half the load-bearing capacity of a 1 metre thick section.

required to improve on this. Nevertheless, superior longevity of SRPC concrete, as opposed to OPC concrete, in clay environments is clearly indicated.

Examination of concrete buried in clay

Samples of a 2 x 2 x 2 m block of concrete, which has been buried in clay since 1942, have been studied to assess the extent of chemical interaction after 43 years of exposure.

The techniques which have been used include X-ray diffraction, electron and optical microscopy, thermogravimetry and differential scanning calorimetry. The concrete contains two aggregates: a sand with particle size ~ 1 mm and a coarse aggregate (up to 20 mm) which is mainly limestone with some flint and feldspar. The concrete shows two reaction zones as indicated in Fig. 2. The outermost layer is most likely the result of carbonation and sulphate attack and is very thin ($\lesssim$ 1 mm). The second zone (~ 10 mm thick) is rich in sulphur (Fig. 3) and is probably the result of reaction with sulphate from the clay. The material in this zone is mechanically sound and is not as friable as the alteration zone produced by artificially accelerated sulphate attack.

The variation of aqueous chemistry with depth into the concrete was studied by taking slices from a specimen and equilibrating each with a ten-fold excess of water (by weight) for 20 days. The composition of the aqueous phase, as a function of depth, is shown in Fig. 4. The magnitudes of the concentrations and their variation with depth are difficult to interpret because of the large number of possible chemical reactions which may be participating in the overall equilibrium. For example, the pH and concentrations of Ca and Si in equilibrium with a slice from 1 cm below the surface are consistent with the C-S-H gel having a Ca/Si ratio close to 1.7 (Fig. 6) as expected in unaltered cement. On the other hand, the results from 2 mm below the surface indicate that in this region the Ca/Si ratio in the gel is close to its minimum value of 0.8 (Fig. 6) and that interaction with groundwater has removed Ca from the gel. These experiments therefore also indicate an alteration zone about 1 cm in thickness.

From equation (3) we would predict that after 40 years in clay the thickness of the alteration zone would be between 1 and 9 cm depending on

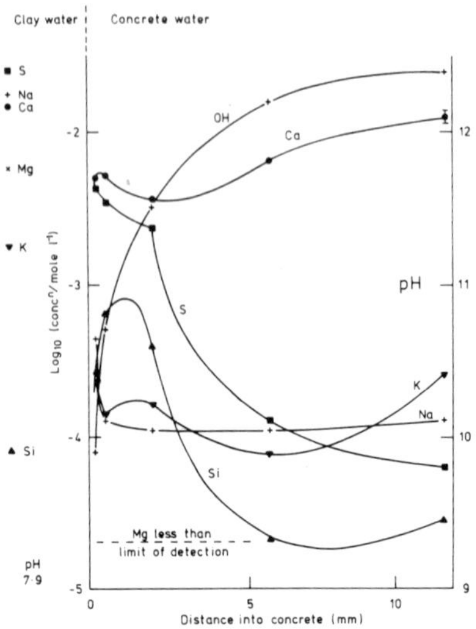

Fig. 4 The composition of the aqueous phase when slices, taken from different depths below the surface of concrete (buried for 43 years in clay) are equilibrated with a ten-fold excess of water by weight. The composition of the interstitial water extracted from the clay is shown on the left.

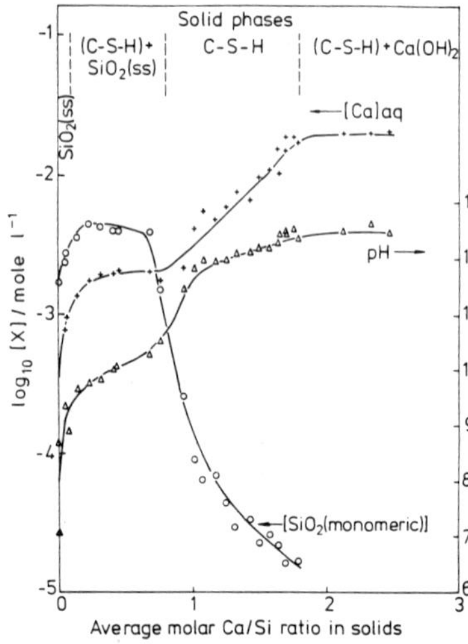

Fig. 6 The composition of the aqueous phase when solids in the $CaO-SiO_2-H_2O$ system, having various Ca/Si ratios, are equilibrated with water (ref. 5). X denotes Ca or Si.

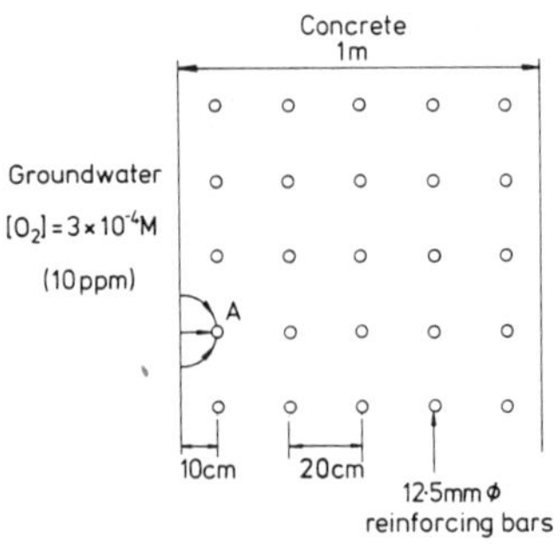

Fig. 5 The corrosion of an array of steel reinforcing bars in concrete controlled by the inward diffusion of O_2 dissolved in the groundwater.

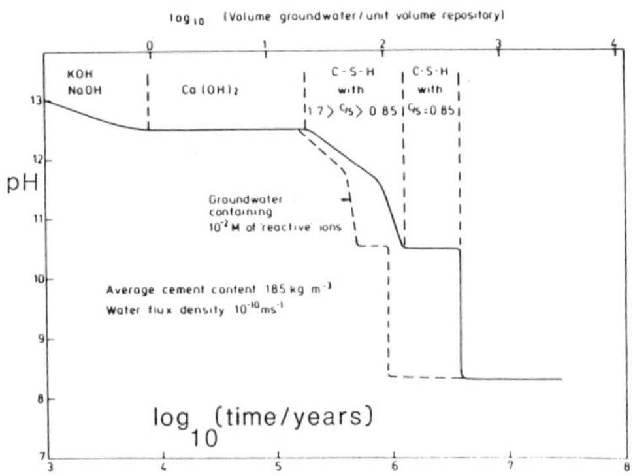

Fig. 7 The estimated time dependence of pH within a repository of radius 20 m containing 185 kg m^{-3} cement and situated in a groundwater flux density of 10^{-10} ms^{-1} (3.2 mm y^{-1}). The phases which control various pH regions are indicated. The broken curve is for a groundwater containing 10^{-2} mole ℓ^{-1} of 'reactive' species, e.g. Mg^{++}.

the tricalcium aluminate content of the cement. Since we have not yet been able to determine this, we conclude that equation (3) gives either a correct estimate or an overestimate of the rate of sulphate attack.

Corrosion of reinforcement

The potential effects of reinforcement corrosion in concrete are reduction in load-bearing capacity, through loss of reinforcement, and cracking of the concrete by the extra volume of the corrosion products. Corrosion of the reinforcement in a repository will be very slow (in comparison with corrosion in reinforced concrete structures above ground) and controlled by oxygen supply to the steel surface. We have estimated the maximum corrosion rate which oxygen, diffusing from the clay groundwater through the concrete to the steel, could sustain for a reinforcement array as shown in Fig. 5. Assuming an oxygen diffusion coefficient of 10^{-9} m^2 s^{-1} and oxygen concentration in the groundwater equal to 3 x 10^{-4} mole ℓ^{-1}, the maximum rate of corrosion of the first layer of reinforcement is estimated to be 10 μm y^{-1} [3]. The resulting degradation sequence depends on whether the corrosion products will crack the concrete at this low corrosion rate. Unfortunately, there are no data available which can be used to predict whether cracking will occur or not. Consequently, we have estimated [3] the lifetime both on the assumption that no cracking occurs, in which case the result is 3000 y, and on the assumption that corrosion of 50% of a reinforcing bar is capable of extensively cracking the concrete between it and the surface of the wall, in which case the result is 900 years. These estimates are compared with those expected from sulphate attack in Table II.

TIME DEPENDENCE OF pH IN THE REPOSITORY

The alkaline environment in cement and concrete is generated, in the long term, by calcium compounds which form Ca^{++} and OH^- ions in aqueous solution. The most abundant compounds are in the $CaO-SiO_2-H_2O$ (C-S-H) system and Fig. 6 shows how the composition of the aqueous phase depends on the composition of the solids, expressed by the Ca/Si atom ratio. Most cements have Ca/Si between 2 and 3 and therefore, from Fig. 6, the aqueous phase is essentially in equilibrium with solid $Ca(OH)_2$ at pH = 12.5. (In some materials the pH may be even greater due to NaOH and KOH in solution.)

When the cement is part of a concrete repository buried in the ground the pH will tend to fall with time as a result of;

(a) loss of Ca from the repository by diffusion or advection,

(b) reaction with groundwater constituents coming into the repository (e.g. Mg^{++}, CO_3^{--}),

(c) slow reactions with other repository constituents (e.g. SiO_2 in aggregates).

We have estimated [6] the way in which pH will be reduced with time for processes (a) and (b) above. The repository was approximated as a sphere of radius 20 m with an average cement content of 185 kg m^{-3} (this is equivalent to half its volume being occupied by a good quality concrete). The groundwater flux density was assumed to be 10^{-10} ms^{-1}, which is a likely upper bound for a typical repository (e.g. sited in an argillaceous formation with a 1% hydraulic gradient [7]). The Ca/Si ratio in the repository was calculated as a function of time using the data in Fig. 6, for the composition of the aqueous phase, and assuming loss of Ca by advection in the flowing groundwater. The result is shown in Fig. 7 which indicates that, in this case, the pH is expected to be above 10.5 for 4 x 10^6 years. If the groundwater is assumed to contain 10^{-2} moles ℓ^{-1} of dissolved species which can react with the Ca-containing phases the pH dependence is estimated to follow the broken line and the timescale for pH to fall below 10.5 is reduced to 0.9 x 10^6 years.

These timescales are far in excess of that predicted for the engineering lifetime of the repository walls. However, the pH buffering capacity does not rely on structural integrity of the material; only on its continued presence in the repository.

The calculations described above have also assumed sufficient diffusive transport within the repository to maintain the contents at a uniform pH. This may not be true in reality since the pH will tend to be reduced preferentially on the side of the repository facing the incident water flow. If there is negligible diffusion within the repository then a similar calculation [6] estimates that the pH = 10.5 front moves into the repository at a rate of 1 m every 1.5 x 10^5 years (assuming a groundwater flux density of 10^{-10} ms^{-1} and no 'reactive' species).

CONCLUSIONS

An engineering lifetime in excess of 1000 years is estimated for a 1 m thick reinforced concrete section in clay provided that sulphate-resisting cement is used in the concrete. The life-limiting process may be either sulphate attack or corrosion of the reinforcement.

Examination of concrete buried in clay for 43 years has revealed an alteration zone, which is rich in sulphur, approximately 1 cm in depth. This is not greater than the depth estimated from the empirical model on which the estimate of engineering liftime is based.

The pH in the repository will be maintained at a high level by the cement for a much longer time than the engineering lifetime. For a representative case it is estimated that the pH will remain above 10.5 for about 10^6 years.

ACKNOWLEDGEMENT

The authors wish to thank NIREX for financial support.

REFERENCES

1. 'NIREX Plaintalk', November 1983, p. 5, NIREX, Harwell, U.K.
2. C.A. Langton and D.M. Roy, Mat. Res. Symp., 26, 543 (1984).
3. A. Atkinson and J.A. Hearne, U.K. Atomic Energy Authority Report AERE-R-11465 (1984).
4. F.M. Lea, The Chemistry of Cement and Concrete, 3rd Edition, (Edward Arnold Ltd., London, 1970).
5. W.H. Harrison and D.C. Teychenné, Sulphate Resistance of Buried Concrete: Second Interim Report on Long Term Investigation at Northwick Park, (Building Research Establishment, HM Stationery Office, 1981).
6. S.A. Greenberg and T.N. Chang, J. Phys. Chem., 69, 182 (1965).
7. A. Atkinson, U.K. Atomic Energy Authority Report AERE-R11777 (1985).
8. A. Atkinson, Nuclear and Chemical Waste Management, 5, 203 (1985).

EFFECTS OF ROCKS AND BACKFILL MATERIALS ON WASTE GLASS LEACHING

K. ISHIGURO, N. SASAKI, H. KASHIHARA AND M. YAMAMOTO
Power Reactor and Nuclear Fuel Development Corporation, Tokai-mura, Japan

ABSTRACT

Extensive studies have been made on the interactions between a waste glass and repository materials under static conditions. One of the PNC reference glasses was leached in the solution prepared from water in contact with crushed granite, tuff, diabase and backfill materials such as bentonite and zeolite. The leachant solutions except for some bentonite solutions reduced the glass leach rate compared with that measured in distilled water. The extent of the reduction was a function of silicon concentration in solution. The bentonite solutions enhanced the glass dissolution rate by a factor of 2 to 3 at low bentonite/water ratios but the effect was found to be less important at high bentonite/water ratios and in the long-term experiment. Addition of granite and zeolite to the bentonite solutions decreased the leach rate below the value measured in distilled water.

INTRODUCTION

The current concept for isolating nuclear wastes in deep underground repositories relies on the multiple barrier system [1]. However, very little dada are available to evaluate the leaching performance of waste forms buried in deep rock formations with other waste package materials. In order to model the release rates of radionuclides from the repository and to predict its long-term behavier, it is important to evaluate a source term for the multiple waste package system. Recently several multicomponent tests including some field tests [2]-[4] have been made. Actual or synthetic groundwaters seem to reduce the leach rate of waste glasses compared with distilled water. The presence of iron metal or bentonite in leachants accelerated the waste glass leaching [4],[5]. Currently it is thought that the enhanced leaching is due to the sorptive properties of the colloidal particles in the solutions [5],[6]. However, further extensive studies are necessary to understand the leach mechanism and to evaluate the long-term performance of waste glasses.

In the present study, experiments have been made on the interaction between a waste glass and some repository materials. The materials used are selected from some typical rocks and candidate backfill materials. This study is the first step towards fully integrated multicomponent tests. Simple interaction systems are used for basic understanding.

MATERIALS AND METHODS

Samples

The glass used was P0500, one of the PNC reference glasses, containing about 30 wt% simulated waste oxides. Its composition is given in Table 1. Samples were cut from cast bar into monolithic blocks with dimensions of 1 cm x 1 cm x 1 cm and finely polished with cerium oxide on all faces.

Leachant solutions were prepared from the water in contact with crushed rocks($<$ 60 μm) and backfill materials. Crushed granite (A: from the Stripa mine, B: from Inada), diabase (from Shimokawa), Tuff (From Ohya) and backfill materials, such as Na-bentonite, zeolite, ie mordenite (from Maji) and silica sand were used in the present study. The water was in contact with those materials in a Teflon container for 28 days at 98 °C and was used as leachant solutions after centrifuging or filtrating. Some bentonite fluids are used as such without those separation procedures. The matrix of leachant preparations is given in Table 2, and their compositions given in Table 3.

Leaching Conditions

All leaching experiments were performed at 98 °C for the period up to 28 days except for several runs up to 364 days in static conditions. Sample glasses were placed on the net in Teflon (PFA) containers. A sample surface area to leachant volume ratio was adjusted to 0.1 cm^{-1} and the nominal volume of leachant was 60 ml. On completion of the tests, the samples were rinsed with distilled water and then dried at 120 °C for 2 hours.

Table 1 Composition of P0500 glass used in this study (wt%)

SiO_2	43.2	Cr_2O_3	0.5	PdO	0.5
B_2O_3	14.1	Rb_2O	0.2	TeO	0.3
Li_2O	3.0	Cs_2O	1.0	Y_2O_3	0.3
Na_2O	8.4	SrO	0.4	La_2O_3	0.6
K_2O	2.0	BaO	0.7	CeO_2	1.1
CaO	2.0	ZrO_2	2.0	Pr_6O_{11}	0.3
ZnO	2.0	MoO_3	2.0	Nd_2O_3	2.5
Al_2O_3	3.5	MnO_2	0.3	Sm_2O_3	0.1
Fe_2O_3	7.4	RuO_2	1.0		
NiO	0.5	Rh_2O_3	0.2		

Table 2 Matrix of leachant preparations

Materials	Contact ratio with water[1)]	Centrifuging(C)/Filtrating(F)
Granite(A)	20g/60ml	[C]
Granite(B)	20g/60ml	[C]
Tuff	20g/60ml	[C]
Diabase	20g/60ml	[C]
Bentonite	0.3, 3, 30g/60ml	
Bentonite(C)	0.3g/60ml	[C]
Bentonite(F)	0.3g/60ml	[F] (0.05 μm filter)
Zeolite	20g/60ml	[C]
Silica Sand	20g/60ml	[C]
Bentonite/Silica	0.3g/20g/60ml	[C]
Bentonite/zeolite	0.3g/20g/60ml	[C]
Bentonite/Granite(A)	0.3g/20g/60ml	[C]

Note: 1) Contact time and temperature prior to experiments: 98°C - 28 days

Sample Analyses

Total mass losses were evaluated by mass loss measurements. Leachates were analyzed by inductively coupled plasma (ICP) optical emission spectroscopy and atomic absorption spectrophotometer. Sample glasses after leaching were analyzed using a scanning Auger micro-probe analyzer.

Table 3 Compositions of leachant solutions (mg/l)

Leachant Solution	Si	Na	K	Mg	Ca	Al	Fe	S	Sr	pH
Granite(A)	24.3	79.0	43.9	—	34	0.5	0.3	53	0.03	7.8
Granite(B)	30	24.0	12.6	0.1	3.6	1.8	0.3	0.9	0.01	9.5
Tuff	285	133	33.0	7.8	23	41.4	41	84	0.03	6.0
Diabase	8.3	57.5	0.6	0.2	110	0.1	0.3	41	0.03	6.0
Bentonite(C)	820	66.8	2.5	72	11	829	48	6.4	0.6	7.7
Bentonite(F)	8.6	35	0.4	0.1	0.2	0.2	0.3	5	0.005	
Zeolite	200	11.3	4.7	2.1	4.1	22.6	3.1	16	0.1	7.2
Silica Sand	41	—	—	—	—	—	—	—	—	7.3
Bentonite/Silica	670	92	1.8	—	0.1	382	34	6.4	0.4	9.2
Bentonite/Zeolite	145	231	7.9	12	10	1.9	0.6	172	0.1	3.5
Bentonite/Granite	270	145	46.9	21	18	116	16	64	0.1	8.5

RESULTS

Leaching in Rock water Systems

Table 4 summarizes the results of 28 day leaching in various rock waters. The rock waters reduced the extent of leaching in all cases.

Table 4 Summary of the results of 28 day leaching (98 °C - 28 days)

	Normalized Mass Loss (x 10^{-5} g/cm^2)				
Leachant Solution	B	Li	Cs	Mo	Mass Loss
Granite(A)	4.2	D.L.	2.8	—	3.4
Granite(B)	D.L.	D.L.	2.9	D.L.	1.2
Tuff	47.1	53.7	18.3	18.8	11.7
Diabase	137.9	117.0	93.6	129.5	50.5
Distilled Water	161.6	174.6	131.2	155.6	96.5

Note: D.L.: below the detection limit

Leaching in Backfill Water Systems

Total mass losses of the glass exposed to various bentonite fluids are given in Fig. 1. The magnitude of the glass/bentonite reaction depended on the bentonite/water ratio in the fluids. The bentonite solutions enhanced the glass dissolution rate by a factor of 2 to 3 at low bentonite/water ratios. At high bentonite/water ratios, however, the glass leach rate was about 10 times lower than in distilled water. The bentonite solution filtered through 0.05 µm filter prior to the leaching experiment provided a similar glass dissolution rate compared with that in distilled water.

The effect of the additions of other materials to the bentonite solutions are shown in Fig. 2. The enhanced leaching was lowered or eliminated in the granite/bentonite solution and the zeolite/bentonite solution. Silica sand addition did not have a significant effect on the leach rate.

Long-term Leaching Results

Long-term leach tests were conducted in distilled water, the granite water and the bentonite solution. Table 5 presents the results from these tests. Total mass losses were in the order: Bentonite solution > Distilled water > Granite water, which corresponded to that of the short term experiment up to 28 days. It may be noted that the leaching kinetics in the bentonite solution, which provided the largest mass loss, tended to reach a saturation. While the leach rate in distilled water was relatively high even after 182 day leaching.

Table 5 Results of the long-term leaching experiment (98 °C)

Leachant Solution	Total Mass Loss (x 10^{-5} g/cm^2) 28d	91d	182d	364d
Distilled Water	97	123	229	314
Granite(A)	3	—	—	55
Bentonite(C)	221	368	382	397

Solid State Analysis

Silicon depth profiles through altered layers of the leached glasses are shown in Fig. 3 for different leachants. The thickness of altered layers based on boron depletion depth corresponded to the total weight losses. The silicon depletion to about one half its original value was observed for the samples exposed to distilled water, the diabase water and the bentonite solution. For the samples in the tuff, granite and zeolite waters, which leached at relatively small rates, their silicon depletions were negligibly small.

DISCUSSION

Effects of Rocks on Waste Glass Leaching

The present study provides reduced leach rates of the waste glass in the rock waters in all cases. Currently, two different mechanisms, the barrier effect of altered layer[7] and the saturation effect of leachates[8], are presented for the supression of waste glass dissolutions. The barrier effect of the altered layers seems to be less important from the results of surface analyses (Fig. 3) because the thickness of the altered layers increased with the increase in the average leach rates.

Groundwaters and synthetic silicate waters are found to reduce the leach rate of waste glasses because of their relatively high silicon concentration[9]. The relation between mass loss and initial silicon concentration in the rock leachants are shown in Fig. 4. Silicon concentrations after

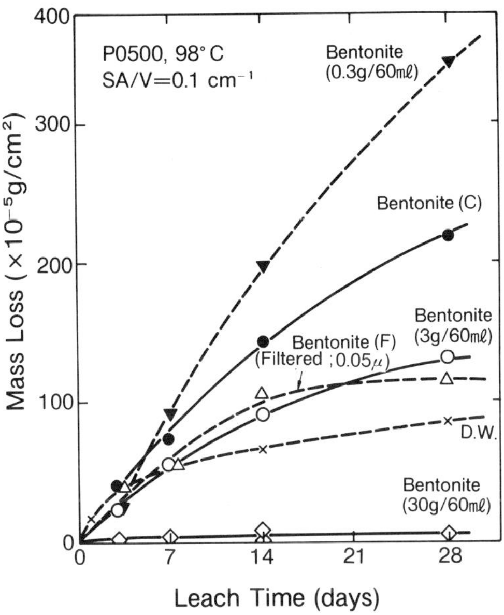

Fig. 1 Mass losses as a function of time for P0500 glass leached in the bentonite fluids at 98 °C. D.W.:distilled water, (C):centrifuged

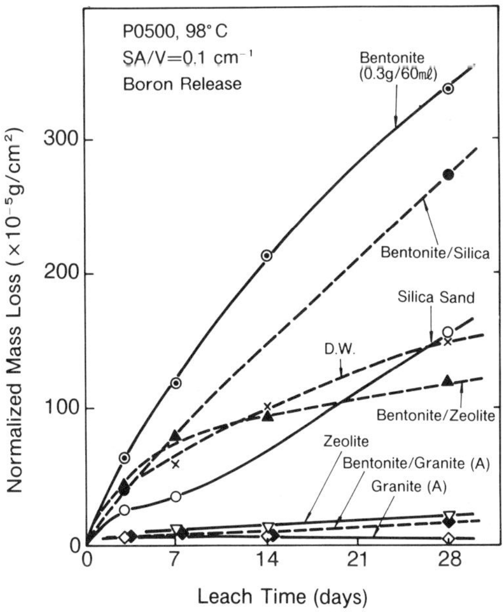

Fig. 2 Normalized releases of boron from P0500 glass leached in the rock/backfill leachants at 98 °C.

filtrating the solutions (0.05 μm filter) are plotted because colloidal silicon dose not contribute. The extent of the leaching supression generally depends on silicon concentration in the leachants.

The composition of the rock waters depends on rock/water reactions in their preparation. The rock waters prepared for the present study may or may not be saturated with the rock components. Therefore, the leach rates shown in Table 4 do not seem to be inherent in the rock types used.

The leaching inhibition was shown to be effective in one year leaching for the granite water. It is noted that silicon concentration in the granite leachate after one year is only about 40 ppm, which is comparable to the silicon solubility from quartz in water[10]. This suggests that the relatively low leach rate in the granite water is due to silicon saturation effects.

Effects of Backfill Materials on Waste Glass Leaching

An acceleration of leaching was observed in the bentonite solutions at low bentonite/water ratios. Similar results were reported by other workers[2]-[4],[11] and it is thought that the enhanced leaching is due to the sorptive and ion exchange properties of the bentonite. In the present study, the filtered solution (Bentonite[F]) did not accelerate the glass leaching compared with not filtered solutions. This suggests that colloidal particles in the solutions exert an influence on the leaching. Table 6 shows a comparison of the leachates (Bentonite[C]) before and after filtrating. Some cations released from the glass such as Cs, Sr and Ca appear to be adsorbed on the bentonite in the solution. These findings support the leaching acceleration mechanism, currently discussed[12], that bentonite prevents the saturation of released cations from the waste glass. However, it is not clear from the present study whether bentonite colloids can provide a silicon sink.

These effects of bentonite appeared to be less important at high bentonite/water ratios, where leaching could be controlled by mass transport through the bentonite[13]. The enhanced leaching by bentonite was lowered in the long-term experiment (Table 5) or eliminated in the mixed bentonite systems (Fig. 2). These effects may be explained by the saturation of sorptive sites of bentonite. In the granite/bentonite system, cations such as K from granite can saturate the ion exchange sites[14] prior to leaching. Besides the saturation of sorptive sites, leaching supression is considered to be due to the silicon saturation effect for the zeolite/bentonite system shown in Fig. 4.

Zeolite water reduced the leach rate of the glass. In the presence of zeolite in the solution, however, enhanced leaching is considered to take place because of its sorptive property as well as bentonite.

Solution pH was 3.5 to 10 in all cases. PNC reference glasses provide a similar leach rate in the pH range of 4 to 10 [15]. Therefore, differences in solution pH can not explain the leaching behavior obtained here.

Table 6 Leach solution data from P0500 glass before and after filtrating (0.05 μm) for the bentonite solution(C). (All concentrations are in ppm)

	Si	B	Na	Ca	Cs	Sr	Mo
Leachate Solution	1100	18	142	15	4.2	1.74	5.2
Filtered Solution	50	17	70	0.2	0.08	0.005	5.0

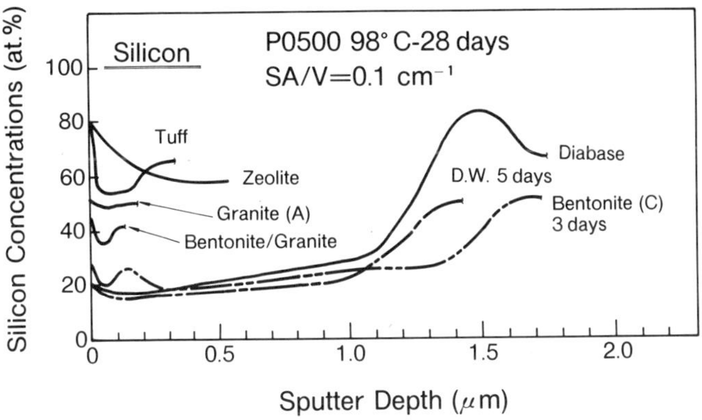

Fig. 3 Silicon depth profiles through altered layers of leached glasses for different leachant solutions, obtained by Auger electron spectroscopy. Vertical scale is normalized to 100% for detected elements except for oxygen.

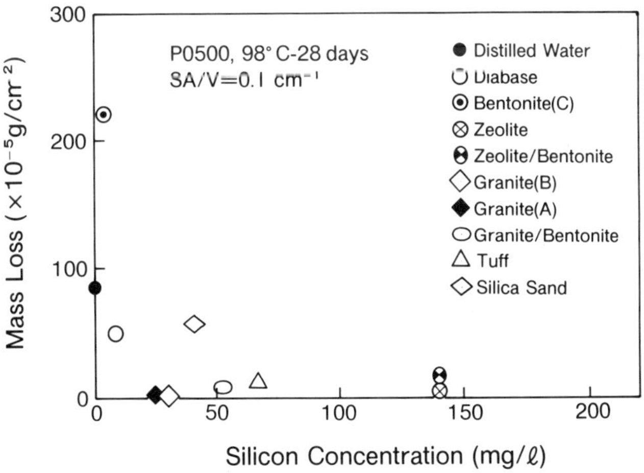

Fig. 4 Relation between mass losses and silicon concentrations in the leachant solutions. (98 °C - 28 days)

ACKNOWLEDGEMENTS

The authors gratefully aknowledge the experimental assistance of T. Hosoya for leach test and Auger analyses by H. Sonobe.

REFERENCES

1. KBS Report(1978), Karnbranslesakerhet, Stockholm, Sweden. S.C. Slate et al., PNL-3838, Pacific Northwest Laboratory, Richland, Washington.
2. L.L. Hench, L. Werme and A. Lodding, in Scientific Basis for Nuclear Waste Management V, edited by W. Lutze (Elsevier Science Publishers, New York, 1982) pp. 153-162.
3. D.E. Clark and C.A. Maurer, in Scientific Basis for Nuclear Waste Management V, edited by W. Lutze (Elsevier Science Publishers, New York, 1982) pp. 71-82.
4. H. Christensen et al., in Advances in Ceramics, Vol.8, Nuclear Waste Management, edited by G.G. Wicks and W.A. Ross, Am. Ceram. Soc., Ohio, pp. 346-357.
5. G.L. McVay and C.O. Buckwalter, PNL-SA-10474, Pacific Northwest Laboratory, Richland, Washington.
6. F. Lanza and C. Ronsecco, in Scientific Basis for Nuclear Waste Management V, edited by W. Lutze (Elsevier Science Publishers, New York, 1982) pp. 125-133.
7. R.M. Wallance and G.G. Wicks, in Scientific Basis for Nuclear Waste Management VI, edited by G.L. Brookins, (North-Holland, New York, 1983) pp. 23-28.
8. L.A. Chick and L.R. Pederson, in Scientific Basis for Nuclear Waste Management VII, edited by G.L. McVay,(North-Holland, New York, 1984) pp. 635-642.
9. G.L. McVay and C.O. Buckwalter, J. Am. Ceram. Soc. 66, 170 (1983).
10. A.E. Hughes et al., Nuclear Technology, 61, 496 (1983).
11. P.V. Iseghem et al., in Scientific Basis for Nuclear Waste Management V, edited by W. Lutze (Elsevier Science Publishers, New York, 1982) pp. 219-227.
12. J.E. Mendel, compiler, Pacific Northwest Laboratory Report PNL 5157 (1984) Richland, Washington
13. T.H. Pigford et al., in A Study of the Isolation System for Geologic Disposal of Radioactive Wastes, National Research Council, (National Academy Press, Washinton, D.C. 1983).
14. A. Inoue, Clays Clay Miner., 31, 81-91 (1983).
15. N. Kawanishi et al., PNCT831-80-01 (1980), pp. 140-150.

CRYSTALLIZATION BEHAVIOR OF A FERRI-SILICATE α-WASTE GLASS

A.D. STALIOS AND R. DE BATIST*
Materials Physics Department, S.C.K./C.E.N., B-2400 Mol, Belgium
*Also Rijksuniversitair Centrum Antwerpen, B-2020 Antwerpen, Belgium

ABSTRACT

The crystallization behavior of a ferri-silicate α-waste glass was studied by means of several experimental techniques. The main crystal phase showed a plate-like morphology and was identified as a monoclinic pyroxene. The Johnson-Mehl-Avrami equation was used for the determination of the kinetic parameters of the process. Following both isothermal and non-isothermal techniques, the Avrami exponent, n, was found to be nearly one. The activation energy for crystal growth is Eg = 356 kJ mole^{-1}. The crystallization process is governed by instantaneous nucleation and diffusion controlled two dimensional growth.

INTRODUCTION

In the past two decades a large number of papers have been published dealing with crystallization kinetics of silicon-oxide based glassy systems [1-8]. According to the basic theory of reaction kinetics in processes of nucleation and growth [9,10], two approaches have been used mainly for determining the kinetic parameters of the crystallization process.

Non-isothermal differential thermal analysis has been used by various authors [1,4-8] since it is a fast dynamic experimental method, which is considered to give reasonably accurate results. The second approach is an isothermal one and is based on the determination of the relative fraction of crystallized matter as a function of time (t) and its use for the calculation of the growth rate and the kinetic parameters. Quantitative X-ray diffraction [11,12] is one of the experimental tools used for the determination of the relative fraction of crystallized matter.

In this paper the crystallization kinetics of a base glass WG119 related to the operation of the high temperature incinerator for α-contaminated waste installed at the Waste Department of S.C.K./C.E.N. Mol has been investigated [13]. The base glass has the following composition (in w/o):

SiO_2 : 56.01; Fe_2O_3 : 22.57; Na_2O : 5.26; K_2O : 8; CaO : 4.75; MgO : 3.42

A combination of non-isothermal differential thermal analysis (D.T.A) and X-ray diffraction analysis methods was employed to investigate the kinetics of crystallization of the glass.

Scanning electron microscopy (S.E.M.) as well as energy dispersive X-ray analysis (EDXA) was used furthermore on a series of isothermally (710°C) treated specimens to determine the evolution of the composition of the crystallites as well as their growth habit in the amorphous matrix.

Very few authors [14,15] dealt till now with "basaltic glass ceramic" compositions comparable to the base glass used in this study. These authors used natural basalt which was melted to yield a black vitreous product which upon reheating recrystallized into a fine grained glass ceramic. The structure of this ceramic was identified to be magnetite and pyroxene. This pyroxene is believed to be a diopsidic augite, basically $CaMgSi_2O_6$ with some $MgSiO_3$, $CaFeSi_2O_6$ and $NaFeSi_2O_6$ components likely to be present in solid solution.

CRYSTALLIZATION KINETICS

The Johnson-Mehl-Avrami [9,10] transformation equation (1) is used for describing crystallization reactions under isothermal conditions:

$$X = 1 - \exp [-(Kt)^n] \quad (1)$$

where X is the fraction of material crystallized at the time t, n is the Avrami exponent, depending on the morphology of the crystallites. K has the dimension of a reaction rate (frequency factor) and follows an Arrhenian temperature dependence in a relatively narrow range of temperatures:

$$K^n = K_o \exp [-E_e/RT] \quad (2)$$

where Ee = mEg
Eg = activation energy for crystal growth
m = constant depending on the nucleation and growth mechanisms and on the dimensionality of the crystals (see Table I).

All methods used for interpreting D.T.A. data from a non-isothermal crystallization process are based on equations (1) and (2) and all assume a constant heating rate β [3].

In ref. [4] the following equation (3) was proposed for the calculation of Ee:

$$\ln(T.\Delta T) = \ln [f(X)] + \ln K_o n + \ln \frac{E_e}{R} - \frac{E_e}{RT} \quad (3)$$

where ΔT is the temperature difference between the transforming glass sample and the inert standard used in the D.T.A. run.

The function f(X) is given by:

$$f(X) = (1-X) [-\ln(1-X)]^{(n-1)/n} \quad (4)$$

As shown in the same ref. [4], the function f(X) may be considered a constant for $0.25 < X < 0.75$.

Table I. Values of n and m for various kinds of nucleation and growth processes [16].

	First order nucleation $m + 1 = n$		Instantaneous nucleation $m + 0 = n$	
	linear growth	diffusion control growth	linear growth	diffusion control growth
3-dimens. (spheres)	3+1 = 4	$\frac{3}{2}+1 = \frac{5}{2}$	3+0 = 3	$\frac{3}{2}+0 = \frac{3}{2}$
2-dimens. (plates)	2+1 = 3	$\frac{2}{2}+1 = 2$	2+0 = 2	$\frac{2}{2}+0 = 1$
1-dimens. (needles)	1+1 = 2	$\frac{1}{2}+1 = \frac{3}{2}$	1+0 = 1	$\frac{1}{2}+0 = \frac{1}{2}$

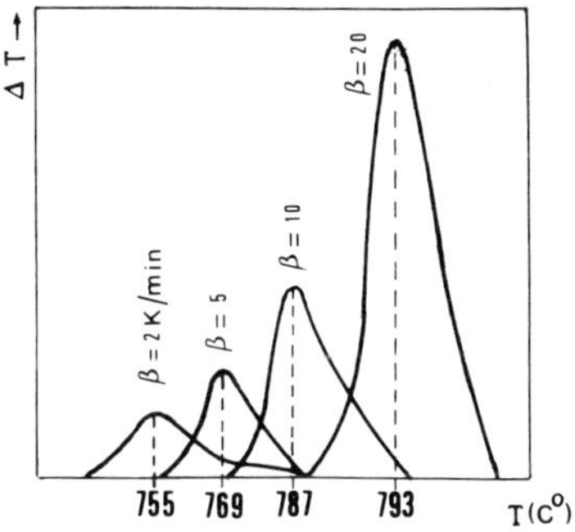

Fig. 1. Schematic representation of DTA plots for various heating rates (β).

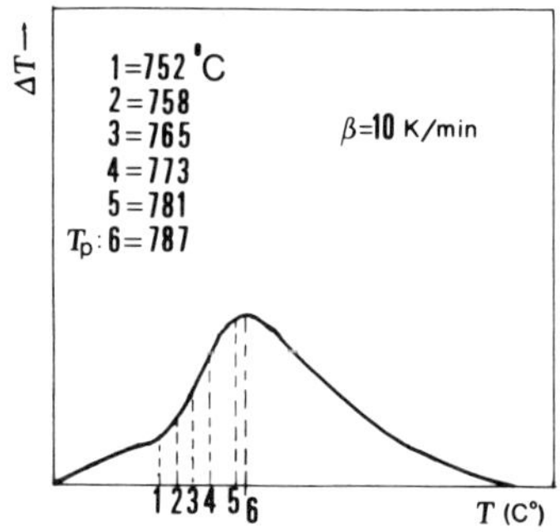

Fig. 2. Single scan D.T.A. plot for the determination of n and Ee.

In ref. [1], eq. (5) was derived for the calculation of the activation energy Ee:

$$\ln (T_p^2/\beta) = E_e/RT_p + \text{constant} \qquad (5)$$

T_p is the peak temperature of the D.T.A exotherm. In ref. [5] eq. (6) was derived, also for the calculation of E_e:

$$\ln \left(\frac{T_p - T_o}{\beta}\right) = \frac{E_e}{RT_p} - \ln K_o \qquad (6)$$

T_o is the starting temperature of the D.T.A. runs.

The results of applying the above-mentioned eqs. (1), (3), (5) and (6) will be discussed in §.IV.

EXPERIMENTAL

1. Differential thermal analysis (D.T.A.)

D.T.A. analysis was used for rapidly obtaining information concerning the kinetic parameters for crystal growth. The glass specimens were annealed for 2 h at 485°C, granulated and about 70 mg of the fraction < 278 mesh was used as sample material for D.T.A. Sapphire was used as reference. The base line was adjusted [17] by adapting the specimen holder position. The heating rates were checked and found constant in the temperature range between 100 - 800°C.

A Du Pont 990 thermal analyser was used with Pt holders. Four different heating rates were used from 2 K min^{-1} to 20 K min^{-1}. A schematic representation of the shift of the exothermal crystallization peak as a function of the heating rate (multiple scan technique) is illustrated in Fig. 1.

In Fig. 2 a single scan exotherm is illustrated taken with a heating rate of 10 K min^{-1}. The five temperatures (1-5) indicated, previous to the peak temperature $T_{p,6}$ (787°C) as well as the corresponding ΔT values (dotted lines) are used in the application of eq. (3) (see Fig. 5).

Note in Fig. 1 that the change in the shape of the exothermal crystallization peaks with changing heating rates is due to the change of the value of the Avrami exponent, n. The higher the value of n the narrower is the peak [18].

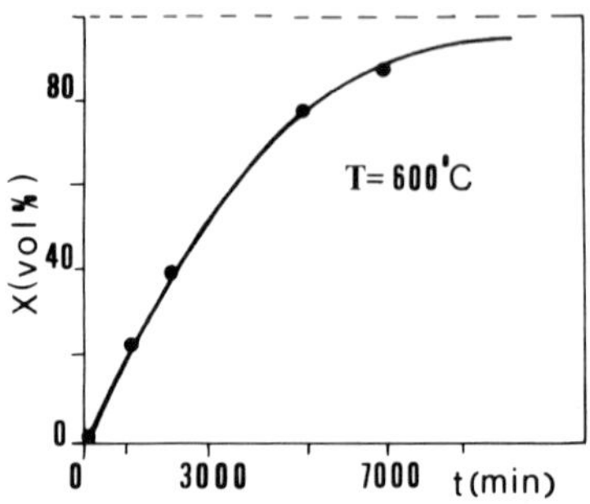

Fig.3. Percent crystallization, X v. time for WG119, 600°C.

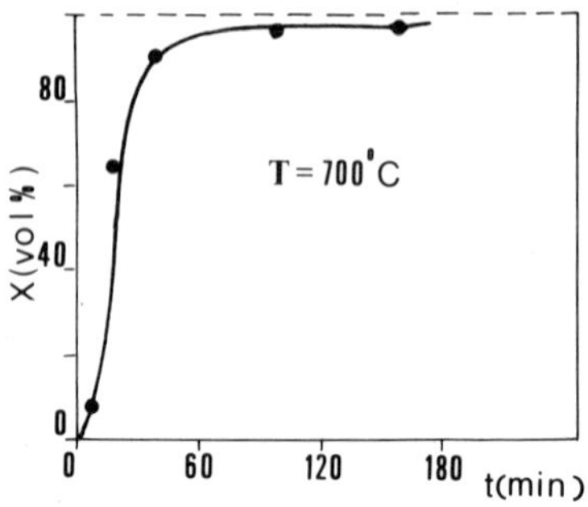

Fig.4. Percent crystallization, X v. time for WG119, 700°C.

The determination of the activation energy Ee and of the Avrami exponent, n, based on the mathematical models of the crystallization kinetics cited in §.II. will be presented in §.IV.

2. X-ray diffraction analysis

X-ray powder diffraction patterns were obtained from each of the glass specimens treated for times between 1200 min and 8400 min at 600°C and for times between 10 min and 160 min at 700°C. The uniformly granulated glass powders (< 30 μm) were mixed with a constant amount of a crystalline internal standard, and mounted in specially prepared holders. The ratio between a prominent peak of the internal crystalline fine powdered standard and the (22$\bar{1}$) peak of the main crystallizing phase (diopside: $FeCaSi_2O_6$) was used as a measure of the relative fraction of crystal, X, [11].

Figs. 3 and 4 are plots of the relative percentage of crystal versus heat treatment time at the two above-mentioned crystallization temperatures. Note that the rate of approach of full crystallinity increases strongly with temperature.

The isothermal determination of the Avrami exponent, n is also presented in §.IV.

RESULTS AND DISCUSSION

In Fig. 5 the eq. (3) is used and Ee is calculated to be 352 kJ mole^{-1}. Figs. 6 and 7 show plots of ln (T_p^2/β) and ln $[(T_p-T_0)/\beta]$ versus T^{-1}, derived from eqs. (5) and (6), respectively. From the slope of the plot in Fig. 6 the value of Ee was calculated to be 356 kJ mole^{-1} and from that of Fig. 7, Ee = 360 kJ mole^{-1}. The least squares fitting method was used for all the linear plots.

The average value of Ee is in fair agreement with the activation energy reported already in the literature [2], Ee = 415 kJ mole^{-1}. The lower value of the activation energy calculated from our results is most probably due to the higher amount of Fe_2O_3 contained in the base glass WG119 used (22.6 w/o), in comparison with ref. [2] where 10 and 13 w/o of Fe_2O_3 were used, leading to activation energies of 502 kJ mole^{-1} and 415 kJ mole^{-1}, respectively. This value of Ee = 356 kJ mole^{-1} also agrees well with the activation energy for viscous flow, En = 314 kJ mole^{-1}, determined from viscosity measurements between 800°C and 1200°C.

Augis and Bennett [1] proposed also the following eq. (6a) for the calculation of the Avrami exponent, n:

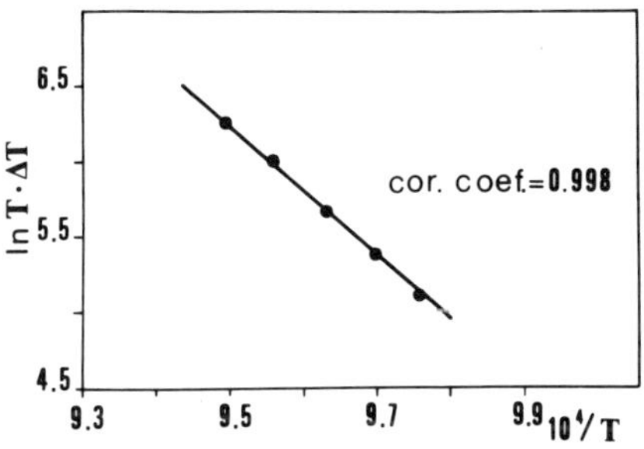

Fig. 5. Piloyan-Borchart plot for the determination of Ee.

$$n = \frac{2.5}{T_{FWHM}} \cdot \frac{T_p^2}{E_e/R} \quad (6a)$$

T_{FWHM} being the full width at half maximum of a D.T.A. exothermal crystallization peak (Fig. 2). n is found to be 1.02.

For the analysis of the isothermal experiments performed at 600°C and at 700°C, use is made of the logarithmic form of eq. (1):

$$\ln[-\ln(1-X)] = n\ln K + n\ln t \quad (7)$$

In Fig. 8 a plot of ln[-ln(1-X)] versus lnt is illustrated for the crystallization temperature of 600°C and results in a value of n = 1.0.

In Fig. 9 the same kind of plot for crystallization at 700°C is given. Due to the pronounced experimental error at small values of X a reduced weight was given to the first (X,Y) pair of the plot. This yielded a value for n of 0.9.

The value of the constant m was thus derived from Table I to be equal to n, i.e. n = m = 1; hence Ee = Eg.

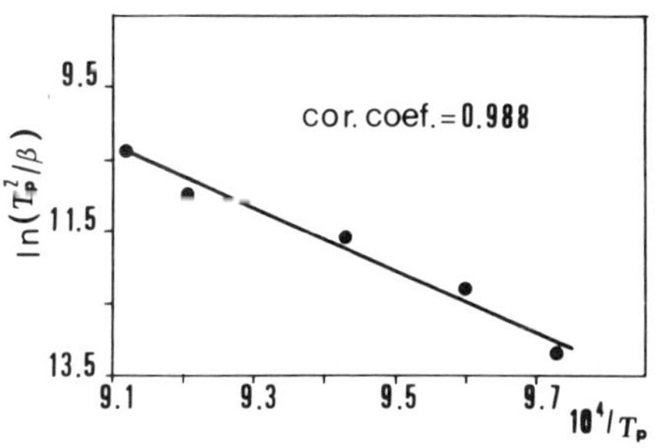

Fig 6. Chen plot for the determination of Ee.

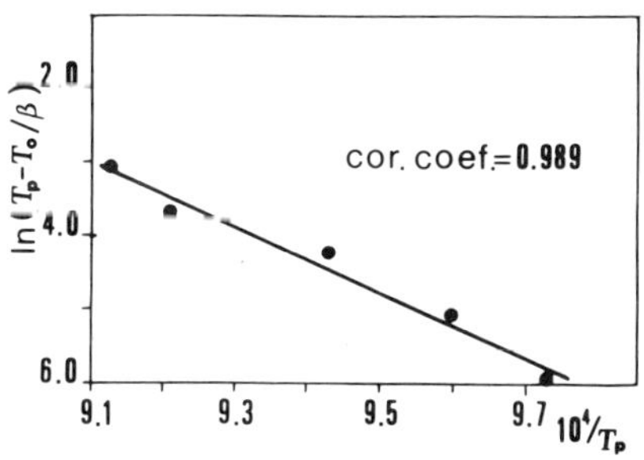

Fig 7. Augis-Bennett plot for the determination of Ee.

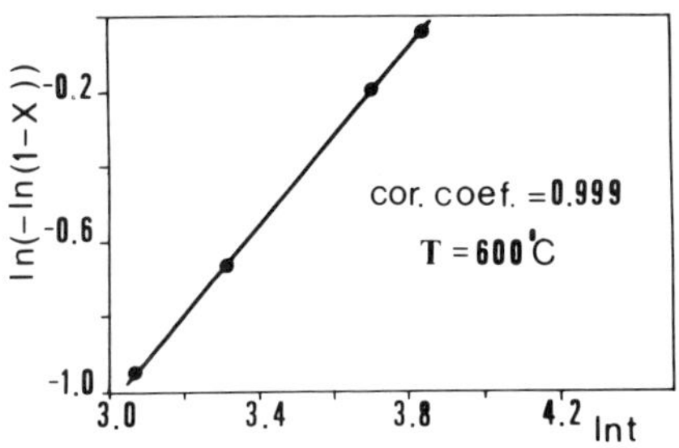

Fig 8. Determination of the Avrami exponent n at 600°C.

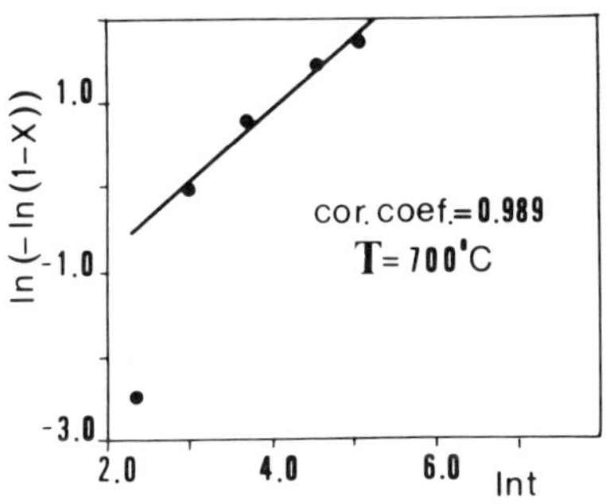

Fig. 9. Determination of the Avrami exponent n at 700°C.

BULK CRYSTALLIZATION: X-RAY DIFFRACTION, SCANNING ELECTRON MICROSCOPY (S.E.M.)

Single crystal electron diffraction and X-ray diffraction of glass specimens annealed in the temperature range of 600°C to 800°C revealed the presence of traces of $MgFe_2O_4$ spinel and of a main monoclinic crystal phase having the general pyroxenic formula:

$$X_{1-p}Y_{1+p}Si_2O_6$$

where $X = Ca^{+2}$, Na^{+1}, $Y = Mg^{+2}$, Fe^{+2}, Fe^{+3} and the sum of X and Y valences must be 4. This corresponds wih diopside ($FeCaSi_2O_6$) with some $CaMgSi_2O_6$ and $NaFeSi_2O_6$ admixtures in solid solution, see also ref. [14,15].

A series of isothermally treated specimens of WG119 (710°C) were used for (S.E.M.) observation. Heat treatment times from 20 min. to 180 min were applied. The glass specimen was first treated at 550°C for 3 h and further at 710°C.

Energy dispersive X-ray analyses (EDXA) were carried out on the glass matrix and on various crystals of each specimen.

In Fig. 10 a small magnification (625 x) of a characteristic transformed region is demonstrated, giving a more macroscopic view of the crystallization.

A higher magnification (5000 X) of the same region is shown in Fig. 11. This allows us to better observe the shape and size of the crystalline precipitates as well as their distribution in the amorphous matrix.

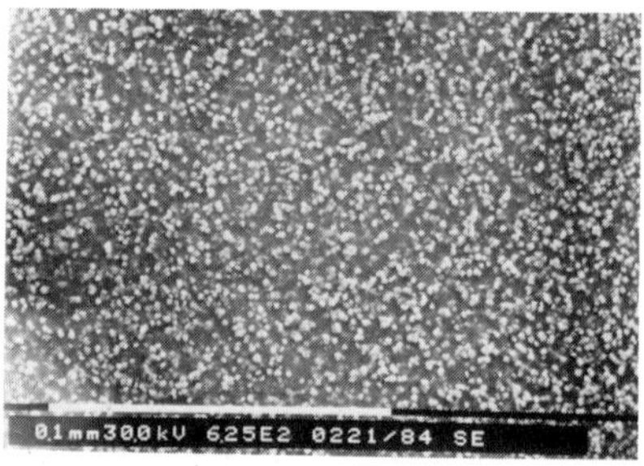

Fig. 10. Heat treatment at 710°C for 20 min. of WG119.

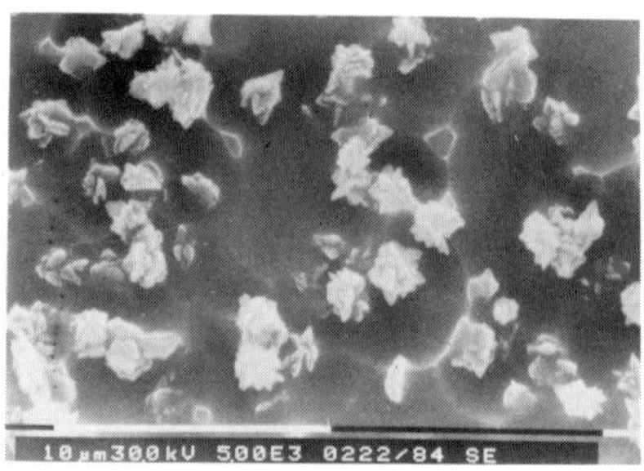

Fig. 11 Heath treatment at 710°C for 20 min. of WG119.

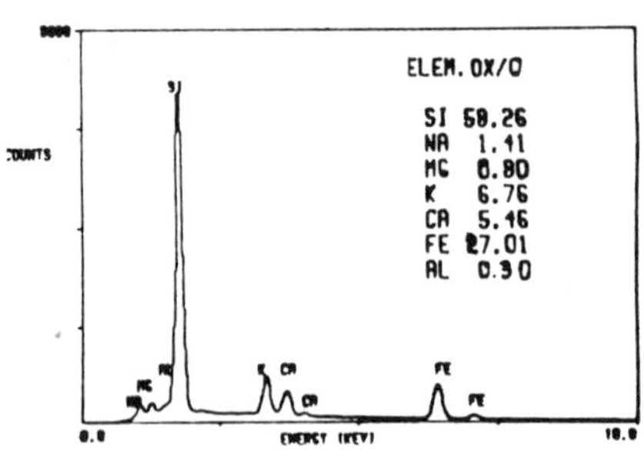

Fig. 12. EDX analysis of the glass matrix. (710°C - 20 min.).

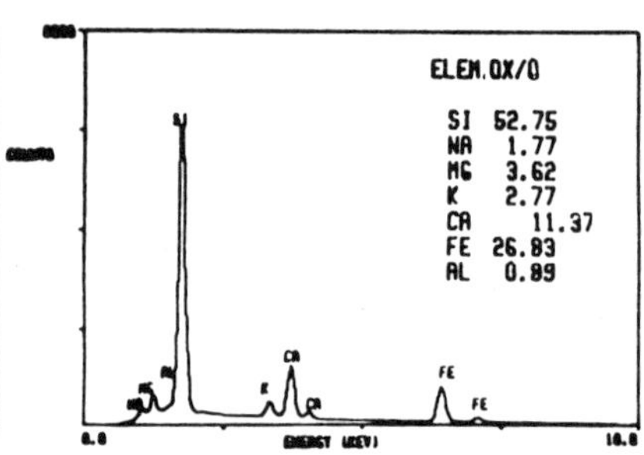

Fig.13. EDX analysis of crystals. (710°C - 20 min.).

Fig. 12 illustrates the mean of two analyses of the composition of the glass matrix in the vicinity of the precipitates shown in Fig. 11. The composition calculated from the integration of the characteristic X-ray peaks of each element is indicated as weight percentage of the oxide (ox/o).

Fig. 13 gives the mean EDXA from three different crystals of Fig. 11. The calculated composition is also indicated as ox/o.

In the two EDXA spectra of Figs. 12 and 13 one can easily observe the difference in the K/Ca ratio between the crystalline phase and the glassy matrix. Since the crystal phase has a composition different from the glassy matrix, the progressive build up of a boundary layer through which the constituents that form the precipitating phase must diffuse is to be expected [2]. Figs. 14, 15 and 16 present characteristic crystallized regions of three different WG119 glass specimens treated at 710°C during 45 min, 80 min. and 180 min respectively. The density of the crystallites increases clearly with longer heat treatments whereas the change in the dimensions is very small. Concerning the shape (morphology) of the crystals one can clearly observe the plate like form of the crystal phase which seems also to be conserved. The EDXA spectra of Figs. 17, 18 and 19, corresponding to Figs. 14, 15 and 16, respectively, show the evolution of the composition of the crystal phase (pyroxene) with increasing periods of heat treatment: a strong decrease of the K/Ca ratios whereas the other elements do not present any observable change: Fig. 17: K/Ca = 0.24, Fig. 18: K/Ca = 0.11, Fig. 19: K/Ca = 0.07. A reduction of the concentration of Ca was observed in the glass matrix, Fig. 12 (see also [14,15] for other basaltic glass compositions).

Fig. 14. Growth treatment at 710°C for 45 min. of WG119.

Fig. 15. Growth treatment at 710°C for 80 min. of WG119.

Fig. 16. Growth treatment at 710°C for 180 min. of WG119.

Fig. 17. EDX analysis of crystals. (710°C - 45 min.).

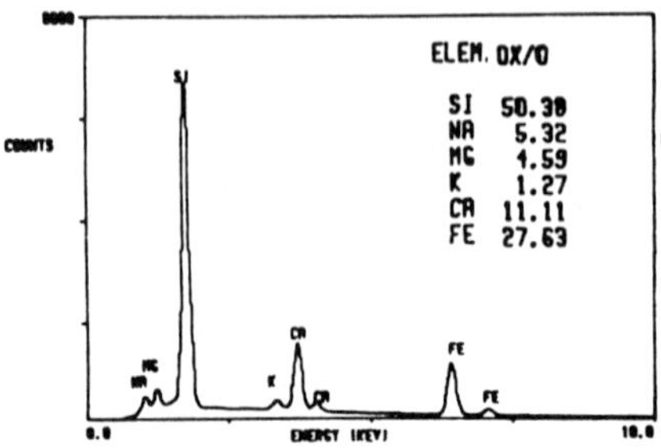

Fig. 18. EDX analysis of crystals. (710°C - 80 min.).

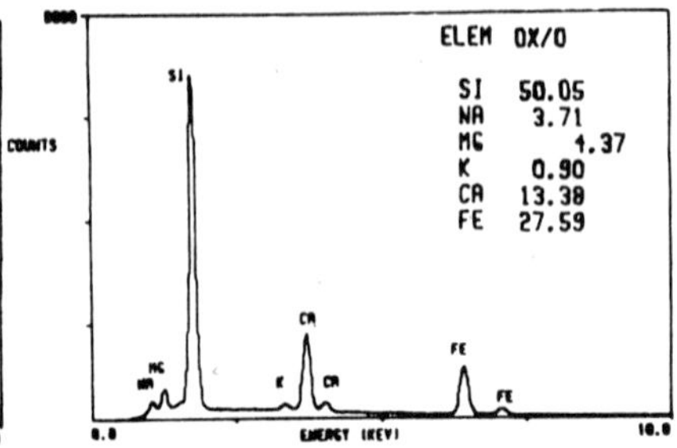

Fig. 19. EDX analysis of crystals. (710°C - 180 min.).

The observation of a plate-like morphology taken together with the fact that n=1 implies, according to Table I, that the crystallization process is governed by instantaneous nucleation and diffusion controlled two dimensional growth. Such a mechanism is also compatible with the observed changes in the alcaline earth concentration in the matrix.

ACKNOWLEDGEMENT

The authors acknowledge the contributions of W. Timmermans and L. Knaepen for the D.T.A. and SEM measurements as well as P. Van Iseghem for useful discussions.

REFERENCES

1. J.A. Augis and Bennet, J. of Therm. Analysis 13, 283 (1978).
2. P.S. Rogers, Minerological Magazine 37, n° 291, 741 (1970).
3. H. Yinnon and D.R. Uhlmann, J. of Non-Cryst. Solids 54, 253 (1983).
4. G.O. Piloyan, Nature 212, 1229 (1966).
5. H.S. Chen, J. of Non-Cryst. Solids 27, 257 (1978).
6. K. Matusita, T. Komatsu and R. Yokota, J. Mat. Science 19, 291 (1984).
7. K. Matusita and S. Sakka, Phys. Chem. of Glasses 20, 81 (1979).
8. A. Marotta, A. Buri and F. Branda, Thermochem. Acta 40, 397 (1980).
9. W.A. Johnson and R.F. Mehl, Amer. Inst. Min. Eng. 135, 416 (1939).
10. M. Avrami, J. Chem. Physics 7, 1103 (1939).
11. L.J. Shelestak, R.A. Chavez and J.D. Mackenzie, J. of Non-Cryst. Solids 27, 83 (1978).
12. V. Kouppa, Phys. Chem. of Glasses 20, n° 4, 85 (1979).
13. N. Van de Voorde et al., STI/TUB/649 2, Proceedings of an international conference on radioactive waste management, (Seattle, U.S.A., May 1983), edited by I.A.E.A., p. 347.
14. G.H. Beall and H.L. Rittler, Am. Cer. Soc. Bulletin 55, n° 6, 579 (1976).
15. A.L. Chick, R.O. Cokken and L.E. Thomas, Am. Cer. Soc. Bulletin 62, n° 4, 505 (1983).
16. L. Mandelkern, Crystallization of polymers, (Mc Graw Hill Book Company, New York, 1964), p. 215.
17. A. Geodakyan, S.V. Stepayan, Fisika i Khimiya Stekla 8, n° 5, 622 (1982) (Plenum Publishing Corporation, 1983).
18. J. Burke, La Cinétique des changements de phase dans les métaux, (Masson, Paris, 1968) p. 209.

WEATHERED BASALT GLASS: A NATURAL ANALOGUE FOR THE EFFECTS OF REACTION PROGRESS ON NUCLEAR WASTE GLASS ALTERATION

B. GRAMBOW*, M.J. JERCINOVIC**, R.C. EWING** and C.D. BYERS***
* Hahn-Meitner Institut, Glienicker Strasse 100, 1000 Berlin 39, F.R.Germany
**Department of Geology, University of New Mexico, Albuquerque, NM 87131 USA
***Argonne National Laboratory, 9700 South Cass Avenue, Argonne, IL 60439 USA

INTRODUCTION

Laboratory experiments alone cannot be used to verify models for the long-term release of radionuclides from nuclear waste glasses. Basalt glasses have been proposed as an appropriate natural analogue for the long-term validation of release models [1]. Their analogous behavior has been demonstrated in laboratory experiments in which both types of glasses display similar reaction rates, alteration products and surface layer morphologies [2,3]. This paper illustrates how empirical data from natural occurrences of basalt glass can be interpreted by a model which is developed to describe the reaction progress of the corrosion of nuclear waste form borosilicate glass [4,5].

SAMPLE DESCRIPTION

The samples include: nine from subglacial volcanoes in north-central British Columbia, Canada; three from pillow basalt sequences in Iceland; four from drill cores of the Deep Sea Drilling Project (DSDP). Table 1 summarizes important characteristics of the samples.

The British Columbia samples are from three separate basalt volcanoes which erupted beneath glacial ice during the late Pleistocene (10 to 15 thousand years ago). #A5B and #A11B are from Ash Mountain, a cone of mostly glassy fragmental deposits (hyaloclastite). #AL14 and #AL18 are from Southern Tuya, a similar but smaller volcano. #A5B and #AL14 represent low vesicularity hyaloclastite. Samples #A11B and #AL18 are from coarse-grained, highly vesicular hyaloclastite deposits. Glass shards in each sample range from tens of microns to millimeters in size. The glasses show evidence of alteration in the form of etch pits on surfaces and the formation of "palagonite", an x-ray diffraction amorphous material which replaces the glass on exposed surfaces. Well-crystallized nontronite is found as a minor constituent filling pore spaces in some samples. Samples #TB1, #TB2, #TB5 and #TBW are from Tuya Butte, a flat-topped volcano composed primarily of outward dipping hyaloclastite with a caprock of subaerially erupted basalt. All of the Tuya Butte samples are from the hyaloclastite unit. These deposits consist of moderately palagonitized, low-vesicularity glass shards (tens of microns to millimeters in size) cemented by zeolites and clays. All samples from British Columbia were altered by meteoric water (glacial melt or rain water) at relatively low temperatures.

Sample #T-2 is from an outcrop of pillow basalt in the Tungufell area of south-central Iceland, 80 km east of Reykjavik. The outcrop is part of a series of Plio-Pleistocene (.7 to 3.1 m.y.) lavas and hyaloclastites. The outcrop consists of small (half-meter sized) pillows and associated interpillow hyaloclastite. Pillow rims are characterized by centimeter thick rinds of low-vesicularity, sideromelane glass. #T-2 contains pillow rim material and interpillow hyaloclastite. Vesicles range from tens of microns to more than a millimeter in size. The glass is partially replaced by palagonite. Palagonite rinds (ranging from a few microns to several hundred microns in thickness) are found on most vesicle walls, fracture surfaces, and hyaloclastite-shard edges. Pore spaces within the interpillow hyaloclastite and most vesicles and fractures within the pillow rims are cemented by fine-grained, brown clay (Fig. 1a), some vesicles are cemented with zeolite, which forms a layer of colorless, subhedral crystals 50 to over 100 microns thick. Zeolite containing vesicles also have thin films of clay which are seen both beneath and covering the zeolite layer. The relationships between vesicle size, palagonite rind thickness and cement type are illustrated in Fig. 1b. Electron microprobe analyses of glass and palagonite of #T-2 show a relative enrichment in the alteration rind of K_2O by a factor of three, indicative of alteration by seawater.

Reyk.-1 and Reyk.-2 were collected from Reykjanes Point, 50 km southwest of Reykjavik, on the southwest coast of Reykjanes Peninsula. The site lies on the axis of Reykjanes Ridge, the northern extension of the Mid-Atlantic Ridge. Pillow basalts, erupted beneath the sea, are now exposed along wave eroded cliffs. The pillows exhibit centimeter thick, highly vesicular glass rims and ash-sized interpillow hyaloclastite and are probably late intraglacial in age (.2 to .5 m.y.). Reyk.-1 is primarily pillow basalt rim and Reyk.-2 is mostly interpillow hyaloclastite. Vesicles are numerous and are mostly a few hundred microns in size (up to 2 mm). Orange to red palagonite replaces the glass on vesicle, fracture and hyaloclastite-shard surfaces. The palagonite rinds are generally a few tens of microns thick. Clay and iron oxide cement is common with the hyaloclastite material, but occurs only rarely in vesicles. These deposits were undoubtedly altered in seawater shortly after eruption. Changes in sea level during the Pleistocene may have resulted in a complex hydrologic history, varying between seawater and meteoric water alteration.

Deep-sea samples include material from DSDP sites 409, 410A, 413 and 417D. Site 409 (#49,409,13-1) is on Reykjanes Ridge just southwest of Iceland and is a hyaloclastite cemented with fossiliferous clay. Glass shards are 50 to 2,000 microns in size and have palagonite rinds up to 200 microns thick. The basalt at site 409 is 2.4 m.y. old [6]. Sample #49,410,2-1 is located near FAMOUS [19], is 9.5 m.y. old [6], and is a hyaloclastite. The glass shards are 100 to 5000 microns in size with palagonite rinds of approximately 40 microns thick, set in a microfossil-rich, calcite cement. Some zeolites (phillipsite) are present in the vesicles. Site 413 (#49,413,2-1), also near FAMOUS, is 3.4 m.y. old [6] and is primarily calcareous, microfossil-rich sediment with scattered pieces of glass and basalt fragments (50 to 1500 microns in size). The glass is completely replaced by palagonite and zeolite. Site 417D (#51,417,41-3 and #52,417,60-6) is located south of the Bermuda Rise. These deposits are 119 m.y. in age [21]. #51,417,41-3 is hyaloclastite (mm-sized shards) with no fresh glass remaining (replaced by chlorite). #52,417,60-6 is a calcite-cemented hyaloclastite and associated pillow rim. All glass is replaced by clay. Glass shards are up to 2000 microns in size. Fresh glass has been reported in samples from site 417D [8] with limited palagonitization. Fresh glass also remains in samples of similar age from site 418A [9]. All DSDP samples are altered by seawater and lie beneath ocean sediments.

DESCRIPTION OF THE BASALT GLASS/WATER REACTION MODEL

A general glass/water reaction model [4,5] is used to calculate the long-term reaction of basalt glass with seawater or meteoric water in the natural environment. The reaction is considered to be a surface reaction. The palagonite rim is interpreted as a surface layer of sparingly soluble reaction products, and the reaction occurs between palagonite and the unleached glass. The forward rate of the reaction is limited by the desorption of activated silanol groups from the pristine glass surface. The

Table 1. Summary of observed alteration features in natural, basalt glasses used in this study. Other studies have cited additional phases [8,12,13].

Sample	Age (m.y.)	Rind thickness (um)	Alteration phases*
British Columbia			
A5B	.01 to .02	8	P
A11B	.01 to .02	2	P,N,S(?)
AL14	.01 to .02	12	P,N
AL18	.01 to .02	3	P,N
TB1	.01 to .02	50	P,S,Ph,Cb,C
TB2	.01 to .02	42	P,N,Ph,Cb
TB5	.01 to .02	50	P,N,Ph,Cb
TBW	.01 to .02	138	P,S,Ph,Cb,An,C
Iceland			
T-2	.7 to 3.1	40	P,S,Cb
Reyk.-1	.2 to .5	20	P
Reyk.-2	.2 to .5	40	P
Deep Sea Drilling Proj.			
49,409,13-1	2.4	200	P
49,410,2-1	9.5	40	P,Ph
49,413,2-1	3.4	700+	P
51,417,41-3	119	1,000+	C,Cl
52,417,60-6	119	1,000+	C,Cl

* Key to phase abbreviations: P = palagonite; N = nontronite; S = Fe-saponite; C = calcite; Cb = chabazite; Ph = phillipsite; An = analcime; Cl = chlorite.

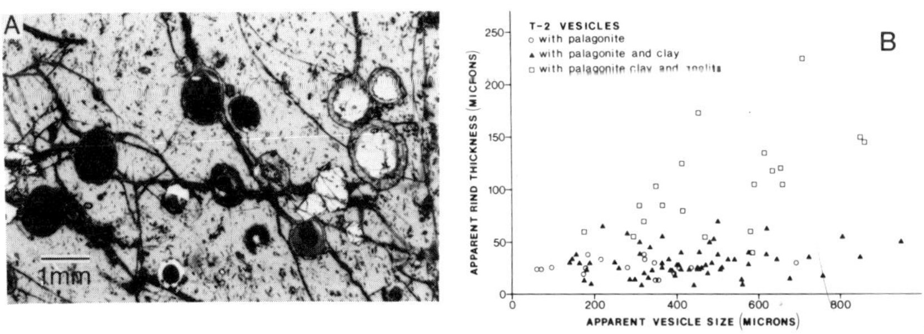

Figure 1. a) Photomicrograph of #T-2. Some vesicles are cemented with brown clay, while others contain zeolites. The orange-yellow (dark-gray in this photograph) ring in some vesicles is palagonite. The width of the field of view is approximately 9 mm. b) Relationship between apparent rind thickness and apparent vesicle size. Zeolites are associated with relatively thick palagonite rinds.

rate of reaction becomes considerably lower than the forward rate when due to the accumulation of dissolved silica in solution or high initial silica concentrations in natural waters, the glass surface becomes stabilized by the condensation of surface silanol groups. At saturated or supersaturated conditions, the reaction does not stop, but is controlled by a constant final rate of reaction, governed by the formation of alteration products.

The rate of glass dissolution critically depends on all reactions in the system which consume or produce soluble silica. Reactions considered for modelling basalt glass/water reactions are the hydrolysis of orthosilicic acid at high pH and the formation of clays, zeolites and SiO_2. For the case of the reaction of seawater with basalt the following additional controls on silica concentration were considered: (1) an initial seawater concentration of 0 to 300 uM of silica and (2) dissolution of biogenic silica (i.e. radiolaria in deep sea sediments). With the help of the computer code PHREEQE [10] the reaction path is calculated for an idealized closed system assuming congruent glass dissolution, precipitation of solid alteration products, and accumulation of soluble glass constituents in solution. This reaction path represents an array of steady state open systems. The steady state phase assemblage is governed by the ratio of the reaction rate to the flow rate of water. Potential, precipitated phases for the purpose of these calculations are taken from the modelling of the experiments on synthetic basalt glass [11]. Experimentally observed alteration products such as hydrotalcite, pyroaurite and illite are excluded, as they are not typical of alteration products of naturally altered basalt glass. The model is used to interpret the observed assemblages of alteration products of basalt glass and palagonite rind thickness as a function of age and environment.

APPLICATION OF THE MODEL

Alteration Products: A comparison of empirical observations and calculated phases

The calculated sequence of phase formation as a function of reaction progress and starting solution composition is shown in Fig. 2. The "phase" SiO_2-glass refers to the saturation concentration with respect to the glass surface. The calculated redissolution of initially formed phases is a result of the formation of more stable phases (nontronite or ferrihydrite) or the consumption of a major reactant (Si in the case of phillipsite in seawater at 8 mg/l initial silica). In general, the phases that are calculated to form as a result of solubility limits are actually observed in the natural samples. The observations support the calculations which indicate a similarity in the clay phases and zeolites that form in either seawater or fresh water reactions (see Table 1 and Fig. 2). The only major difference in behavior between the two environments is the calculated absence of Mg-rich saponite in samples altered in fresh water. This is consistent with the observation that only Fe-saponite was observed in fresh-water samples from British Columbia, while Mg- and Fe-saponite form in the marine environment [12].

In order to compare the calculated sequence of precipitated phases with those observed, we approximate the composition of palagonite with that of nontronite (Nontronite: $Si_{3.84}Al_{.94}Fe_{.9}Mg_{.33}O_{10}(OH)_2Ca_{0.1}Na_{0.05}K_{0.2}$; Palagonite: (the formula is based on #T-2) $Si_{3.46}Al_{.93}Fe_{1.15}Ti_{.14}Mg_{.30}O_{10}(OH)_2Ca_{.34}Na_{.05}K_{.02}$). No thermodynamic consideration is given to the fact that the palagonite is x-ray diffraction amorphous. Thus, the calculated sequence is: nontronite-saponite-phillipsite-chabazite, and it agrees with the observed sequence: palagonite-smectite(saponite)-phillipsite-gyrolite-chabazite-analcime as reported in the literature [13] and in part observed in these samples. There are, however, discrepancies. The calculations predict that in reactions with seawater, Mg-saponite forms after Fe-saponite. In contrast, it is reported [12,13] that the Fe-saponite is a late-forming product. The sequence of clay mineral formation depends on the Fe concentration in

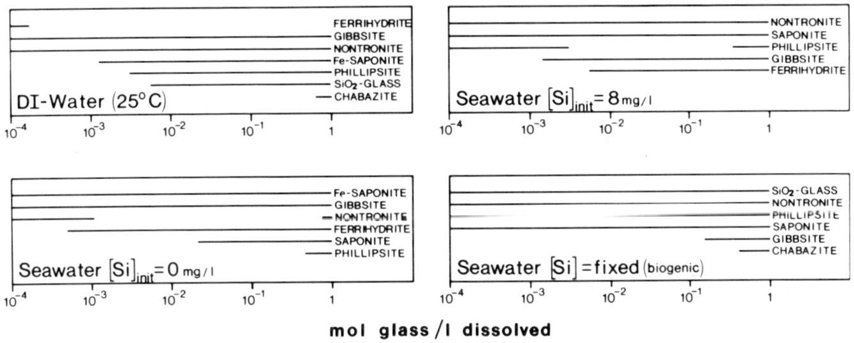

Figure 2. Calculated sequence of phase formation as a function of reaction progress in a closed system.

solution. Additional calculations show that if less soluble lepidocrosite is used instead of ferrihydrite to control the Fe concentration, the observed sequence of precipitation can be modelled.

According to the calculations, zeolites (phillipsite followed by chabazite) should form when large amounts of glass are dissolved in small solution volumes (i.e. low flow rates or static conditions). Chabazite forms only at relatively high values of pH, which are only achieved in seawater when most of the Mg is used in the formation of clays. Zeolites are present in the vesicles which are connected by the smallest cracks to potential water sources (Fig. 1a). Similarly, the zeolite rich samples from Tuya Butte, British Columbia, are well-cemented (restricted flow rate) and the residence time of any given volume of water is higher than that of more porous samples of the same age. Additionally, the calculations show that a critical concentration of silica in solution is necessary for the formation of phillipsite (see ref. 11]. At low initial concentrations of silica, phillipsite forms at relatively high reaction progress, but in the case of silica concentrations controlled by the presence of siliceous tests of radiolaria, the phillipsite forms much earlier. The early formation of phillipsite indicates that it is in equilibrium with the surrounding solutions. In the case of glass dissolution in the presence of a sedimentary column on the deep ocean floor, abundant biogenic silica is available and phillipsite will form at low reaction progress. The phillipsite is maintained throughout the reaction progress of basalt glass dissolution at the expense of the dissolution of biogenic silica. This explains the observed positive correlation of the dissolution of biogenic silica with the occurrence of phillipsite [14]. Observations suggest that phillipsite continues to grow for a million years with no observed dissolution (etching) [13,14]. In contrast, in shallow ocean water the silica concentration is lower [15]; phillipsite does not precipitate or, once formed, is redissolved as evidenced by etching on crystal faces [14].

Supersaturated Phases

The present calculations are based on a model that does not consider kinetically controlled non-equilibrium precipitation of phases from supersaturated solutions. Therefore, for long-term extrapolations, it must be determined whether the phases used in the calculations are thermodynamically stable or absent because of kinetic barriers. As an example, the calculations for basalt glass/seawater reactions indicate the precipitation of chlorite and celadonite (at the expense of saponite). However, observations show that chlorite is rare as a low temperature alteration product of basalt glass. In nature, saponite forms first and celadonite forms (if at all) much later [12]. This illustrates that celadonite formation is kinetically hindered and that supersaturation with respect to celadonite may persist for millions of years (although there is some evidence to suggest a saponite-celadonite transition [12]). Thus, at low temperatures, it is possible to exclude celadonite and chlorite from the list of potential alteration products used in the computational simulation of solution concentrations which are controlled by precipitation reactions.

Rates of Reaction: As inferred from palagonite thickness and age

The long-term prediction of the degree of alteration of nuclear waste form glasses clearly requires information on the final corrosion rate, k, [4]. The degree of alteration (palagonite rind thickness as a function of age) provides a means to extract this type of information from naturally occurring glasses. Various authors [8,16,17,18] have used the palagonite rind thickness/age as a means to determine the time dependence of the corrosion reaction, but no consistent picture of the effect of environment on the reaction rate for basalt glass has emerged. Nevertheless, it is generally agreed that the reaction of basalt glass in seawater is a low temperature reaction that continues as long as circulation of water continues. Table 2 provides a summary of data from the literature on the "apparent" reaction rates.

Table 2. Summary of "apparent" reaction rates of basalt glass in aqueous environments at approximately 3°C, as calculated from palagonite rind thickness.

Rate (um/1000yr)	Environment	Age (m.y.)	Reference or Sample #
3	seawater/open ocean	<.1	[17]
20	seawater/open ocean	0.01-0.001	[19]
0.10	seawater/sediment cover,DSDP 409	2.4	#49,409,13-1*
0.004	seawater/sediment cover,DSDP 410	9.5	#49,410,2-1*
>0.2	seawater/sediment cover,DSDP 413	3.4	#49,413,2-1*
>0.007	seawater/sediment cover,DSDP 417D	119	#51,417,41-3* #52,417,60-6*
no glass	seawater/sediment cover,DSDP 417A	119	[8]
0.003	seawater/sediment cover,DSDP 417D	119	[8]
0.15	seawater/sediment cover,DSDP 332	1	[16]
0.15	seawater/sediment cover,DSDP 395	2	[16]
0.04-0.2	seawater/shallow marine	.2-.5	Reyk.-1-2*
0.01-0.06	seawater(?)/mixed, subaerial	.7-3.1	T-2*
100	seawater/laboratory exper. (pH8.8)	.000001	[11]
0.1-1.2	fresh water/cemented	0.01-0.02	A5B*,A11B* AL14*,AL18*
2-14	fresh water/highly cemented	0.01-0.02	TB1-5*,TBW*

*Samples used in this study.

The striking result of the compilation in Table 2 is that there is not a common or consistent rate of reaction for basalt glass corrosion. In seawater/glass reactions the apparent rate may vary by five orders of magnitude. Glass samples may be preserved for 100 million years, or they may be completely altered in a few million years. In the following discussion, we attempt to reconcile this large variation in reaction rate by making a distinction between the forward rate, k+, and final rate of reaction, k, [4] in the various geologic environments. The data in Table 2 are plotted in Figure 3 to illustrate: (1) the range of values determined for the forward rate (3-20 um/1,000 yrs) and (2) the value for a final rate (0.1 um/1,000 yrs) at Si-saturation. Note, that the difference in the silica content of the solution allows one to determine whether the reaction is proceeding at the forward or final rate.

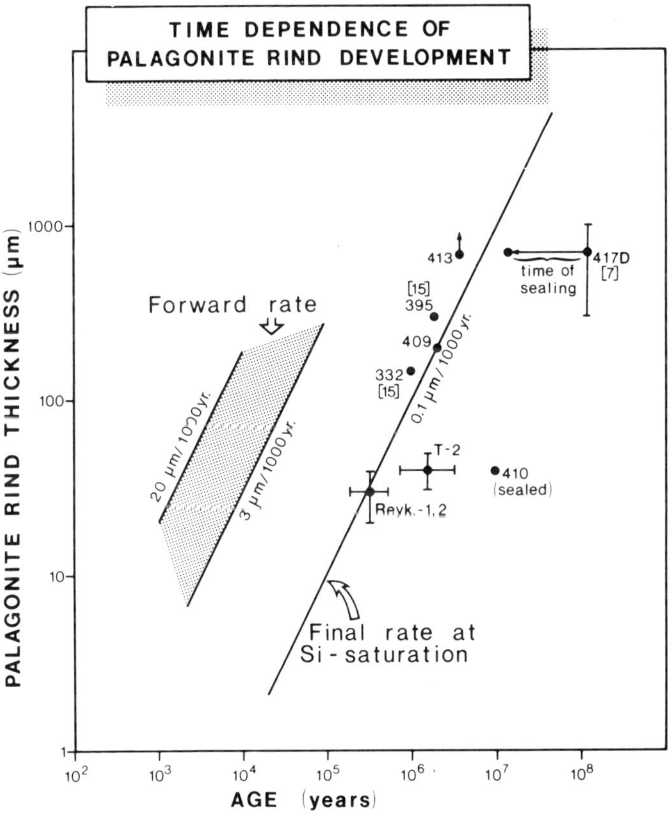

Figure 3. "Apparent" rates of palagonite rind formation (taken from Table 2) are plotted to indicate their relationship to the forward rate and final rates of reaction.

Alteration in Seawater

The Si concentration in open seawater varies between 0.4 and 4 mg/1, whereas the Si-content of interstitial waters of deep-sea sediments is in the range of 5-25 mg/l [20]. In addition, ocean sediments usually contain biogenic silica as a potential source for silica consuming reactions. The generally lower reaction rate of basalt glass in oceanic basalts covered by sediments as compared to the higher rate of relatively young dredge samples on the open-ocean floor is probably a result of this difference in silica concentration. The rate for dredge samples (sediment absent; pH=8.2) of 3-20 um/1000 years (Table 2, Fig. 3) is interpreted as the forward rate of reaction. Higher rates measured in synthetic basalt/seawater laboratory experiments may result from the higher pH (8.8). In contrast, in the presence of sediments with silica-saturated, interstitial water, palagonite growth may be described by the final rate of reaction and is calculated within the narrow range of 0.02-0.2 um/1000 years. The apparently low rate of the 119 million year old samples in DSDP 417D and the 9.5 million year old sample in DSDP 410 require special comment. The rates become higher for 417D if we only consider the reaction with seawater to have occurred during the first 13±5 million years of the life of the sample. This is based on dating of the alteration products [21] and the suggestion that it only requires approximately 10 m.y. to seal oceanic crust with circulating water [21,22]. The resulting calculated rate is 0.02-0.06 um/1000 years. In the case of the sample at DSDP 417A [8], complete alteration has occurred because the site is a topologic high and was free of sediments for about 20 m.y. [23]. For DSDP 410, the sample is enclosed in an undated, calcite matrix which has sealed the sample; therefore, the rate may be much higher than calculated (Table 2, Fig. 3).

The rates of reaction of the basalt hyaloclastites in Iceland have a similar final rate of reaction as that found for the sedimentary environment at the ocean floor. This is indicative of silica saturated conditions, which is reasonable in light of the fact that the samples are well-cemented and flow rates would be low. The interpretation of a constant rate controlling the reaction is supported by the constant layer thicknesses observed in vesicles (zeolites absent) of all sizes in the sample (Fig. 1b).

Thus, the same final rate of reaction may apply to samples as young as 200,000 years and as old as 13 million years. The reaction does not stop, as long as the sample is not sealed from water.

Alteration in Glacial Melt or Meteoric Water

For samples altered in a fresh water environment, it is more difficult to assess the conditions of water contact (the water composition and the time of contact are not known explicitly). Nevertheless, the alteration products can be used to estimate the alteration environment, extent of reaction progress or flow regime (Fig. 2). The formation of chabazite occurs only at high reaction progress (almost stagnant conditions), whereas silica saturation is likely to occur already at much higher flow rates. Therefore, the occurrence of chabazite in the Icelandic and Tuya Butte samples can be used as an indication for silica-saturated conditions. On the other hand, the apparent rate (as calculated from palagonite rind thickness) is even higher for vesicles with zeolites, than for other vesicles in the same sample (Iceland) or for different samples in the same type of environment (British Columbia). This suggests that silica-saturation probably governs the reaction behavior of the non-zeolitized samples, but the extent of reaction progress is not great enough to cause the formation of chabazite. The higher reaction rate (thicker rinds) is probably caused by the higher pH, a result of higher reaction progress (as inferred from the presence of chabazite). This interpretation illustrates the paradoxical effect of permeability on reaction rate. Because of the higher pH, glass in a low permeability environment may be more altered than glass in a highly permeable environment.

The interpretation of a rate of 0.1-1.2 um/1000 years as a final rate of reaction in fresh water can only be true if the flow rate of water within the sample is not higher than a limiting value. This value may be calculated as follows: According to Fig. 2, the steady-state "concentration" of glass per unit volume (reaction progress) must be higher than 6 x 10^{-3} mol/l (NL x S/V = 360 g/m^3 in order to maintain silica saturation and about a factor of 100 higher than this value for the formation of chabazite). NL = weight of glass reacted per surface area; S/V = sample surface area to solution volume ratio. With a rate of 0.5 um/1000 years, NL/t = 3.8 x 10^{-6} g/m^2/d. Hence, the maximum flow rate is 1 x 10^{-8} m/d. In the zeolitized region of the samples, the flow rate is at least a factor of 100 lower. For a vesicle of 200 um diameter (without zeolites), this corresponds to a residence time for water of at least 6 years, as compared to 600 years for zeolite containing vesicles.

IMPLICATIONS FOR THE PREDICTION OF NUCLEAR WASTE GLASS LONG-TERM BEHAVIOR

The most important conclusion from this work is that the glass/water reaction model [4,5,24] can used to model the long-term reaction of basalt glass with water in the natural environment. This suggests that the application of a similar approach to the evaluation of the long-term behavior of borosilicate glasses will be successful.

Specific implications include:

1. There are similarities and differences in the phases that form in nature as compared to those observed in experiments (either on synthetic basalt glass or nuclear waste form borosilicate glass). For example, saponite is formed both in experiments and nature. Although saponite is metastable with respect to celadonite, it persists for millions of years and can therefore be used for modelling with fair confidence. This emphasizes how important the formation of metastable phases (and the absence of thermodynamically stable phases) can be in controlling solution compositions. Some phases that form in experiments, such as illite, are however, absent in nature, indicating that more stable phases are formed or that the environment is different. More studies are required to provide a consistent database which either empirically or theoretically considers the kinetics of precipitation.

2. Depending on environment, the rate may vary by five orders of magnitude. Even in a saturated environment the glass/water reaction does not cease. A comparison of basalt glass in sediment covered ocean floor (Si-saturated conditions) with open-ocean dredge samples (non-saturated conditions) shows that the rate of alteration is a factor of 100 lower under silica-saturated conditions. A similar relative decrease from the initial to the final rate has been observed for nuclear waste glass [4,25]. The final rate of reaction for nuclear waste glasses is higher (300 um/1000 years, SRL 131 at 70°C <u>vs</u>. 0.1 um/1000 years, basalt glass at 3°C) due to differences in glass composition and the temperature of reaction.

3. Basalt glasses confirm the effect of fracturing on nuclear waste glasses. The surfaces of the smallest cracks are altered to the same degree as that observed at more exposed surfaces. The degree of alteration is proportional to the total surface area.

4. Under certain conditions, basalt glasses can be preserved for periods of time up to hundreds of millions of years, while at other conditions the glass may be completely altered much earlier. The long-term stability of basalt glass is related to low temperatures and effective sealing of the glass against circulating water. Both of these factors should be an important part of any repository design.

ACKNOWLEDGMENTS

This work was supported as part of the JSS Project (B. Grambow) and by the U.S. Nuclear Regulatory Commission, through Argonne National Laboratory, contract No. 60-84-254, FIN No. A2254-4 (R. Ewing). The authors are grateful for samples provided by the U.S. Ocean Drilling Project.

REFERENCES

1. G. Malow, W. Lutze and R.C. Ewing, J. of Non-Crystalline Solids, 67, 305 (1984).
2. C.D. Byers, M.J. Jercinovic, R.C. Ewing and K. Keil, in Scientific Basis for Nuclear Waste Management VIII, edited by C.M. Jantzen, J.A. Stone and R.C. Ewing (Materials Research Society, Pittsburgh, 1985), p. 583.
3. W. Lutze, G. Malow,R.C. Ewing, M.J. Jercinovic and K. Keil, Nature, 314, 252 (1985).
4. B. Grambow, in Scientific Basis for Nuclear Waste Management VIII, edited by C.M. Jantzen, J.A. Stone and R.C. Ewing (Materials Research Society, Pittsburgh, 1985) pp. 15-27.
5. B. Grambow, Ph.D. thesis, Freie Universitat, Berlin, 1984.
6. B.D. Luyendyk, J.R. Cann, et al., Initial Reports of the Deep Sea Drilling Project, 49 (1978).
7. T. Donnelly, J. Francheteau, W. Bryan, P. Robinson, M. Flower, M. Salisbury, et al., Initial Reports of the Deep Sea Drilling Project, 51, 52, 53 (1979).
8. T. Juteau, Y. Noak, H. Whitechurch and C. Courtois, Initial Reports of the Deep Sea Drilling Project, 51, 52, 53 (1979).
9. H. Staudigel and S.R. Hart, Geochimica et Cosmochimica Acta, 47, 337 (1983).
10. D.L. Parkhurst, D.C. Thorstensen and L.N. Plummer, Water-Resources Investigations 80-96 (U.S. Geological Survey, Reston, Virginia, 1980). Extension to run precipitation controlled reactions by P. Offermann, HMI; updated with MINTEQ data base by W. Howden and K. Kruppka, Battelle PNL.
11. J.L. Crovisier, B. Fritz, B. Grambow and J.P. Eberhart, this volume.
12. R.G. Pritchard, J.R. Cann and D.A. Wood, Initial Reports of the Deep Sea Drilling Project, 49, 709 (1978).
13. J. Honnorez, in Natural Zeolites: Occurrence, Properties, Use, edited by L.B. Sand and F.A. Mumpton (Pergamon Press, Oxford, 1978) pp. 245-258.
14. M. Kastner, in The Sea, Volume 7, The Oceanic Lithosphere, edited by C. Emiliani, (John Wiley and Sons, New York, 1981) pp. 915-980.
15. R. Wollast, in the The Sea, Volume 5, edited by E.D. Goldberg (John Wiley and Sons, New York, 1976) pp. 359-292.
16. Y. Noak, Sciences Geologiques (Universite Louis Pasteur de Strasbourg Institut de Geologie, 1983) 36, 111-116.
17. H. Hekinian and M. Hoffert, Marine Geology, 19, 91 (1975).
18. M. Morgenstein and Th.J. Riley, Asian Perspectives, 27, 145 (1974).
19. W.B. Bryan and J.G. Moore, Geological Society of America Bulletin, 88, 556 (1977).
20. J.M. Gieskes, H. Reese et al., Initial Reports of the Deep Sea Drilling Project, 51, 52, 53, 747 (1979).
21. H. Staudigel and S.R. Hart, Initial Reports of the Deep Sea Drilling Project, 83 (1985).
22. T.W. Donnelly, R.A. Pritchard, R. Emmermann and H. Puchelt, Initial Reports of the Deep Sea Drilling Project, 51, 52, 53 (1979).
23. T.W. Donnelly, G. Thompson and M.H. Salisbury, Initial Reports of the Deep Sea Drilling Project, 51, 52, 53 (1979).
24. B. Grambow, W. Lutze, R. Ewing, Nature, submitted.
25. E. Freude, B. Grambow, W. Lutze, H. Rabe and R.C. Ewing, in Scientific Basis for Nuclear Waste Management VIII, edited by C.M. Jantzen, J.A. Stone and R.C. Ewing (Materials Research Society, Pittsburgh, 1985), p. 99.

DISSOLUTION OF BASALTIC GLASS IN SEAWATER : EXPERIMENTS AND THERMODYNAMIC MODELLING

J.L. CROVISIER,* B. FRITZ,** B. GRAMBOW, *** and J.P. EBERHART *

* Laboratoire de Cristallographie - Université Louis Pasteur
1 rue Blessig, 67084 Strasbourg (France).

** Centre de Sédimentologie et de Géochimie de la Surface (CNRS)
1 rue Blessig, 67084 Strasbourg (France)

*** Hahn-Meitner-Institut für Kernforschung - Glienickerstr.100
1000 Berlin 39 (FRG).

INTRODUCTION

The reaction of basalt glasses with natural waters were proposed as analog systems to the long-term alteration of nuclear waste glasses in the geosphere [1].

Recent application of geochemical models seems promising in order to interpret experimental data for leaching of nuclear waste glasses in aqueous environments [2]. However, the thermodynamic data base of such models usually containes idealized mineral compositions, while various compositions (phyllosilicates, zeolites etc.) are observed both in nature and in laboratory experiments. Moreover, due to the presence of metastable phases, it is an open question, wether it is possible to use such calculations also for long-term predictions of the behavior of nuclear waste glasses in an actual repository.

In this paper we study the formation of alteration products from seawater/-basalt glass laboratory experiments as an analog system to nuclear waste glass leaching. We used solid solution models [3,12,13] to calculate the stability constants for the experimentally determined alteration products as required for the geochemical reaction path model. Comparative calculations were performed both for the experiments and the naturally altered deep sea basalt glasses.

EXPERIMENTAL AND MATERIAL

Static leaching tests were performed using an artificial tholeiitic glass, artificial and natural seawater. The composition of glass is : SiO_2 (50.76), Al_2O_3-(13.62), Fe_2O_3(4.08), FeO(8.18), CaO(10.72), MgO(7.35), Na_2O(2.48), K_2O(0.18), TiO_2(1.66), P_2O_5(0.10) and MnO(0.16) in weight %.

The compositions of seawater are given in Table 3. The tests were performed with Surface/Volume ratio=7m^{-1} at 60, 50, 25 and 3°C during periods of 1 to 600 days. The experimental procedure is given in CROVISIER et al. [4].

EXPERIMENTAL RESULTS

A) <u>Observed alteration products.</u>

The glass surface chemistry was followed by photoelectron spectroscopy and the minerals formed on the surface were determined by electron microscopy [4,5].

The sequence of phases formed at 50°C on the surface of the glass is shown on Fig.1.

1) Formation of an iron-rich layer, tentatively identified as goethite (FeOOH).
2) Formation of hydrotalcite crystals ($Mg_6Al_2CO_3(OH)_{16}4H_2O$), forming a one micrometer thick layer after 10 days.
3) After 120 days, hydrotalcite is covered by an amorphous product containing silica, magnesium and aluminium. A fibrous layer of Mg-Fe hydroxides develops between pristine glass and hydrotalcite crystals.
4) After 600 days the chemical composition of the external part of the layer is close to saponite (Mg-Al phyllosilicate), while hydrotalcite crystals begin to be destroyed. A poorly crystallized phyllosilicate appears in the fibrous layer, whose composition is : MgO(20.5), Al_2O_3(25.5), SiO_2(19.7), CaO(0.7) and Fe_2O_3(33.6) in weight %.

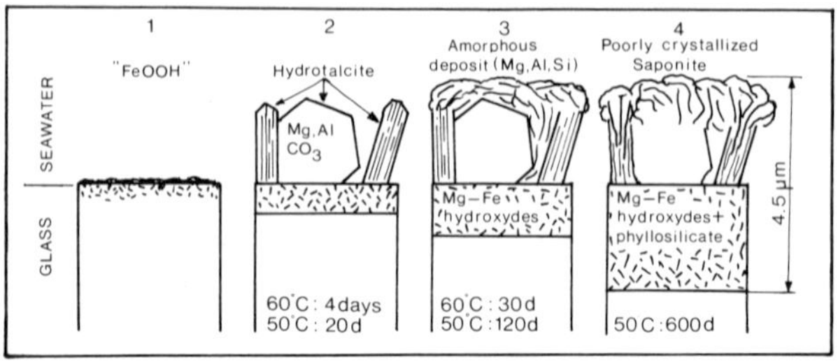

Fig.1 After CROVISIER et al. [6] Successives steps for the formation of the alteration layer at 50°C.

At 60°C the same hydrotalcite forms. The first crystals were identified after 4 days. Further evolution of the alteration layer is identical.
At 25°C the first observed phase is an amorphous iron hydroxide. Later (240 days), Pyroaurite ($Mg_6Fe_2CO_3(OH)_{16}4H_2O$) was observed in some experiments while in others (depending on pH, see CROVISIER et al. [5]) a poorly crystallized Fe-rich saponite was formed.
At 3°C an iron-rich layer forms on the surface during the first months. After 600 days illite (K-Al phyllosilicate) was identified, forming a 0.1 micrometer thick layer on the surface.

B) Analysis of leaching mechanism.

Observation of ultrathin sections by electron microscopy shows a sharp interface between the pristine glass and the leached surface layer. No chemical gradient was noted in the external part of the glass which strongly suggests a reaction mechanism governed by a congruent dissolution of the glassy matrix and subsequent precipitation (or immobilization) of sparingly soluble elements within the surface layer (CROVISIER et al. [7]). Confirmation of this mechanism, by studying the relative release of all glass constituants into solution, was not possible, because respective solution data were not available due to the high ionic concentrations of the major constituents in the initial seawater and the low S/V ratio during experiments. However, the linear dependance of the release of silica on time at 60°C (Fig.2a), indicates a breackdown of the glassy network during the early stages of interaction. This supports an interpretation of the leaching process as controlled by the dissolution rate of the glassy network. Selective alkali extraction (if any) plays only a minor role in the overall mass balance. This reaction mechanism for a basaltic glass is similar to that identified for the reaction of nuclear waste glasses with aqueous solutions [8,9,10]. For the latter glasses it was found, that initially the rate is constant with time, while for longer times of reaction in static leach tests, the rate decreases with time by saturation of silica in solution. It is the purpose of the following section to identify the time dependance of the dissolution rate at various temperatures.

C) Dependance of leaching on time.

The interpretation of leach data of basaltic glass/seawater reaction in terms of its time dependance has to face certain experimental difficulties. In solution, only Si and Al accumulation could be used to measure the reaction progress, as the more soluble elements Na, Ca, K, and Mg are already in excess in the seawater. However, both Si and Al may be incorporated into secondary alteration products. Therefore, at 50 and 90°C additional determinations of reaction progress was performed by measuring the surface layer thickness and the mass loss of the samples after removing the surface layer completely by mechanical means. Table 1 contains the results of these experiments. At 50°C fair agreement is obtained between the Si - data and the surface layer thickness and also at 72 days with the mass loss data. In contrast, at 90°C the accumulated amount of Si in solution is much lower than expected from the reaction progress determined from surface layer thickness or mass loss data. This is likely a result of the formation of secondary alteration products. Hence, at 60, 50°C and possibly also at lower temperature, the silica concentrations may be used in short time experiments as an indicator for the degree of matrix dissolution, while at higher temperature this is not possible.

	50°C		90°C	
	28d	72d	28d	72d
Si	1.8	2.5	0.2	0.2
Al	-	1.3	-	1.4
Na	-	-	-	-
Ca	-	-	-	-
Mg	-	-	-	-
Layer thick.	1.5	6.0	3.0	3.5
Mass loss	-	1.8	1.1	3.0

Table 1. Results from MCC-1 (MCC 1981) static leach tests performed in seawater, expressed as normalized mass loss (G/m2).

Fig.2a-d shows the time dependence of silica release at 60, 50, 25 and 3°C. Despite of the scatter of data, early time data at 50°C and all data at 60, 25 and 3°C were interpreted as controlled by the forward rate of reaction in terms of a linear time law. The resulting rates were plotted in an Arrhenius diagram in Fig.2e. An activation energy of 65 KJ/mol was determined from this plot, which is close to that found for the forward rate of reaction for nuclear waste glasses, 75 KJ/mol [2], quartz and amorphous silica, 72 and 63 KJ/mol, resp. [11].

For times longer than 3 month data at 50°C indicate, that silica in solution has become saturated. This goes along with a decrease of the overall rate of reaction after 2 month (indicated by the decrease of surface layer growth rate with time, [4], when saponite is formed.

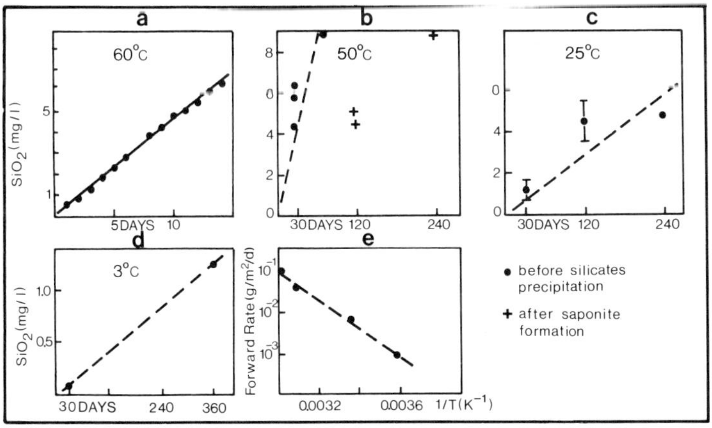

Fig.2 Effect of time and temperature on the leaching rate.

CALCULATION OF EQUILIBRIUM CONSTANTS

Hydrotalcite and pyroaurite solubility products were calculated considering these minerals as ideal solid-solutions of well known end members. The following formula has been used :

$$\text{Log } K = \sum_i (x_i \log K_i + x_i \log X_i)$$

where x = number of moles of a given end member in the solid-solution.
and X = molar fraction of this end member in the solid solution.

Example : Hydrotalcite = Brucite + Gibbsite + Magnesite

$$Mg_6Al_2CO_3(OH)_{16}nH_2O = 5(Mg(OH)_2) + 2(Al(OH)_3) + MgCO_3$$

The equilibrium constants for natural and experimentally formed phyllosilicates have been calculated by means of the solid-solution model described by FRITZ [3,12,13] with 27 end members (ideal phyllosilicates) selected for these calculations. The results are given in Table 2.

RESULTS OF REACTION PATH CALCULATIONS

Reaction path calculations were performed with the help of the computer codes DISSOL (FRITZ [3]) and PHREEQE (PARKHURST et al. [14]) Alkalinity was assumed to be constant at the various temperatures (no attempt was made to measure the alkalinity at test temperature).

Modelling was performed for 60, 25 and 0°C. As an example some results are shown for 60°C (Fig.3) :

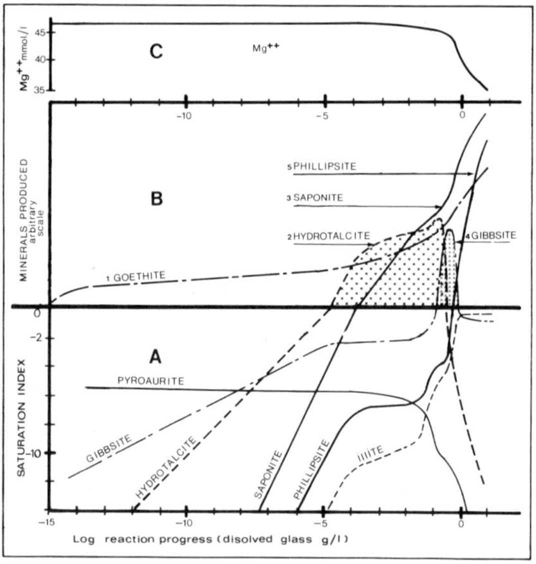

Fig.3 Reaction path calculated for 60°C.

The reaction progress (dissolved glass : g/l) is plotted in part A of the figure vs. the state of saturation of the resulting solution (saturation index=log IAP M/K where IAP is the ion activity product of the mineral M while K is the respective solubility product).

In part B the minerals produced are plotted vs. the reaction progress. All the minerals given in Table 2 were introducted in the codes during calculations, but only some of them are plotted (for clearness) on the figure.

The first mineral formed is goethite followed by hydrotalcite (after 1.6 x 10^{-5}g of glass have reacted). Later saponite then gibbsite and K-phillipsite form at high values of reaction progress. Hydrotalcite dissolves when Mg^{++} concentration decreases in solution (part C of the figure) , which confirms the transitional character of this phase as pointed out by electron microscopy [4,22].

Table 2. Minerals used for calculations with respective equilibrium constants.

NAME	SOURCE	DISSOLUTION EQUATION	Log K (T) 0°C	25°C	60°C
HYDROXYCARBONATES					
Hydrotalcite	1	$(Mg_6Al_2CO_3(OH)_{16}n\ H_2O) + 6H^+ \rightleftharpoons 6Mg^{++} + 2Al^{+++} + CO_3^= + 6\ H_2O + 10\ OH^-$	-47.86	-50.77	-55.08
Pyroaurite	1	$(Mg_6Fe_2CO_3(OH)_{16}n\ H_2O) + 6\ H^+ + 2e^- \rightleftharpoons 6Mg^{++} + 2Fe^{++} + CO_3^= + 6\ H_2O + 10\ OH^-$	-27.74	-31.39	-36.64
OXIDES-HYDROXIDES					
Amorphous Iron	2	$Fe(OH)_3 + e^- + 3H^+ \rightleftharpoons Fe^{++} + 3\ H_2O$	19.92	17.92	15.62
Artinite, Brucite, Goethite, Gibbsite see Fritz (1981)					
PHYLLOSILICATES					
Illite I (formed during experiments)	1	$(Si_{3.2}Al_{2.2}Mg_{.1}Fe_{.5}O_{10}(OH)_2K_{.6}Ca_{.15}) + .8H_2O + 9.2H^+ \rightleftharpoons .15Ca^{++} + .6K^+ + .5Fe^{+++} + .1Mg^{++} + 2.2Al^{+++} + 3.2H_4SiO_4$	13.50	10.26	6.85
Illite II (average marine)	3	$(Si_{3.43}Al_{2.24}Mg_{.38}O_{10}(OH)_2\ K_{.8}) + 1.72\ H_2O + 8.28H^+ \rightleftharpoons .8K^+ + .38Mg^{++} + 2.24Al^{+++} + 3.43\ H_4SiO_4$	16.40	13.38	9.90
Montmorillonite	3	$(Si_{3.83}Al_{1.84}Mg_{.38}O_{10}(OH)_2K_{.4}) + 3.32\ H_2O + 6.68\ H^+ \rightleftharpoons .4K^+ + .38Mg^{++} + 1.84Al^{+++} + 3.83\ H_4SiO_4$	9.26	7.10	4.58
Nontronite	4	$(Si_{3.84}Al_{.94}Fe_{.9}Mg_{.33}O_{10}(OH)_2Ca_{.105}Na_{.05}K_{.2}) + 3.36\ H_2O + 6.64\ H^+ \rightleftharpoons .2K^+ + .05Na^+ + .105\ Ca^{++} + .33Mg^{++} + .9Fe^{+++} + .94Al^{+++} + 3.84\ H_4SiO_4$	1.56	-0.16	-2.19
Saponite	3	$(Si_{3.67}Al_{.33}Mg_3O_{10}(OH)_2Na_{.33}) + 2.68\ H_2O + 7.32\ H^+ \rightleftharpoons .33Na^+ + 3Mg^{++} + .33Al^{+++} + 3.67\ H_4SiO_4$	34.00	30.54	26.74
Saponite-Fe	5	$(Si_{3.67}Al_{.8}Fe_{.46}Mg_{1.52}O_{10}(OH)_2Ca_{.11}Na_{.11}K_{.17}) + 2.68H_2O + 7.32H^+ \rightleftharpoons .17K^+ + .11Na^+ + .11Ca^{++} + 1.52Mg^{++} + .46Fe^{+++} + .8Al^{+++} + 3.67\ H_4SiO_4$	15.52	14.78	11.84
ZEOLITES					
Analcime see Fritz (1981)					
Clinoptilolite	6	$(Na_{.39}K_{.91}Ca_{.26}Mg_{.11}Al_{2.08}Si_{9.93}O_{24}.10H_2O) + 5.72\ H_2O + 8.28H^+ \rightleftharpoons .39\ Na^+ + .91K^+ + .26Ca^{++} + .11Mg^{++} + 2.08\ Al^{+++} + 9.93\ H_4SiO_4$	-14.28	-13.15	-12.33
Phillipsite-Na	6	$(Na_{.6}K_{.46}Ca_{.07}Mg_{.04}Al_{1.12}Si_{2.84}O_8.3.73H_2O) + 4.64H^+ \rightleftharpoons .6Na^+ + .46K^+ + .07Ca^{++} + .04Mg^{++} + 1.12Al^{+++} + 2.84H_4SiO_4 + .37\ H_2O$	5.53	4.74	3.75
Phillipsite-K	7	$(Na_{.46}K_{.6}Ca_{.07}Mg_{.04}Al_{1.12}Si_{2.84}O_8 3.73H_2O) + 4.64H^+ \rightleftharpoons .6Na^+ + .46K^+ + .07Ca^{++} + .04Mg^{++} + 1.12Al^{+++} + 2.84H_4SiO_4 + .37H_2O$	4.97	4.26	3.19

1)This work. 2)Wagman et al.[15], ΔH_r calculated according to Baes and Mesmer [16]. 3)This work, Composition given in [17]. 4)This work. Composition given in [18]. 5)This work. Composition given in [19]. 6)Log K(T) calculated by means of SUP CRT program of H.Helgeson (data of Kosiur [17]). 7)Proposed composition. Log $K_{25°C}$ calculated according to Tardy and Garrels [20], ΔH_r assumed to be the same as ΔH_r of phillipsite-Na.

Hydrotalcite has never been described as an alteration product of basaltic glass in a natural environment. This either results from its transitionary character or as will be shown below, by its high solubility at typical low temperatures. Conversely, the sequence : saponite-phillipsite has been observed in the North West Pacific Ocean (DSDP Leg 55) as the major alteration products of basaltic glass (KARPOFF et al. [23]). The present calculations confirm that saponite is the major reaction product, followed by phillipsite at high reaction progress.

The same run has been performed for 25°C assuming the formation of $Fe(OH)_3$ instead of goethite, which was not observed during the experiments. The results are plotted on Fig.4A.

Hydrotalcite cannot form due to its high solubility at that temperature. Pyroaurite, observed during experiments, is close to saturation where saponite forms (log IAP/K=-0.04).Calculations show that pyroaurite is not stable: if we change its log K (-31.44 instead of -31.39) in order to reach saturation,the mineral forms effectively but becomes undersaturated very quickly due to Mg consumption by saponite. Fe-saponite is the major alteration product : at the end of simulation, 12.5g of glass were disolved ; we calculate that 11g of Fe-saponite were formed while only 0.9g of amorphous iron, 0.9g of gibbsite and 0.4g of saponite were precipitated. No zeolite appeared. The composition of the solution at the end of simulation is given in Table 3.

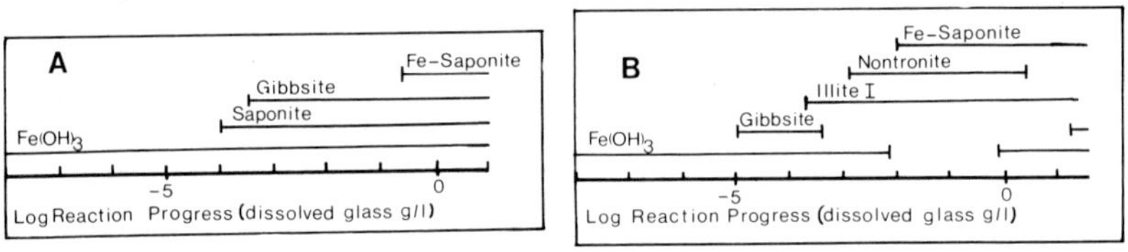

Fig.4 Schematic representation of the calculated sequences for 25°C (part A) and 0°C (part B).

At 0°C (Fig.4B) the first mineral formed is $Fe(OH)_3$. Gibbsite appears but redisolves when the amount of illite increases. Later, nontronite and Fe-saponite form. At the end of simulation, 25g of glass were dissolved to form 23g of Fe-saponite, 1.6g of amorphous iron, 1.3g of gibbsite and 1.2g of illite ; see Table 3 for the solution composition.

These calculations reproduce the experimentally observed sequence for the short term : iron hydroxide and illite. Illite formation as an alteration product of basaltic glass in natural environments is however an open question. Geologists consider this mineral as a detrical remnant. It must be noted in this respect that the average marine illite (Illite II in Table 2) is not calculated to be formed and that illite 1 (observed during laboratory work) dissolves at the end of simulation. Other calculations have been performed considering an additional source for silica (i.e.silica rich sediments present in Deep Oceans). In that case Illite 1 dissolves very quickly and a K-rich Phillipsite forms as observed in natural environments. Then, according to these calculations, phillipsite should not precipitate as a result of the glass/seawater reaction alone if equilibrium is achieved between the formed clay phases (nontronite, saponite) and the solution. However, in the case the solution remains supersaturated with respect to these phyllosilicates phillipsite can be formed.

Nontronite and Fe-rich saponite are commonly associated with volcanic glass as low temperature alteration products, as described for ex. by HONNOREZ [24], KARPOFF [18] or DESPRAIRIES [19].

Hence the discrepancy of observed and predicted alteration products and comparison with what is found in nature may yield important information on the state of disequilibrium of the overall system and on the way a single reaction (glass/water system) may be tied to the overall reaction system. However, no data are available yet concerning the degree of saturation of this natural system.

Table 3 shows the compositions calculated for the resulting solutions.We can see that calcium concentration increases while magnesium concentration decreases as a result of glass dissolution and secondary products formation.In this respect our calculations are in good agreement with what is found in nature [25].

Table 3. Composition of the solutions

INITIAL SOLUTIONS (moles/KGM H_2O)			RESULTING SOLUTIONS (moles/KGM H_2O)**			
EL^T	1	2	EL^T	1.1(60°C)	1.2(25°C)	2.1(0°C)
			DG*	6.84	12.46	25.12
			Al	0.61E-06	0.84E-06	0.87E-07
K	0.92E-02	1.06E-02	K	0.41E-02	0.49E-02	0.36E-04
Na	0.43	0.49	Na	0.43	0.43	0.50
Ca	0.91E-02	1.06E-02	Ca	2.14E-02	3.04E-02	5.30E-02
Mg	0.46E-01	0.55E-01	Mg	0.35E-01	0.23E-01	0.15E-01
			Fe	1.44E-16	1.28E-14	8.54E-15
			SiO_2	0.18E-03	0.25E-05	0.24E-03
S	0.24E-01	0.29E-01	S	0.24E-01	0.24E-01	0.29E-01
Cl	0.49	0.57	Cl	0.49	0.49	0.57
Alkal.	0.26E-02	0.24E-02	Alkal.	0.71E-04	0.18E-02	0.12E-01
pH(25°C)	8.88	8.22	pH **	7.05	8.81	9.05
eH(25°C)	400 m V	500 m V	eH **	399 m V	404 m V	534 m V

1) Artificial seawater used during experiments [4,5]
2) Average marine seawater [21]
1.1) Simulation of glass dissolution at 60°C using solution 1
1.2) Simulation at 25°C using solution 1
2.1) Simulation at 0°C using solution 2 (The results obtained with composition 1 are not basically differents).
* Dissolved Glass (G/KGM H_2O)
** Alkalinity, pH and eH are given for the corresponding temperature.

The reaction progress, at which certain phases precipitate may be correlated with the time it takes to reach this state of reaction experimentally [2].

The Table 4 summarizes the results of this comparaison for the time solubility limits are just achieved and some later time, when the phases are large enough to be seen by electron microscopy. As can be seen fair agreement of experimental findings with the calculations is achieved.

Table 4. Comparison of the experimental time necessary to form secondary alteration products with the respective time calculated via the computer codes. (RS=just reaches solubility limit) ; layer thickness of phases calculated assuming homogeneous precipitation with a density of $2g/cm^3$.

Temp.	Phase	Calculated time	Observation
0°C	Illite	RS: 30 d 0.06 µm: 560 d	-- 0.1 µm at 600 d
25°C	Saponite	RS: 2 d 0.5 µm: 480 d	-- 0.2 µm at 240 d
60°C	Hydrotalcite	RS: 0.3 d 0.02 µm: 4 d	-- first observed : 4d

CONCLUSIONS

As for nuclear waste glasses the mechanism of the reaction of basalt glass with seawater is controlled by the congruent dissolution of the glass matrix and subsequent precipitation of secondary products.

When alteration products of non-ideal compositions are formed their stoichiometry and their stability constants must be known in order to model the reaction path.

The observed period of time required to form certain secondary alteration products fits quite well with the respective time calculated with the reaction path model. The relationship between reaction progress and time has to be studied experimentally.

The presence of alteration phases in disequilibrium with the rest of the system and the competition of the glass/water reaction with other dissolution/precipitation reactions complicates the modelling of the reaction of basaltic glasses in a natural environment.This may also be important if an overall nuclear waste glass groundwater/rock system is to be modelled.

1 LUTZE W., MALOW G., EWING R.C., JERCINOVIC M.J. and KEIL K., Nature 314, 252 (1985).

2 GRAMBOW B., in : Scientific Basis for Nuclear Waste Management (VIII) Elsevier Science Publ. Co (in Press) (1984).

3 FRITZ B., Sci.Géol.Mém.Univ.Louis Pasteur.Strasbourg 65, 197 (1981).

4 CROVISIER J.L., THOMASSIN J.H., JUTEAU T., EBERHART J.P., TOURAY J.C. and BAILLIF P., Geoch.Cosmoch.Acta.47, 377 (1983).

5 CROVISIER J.L., EHRET G., EBERHART J.P. and JUTEAU T., Sciences Géol.Mém.36, 187 (1983).

6 CROVISIER J.L.,THOMASSIN J.H.,JUTEAU T.,TOURAY J.C.and EBERHART J.P., Water.Rock.Interaction IV.Misasa (Japon) Extended Abstracts, 99 (1983).

7 CROVISIER J.L., EHRET G. and EBERHART J.P., Rivista della Staz.Sper. Vetro, 149 (1984).

8 GRAMBOW B., XIII Int. Congress on Glass, Hamburg, Glastech. Ber. Sonder band. 4.9, 7 (1983).

9 NOGUES J.L., Thèse Docteur Ing. Montpellier 322 (1984).

10 STRACHAN D.M., Nuclear and Chemical Waste Management,4, 177 (1983).

11 RIMSTIDT J.D. and BARNES H.L., Geoch.Cosmoch.Acta.44, 1683 (1980).

12 TARDY Y. and FRITZ B., Clay Minerals. 16, 361 (1981).

13 FRITZ B., in: The Chemistry of Weathering. NATO ASI Series (C) D.Reidel Publishing Company 19 (1985).

14 PARKHURST D.L., THORSTENSEN D.C. and PLUMMER L.N., Water-Resources Investig.80-96 U.S.Geol.Surv.Reston Virginia (1980).

15 WAGMAN D.D. et al., N.BS.Technical Note 270-4 (1969).

16 BAES C.F. and MESMER R.E., Am.J.Sci.281, 935 (1981).

17 KOSIUR D.R., Thesis Univ.California Los Angeles 148 (1978).

18 KARPOFF A.M., Init.Rep. of DSDP,LXXIII Washington (US Govern.Print.Off.) 515 (1984).

19 DESPRAIRIES A., Init.Rep. of DSDP, LX Washington (US Gov.Print.Off.) 455 (1981).

20 TARDY Y. and GARRELS R.M., Geoch.Cosmoch.Acta,38 1101 (1974).

21 NORDSTROM et al., Am.Chem.Soc.ACS Symp.Series 93.Washington D.C. 857 (1979).

22 THOMASSIN J.H. et TOURAY J.C., Bull.Mineral.105, 312 (1982).

23 KARPOFF A.M., PETERSCHMITT I. and HOFFERT M., Init.Rep. of DSDP, LV, Washington (US Gov.Print.Off.) 463 (1980).

24 HONNOREZ J., in : "The Sea" EMILIANI C. Ed.Wiley Interscience 525 (1980).

25 GIESKES J.M. and LAWRENCE J.R.,Geoch.Cosmoch.Acta.45,1687 (1981).

PART IV

Waste Form Performance— Ceramics and Spent Fuel

THE EXTENDED DEFECT AS A MECHANISM FOR THE IMMOBILIZATION OF HLW SPECIES IN ZIRCONOLITE, PEROVSKITE AND HOLLANDITE

T.J. WHITE*
School of Science, Griffith University, Nathan, Brisbane, QLD 4111, Australia

ABSTRACT

High resolution electron microscopy, analytical electron microscopy and selected area electron diffraction are used to establish the crystallo-chemical mechanisms by which simulated high level waste enters Synroc. It is shown that waste species are incorporated largely as a non-continuous solid solution. In other words, simple isomorphic substitution of waste elements in place of matrix cations, i.e. the classical continuous solid solution, does not operate (or is only significant at very low waste levels of rad-waste). Usually, substitution is accompanied by the appearance of coherent extended defects which can take the form of antiphase boundaries, unit cell twinning (both mimetic and polysynthetic), or cation ordering. In this manner, a number of closely related phases appear in response to variations in waste composition and loading. It is believed that the mechanism of structural modification will endow Synroc with considerable flexibility to respond to chemical changes in the wastestream. Other ceramic materials currently being evaluated as possible wasteforms (viz. sphene glass-ceramics and APO_4 phosphates), are also likely to possess this capacity.

INTRODUCTION

Synroc is the generic name for a number of titania based wasteforms designed to immobilize the radioactive products of commercial light water reactors [1]. These ceramics are "synthetic mineral" assemblages which consist of several oxide structure types including fluorites, perovskites, hollandites, "reduced" rutile and magnetoplumbite [2]. Most of the components of high level waste (HLW) enter these phases in solid solution, but some are reduced to the metallic state and form alloys [3].

To date, the chemical partitioning of waste species amongst Synroc phases has been determined on the basis of electron microprobe data. This technique suffers from the limitation that analyses can only be collected from crystals somewhat larger than 1 μm^3, and crystals of these dimensions are only found in Synroc specimens hot pressed close to the solidus (> 1200°C) [4]. During actual fabrication, calcine powders are hot pressed at lower temperatures (1100 - 1170°C). It is not clear if the solid state chemistry operating under these two preparative regimes will yield identical phase assemblages, or whether, at the higher temperature elemental partitioning into grain boundaries [5] will be enhanced. Consequently, conventional methods of microstructural evaluation, i.e. scanning electron microscopy combined with X-ray energy dispersive spectroscopy are unable to probe individual grains of the final Synroc product, and the data obtained from such studies should be viewed circumspectly. Furthermore, the composition of several Synroc minerals are similar, so chemical analyses alone are insufficient for definitive phase and structure identification [6].

In this article, we describe the use of high resolution electron microscopy to explore directly, at the unit cell and atomic level, the

*Current address: Materials Division, Australian Atomic Energy Commission, Lucas Heights Research Laboratories, Private Mail Bag, Sutherland NSW 2232, Australia.

crystallographic changes which accompany the incorporation of simulated radionuclides into the structural aristotypes of Synroc. Using these insights it is possible to more accurately predict the response of the phase assemblage to differences in wastestream composition, and in the future, permit more realistic modelling of radiation damage in ceramics.

CRYSTAL STRUCTURES OF THE ARISTOTYPES: ZIRCONOLITE, PEROVSKITE, HOLLANDITE

In the original Synroc strategy it was envisaged that all the chemical species in HLW could be "dissolved" (by isomorphic substitution) in three titanate phases - zirconolite, perovskite and hollandite. Although substantially correct, it is now recognised that this scenario is an oversimplification. In fact, the Synroc phase assemblage is dependent upon waste loading, and a number of closely related, but distinct phases appear in response to variations in radwaste composition and concentration (Table I). Crystallographically, it is simplest to regard these synthetic minerals as arrays of TiO_6 octahedra [7]. These share edges and vertices in a number of motifs to create large interstices (which prototypically contain Ca, Zr and Ba), that are suitable for the inclusion of all the nuclide species (not reduced to metal).

TABLE I
APPROXIMATE PROPORTIONS OF STRUCTURE-TYPES (WT%) IN SYNROC C

Fluorites	30%	Perovskite	15%	Magnetoplumbite	≤ 5%
Hollandites	30%	"Reduced" Rutile	15%	Alloy	≤ 5%

The nominal formulation of zirconolite is $CaZrTi_2O_7$, though a phase of integral stoichiometry probably cannot be prepared [8]. It is useful to consider zirconolite formally as a layer structure, in which TiO_6 octahedra are arranged in rings of three and six to form continuous (001) sheets as shown in Figure 1(a) [9]. These layers, which possess the so-called hexagonal tungsten bronze (HTB) type topology, are separated by rows of calcium and zirconium atoms. The former are located in distorted cubic polyhedra, whilst the latter are co-ordinated to seven oxygens in what are best described as face-capped octahedra [10]. Two TiO_6 octahedral layers, stacked along [001], are required to yield a crystallographic repeat.

In $CaTiO_3$ perovskite, TiO_6 octahedra share every vertex to form a continuous network (Figure 1(b)). The large calcium interstice has 12 equidistant oxygens in the disposition of a cuboctahedron. Strictly though, the octahedra are rotated slightly, collapsing the structure and reducing the symmetry from cubic to orthorhombic [11]. Consequently, the cuboctahedra are distorted [12] in a manner which furnishes the calcium atom with eight nearest and four second nearest oxygen neighbours.

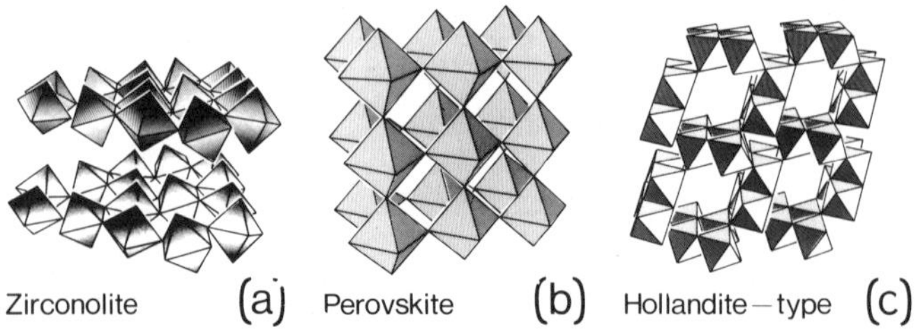

Fig. 1 The octahedral motifs of the structural aristotypes that occur in Synroc

Controversy and misunderstanding surround both the precise structure and stoichiometric representation of hollandite.* This has been clarified by Kesson and White [14], who reaffirmed the general formulation proposed by Sabine and Hewat [15], viz. $Ba_x(Al,Ti)_{2x} Ti_{8-2x}O_{16}$, $1.09 < x < 1.14$. Hollandite consists of channels generated by corner and edge-sharing oxygen octahedra that are occupied by Al^{3+}, and trivalent or tetravalent titanium (Figure 1(c)). Ba^{2+} ions are placed in the larger tunnels entering 8-coordinate parallelelepipedons. In detail, the situation is somewhat complex, since the barium cations may fill either of two sites within each tunnel polyhedron, and only partial occupancy of these sites is required.

EXPERIMENTAL

Synthetic minerals as constituents of Synroc, and also as specially prepared "single phase" assemblages were studied by selected area electron diffraction (SAD) and high resolution electron microscopy (HREM). The methods of synthesis are detailed elsewhere [16,17,18]. The specimens were prepared for TEM analysis by crushing under ethanol and depositing a drop of the suspension on holey carbon films [2].

ZIRCONOLITE-ZIRKELITE INTERGROWTHS

The fluorite-related oxides exhibit the greatest diversity of structural modification, engineered simply by varying the offset between HTB layers. These are displaced with respect to each other by an interlayer stacking vector (SV) which is defined as the projected distance, viewed down the (pseudo) six-fold axis, between crystallographically similar atoms in adjacent layers (Figure 2(a)). The angle between successive stacking vectors is

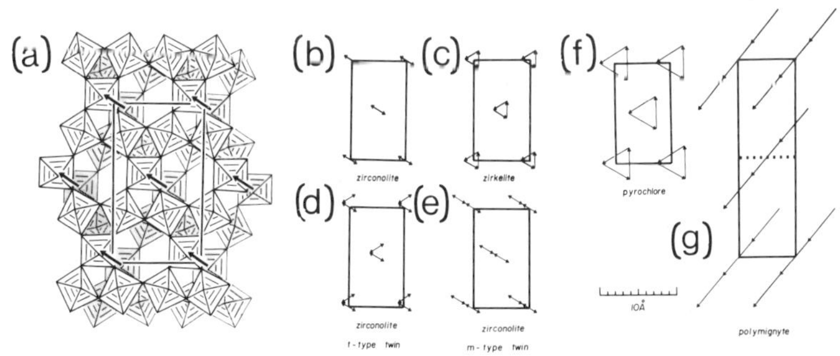

Fig 2 (a) Two HTB layers offset by an interlayer SV (arrowed)
(b)-(g) Displacement vector diagrams for zirconolite and its polytypes plotted on the common a-b plane

the intermodular stacking angle (SA) and can take any $-n\pi/3$ value [19]. The sequence of SVs can be usefully summarised on a stacking vector diagram, constructed by plotting SVs on the C-centred base of the zirconolite unit cell (Figure 2(b)). In zirconolite itself successive modules are related to each other by a SA of -180°. The orientation relationships are such that the

---

*Actually, the hollandite type compound, $BaMn_8O_{16}$ is not compositionally similar to Synroc hollandite. A close natural analogue for the Synroc species is priderite, $(Ba,K)_{1.6}(Ti,Fe)_8O_{16}$ [13].

insertion of a SA of 120° results in a triad or t-type twin (Figure 2(d)), while a SA of -360°results in a monad (m-type) twin (Figure 3(e)). If t-type boundaries are inserted in a completely regular fashion into the aristotype (i.e. the twinning is said to be mimetic), then a new phase called zirkelite [20] is produced (Figure 2(c)). Clearly, the close structural relationship between zirconolite and zirkelite makes it eminently feasible that coherent intergrowth should occur between the phases. This is indeed the case, as illustrated by the HR electron image in Figure 3, where completely coherent intercalation occurs in a mixed crystal of approximate composition $Ca_{0.8}Th_{0.2}ZrTi_{0.8}Fe_{0.2}O_7$. In Synroc, large recognisable supercells of zirkelite are not usually apparent, but frequent introduction of twin boundaries does occur (Figure 4). These are best described as small portions of intergrown zirkelite, that provide suitable sites for the insertion of radwaste. Zirconolite-zirkelite intergrowths predominate in Synroc with lower waste loadings (i.e. ~ 10 wt%). If the waste is rich in rare earth elements, or a 20 wt% loading is used, then another fluorite-related structure, pyrochlore appears. It is also constructed of HTB layers and its SV diagram is shown in Figure 2(f). At still higher waste loadings (40 wt%) polymignyte becomes the dominant phase - zirconolite is seldom observed. Polymignyte also consists of HTB layers but in this instance they are sheared in a manner which results in a doubling of the long-axis of the c-centred basal cell (Figure 2(g)).

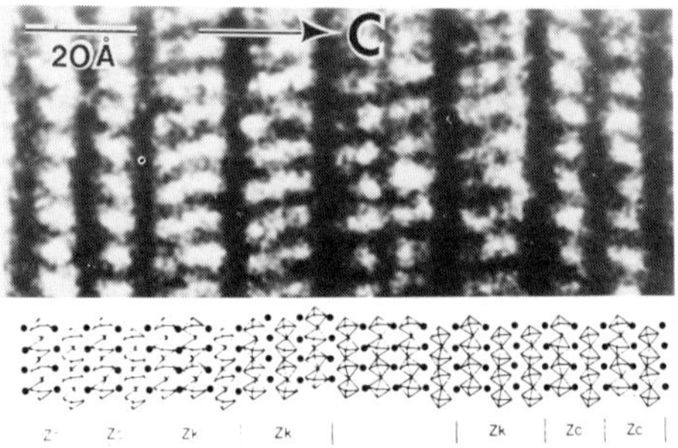

Fig. 3 Coherent intergrowth between zirconolite (Zc), zirkelite (Zk) and a higher polytype. Approximate composition of the crystal was $Ca_{0.8}Th_{0.2}ZrTi_{0.8}Fe_{0.2}O_7$

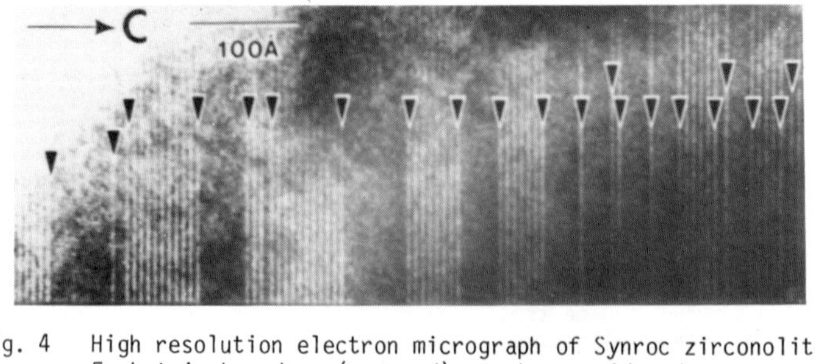

Fig. 4 High resolution electron micrograph of Synroc zirconolite. Each twin boundary (arrowed) can be considered as a narrow lamellar of zirkelite.

Thus, zirconolite may be regarded as the structural aristotype of a group of minerals composed of HTB layers. Providing altervalent cations are available (usually as processing contaminants), to ensure electrostatic charge balance, these layers can be stacked in a number of orientations which permit the incorporation of rare earth elements and actinides at variable concentrations.

PEROVSKITE POLYTYPES

Polytypes based upon the simple perovskite aristotype are observed very rarely in Synroc [12], and are apparently favoured by simultaneously high local concentrations of Al, Nd and Fe (Figure 5(a)). The polytypes are produced by the mimetic twinning of perovskite on the (approximately) close-packed atom planes. Figure 5(b) shows the most common structure (idealised to perfect cubic-closest-packing) and its high resolution image.

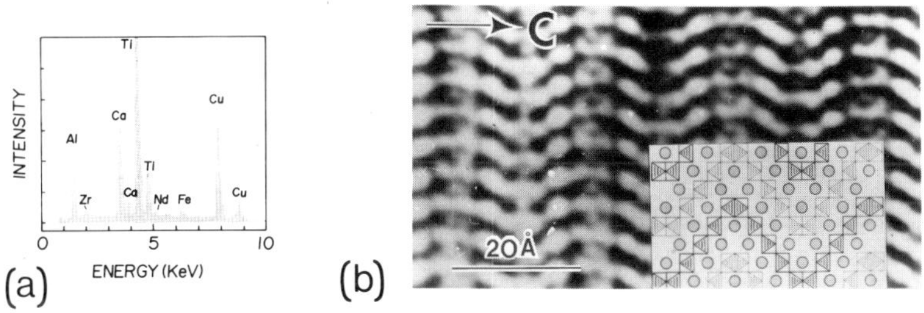

Fig. 5 (a) Chemical constitution as determined by energy dispersive analysis of X-rays and (b) high resolution electron micrograph of a perovskite polytype in Synroc

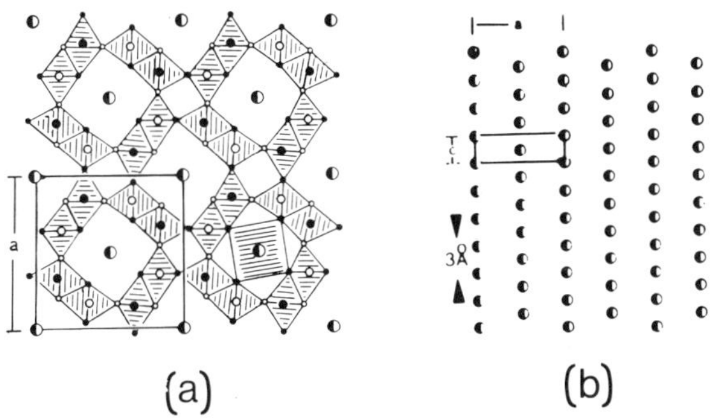

Fig. 6 (a) $(Ti,Al)O_6$ octahedral motif of hollandite projected along the tunnel direction. Large cations, (Ba,Cs), enter the (2x2) octahedral channels.
(b) Arrangement of large cations viewed at right angles to the tunnel, and assuming a statistical distribution over all available sites. For clarity the octahedra are omitted.

HOLLANDITE SUPERSTRUCTURE MOSAICS

In Figure 6 the relationship between the $(Al,Ti)O_6$ octahedral framework of hollandite, and the tunnel cation sites is shown - it should be noted that the spacings between the latter interstices along c is ~ 3Å. The filling of the tunnel positions is quite complex, with a variety of n x c = n x 3Å (n integral) supercells arising from several cation-vacancy ordering sequences. The diameter of these commensurate superstructure microdomains is usually less than 50Å, and the contribution from a number of supercells of different c-axis repeat, frequently leads to an overall incommensurate (i.e. n non-integral) supercell.

Under favourable conditions (viz. a sufficiently thin crystal and well developed supercells), the disposition of supercell microdomains can be inferred directly from high resolution electron micrographs. With the qualification that the projected charge density approximation be applicable, it is possible to assume direct correspondence between lines of white blobs along [001] and the position of empty tunnel site columns. As noted above, n x 3Å (n integer) spacings between unoccupied tunnel sites should be preset. This effect is illustrated in Figure 7(a), where ordered and disordered sequences of white blobs (i.e. empty tunnel interstices) along the c-axis are apparent, with the most common spacing being ~ 6Å, corresponding to a vacancy-cation-vacancy pattern. Intertunnel correlation along b is, in general, quite limited. An enlargement of some areas and their schematic interpretation in terms of filled and vacant tunnel cation columns, is shown in Figure 7(b). The random displacements of ± 1/2 c between adjacent .... cation-vacancy-cation-vacancy .... sequences, in the lower figure, may be regarded as anti-phase boundaries (APB) of the form (100) n+1/2[001], n > 0. In general, the incorporation of higher concentrations of cesium (i.e. greater waste loadings), leads to longer range supercell ordering (described as greater multiplicity) in the tunnel direction (Figure 8).

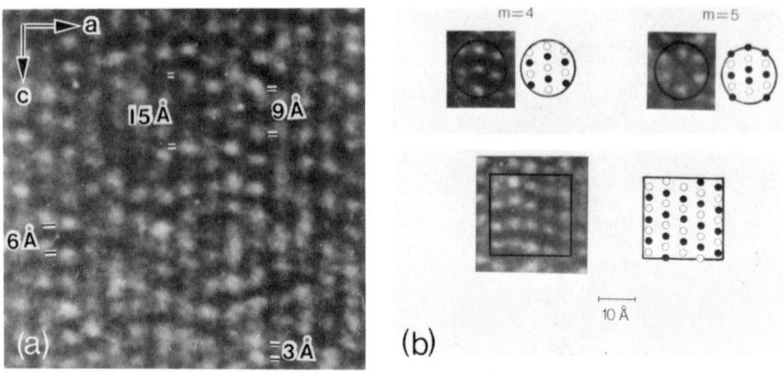

Fig. 7 (a) High resolution micrograph of Synroc hollandite viewed in the same direction as Figure 6(b). White blobs (empty tunnel sites) occur at spacings which are multiples of ~ 3Å.
(b) Enlargement of some areas of this crystal, and their schematic interpretation in terms of occupied (filled circles) and unoccupied (open circles) tunnel sites.

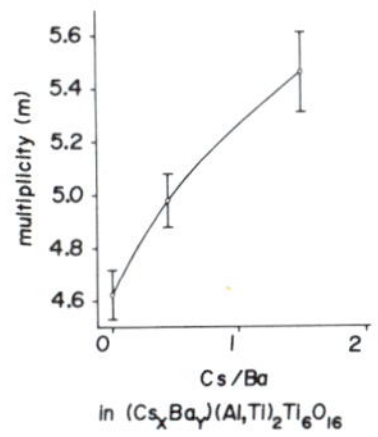

Fig. 8 The number of subcell units per supercell, or the multiplicity (m) of hollandite crystals increases as the Cs/Ba ratio is raised. That is, higher waste loadings lead to specific long range cation ordering sequences along the larger hollandite tunnels.

STRUCTURAL MODIFICATION IN SPHENE AND ABO_4 PHOSPHATES

Sphene-based glass ceramics are being developed for the immobilisation of wastes generated by the recycling of Candu nuclear fuel [21]. The mineral sphene ($CaTiSiO_5$) is able to take up a wide range of waste cations into two of its cation acceptor sites. Substituents for 7-coordinated calcium include alkali earths, some alkalis and rare earth elements. Charge balance for rare earths over the calcium site can be maintained by substituting Al and Fe onto the titanium octahedron. However, linear domains are well known in natural sphene from the examination of X-ray precession photographs [22]. This texture is due to insertion of antiphase boundaries (APB) in the aristotype. Furthermore, the occurrence of APBs is controlled by cation substitutions over the octahedral titanium site. Thus, it appears likely that a HREM investigation of sphene, as a component of the Candu wasteform, would reveal in greater detail the manner in which structural modification assists in the uptake of waste species.

ABO_4 (A = Ca, Zr, rare earth, actinide; B = Si, P, Nb) materials are also being evaluated as potential hosts for HLW [23]. In this case, four closely related structures - monazite, zircon, scheelite and fergusonite - containing the same polyhedra, but with slightly different topologies are known. The formation of these polytypes is controlled by composition, and once again it seems feasible that the ultrastructure of these compounds may possess extended defects or microdomains.

CONCLUSION

The characterisation of Synroc by high resolution electron microscopy, has revealed the uptake of radwaste to be not only a question of inserting radionuclides into the appropriate crystallographic sites of a host matrix as a continuous solid solution. We have shown the phenomenon of structural modification to very often accompany the incorporation of a substantial level of waste. Moreover, it is believed that analogous mechanisms will occur in other ceramic wasteforms.

The realisation that radwaste is not distributed uniformly through a ceramic, but concentrated in extended defects as a non-continuous solid solution, may influence the process of metamictisation. For example, it could provide an explanation for the observation that in partially metamict natural zirconolite, crystalline regions are interspersed with amorphous material [24]. In particular, theoretical models of radiation damage must be carefully constructed to take into account this heterogeneous distribution of waste elements, to ensure that one arrives at a realistic description of the final microstructure.

ACKNOWLEDGEMENTS

This work was undertaken with the support of AINSE and NERDDC postdoctoral fellowships. The encouragement and support of the members of the Griffith University Synroc Research Group is gratefully acknowledged. The author is obliged to Dr S. Kesson for numerous discussions and shared insights on many aspects of this work.

REFERENCES

1. A.E. Ringwood, V.M. Oversby, S.E. Kesson, W. Sinclair, N. Ware, W. Hibberson, A. Major, Nucl. Chem. Waste Manage., 2, 287 (1981).
2. J.A. Cooper, D.R. Cousens, R. Lewis, S. Myhra, R.L. Segall, R.St.C. Smart, P.S. Turner, T.J. White, J. Amer. Ceram. Soc., 68, 64 (1985).
3. R.G. Dosch, T.J. Headley, P.F. Hlava, J. Amer. Ceram. Soc., 67, 354 (1984).
4. A.E. Ringwood, S.E. Kesson, W.G. Ware, W. Hibberson, A. Major, Nature, 278, 219 (1979).
5. R.L. Segall, R.St.C. Smart, P.S. Turner, T.J. White, these proceedings.
6. T.J. White, R.L. Segall, P.S. Turner, Angew. Chem., 97, 369 (1985).
7. J. Campbell, C. Hoenig, F. Bazan, F. Ryerson, M. Guinan, R. Van Konynenburg, R. Kozsa, LLRL Report No. UCKL-53240 (1981).
8. H.J. Rossell, Nature, 283, 282 (1980).
9. B.M. Gatehouse, I.E. Grey, R.J. Hill, H.J. Rossell, Acta Crystallogr., B37, 306 (1981).
10. T.J. White, Amer. Mineral., 69, 1156 (1984).
11. H.F. Kay and P.C. Bailey, Acta Crystallogr., 10, 219 (1979).
12. T.J. White, R.L. Segall, J.C. Barry, J.L. Hutchison, Acta Crystallogr., B41, 93 (1985).
13. W. Sinclair and G.M. McLaughlin, Acta Crystallogr., B38, 245 (1982).
14. S.E. Kesson and T.J. White, submitted for publication.
15. T.M. Sabine and A.W. Hewat, J. Nucl. Mat., 110, 173 (1982).
16. S.E. Kesson and A.E. Ringwood, in Scientific Basis for Nuclear Waste Management, VI, edited by G.C. McVay (Elsevier Science Publishers, New York, 1984), pp 507-512.
17. S.E. Kesson, W.J. Sinclair, A.E. Ringwood, Nucl. Chem. Waste Manage., 4, 259 (1983).
18. S.E. Kesson and A.E. Ringwood, Nucl. Chem. Waste Manage., 2, 53 (1981).
19. T.J. White, R.L. Segall, J.L. Hutchison, J.C. Barry, Proc. R. Soc. Lond., A392, 343 (1984).
20. F. Mazzi and R. Munno, Am. Mineral., 68, 262 (1983).
21. E.R. Vance, P.J. Hayward, C.D. Cann, S.L. Mitchell, M.A.T. Stanchell, D.J. Wren, Glass Tech., 25, 232 (1984).
22. J.B. Higgins and P.H. Ribbe, Amer. Mineral., 61, 878 (1976).
23. A.T. Aldred, in Geochemical Behaviour of Disposed Radioactive Waste, edited by G.S. Barney, J.D. Navratil and W.W.S. Schultz (American Ceramic Society, Columbus, 1984) pp 305-314.
24. R.C. Ewing and T.J. Headley, J. Nucl. Mat., 119, 102 (1983).

INTERGRANULAR FILMS AND PORE SURFACES IN SYNROC C

S. MYHRA, R.L. SEGALL, R. ST. C. SMART, P.S TURNER AND T.J. WHITE*
School of Science, Griffith University, Nathan, Brisbane, QLD 4111, Australia

ABSTRACT

High resolution electron microscopy and scanning electron microscopy were used to determine the distribution of intergranular films and microvoids in Synroc C. Diffraction contrast derived from these films, which were 1-3 nm thick, showed then to be ill-defined crystallographically, and they may be described as glassy. Pores were usually several micrometers in extent and occurred principally in rutile-rich areas. The chemical composition of these structural features was obtained using analytical transmission electron microscopy, secondary ion mass spectrometry, x-ray photoelectron spectroscopy and scanning Auger microscopy. Within intergranular films, elemental enhancement of cesium, sodium, potassium and aluminium, and possibly silicon and molybdenum was observed. Enhancement of cesium, sodium, phosphorous, aluminium and silicon was found at triple points. Fracture faces preferentially expose grain boundaries, and dissolution of these surfaces proceeds rapidly at ambient temperatures. During the fabrication of Synroc C, microvoids trap cesium vapour, and after cooling this condenses onto pore surfaces. Recognition of the (simulated) waste species which segregate at grain boundaries and pores, permitted the reinterpretation of published leach data for monolithic and crushed Synroc C.

INTRODUCTION

The capacity of a material to retain (simulated) radwaste during severe hydrothermal testing, is the most significant short term scientific criterion used during the evaluation of a waste form. The current paper is an appraisal of the contribution that waste elements, which have undergone segregation into intergrannular films# and pores, make towards the overall dissolution rate of HLW species from Synroc [1]. Such an assessment must consider the composition and structure of films; their inherent resistance to hydrothermal attack; the distribution of pores in the multiphase assemblage; and the connectivity of resulting diffusion pathways.

In this report, we present chemical analyses for these "non-crystalline" structural features, that were derived from transmission electron microscopy and surface analytical techniques. These results, together with published data on similar materials, permit the interpretation of the early leaching characteristics of alkali elements relative to other species.

EXPERIMENTAL METHODS

Synroc C specimens were obtained from the Lucas Heights Research Establishment and the Research School of Earth Sciences at the Australian National

---

* Present address: Materials Division, Australian Atomic Energy Commission, Lucas Heights Research Laboratories, Private Mail Bag, Sutherland NSW 2232, Australia.

The term "intergranular film" is widely used [2], although in the case of the multiphase ceramic Synroc C, the term interphase film might be more suitable, since films have always been found on boundaries between grains of different phase [3].

(HRTEM) by either argon ion-beam milling, or crushing under ethanol in an agate mortar and pestle, and depositing a drop of the suspension on a holey carbon film. Non-crystalline layers were identified by lattice fringe or defocus imaging [2], in a TEM operated at 100 kV and utilising a conventional tungsten filament. X-ray energy dispersive spectra (EDS) of glassy triple points and intergranular films were obtained using a 100 kV microscope fitted with a high-brightness field emission gun (FEG). This instrument permitted analyses from regions a few nanometres in diameter.

Surface analytical measurements were initially conducted on fresh fracture faces. These were subsequently treated by dipping in doubly distilled water, or by argon ion beam etching. Secondary ion mass spectrometry (SIMS) was carried out by bombarding a 2 x 2 mm rastered area of the specimen with 1 kV Ar ions focused to a beam ~ 250 μm in diameter. Secondary ions emanating from the sample were collected and analysed using a quadrupole mass spectrometer. Provided mixing of surface species is avoided, by using low ion currents (10^{-7}-10^{-8}A cm^{-2})[13], this method is capable of monolayer depth resolution. Its lateral resolution can be no better than the size of the rastered area.

The basis of x-ray photoelectron spectroscopy (XPS) is that x-rays promote the ejection of outer shell electrons from the first few monolayers. These electrons are then energy analysed in order to generate a spectrum. XPS has a depth resolution of ~ 2 nm and lateral resolution in the millimetre range. In addition to providing information about elemental abundances with a sensitivity of 0.1 at%, it can also yield data regarding the chemical environments of atomic species. XPS may be combined with ion beam etching to yield depth profiled chemical analysis.

Scanning Auger Microscopy (SAM) uses a focused beam of monoenergetic electrons (typically 10 keV) to induce expulsion of Auger electrons from near surface atoms. SAM offers high lateral resolution (≥ 100 nm in favourable cases) for imaging and point analyses. When combined with scanning electron microscopy (SEM) operated in the secondary electron mode, it provides direct correlation between surface features and composition.

TRANSMISSION ELECTRON MICROSCOPY

As reported previously [3], intergranular films about 2 nm wide are commonly observed between crystallites of different phases, whereas more nearly coherent boundaries occur at the juncture of grains of the same phase. It is believed that Synroc samples cleave predominantly along crystal boundaries, rather than by transgranular fracture. Confirmation of this was obtained from TEM images of crushed samples, which demonstrated the clear association of cracks and intergranular films. Figure 1 illustrates separation along an amorphous film between a perovskite grain and a phase possessing the magnetoplumbite-type structure. We note that, rather than breaking along its midline, the intergranular film has adhered to one of the crystalline phases. Cracks have not been observed within single grains of the matrix minerals, although fractures of this kind are likely to propagate rapidly over large distances, and thus their recognition would be difficult.

Regions 10-20 nm in diameter, and showing no diffraction contrast - presumably due to their amorphous state - were observed at trigranular junctions in some samples of Synroc C [3]. In other specimens, distinct triple point regions could not be detected. Although an unequivocal statement is unwarranted, the occurrence of these structural features appears to be dependent upon the presence of silicon as an impurity. Using the 2 nm probe of the analytical TEM (FEG), the amorphous nature of these regions was confirmed, and their major chemical constituents determined to be aluminium,

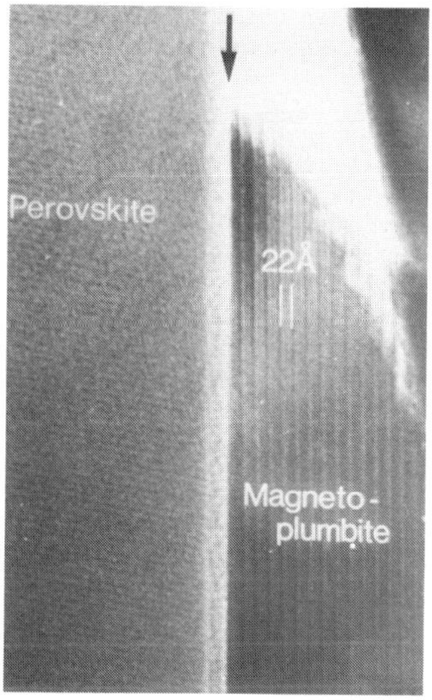

Fig. 1. High resolution transmission electron micrograph of an intergranular film lying at the interface between a perovskite and magnetoplumbite-type crystal. Cleavage has begun, with the crack (arrowed) running between the crystallites, and the film adhering to the perovskite grain.

phosphorous, cesium, sodium and silicon (Figure 2). (Titanium and calcium were also detected, most probably through fluorescence from neighbouring grains.) The number of triple points examined in detail was small, but it is believed these results are representative of such features. The TEM (FEG) instrument was also used to assess the cesium content of interphase layers. Again, only a few cases could be thoroughly investigated. However, regardless of the presence or absence of silicon, cesium and sodium were usually detected in intergranular films.

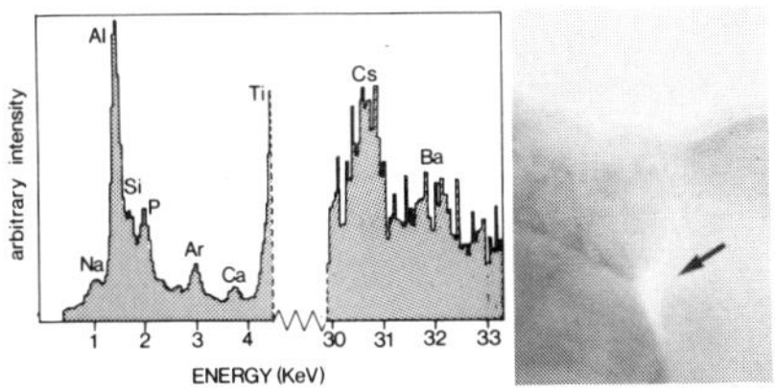

Fig. 2. Bright field transmission electron microscope image of a glassy triple point in Synroc C. Energy-dispersive x-ray analysis for the region show it to be rich in silicon, aluminium, phosphorous, sodium and cesium.

SIMS AND XPS ANALYSIS OF FRACTURE SURFACES

SIMS and XPS techniques collect data from large areas of sample surface, and so do not distinguish between chemical species arising from grains, pores and intergranular films. Analyses reflect the compositional average of the surface under examination.

Mass spectrometry of freshly fractured Synroc C surfaces invariably exhibited well defined cesium, sodium, potassium and silicon peaks (Figure 3(a)), which were reduced substantially after removal of ~ 5 nm surface layer (Figure 3(c)) by ion milling. Other elements such as aluminium, calcium, titanium, strontium, zirconium, molybdenum and barium showed no such enhancement. (Note that both XPS and SIMS are very insensitive to phosphorous, so failure to detect this element is inconclusive.) The leach resistance of elements residing in intergranular layers exposed by fracture, can be assessed by dipping the specimen in doubly distilled deionised water at ambient temperatures for five minutes. As evidenced by Figure 3(b), these conditions are sufficient to remove the surface features enhanced in alkali elements and silicon. Further contact with water does not significantly alter the concentration of any element. An indication of the width of these layers may be obtained by SIMS depth profiling. The amplitudes of the alkali element signals typically reached about four times the bulk level within the first two monolayers, declining to bulk levels at a depth of 1 nm (Figure 4).

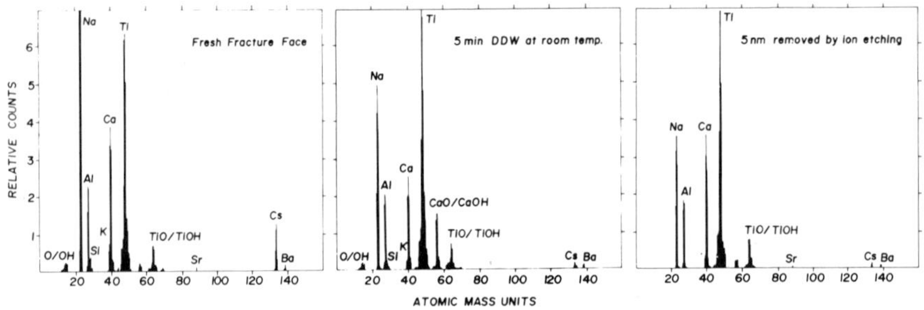

Fig. 3. Secondary ion mass spectrographs of
(a) a fresh fracture surface of Synroc C,
(b) a fracture surface treated for 5 minutes in doubly distilled water at ambient temperatures,
(c) the same surface after 5 nm has been removed by ion milling.

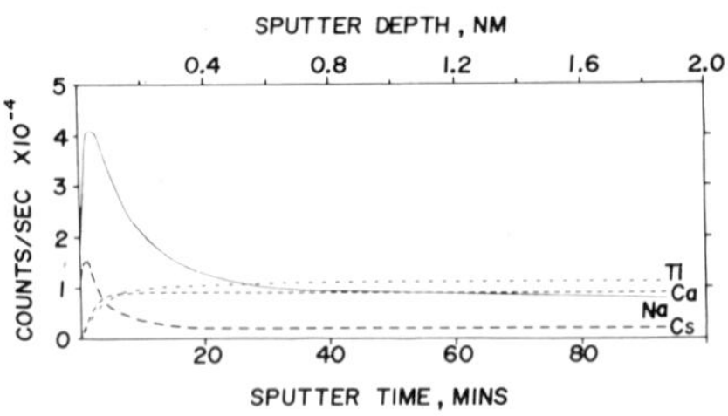

Fig. 4. SIMS depth profile of a fresh fracture face of Synroc C, illustrating enhancement of cesium and sodium.

XPS of fresh fracture faces in Synroc C yielded data in general agreement with the results obtained from SIMS (Figure 5). The concentration of alkali elements on these surfaces, as determined by XPS, was ~ 2.5 times the bulk level. When it is borne in mind that the XPS signal is averaged over a greater depth (2 nm) than that utilised for SIMS analyses (0.3 nm), it is not unexpected that a qualitatively less dramatic enhancement is observed when the former technique is employed. XPS also indicated segregation of aluminium and a little molybdenum into intergranular regions. All other elements (viz. calcium, zirconium, barium, titanium and oxygen) were depleted with respect to bulk concentrations, and, as the surface layers were removed by ion etching, the relative percentages of these species rose to expected matrix levels. XPS analyses of fracture surfaces after short immersions in doubly distilled water at room temperature, demonstrated the rapid removal of alkalis and aluminium from intergranular films by this treatment.

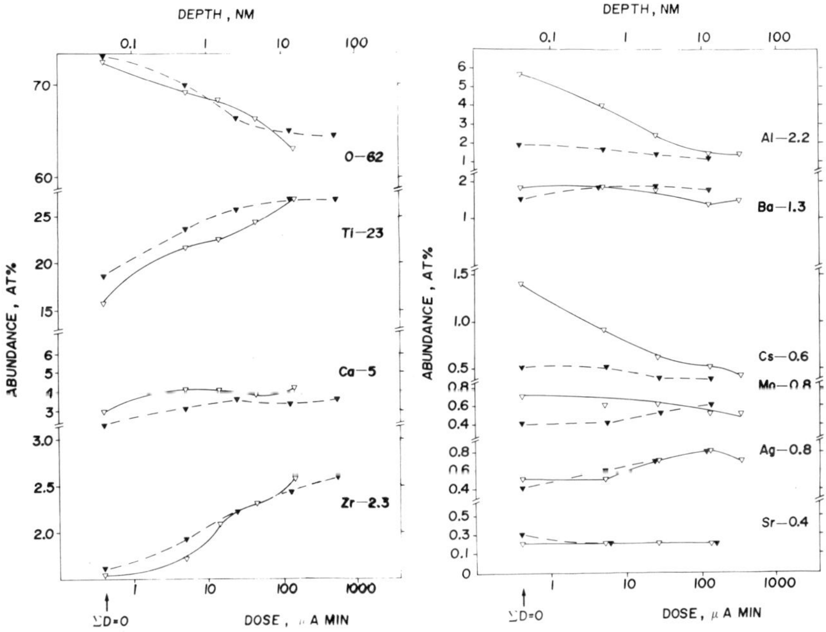

Fig. 5. XPS depth profiles for a fracture surface of Synroc C before (▽) and after (▼) treatment in doubly instilled water at room temperature. Expected bulk concentrations of elements are shown on the right hand side of each figure.

SAM OF MICROVOIDS

Cleavage of Synroc monoliths not only exposes intergranular films, but also microvoids that were usually 10-20 μm in width. SAM of these latter features was carried out in the secondary electron and cesium Auger mapping modes. Figure 6(a) presents the secondary electron image of a pore exposed by fracture. The cesium map of this initial surface (Figure 6(b)) shows a strong cesium signal over a substantial fraction of the cleavage face, with particularly high intensity within the interior of the microvoid. After ion etching to a depth of 15 nm, the cesium-rich surface layer is effectively removed (Figure 6(c)).

Fig. 6. (a) Scanning electron micrograph, taken in the secondary electron mode, of a microvoid in a fracture surface of Synroc C.
(b,c) Cesium maps of the same area before and after removal 15 nm by ion etching.

Current fabrication techniques for Synroc C produce material whose density is routinely greater than 98% of theoretical density. Microvoids, which are responsible for a reduction of density, nearly always occur within rutile-rich areas [4]. Rutile, that is added in "excess", and accumulates by the oxidation of titanium metal that is added to maintain a low oxygen fugacity during fabrication, contains no waste elements [2] and is very leach resistant [5]. Volatile HLW species, such as cesium, are trapped in these voids during hot pressing, condense on pore surfaces, and are thus encapsulated within the rutile regions. Provided that Synroc is of high density, and no significant cracking occurs, the contribution to alkali element leach rates from this source will be limited by diffusion along grain boundaries which terminate within a pore.

THE CORRELATION OF MICROSTRUCTURAL FEATURES TO SHORT TERM LEACHING PHENOMENA

The fact that substantial "wetting" of crystallites by amorphous films occurs, and alkali ions are rapidly removed from such layers by brief exposure to water at ambient temperatures, provides the basis for a ready explanation of short term dissolution characteristics. Higher initial cesium leach rates, a feature of all Synroc C samples, are largely the result of H_3O^+ ion exchange of Cs^+ from intergranular films and pore surfaces, which intersect the exterior of a bulk specimen [6]. In the longer term, cesium losses will have contributions from both hollandite dissolution and diffusion limited leaching from grain boundaries [6]. Near surface, cesium containing microvoids, in which intergranular films terminate, will also augment the overall dissolution rate of Cs^+.

Coles and Bazan [7] conducted dissolution experiments on both monolithic and crushed Synroc C samples. After normalising the surface areas, the leach rates for strontium, calcium and barium, obtained from both type of specimen agreed closely. But cesium dissolution from the crushed sample was consistently more rapid by a factor of 5 (Figure 7). The continuing higher loss rate for cesium from crushed material may be attributed to the relatively poor quality of the Synroc (95% TD) employed by these workers. Such a a specimen must have been very porous, and for the powdered sample, near surface microvoids would provide a large reservoir of cesium, which could be accessed via the rapid diffusion paths furnished by intergranular films. It is noteworthy that in this poorly densified Synroc, barium leach rates were always less than those of cesium. As both waste species reside in hollandite, this comparative result lends further support to the notion that a portion of the cesium occurs in less durable regions of the microstructure. It is pertinent that Oversby and Ringwood [8], who studied Synroc with densities ⩾ 99% TD, observed similar leach rates for cesium and barium (Figure 8).

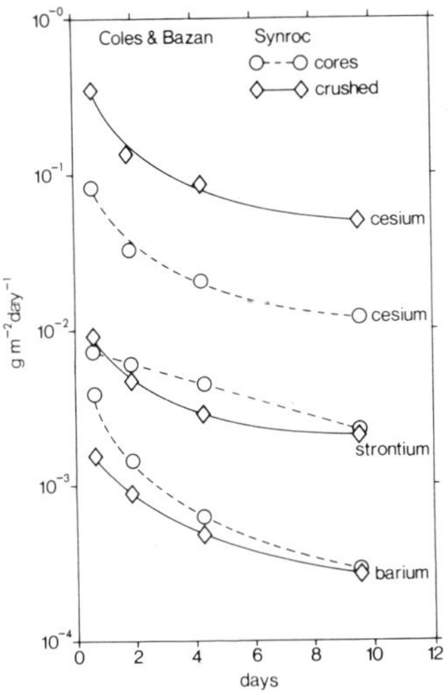

Fig. 7. Leach data for cored and crushed Synroc C of low density (~ 95% TD). Adapted from Coles and Bazen [7].

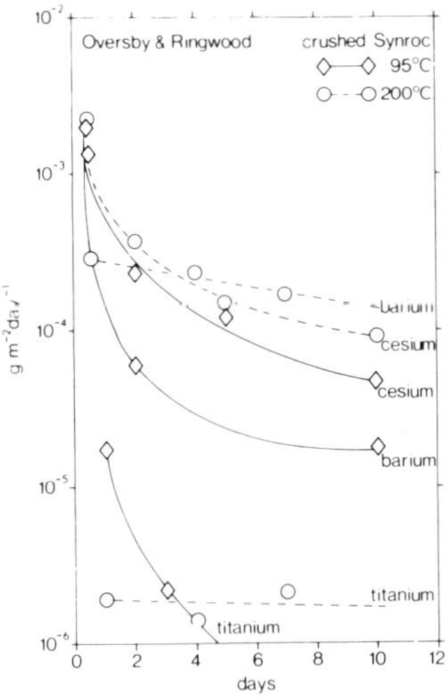

Fig. 8. Leach data for fully dense Synroc C at 95°C and 200°C. Adapted from Oversby and Ringwood [8].

Coles and Bazan also examined the effect of leaching temperature (25°C versus 75°C), and found this parameter was only significant in terms of cesium dissolution from crushed Synroc (Figure 9). At the higher temperature cesium losses were approximately an order of magnitude greater than at 25°C. The leach rates for species which do not segregate into grain boundaries, viz. barium and strontium, showed no temperature dependence. Oversby and Ringwood carried out similar experiments at 95°C and 200°C, but unlike Coles and Bazan found the hydrothermal durability of their specimens not to be strongly contingent upon leaching temperatures (Figure 8). Thus, it is apparent that in porous material, slightly elevated temperatures accelerate the diffusion of cesium out of microvoids along grain boundaries.

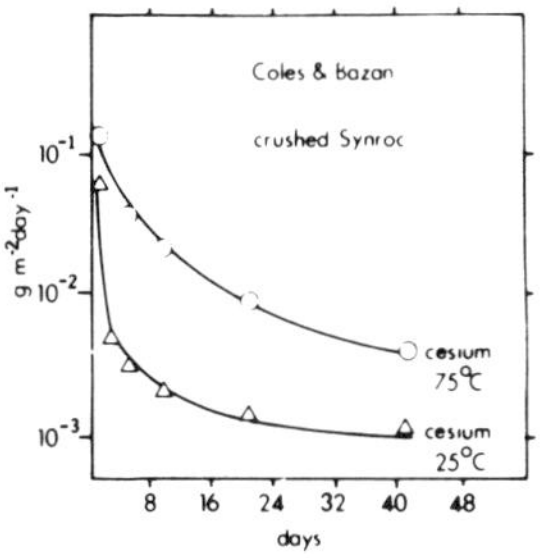

Fig. 9. Temperature dependence of cesium leach rates in porous Synroc C. Adapted from Coles and Bazan [7].

ACKNOWLEDGEMENTS

The authors are especially grateful to T. Chan and J.B. Metson (Surface Science Western, Ontario), for carrying out the scanning Auger microscopy presented in this paper. Specimens were provided by members of the Australian Atomic Energy Commission Synroc Research Group and Dr S. Kesson of of the Research School of Earth Sciences (ANU). Peter S. Turner is appreciative of the support and assistance given by the staff of the Centre for Solid State Science at Arizona State University, whilst undertaking TEM/FEG experiments. T. White was supported by AINSE and NERDDC postdoctoral fellowships.

REFERENCES

1. A.E. Ringwood, V.M. Oversby, S.E. Kesson, W. Sinclair, N. Ware, W. Hibberson, A. Major, Nucl. Chem. Waste Manage., 2, 287 (1981).

2. D.R. Clarke, Ultramicroscopy, 4, 33 (1979).

3. J.A Cooper, D.R. Cousens, R. Lewis, S. Myhra, R.L. Segall, R. St. C. Smart, P.S. Turner, T.J. White, J. Amer. Ceram. Soc., 68, 64 (1985).

4. R.L. Segall, S. Myhra, R. St. C. Smart, P.S. Turner, Final Report NERDDC No. 80/0049, Griffith University, Synroc Group, 1984.

5. S. Forberg, T. Westermark, H. Larker, B. Widell, in Scientific Basis for Nuclear Waste Management I, edited by G. McCarthy (Plenum Press, New York, 1979) pp. 201-205.

6. S. Myhra, R.L. Segall, R. St. C. Smart, M. Stephenson, P.S. Turner, T.J. White, International Seminar on International Waste Management for Final Disposal, 10-14 June, 1985. KFA/PTB Juelich.

7. D.G. Coles and F. Bazan, Nucl. Tech., 56, 226 (1982).

8. V.M. Oversby and A.E. Ringwood, Radioactive Waste Manage., 2, 223 (1982).

REDOX PROCESSES AND UO_2(s)SOLUBILITY: THE DETERMINATION OF THE UO_2^{2+}/U^{4+} REDOX POTENTIAL AT 25°C in $HClO_4$ MEDIA OF DIFFERENT STRENGTH

JORDI BRUNO, INGMAR GRENTHE and BO LAGERMAN
Royal Institute of Technology, Dept. of Inorganic Chemistry
S-100 44 Stockholm 70, Sweden

SUMMARY

The redox potential of the couple UO_2^{2+}/U^{4+} has been determined at 25°C in various $HClO_4$ media by direct potentiometric determinations in 0.25, 0.5, 1, 2 and 3 M $HClO_4$ media. Some spectrophotometric measurements of the equilibrium:

$$Cu(s) + UO_2^{2+} = U^{4+} + Cu^{2+} + H_2O$$

in 0.5,0.75,1.0,1,25,1.5,2.0 and 3.0 M $HClO_4$ media have also been made.

The results have been extrapolated to a common standard state (the pure water state) by using two different methods, the Brønsted-Guggenheim-Scatchard specific interaction theory (SIT) [11] and the more elaborated Pitzer approach [10].

The values obtained at the infinite dilution state are $\log K_0 = 8.8 \pm 0.1 (E_0 = 260 \pm 3 mV)$ by using the SIT approach and $\log K_0 = 8.4 \pm 0.1 (E_0 = 248 \pm 3 mV)$ by using the Pitzer approach.

The extrapolation methods are compared and their reliability in treating electrolyte mixtures containing ions of high charge (U^{4+}) is discussed.

INTRODUCTION

The standard potential for the UO_2^{2+}/U^{4+}couple i.e. the reaction

$$UO_2^{2+}(aq) + 4H^+(aq) + 2e^- = U^{4+}(aq) + H_2O \qquad (1)$$

is an important thermodynamic quantity, nevertheless its value is not precisely known. The standard potential is used to calculate other thermodynamic quantities, notably the standard free energy of formation of U^{4+}(aq), cf. Fuger and Oetting [1]. It is also a critical datum in the modelling of uranium solubility in natural waters, as the solubility of UO_2(s) is highly dependent on the redox state.

Fuger and Oetting have discussed some experimental determinations of the standard UO_2^{2+} / U^{4+} redox potential made by Sobkowski and Minc [2], and by Nikolaeva [3], these and other previous determinations are summarized in Table 1. The reviewers have in our opinion underestimated the experimental shortcomings, furthermore they have not described and discussed the procedure used for the extrapolation of the experimental data to the infinite dilution reference state.

We will divide this report in three sections and discuss:

- the quality of previous experimental data

- our own determination of the standard UO_2^{2+}/U^{4+} potential
- the extrapolation methods to obtain the standard potential in the infinite dilution state.

Table 1

Authors	year	medium	E mV	ref
Luther, Michie	1908	0.5 M H_2SO_4	281(419)	4*
Tittlestadt	1910	0.5 M H_2SO_4	264(404)	5*
Taylör, Smith	1944	⇒ 0	334	6
Kraus, Nelson	1949	⇒ 0	310	7
Sobkowski	1960	⇒ 0	329	2
Grenthe, Varfeldt	1969	4M($NaClO_4$,$HClO_4$)	268	8**
Nikolaeva	1973	⇒ 0	273	3

* Corrections for complexation of U^{4+} and UO_2^{2+} with sulphate have been made (uncorrected values within parenthesis).

** Estimation of the liquid junction potential has been made.

Comments on previous experimental determinations of the UO_2^{2+}/U^{4+} standard potential

The measurements made in H_2SO_4 media have to be corrected for the sulphate complexation. These corrections have been made by using the constants compiled in ref [13]. The corrected standard potentials have fairly large errors (±20 mV), mainly due to uncertainties in the equilibrium constants for the protonation of SO_4^{2-} and for the formation of sulphate complexes. The error quoted by Kraus and Nelson [7], ±30 mV, seems mainly to be due to shortcomings in the measuring procedure.

Many authors have observed that the equilibrium potential is established very slowly. This is often the case for processes involving a two-electron transfer with major structural reorganization. The exchange current density at the inert electrode is then low, which means that the measured potential may be influenced by other redox couples and the surface state of the electrode.

In order to decide whether one measures an equilibrium potential or a "mixed" potential, one should both ascertain that the absolute value of the measured emf is independent of the material used in the inert electrode, and that the electrode gives a Nernstian response to changes in the concentrations of the red. and ox.forms.

Sobkowski and Minc [2] have tested two different electrode materials, smooth platinum and platinum black, electrolytically deposited. However, they have not ascertained that the electrodes give a Nernstian response, since only one $[UO_2^{2+}]/[U^{4+}]$ ratio was measured at each perchlorate level.

Nikolaeva [3] tested three electrodes (one polished Pt and two Pt-black) and varied the ratio UO_2^{2+}/U^{4+} from 3:4 to 30:1. The different electrodes gave the same readings and the author claims that a Nernstian response was obtained.

In order to avoid excessive hydrolysis of U^{4+} (aq), the redox potential measurements must be made at fairly high acidity. This may cause problems with the liquid junction potentials between the test solution and the reference half-cell if a cell with liquid junctions is used. Sobkowski et al. have avoided this problem by using a cell with liquid junction but with the same concentrations of H^+ throughout. The liquid junction will in this way only be influenced by the changes of the concentrations of UO_2^{2+} (aq) and U^{4+}(aq) in the test solution. We have evaluated this effect and it should be 1 mV, or less.

Nikolaeva [3] has used a cell where the test solutions contain $HClO_4$ as the ionic medium and where the reference half-cell is

Tl(Hg) TlCl saturated KCl

This choice is unfortunate, as $KClO_4$ will precipitate in the liquid junction and the diffusion potential will be very large because of the large differences in the concentrations of H^+ between test solution and reference half-cell. It is unlikely that this liquid junction potential is reproducible and the value of the measured UO_2^{2+}/U^{4+} may thus be in error.

The measurements, made by Nikolaeva [3], at t> 25°C may also be influenced by thermodifussion potentials because the test solution and the reference solution are not then at the same temperature.

Grenthe and Varfeldt [8] made a determination in connection with a study on the complex formation between uranium (U^{4+} and UO_2^{2+}) and fluoride. They used an Ag, AgCl reference electrode connected to the solution through a salt bridge. The arising junction potential as a result of the variation in UO_2^{2+} and U^{4+} concentrations in the test solution was estimated to be of the same magnitude as in Sobkowski's case (1.5 mV). Nevertheless, the determination of the magnitude of the E_0 of the semicouple Ag, AgCl is more problematic since no data in 4.0 M perchlorate is available. The error on the value extrapolated from measurements in 1.0, 2.0 and 3.0 M $NaClO_4$ by means of the SIT approach is at most ±5 mV.

Experimental determination of the UO_2^{2+}/U^{4+} redox potential as function of the ionic strength

In view of the experimental shortcomings mentioned above, we have made a redetermination of the UO_2^{2+}/U^{4+} potential by using a cell without liquid junction of the following type:

glass	$H^+(C_1)$, $UO_2^{2+}(C_2)$, $U^{4+}(C_3)$	Pt, Au
electrode	$[ClO_4^-] = C_1 + 2C_2 + 4C_3$	Pt, glassy carbon

The glass electrode was measured vs a hydrogen electrode in a separate experiment.

The ratio UO_2^{2+}/U^{4+} was varied a factor of ten at each perchlorate level, in order to establish whether the electrodes gave a Nerstian response, or not. For a more detailed description of the experimental technique the reader is referred to [9]. Four different electrode materials were tested, Pt, Au, Pd-black and glassy carbon. The electrode readings agreed within 0.3 mV, or less. The Nernstian response is depicted in Figure 1.

The concentration of H^+ was varied from 0.23 m to 3.5 m and the reversibility tests outlined above were made at each acidity level investigated.

The following results were obtained:

Table 2

I (m)	$[H^+]$(m)	E		log K	
0.331	0.239	305	3	10.3	0.1
0.643	0.487	306	2	10.3	0.1
1.21	1.05	299.0	0.7	10.11	0.02
2.37	2.13	279.2	0.5	9.44	0.02
3.89	3.43	268.3	0.7	9.07	0.02

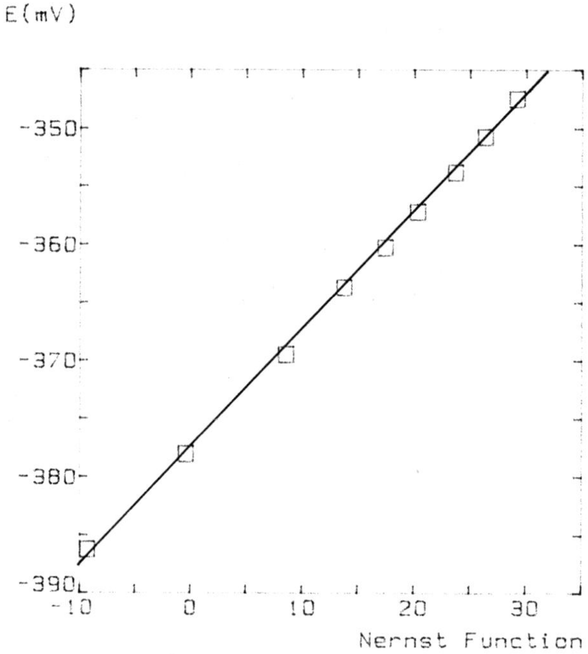

Fig. 1.

In view of the experimental difficulties often encountered in emf investigations of two-electron transfer processes we believe that it is important to determine the UO_2^{2+}/U^{4+} redox potential by an independent experimental method. We have determined the equilibrium constant for the reaction

$$Cu(s) + UO_2^{2+}(aq) + 4H^+(aq) = Cu^{2+}(aq) + U^{4+}(aq) + 2H_2O \qquad (2)$$

from which the value of the UO_2^{2+}/U^{4+} redox couple can be calculated.

The equilibrium constants were obtained by equilibrating solutions of known H^+ and UO_2^{2+} concentrations with Cu(s) in powder form. The time to reach equilibrium varied from 3 to 12 days, depending on the acidity of the solution. The concentrations of UO_2^{2+}, U^{4+} and Cu^{2+} respectively were determined spectrophotometrically at 412, 647 and 826 nm. The error in these measurements is about 3%. This error can be reduced considerably by using other analytical methods (e.g. titrimetric) and a more precise investigation will be carried out in the near future.

The results plotted in Fig. 2 show that the agreement between the two methods is satisfactory.

Thus, we have established that one can make precise emf measurements of the UO_2^{2+}/U^{4+} redox potential and that true equilibrium can be obtained also in emf measurements.

EXTRAPOLATION TO THE PURE WATER REFERENCE STATE

In order to compare the experimental results of different investigators and our determinations at different ionic strength, we need a method for the estimation of activity factors of the reactants and the products in equilibrium (1). We have used the Brønsted, Guggenheim, Scatchard (SIT)[11] and the Pitzer[11,12] specific interaction for this purpose.

The Specific Ion Interaction Theory (SIT) [11]

The Brønsted, Guggenheim, Scatchard SIT [11] approach assumes that the activity coefficient of a single ion i is the summation of two terms. The Debye-Hückel term which takes into account the pure long range electrostatic component for the nonideal behaviour of ions in solution and a second term $\Sigma\varepsilon(i, j, I)m_j$ which accounts for the short range interactions, but only between ions of opposite charge. In the second term one often assumes that ε is concentration independent. This is an approximation which is fairly good for ions of low charge, but not for ions of high charge cf. Figure 3. This second term is assumed to be a linear function of m, the molality of the main component of the solution

$$\log\gamma_i = -Z^2D + \Sigma\varepsilon(i,j,I)m_j \qquad (3)$$

where $D = 0.5109\sqrt{I}/(1+1.5\sqrt{I})$ (3a), is the Debye-Hückel term.

By applying this to the equilibrium expression for the reaction

$$UO_2^{2+}(aq) + H_2(g) + 2H^+(aq) = U^{4+}(aq) + H_2O \qquad (4)$$

we obtain

$$K_0=K_c\gamma(U^{4+})/\gamma(UO_2^{2+})\gamma^2(H^+) \qquad (5)$$

hence

$$\log K_c - 10D = \log K_0 - m(ClO_4^-)[(\varepsilon(U^{4+},ClO_4^-)-\varepsilon(UO_2^{2+},ClO_4^-)-2\varepsilon(H^+,ClO_4^-)] \quad (6)$$

According to the SIT this should be a linear function of $m_{ClO_4^-}$ with log K as the intercept on the (log K_c-10D) axis and with $[\varepsilon(U^{4+},ClO_4^-)-\varepsilon(UO_2^{2+},ClO_4^-)-2\varepsilon(H^+,ClO_4^-)]$ as the slope, cf. Figure 2. logK_c-10D is an approximately linear function of $m_{ClO_4^-}$. However, there is a large discrepancy between our data and those of Sobkowski, particularly at high ClO_4^- molalities. (see Fig. 2)

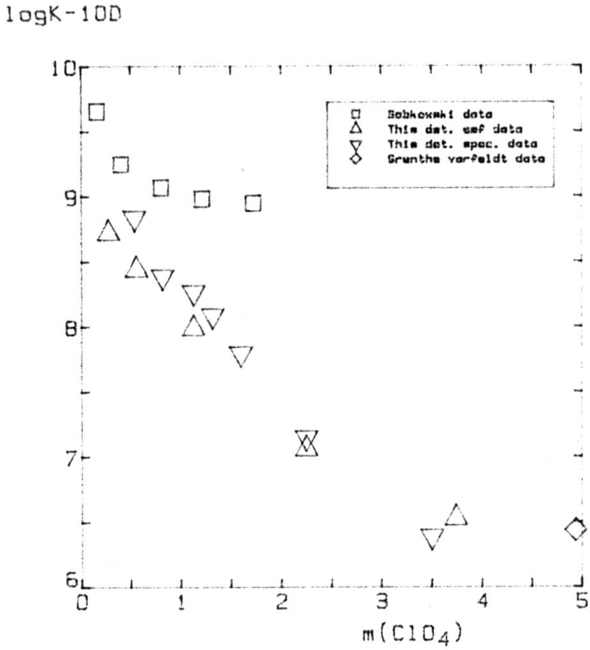

Fig 2

The result of the linear fitting gives log K_0=8.8±0.1 (see Fig 2) and $\varepsilon(UO_2^{2+},ClO_4^-) + 2\varepsilon(H^+, ClO_4^-) - \varepsilon(U^{4+}, ClO_4^-)= -0.64$

Since the interaction coefficients

$\varepsilon(UO_2^{2+}, ClO_4^-)= 0.46\pm0.01$ and $(H^+ ,ClO_4^-)= 0.14\pm0.01$ are well known from activity data [12], we obtain $\varepsilon(U^{4+} ,ClO_4^-)=1.4$

The same treatment of the data obtained from the system U/Cu gives

log K_0= 9.2 ±0.2 and $\varepsilon(U^{4+} , ClO_4^-)= 1.6$.

The Pitzer approach [10]

The Pitzer method is a more developed specific interaction approach than the SIT and takes the interactions between ions of the same charge sign into account. It uses a virial expansion which permits the calculation of fairly accurate activity factors also at higher ionic strengths. The dependence of the second virial coefficient B_{kj} (corresponding approximately to ε in SIT) on ionic strength is also taken into account in this approach: $B_{kj}=2(\beta^{o}+\beta^{1}f1))$ where f1 is a funcion of the ionic strength.

Because of the lack of data on the parameters for the interactions between ions of the same charge sign, they will not be included in the treatment developed below. The effect of these interactions is, at the I range studied, generally small and can be neglected.

Hence, for the activity coefficient of a positive ion i we have the expression [10] [11].

$$\ln\gamma_i = Z_i^2 D+2\Sigma m_j[B_{ij}+\Sigma m_k Z_k C_{ij}] + \Sigma_k\Sigma_j m_k m_j[Z_i^2 B_{kj} + Z_i C_{kj}] + + \Sigma_k\Sigma_j[Z_1^2 B'_{kj} + Z_1 C_{kj}] \qquad (7)$$

which applied to the reaction (4) gives:

$$\ln K_o = \ln K_c + 10D + 4mClO_4^- \beta^{o}(U^{4+}) - \beta^{o}(UO_2^{2+}) - 2\beta^{o}(H^+) + 2m(ClO_4^-)f1[\beta^1(U^{4+}) - \beta^1(UO_2^{2+}) - 2\beta^1(H^+)] + 2[m(U^{4+})m(ClO_4^-) + m(UO_2^{2+})m(ClO_4^-) + m(H^+)m(ClO_4^-)][C(U^{4+}) - C(UO_2^{2+}) - 2C(H^+)] + 10f2[m(U^{4+})m(ClO_4^-)\beta^1(U^{4+}) + m(H^+)m(ClO_4^-)\beta^1(H^+)] \qquad (8)$$

where:

B_{kj} = The second virial coefficient, $B_{kj}= 2(\beta^{o}+\beta^{1}f1)$, C_{ij}= The third virial coefficient, B'_{kj}=The derivative of the second virial coefficient, k= all cations, j=all anions.

D is the Debye-Huckel term, $D = 0.392[\sqrt{I}/(1+1.2\sqrt{I})+(2/1.2)\ln(1+1.2\sqrt{I})]$ (9)

Note the difference with the D-H term in the SIT, eq.3a.

$\beta^{o}(UO_2(ClO_4)_2) = 0.6113$, $\beta^{o}(HClO_4) = 0.1747$, $\beta^1(UO_2(ClO_4)_2 = 2.144$

$\beta^1(HClO_4) = 0.2931$, $C(UO_2(ClO_4)_2) = 0.00767$, $C(HClO_4) = 0.00410$

(All data in ref. [10])

The results obtained from the treatment from the emf data are $\log K_o = 8.1\pm0.1$,

and the parameters: $\beta^{o}(U^{4+}) = 1.79$; $\beta^1(U^{4+}) = 1.49$; $C(U^{4+}) = -0.244$

By using these parameters with the data obtained from the system U/Cu we obtain $\log K_0$=8.5±0.2.

The Pitzer parameters are not very precise because we have only data at five different ionic strengths. At this stage they should only be considered as fitting parameters for the nonlinear behaviour. It is highly desirable to determine them from independent activity coefficient data.

DISCUSSION AND CONCLUSIONS

The reversibility of redox reaction (1), involving two electron transfer and bonding rearrangements between UO_2^{2+}/U^{4+}, has been questioned. However, from the fact that we have obtained Nernstian responses in all titrations run at different ionic strength, and by using different inert electrode materials (Pt, Au, Pd, graphite) we can conclude that the redox process is reversible. The fact that equilibrium measurements with the semicouples UO_2^{2+}/U^{4+} and $Cu(s)/Cu^{2+}$ give the same results (within the experimental error) is an additional proof. We have also observed a relation between the proton concentration and the time for the attainment of equilibrium in both reactions (1) and (2). This may explain the difference between emf and spectrophotometric data at the lowest $HClO_4$ level (0.5 M).

By working at high $HClO_4$ concentrations we have minimized the U(IV) hydrolysis. Furthermore, since we used a diffusion free reference system (GE) another cause of experimental uncertainty was eliminated.

Even if we have obtained reliable redox data in the range 0.5 to 3 M $HClO_4$, we must discuss the extrapolation of these data to the pure water state. Two well established approaches have been used for this purpose. In the SIT approach a linear relationship between logK-10D and $m(ClO_4^-)$ is assumed. From the experimental data collected and plotted in Fig. 2 one observes a non linear behaviour. In Fig. 3 we have plotted ε vs I for some of the salts in the redox reaction, $HClO_4$ and $UO_2(ClO_4)_2$, also for $ThCl_4$ and $Th(NO_3)_4$ as models for $U(ClO_4)_4$, they are the only available data [10] for 1:4 type salts. One observes that while $\varepsilon(HClO_4)$ and $\varepsilon(UO_2(ClO_4)_2)$ are nearly concentration independent and thus behave in the way predicted by the SIT, $\varepsilon(ThCl_4)$ and $\varepsilon(Th(NO_3)_4$ increase at lower I. This could be the reason for the deviations from linearity observed at low $m(ClO_4^-)$.

The behaviour of $\varepsilon(ThCl_4)$ and $\varepsilon(Th(NO_3)_4)$ deviates strongly from that of the interaction coefficients of 1:1, 2:1 and 3:1 type of salts. If this is typical for high charged ions (Z>3) one can question the validity of the SIT approach to describe the activity coefficients of these species.

Hence, we tried to use the more elaborated Pitzer specific interaction approach, which include parameters for the variation of the interaction coefficients with ionic strength. The problem in this case is that one does not have accurate values of the Pitzer parameters, which instead have to be determined from equilibrium constant data at various ionic strengths. This method is laborious and in general not very precise.

Fuger and Oetting [1] proposed a value of 273±5mV for the normal potential of the UO_2^{2+}/U^{4+} couple. We obtain values which are considerably lower, 260 mV using the SIT, and 248 mV using the Pitzer approach. The precision of the data are satisfactory but additional information on the mean activity coefficients of some $M(ClO_4)_4$ salts are needed in order to improve the accuracy of the E_0 value.

Fig 3

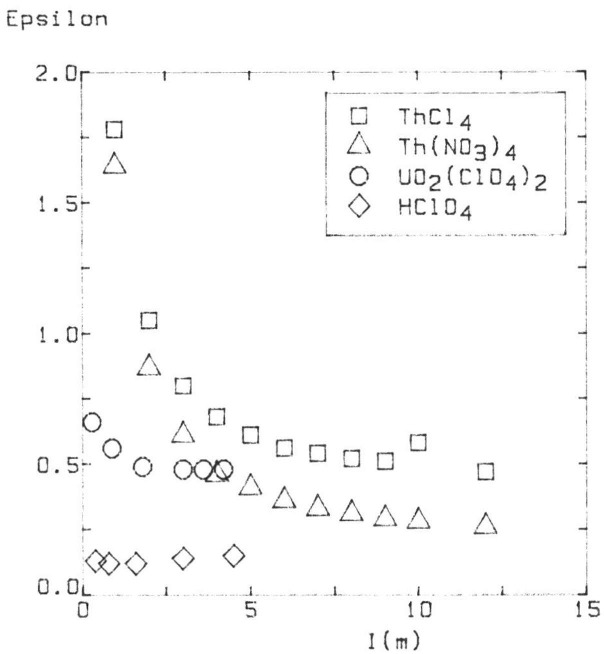

Acknowledgements

Anders Molin is acknowledged for carrying out the spectrophotometric analysis in the U/Cu system.

REFERENCES

1. J. Fuger and F.L. Oetting, The Chemical Thermodynamics of Actinide Elements and Compounds-II. The Actinide Aqueous Ions. Intl. Atomic Energy Agency, Vienna, pp 16-60, 1976
2. J. Sobkowski and S. Minc, J. Inorg. Nucl. Chem. 23,81 (1961)
3. N.M. Nikolaeva, Izv. Sib. Otd. Akad. Nauk SSSR, Ser. Khim 1, 14 (1974)
4. R. Luther and A.C. Michie, Z. Electrochem. 14, 826 (1908)
5. N. Tittlestadt, Z. Phys.Chem. 72, 257 (1910)
6. J.K. Taylor and E.R. Smith, Report A 1972, 1944
7. K.A. Kraus and F. Nelson, J. Amer. Chem. Soc. 71, 2517 (1949)
8. I. Grenthe and J. Varfeldt, Acta. Chem. Scand. 23, 988 (1969)
9. J. Bruno and I. Grenthe, Environmental and Toxicological Chemistry, 10, 257 (1985)
10. K. Pitzer in Activity coefficients in electrolyte solutions Part I, edited by R.M. Pytkowicz (CRC Press, Florida, 1979) pp. 157-208,

11. M. Whitfield in Activity coefficients in electrolyte solutions Part II, edited by R.M. Pytkowicz (CRC Press, Florida, 1979) pp.153-299.
12. L. Ciavatta, Anali di Chimica. Soc. Chim. It. 1980, pp. 551-567.
13. J.J. Katz and G.T. Seaborg, The Chemistry of the Actinide Elements, 1st ed. (Methuen & Co. Ltd., London, 1957) pp. 184-186.

OXIDATION OF CANDU UO_2 FUEL BY THE ALPHA-RADIOLYSIS PRODUCTS OF WATER

D.W. SHOESMITH, S. SUNDER, L.H. JOHNSON AND M.G. BAILEY
Atomic Energy of Canada Limited
Whiteshell Nuclear Research Establishment
Pinawa, Manitoba, Canada R0E 1L0

ABSTRACT

The oxidation of CANDU fuel (UO_2) by the alpha-radiolysis products of water has been investigated using electrochemical and X-ray photoelectron spectroscopic experiments. Experiments with O_2 and H_2O_2, two of the expected products of radiolysis of water, indicate that the rate of oxidation of UO_2 by H_2O_2 is about 200 times faster than by dissolved oxygen. Oxidation by both H_2O_2 and O_2 shows pH dependence. Possible reaction paths for the oxidation of UO_2 by radiolysis products are discussed.

INTRODUCTION

Canadian nuclear power reactors (CANDU™) use natural UO_2 as fuel. The used fuel from these reactors is mainly UO_2 (> 98%), the remainder being fission products of uranium and other actinides formed in the reactors. If the used nuclear fuel is to be disposed of in an underground nuclear waste vault, the interaction of the uranium dioxide waste form with the groundwater could be crucial in determining the release of the radionuclides within the fuel, as over 90% of the radionuclide inventory is located within the UO_2 grains. Consequently, much work has been done to understand the dissolution and oxidation mechanisms of UO_2 in water [1-7]. UO_2 is very insoluble in water under reducing conditions, but under oxidizing conditions its solubility increases. Oxidizing conditions at the fuel surface could be produced by the radiolysis of water. It can be argued that the effects of gamma radiolysis of water on UO_2 fuel dissolution are of little significance for the nuclear waste management program, since "gamma fields" will have decreased to insignificant levels at the time of waste-container failure. Alpha radiolysis is more likely to be an important factor in the long-term stability of the disposed UO_2 fuel as "alpha fields" remain high enough to be of significance for more than 100 000 years. Alpha radiolysis of water will produce a mixture of molecular (e.g., H_2O_2) and radical (e.g., OH) oxidants and reductants [8]. The effects of the various oxidants (O_2, H_2O_2, OH) formed by alpha radiolysis, or introduced by other sources (e.g., atmospheric O_2 in groundwater), will be interrelated, as indicated in Figure 1, i.e., the presence of one means that there is a possibility that the other two will be involved in reactions with the UO_2. Therefore, we have investigated the oxidation of CANDU fuel by oxygen and hydrogen peroxide to delineate the mechanism of oxidation of the fuel by the radiolysis products of water. Here we discuss these results in addition to the experiments carried out on UO_2 oxidation by the direct alpha radiolysis of water.

EXPERIMENTAL

All experiments were performed on UO_2 electrodes constructed from unirradiated UO_2 fuel pellets [1,4]. Generally, experiments were performed in 0.1 $mol \cdot dm^{-3}$ $NaClO_4$. All potentials were measured, and are quoted, against the saturated calomel electrode. The potential profile used in our

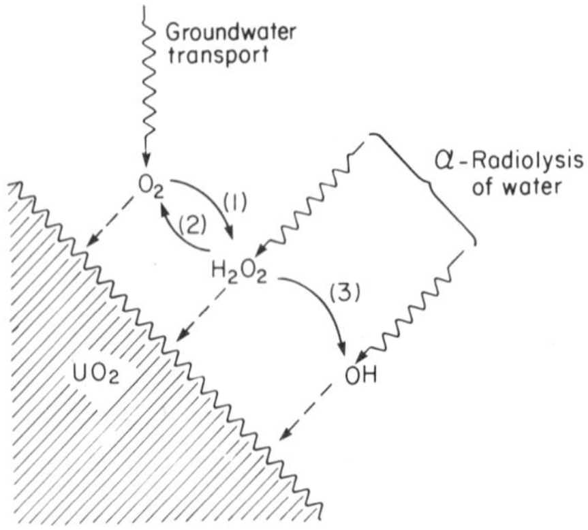

Figure 1. Source of Potential Oxidants and Their Chemical Interrelationships.

corrosion and electrochemical experiments is shown in Figure 2. In region 1, the electrode is cathodically cleaned at -2.0 V [1]. The UO_2 is then allowed to corrode on open-circuit in the presence of dissolved O_2, H_2O_2 or radiolytically decomposed water, and the corrosion potential is followed as a function of time (region 2). The electrode can then be either cathodically stripped, using a voltammetric sweep from the corrosion potential to the initial reducing potential (region 3), or removed for surface analysis by X-ray photoelectron spectroscopy (XPS). The details of our XPS procedures and analyses [1,2] and the cell used in the alpha-radiolysis experiments [9] have been described previously.

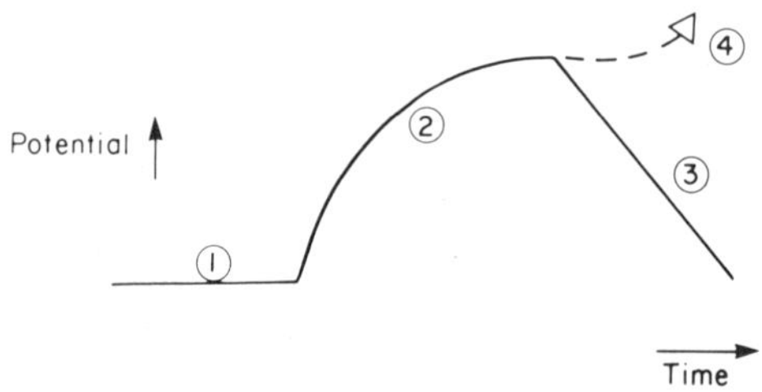

Figure 2. Potential-time profile used in corrosion and electrochemical experiments: (1) electrode cathodically cleaned; (2) open-circuit corrosion; (3) surface film cathodically stripped; (4) electrode removed for surface analysis.

RESULTS AND DISCUSSION

Dissolved Oxygen

Figure 3 shows the UO_2 corrosion potential recorded in air-saturated 0.1 mol•dm^{-3} $NaClO_4$ (pH = 9.5). The potential rises rapidly (in ∼ 1 min) to ∼ -0.400 V from -2.0 V, the potential at which the electrode was cathodically cleaned. Over the next six to eight hours, the corrosion potential rises to ∼ +0.030 V and a film is observed to grow on the UO_2. This film thickens slowly over a period of eight hours, achieving a limiting thickness of 6±1 nm. For longer times, the corrosion potential slowly drifts to ∼ +0.090 V, but no further film thickening is observed. The final potential achieved varies somewhat from electrode to electrode (from ∼ 0.050 V to ∼ 0.090 V), but is reproducible on the same electrode to within ∼ 20 mV. Our XPS data indicate that the surface composition is close to $UO_{2.33}$ after ∼ 10 hours of immersion [10]. This composition is generally achieved prior to the onset of dissolution [1]. Thus, UO_2 dissolution in the presence of dissolved oxygen is preceded by the film-formation process

$$UO_2 + 0.33H_2O \rightarrow UO_{2.33} + 0.66H^+ + 0.66e \quad . \qquad (1)$$

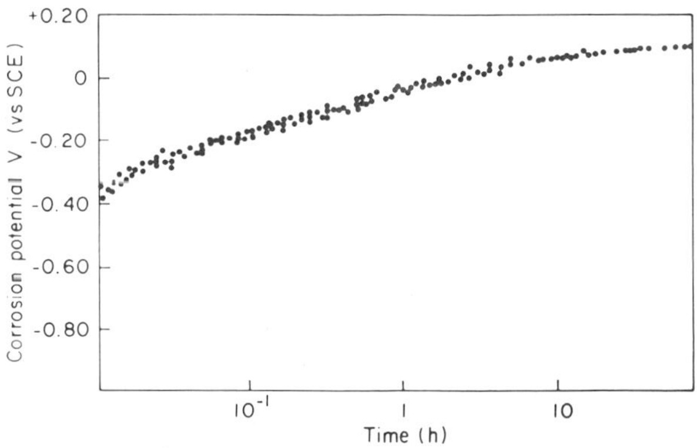

Figure 3. Corrosion potential of the UO_2 electrode as a function of time in air-saturated 0.1 mol•dm^{-3} $NaClO_4$; pH = 9.5; data points from thirteen separate experiments.

If one attributes the slow drift in corrosion potential (from ∼ +0.030 V to ∼ +0.090 V), for times greater than ∼ 8 h, to a slow change in composition of the UO_2, one can envisage the film-formation reaction as composed of two stages, i.e.,

$$UO_2 \rightarrow UO_{2+x} \rightarrow UO_{2.33} \quad . \qquad (2)$$

The corrosion potential, under steady-state dissolution conditions, shows three regions (see Figure 4). In region 1, for pH $\leq$ 2, the potential appears to be pH-dependent, and the steady-state potential is achieved in $\leq$ 1 h, compared with ∼ 10 h for higher pH. In region 2, for 2 $\leq$ pH $\leq$ 10,

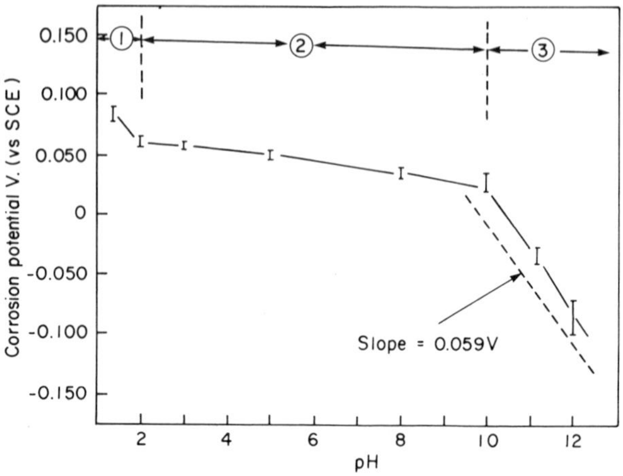

Figure 4. Corrosion potential of UO_2 as a function of pH in air-saturated 0.1 mol•dm^{-3} $NaClO_4$; pH adjusted with $HClO_4$ or NaOH.

the corrosion potential is almost independent of pH, suggesting no involvement of protons in the potential-determining reaction. However, in region 3 (pH $\geq$ 10), the corrosion potential is pH-dependent, indicating a change in the potential-determining reaction to one involving protons. The slope of the corrosion potential vs pH plot is close to 0.059 V (dashed line in Figure 4), suggesting the involvement of one proton.

For low pH values in region 1, our experiments [10] indicate a different mechanism for the oxidative dissolution of UO_2 from that observed in neutral and alkaline solutions. The surface film formed is different from that formed at higher pH values and it is relatively thin, $\sim$ 1 to 2 nm compared with $\sim$ 6 nm at pH = 9. In the neutral pH region, electrochemical measurements show that the UO_2 dissolution reaction is independent of pH [1,11]. However, it is not possible on the basis of the evidence presented here to determine which reaction is controlling the reaction kinetics, the corrosion potential almost certainly being a mixed potential in this region. In alkaline solutions, there is a dependence of the anodic dissolution on pH [5,12], the dissolution current decreasing with increasing pH. Consequently, if the anodic reaction is potential-determining, the corrosion potential would be expected to shift in an anodic direction, which is contrary to the data of Figure 4. On the other hand, the reduction of oxygen at a UO_2 surface shows a first-order dependence on the hydrogen-ion concentration, which would account for the 0.059-V cathodic shift of the corrosion potential, if the oxygen reduction reaction is potential-determining.

These results suggest that the mechanism of UO_2 corrosion in the presence of dissolved oxygen changes from neutral to alkaline solutions. In neutral solutions, the anodic reaction, which is a combination of the film-formation reaction, (1), and the film-dissolution reaction

$$UO_{2.33} + 0.66H^{+} \rightarrow UO_2^{2+} + 0.33H_2O + 1.33e \quad , \qquad (3)$$

is pH-independent, the overall reaction being the sum of reactions (1) and (3), i.e.,

$$UO_2 \rightarrow UO_2^{2+} + 2e \quad . \qquad (4)$$

The subsequent hydrolysis of UO_2^{2+} does not confer a pH dependence on the overall dissolution reaction, indicating an irreversible process. The corrosion reaction appears to be jointly controlled by the kinetics of both the anodic (UO_2 dissolution) and cathodic (O_2 reduction) reactions. In alkaline solutions, a step involving one proton appears to be potential-determining. One possibility is a reaction involving the formation of a superoxide (HO_2^-) species [13], i.e.,

$$O_2 + H_2O + 2e \rightleftharpoons HO_2^- + OH^- \quad . \qquad (5)$$

Hydrogen Peroxide

Figure 5 shows the corrosion potential of UO_2 as a function of time after the addition of H_2O_2. The two concentrations shown are close to those of dissolved oxygen in air-saturated ($[O_2] = 2.66 \times 10^{-4}$ mol•dm^{-3}) and oxygen-saturated ($[O_2] = 1.27 \times 10^{-3}$ mol•dm^{-3}) solutions. Cathodic stripping voltammetry shows that surface oxidation of UO_2 occurs, via a reaction similar to reaction (1). Our XPS data show a composition close to $UO_{2.33}$ is achieved at these concentrations of H_2O_2. Hence, the overall oxidation process in H_2O_2 at these concentrations is the same as that in dissolved O_2. However, the rate of oxidation is faster for both stages of film growth, reaction (2), the corrosion potential achieving a value of

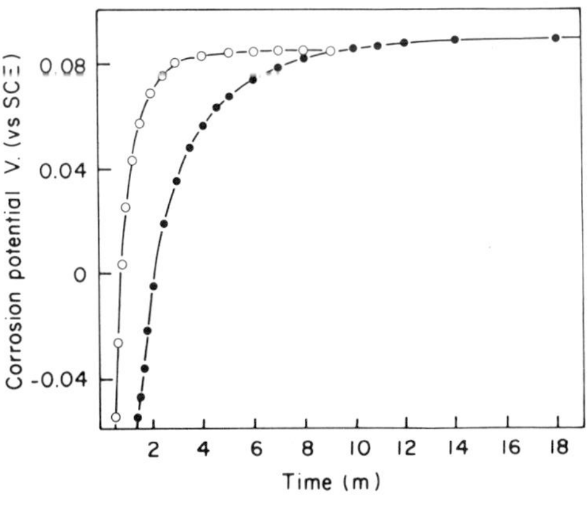

Figure 5. Corrosion potential of UO_2 as a function of time in 0.1 mol•dm^{-3} $NaClO_4$ containing H_2O_2; ● $[H_2O_2] = 2.66 \times 10^{-4}$ mol•dm^{-3}; ○ $[H_2O_2] = 1.27 \times 10^{-3}$ mol•dm^{-3}.

+0.085 V in ∽ 5 min in 10^{-3} mol•dm^{-3} H_2O_2, compared with ∽ 24 h for a similar concentration of dissolved oxygen. This may be due to a reaction involving OH radicals, as the decomposition of H_2O_2 can lead to the formation of radical intermediates [14], i.e.,

$$H_2O_2 + e \rightarrow OH + OH^- \qquad (6)$$

$$H_2O_2 \rightarrow HO_2 + H^+ + e \qquad (7)$$

which react in a catalytic fashion with H_2O_2, yielding water and oxygen and regenerating the radical intermediates,

$$H_2O_2 + OH \rightarrow H_2O + HO_2 \qquad (8)$$

$$H_2O_2 + HO_2 \rightarrow O_2 + H_2O + OH \ . \qquad (9)$$

Reactions (6) and (7) suggest that, providing the reaction mechanism does not change with concentration, the shift in potential with $[H_2O_2]$ for the two reactions will be equal, but of opposite sign. Consequently, the potential would not be expected to change with H_2O_2 concentration, as observed by us at higher concentration [10]. The invariance of the UO_2 potential at $[H_2O_2]$ ∽ 5 x 10^{-2} mol•dm^{-3} suggests that a similar process is occurring on UO_2 in this concentration range. The potential stabilizes in a range where the surface composition is close to $UO_{2.33}$ within approximately 5 min, indicating that the surface oxidation $UO_2 \rightarrow UO_{2+x}$ is rapid and proceeds via a different mechanism from that observed in dissolved oxygen.

If the discharge reaction (6) occurs on the surface of the UO_2, then the oxidation of UO_2 (containing U^{IV}) to UO_{2+x} (containing U^{IV} and U^{VI}) by the OH radical

$$U^{IV} + 2OH \rightarrow U^{VI} + 2OH^- \qquad (10)$$

(where U denotes a lattice species) must be fast enough to compete with reaction (7). An alternative reaction pathway is that H_2O_2 dissociates chemically on the UO_2 surface, i.e.,

$$H_2O_2 \rightarrow 2OH \qquad (11)$$

followed by reaction (10). Once the potential becomes independent of time (see Figure 5), its value appears to be determined by the balance of reactions (6) and (7). This does not preclude the occurrence of reaction (3), and it remains to be demonstrated what the balance is between peroxide decomposition and oxide dissolution.

For $[H_2O_2] \geq 10^{-2}$ mol•dm^{-3}, the UO_2 electrode potential shifts to more positive values, suggesting that further oxidation of UO_2 is occurring. This is confirmed by our XPS results [10], which show that the surface composition of the electrode progresses through the sequence

$$UO_{2.33} \rightarrow UO_{2.5} \rightarrow UO_{2.67} \ . \qquad (12)$$

These observations agree with our electrochemical results [1,5], and are consistent with the rotating-disk results of Needes and Nicol [15], which show that the formation of new surface phases on the UO_2 leads to a shift from predominantly peroxide decomposition to the oxidative dissolution of UO_2 as the electrode potential is increased.

Alpha Radiolysis

As reported earlier [9], corrosion-potential measurements on a UO_2 electrode placed within ∿ 30 μm of a gold-plated alpha source (686 μCi*) developed a potential of ∿ 0.075 V over a period of ∿ 20 h. The shape of the potential-time curve reflected the buildup of radiolysis products as a function of time. The potential is similar to that achieved in H_2O_2 and O_2-containing solutions, demonstrating that oxidizing conditions are achieved at the UO_2 surface as a consequence of water radiolysis. Weaker alpha sources (4.7 μCi and 100 μCi), which produce fluxes close to those expected to occur at the surface of irradiated fuel after storage for 500 years, showed little tendency to oxidize the UO_2 surface.

The oxidizing conditions created by alpha radiolysis are confined to the gap between the UO_2 electrode and the alpha source [9]. When the electrode separation is increased from 30 μm to 750 μm (after 21.5 h), the radiolysis products are swept from the gap to the bulk of solution, and the potential instantly falls to less than 0.0 V, finally reaching a value of -0.040 V. The value of -0.040 V is as negative as the potential will go, since the formation of UO_{2+x} (the first stage of reaction (2)) is irreversible.

An explanation, similar to that advanced above to explain the behaviour of UO_2 in the presence of hydrogen peroxide, can be used to explain the oxidation due to alpha radiolysis. After an initial oxidation of the UO_2 to UO_{2+x}, the predominant reaction occurring on the UO_2 surface appears to be the decomposition of hydrogen peroxide (via reactions (6), (7), (8) and (9)), which would explain why the positive potential of 0.075 V can only be maintained in the presence of the radiolysis products [9]. It is possible that this reaction is accompanied by oxide dissolution (3). The instantaneous drop in potential on separating the electrode and the alpha source suggests that oxidation beyond the $UO_{2.33}$ stage is not occurring, at least for times up to ∿ 30 h. For times $\geq$ 30 h with alpha radiolysis, the potential has been observed to go more positive, a value of ∿ 0.115 V being achieved after ∿ 50 h. When the gap was subsequently opened, the potential decreased slowly, the decrease being accelerated by degassing the solution with argon. After ∿ 16 h of degassing, with the gap open to > 750 μm, the potential finally stabilizes at 0.065 V. The decrease in potential on degassing can be attributed to the removal of radiolytically produced oxygen. The observation of a potential as positive as 0.115 V with the gap closed to ∿ 30 μm suggests that, in this case, oxidation has proceeded beyond the $UO_{2.33}$ stage, via reaction (12). This oxidation process appears to be irreversible since the final potential on opening the gap and degassing the solution is as positive as 0.065 V. It is not clear at this stage whether this increased oxidation is due to a high surface concentration of hydrogen peroxide produced radiolytically, or a more facile oxidation of UO_2 due to a variation in surface reactivity.

SUMMARY AND CONCLUSION

The oxidation of UO_2 in water is 200 times faster by H_2O_2 than by dissolved oxygen at the same concentration, and the mechanisms of oxidation change with pH. The oxidation of UO_2 by direct radiolysis products is dependent upon the concentration of the radiolysis products and suggests a dose-rate effect. The corrosion potential of UO_2 changes from nominally reducing to oxidizing conditions with the increasing strength of the alpha source.

*1Ci = 37 MBq

REFERENCES

1. S. Sunder, D.W. Shoesmith, M.G. Bailey, F.W. Stanchell and N.S. McIntyre, J. Electroanal. Chem. 130, 163 (1981).

2. N.S. McIntyre, S. Sunder, D.W. Shoesmith and F.W. Stanchell, J. Vac. Sci. Technol. 18, 714 (1981).

3. S. Sunder, D.W. Shoesmith, M.G. Bailey and G.J. Wallace, in Proc. Int. Conf. Radioactive Waste Management, Edited by M.A. Feraday (Canadian Nuclear Society, Toronto, 1982), pp. 398-405.

4. L.H. Johnson, D.W. Shoesmith, G.E. Lunansky, M.G. Bailey and P.R. Tremaine, Nucl. Tech. 56, 238 (1982).

5. S. Sunder, D.W. Shoesmith, M.G. Bailey and G.J. Wallace, Electroanal. Chem. 150, 217 (1983).

6. D.W. Shoesmith, S. Sunder, M.G. Bailey and D.G. Owen, Thin Films Sc. Tech. 4, 125 (1983).

7. D.W. Shoesmith, S. Sunder, M.G. Bailey, G.J. Wallace and F.W. Stanchell, Applic. of Surf. Sci. 20, 39 (1984).

8. W.G. Burns, A.E. Hughes, J.A.C. Marples, R.S. Nelson and A.M. Stoneham, AERE-R10189 (1981).

9. M.G. Bailey, L.H. Johnson and D.W. Shoesmith, Corros. Sci. 25, 233 (1985).

10. S. Sunder, D.W. Shoesmith, G.J. Wallace and M.G. Bailey, unpublished data.

11. M.J. Nicol and C.R.S. Needes, Electrochimica Acta 20, 585 (1975).

12. M.J. Nicol and C.R.S. Needes, The Mechanism of the Anodic Dissolution of Uranium Dioxide, Nat. Inst. Met., Repub. S. Afr. Rep. No. 7079 (1973).

13. E. Yeager, Electrochimica Acta 29, 1527 (1984).

14. R. Genscher and H. Genscher, Z. Phys. Chem. 6, 178 (1956).

15. C.R.S. Needes and M.J. Nicol, A Study of Some Redox Reactions at a UO_2 Surface, Nat. Inst. Met. Repub. S. Afr. Rep. No. 7073 (1973).

DISSOLUTION OF USED CANDU FUEL AT VARIOUS TEMPERATURES AND REDOX CONDITIONS

S. STROES-GASCOYNE, L.H. JOHNSON, P.A. BEELEY AND D.M. SELLINGER
Atomic Energy of Canada Limited, Whiteshell Nuclear Research Establishment, Pinawa, Manitoba, ROE 1LO, Canada.

ABSTRACT

Samples of used UO_2 fuel exposed to air-saturated water at 25°C for eight years have been examined using electron microscopy and X-ray diffraction techniques. The results, in conjunction with solution analysis data, show evidence for a $UO_3 \cdot 2H_2O$ precipitate on the fuel surface, and confirm the importance of grain-boundary leaching in controlling Cs-137 release. Studies of used-fuel dissolution under both oxidizing and reducing conditions at 150°C show significantly lower radionuclide release to solution for reducing conditions, illustrating the potential importance of redox chemistry in determining the long-term stability of used fuel after disposal.

INTRODUCTION

As part of a program to assess the concept of deep geological disposal of nuclear fuel wastes in plutonic rock of the Canadian Shield, the leaching and dissolution properties of used CANDU UO_2 fuel have been studied. Previous reports from this program have discussed fuel dissolution in oxidizing synthetic groundwaters at 25°C [1,2], dissolution at 150°C under both oxidizing and reducing conditions [3,4], and the relationship between fuel irradiation conditions and the short-term leaching of Cs-137 and I-129 [5,6,7].

In this paper we describe solution and surface analysis results from experiments involving the exposure of used UO_2 fuel samples to air-saturated solutions at 25°C for eight years. These studies help to clarify the role of surface precipitates in fuel dissolution, and illustrate that micro-analysis techniques can provide valuable supporting data for the interpretation of the dissolution mechanism of used fuel.

Other results reported here were obtained from a series of 10- to 30-day used-fuel dissolution experiments at 150°C in distilled water and in a synthetic saline groundwater under both oxidizing and reducing conditions. These experiments, using low linear power rating (LPR) fuel, differ from previous studies performed under similar conditions [3,4,8] that used higher LPR fuel. The LPR is shown to have a significant effect on the dissolution properties of the fuel, and the present results differ considerably from those of the earlier experiments.

DISSOLUTION EXPERIMENTS AT 25°C

Experimental

The experiment, which began in 1977, has been described in detail elsewhere [1]. It involved exposure of 50-mm sections of used-fuel elements, with attached cladding, to deionized distilled water and tap water. The major difference between the two solutions is the carbonate concentration, which is < 2 mg/L in the distilled water and approximately

50 mg/L in the tap water. Experiments were performed in duplicate. Fuel elements were sectioned using a silicon carbide cut-off wheel with no lubricant. The resultant fuel-element sections were leached in polypropylene bottles containing 100 mL of leachant, and at the end of the leaching interval (ranging from 1 to 90 days), the fuel samples were transferred to bottles containing fresh leachant. Two mL of concentrated HNO_3 were added to the bottle containing the leachate, to ensure dissolution of particulate material and desorption of radionuclides from the vessel walls prior to radiochemical analysis. Early in 1985, the fuel samples were removed from the leaching vessels, and 3-mm-thick sections were sliced from the ends to permit scanning electron microscope (SEM) studies of the leached surfaces. Upon completion of this work, using a shielded SEM in the Hot Cell Facility, samples of the precipitate covering the fuel surface were removed using adhesive tape, for subsequent examination by SEM/energy dispersive X-ray (EDX) and X-ray diffraction (XRD) analyses.

Results and Discussion

Fractional release rates for Cs-137, Sr-90, Pu-239+240 and uranium are shown in Figure 1 for the fuel samples leached in distilled water. Each data point is an average value for the duplicate samples. Results for the tap water were similar, except that uranium releases were approximately one order of magnitude higher [9].

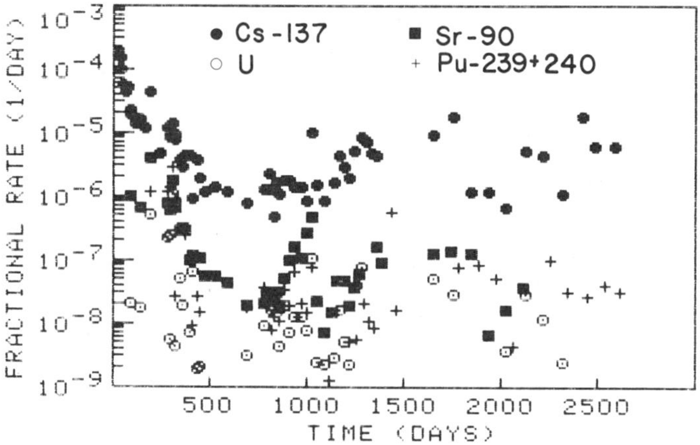

FIGURE 1: Fractional Release Rates of Cs-137(●), Sr-90(■), Pu-239+240(+) and Uranium (o) in Distilled Water at 25°C

Although there is a large degree of scatter in the data, it is apparent that Cs-137 release rates are about two orders of magnitude higher than release rates of the other species. The results suggest a very high degree of preferential leaching of Cs-137 from the fuel.

It is also clear from the data that Sr-90 and Pu-239+240 have very similar release rates. The higher rate of uranium release in tap water is probably due to the complexation of uranium (in the form of UO_2^{2+}) by carbonate. It should be emphasized that the data represent total release from the fuel, including adsorbed and colloidal species.

Scanning electron micrographs of the surface of the fuel sample that was exposed to distilled water show that it is partially covered by a yellow precipitate. EDX analysis of this material shows the presence of only uranium, cesium and silicon, with uranium by far the most abundant element [9]. The EDX detection limit is approximately 1%; the presence of cesium, therefore, suggests that it is enriched in the precipitate relative to the amount originally present in the fuel (0.05 wt.%). No plutonium was detected in the precipitate. Since the Pu concentration in the fuel is ∽ 0.4%, this suggests no significant Pu enrichment in the precipitate layer. Similar analysis of the material from the surface of the sample exposed to tap water indicated that there was relatively little precipitate. No cesium was detected, and few uranium-rich particles were observed other than particles of fuel pulled from the surface during sample preparation.

Several attempts were made to remove sufficient precipitate from the surface of the fuel sample exposed to distilled water to obtain an XRD pattern. A sample obtained using plastic adhesive tape gave peaks at 0.725-0.735, 0.319 and 0.253 nm. These peaks are characteristic of schoepite ($UO_3 \cdot 2H_2O$), or closely related compounds. The presence of one of the characteristic schoepite peaks at 0.366 nm could not be confirmed due to a broad, high-intensity reflection in this region from the plastic tape.

The XRD and EDX data strongly suggest that the precipitate is composed of $UO_3 \cdot 2H_2O$, or a related hydrate, which have been shown to occur in similar conditions [10], with a cesium uranate phase as a minor constituent. The formation of a cesium uranate compound during dissolution of used fuel in distilled water would explain the results of Wilson and Oversby [11], who observed that, in static dissolution experiments, the Cs-137 concentration in solution decreased by over an order of magnitude during their 250-day experiment. This did not occur in their tuff groundwater (containing ∽ 120 mg/L carbonate), a result consistent with the EDX results reported here for the samples exposed to tap water. It is likely that a cesium uranate would not be stable in a solution containing significant amounts of sodium, calcium or carbonate.

The presence of a solid phase containing uranium on the leached surface suggests that the solubility of this phase should control the uranium concentration in solution. The distilled water leachates were typically observed to have a pH of approximately 8. Assuming that $UO_3 \cdot 2H_2O$ is the solid phase, uranium concentrations should be 10^{-5} - 10^{-4} mol/L for these conditions [12]. The actual concentrations measured in solution are shown in Figure 2, for various solution contact times, illustrating that uranium concentrations are well below saturation, the majority of solutions being undersaturated by 1-2 orders of magnitude. This is surprising since the da°ta plotted include adsorbed and colloidal uranium; thus, the lowest values in Figure 2 probably represent the true dissolved uranium concentrations. Contact time apparently has no effect on the degree of undersaturation. It is surprising to find substantial formation of a precipitate of such apparent high solubility occurring in a situation where undersaturation is maintained by periodic replacement of the leaching solution. The degree of scatter in the uranium data is very large, and this result suggests that spallation of the precipitate occurs to varying degrees when the fuel is transferred to a new leaching vessel. The lower concentrations, 10^{-7} - 10^{-6} mol/L, may therefore reflect the true solubility of the precipitate. Similar uranium concentrations have been observed in other fuel-dissolution experiments performed with distilled water [11,13].

In the case of the fuel samples exposed to tap water, the absence of any significant amount of uranium-rich precipitate on the surface is consistent with the dissolution mechanism proposed by Shoesmith et al., [14]. In solutions containing $\gtrsim$ 60 mg/L carbonate, U(VI) remains in

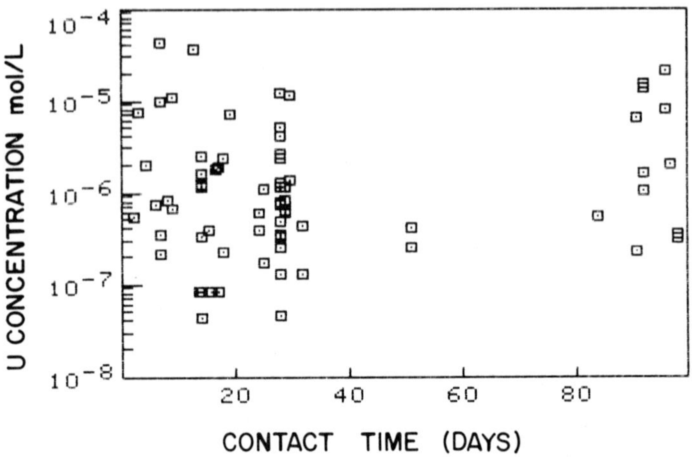

FIGURE 2: Uranium Concentrations Measured in Distilled Water Leachates at Various Exposure Times

solution as the carbonate complex. In such a solution, the fuel probably maintains a surface composition near U_3O_7.

SEM examination of fuel fracture surfaces exposed to the leachant for the duration of the experiment shows microstructural features consistent with the observed preferential leaching of Cs-137. In the central region of the fuel where large numbers of fission-gas bubbles formed on grain boundaries, gas tunnels are visible through which the leachant may have penetrated to dissolve the cesium.

Conclusions

The most important result of this experiment is the observation that, even after eight years of exposure to solution, the rate of Cs-137 leaching from used fuel remains much higher than the release rates of other radionuclides. The SEM results suggest that the reason for this is the release of Cs-137 from grain boundaries. As discussed in previous studies and in the second half of this paper, the degree and duration of preferential leaching are highly dependent on the LPR of the fuel. With a high LPR fuel of the type used in this study, it is conceivable that a substantial fraction of the inventory of Cs-137, Cs-135 and I-129 [6,7] stored at grain boundaries (approximately 7% of their total inventory in CANDU fuel) could be leached in a relatively short time, perhaps hundreds or thousands of years.

This study also illustrates that a combination of surface chemistry techniques and solution analysis provides valuable data for an understanding of the dissolution mechanism of used fuel. Further studies of the type performed by Bruno et. al. [13], which examined the relationship between thermodynamic data and actinide solution concentration data in dissolution experiments, in conjunction with surface analysis of the type reported here, will be required to develop predictive models for used-fuel dissolution.

DISSOLUTION EXPERIMENTS AT 150°C

Experimental

The basic approach used in the experiments performed at 150°C has been described [3,4], and is outlined here briefly to highlight minor differences in approach. The fuel samples used in the experiments were taken from two intermediate ring elements in Bruce fuel bundle F21271C, irradiated at an average LPR of 35-39 kW/m to a burnup of ~ 900 GJ/kg U. The fission-gas releases for the elements were ~ 0.1%. Fuel-element sections, ~ 20 mm long and weighing ~ 25 g, were cut as described in the first part of this paper.

A 1-L titanium autoclave, equipped with a dip tube for solution sampling, a thermocouple well, and a gas sampling port, was used in all experiments. A new glass liner was used for each experimental run, to prevent cross-contamination from previous tests. In some experiments, the fuel was placed in a porous zirconium basket (5-μm pore size) to prevent fuel particles from being removed during solution sampling.

Experiments performed under oxidizing conditions used 300 mL of air-saturated, deionized distilled water or WN-1 saline solution (see Table I for composition), with air as the cover gas in the autoclave. For reducing runs, the solutions were degassed for 3 h with Ar-3% H_2 before being forced through the inlet valve into the autoclave (also degassed) containing the fuel sample. Solution samples of 15 mL were removed at 150°C throughout the 10- to 30-day experiments, their pH was measured, and they were analyzed for Cs-137 by γ-spectrometry, for U-238 by instrumental neutron activation analysis (INAA), for Sr-90 and Tc-99 by chemical separations followed by β-counting, and for Pu-239+240 by α-spectrometry. The solution volume remaining at the end of a run was approximately 150 mL. A final solution sample was taken at 25°C one day after completion of the experiment. The liners and baskets were rinsed with 5% HNO_3 for one week and the solutions analyzed to determine plateout activity. In some cases, this was followed by a one-week rinse with 1 mol/L NaOH, to ensure complete removal of any radionuclide that may have diffused into the glass, followed by radiochemical analysis of the rinse solution.

TABLE I

COMPOSITION OF WN-1 SALINE SOLUTION (mg/L)

	As prepared (25°C)	After reaction with glass for two weeks (150°C)
K	14.1	16.3
Na	1940	1840
Fe	0.28	0.43
Sr	24.6	26.0
Ca	2190	1610
Mg	42.0	49
Si	2	63
Cl	6240	5850
SO_4	1100	90
HCO_3	13	4
pH (25°C)	6.9	6.5

Results and Discussion

The results of the experiments are summarized in Table II. For U, Pu-239+240 and Tc-99, average concentrations after the first day of the experiments are given because the concentrations were relatively constant with time, probably indicating that saturation had been achieved for these elements. For Cs-137 and Sr-90, the final concentration is given, because in some cases concentrations continued to increase throughout the duration of the experiment, and because solubility control of the concentrations of these elements in short-term tests is considered less likely than for elements such as uranium and plutonium. The final column in Table II gives the quantities of radionuclides found in solution, as fractions of the initial inventory in the fuel. Solution pH (at 25°C) generally increased from 6 to 7.5 during distilled-water runs and remained constant at 6.5 to 7.5 in saline solution runs.

TABLE II

RADIONUCLIDE CONCENTRATIONS IN DISSOLUTION EXPERIMENTS AT 150°C

Experiment No.	Duration of Experiment (d)	Redox Condition	Leaching Solution	Radionuclide	Average † Concentration (mol/L)	Fraction of Inventory in Solution**
DO	17	oxidizing	Distilled	U	1.4×10^{-5}	5×10^{-5}
			Water(DW)	Cs-137	3.2×10^{-7}	1.7×10^{-3}
				Sr-90	3.2×10^{-8}	4×10^{-4}
				Pu-239+240	4×10^{-9}	1.5×10^{-6}
DR1	28	reducing	DW	U	8×10^{-8}	2.9×10^{-7}
				Cs-137	3.2×10^{-8}	1.8×10^{-4}
				Sr-90	8.1×10^{-9}	1.1×10^{-4}
				Tc-99	$< 7 \times 10^{-9}$ ††	--
				Pu-239+240	7×10^{-10}	5×10^{-7}
DR2	28	reducing	DW	U	2×10^{-7}	7.2×10^{-7}
				Cs-137	1.1×10^{-8}	1.4×10^{-4}
				Tc-99	2×10^{-8}	1×10^{-4}
				Pu-239+240	5×10^{-10}	3×10^{-7}
WO1	10	oxidizing	WN-1	U	2.5×10^{-6}	8×10^{-6}
				Cs-137	1.7×10^{-6}	9×10^{-3}
				Sr-90	2.2×10^{-7}	1.8×10^{-3}
				Tc-99	7×10^{-8}	3.5×10^{-4}
WO2	10	oxidizing	WN-1	U	9×10^{-6}	3.2×10^{-5}
				Cs-137	1.0×10^{-6}	5.5×10^{-3}
WR	30	reducing	WN-1	U	$< 3 \times 10^{-7}$ *	--
				Cs-137	1.0×10^{-7}	5.5×10^{-4}
				Sr-90	6.1×10^{-8}	5.1×10^{-4}
				Tc-99	2×10^{-8}	1×10^{-4}

† for U, Pu and Tc, the average concentration after the first day is given; for Cs and Sr the final concentration is given.
†† detection limit, Tc detected in only one sample.
* detection limit high due to interference from Na in INAA.
** based on concentrations given in the previous column.

The results in Table II generally show lower releases of all radionuclides under reducing conditions. This is illustrated quite clearly by the data in Figure 3 for experiments in distilled water (DO and DR1).

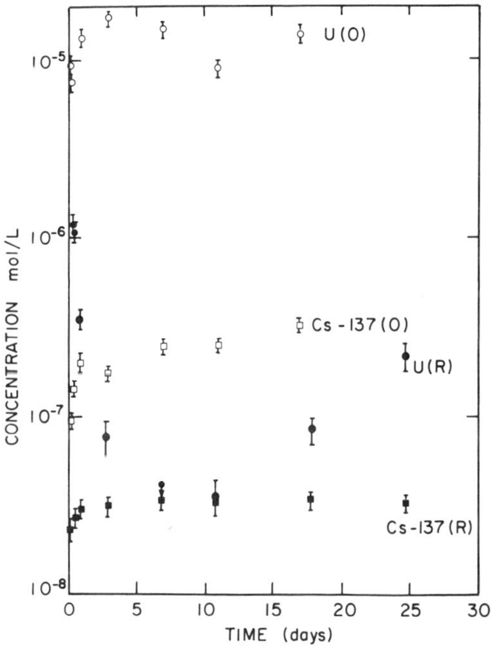

FIGURE 3: Concentrations of Uranium and Cs-137 in Distilled Water in Contact with Used Fuel Under Oxidizing (O) and Reducing Conditions (R).

Error bars reflect the combined uncertainties associated with the radiochemical analysis and the determination of the volume of leachant removed from the autoclave. Under reducing conditions, the uranium concentration decreased from > 10^{-6} mol/L to near the detection limit (~ 3.5 x 10^{-8} mol/L) within the first few days of the experiment. This is assumed to be due to the reduction of U(VI) to U(IV), followed by precipitation of UO_2, which has been shown to occur within the same time frame under similar conditions [4,15]. In contrast, the results from the oxidizing run show the uranium concentration stabilizing at 10^{-5} mol/L, possibly reflecting solubility control by $UO_2(OH)_2$, or an amorphous U(VI) oxide [16].

The release of Cs-137 and Sr-90 was similarly influenced strongly by the redox condition. In the oxidizing run, the Cs-137 concentration continued to increase with time at a significant rate, whereas in both reducing runs the Cs-137 concentration became constant after a few days. Agreement between duplicate experiments was generally good. Concentrations of Pu-239+240 were very similar for experiments in distilled water under both oxidizing and reducing conditions, and may reflect the solubility of a plutonium hydrous oxide [13].

In the WN-1 solution, fission-product and uranium concentrations were again higher under oxidizing conditions. Some enhancement of fission-product release in WN-1 solution relative to distilled water was evident, for both oxidizing and reducing conditions. Sr-90 and Cs-137 concentrations became constant under reducing conditions after 10 to 25 days. No Pu-239+240 data were available because of difficulties in analytical procedures due to the salinity of the solutions.

The results of these studies suggest that, after a period of less than ten days, the release of Cs-137, Sr-90 and uranium is controlled by the dissolution rate of the UO_2 grains, which is a strong function of the redox condition. Under reducing conditions, fission-product concentrations became constant, suggesting very slow dissolution compared with oxidizing conditions. Alternative possibilities are that secondary phases form, containing Cs-137, Sr-90 and uranium, which are highly insoluble under reducing conditions, or that sorption of these elements on the vessel walls occurs to a much greater degree under reducing conditions. Analysis of rinse solutions from the glass liners offers no support for the latter suggestion. It therefore seems likely that the high stability of UO_2 under reducing conditions is the only credible explanation for the results.

The present results, however, are inconsistent with those of previous studies [4,8] that showed Cs-137 and Sr-90 concentrations increasing with time under both oxidizing and reducing conditions at 150°C. The difference between the two results is thought to be due to the higher LPR of the fuel used in the earlier studies. Calculations using the ELESIM code [17] indicate that fuel samples used in the previous tests reached a centreline temperature of 1400 to 1500°C during irradiation, exceeding the threshold temperature of ~ 1300°C, above which significant grain growth and the formation of interconnected gas porosity occur. The fuel used in this study reached a maximum centreline temperature of only ~ 1150°C. The result of this is that, in the reducing experiments reported here, dissolution of the UO_2 grains in which the fission products are trapped is the only significant mode of release after the first few days of the experiment, when gap release occurs [5]. With fuel of higher LPR, interconnected fission-gas tunnels at grain boundaries allow water to percolate through the central regions of the fuel, permitting continuing release of fission products, despite the stabilization of the UO_2 by the reducing conditions.

Another result that is inconsistent with the previous 150°C dissolution studies [3,4] is the concentration of uranium in distilled water under oxidizing conditions. Previous results showed uranium concentrations at 2-4 x 10^{-7} mol/L, whereas values exceeding 10^{-5} mol/L were obtained here. The major difference between the two studies was the use of an unlined titanium vessel previously, whereas the experiments here used a glass-lined vessel. Based on measurements of the amount of uranium desorbed from the glass liner and the titanium in the two tests, in addition to the uranium concentrations in solution during and after the tests, it is apparent that considerably more uranium was adsorbed on the titanium than on the glass at 150°C, and this may account for the difference between the two results. Further work will be required to confirm this.

Conclusions

Releases of Cs-137, Sr-90 and uranium from used fuel in both saline solution and distilled water at 150°C are significantly lower under reducing conditions than under oxidizing conditions. Although the results clearly need confirmation in long-term tests, they indicate that, in spite of this strong redox dependence, the effect may be masked in experiments with some UO_2 fuels due to preferential leaching of fission products from interconnected porosity. Fuel microstructure is clearly an important factor in the interpretation of used-fuel dissolution experiments. Future studies will examine this aspect further, in addition to investigating the effect of reducing agents likely to be present in the vault environment, such as Fe^{2+}, on the release of fission products and actinides from used UO_2 fuel.

Acknowledgements

We thank A.P. Snaglewski for assistance with the dissolution experiments, F.W. Stanley for the SEM studies of the used fuel, L. Brown for the EDX analysis and A.M. Duclos for performing XRD analysis.

REFERENCES

1. T.T. Vandergraaf, Atomic Energy of Canada Limited Technical Record, TR-100* (1980).

2. L.H. Johnson, Atomic Energy of Canada Limited Report, AECL-6837 (1982).

3. L.H. Johnson, K.I. Burns, H.H. Joling and C.J. Moore, Atomic Energy of Canada Limited Technical Record, TR-128* (1981).

4. L.H. Johnson, H.H. Joling, in Scientific Basis for Nuclear Waste Management, S.V. Topp (ed.) p. 321 (1982).

5. L.H. Johnson, K.I. Burns, H.H. Joling and C.J. Moore, Nucl. Technol. 63,470 (1983).

6. L.H. Johnson, S. Stroes-Gascoyne, J.D. Chen, M.E. Attas, D.M. Sellinger and H.G. Delaney, in Proceedings of the ANS Topical Meeting on Fission Product Behaviour and Source Term Research, Snowbird, Utah, July, 1984 (in press).

7. L.H. Johnson, H.H. Joling, Atomic Energy of Canada Limited Technical Record, TR-280* (1984).

8. L.H. Johnson (unpublished data).

9. L.H. Johnson, in Proceedings of the Nineteenth Information Meeting of the Canadian Nuclear Fuel Waste Management Program, Atomic Energy of Canada Limited Technical Record, (in press).

10. R. Wang, Battelle Pacific Northwest Laboratory Report, PNL-3566 (1981).

11. C.N. Wilson, V.M. Oversby, in Proceedings of the Waste Management 85 Conference, Tucson, Arizona, March, 1985 (in press).

12. K.M. Krupka, D. Rai, R.W. Fulton and R.G. Strickert, in Scientific Basis for Nuclear Waste Management, C.M. Jantzen, J.A. Stone and R.C. Ewing (eds.) (1985.

13. J. Bruno, R.S. Forsyth and L.O. Werme, in Scientific Basis for Nuclear Waste Management, C.M. Jantzen, J.A. Stone and R.C. Ewing (eds.) (1985).

14. D.W. Shoesmith, S. Sunder, M.G. Bailey, G.J. Wallace and F.W. Stanchell, Application of Surface Science 20, 39 (1984).

15. B. Bunji and B. Zogovic, in Proceedings of the Second U.N. International Conference on the Peaceful Uses of Atomic Energy, 3, 350 (1958).

16. A.A. Nikitin, E.I. Sergeyeva, I.L. Khodakovskiy and G.B. Naumov, Geokhimiya 3, 297 (1972).

17. M.J.F. Notley, Nucl. Technol., 44, 445 (1979).

* Unrestricted, unpubished report, available from SDDO, Atomic Energy of Canada Limited Research Company, Chalk River, Ontario K0J 1J0

THE CORROSION OF SPENT UO_2 FUEL IN SYNTHETIC GROUNDWATER

R.S. FORSYTH* and L.O WERME**
* Studsvik Energiteknik AB. Sweden
** The Swedish Nuclear Fuel and Waste Management Co, Sweden

ABSTRACT

Leaching of high burnup BWR fuel for up to 3 years showed that both U and Pu attain saturation rapidly at pH 8.1, giving values of 1-2 mg/l and 1 µg/l respectively. The leaching rate for Sr-90 decreased from about 10^{-5}/d to 10^{-7}/d but was always higher than the rates for U, Pu, Cm, Ce, Eu and Ru. Congruent dissolution was only attained at pH values of about 4.

The hypothesis that alpha radiolytic decomposition of water is a driving force for UO_2 corrosion even under reducing conditions has been examined in leaching tests on low burnup (low alpha dose-rate) fuel. No significant effect of alpha radiolysis under the experimental conditions was detected.

INTRODUCTION

In Sweden, the current main option for management of the back-end of the fuel cycle is the direct disposal of spent nuclear fuel in granitic bed-rock. As a part of the extensive SKB program for research and verification within this option, studies are being performed of the corrosion of spent fuel in granitic groundwater in order to measure dissolution and leaching rates of radioactive species, and to determine the mechanisms for corrosion. This paper presents the latest results arising from this work.

HIGH BURNUP FUEL: SEQUENTIAL AND STATIC LEACH TESTS

Experimental

All specimens used in these experiments consisted of 20 mm long segments of fuel and clad cut out from a well-characterized BWR fuel rod irradiated to a burnup (rod maximum) of 42 MWd/kg U.

Each fuel specimen, containing about 16 g UO_2, was suspended in a spiral of platinum wire in 200 ml of the leachant in a 250 ml Pyrex flask. All leach tests were performed at 20-25°C, the ambient temperature of the hot cell.

On completion of the planned contact period, the specimen was removed from the flask and transferred to a new flask with new leachant in the case of sequential leaching, or discarded in the case of the static tests (or exposed to low pH leachant).

After removal of the leachant from the flask the pH was measured on one aliquot and then two 10 ml aliquots were centrifuged through membrane filters (Aminco Corp, USA) with apertures of 1.5-2 nm. The material retained on these membranes has been defined arbitrarily here as "colloidal". Both centrifugate and membrane were subjected to radiometric and radiochemical analysis. The empty flask, after a rapid rinse with deionized water, was exposed for a few days to 200 ml of 5M HNO_3/0.5M HF to desorb any activity on the vessel walls. This rinse solution was also analyzed.

The experiments discussed in this section, three series each with a fuel/clad specimen exposed to sequential leaching and two series of static tests - each with 8 fuel/clad specimens, were all performed under so-called oxidizing conditions. Other test parameters are given in TABLE I.

Table I
Summary of leach test conditions.

Series	Leachant	Procedure	Contact period (days)
3.1	DIW	SEQU	7,7,14,63,91,83,182,159,218,174
3.2	GW	SEQU	"-
3.3	"	SEQU	7,7,14,63,91,182,182,159,218,174
3.4	"	STAT	14d Preleach. 14:77:168:350:532:909
3.5	"	STAT	" " 14:77:168:350:691:1083

The leachant used for Series 3.1 was deionized water while a synthetic groundwater whose composition is shown in Table II was used for the other tests. The deionized water, which was also used in preparation of the groundwater, was de-aerated by bubbling with nitrogen prior to use. However, as mentioned above, each leachant was in contact with about 50 ml of air during leaching.

Table II
Composition of the synthetic groundwater.

Species	HCO_3^-	SiO_2	SO_4^{2-}	Cl^-	Ca^{2+}	Mg^{2+}	K^+	Na^+
pm	123	12	9.6	70	18	4.3	3.9	65

pH: 8.0 - 8.2 Ionic strength: 0.0085.

Results and discussion

At previous MRS symposia in this series, results obtained during the first 18 months of leaching have been reported (1) and a tentative model of the solubilities of actinides (U, Pu) and lanthanides (Ce, Eu) as a function of pH has been described (2). The results now available are in general agreement with those reported previously from the earlier phases of the program.

In the following, the experimental results are often reported as Fractions of the Inventory in the Aqueous Phase (FIAP values) - the sums of the fractions in the centrifugate and retained on the membrane filter. The term Apparent Leach Rate (ALR) in units of FIAP/day, is used as the corresponding definition of rate. However, only concentrations in the centrifugate are used when discussing solubility.

The concentrations of uranium in the groundwater leachant centrifugates at pH 8.1 in the experimental Series 3.2 - 3.5 are shown in Figure 1a. The results show appreciable scatter but concentrations can be seen to level out rapidly and attain saturation at a level of 1-2 mg/l. The concentration levels in deionized water - Series 3.1 - were around or below the detection limit of about 50 µg/l. The concentrations of plutonium in the centrifugates, presented in Fig 1b, show a marked decrease with contact time.

For the nuclides with the highest values of ALR, Cs-137 and Sr-90, no such saturation effects have been found, since the cumulative amounts leached from the fuel specimens during a given contact time appear to be independent of whether the leaching was performed in several stages (sequential leaching) or in one (static leaching). It can also be noted that due to their low inventories in the fuel, the concentrations of the chemical elements cesium and strontium in the leachants are also low: even those leachants which contained most Cs-137 - the first contact fuel/water, when cesium released from the fuel matrix during reactor operation is rapidly dissolved - had a concentration of only 2 mg cesium/l, while for elementary strontium, even

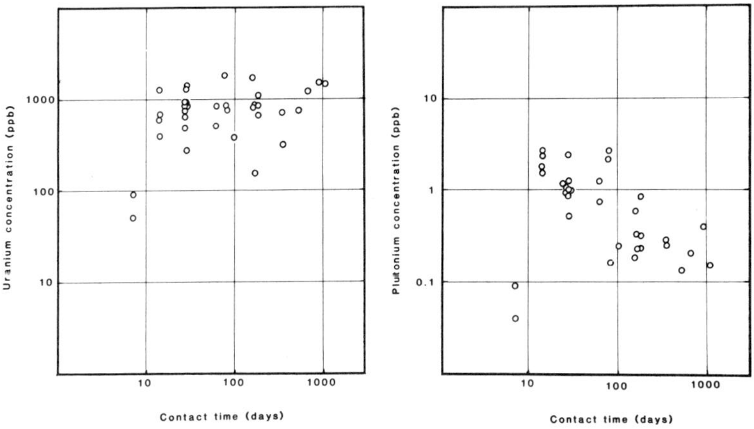

a) Uranium b) Plutonium

Fig 1. Measured concentrations in leach solution centrifugates as a function of contact time.

after 1000 days of leaching, the solution concentration corresponded to only 50 µg strontium/l. Thus, while values of leach rate for U and Pu can be misleading, since due to saturation the values are always inversely proportional to the contact time, the values for Cs and Sr are meaningful.

Fig 2 presents the values of ALR for Cs-137 and Sr-90 for all experiments performed in groundwater (pH 8.1) and in deionized water. (For a particular series of static tests, the differential leach rates for each extension of the contact time are plotted). Only the results for the sequential and static leaching of high burnup BWR fuel will be discussed here.

The results are very similar to those published earlier by us and by other workers (1, 3, 4, 5). They show the initial rapid dissolution of Cs-137 (instant release fraction) which over the first 2 weeks corresponded to about 1% of the Cs-137 inventory in the fuel, and agreed well with the reported release (of 0.7 %) of fission gas from the whole fuel rod during reactor operation.

For both nuclides, after the first 2 weeks, the apparent leach rates were inversely proportional to the contact time over the next 3 years, Sr-90 attaining a value of about 10^{-7}/d at the end of the period. The values of Cs-137 were always higher than the corresponding values for Sr-90 by a factor of 3-5, suggesting perhaps grain boundary attack. This effect has also been observed by Vandergraaf and Johnson (4, 5).

It is also of interest to note that the leaching behaviour of Cs-137 and Sr-90 in deionized water was very similar to that in the groundwater, although the measured uranium solubility in DIW was found to be very low (about 50 ppb).

The ALR values for U were about one, and for Pu, two orders of magnitude lower than the corresponding values for Sr-90. (The limited validity of these figures mentioned above should be kept in mind.) Values of ALR for Cm-244, Ce-144 and Eu-154 were apparently of the same order as those for Pu, but since these nuclides were usually only measurable on the membrane filters, and contamination with small fuel particles can be suspected, it is possible that their real ALR values are an order of magnitude lower.

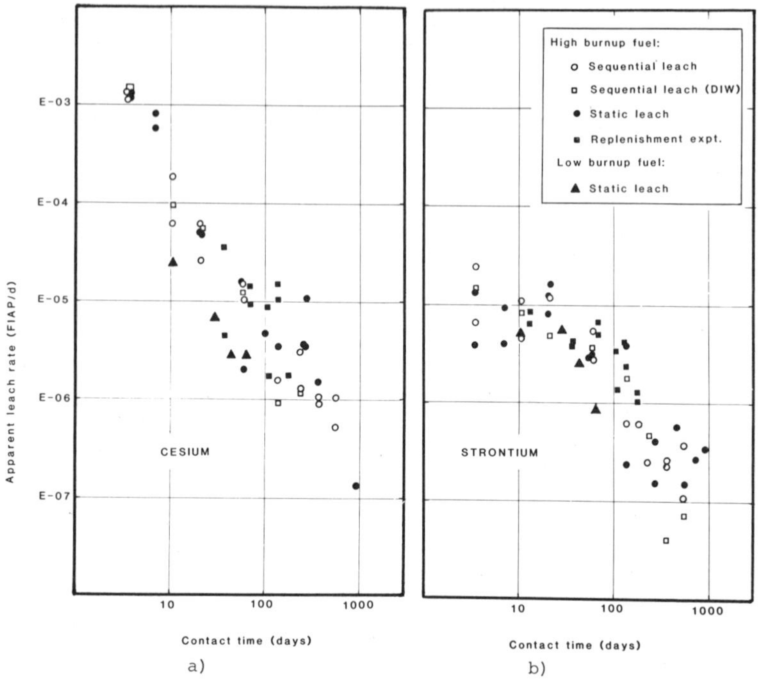

Fig 2. Values of Apparent Leach Rate for a) Cs-137 B) Sr-90 as a function of contact time (chronological midpoint of contact period).

HIGH BURNUP FUEL: LOW pH TESTS.

In order to explain the marked differences in the leaching behaviour of the various nuclides reported above, two main hypotheses can be advanced:

1) that in the groundwater used here (pH 8.1, 123 ppm HCO_3^-), uranium saturation is attained rapidly (1-2 mg/l). Other actinides and lanthanides are released congruently to the solution but are precipitated and/or adsorbed on available surfaces. Cs and Sr are preferentially leached from the fuel: cesium partly as an instant release fraction comparable in size to the release of fission gases from the fuel during reactor operation, and then, as for strontium, from enriched inclusions or zones in the fuel matrix, perhaps from grain boundaries.

2) that Cs is leached as in the first hypothesis above, but the leaching rate for Sr represents the dissolution rate of the fuel matrix, and although dissolution occurs congruently, the amounts in solution of the actinides and lanthanides are determined by solubility limitations.

The second hypothesis implies that after long exposures to groundwater, as in the static tests in Series 3.4 and 3.5, significant quantities of U, Pu and lanthanides would have entered and then left the solution by precipitation or adsorption, or been retained on the fuel surfaces as alteration products.

Analysis of the more than 100 flask rinse solutions examined has indicated that this material does not appear to have deposited on the flask walls, since their contents correspond to the dissolution of small fragments of fuel which have been lost during handling.

Experimental

As a practical check on the second hypothesis, the 6 fuel/clad specimens from the 3.4 and 3.5 series with the longest contact times in the pH 8.1 groundwater (350-1083 days) were subsequently contacted for 20 days with 200 ml of new groundwater which had pH values adjusted to lower values by means of HCl.

It was expected, of course, that the dissolution of the matrix would increase at low pH, but if the redeposited material or alteration products were also soluble, then the ratios of actinides and lanthanides to strontium in the leachants would be enhanced compared with those in the fuel iself.

Results

Since the apparent leaching rate for Sr-90 decreases with increasing contact time as shown in Fig 2b, the results are expressed (in Table III) as rounded ratios to the Sr-90 values.

Table III
Normalised apparent leaching rates in low pH tests.

pH	2.0	3.2	3.4	4.0	4.1	5.5
Initial contact (d)	532	1083	909	350	691	350
Sr-90: ALR (x E-06)	25.8	14.0	16.4	9.6	10.7	10.3
Sr-90	100	100	100	100	100	100
U	100	80	120	90	40	4
Pu	80	130	12	100	60	40
Cm	260	90	230	170	140	60
Ce-144	150	120	110	180	100	65
Eu-154	110	80	100	130	100	65
Cs-137	70	90	80	80	60	70

Although the results for Cm appear to show some enhancement, this is thought to be a combination of experimental error and variation in local Cm inventories, and it can be concluded that the experiments give no support to the hypothesis of appreciable adsorption or precipitation of actinides and lanthanides on the fuel/clad surface.

However, the results do show that congruent dissolution is attained at pH 4 and below, although significant fractions of several leached species were retained on the membrane filters. Another interesting feature is that at low pH, the Sr-90 dissolution rate exceeded the Cs-137 rate, indicating, perhaps, different attack sites on the fuel surface than those at high pH.

HIGH BURNUP FUEL: REPLENISHMENT EXPERIMENT

The experimental finding that uranium saturation at 1-2 mg/l was attained rapidly in the groundwater is, of course, of significance for safety analysis, and it was decided to check the finding in a further experiment series.

Experimental

The 6 experiments were all performed under the same oxidizing conditions used in the main program, i.e., there was 50 ml of air in contact with the 200 ml of the synthetic groundwater leachant (pH 8.1). Polypropylene flasks were also compared with Pyrex flasks. The fuel fragments were suspended in the leachant on Pt trays. Details of the experiments are given in Table IV.

Table IV
Replenishment experiment: Summary of test conditions

Experiment	Flask	Fuel Specimen		Wt UO_2 (g)
3.13	Polypropylene	Fuel/clad segment		16.6
3.14	"	Fuel fragments >4 mm		5.7
3.15	"	"-	3.15-4 mm	5.7
3.16	Pyrex	Fuel/clad segment		16.6
3.17	"	Fuel fragments >4 mm		5.8
3.18	"	"-	2-3.15 mm	3.9

After initial filling, the flasks were allowed to stand for 27 days in the hot-cell at ambient temperature, following which 40 ml of the leachants were removed (and subjected to the same procedures as previously) and immediately replaced by 40 ml of fresh groundwater. This was repeated four times after different contact periods. On the basis of the previous results, all solutions should have shown uranium saturation at the 1-2 mg/l level.

Results

The measured uranium concentrations in the centrifugate samples from the various replenishment tests are presented in TABLE V.

Table V
Uranium concentrations (ppb) as a function of total contact time.

Contact time (d)	27	47	91	117	145	211
3.13	4 950	11 500	10 000	9 000	8 000	2 450
3.14	650	1 300	1 900	1 600	1 450	340
3.15	800	1 100	1 600	1 450	1 700	875
3.16	4 150	6 000	7 500	5 500	3 900	2 400
3.17	600	950	900	1 250	1 000	700
3.18	900	2 050	1 750	1 900	1 050	850

The results for the fuel fragment experiments show fairly good agreement with the previously established saturation concentration but the high (4-5 mg/l) initial U concentrations found for the fuel/clad segment experiments, increasing over the first few months and then decreasing approximately at the same rate as dilution, were unexpected.

Fig 3 presents the FIAP values for Cs, Sr, U and Pu for the experimental series. Note that the abscissa represents the number of replenishment stages, although the time for each contact period is also given.

To simplify the figure, the results have been averaged for the 3 groups shown in the legend, and the dashed line represents the slope of successive 20% dilutions.

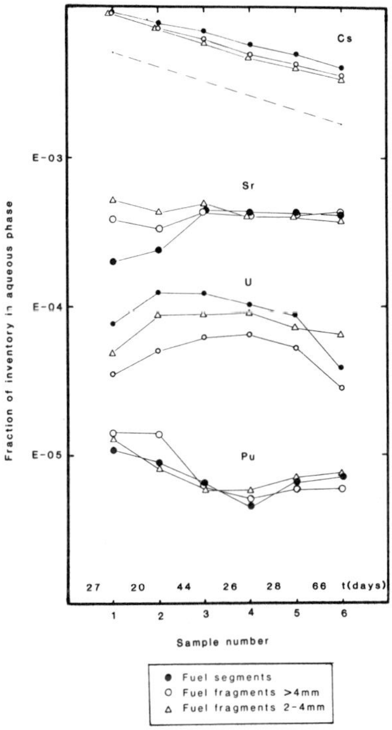

Fig 3. Replenishment experiments: Variation of leach fraction with successsive replenishment.

The Cs curves display, as expected, the initial release fraction of about 1% of inventory during the first 27 day contact period. The concentrations then almost follow the 20% successive-dilution curve, but show some further Cs leaching, particularly for the fuel/clad segments. FIAP values for fuel/clad segments and fuel fragments are very similar suggesting that only a minor part of the cesium released from the fuel during reactor operation was deposited on the inner surface of the clad.

The Sr results also show good agreement with the earlier results from the program. During the initial contact periods, the ALR values were higher for fuel fragments than for fuel/clad segments, probably reflecting easier access of the leachant to the fuel surface, but calculation of the cumulative leach fractions at 211 days showed all specimens to have released about the same fraction ($\sim 8 \cdot 10^{-4}$) of inventory. (The derived values for ALR for the two fuel/clad segments are plotted in Fig 2b.)

As mentioned above, the U results are somewhat surprising. For fuel fragment leaching, the leach fractions agree well with previous work, being an order of magnitude lower than those for Sr-90, and the measured uranium concentrations were in the range 1-2 mg/l. Leaching of the 2 fuel/clad segments, however, showed increasing dissolution of U over the first 91 days, attaining U concentrations of the order of 10 mg/l, which then decreased roughly following the 20% successive dilution curve. This suggests the same effect of supersaturation and precipitation observed by Johnson (5). A tentative explanation for the initial high concentrations is that the cut fuel surfaces of the two segment specimens had experienced some oxidation during the two years of air storage prior to leaching.

The Pu results also show a somewhat high initial concentration, but appear to attain saturation after about 91 days at a level of a few μg/l.

LOW BURNUP FUEL: ALPHA RADIOLYSIS EXPERIMENT

The corrosion of spent fuel in deep granitic groundwater, which is highly reducing, is expected to occur several orders of magnitude more slowly than under the oxidizing conditions employed in most leaching experiments. It has been postulated, however, that local oxidizing conditions, caused by alpha radiolytic decomposition of water, may be created in a thin aqueous film adjacent to the fuel surface, and that a continuous dissolution of the fuel matrix may nevertheless occur in this way.

In experiments where leaching of the high burnup BWR fuel was performed in the synthetic groundwater on which reducing conditions had been imposed by means of H_2/Ar bubbled over a Pd catalyst suspended in the leaching vessel, the leaching rates were shown to be reduced by about a factor of 20 compared with oxidizing conditions (1). It is possible, of course, that these lower leaching rates represent fuel dissolution by the alpha radiolysis effect, and postulated conditions for this have been discussed in the context of calculations of the radiolysis induced by radiations from spent fuel (6).

In the following section, results are presented from leaching tests on fuel irradiated to a low burnup, such that the actinide content, and hence the alpha particle dose to the leachant, was much lower than in the fuel considered previously in this paper.

Experimental

A short fuel rod was irradiated for one reactor cycle (17 days) in the Studsvik R2 test reactor at a linear power of about 22 kW/m up to a burnup of about 0,5 MWd/kg U. After a few weeks cooling the fission product radioactive inventories were similar in size if not in isotope to those in the high burnup fuel, while the inventory of alpha emitters was much less. TABLE VI compares the inventories of the two fuels at their respective reference dates prior to leaching.

Table VI
Comparison of radionuclide inventories for the two fuel types.

Nuclide	Bq/gU	
	High burnup	Low burnup
Sr-89	-	4.06 E 09
Sr-90	2.46 E 09	-
Zr-95	-	3.84 E 09
Ru-103	-	2.79 E 09
Ru-106	1.31 E 09	-
Cs-137	4.43 E 09	5.13 E 07
Ba-140	-	2.11 E 09
Ce-141	-	5.12 E 09
Ce-144	2.02 E 09	1.53 E 09
Nd-147	-	6.48 E 08
Eu-154	2.14 E 08	-
Pu-239/240	3.38 E 07	5.33 E 05
Pu-238/Am-241	1.52 E 08	-
Cm-242	6.76 E 07	-
Cm-244	1.52 E 08	

14 segments of fuel/clad (20 mm long) were cut out from the rod and exposed to the synthetic groundwater (pH 8.1) for 3 weeks under oxidizing conditions as a pre-leach to remove the cesium instant release fraction and to dissolve up any uranium oxidixed by the cutting procedure. The leachants were analyzed by the standard procedure. The fuel specimens were then transferred to 14 new leaching flasks where 6 of them, in pairs, were leached again under oxidizing conditions for periods of 2, 4 and 8 weeks. The remaining 8 specimens were leached, under the reducing conditions described above, for the same periods. Again, the final leachants were subjected to analysis.

RESULTS

The results are summarized in TABLES VII and VIII.

Table VII
Uranium concentrations in centrifugates (ppb).

Contact time (d)	22*	15	28	55
Oxidizing conditions	6380	3700	6250	8250
Reducing conditions		34	37	21

*Pre-leach.

Table VIII
Comparison of ALR values under oxidizing and reducing conditions.

	ALR (FIAP/d) (x E-06)			
Contact time (d)	22	15	28	55
	Oxidizing conditions			
U	5.1	3.9	3.7	2.5
Cs-137	23.6	6.8	4.9	3.9
Sr-89	5.2	5.6	4.1	2.6
I-131	4.1	4.1	4.1	2.6
Ba-140	3.9	4.4	2.7	1.4
Pu	1.0	1.0	0.6	0.4
	Reducing conditions			
U		0.04	0.03	0.04
Cs-137		1.6	0.8	0.7
Sr-89		0.36	0.24	0.21
I-131		1.7	ND	ND
Ba-140		0.53	0.39	0.22
Pu		0.05	0.02	0.07

Inspection of the results leads to the following conclusions:

a) The measured U concentrations under oxidizing conditions were much higher than expected and even after 55 days had probably not attained saturation.

b) Cs, Sr, I and Ba appeared to dissolve congruently with the UO_2 matrix, while Pu and the lanthanides (not reported here) showed lower values of leach rate.

c) Although there are some differences between nuclides, and depending on whether the values for centrifugate concentration or FIAP are used, the decreases in leaching rates for leaching under reducing conditions were of the same order as those obtained with the high burnup fuel.

It seems unlikely that the presence (or absence) of alpha radiolytical effects can be advanced as the cause of these observations. The explanation is probably to be found in differences in microstructure and stoichiometry between the high burnup and low burnup fuel types.

CONCLUSIONS

1. The results from the Swedish program on sequential and static leaching of high burnup BWR fuel extended to 3 year contact times is in general agreement with the earlier results from the program and with those of other workers.

2. Although with the high burnup fuel, uranium saturation in the groundwater (pH 8.1) appears to be attained at a level of 1-2 mg/l, there is some evidence that this level can be raised if the fuel is air-stored for prolonged periods prior to leaching.

3. For high burnup fuel, congruent dissolution was attained at pH values lower than 4.

4. Experiments designed to test the hypotheses that the Sr-90 leach rate represents the dissolution of the fuel matrix, and that the concentrations of the actinides and lanthanides are much lower due to solubility limitations gave indecisive results. SEM and EPMA examination of fuel before and after leaching will probably be needed to give pertinent data.

5. No evidence that alpha radiolytic decomposition of leachant is a significant factor for fuel leaching was obtained in a comparison of leaching rates under oxidizing and reducing conditions of two fuel types which gave alpha doses to the leachant differing by a factor of about 1000.

6. Differences in leaching behaviour between these two fuel types may be due to differences in fuel morphology and stoichiometry.

ACKNOWLEDGEMENTS

We gratefully acknowledge the contributions of the personnel of Studsvik's Hot Cell Laboratory and Analytical Laboratory.

REFERENCES

1. R.S. Forsyth, K. Svanberg and L.O. Werme, in Scientific Base for Nuclear Waste Management - VII edited by G.K. McVay (Elsevier Science Publishers, New York, 1984) p. 179

2. J. Bruno, R.S. Forsyth, and L.O. Werme in Scientific Base for Nuclear Waste Management VIII (Elsevier Science Publishers, New York, 1985)

3. U.B Eklund and R S Forsyth, KBS Technical Report No 70, SKB, Stockholm, Sweden 1978.

4. T.T. Vandergraaf, Report TR-100, AECL Whiteshell Nuclear Res. Est., Pinawa, Manitoba, Canada, 1980.

5. L.H. Johnson, Report AECL-6837, AECL Whiteshell Nuclear Res. Est., Pinawa, Manitoba, Canada, 1982.

6. H. Christensen and E. Bjerkbakke, Alpha-radiolysis of aqueous solutions, paper presented at this symposium, 1985.

DERIVATION OF A WASTE PACKAGE SOURCE TERM FOR NNWSI FROM THE RESULTS OF LABORATORY EXPERIMENTS

VIRGINIA M. OVERSBY* AND CHARLES N. WILSON**
*Lawrence Livermore National Laboratory, P.O. Box 808, Livermore, CA 94550
**Westinghouse Hanford Company, P.O. Box 1970, Richland, WA 99352

ABSTRACT

Results are presented for the dissolution of Turkey Point pressurized water reactor (PWR) spent fuel in J-13 well water at ambient hot cell temperatures. These results are compared with those previously obtained on Turkey Point fuel in deionized water, on H. B. Robinson PWR fuel in J-13 water, and by other workers using various fuels in dilute bicarbonate groundwaters. A model is presented that represents the conditions under which maximum dissolution of spent fuel could occur in a repository sited at Yucca Mountain, Nevada. Using an experimentally determined upper limit of 5 mg/l for uranium solubility in J-13 water, a fractional release rate of 6.4×10^{-8} per year is obtained by assuming that all water entering the repository carries away the maximum amount of uranium.

INTRODUCTION

The Nevada Nuclear Waste Storage Investigations Project (NNWSI) is investigating the suitability of the volcanic rocks at Yucca Mountain, Nye County, Nevada, for potential use as a disposal site for high-level nuclear waste. The repository horizon under study is the Topopah Spring Member of the Paintbrush Tuff. At the potential repository level, the tuff is densely welded and devitrified, and lies approximately 200 to 400 meters above the water table in the unsaturated zone [1].

Laboratory tests on spent fuel are being conducted to determine the release rates of radionuclides from waste packages emplaced in a repository at Yucca Mountain. Four specimen types are used to represent a range of fuel conditions that might occur: (1) fuel rod sections split open to expose bare fuel particles to the liquid phase, (2) rod sections with a 2.5-cm long by 150-μm wide slit through the cladding, (3) rod sections with two 200-μm laser-drilled holes through the cladding, and (4) undefected rod segments. The last three specimen types are fitted with water-tight end fittings containing a vent in the upper fitting to aid water entry through the defects. The undefected specimen type provides a control to show the portion of radionuclide release originating from residual contamination on the cladding exterior surfaces. The details of the test procedure are given in Reference 2.

The first series of tests used Turkey Point PWR fuel discharged in November 1975, with a burnup of approximately 27 MWd/kgM and estimated fission gas release of 0.3%. These tests were conducted in deionized water (DIW) with periodic sampling of the solution phase and rods contained in the test vessel to monitor the progression of radionuclide "plate-out." The tests were run for 8 months, terminated, and restarted for a further four-month testing period in fresh DIW. At the termination of both testing periods, the test specimens and test vessels were rinsed in DIW, and the test vessels were subjected to an acid strip procedure using 8M nitric acid. Uranium and the radionuclides recovered in each phase of the testing were separately measured so that a complete accounting of these species could be made. The results from this test series (DIW/TP) are contained in Reference 3.

The second series of tests is being conducted in J-13 well water, a dilute bicarbonate groundwater that is believed to be representative of the water to be encountered at the potential repository horizon [4]. Spent fuel from the H. B. Robinson reactor, discharged in May 1974, with a burnup of approximately 31 MWd/kgM and 0.2% gas release, was used in the first part of the Series 2 tests (J-13/HBR). Turkey Point fuel from the same pins that provided material for the DIW/TP Series 1 tests was used in the second part of the Series 2 tests (J-13/TP). All tests are conducted under air at ambient hot cell temperatures. Results for the H. B. Robinson fuel in J-13 water through the first termination are given in Reference 5. A summary of selected results from the Turkey Point tests in J-13 water through the first termination is given below. The restarted tests for both fuels are still in progress.

RESULTS OF TESTING TURKEY POINT FUEL IN J-13 WELL WATER

Figure 1 shows the uranium concentration in unfiltered solutions for each of the four specimen types as a function of time. Samples filtered through 0.4-μm and 1.8-nm filters gave identical concentrations, showing that all of the uranium was in true solution. Fuel contained in cladding with induced defects produced much lower concentrations of uranium in solution than did bare fuel. In particular, much lower uranium concentrations were produced by the specimens with the laser-drilled holes. Reduced solution concentrations for the actinides Pu, Am, and Cm were also observed with the slit and the laser-drilled specimens relative to the bare fuel specimens. Similar results were also obtained in the DIW/TP and J-13/HBR test series, indicating that some control on the dissolution rate was provided when the fuel was contained in its original tight geometry in the defected cladding.

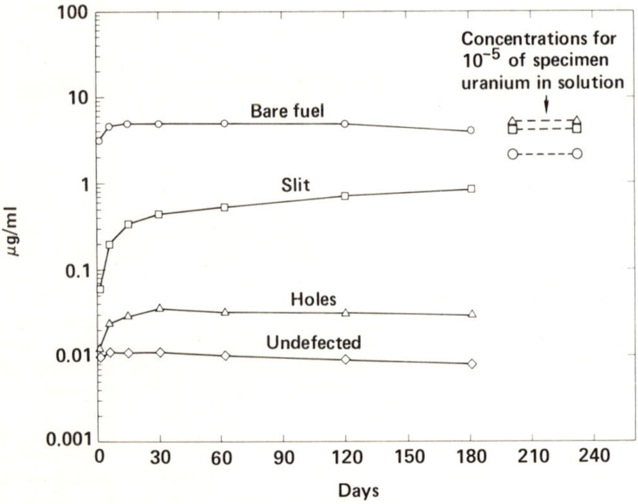

FIGURE 1. Uranium Concentrations (Unfiltered) for the First 181 Days for Tests Using Turkey Point Fuel in J-13 Water.

About half of the total uranium was recovered in periodic solution samples plus the final solution sample and half in the rinse and strip solutions. In contrast, for the other actinides only 12 to 18% was in solution and the bulk of the material was recovered in the acid strip. The initial release of material from the Turkey Point sample appears to be dominated by dispersion of fine particles of fuel, possibly produced during the sample preparation. The smallest of these appear to have rapidly dissolved while the larger settled rapidly to the bottom of the test vessel, since there is no evidence in the filtration data for any particulate matter in suspension. If this explanation is correct, there should be less uranium and actinides in the vessel rinse and strip samples on termination of the second portion of the tests. In the tests using Turkey Point fuel in deionized water the amount of material recovered in the rinse and strip on second termination was only 1/4 of that found in the first termination [3], suggesting much less dispersion of fines on restarting of the test.

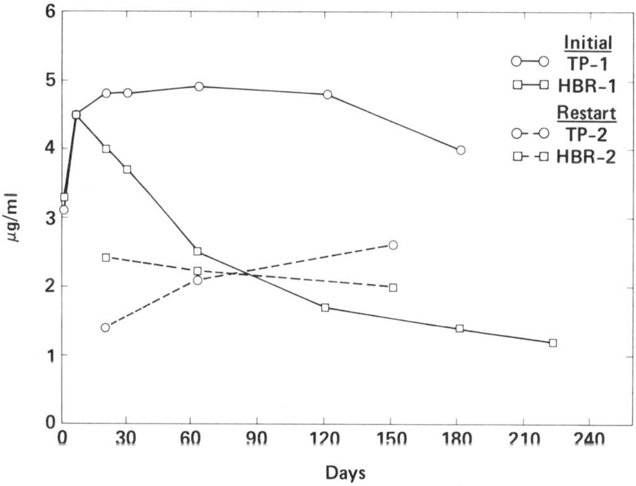

FIGURE 2. Uranium Concentration for Turkey Point and H. B. Robinson Bare fuels Tested in J-13 Water.

Figure 2 compares the uranium concentration observed with the Turkey Point bare fuel in J-13 water to that observed with the H. B. Robinson bare fuel in J-13 water. The Turkey Point fuel produced a nearly constant concentration of 4.8 mg/l for 4 months, but decreased to 4.0 mg/l at 180 days. The H. B. Robinson fuel produced a maximum concentration of 4.5 mg/l after 6 days of testing, followed by a steady decrease to 1.4 mg/l at 181 days and 1.2 mg/l at 223 days. When the tests were restarted, the uranium concentrations were 2.4 mg/l for H. B. Robinson and 1.4 mg/l for Turkey Point fuels at 20 days; 2.2 mg/l and 2.1 mg/l, respectively, at 2 months; and 2.0 mg/l and 2.6 mg/l, respectively, at 5 months. The higher initial solution concentrations reached over both bare fuel types is thought to be due to a thin oxidized layer on the surface of the fuel formed by air exposure of the fuel after the fuel rods were sectioned. In this hypothesis, as the more highly oxidized surface layer dissolves and a steady-state condition is approached with an intermediate oxidation state or "plate-out" phase, uranium concentration decreases. The time between rod sectioning and testing was approximately 5 years for the Turkey Point fuel versus a few months for the H. B. Robinson fuel. The more rapid decrease in uranium concentration over the H. B. Robinson fuel is thought to be due to the shorter time between rod sectioning and testing resulting in less of the higher oxidation state. After the first exposure to J-13 water, the surface oxidation states of the two fuels may be similar, since they exhibited more similar solubility behavior when the tests were restarted.

Data for total release of actinides, technetium, and cesium during the initial six-month testing period for the bare fuel samples are given in Table I. Total release of Pu, Am, and Cm was congruent and lower than that found for uranium. The preferential release observed for uranium relative to the actinides may be due to leaching of oxidized uranium from the surface of the fuel. If this is the case, the total release of uranium measured on termination of the restarted tests should be more similar to the actinide release.

TABLE I: RELEASE DATA FOR TURKEY POINT BARE FUEL IN J-13 WATER

	Uranium (μg)	$^{239+240}$Pu (nCi)	^{241}Am (nCi)	^{244}Cm (nCi)	^{99}Tc (nCi)	^{137}Cs (μCi)
Solution Samples	350.5	63	138	138	15[a]	1010
Rod Samples	15.4	22	42	38	n.m.	2.7
Final Solution	1000	114	243	250	53	3090
Water Rinse	366	63	111	104	11.6	786
Acid Strip	960	1140	2180	1610	6.1	138
TOTAL	2691.9	1401	2714	2140	86	5027
Fractional Release (parts in 100,000)	11.67	7.30	6.87	7.44	32.4	295
% in Solution	50.2	12.6	14.0	18.1	79.1	81.5

[a] Includes 8 nCi in solutions on which Tc was actually measured and an estimate of 7 nCi removed in other solutions.

Technetium is expected to be very soluble under the conditions of these tests where there is free access of air to the solutions. It is, therefore, reasonable to assume that the technetium recovered from the sample rinse and acid strip solutions represents undissolved fine particles of spent fuel. If this is the case, we should be able to predict the uranium recovered in these samples by assuming they contain an inventory of technetium and uranium that is representative of the average in the fuel sample. The 17.7 nCi of Tc found in these samples for the bare fuel test implies that 1540 μg of U should be in the rinse and acid strip samples; 1326 μg was recovered in these samples, in good agreement with the prediction. A similar calculation using the technetium data from the H. B. Robinson bare fuel test gives the predicted amount of uranium as 3585 μg as compared with 3300 μg actually recovered. This evidence supports the arguments given above that the sample preparation method for bare fuel samples causes fine particles of fuel to become dislodged. Thus, the first run of a test using bare fuel will somewhat overestimate the "dissolution" rate of the fuel. The slit defect and laser puncture sample results should be much less affected by sample preparation.

The data for technetium in the solution samples may be compared with the uranium recovered from solution samples plus that calculated to have precipitated from solution to obtain an estimate of the degree of preferential dissolution of technetium relative to the bulk of the matrix. Using a peak uranium concentration of 4.8 mg/l and a final value of 4.0 mg/l for the first run of the Turkey Point bare fuel tests gives an estimate of 200 μg of uranium precipitation. If this uranium is added to that recovered from solution samples, we estimate that 1550 μg of uranium were once in solution. Assuming the average fuel inventory ratio of technetium to uranium, this corresponds to 17.8 nCi of technetium. The amount of technetium in solution was 68 nCi, indicating a preferential dissolution factor of 3.8 for technetium in the first run of the test. A similar calculation for the H. B. Robinson bare fuel test gives a factor of 8.7 for preferential technetium dissolution. Microstructural examination of cathodic vacuum etched ceramographic sections from both fuel types indicated a finer pre-irradiation grain size for the H. B. Robinson fuel (6 μm for H. B. Robinson versus 25 μm for Turkey Point). Partial diffusion of technetium from the oxide fuel matrix to the grain boundaries likely occurs during irradiation. The higher preferential dissolution factor for technetium from the H. B. Robinson fuel correlates well with the higher grain boundary fractional technetium inventory expected for the finer grain fuel.

Total measured fractional release for the initial run of the J-13/TP tests are compared to that of the J-13/HBR and DIW/TP tests in Table II. Actinide releases from both bare fuel tests in J-13 water were nearly identical. The slightly lower fractional uranium release with the H. B. Robinson fuel may be explained by precipitation (as suggested by Figure 2) onto the bare fuel bed covering the bottom of the vessel where it would not necessarily be completely recovered in the rinse or acid strip solutions. Higher actinide and ^{99}Tc release in the DIW/TP tests have been correlated to extensive grain boundary dissolution, which was not observed in the J-13 water bare fuel tests [5]. Higher fractional release from Turkey Point fuel (both J-13 and DIW) slit and holes defect specimens relative to the corresponding J-13/HBR tests may be related to the longer pretesting exposure to air for the Turkey Point fuel.

The release of cesium from the bare fuel was comparable to the amount of fission gas release for the fuel, in agreement with the results of Johnson et al. for CANDU fuel [6]. The cesium release for the H. B. Robinson bare fuel was nearly four times the estimated gas release. Forsyth et al. [7] have found cesium release for boiling water reactor (BWR) high burn-up fuel to be 1.5 times the fission gas release. Variations in reported cesium release to fission gas release ratios may be due to a number of factors, including specimen preparation and storage conditions, and irradiation parameters for different fuel types. Calculation of the preferential release factor for Cs based on U that once was in solution (as done above for Tc) gives a factor of 36 for Turkey Point fuel versus 358 for H. B. Robinson fuel. As with technetium, the greater fractional Cs release with the H. B. Robinson fuel may be correlated to the finer grain size in this fuel.

The summary of release data for the first run of all sample types in both series of tests (Table II) shows that there is a larger total release from the bare fuel in DIW than in J-13. This is thought to reflect a difference in the amount of fine particles dispersed at the start of the test and not a true difference in "solubility" of the fuel in the two water types. The cesium release for Turkey Point bare fuel is similar in both waters and is probably a true reflection of the amount of readily soluble Cs in the fuel. The release of technetium appears to depend on the water type, with much lower release occurring in the J-13 water. Total release of actinides from the slit defect samples is generally two orders of magnitude less than that of the bare fuel, and the release from the hole defect specimens is another order of magnitude

TABLE II: SUMMARY OF TOTAL MEASURED FRACTIONAL RELEASE DATA
(Units: parts per 100,000 of the test specimen inventory)

Nuclide	Series[a]	Bare Fuel	Slit	Holes	Control
Uranium	J-13/HBR	5.63	0.05	0.005	0.009
	J-13/TP	11.7	0.65	0.030	0.011
	DIW/TP	21.2	0.07	0.032	0.010
$^{239+240}Pu$	J-13/HBR	7.04	0.01	0.003	0.002
	J-13/TP	7.30	0.11	0.022	0.006
	DIW/TP	22.4	0.20	0.042	0.021
^{241}Am	J-13/HBR	7.76	0.01	0.002	0.003
	J-13/TP	6.87	0.12	0.023	0.004
	DIW/TP	17.2	0.15	0.019	0.008
^{244}Cm	J-13/HBR	6.54	0.01	0.001	0.002
	J-13/TP	7.44	0.13	0.026	0.005
	DIW/TP	21.5	0.35	0.026	0.006
^{237}Np	J-13/HBR	<6.4	---	---	---
	J-13/TP	<7.4	---	---	---
	DIW/TP	18	0.2	---	---
^{137}Cs	J-13/HBR	739	633	400	0.18
	J-13/TP	295	137	7.6	0.13
	DIW/TP	230	110	50	0.04
^{99}Tc	J-13/HBR	22.2[b]	2.4	<1.5	---
	J-13/TP	32.4	<13.7	<2	---
	DIW/TP	152	8.1	---	---

[a] J-13/HBR -- H. B. Robinson fuel tested in J-13 water [5].
J-13/TP -- Turkey Point fuel tested in J-13 water.
DIW/TP -- Turkey Point fuel tested in deionized water [3].
[b] Includes an estimate of 22.5 nCi for Tc removed in solutions on which direct measurement was not made.

lower. Cesium release is only slightly lower for the defected samples in comparison to the bare fuel, an indication of the readily accessible nature of the rapid release component and the high solubility of Cs.

UPPER BOUND TO RELEASE RATES FROM SPENT FUEL UNDER NNWSI CONDITIONS

Based on our data and that of others [6,7,8], we have developed a model to estimate the upper bound to the release rate of radionuclides from spent fuel under NNWSI conditions. General corrosion rates for the stainless steel containment barrier are sufficiently low that the container will remain substantially intact for more than 10,000 years [8]. Localized corrosion mechanisms, such as stress corrosion cracking or crevice corrosion, may cause containers to develop breaches that allow water to enter the container. The most likely location for the breach to occur is at or near the final closure weld [8].

Figure 3 shows a schematic representation of a spent fuel container in a vertical emplacement hole. The hole is assumed to be lined with a carbon steel liner that allows water to collect in the hole. In the absence of the liner, water would be expected to drain away and not fill the hole [9]. If we assume that the container has a breach large enough to allow free access of water to the interior and that the liner is intact, approximately 1800 liters of water could accumulate in the hole. The weight of fuel in the container is 3140 kg, giving a fuel-to-water ratio of 1.74 kg/l. Our tests using Turkey Point bare fuel had a ratio of 0.11 kg/l, and those using H. B. Robinson fuel had a ratio of 0.33 kg/l. The solution concentration of uranium for both fuels on the restarted tests was about 2 mg/l, suggesting that solubility of uranium in the system was controlling the solution concentration and the dissolution of the matrix after initial dissolution of any oxidized surface films and equilibration of the system.

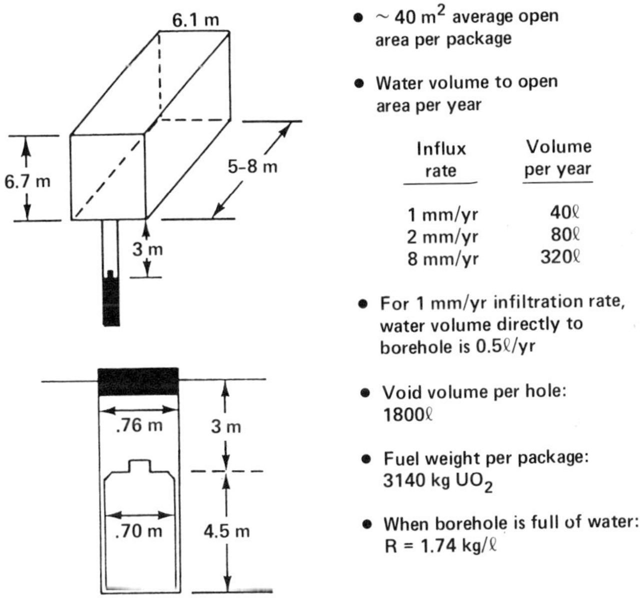

FIGURE 3. Water-to-Fuel Ratio and Vertical Emplacement Configuration.

Forsyth et al. [7] used a fuel-to-water ratio of 0.08 kg/l, similar to our Turkey Point tests, and found a concentration of about 1 mg/l in water with similar bicarbonate concentration to J-13. Johnson [10] found uranium concentrations of 1.2 mg/l to 4.5 mg/l in "KBS" water with bicarbonate 2.5 times higher than J-13, and 0.2 mg/l to 3 mg/l in granite groundwater with bicarbonate half of the J-13 value. Both sets of experiments used a fuel-to-water ratio of 0.72 kg/l. Taken together, the data strongly suggest that the ambient temperature solubility of uranium over spent fuel in dilute bicarbonate solutions is 5 mg/l or less. We will use this value as the matrix solubility control for dissolution of spent fuel in our model calculations.

Control of matrix solubility at less than 5 mg/l of uranium does not necessarily imply that matrix dissolution ceases when solution concentrations reach that value. The dissolution might still continue, accompanied by precipitation of a secondary uranium-bearing phase. In our tests, technetium can be used to monitor whether dissolution of the matrix continues once the uranium in solution has reached the solubility limit. Table III shows data for technetium in solution for the first run of both the H. B. Robinson and Turkey Point bare fuel tests. Solution concentrations are constant to within the precision of the measurements for both tests. Since solution samples have been taken at 2, 4, and

6 months in each case, we would expect decreasing concentrations of technetium if no further dissolution of the fuel matrix occurred when fresh J-13 water was added to replace the volume removed during sampling. The fact that solution concentrations are essentially constant implies some matrix dissolution has occurred in the added water. For H. B. Robinson, the total technetium increase from day 63 to day 223 was 27 nCi; for Turkey Point the increase from day 62 to day 181 was 8 nCi. Applying the preferential dissolution factors calculated above, this would correspond to 249 and 182 μg of uranium dissolution respectively. In the H. B. Robinson test, a total of 60 ml of new water was added between day 63 and day 223. If this water had dissolved 5 mg/l of uranium, a total of 300 μg of uranium should have dissolved. In the Turkey Point test, 40 ml of water was added, implying that 200 μg of uranium should have dissolved. In both cases the results based on assuming that uranium dissolved until the new water reached 5 mg/l concentration are close to, and slightly higher than, those based on assuming that technetium was added to solution with the average preferential release factor remaining constant throughout the test. We would expect the preferential release factor to decrease with time as the grain-boundary inventory is depleted, and eventually to observe congruent technetium and uranium release. Once congruent release is achieved we will be able to base our calculations on measured matrix dissolution rates and will revise the release rate model accordingly.

TABLE III: CONCENTRATION OF TECHNETIUM IN SOLUTION FOR TURKEY POINT AND H. B. ROBINSON TESTS IN J-13 WATER, In pCi/ml.

H. B. Robinson Bare Fuel		Turkey Point Bare Fuel	
Sampling Day	[Tc]	Sampling Day	[Tc]
63	450	62	198
181	468	120	203
223	450	181	212

Our initial model also assumes that the oxide fuel matrix is not substantially degraded by oxidation or other mechanisms that are not observed in short-term laboratory leaching tests and that all actinides are therefore released congruently. Under the assumptions of the model, the rate at which the matrix of the spent fuel will dissolve depends linearly on the rate of delivery of water to the fuel. The infiltration rate at Yucca Mountain has been estimated to be less than 1 mm/yr [9]. If we assume a flow rate of 1 mm of water per year through the repository horizon and that all of the water that intersects the emplacement drift excavation area is able to collect into the package emplacement holes, we can calculate a rate of influx of new water into the borehole. The flow rate calculated by this method will be far higher than that expected to actually occur in the unsaturated zone, since water flow should be channeled around rather than through the large openings in the rock [11].

Figure 3 shows a model of the emplacement drift with a vertical package borehole. The excavation area is 40 m^2 per package, corresponding to a delivery rate of 40 l/yr of new water to the borehole. Dissolution of U to give a concentration of 5 mg/l would result in a fractional dissolution rate of 6.4×10^{-8} per year. Using a borehole plus container void volume of 1800 l and an inflow rate of 40 l/yr, it would take 45 years for the borehole to fill. During this time, no release would occur under the model assumption of an intact borehole liner (inflow opening near the top of the liner). After the borehole plus container are filled with water, new water is assumed to displace U-saturated water and to itself reach saturation. This would produce a long-term release rate of 6.4×10^{-8} per year for those nuclides whose release is controlled by matrix solubility.

The time of 45 years to fill the void volume is much longer than our longest test interval of 223 days. The present ratio of model extrapolation time to experiment length is 74; by conducting a 3 year test this ratio can be decreased to 15, and a 5 year test reduces the ratio to 9.

There are four types of release that are not accounted for by the model discussed above. First, ^{14}C present in the Zircaloy can be released because of contact with air as soon as the container is breached. Measurements at 275°C on Turkey Point fuel showed an initial release of about 0.3% of the calculated ^{14}C inventory [12]. This material is thought to come from the oxidized layer on the cladding surface. Further release was very low. The pulsed ^{14}C release will be controlled by the container failure rate. A container failure rate of 0.1% per year would control the release rate of ^{14}C to 0.3 parts in 100,000 per year.

The second component of release from non-fuel-matrix sources is the release of activation products contained in the cladding matrix and in fuel assembly parts. The rate of corrosion of Zircaloy under expected Yucca Mountain conditions is being determined to provide data on the release rate of ^{14}C and other activation products. The release rate of activation products from other metal components will be determined using the corrosion rates obtained as part of the metal containment barrier testing program.

The third component of the release rate model is that for the gap/grain boundary components that are released initially at a rapid rate, but then more slowly until finally controlled by the matrix fuel solubility. The highest initial release rate is for cesium, and the amount of release is similar to the fission gas release for the fuels. However, after the 300-year minimum 10 CFR 60 required [13] containment period, ^{137}Cs and ^{134}Cs will have essentially decayed away, and the only remaining radioactive ^{135}Cs isotope accounts for only about 0.01% of the total 300-year curie activity of a spent PWR fuel assembly [14]. We will assume that the average gap/grain boundary inventory is 0.5% and that all of this material is released in the first year following breach of the cladding. For a cladding breach rate of 0.1% per year, the release rate from the repository ensemble is dominated by the pulsed release for the 1000-year period over which the pins are being breached. The rate would be 5.55×10^{-6} per year. After this component is exhausted, the release rate would rapidly decrease to the much lower rate controlled by matrix dissolution. Release rates for technetium are expected to be intermediate between those of cesium and the matrix.

A fourth component of the release rate model to be developed is the possibility of slow degradation of the oxide fuel matrix by oxidation. Such oxidation of the UO_2 spent fuel matrix phase would provide a mechanism for accelerated release of soluble long-lived radionuclides such as ^{99}Tc, ^{135}Cs, ^{129}I, ^{237}Np, ^{79}Se and ^{14}C. The kinetics for oxidation of the spent fuel matrix are known to be extremely slow at post-containment repository temperatures (<130°C) and may not be significant; however, studies have been initiated to better characterize the low temperature oxidation behavior of spent fuel.

In order to improve the model described above, several types of data are needed. The most important are (1) the dependence of uranium solubility on temperature, (2) the rate of failure of the containment barrier, (3) the rate of breach of cladding and (4) the rate of oxidation of the UO_2 fuel matrix. Experiments and tests are in progress to improve our understanding of all of these processes and to provide the needed data.

REFERENCES

1. S. Sinnock, Y. T. Lin, and J. P. Brannen, SAND84-1492, Sandia National Laboratories, Albuquerque, NM (1984).
2. C. N. Wilson, HEDL-TC-2353-3, Hanford Engineering Development Laboratory, Richland, WA (1984).
3. C. N. Wilson, HEDL-TME-84-30, Hanford Engineering Development Laboratory, Richland, WA (1985).
4. V. M. Oversby, UCRL-53629, Lawrence Livermore National Laboratory, Livermore, CA (1985).
5. C. N. Wilson, and V. M. Oversby, UCRL-91464, Lawrence Livermore National Laboratory, Livermore, CA (1985).
6. L. H. Johnson, K. I. Burns, H. Joling, and C, J. Moore, Nucl. Technol 63, 470 (1983).
7. R. S. Forsyth, K. Svanberg, and L. Werme, in Proceedings of the Third Spent Fuel Workshop, L. Werme editor, KBS Technical Report No. 83-76, Swedish Nuclear Fuel Supply Co., 1983 (issued March 1984).
8. V. M. Oversby, and R. D. McCright, UCRL-91257, Lawrence Livermore National Laboratory, Livermore, CA (1984).
9. P. Montazer, and W. E. Wilson, USGS-WRI-84-4345, U.S. Geological Survey Water-Resources Investigations, Lakewood, CO (1984).
10. L. H. Johnson, AECL-6837, Atomic Energy of Canada Limited, Pinawa, Manitoba, Canada (1982).
11. J. A. Fernandez, and M. D. Freshley, SAND83-1778, Sandia National Laboratories, Albuquerque, NM (1984).
12. R. A. Van Konynenburg, C. F. Smith, H. W. Culham, and C. H. Otto, Jr., UCRL-90855, Rev. 1, Lawrence Livermore National Laboratory, Livermore, CA (1984).
13. Code of Federal Regulations, 10CFR60, Title 10, Chapter 1, Part 60 (June 30, 1983).
14. C. W. Alexander et al., ORNL/TM-6008, Oak Ridge National Laboratory, Oak Ridge, TN (October, 1977).

---

*Work performed under the auspices of the U.S. Department of Energy by the Lawrence Livermore National Laboratory under contract number W-7405-ENG-48.

INFLUENCE OF HIGHER WASTE LOADING ON CRYSTALLIZATION OF TITANOSILICATE GLASS-CERAMICS

P.J. HAYWARD, E.R. VANCE, C.D. CANN AND I.M. GEORGE
Atomic Energy of Canada Limited, Whiteshell Nuclear Research Establishment, Pinawa, Manitoba, ROE 1LO, Canada

ABSTRACT

Sphene ($CaTiSiO_5$) is obtained as the main crystalline phase in Na_2O-Al_2O_3-CaO-TiO_2-SiO_2 glass-ceramics containing up to ~ 15 wt% of simulated nuclear fuel recycle waste, by maintaining the precursor glass at ~ 1050°C for sufficient time to allow nucleation and crystal growth. The implications for production of an engineering-scale waste form are discussed.

INTRODUCTION

Glass-ceramics with sphene ($CaTiSiO_5$) as the major crystalline phase are being developed in the Canadian Nuclear Fuel Waste Management Program as candidate materials for immobilizing the high-level wastes (HLW) arising from possible future recycling of irradiated CANDU™ UO_2 fuel. The potential advantages of the glass-ceramics include their excellent durability, their relative ease of fabrication, and their ability to incorporate a significant fraction of the waste ions in the more durable crystalline phase [1].

The glass-ceramics are prepared by controlled crystallization of precursor melts in the system Na_2O-Al_2O_3-CaO-TiO_2-SiO_2-X (X = waste ions and/or minor processing additives). This paper examines crystallization during slow cooling and subsequent reheating of precursor melts containing up to 25 wt% simulated HLW. Prior to this work, it was uncertain whether slow rates of cooling, anticipated for the production of an engineering-scale waste form of several hundred kg, would produce crystallization of alternative or additional phases during cooling. Previous laboratory experiments (see Ref. 1) have typically involved approximately 0.2 kg melts, and reheating of quickly cooled glasses to achieve crystallization of sphene. Also, since previous work [1] has shown that the presence of several wt% of certain simulated radionuclides, particularly U and rare earth (RE) ions, can promote crystallization of phases other than sphene, it was necessary to establish the maximum waste loading that the glass-ceramics could accommodate while still forming sphene as the major crystalline phase.

EXPERIMENTAL

The composition of the simulated HLW (SHLW), given in Table I, was based on the recycling of 10-year cooled Pickering UO_2 fuel, irradiated to a burn-up of 650 GJ/kg U, and with 99.9% (U+Pu) extraction [2]. Elements present at less than 0.1 wt% were omitted; Pd was substituted for Rh and Ru; La was substituted for Pm, Sm, Eu, Gd and Tb; Ce was used to simulate Pu, Am, Np and Cm; Mo was substituted for Tc. The SHLW was prepared by milling of pre-mixed nitrates and oxides, followed by calcination in air at 580°C for three hours.

The composition of the base glass-ceramic was (wt%): Na_2O, 6.33; Al_2O_3, 8.05; CaO, 14.32; TiO_2, 18.30; SiO_2, 53.00. A 400-g batch of this glass was prepared by mixing appropriate quantities of oxides and carbonates, followed by melting the mixture in air at 1350°C for 1 h in a Pt-10% Rh crucible, fritting in deionized water and pulverizing to < 150 μm powder. Additions

Table I.
Calculated Composition of Simulated High-Level Waste After Calcination

Oxide	wt%	Oxide	wt%
PdO	11.85	ZrO_2	8.95
UO_3	12.03	Cs_2O	7.72
La_2O_3	6.32	SrO	2.13
CeO_2	7.54	BaO	4.12
Nd_2O_3	11.14	SnO_2	0.31
Pr_2O_3	1.36	Fe_2O_3	3.41
Y_2O_3	1.33	Cr_2O_3	0.92
MoO_3	18.09	NiO	2.78

of 10, 15, 20 and 25 wt% of calcined SHLW were then made to portions of the base glass frit, to give compositions designated W-10, W-15, W-20 and W-25, respectively. After thorough mixing, the SHLW-doped frits were remelted in covered graphite crucibles for 1 h at 1400°C in a nitrogen atmosphere. The melts were cast on a steel plate and annealed at 725°C before cooling to room temperature.

A portion of each specimen was crushed and sieved to size fractions of < 53 μm, 75 to 106 μm, 300 to 350 μm, 500 to 600 μm, 1000 to 1180 μm and > 2360 μm. Differential Thermal Analysis (DTA) was performed on these sieved fractions, using the method and equipment described previously [3]. X-ray diffraction (XRD) measurements, and scanning electron microscopy (SEM) or transmission electron microscopy (TEM), combined with energy dispersive X-ray (EDX) analyses, were also made using previously described [3] facilities.

RESULTS AND DISCUSSION

Crystallization During Reheating of Precursor Glasses

As found in previous work (see Ref.1), all the glasses separated into two amorphous phases during the relatively rapid cooling from the melt to form an irregular (CaO, TiO_2)-rich droplet phase in a continuous SiO_2-rich matrix. A metallic phase was also found to have segregated from each melt, forming a discontinuous layer on the base and sidewalls of each crucible. In addition, droplets of this phase, ranging from in size 1 μm upwards, were observed by SEM in the interiors of the samples. Subsequent SEM/EDX analyses showed that the metallic phase consisted mainly of Mo, with some Pd, Ni and Fe. Figure 1(a) shows a representative SEM photomicrograph of the interior of the W-10 glass.

DTA results for the < 53 μm and > 2360 μm size fractions of the four glasses are listed in Table II, together with the phases crystallized

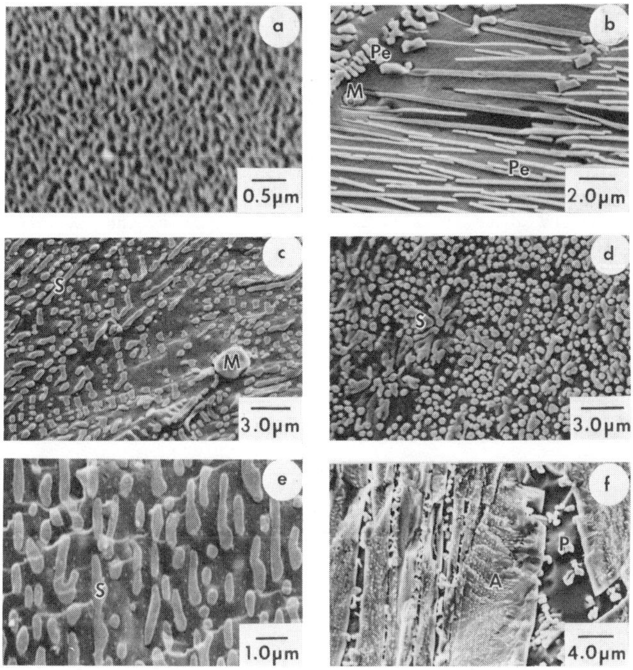

Figure 1. SEM micrographs of:
(a) W-10, phase-separated precursor glass;
(b) W-10, cooled from 1400°C to 500°C at 5°C/min., showing perrierite (Pe) and metallic droplets (M);
(c) W-10, cooled as in (b), then reheated to 1050°C and held for 6 h. S = sphene, M = metallic phase;
(d) W-10, cooled from 1400°C to 1050°C at 5°C/min., held at 1050°C for 3 h. S = sphene;
(e) W-15, same cooling and heating schedule as in (c). S = sphene;
(f) W-25, same cooling and heating schedule as in (c). A = anorthite, P = perovskite.

after heating samples to 1100°C at 10°C/min in the DTA apparatus (identified using XRD). The minimal difference between T_c values for the fine and coarse fractions of the glasses indicates bulk nucleation of the crystal phases, as discussed previously [3,4]. However, it is likely that nucleation was produced either by the presence of precipitated droplets of metals such as Pd [4], or by the initial crystallization of the CaO-, TiO_2-rich glass phase to produce fluorite- or pyrochlore-structured phases such as $(U,RE)O_{2+x}$ or $Ca(U,RE)Ti_2O_{7+x}$ $(0 < x < 1)$ [3,5], followed by further reaction of the crystalline phases with residual glass to produce sphene and/or perrierite-structured $(\underline{Ca},RE,U)_2Ti_2Si_2O_{11}$. For the two traces that gave more than one exothermic crystallization peak, no simple correlation was found between an individual peak and the crystallization of a new phase. The phase listed as perrierite-structured $(Ca,U,RE)_2Ti_2Si_2O_{11}$ in Table II was shown by subsequent TEM/EDX analysis to depart considerably from this idealized stoichiometry, although giving essentially the perrierite XRD pattern. The precise nature of this phase, hereafter referred to as perrierite, is discussed below.

The "ideal" stoichiometry of the mineral perrierite has been reported as $Ce_2Si_2Ti_2O_{11}$ [6]. The phase observed in the present work gave an XRD pattern similar to that of perrierite. However, TEM/EDX analyses indicate that the Ce site is mainly occupied by Ca, while the Ti/Si molecular ratios are closer to 2 than to unity.

This phase has previously been observed in titanosilicate glass-ceramics doped with individual RE ions only [7]. In the present work, the analyses indicate a stoichiometry close to $(Ca,M)Ti_2SiO_{7+x}$, where M represents U, RE and other impurity ions. Attempts to synthesize the perrierite-structured phase are in progress.

Table II. DTA and XRD Results for Waste-Loaded Glasses During Reheating to 1100°C at 10°C/min.

Glass	Size	T_g (°C)	M_g (°C)	T_c (°C)	Phases Crystallized
W-10	<53 µm	685	733	860	S
W-10	>2360 µm	691	739	886	S
W-15	<53 µm	698	744	873	S + trace F
W-15	>2360 µm	693	744	858,889	S + trace F
W-20	<53 µm	695	725	882	S + F
W-20	>2360 µm	700	735	892	S + F
W-25	<53 µm	680	729	841,977	S + F + trace Pe
W-25	>2360 µm	695	735	847	Pe + tr.S

T_g = transformation temperature (± 6°C), M_g = softening point (± 4°C), T_c = peak temperature of crystallization exotherm.
S = sphene, F = fluorite-structured $(U,RE)O_{2+x}$, Pe = perrierite-structured $(Ca,U,RE)_2Ti_2Si_2O_{11}$. Most abundant phases underlined.

Crystallization During Slow Cooling and Reheating

Devitrification of a precursor melt may also occur during cooling from the molten state, so that reheating from lower temperatures to achieve crystallization may become redundant. Devitrification will be most likely if the cooling rate in the critical range 1050°C to 1200°C is low [8], as could be the case for the cooling of a large waste form after casting. Thus a series of controlled cooling experiments, with or without a subsequent heating stage, were performed on ~ 25 g samples of the waste-loaded glasses in alumina crucibles and an N_2 atmosphere. These experiments involved:

(a) cooling from 1400°C to room temperature at 5°C/minute,

(b) cooling from 1400°C to 500°C at 5°C/minute, followed by reheating to 1050°C at 10°C/minute and holding at 1050°C for 6 hours before cooling in the shut-down furnace, and

(c) cooling from 1400°C to 1050°C at 30°C/minute, holding at 1050°C for

3 hours, before cooling in the shut-down furnace.

After completion of each schedule, a 5 mm diameter central core was taken from each sample and studied using XRD, SEM and TEM. The crystalline phases identified by XRD for each sample are listed in Table III.

The results show that the major phase produced by cooling from 1400°C at 5°C/minute was perrierite (see Figure 1(b)). However, for glasses with waste loadings of $\leq$ 15 wt%, conversion to a sphene-dominated glass ceramic could be achieved by arresting the cooling and holding at 1050°C, or, after cooling to lower temperatures, by reheating to 1050°C and holding (see Figures 1(c) to 1(e)). At higher waste loadings, however, the new phase assemblages were dominated by perrierite-, perovskite- and pyrochlore-structured phases (see Figure 1(f)), in which the solubilities of RE and U ions are presumably greater than those in the $CaTiSiO_5$ structure. Partial devitrification of the residual glass to form anorthite also occurred.

Table III. Crystalline Phases Produced by Various Cooling Schedules in Waste-Loaded Glasses

Heat Treatment	W-10	W-15	W-20	W-25
a (controlled cooling)	Pe+Pk+An+Wo+R	Pe+Pk+Wo	Pe+Pk+Wo	Pe+Py+Pk+An
b (cooling+reheating)	S	S+minor F	Pk+F+An	An+Py+Pk
c (interrupted cooling)	S+trace F	S+minor F	Pe+Pk+ F,S	Pe+traces Pk F,Wo,S

Most abundant phases are underlined.
S = sphene; Pk = perovskite-structured $(Ca,Na+RE)TiO_3$; Pe = perrierite (see Appendix 1); R = rutile, TiO_2; F = fluorite-structured $(U,RE)O_{2+x}$; Py = pyrochlore-structured $Ca(U,RE)Ti_2O_{7+x}$; Wo = wollastonite, $CaSiO_3$; An = anorthite, $CaAl_2Si_2O_8$.

Waste Ion Partitioning Between Glass and Crystal Phases

The glass and the major crystalline phases in selected samples of heat-treated glasses were analyzed using TEM-EDX on ion-thinned specimens. The EDX spectra were analyzed for elemental concentrations using the thin-film standardless program ZAPTEM [3]. No correction was made to the analyses for Na^+ migration from the area irradiated by the electron beam. The precision obtained was approximately $\pm$ 10% for major elements and approximately $\pm$ 25% for minor elements (< 2 mole %). Table IV shows the results obtained for W-10 after cooling from 1400°C to room temperature at 5°C per minute, and after cooling from 1400°C to 500°C at 5°C/min followed by reheating to 1050°C and holding for six hours.

It is difficult to estimate the degree of partitioning of the waste ions between the glass and crystalline phases from these results because the volume fraction of crystalline material is considerably higher in the sphene glass-ceramic than in the perrierite glass-ceramic (see Figures 1(c) and 1(b)). However, both sphene and perrierite acccept the major part of the RE^{3+}, Y^{3+} and Zr^{4+} contents of these compositions, while Ba^{2+} and Cs^+ are too large to be easily accommodated in the crystal structures, and partition strongly into the glass phases. Uranium (presumed to be in the U^{4+} state) is preferentially incorporated into perrierite but not into sphene; possibly the level of RE substitution in the sphene structure and/or a lack of charge balancing cations (such as Na) prevent further substitution of U for Ca. Thus, the mean analyses for sphene were fairly close to the idealized $CaTiSiO_5$ formula. As mentioned previously, the major parts of Mo, Fe, Ni and Pd were present in these samples as separate alloy phases.

Table V shows TEM/EDX analyses of the perrierite-structured and pyrochlore-structured phases and of the residual glass in samples of W-25 cooled from 1400°C to room temperature at 5°C/minute. The mean analysis of the perrierite-structured phase is close to that given in Table IV, except for a higher degree of RE substitution for Ca, as expected from the higher waste loadings. The pyrochlore-structured phase approximates a stoichiometric composition of $(Ca^{2+},Sr^{2+},Sn^{2+};RE^{3+},Y^{3+};\ U^{4+},Zr^{4+})_2Ti_2O_7$, except for an excess of Ti, and contains a significant fraction of the Cs and Ba.

Table IV.
Mean TEM/EDX Analyses (Mole %) of Major Phases in W-10 Samples After Different Cooling Schedules.

Element	W-10, cooled at 5°C/min		W-10, cooled at 5°C/min reheated to 1050°C	
	Glass (4 analyses)	Perrierite (6 analyses)	Glass (7 analyses)	Sphene (6 analyses)
Al_2O_3	14.0	0.0	10.0	0.0
SiO_2	57.8	27.5	77.0	33.3
CaO	17.0	21.0	6.2	30.0
TiO_2	7.5	46.8	2.2	34.5
FeO	0.17	0.00	0.39	0.00
SrO	0.35	0.38	0.39	0.16
Y_2O_3	0.10	0.28	0.10	0.12
ZrO_2	0.44	1.92	0.54	0.85
UO_2	0.24	1.31	0.53	0.16
MoO_2	0.14	0.23	0.22	0.18
SnO	0.19	0.24	0.21	0.17
Cs_2O	0.18	0.02	0.23	0.04
BaO	0.18	0.01	0.23	0.05
Ce_2O_3	0.12	0.31	0.06	0.13
La_2O_3	0.02	0.42	0.05	0.07
Nd_2O_3	0.00	0.43	0.05	0.16
TOTAL	98.43	100.85	98.40	99.89

SUMMARY: IMPLICATIONS FOR WASTE FORM PRODUCTION

This work suggests that up to 15 wt% of HLW can be incorporated in a sphene-based glass-ceramic monolith, by either inter-rupting the cooling of the cast waste form and holding it at ~ 1050°C for up to 3 hours, or by reheating the cooled waste form to ~ 1050 C for up to 6 hours. At higher waste loadings, the crystalline phases formed are those in which RE and U are essential (rather than substituted) ions, such as perrierite-, pyrochlore-, perovskite- or fluorite-structured phases.

Further work is required to assess the optimum melting atmosphere to minimize volatilization of Cs, Tc, Ru and Mo from melts containing HLW, and to encourage multivalent ions, such as Ce and U, to adopt a lower valence state (which should, on crystal chemical arguments [3], promote their incorporation into the Ca site in sphene during crystallization). The melting conditions used in the present experiments caused formation of a Mo/Ni/Fe/Pd alloy layer from oxide reduction and gravity segregation. Experiments using CO/CO_2 atmospheres of controlled oxygen fugacity are in progress to address this problem.

Table V.
Mean TEM/EDX Analyses (Mole%) of Major Phases in W-25 Samples After Crystallization During Cooling.

Element	W-25, cooled from 1400°C to 500°C at 5°C/minute		
	Glass (4 analyses)	Perrierite (4 analyses)	Pyrochlore (5 analysis)
Al_2O_3	17.5	0.0	0.0
SiO_2	57.4	25.3	1.5
CaO	14.7	17.8	14.6
TiO_2	4.2	49.3	60.3
FeO	0.04	0.35	0.00
SrO	0.65	0.51	1.07
Y_2O_3	0.11	0.39	1.26
ZrO_2	0.44	2.62	7.34
UO_2	0.20	1.69	8.27
MoO_2	0.16	0.20	0.78
SnO	0.20	0.25	1.08
Cs_2O	0.33	0.10	0.44
BaO	0.40	0.02	0.17
Ce_2O_3	0.15	1.45	0.75
La_2O_3	0.13	1.44	0.47
Nd_2O_3	0.10	1.46	1.64
TOTAL	96.71	102.88	99.67

ACKNOWLEDGEMENTS

Thanks are due to A.M. Duclos and J. Snider for XRD measurements, P. Richardson for TEM/EDX assistance, and J.C. Tait and A.G. Wikjord for helpful discussions.

REFERENCES

1. P.J. Hayward, this volume (1985).

2. L.J. Clegg and J.R. Coady, "Radioactive Decay Properties of CANDU Fuel. Vol.1: The Natural Uranium Fuel Cycle. Part 3: Reprocessed Fuel. Atomic Energy of Canada Limited Report, AECL-4436/2 (1977).

3. E.R. Vance, P.J. Hayward, C.D. Cann, S.L. Mitchell, M.A.T. Stanchell and D.J. Wren, Glass Tech., 25, 232 (1984).

4. D. Liu, G.R. Piercy, G.R. Purdy and P.J. Hayward, Phys. Chem. Glasses, in press.

5. P.J. Hayward, W.H. Hocking, S.L. Mitchell and M.A.T. Stanchell, Nucl. Chem. Waste Management, 5, 27 (1984).

6. R.S. Mitchell, Amer. Mineral. 51, 1394 (1966).

7. E.R. Vance, P.J. Hayward and I.M. George, submitted to Phys. Chem. Glasses.

8. J.C. Tait, P.J. Hayward and H.E.C. Rummens, submitted to Phys. Chem. Glasses.

REVIEW OF PROGRESS IN THE DEVELOPMENT OF SPHENE-BASED GLASS-CERAMICS

P.J. HAYWARD
Geochemistry & Applied Chemistry Branch, Whiteshell Nuclear Research Establishment, Pinawa, Manitoba, ROE 1LO, Canada

ABSTRACT

Since 1980, considerable research has been invested in the development of sphene-based glass-ceramics as hosts for possible future CANDU™ fuel recycle wastes. This paper reviews the fabrication of the glass-ceramics, their detailed microstructure, the partitioning of waste ions between the glassy and crystalline phases, and their resistance to leaching in deionized water and synthetic brines.

INTRODUCTION

Glass-ceramics based on crystalline sphene ($CaTiSiO_5$) are being studied in the Canadian Nuclear Fuel Waste Management Program as possible alternatives to glasses for the immobilization and deep geological disposal of high-level wastes (HLW) that would arise if used CANDU™ fuel were recycled[3]. These materials were selected for development because of the relatively simple process technology involved, and because of the well-documented persistence of sphene as a naturally occurring mineral in many geochemical environments, with an ability to take a wide variety of ions into solid solution [4,5]. Recent leaching experiments [6] and thermodynamic calculations [6,7] have demonstrated that sphene should be stable in the Ca-, Na-, Cl-rich brines found at depth in the Canadian Shield [8].

The aluminosilicate glass matrix remaining after sphene crystallization is a highly durable material [1,2,9,10] for immobilizing those waste ions that do not partition into the sphene phase. Further, in the event of dilation and eventual amorphization of the sphene lattice from α-recoil damage and fission product transmutation, slow viscous flow in the glass matrix could minimize or delay the buildup of internal stresses.

FABRICATION BY MELT CASTING

Studies of the formation of sphene glass-ceramics have concentrated on detailed examinations of a relatively small range of compositions, in order to understand the sequence of reactions that occur during controlled crystallization of the precursor glasses.

Melting of Precursor Glasses

The glass-ceramics may be prepared by conventional means, i.e. by melting compositions in the system Na_2O-Al_2O_3-CaO-TiO_2-SiO_2-X (X = waste oxides or minor additives) at temperatures > 1250°C, followed by casting and crystallization during controlled heating at 900-1050°C. The melts are extremely fluid [11], as shown in Figure 1, but are prone to crystallization below ~ 1200°C. They are also corrosive to many refractories. However, tests by the British Glass Industry Research Association [12] have established that Cr_2O_3-based refractories and SnO_2-based electrode materials offer sufficient corrosion-resistance to the melts to make electric melting a feasible production method.

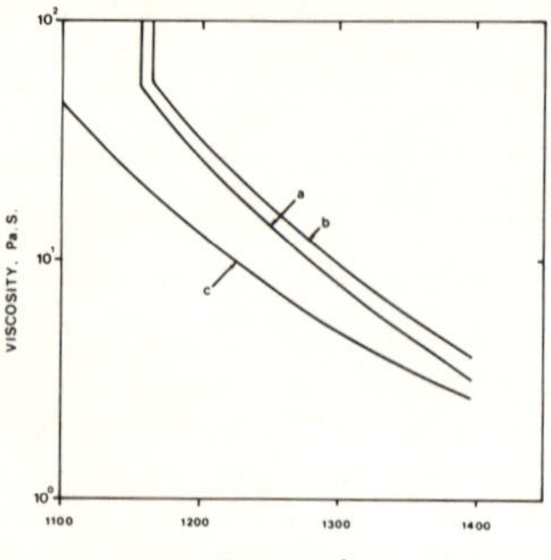

Figure 1. Viscosity versus temperature plots for (a) a titanosilicate melt, (b) a titanosilicate melt + 5 wt% HLW, and (c) a borosilicate melt (mole %: Na_2O, 16.7; B_2O_3, 16.7; SiO_2, 66.7).

Redox Influence During Melting

The melting of waste-loaded compositions requires an atmosphere of low oxygen fugacity to promote the formation of RuO_2, TcO_2 and MoO_2 rather than the volatile RuO_4 (b.p. 100°C), Tc_2O_7 (sublimes below 400°C) and MoO_3 (sublimes at ∼ 1155°C). Another benefit of low oxygen fugacity is the encouragement of lower valence states in multivalent ions such as U and Ce, aiding their partitioning into the more durable sphene phase during glass-ceramic crystallization [13,14]. Further, high temperature decomposition reactions that produce foaming in the melts, such as

$$U^{6+} + 6SiO^- \rightarrow U^{4+} + 4SiO^- + Si-O-Si + \frac{1}{2}O_2\uparrow$$

are eliminated.

Atmospheres of 5% H_2/95% N_2 and of air equilibrated with excess graphite have both been shown to achieve all the desired reductions. However, they also have the unwanted effect of reducing MoO_3, Fe_2O_3 and NiO to the metals and partially reducing Ti^{4+} to Ti^{3+} (characterized by a deep blue or black colour in the precursor glasses). Thus, further optimization of the melting atmosphere, at a higher oxygen fugacity (e.g. a CO_2 atmosphere), is required.

Crystallization During Reheating

Provided that a typical precursor melt containing HLW is cooled sufficiently rapidly (≥ 10°C/min.) through the temperature range 1200-800°C, crystallization is avoided, and a 2-phase unmixed glass is formed by spinodal decomposition [13,15,16]. The glass is converted to a glass-ceramic by subsequent reheating to 900°-1050°C. The phase-separated glass (Figure 2(a)) comprises a (CaO, TiO_2)-rich phase, close to the $CaTiSiO_5$ composition, and an SiO_2-rich phase. In general, the HLW ions partition between the two phases. However, precipitation of droplets of metallic Pd and Rh, and of a Ru-rich phase occurs, and these can initiate sphene crystallization during reheating by serving as nucleii for the epitaxial crystallization of a Ca-deficient perovskite ($CaTiO_3$) in the (CaO, TiO_2)-rich glass phase, as shown in Figure 2b [15]. The perovskite is unstable with respect to sphene in the presence of excess SiO_2, and acts as a heterogeneous nucleus for sphene formation (Figure 2(c)). The sphene grows as elongated or dendritic crystals to the margins of the CaO-, TiO_2-rich glass phase (Figure 2(c)). At higher temperatures, growth continues across the SiO_2-rich glass phase to produce a microstructure of elongated crystals in an aluminosilicate glass matrix (Figure 2(d)).

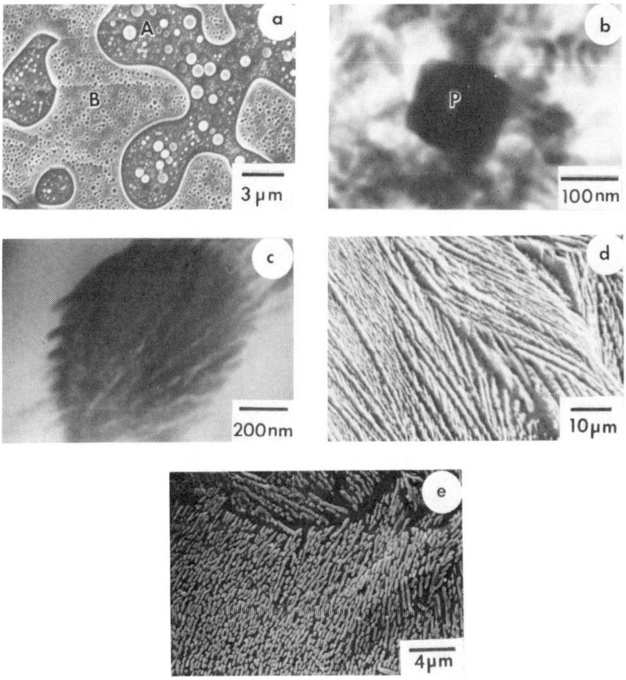

Figure 2. (a) SEM micrograph of phase-separated precursor glass, showing (CaO, TiO_2)-rich and SiO_2-rich regions (A and B, respectively). Note droplets of B in A and A in B;
(b) TEM micrograph showing epitaxial crystallization of perovskite around a Pd nucleus (P);
(c) TEM micrograph showing sphene crystals growing in CaO-, TiO_2-rich glass phase;
(d) SEM micrograph of fully crystallized sphene glass-ceramic, showing elongated crystal growth.
(e) SEM micrograph of bulk-nucleated sphene glass-ceramic containing 10 wt% simulated HLW.
[Figures 2(b) and 2(c) from ref. 15].

In the absence of Pd, Rh or Ru, or of simulated HLW, the crystallization of the precursor glasses proceeds from surface or internal heterogeneous nucleating sites only. Common nucleating oxides for other glass systems, such as Cr, Zr and Mo, or the boundaries between the unmixed glass phases, have no significant nucleating effect [15].

As the waste content of the precursor glass is increased, additional phases crystallize during reheating. The results for individual additions of U, La, Nd, Ce, Y, Cs and Sr [13,16,17] are summarized in Table 1. In general, at higher concentrations of U and rare earth (RE) ions, the major phases to form are those which can accommodate a greater degree of ionic substitution than is possible in the sphene structure.

Table I.
MAJOR PHASES TO CRYSTALLIZE DURING HEATING OF INDIVIDUALLY DOPED PRECURSOR GLASSES TO 1100°C at 10°C/MINUTE.

Dopant Ion and Ionic Radius (nm)	Corresponding Waste Ions and Ionic Radii (nm)	Mole % Addition of Dopant Ion			
		1	2	3	4
La^{3+} (0.113)	Np^{3+} (0.110)	S	S+A	S+Pk+A	
Ce^{3+} (0.109)	Pu^{3+} (0.109)	S	S		S+Pe+Pk
Nd^{3+} (0.106)	Cm^{3+} (0.106) Am^{3+} (0.104)	S	S	S+Pk	S+A
Eu^{3+} (0.103)	Am^{3+} (0.104) Gd^{3+} (0.102)	S		S+Pk	
Y^{3+} (0.098)	Gd^{3+} (0.102) Tb^{3+} (0.098)	S	S		S+Py
Ce^{4+} (0.088)	Pu^{4+} (0.088)		S+F+Py	S+F	
U		S	S+Py		Py+S
Th^{4+} (0.108)		S	S+F		S+F

S = sphene, $CaTiSiO_5$; Pk = perovskite-structured $(Ca,Na+RE)TiO_3$;
Pe = perrierite-structured $(Ca,U,RE)_2Ti_2Si_2O_{11}$;
A = apatite-structured $Ca_2(RE)_8(SiO_4)_6O_2$;
Py = pyrochlore-structured $Y_2Ti_2O_7$ or $Ca(U^{4+},Ce^{4+})Ti_2O_{7\pm x}$;
F = fluorite-structured $(U^{4+},Th,RE)O_{2+x}$. Oxidation state of U and Mn unknown. Ionic radii from reference 18 (6-fold coordination assumed). Phases listed in order of abundance.

Experiments with titanosilicate glasses containing 0 to 5 wt% of simulated HLW, prepared under oxidizing conditions [11] and with similar glasses containing 10 to 25 wt% simulated HLW, prepared under reducing conditions [14], have shown that the crystallization behaviour is surprisingly simple for compositions with $\leq$ 15 wt% HLW. When such glasses are reheated to ~ 1100°C, the only crystalline phase in the final product is sphene, the residual material being an aluminosilicate glass. The microstructure of a typical glass-ceramic (Figure 2e) suggests a bulk nucleation mechanism. Here, sphene may form via a precursor crystalline phase, which is itself bulk-nucleated in the initial glass [10,15] and then reacts with the glassy phase to produce sphene plus the final aluminosilicate glass.

The partitioning of waste ions between sphene and the glass phase has been studied by Transmission Electron Microscopy combined with Energy Dispersive X-ray Analysis, (TEM/EDX), by Scanning Auger Microscopy (SAM) and by Photoluminescent Spectroscopy (PLS). These results are summarized in Table 2. Unfortunately, only the PLS technique can detect realistic concentrations ($\leq$ 1 wt%) of individual ions. With higher dopant concentrations, the glass-ceramics may contain additional crystalline phases that can also serve as hosts for the dopant ion, making the partitioning results more difficult to interpret. In general, RE ions preferentially partition into sphene, whereas Sr^{2+} partitions equally into the sphene and glass phases. Cesium, which has an ionic radius too large to be easily accommodated in the sphene structure, partitions strongly into the glass phase. The preference of U for the glass phase probably results from high $U^{6+}:U^{4+}$ ratios in these air-melted glasses [13].

Table II.
SELECTED ION PARTITIONING RESULTS FOR SPHENE GLASS-CERAMICS

Ion and Concentration	Ion Partitioning Sphene : Glass			Technique	Reference
	Sphene	:	Glass		
La^{3+}, 2 mole %	> 10	:	1	SAM	19
La^{3+}, 0.1 mole %	> 2	:	1	TEM/EDX	16
Ce^{3+}, 0.1 mole %	> 6	:	1	TEM/EDX	16
Eu^{2+}, 5 wt%	> 10	:	1	SAM	20
Eu^{3+}, Sm^{3+} and Gd^{3+}, 0.01 to 1 wt%	∿ 100% in sphene, not detected in glass.			PLS	20
U^{6+}, 0.5 mole %	∿ 1	:	3	TEM/EDX	16
Sr^{2+}, 0.1 to 2 mole %	∿ 1	:	1	TEM/EDX	16
Cs^{+}, 0.1 to 2 mole %	1	:	>40	TEM/EDX	16
Ni^{2+}, 0.01 to 1 wt%	strongly partitioned into sphene.			PLS	20

Crystallization During Slow Cooling

Crystallization of the precursor glass may also occur during cooling from the melt, especially if the cooling rate in the critical range 1050-1200°C is low, as might be expected for an engineering-scale wasteform. However, it has been shown that a monolithic glass-ceramic waste form, containing up to 15 wt% simulated HLW, and with sphene as the predominant crystalline phase, can be produced by either reheating the slowly cooled glass precursor to ∿ 1050°C and holding at this temperature for up to 6 h, or by interrupting the slow cooling by holding the glass precursor at 1050°C for up to 3 h [14].

FABRICATION BY SINTERING

Sintering techniques have also been used to fabricate a glass-ceramic waste form. The starting materials may be either a finely ground glass frit, or the milled product of sol-gel preparation techniques [21]. In the former case, an undoped precursor glass "frit" is prepared by melting in an inactive facility (e.g. a conventional frit melter), followed by milling and blending with suitable binders and plasticisers prior to pressing and sintering at temperatures of 850 - 900°C. The HLW may be added during the milling stage, in concentrations of up to ∿ 10 wt%. An alternative approach is to presinter the pressed body at ∿ 720°C, to produce a partially sintered glass compact with an open porosity of ∿ 25%. This compact is used to absorb liquid HLW, and is then dried and fired by slow heating to ∿ 900°C. During the firing, the glass compact initially consolidates by liquid-phase sintering at 720 - 750°C, to give a body containing closed pores only (typically 90 to 95% dense). At temperatures above 750°C, crystallization occurs, so that most of the encapsulated wastes are dissolved and redistributed between the final glass and sphene phases.

The sol-gel approach involves the premixing of acidic nitrate solutions of Na, Al and Ca, together with tetraethylorthotitanate and colloidal silica, to give a gel in which the ions are mixed on a molecular scale. After drying and calcination, the gel is milled to produce a highly reactive powder, which may then be processed in a similar manner to the milled frit.

Both these techniques offer the promise of yielding a durable waste form at relatively low temperatures, thus minimizing radionuclide volatilization and eliminating refractory corrosion in high-temperature melters. However, problems have been encountered with migration of the liquid wastes during drying of the pre-sintered powder compacts, so that freeze-drying may be required to prevent a grossly uneven distribution of HLW ions in the final glass ceramic. Also, high localized concentrations of HLW ions have been found to prevent satisfactory sintering. Further work is in progress to optimize both techniques.

FABRICATION USING ION-EXCHANGE STARTING MATERIALS

Sphene glass-ceramics have also been produced from the melting and subsequent crystallization of suitable pre-mixed inorganic ion-exchange materials. A method in which a bed of pre-mixed zeolite (mordenite), sodium titanate ($NaTi_2O_5H$), hydroxyapatite ($Ca_5(PO_4)_3OH$) and wollastonite ($CaSiO_3$) is used to decontaminate alkaline or neutral waste solutions has been described [22]. After the bed is exhausted, it is dried and then heated to ~ 1350°C to melt the bed components in their Inconel container, followed by cooling and reheating to 900-1050°C to produce a sphene glass-ceramic. The advantage of this technique is that it affords a large volume reduction of the original waste solution, the latter emerging from the bed as a low-level waste solution. However, it is unsuitable for use with wastes containing actinides or RE's, since these ions have been found to precipitate as insoluble hydroxides, as a result of pH increase during ion exchange; blockage of flow through the bed eventually occurs. Thus the method is probably restricted to neutral or alkaline wastes.

LEACHING PROPERTIES

Previous leaching studies [10,23] have shown that sphene glass-ceramics may be approximately considered as a mixture of sphene and aluminosilicate glass, with independent leaching behaviours. Aluminosilicate glass leaching has already been the subject of considerable study [1,2,24,25]. Consequently, recent experiments have focussed on the leaching of natural sphenes and doped sphene ceramics [6,26,27] to establish whether thermodynamic predictions of sphene stability are also applicable to synthetic sphenes containing substituted waste ions.

Leaching experiments in synthetic brine, using natural sphene and a sphene ceramic [6], have confirmed the thermodynamic stability of sphene in Ca-Na-Cl brines at temperatures in the range 25°-100°C [6,7,9]. Further experiments were performed using a sphene ceramic containing simulated HLW ions in solid solution and brines of varying strengths (representing groundwaters from 300 to 700 m depth in the Canadian Shield) at temperatures in the range 25°-100°C [27]. The results confirmed that there are major differences between the leaching mechanisms in deionized water and brine: they also showed that the behaviour in brines was (a) strongly dependent on temperature, with little or no reaction below 60°C, and (b) insensitive to Ca, Na and Cl concentrations in the brines above a minimum value. Little release of any of the impurity ions, except for U, was found; the latter formed isolated aggregates of uranophane ($Ca(UO_2)_2(SiO_3OH)_2$) crystals at the specimen surfaces. The U mobility was attributed to the presence of a minor unidentified U^{6+}-rich phase (observed using XRD) in the original ceramic, and to the formation of soluble U complexes with carbonate anions in solution. The leaching experiments are being repeated under an argon atmosphere with ceramics that have been fired at lower oxygen fugacities to maximize the U^{4+}/U^{6+} ratio, and, thus, to allow for U incorporation in the sphene structure.

Leaching studies, including Secondary Ion Mass Spectrometry (SIMS) depth profiling, have been made on doped sphene glass-ceramics that have been leached for up to 360 days in deionized water and concentrated brines [10,23]. In each case, leaching was predominantly confined to the glass phase, and was most severe with deionized water; leach rates for the glass phase were $\sim 10^{-10}$ kg.m^{-2}.s^{-1} during one year of leaching. The brines had a passivating effect on leaching and surface alteration phenomena, partly through precipitation of new phases on the specimen surfaces.

In a separate experiment, a sphene glass-ceramic containing U, La, Cs and Sr was leached for 485 days at 100°C in synthetic brine and in deionized water, while being subjected to a dose rate of 400 R.h^{-1} from a used fuel gamma ray source [28]. Leach rates of $\sim 5 \times 10^{-11}$ and $\sim 2 \times 10^{-11}$ kg.m^{-2}.s^{-1}, respectively, were calculated from the weight change measurements, and the observed surface alteration phenomena resembled those seen after similar leaching in the absence of a radiation field. The results indicate that gamma-radiolysis effects in the leachant do not significantly alter the durability of the glass-ceramic.

Using 3 MeV Ar^+ bombardment to simulate alpha decay, it has been deduced [29] that a fluence of $\sim 7 \times 10^{18}$ α/g is required to produce complete metamictization of sphene. A similar conclusion was reached based on a study of metamict natural samples [30]. The durability of metamict sphene in deionized water at 100°C was lower than that of undamaged sphene by a factor of ~ 5. However, no corresponding effect in leach rates was found for a sphene glass-ceramic after similar irradiation.

INTERACTIVE SYSTEMS TESTS

A representative sphene glass-ceramic, containing 5 wt% simulated HLW, has been included in the Materials Interface Interaction Tests (MIIT) in the salt dome at Carlsbad, New Mexico. In these tests, which are coordinated by the U.S. Savannah River Laboratory, cored circular slices ("pineapple slices") of waste forms, buffers and container materials will be stacked in contact with each other, with a cylindrical heater at the centre of each stack, and subjected to in-situ leaching for up to five years. A detailed materials evaluation program is planned for the end of this period. Glass-ceramic/groundwater/buffer/rock interactions are also being studied in laboratory experiments at WNRE.

CURRENT AND FUTURE RESEARCH

Current research involves studies of the influence of gamma-ray radiolysis of groundwaters on the leaching behaviour of HLW-doped glass-ceramics, the preparation of active samples for leach testing, and the formation and properties of Np- and Tc-doped glass-ceramics. Other topics to be studied further include volatilization during melting, glass-ceramic mechanical properties, natural analogs and the development of mathematical leaching models. A demonstration facility for production of sphene glass-ceramics on a small engineering scale is being considered.

REFERENCES

1. W.F. Merritt, Atomic Energy of Canada Limited Report, AECL-5317 (1976).
2. T.W. Melnyk, F.B. Walton and L.H. Johnson, Nucl. Chem. Waste Management 5, 49 (1984).
3. T.E. Rummery and E.L.J. Rosinger, Atomic Energy of Canada Limited

Report, AECL-7787 (1983).
4. W.A. Deer, R.A. Howie and J. Zussman, Rock-Forming Minerals, 2nd ed. Vol.1A, Orthosilicates (Longmans, London and New York, 1982).
5. E.R. Vance and D.K. Agrawal, Nucl. Chem. Waste Management 3, 229 (1982).
6. P.J. Hayward, F.E. Doern, E.V. Cecchetto and S.L. Mitchell, Can. Mineral, 21, 611 (1983).
7. H.W. Nesbitt, G.M. Bancroft, W.S. Fyfe, S.N. Karkhanis, A. Nishijima and S. Shin, Nature 289, 358 (1981).
8. P. Fritz and S.K. Frape, Chem. Geol. 36, 176 (1982).
9. P.J. Hayward and E.V. Cecchetto, in Scientific Basis for Nuclear Waste Management 4, edited by S.V. Topp (Elsevier Science Publishers, New York, 1982), p.91.
10. P.J. Hayward, W.H. Hocking, S.L. Mitchell and M.A.T. Stanchell, Nucl. Chem. Waste Management 5, 27 (1984).
11. J.C. Tait, P.J. Hayward, H.E.C. Rummens and D.L. Wilkin, submitted to Phys. Chem. Glasses.
12. T.S. Busby and M.P. Woods, British Glass Industry Research Association, unpublished work (1985).
13. E.R. Vance, P.J. Hayward, C.D. Cann, S.L. Mitchell, M.A.T. Stanchell and D.J. Wren, Glass Techn. 25, 232 (1984).
14. P.J. Hayward, E.R. Vance, C.D. Cann and I.M. George, this volume (1985).
15. D. Liu, G.R. Piercy, G.R. Purdy and P.J. Hayward, Phys. Chem. Glasses, in press.
16. P.J. Hayward, E.R. Vance, C.D. Cann and S.L. Mitchell, in Advances in Ceramics 8, (American Ceramic Society, 1984), p.291.
17. E.R. Vance, P.J. Hayward and I.M. George, submitted to Phys. Chem. Glasses.
18. E.J.W. Whittaker and R. Muntus, Geochim. Cosmochim. Acta 34, 945 (1970).
19. W.H. Hocking, P.J. Hayward, D.G. Watson and G.C. Allen, Phil. Mag. A 49, (5), 637 (1984).
20. D. Simkin, J.C. Tait, P.J. Hayward, W.H. Hocking and E.R. Vance, unpublished work.
21. E.R. Vance, P.J. Hayward, J.C. Tait, I.M. George and A.A. Carmichael, unpublished work.
22. R.A. Speranzini and P.J. Hayward, in Advances in Ceramics 8, American Ceramic Society, 1984), p.273.
23. P.J. Hayward, W.H. Hocking, F.E. Doern and E.V. Cecchetto, in Scientific Basis for Nuclear Waste Management V, edited by W. Lutze (Elsevier Science Publishers, New York, 1982), p.316.
24. K.B. Harvey and C.D. Litke, J. Amer. Ceram. Soc., 67, 553 (1984).
25. J.C. Tait and D.L. Mandolesi, Atomic Energy of Canada Limited Report, AECL-7803 (1983).
26. J.B. Metson, G.M. Bancroft, S.M. Kanetkar, H.W. Nesbitt, W.S. Fyfe and P.J. Hayward, in Scientific Basis for Nuclear Waste Management V, edited by W. Lutz (Elsevier Science Publishers, New York, 1982), p.329.
27. P.J. Hayward, D.G. Watson, A.K. McIlwain, I.M. George and S.L. Mitchell, submitted to Nucl. Chem. Waste Management.
28. J.C. Tait, K.B. Harvey, P.J. Hayward, unpublished work.
29. E.R. Vance, F.G. Karioris, L. Cartz and M.S. Wong, in Advances in Ceramics 8, (American Ceramic Society, 1984), p.62.
30. E.R. Vance and J.B. Metson, Phys. Chem. Minerals, in press.

RADIATION DAMAGE IN NATURAL TITANITE BY CRYSTAL STRUCTURE ANALYSIS

M.E. FLEET AND G.S. HENDERSON
Department of Geology, University of Western Ontario, London, Ontario
N6A 5B7, Canada

ABSTRACT

Study of radiation damage in natural titanite ($CaTiSiO_5$) permits estimation of the long-term performance of titanite-bearing glass ceramics in the immobilization of radioactive waste. Single-crystal X-ray structure analyses have been made on two geologically old, U,Th-bearing titanites (sphenes), on an undamaged gem-quality titanite and on their annealed equivalents. The alpha-damaged titanite crystal structure is characterized by a marked increase in magnitude of the apparent thermal parameters of all structural positions, reflecting a general increase in positional disorder. The ratio of observed structure factors (F_O-damaged/F_O-annealed) may be used to estimate the overall structural damage.

INTRODUCTION

In the Canadian Nuclear Fuel Waste Management Program, glass ceramics containing micron-sized crystals of synthetic titanite ($CaTiSiO_5$) embedded in an aluminosilicate glass matrix are being studied as possible alternatives to glasses for the immobilization of radioactive waste [1, 2]. The titanite structure is capable of incorporating in solid solution a wide variety of impurity ions, including actinides and the stable isotopes corresponding to fission products [3]. Natural titanite (previously called sphene) has a very broad field of stability in all ages of rocks of the earth's crust: it occurs as an accessory mineral in acid and alkalic igneous rocks, in low and medium grade metamorphic rocks, in low temperature vein systems, and (rarely) as a detrital mineral in sediments [3]. Its chemical stability in possible waste repository environments in the Canadian Shield is supported by thermodynamic calculations [4].

Natural titanite has the formula $Ca(Ti_{1-x}M_x)SiO_{5-x}(OH,F)_x$, with x = 0.0-0.35 and M = Al + Fe^{3+} [e.g., 5], and contains trace and minor amounts of Na, Mn, Sr, Ba, Y, rare earth, U and Th substitutions for Ca and Nb and Ta substitutions for Ti. Thus information on the susceptibility of the titanite structure to long-term, low-level radiation damage may be obtained from study of geologically old, U- and Th-bearing titanites. Natural titanites exhibit all stages of radiation damage from undamaged to largely metamict [5, 6, 7]. Radiation damage is accompanied by an increase in unit-cell dimensions and a decrease in density. The undamaged titanite structure is recovered by annealing at 800-1100°C for 3 hours [5, 6]. Vance and Metson [7] report that a fluence of about 1.5 x 10^{18} alphas/g results in moderate structural damage, which is largely recovered by annealing at approximately 800°C for just 1 hour. In the present study, the nature of the radiation damage in titanite is examined in detail by single crystal X-ray structure analysis of three natural titanites and their annealed equivalents.

Titanite is monoclinic with space group $P2_1/a$ for x ~ 0 and A2/a for x $\geqslant 0.03$. Most natural specimens have a moderate Al,Fe^{3+} substitution for Ti and therefore have the A2/a space group. Synthetic $P2_1/a$ titanite transforms to the A2/a structure above 220°C at 1 bar [8]. The basic titanite structure (Fig. 1) has chains of corner-shared TiO_6 octahedra parallel to the a-axis, cross-linked by isolated SiO_4 tetrahedra [8, 9, 10, 11]. Because two of the SiO_4 tetrahedral oxygens are from the same TiO_6 chain, individual octahedra are canted from the b-axis direction in a zigzag pattern, Ca (and

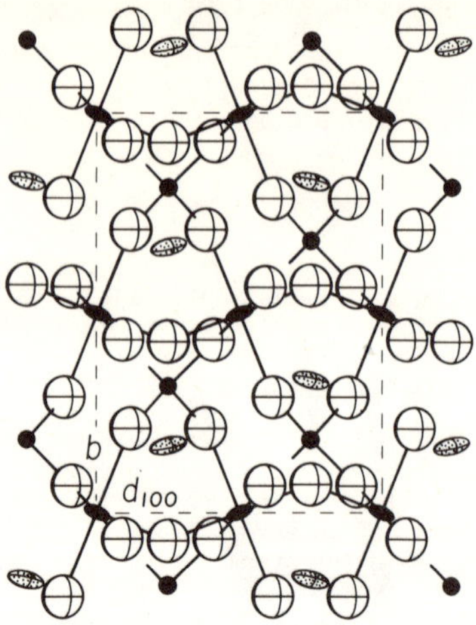

Fig. 1. Crystal structure of titanite in c-axis projection: Ca- stippled ellipses; Ti- solid ellipses; Si- solid circles.

substituted U,Th) is accommodated in an irregular 7-fold coordinated polyhedron. The $P2_1/a$ structure results from an ordered displacement of Ti from the geometrical centre of its coordination octahedron. The Ti^{4+} cations are shifted in the +a-axis direction and in the -a-axis direction in alternate TiO_6 chains [10, 8].

EXPERIMENTAL

Eight natural and one synthetic titanites and their annealed equivalents were thoroughly characterized by electron microprobe and single-crystal X-ray precession camera analysis. Data for the titanites selected for single crystal structure analysis are presented in Table I. Electron microprobe totals for damaged titanites were consistently lower by up to 5 wt. % than expected [5]. Structural formulae are therefore reported as (OH,F)-free. All the natural and annealed natural samples had space group A2/a. Moderately damaged titanites characteristically yield X-ray diffraction patterns with broad Bragg reflections. Diffuse scattering in the vicinity of Bragg reflections has a weak preferential radial (ω) distribution in reciprocal space but there is no tendency for it to be concentrated along lattice rows or lattice planes. Furthermore, the intensity maxima for this diffuse scattering are not apparently displaced from the reciprocal lattice points as in the theory of Krivoglaz [16] for severely distorted crystals. Background in between Bragg reflection positions is enhanced, in the manner of scattering from an X-ray amorphous component. Compared to the diffraction patterns of annealed titanites, the decrease in intensity and increase in peak broadening of Bragg reflections is greatest for reflections with $h \gg k,\ell$. Also, compared to results for annealed or undamaged titanites, Bragg reflection intensity decreases markedly with increasing $\sin\theta/\lambda$. The compositional range for the presence of diffuse P-lattice reflections (h00 relplane scattering [10]) is extended to x = 0.11 (1.9 at. % Fe, 8.8 at. % Al) [5, Table 2]. Moreover, diffuse P-lattice reflections are present in quenched annealed titanites with x = 0.03-0.11.

Single crystals for X-ray structure analysis were approximately equidimensional polyhedral fragments 0.003-0.005 mm^3 in volume. Each single

Table I. Chemical Composition for Natural and Annealed Titanites

	1	1A	2	2A	3
SiO_2	30.8	31.7	30.8	30.8	31.4
TiO_2	27.4	27.8	35.5	35.1	36.1
Nb_2O_3	0.5	0.4	0.2	0.2	0.0
Y_2O_3	0.9	0.9	0.4	0.3	0.0
Al_2O_3	6.6	7.7	1.6	2.0	2.3
Fe_2O_3	0.8	0.7	1.7	1.9	0.8
MnO	0.1	0.1	0.2	0.2	0.0
MgO	0.4	0.6	0.0	0.0	---
CaO	26.9	27.5	27.4	28.2	29.4
Na_2O	0.0	0.6	0.0	0.0	---
Total	94.4	98.0	97.8	98.7	100.0

1. JB401. Natural titanite from Cardiff mine, Ontario [7], moderately damaged: $(Ca_{.93}Y_{.02})(Ti_{.67}Al_{.25}Fe_{.02}Mg_{.02})SiO_5$

1A. JB401A. JB401 annealed in air at 1000°C for 1 d, air-quenched: $(Ca_{.92}Na_{.04}Y_{.02})(Ti_{.66}Al_{.29}Fe_{.02}Mg_{.02})SiO_5$

2. C-80-7. Natural titanite from North Trout Lake Batholith, Favourable Lake area, Ontario, 240 ppm U, 160 ppm Th, moderately damaged: $Ca_{.96}(Ti_{.86}Al_{.06}Fe_{.04})SiO_5$

2A. C-80-7A. C-80-7 annealed with restricted access to air at 1000°C for 5 d, air-quenched: $Ca_{.98}(Ti_{.86}Al_{.08}Fe_{.05})SiO_5$

3. A6. Natural titanite from Switzerland [7, Table 1, No. 4], undamaged: $Ca_{1.00}(Ti_{.87}Al_{.09}Fe_{.02})SiO_5$

3A. A6B. A6 annealed in air at 1000°C for 5 d, air-quenched: Not analysed

crystal was selected following a survey of the diffraction quality of several larger grains and size-reduced by grinding with 600 abrasive paper lubricated with a light oil. Unit-cell parameters (Table II) were obtained in the manner of [12, 13]. X-ray intensity data were collected on an Enraf-Nonius CAD-4F single crystal diffractometer system with Mo $K\alpha$ radiation [12], over a hemisphere of reciprocal space out to $2\theta = 80°$ (the limit of useful data for moderately damaged titanites). Raw data were reduced in the manner of [12, 13] (with $\mu = 36.2$ cm^{-1} for titanite 1, etc.), and averaged to give about 1140 non-equivalent reflections for each data set.

Crystal structure refinement in space group A2/a was made after [12, 13], using anisotropic thermal parameters for all six structural positions. Starting parameters were taken from [8, 9]. An isotropic extinction correction was used, and the scale factor was applied to the observed structure factors (F_O). Final positional parameters and equivalent isotropic thermal parameters are given in Table II. For Ca and Ti positions, $B_{11} \gg B_{22}, B_{33}$ for all six structures, and for titanites 1A, 2A, 3 and 3A a significant reduction in residual index (R) was obtained by refining a Ca position disordered along the a-axis (Table II). Electron density maps of the moderately damaged structures did not indicate a significant occupancy

Table II. Unit-Cell Parameters, Positional and Thermal Parameters and R for Natural and Annealed Titanite Structures

		1	1A	2	2A	3	3A
a (Å)		7.051(2)	7.033(1)	7.058(3)	7.0504(7)	7.0684(6)	7.0539(5)
b		8.724(3)	8.6753(4)	8.749(2)	8.7027(4)	8.6943(5)	8.7060(6)
c		6.565(2)	6.5296(5)	6.582(2)	6.5479(5)	6.5537(4)	6.5510(5)
β (°)		113.95(2)	113.945(8)	113.88(3)	113.769(6)	113.983(5)	113.788(6)
V (Å^3)		369.07	364.08	371.60	367.69	367.99	368.13
Ca	x	1/4	0.237(2)	1/4	0.2340(8)	0.244(2)	0.244(2)
	y	0.1687(2)	0.1673(1)	0.1698(1)	0.1680(1)	0.1674(1)	0.1679(1)
	B	2.50(5)	0.95(7)	2.14(4)	0.63(5)	1.13(4)	1.05(4)
Ti	B	1.81(5)	0.83(2)	1.63(4)	0.73(2)	0.81(2)	0.65(2)
Si	y	0.1831(2)	0.1836(1)	0.1827(2)	0.1826(1)	0.1830(1)	0.1827(1)
	B	1.50(5)	0.52(2)	1.19(4)	0.48(2)	0.40(2)	0.38(2)
O(1)	y	0.0706(5)	0.0711(2)	0.0703(5)	0.0713(2)	0.0708(2)	0.0709(2)
	B	1.8(1)	0.73(4)	1.8(1)	0.73(5)	0.68(4)	0.63(4)
O(2)	x	0.9093(6)	0.9106(2)	0.9105(5)	0.9109(2)	0.9102(2)	0.9109(2)
	y	0.0658(4)	0.0665(1)	0.0655(4)	0.0665(2)	0.0663(1)	0.0664(1)
	z	0.1866(6)	0.1865(2)	0.1870(5)	0.1854(2)	0.1859(2)	0.1854(2)
	B	2.2(1)	0.85(3)	1.96(9)	0.81(4)	0.70(3)	0.70(3)
O(3)	x	0.3826(6)	0.3827(2)	0.3828(5)	0.3813(2)	0.3815(2)	0.3820(2)
	y	0.2094(4)	0.2095(1)	0.2105(3)	0.2107(2)	0.2104(1)	0.2107(1)
	z	0.3983(6)	0.3971(2)	0.3982(5)	0.3970(2)	0.3974(2)	0.3970(2)
	B	1.9(1)	0.77(3)	1.77(9)	0.71(4)	0.61(3)	0.59(3)
R		0.077	0.024	0.070	0.032	0.025	0.023

Titanite samples: 1- JB401; 1A- JB401 annealed; 2- C-80-7; 2A- C-80-7 annealed; 3- A6; 3A- A6 annealed

Ca- z=0; Ti- x=1/2, y=0, z=1/2; Si- x=3/4, z=0; O(1)- x=3/4, z=1/2

B- B_{equiv}; R- residual index for observed reflections ($I \geqslant 3\sigma(I)$)

of interstitial atom positions and the electron density distributions at the oxygen atom positions were not apparently anomalous.

DISCUSSION

The Undamaged Titanite Structure

The structures of the relatively undamaged titanite 3 and the three annealed titanites (1A, 2A, 3A) are all very similar (Table II) and the small differences which do exist can be attributed to composition- largely substitution of Ti by Al,Fe^{3+}. The close agreement between the refined structures of titanites 3 and 3A is particularly gratifying because it suggests that the annealed structure is not adversely affected by possible loss of volatile components (H_2O, F_2) during heat treatment and that it is probably a good representation of the undamaged titanite structure.

Earlier studies on the $P2_1/a \rightarrow A2/a$ transformation [8, 5, 10] have attributed it to an ordered displacement of the Ti^{4+} cation in the general direction of the a-axis, giving one short and one long Ti-O bond, citing

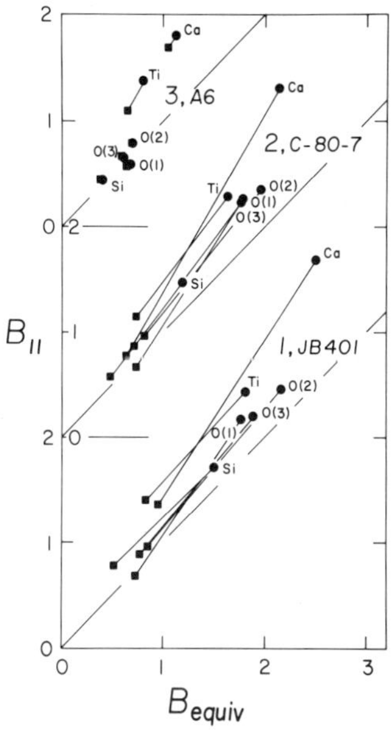

Fig. 2. Refined thermal parameters for damaged titanite (circles) and annealed titanite (squares).

analogy with $BaTiO_3$ and other ferroelectric materials. Substitution of Ti by Al and Fe^{3+} results in a one-dimensional domain structure in low-temperature A2/a titanite. Above the transformation temperature (220°C), the A2/a space group is presumably associated with a structure in which Ti^{4+} is in the centre of its coordination octahedron. The anisotropy of the thermal parameters of Ca and Ti in the present structure refinements (Fig. 2) and of all other room-temperature A2/a refinements [9, 11], together with the presence of h00 relplane scattering, certainly supports the domain structure model. There is no doubt that the Ca position, in particular, is disordered (in the average structure) in the a-axis direction. In our interpretation, the substituent Al^{3+},Fe^{3+} act as blocking cations, below the transformation temperature, perhaps causing a local reversal in the direction of displacement of the Ti^{4+} cation in any one TiO_6 chain. h00 relplane scattering is equally developed in both natural and annealed equivalents when examined at room temperature. This suggests that the domain structure develops quite readily on cooling through the transformation temperature. When $x > 0.11$, with more than 10 % blocking cations, the domain structure is effectively dissipated.

Previous studies have not noted that the Ca^{2+} cation is also displaced from the centre of its coordination polyhedron in the low-temperature A2/a structure (in fact, Ca^{2+} is displaced further than Ti^{4+}). Moreover, the displacement of the Ca^{2+} cation (as indicated by anisotropy of its thermal ellipsoid) also persists above the transformation temperature [8], and this may indicate that the $P2_1/a \rightarrow A2/a$ transformation is somewhat more complicated than previously imagined.

The Radiation Damaged Titanite Structure

The α-particle radiation-damaged titanite crystal structure is characterized by a marked increase in magnitude of the thermal parameters of all six structural positions (Ca, Ti, Si, O(1), O(2), O(3)), no doubt reflecting a general increase in positional disorder (Table II, Fig. 2). Equivalent isotropic B's are 2x to 3x greater for the moderately damaged structures than for the equivalent annealed structures. Also, the apparent relative displacement of the cations is similar to that of the oxygen anions. This is in marked contrast, then, to the structure of damaged natural zircon ($ZrSiO_4$), in which the principal defect appears to be the displaced oxygen ion [14, 15]. The Ca data presented here are somewhat misleading because the Ca position is disordered in the annealed titanite structures (1A, 2A, 3A), which results in smaller thermal parameters.

The positional disorder is also markedly anisotropic, with $B_{11} >> B_{22}, B_{33}$. Fig. 2 shows that this anisotropy is directly related to the anisotropy in the thermal parameters of the equivalent annealed structures. Thus, off-centre displacement of Ca^{2+} and Ti^{4+} causes the damaged titanite structure to relax preferentially in the a-axis direction. It certainly appears as though the cations are displaced along with the oxygen anions. This may indicate that the integrity of the nearest-neighbor coordination polyhedra is generally preserved in the damaged structure.

Estimates of Structural Damage

Radiation damage in the titanite structure consists very largely of displacements of individual atoms (or more likely of atomic clusters) preferentially in the a-axis direction. The present method of structure analysis can interpret such deviations from the ideal structure only as "thermal vibration". Other methods of structure analysis, particularly Huang diffuse scattering [17] and high resolution transmission electron microscopy, may yield more specific information on the structural damage. However, we emphasize that the distribution of diffuse scattering associated with the Bragg reflections is not asymmetric in precession camera films. Statistical broadening of the X-ray scattering source causes diminution of peak intensity with increasing $\sin\theta/\lambda$. An appropriate analogy for metamict titanite is provided by (optical) transform theory. The transform of a periodic step function gives an interference pattern (related to the step periodicity) limited by a diffraction envelop (related to the step width). If the periodicity is retained and the step width broadened the interference pattern has a more limited extent in reciprocal space. If now each step of the step function is replaced by a series of finely spaced steps (which causes the periodicity to be modulated) the peaks in the interference pattern will become broadened and diffuse. Statistical displacement of atoms in real crystals also results in some loss in atomic correlation, which adds an X-ray amorphous component to the damaged structure.

It is well known that metamictization is principally manifested by atomic displacements and that unit-cell volume increases (or density decreases) in proportion to the amount of structural damage. In addition, the present study shows, after Higgins and Ribbe [5], that electron microprobe totals are proportionately decreased by radiation damage. We tentatively attribute this effect to the lower mass density of damaged titanite relative to that of the microprobe standards. The residual index for the structure refinement (R) is also proportionately larger for damaged crystals. This effect is largely attributable to the presence of atomic displacements which cannot be interpreted as thermal vibration in the structure factor expression.

More sensitive estimates of structural damage are afforded by manipulation of the magnitudes of the scaled observed structure factors (F_O). F_O's are directly related to the damaged structure and loss of diffraction coherence may be a more satisfactory definition for metamictization than statistical models of disorder or randomness. This topic will be developed more fully elsewhere. For the present we offer a

Fig. 3. Qualitative estimate of damage in titanite using ratio of observed scattering factors.

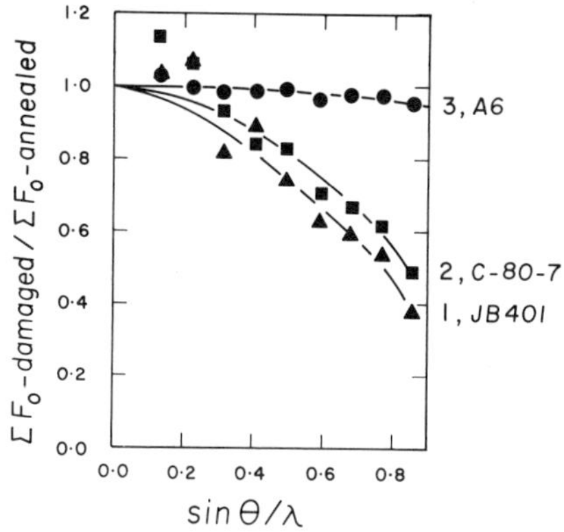

Table III. Damage Estimate from Ratio of Sum of Scaled Observed Structure Factors

Specimen	ΣF_0-damaged / ΣF_0-annealed (%)	Damage (%)	Equivalent U (μg/g)	Age (By)
1, JB401	65	35	500	1.0
2, C-80-7	73	27	290	2.7
3, A6	98	2	<100	<0.1

rudimentary approach to show the potential of this type of synthesis. The F_0 ratio (ΣF_0-damaged/ΣF_0-annealed) has been calculated for concentric shells of sinθ/λ (Fig. 3). The erratic behaviour of the low 2θ data is due to unfavorable statistics (too few lattice points per shell). Clearly this plot is a sensitive qualitative measure of damage. It is now apparent that even the "undamaged" natural titanite (3, A6) is, in fact, slightly damaged. One simple way to make these data more quantitative is to use the ratio of the sum of the scaled F_0's (Table III). Equivalent U values for titanites 1 (JB401) and 3 (A6) are from Vance and Metson [7] and that for titanite 2 (C-80-7) is estimated [7]. The alpha fluence for C-80-7 is greater than that for JB401 but the structure of the former may have partly recovered by metamorphism to low grade conditions (300-500°C, 1-3 kbar). Clearly, the present estimates of damage are minimum values for the F_0 method because the measurable intensity data for the annealed crystals extend beyond the present sinθ/λ limit of data collection (Fig. 3). Also, we have assumed that loss of diffraction coherence is associated very largely with atomic displacement, which may not be true for all stages of damage in metamict materials, and that the ratio of the sum of the F_0's is linearly related to damage. The scattering intensity due to the amorphous component of the damaged structure should not significantly interfere with the measured Bragg intensity because it is distributed isotropically over three dimensions of reciprocal space.

Previous estimates of the significance of radiation damage effects to the performance of titanite-bearing glass ceramics [7, 18] remain essentially unaltered.

ACKNOWLEDGEMENTS

We thank E.R. Vance for damaged titanite materials and helpful discussion, R.L. Barnett for electron microprobe analysis and Atomic Energy of Canada Limited for financial support.

REFERENCES

1. P.J. Hayward and E.V. Cecchetto, in Scientific Basis for Nuclear Waste Management IV, edited by S.V. Topp (Elsevier Science Publishers, New York, 1982), pp. 91-98.
2. P.J. Hayward, F.E. Doern, E.V. Cecchetto, S.L. Mitchell, Canad. Mineral. 21, 611 (1984).
3. W.A. Deer, R.A. Howie, J. Zussman, Rock Forming Minerals 1a, 2nd. ed. (Longman, London, 1982).
4. H.W. Nesbitt, G.M. Bancroft, S.N. Karkhanis, W.S. Fyfe, in Scientific Basis for Nuclear Waste Management III, edited by C. Northrup (Elsevier Science Publishers, New York, 1980), pp. 131-138.
5. J.B. Higgins and P.H. Ribbe, Am. Mineral. 61, 878 (1976).
6. P. Cerny and P. Povondra, Neues Jahrb. Mineral. Monatsch. 1972, 400.
7. E.R. Vance and J.B. Metson, Phys. Chem. Mineral., (in press).
8. M. Taylor and G.E. Brown, Am. Mineral. 61, 435 (1976).
9. R. Mongiorgi and L.R. Di Sanserverino, Mineral. Petrogr. Acta 14, 123 (1968).
10. J.A. Speer and G.V. Gibbs, Am. Mineral. 61, 238 (1976).
11. C.L. Hollabaugh and F.F. Foit Jr., Am. Mineral. 69, 725 (1984).
12. M.E. Fleet, Acta Cryst. C40, 1491 (1984).
13. M.E. Fleet and T.A. Mowles, Acta Cryst. C40, 1778 (1984).
14. L.A. Bursill and A.C. McLaren, Phys. Stat. Sol. 13, 331 (1966).
15. E.R. Vance, L. Efstathiou, F.H. Hsu, Rad. Effects 52, 61 (1980).
16. M.A. Krivoglaz, Theory of X-ray and Thermal-Neutron Scattering by Real Crystals (Plenum, New York, 1969), pp. 249-258.
17. B.C. Larson and W. Schmatz, Phys. Rev. B10, 2307 (1974).
18. E.R. Vance, F.G. Karioris, L. Cartz, M.S. Wong, in Advances in Ceramics 8, edited by G.G. Wicks and W.A. Ross (American Ceramic Society, 1984), pp. 62-70.

STRUCTURAL CHANGES IN ^{238}Pu-SUBSTITUTED ZIRCONOLITE ON RECOVERY FROM THE METAMICT STATE*

F. W. CLINARD, JR., R. J. LIVAK, L. W. HOBBS+ AND D. L. ROHR
Materials Science and Technology Division, Los Alamos National Laboratory
Los Alamos, NM 87545, USA

ABSTRACT

^{238}Pu-substituted zirconolite ($CaPuTi_2O_7$) was held at ≈350 K for a time sufficient to allow self-irradiation damage to convert the material to the metamict condition. Samples were then stored at 875 K, while dimensional and microstructural changes were monitored by dilatometry and transmission electron microscopy. Densification proceeded sporadically over a period of ≈400 days, attaining a final value of 4 vol% as a result of crystallization. The material was found to be friable at an intermediate stage of recovery, apparently as a result of internal strains. Evidence is presented that $CaPuTi_2O_7$ can exist in more than one metamict state, depending on extent of prior damage and thermal history.

INTRODUCTION

The zirconolite phase of the SYNROC [1] ceramic nuclear waste form is a host for actinide isotopes that generate self-irradiation damage by alpha decay. It has been shown [2] that ≈94% of this damage results not from the alpha particle, but from the heavy recoil nucleus.

Transmission electron microscopic (TEM) examination of damage in Pu-substituted [3], Cm-doped [4], Pb-ion implanted [5], and natural [5,6] zirconolite, as well as calculations of damage microstructures [7], have revealed that each recoil event induces a disordered damage track which, if spherical, would be of diameter ≈5 nm. During storage near room temperature these tracks accumulate and overlap until the material becomes metamict, resulting in swelling of 5.4 vol% for the Pu-substituted material [3]. Aging at 575 K results in slower accumulation of damage due to partial annealing, with eventual conversion to a metamict state different from that attained at the lower temperature. When stored at 875 K this material does not become metamict at all, but remains crystalline [3].

Recovery from the damaged state attained near room temperature has been studied by differential thermal analysis [4,8], and has been shown to be correlated with the damage microstructure [8]. The latter work has demonstrated that zirconolite crystallizes from the metamict state at temperatures between 860 and 1035 K, depending on extent of prior damage. As damage accumulates, the amount, size, and degree of interconnectivity of the residual crystalline phase all decrease. This is accompanied by an increase in temperature and abruptness of the conversion, suggesting that recovery is inhibited by these changes [8]. The magnitude of the energy released initially increases roughly in proportion to the irradiation dose, finally attaining a value of 100 J/g. However, at higher doses ($>2\times10^{25}$ α decays/m^3) stored energy gradually decreases, to a value of 50 J/g at 1.6×10^{26} α/m^3. It has been suggested [8] that this reduction is attributable either to redamage effects or to reduction of the internal strain that results from the heterogeneous nature of the transformation to the metamict state.

----

* Work performed under the auspices of the U.S. Department of Energy.
+ Permanent address: Department of Materials Science and Engineering, Massachusetts Institute of Technology, Cambridge, MA 02139.

In the work reported here, recovery processes are further examined by applying dilatometric and TEM techniques used earlier [3] to evaluate metamictization. Results are discussed in terms of stability of this nuclear waste phase under repository conditions.

EXPERIMENTAL PROCEDURE

Samples of ^{238}Pu-substituted zirconolite were fabricated by a technique described elsewhere [3]. In brief, powders of PuO_2, CaO, and TiO_2 in the molar ratio 1:1:2 were ball-milled together, calcined, and cold-pressed into two cylinders of approximately 10 mm diameter and height. The cylinders were air-fired at 1675 K for 16h, and furnace-cooled. The resulting product was characterized by a grain size of ≈20 µm, 2-3 vol % residual PuO_2, a lesser amount of an unidentified phase, and ≈7 vol % porosity.

The cylinders were stored at ambient temperature for 460 days, to a dose of 6.1 x 10^{25} α decays/m^3, to allow self-damage to accumulate until the metamict condition was achieved [3]. Actual storage temperature was estimated to be ≈350 K, the increment in temperature being attributable to the ≈1 MW/m^3 released by radioactive decay.

One cylinder was then placed in a dilatometer at 875 K and its length monitored several times a week for 770 days. Volumetric densification was calculated assuming dimensional changes to be isotropic, an assumption supported by the similarity of swelling results from dilatometric [3] and immersion density [9] measurements.

The other cylinder was broken into a number of pieces and stored at 875 K for periodic TEM examination. This experiment was begun 165 days after the dilatometer run was initiated, in order to allow selection of optimum times for examination by reference to the dilatometry results; as a consequence, storage time of the TEM material prior to heating was 625 days. Samples were prepared by extraction of thin shards from fractured faces, by use of a replication technique [3]. A fresh piece of material was removed from the furnace for each examination. Particles with electron-transparent edges were supported on carbon substrates and examined in a Siemens Elmiskop I microscope at 100 keV. Reduced beam intensities were employed to avoid excessive heating and consequent annealing of self-damage [10]. Samples were examined at 1.7, 6, 129, 180, 204, and 333 days. TEM evaluations were terminated at that point, since little change had been noted between the latter two observations.

RESULTS AND DISCUSSION

Densification

Shrinkage of the dilatometer sample as a function of annealing time at 875 K is shown in Fig. 1. Densification was initially relatively rapid, but by about 15 days had drastically slowed. This condition persisted until ≈130 days (corresponding to a volume change of 0.7 %), at which time shrinkage again became rapid. This stage ended after a total densification of 3.5 vol % had been attained. Dimensional changes at first appeared to have ended at that point, but further shrinkage occurring at ≈400 days brought the density change to 4.0 vol %. No further dimensional changes of significance were observed between that time and termination of measurements at 770 days.

Saturation of swelling occurs in this material at 5.4 vol % [3], and the corresponding shrinkage required to return the material to its original density is 5.1 vol %. However, only 77 % of that change was realized by densification at 875 K. This apparent failure to recover fully may be attributable to either of two factors (or a combination thereof): incomplete recovery to the crystalline state, or microcracking.

The first condition could result either from incomplete annealing at this temperature, or from concurrent redamage. It was shown earlier [3] that holding this material at 875 K from the time of fabrication resulted in a retention of crystallinity - an observation that suggests no significant disordering effect from continuous self-irradiation at that temperature. It is possible, however, that some damage might be retained at 875 K if previously-formed metamict material is present to stabilize the new damage tracks. With respect to microcracking, considerable friability was found in material stored for 129 days (discussed in the next section), but the overall impression from TEM examinations was that microcracking was not sufficient to account for the observed difference between swelling and densification.

Microstructural Changes

TEM examination of material stored for 1.7 days at 875 K revealed only fine, unresolved details characteristic of near-room temperature damage (Fig. 2); these features have been reported earlier [3], and tentatively attributed to intense local strain fields. No crystallites could be resolved after this treatment. Diffraction patterns displayed only broad isotropic correlation maxima characteristic of the metamict condition [3].

The material that was aged for 6 days also showed no crystallization, although densification had proceeded somewhat further at this point (Fig. 1). Diffraction patterns again exhibited the inner correlation ring characteristic of metamict zirconolite.

Storage for 129 days for the most part resulted in formation of a fine dispersion of crystallites of average size ≈10 nm, and an inner correlation ring distinctly sharper than that seen after shorter aging times (Fig. 3). Faint spots on this ring confirmed the presence of these small crystallites. In some areas larger crystallites (≈20nm) were observed, and here the inner ring showed correspondingly more intense spots.

The material aged for 129 days was noticeably friable: large pieces sometimes crumbled during preparation of TEM samples, and shards often showed extensive cracking (Fig. 3). Similar behavior (in this case spontaneous fracture upon annealing) has been observed by Routbort et al. [11] for a Cm-doped glass-ceramic waste form, and attributed to coalescence of helium gas bubbles. This does not seem to be a suitable explanation for friability of Pu-substituted zirconolite, since no such bubbles were observed.

The most likely explanation for the behavior noted here is that at this point in the recovery process, internal stresses have neared the level where spontaneous fragmentation can occur. These stresses result from the greater density of crystallites relative to the remanent metamict material. Since this friability was not noted at shorter or longer aging times, it would appear that stresses are highest at an intermediate point in the recovery process, consistent with the fact that microstructures at the beginning and end of the recovery process are essentially single phase.

At 180 days, densification had reached 1.7 vol %. TEM images (Fig. 4) showed that much of the material had been converted to the crystalline state, with some evidence of elongated grains. Diffraction patterns were characterized by spotty rings, indicating the presence of crystallites of random orientation. This pattern of recovery was also observed in beam-heated natural zirconolite, by Ewing et al. [10].

After aging for 204 days, densification had reached 3.1 vol % and TEM examination showed further crystallization. Average grain size at this point remained very fine, on the order of 30 nm. The final examination, at 333 days, showed no apparent change in crystallite morphology but a sharpening of the diffraction pattern (Fig. 5). The magnitude of densification, 3.5 vol %, indicates that most of the material was in the crystalline state at this point.

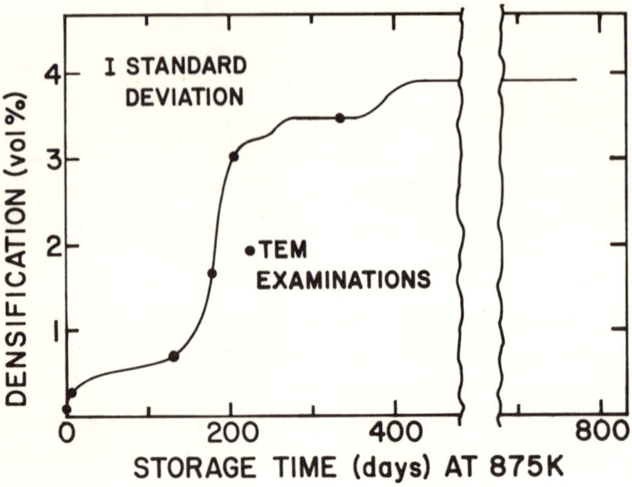

Fig. 1. Densification of originally-metamict $CaPuTi_2O_7$ as a function of storage time at 875 K.

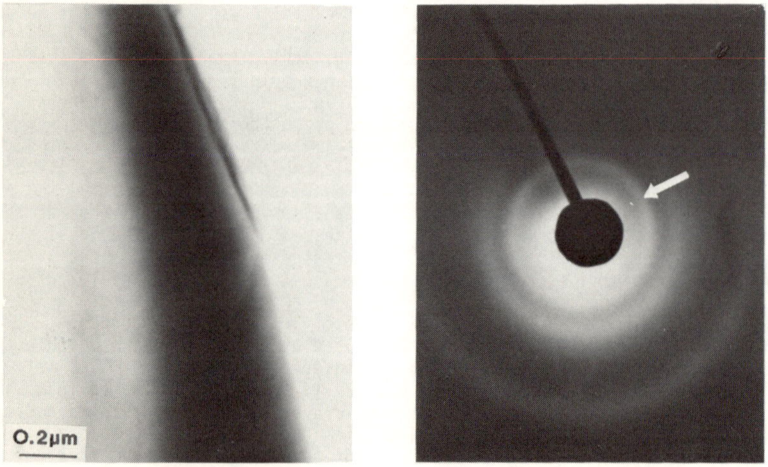

Fig. 2. Bright-field image and diffraction pattern after storage for 1.7 days at 875 K. The arrow indicates the inner correlation ring characteristic of the metamict material. Other diffuse rings are from the carbon substrate.

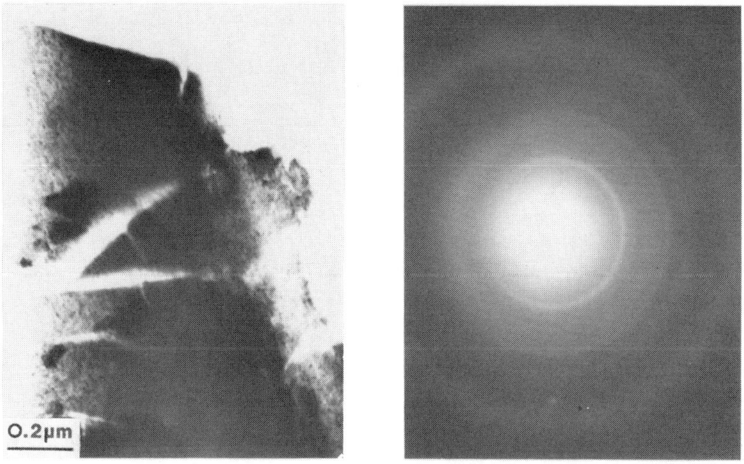

Fig. 3. Bright-field image and diffraction pattern after storage for 129 days at 875 K.

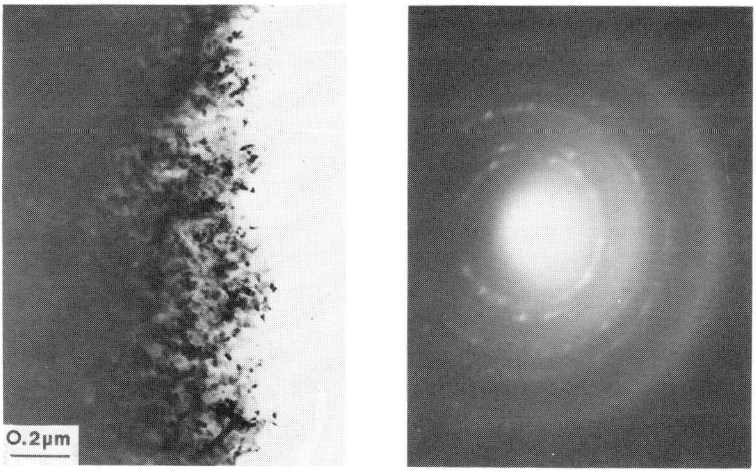

Fig. 4. Bright-field image and diffraction pattern after storage for 180 days at 875 K.

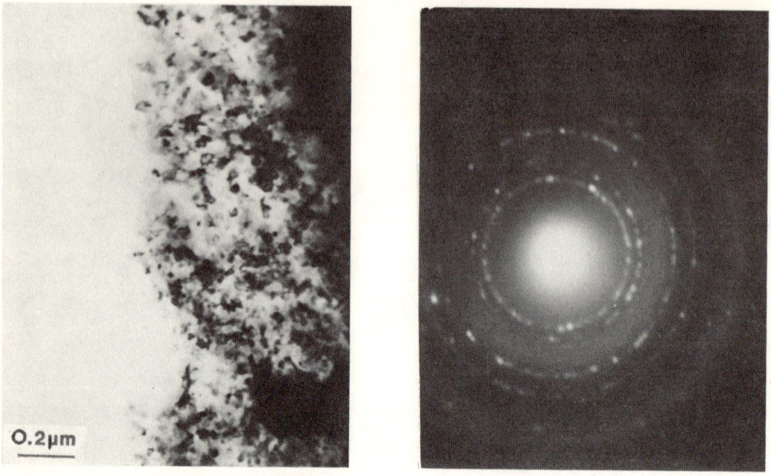

Fig. 5. Bright-field image and diffraction pattern after storage for 333 days at 875 K.

Table I. Width of Inner Correlation Ring in $CaPuTi_2O_7$ as a Function of Damage State (Arbitrary Units).

Sample History	Full Width at Half-Maximum	Full Width near Base
Damaged at ≈350 K for:		
• 460 days	1.00 (±0.23)*	2.13 (±0.38)
• 957 days	0.93 (±0.17)	1.55 (±0.15)
Damaged at 575 K for 430 days	0.71 (±0.23)	1.29 (±0.35)
Damaged at ≈350 K for 625 days, then annealed at 875 K for:		
• 1.7 days	0.60 (±0.14)	1.18 (±0.36)
• 6 days	0.61 (±0.18)	1.20 (±0.25)

* Standard deviations are given in parentheses.

Evidence of Variations within the Metamict State

There is no fundamental reason to assume that the metamict state induced in ^{238}Pu-substituted zirconolite by alpha decay is unique, i.e., a disordered state corresponds to only one characteristic atomic arrangement. In fact, earlier swelling studies [3] showed that when this material is held at 575 K until the metamict state is achieved, the resulting product swells only 80% as much as the value attained on storage at ≈350 K - implying a different atomic arrangement for the two conditions.

We have sought further evidence of variations in the metamict state by analysis of the inner correlation ring in diffraction patterns from this material. Densitometer traces were made across these rings to convert their profiles to peaks, and full widths at half-height and near base (5% of height) were determined. Analyses were made for material subjected to the following conditions:

1. held at ≈350 K for:
   - 460 days
   - 957 days

2. stored at 575 K for 430 days

3. stored at ≈350 K for 625 days, then annealed at 875 K for:
   - 1.7 days
   - 6 days.

Conditions 1 and 2 were attained during an earlier investigation [3]; condition 3 pertains to the present study.

Results of the peak measurements, normalized to unity for the value obtained for full width at half-maximum from material aged at 350 K for 460 days, are shown in Table I. Standard deviations are typically large, but there is evidence that either repeated irradiation of already-damaged material or employment of elevated temperatures (during or after self-irradiation) causes a sharpening of the rings. Since reduced swelling [3] and densification (this paper) accompany a sharpening of the rings, the lower values in Table I appear to be indicative of closer packing of atoms.

Greegor et al. [12] have shown in EXAFS studies of a natural mineral titanate, that the principal difference between the crystalline and metamict forms lies in the degree of misorientation between TiO_2 octahedra. Since zirconolite has a similar structure [13,3], the differences in degree of disorder implied by the results in Table I may also reflect varying misorientations of these octahedra. EXAFS studies of zirconolite are needed to further evaluate its metamict structure(s), and are planned.

Implications for Stability in a Nuclear Waste Repository

The temperature of a nuclear waste form will vary with radioactive isotope decay, waste loading, depth of burial, nature of the surrounding medium, changing geologic conditions, and condition of the waste itself (e.g., whether extensively disordered or microcracked). Although the overall trend in temperature is downward with time, some of these factors could lead to a temperature increase. For a material in the metamict state, a relatively high recovery temperature is desirable to avoid the possibility of fragmentation and consequent increase of the surface area exposed to a leachant. It appears that metamict zirconolite would be structurally stable under small upward temperature excursions, although the effect of extremely long storage times on recovery processes must be taken into account.

CONCLUSIONS

Recovery of Pu-substituted zirconolite from the metamict state is accompanied by densification, which proceeds sporadically at 875 K. The microstructure evolves from electron diffraction-metamict to a fine dispersion of crystallites having random orientations. At an intermediate stage in the recovery process the material becomes friable, apparently as a result of high internal strains. However, this condition does not persist to later stages of recovery. Evidence is presented that more than one metamict structure can exist in this material, depending on degree of self-damage and thermal history.

ACKNOWLEDGMENTS

The authors would like to thank K. M. Axler and M. Gibbs for their valuable assistance in carrying out this study.

REFERENCES

1. A.E. Ringwood,, S.E. Kesson, N.G. Ware, W.O. Hibberson, A. Major, Geochem. J. 13, 141 (1979).

2. K.D. Reeve and J.L. Woolfrey, J. Aust. Ceram. Soc. 16, 229 (1980).

3. F.W. Clinard, Jr., D.E. Peterson, D.L. Rohr, L.W. Hobbs, J. Nucl. Mater. 126, 245 (1984).

4. W.J. Weber, J.W. Wald, Hj. Matzke, in: Materials Research Society Symposium Proceedings Vol. 44, edited by C.M. Jantzen, J.A. Stone, and R.C. Ewing (Materials Research Society, Pittsburg, 1985) (in press).

5. T.J. Headley, G.W. Arnold, C.J.M. Northrup, in: Scientific Basis for Nuclear Waste Management, edited by W. Lutze (Elsevier Science Publishers, New York, 1982) p. 379.

6. W. Sinclair and A.E. Ringwood, Geochem. J. 15, 229 (1981).

7. F.W. Clinard, Jr., D.S. Tucker, G.F. Hurley, C.D. Kise, J. Rankin, op. cit. ref. 4.

8. E.M. Foltyn, F.W. Clinard, Jr., J. Rankin, D.E. Peterson, J. Nucl. Mater. (in press).

9. J.W. Wald and P. Offermann, op. cit. ref. 5.

10. R.C. Ewing, R.F. Haaker, T.J. Headley, P.F. Hlava, in: Scientific Basis for Nuclear Waste Management, edited by S.V. Topp (Elsevier Science Publishers, New York, 1982) pp. 249-256.

11. J.L. Routbort, P. Offermann, Hj. Matzke, Materials Research Society Symposium Proceedings Vol. 15, edited by D.G. Brookins (Elsevier Science Publishers, New York, 1983) pp. 415-422.

12. R.B. Greegor, F.W. Lytle, R.C. Ewing, R.F. Haaker, Nucl. Instrum. and Methods B1, 587 (1984).

13. H.J. Rossell, Nature 283, 282 (1980).

ALPHA-RECOIL EFFECT ON THE DISSOLUTION OF BETAFITE: RAPID NATURAL ANNEALING OF RADIATION DAMAGE WITHIN A METAMICT PHASE

Y. EYAL*, G.R. LUMPKIN** AND R.C. EWING**
*Department of Nuclear Engineering, Technion - Israel Institute of Technology, Haifa 32000, Israel
**Department of Geology, University of New Mexico, Albuquerque, New Mexico 87131, USA

ABSTRACT

Betafite, a radioactive mineral, is the Ti-rich member of the pyrochlore group, $A_{1-2}B_2X_6Y_{0-1}$, (Fd3m, Z = 8), where A = Ca, Na, U, Th, REE (rare earth elements), Y, Ba, Sr, Bi, Pb; B = Nb, Ti, Ta, Zr, Sn, Fe: X = oxygen; Y = O, OH and F. The pyrochlore structure (or structural derivatives, e.g. zirconolite in SYNROC) is a common constituent of polyphase, crystalline radioactive waste forms. Naturally occurring minerals with this structure often occur in the metamict state. Therefore, a principal concern is the effect of alpha-recoil damage on the structure and dissolution behavior of synthetic, analogue, waste-form phases which contain actinides.

We describe the dissolution behavior of ^{238}U, ^{234}U, ^{232}Th, ^{230}Th and ^{228}Th in betafite during leaching in a bicarbonate-carbonate solution. Results indicate enhanced leaching of recently produced alpha-recoil damage as evidenced by the more rapid dissolution rate of ^{228}Th relative to ^{232}Th and ^{230}Th, and only slightly enhanced leaching of older damage as evidenced by a small increase in the dissolution rate of ^{234}U relative to ^{238}U. These experiments demonstrate that individual alpha-recoil tracks are preserved for some time as disordered regions of higher chemical reactivity in already fully-damaged, aperiodic structures. We estimate an annealing time for the alpha-recoil tracks of 2,000 ± 1,300 yr in the metastable, aperiodic structure. Similar alpha-recoil tracks should be formed and annealed in other aperiodic nuclear waste forms, such as borosilicate glass.

SPECIMEN DESCRIPTION

Fragments of betafite (USNM #126235, UNM #083) from Bancroft, Ontario, Canada, with minimum evident alteration were selected. Electron microprobe analyses (EMPA) indicate minor alteration (hydration) along galena-filled veins, large voids, and microfractures which occurred after the specimen was at least partially metamict [1]. The analyses in wt.% (analysis of altered areas in parenthesis) give: Ta_2O_5 1.80 (1.56); Nb_2O_5 37.6 (35.2); TiO_2 15.4 (14.3); ZrO_2 0.08 (0.05); ThO_2 0.33 (0.25); UO_2 26.3 (24.3); Y_2O_5 0.15 (0.13); ΣREE_2O_3 0.70 (0.59); MnO 0.35 (0.36); FeO 3.09 (2.97); CaO 8.46 (7.59); BaO 0.07 (0.05); PbO 0.54 (0.55); Na_2O 0.08 (0.10); F 0.12 (0.08); total = 95.07

(88.08); bulk $H_2O \sim 10.0$, estimated by thermogravimetric analysis (TGA).

It is important to note variations in ThO_2 and UO_2. ThO_2 averages 0.33 wt.% in unaltered and 0.25 wt.% in hydrated regions. In both cases, ThO_2 is homogeneous at a level determined by counting statistics (± 0.05 wt.%). UO_2 averages 26.3 wt.% in unaltered and 24.3 wt.% in hydrated regions. Variations in UO_2 in both areas are above the standard deviation defined by counting statistics (± 0.2 wt.%) by a factor of three. With the high concentrations of UO_2 and ThO_2 and an age of at least 800 million years, the sample has experienced more than 50 displacements per atom (dpa) and is x-ray diffraction amorphous. Samples were annealed in nitrogen at 200°C and 800°C (Fig. 1). Differential thermal analysis (DTA) was completed in an argon atmosphere up to 1150°C (Fig. 2). TGA was performed in air up to 1000°C. Weight loss occurred primarily at 160°C and 380°C, corresponding to endotherms determined by DTA (Fig. 2). Total weight loss on heating to 1000°C was 10.0 wt.%. Using the TGA curve, estimated weight losses for the betafite annealed at 200°C and 800°C are approximately 2.0 and 9.5 wt.%, respectively.

LEACHING EXPERIMENTS

Betafite fragments were ground to fine powder (maximum particle size ∿100 μm). Portions (0.44 to 1.0 g each) of unheated or heat-treated powder were subjected to a series of successive leaching treatments in a bicarbonate-carbonate solution (10 ml of aqueous 0.1 M $NaHCO_3$ and 0.1 M Na_2CO_3 solution) at room temperature (∿22 °C). Each leaching series involved 8 or 9 successive treatments for a total of ∿50 or ∿100 days [2,3]. The radioactivities of ^{238}U, ^{234}U, ^{232}Th, ^{230}Th and ^{228}Th in each solution were determined by alpha spectrometry, preceded by chemical separation and purification of U and Th [2,3].

Initial radioactivities in the mineral were determined by the above method, for a fraction of totally dissolved mineral powder. The measured activity ratios for members of the uranium series are $^{234}U/^{238}U = 1.00 \pm 0.01$ and $^{230}Th/^{238}U = 1.00 \pm 0.04$, indicating a closed system for the last $\sim 10^6$ yrs. The measured absolute ^{238}U activity is 143,500 ± 2,600 disintegrations per minute per gram of betafite. This value is ∿17% lower than that expected from the EMPA, probably due to the presence of non-radioactive impurities (e.g. PbS). Because of the low abundance of Th relative to that of U, accurate values for Th-content and $^{228}Th/^{232}Th$ activity ratio were not determined. However, within the present uncertainty (∿8%), it is experimentally established that ^{232}Th and ^{228}Th are in radioactive equilibrium and that the value of $^{230}Th/^{232}Th$ activity ratio agrees with that expected from EMPA, $^{230}Th/^{232}Th = 244$. We adopt this value for the activity ratio for the present analysis.

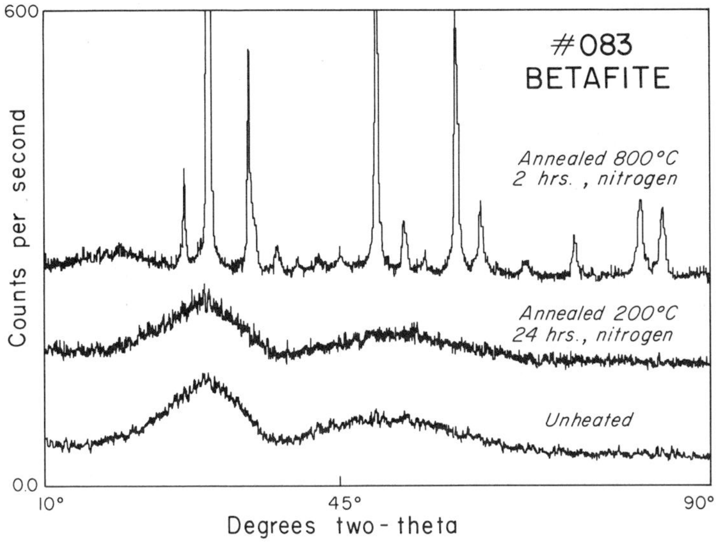

Figure 1. X-ray diffraction patterns for betafite: unheated; annealed at 200 °C in nitrogen; annealed at 800 °C in nitrogen. X-ray patterns are shifted along the vertical axis for comparison.

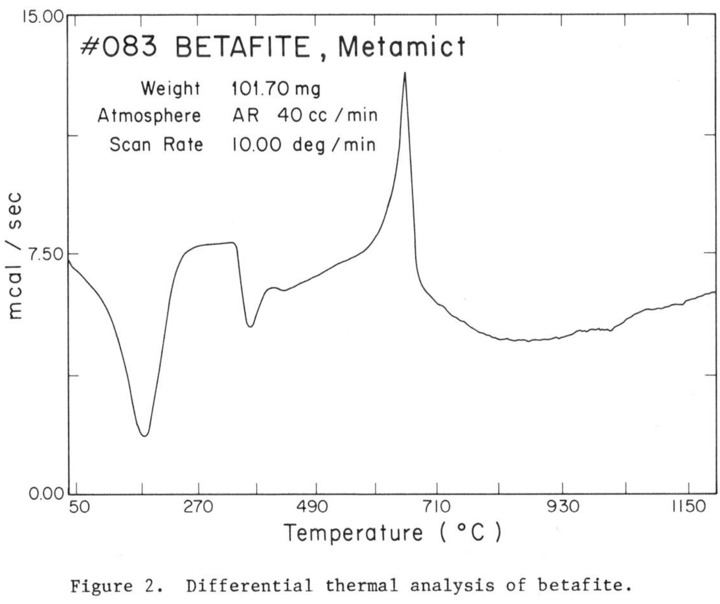

Figure 2. Differential thermal analysis of betafite.

RESULTS AND DISCUSSION

The leaching data are presented in Table 1. The quantity R is the relative leaching rate, given by the ratio of fractions of activities leached. $R(^{234}U/^{238}U)$ values consistently show isotopic fractionation in all leaching series. The enhanced dissolution rate of the long-lived radiogenic ^{234}U nuclide relative to that of its primary isotope, ^{238}U, clearly indicates the continued presence of alpha-recoil tracks even in material that must be considered as fully-damaged and aperiodic (> 50 dpa). Much more pronounced isotopic fractionation is observed for thorium. $R(^{228}Th/^{230}Th)$ values range from ∿8 to almost 100 during the early stages of the dissolution process. Because the leaching of the short-lived ^{228}Th nuclide displays the effect of alpha-recoil damage that is relatively fresh, the much enhanced $^{228}Th/^{230}Th$ fractionation relative to the amount of the $^{234}U/^{238}U$ fractionation indicates that the ambient annealing time for alpha-recoil tracks in betafite must be relatively short. It must be mentioned that the value of $R(^{228}Th/^{230}Th)$ is sensitive to the microdistribution of Th and U, since ^{228}Th is produced by decay of ^{232}Th, whereas ^{230}Th is a member of the ^{238}U-decay chain. Therefore, it is desirable to know the values of $R(^{228}Th/^{232}Th)$. However, because of the low abundance of Th in the mineral, relative to that of U, we were not able to determine accurately the amount of the leached ^{232}Th radioactivity. The activity of each Th source was counted from one to four days. During this period, only a few events were observed which may correspond to alpha particles emitted by ^{232}Th. It appeared, however, that most of these events may be accounted for by the continuous low-energy tail of the relatively intense ^{230}Th alpha-particle peak. Nevertheless, it may be concluded that within a large experimental uncertainty (typically a factor of ∿3), the data are consistent with ^{232}Th and ^{230}Th being leached at nearly the same rate.

It is important to note that $R(^{228}Th/^{230}Th)$ values show considerable variation for early successive leaching treatments in each series and between each of the three series. In addition, the $R(^{228}Th/^{230}Th)$ values pass through a maximum shortly after the initiation of the leaching experiment. At present, there is no direct or simple explanation for this behavior. All three leaching experiments show a significant reduction in the value of $R(^{228}Th/^{230}Th)$ as the dissolution process proceeds. This result may be explained by the rapid depletion of ^{228}Th at grain sufraces in contact with the solution.

Following the procedure of Eyal and Fleischer [4,5], the mean annealing time, τ_A, for alpha recoil tracks in a mineral is given by:

$$\tau_A = \tau_{234} \cdot [R(^{234}U/^{238}U) - 1]/[R(^{228}Th/^{232}Th) - 1] \qquad (1)$$

where τ_{234} is the mean life of ^{234}U (3.53×10^5 yr) and R is the initial relative isotopic leaching rate. If thorium and uranium are distributed uniformly in the mineral phase, then τ_A is also given by

$$\tau_A = \tau_{230} \cdot [R(^{228}Th/^{232}Th) - 1]/[R(^{228}Th/^{232}Th) - 1] \tag{2}$$

where τ_{230} is the mean life of ^{230}Th (1.15×10^5 yr). We assume that the leaching of ^{230}Th is mostly governed by the alpha-recoil damage created by ^{234}U decay, and that it is not influenced by the possible much older nearby damage created by ^{238}U decay (the primary nuclide in the uranium decay chain). However, as mentioned earlier, we were not able to determine accurately the value of $R(^{228}Th/^{232}Th)$. Therefore, we determine τ_A by a combined expression of Equations (1) and (2),

$$\tau_A = \tau_{234} \cdot \frac{R(^{234}U/^{238}U) - 1}{R(^{228}Th/^{230}Th) \cdot \{ \frac{\tau_{234}}{\tau_{230}} \cdot [R(^{234}U/^{238}U) - 1] + 1 \} - 1} \tag{3}$$

By employing Equation (3), we obtain from the leaching data of Table 1 the annealing times given in Table 2. We have obtained the initial relative isotopic leaching rates by integrating the leached activities over the first four leaching treatments (L_1 to L_4 in Table 1) for each experiment; because of the apparent delay in the release of the ^{228}Th into solution, L_1 data alone do not give the true initial relative leaching rates. An average annealing time of τ_A = 2,000 ± 1,300 yr is calculated from the results of all three experiments. The uncertainty is determined from the dispersion in the values of the three separate results.

The leaching data for two heat-treated samples are presented in Table 3. The leaching behavior of the material annealed at 200°C is not much different from that of the unheated sample. Indeed, the anneal at 200°C in nitrogen (Fig. 1) did not lead to any significant x-ray diffraction evidence for recrystallization. For the sample annealed at 800°C, the uranium isotopic fractionation is nearly absent upon leaching. This indicates that almost all alpha-recoil damage has been annealed. This is consistent with the x-ray diffraction pattern which gives sharp diffraction maxima and a pattern for betafite (a_o = 10.2557 ± 0.0005 A). However, heating the powder to 800°C did not eliminate the $^{228}Th/^{230}Th$ fractionation. Similar behavior has been noted with the leaching of heat-treated monazite, thorianite and uraninite [3,5]. This may indicate incomplete lattice repair or that ^{228}Th migrated to the surface during heating. The leaching data of the material annealed at 800°C

TABLE 1. LEACHING DATA FOR UNHEATED BETAFITE*

Leaching Treatment	Cumulative Leaching Time (days)	Activity Leached (dis/min·g^{-1})[a] ^{238}U	^{230}Th	Relative Leaching Rate, R[b] $^{234}U/^{238}U$	$^{228}Th/^{230}Th$
Experiment 1					
L_1	0.018	23.9 ± 1.4	13.0 ± 1.6	1.13 ± 0.03	8.0 ± 2.5
L_2	0.030	5.4 ± 0.7	5.8 ± 0.7	1.16 ± 0.18	31.5 ± 5.3
L_3	0.042	3.9 ± 0.4	3.0 ± 0.5	1.24 ± 0.15	44.2 ± 13.7
L_4	1.03	21.7 ± 1.0	16.9 ± 1.8	1.04 ± 0.04	95.6 ± 4.6
L_5	3.95	16.2 ± 0.9	16.0 ± 1.4	1.07 ± 0.04	13.2 ± 3.6
L_6	11.0	18.9 ± 1.0	17.9 ± 2.1	1.05 ± 0.04	10.4 ± 2.2
L_7	21.0	19.3 ± 0.7	13.5 ± 2.2	1.04 ± 0.03	5.6 ± 1.4
L_8	56.1	19.0 ± 1.0	17.3 ± 1.5	1.14 ± 0.06	4.2 ± 2.6
L_9	101.0	14.2 ± 0.9	12.7 ± 1.1	1.08 ± 0.07	7.3 ± 2.6
Experiment 2					
L_1	0.17	58.5 ± 1.6	26.8 ± 1.2	1.10 ± 0.02	8.2 ± 1.3
L_2	1.06	19.9 ± 1.1	13.4 ± 1.1	1.12 ± 0.05	7.9 ± 2.5
L_3	2.00	15.4 ± 0.9	10.7 ± 0.8	1.08 ± 0.04	11.2 ± 3.5
L_4	3.09	13.6 ± 0.9	9.9 ± 0.8	1.15 ± 0.08	17.5 ± 5.0
L_5	7.1	20.4 ± 1.2	15.4 ± 1.4	1.01 ± 0.06	7.3 ± 3.1
L_6	14.0	26.8 ± 1.2	14.4 ± 1.7	0.97 ± 0.04	3.6 ± 1.4
L_7	24.0	22.3 ± 1.4	14.5 ± 1.7	1.00 ± 0.06	3.3 ± 2.0
L_8	52.1	27.5 ± 2.6	19.0 ± 3.0	0.95 ± 0.09	3.6 ± 2.4
Experiment 3					
L_1	0.012	19.4 ± 1.5	6.4 ± 0.7	1.20 ± 0.03	11.7 ± 3.7
L_2	0.023	4.5 ± 0.3	3.3 ± 0.5	1.03 ± 0.08	21.0 ± 6.7
L_3	0.034	3.1 ± 0.2	2.5 ± 0.4	1.06 ± 0.07	49.9 ± 8.3
L_4	1.13	17.8 ± 0.8	11.9 ± 1.4	1.08 ± 0.05	7.3 ± 1.1
L_5	4.02	13.3 ± 0.7	9.9 ± 0.6	1.01 ± 0.05	4.6 ± 1.7
L_6	11.0	15.0 ± 1.2	9.7 ± 1.2	1.05 ± 0.08	1.7 ± 0.9
L_7	18.0	9.0 ± 1.3	6.7 ± 0.5	0.99 ± 0.05	5.6 ± 1.9
L_8	49.2	17.6 ± 1.8	10.8 ± 0.7	1.00 ± 0.09	4.2 ± 3.4

*The results of three independent experiments are reported separately. Experiments 1, 2 and 3 involved the leaching of 0.5, 0.5 and 1.0 g, respectively, of unheated mineral powder in 10 ml of 0.1 M $NaHCO_3$ and 0.1 M Na_2CO_3 solution at ∿ 22 °C.
[a]Measured radioactivities in units of disintegrations per minute per gram betafite.
[b]Relative leaching rates, R, given by the ratios of fractions of activities leached.

TABLE 2. ANNEALING TIME OF ALPHA-RECOIL DAMAGE IN BETAFITE*

Experiment	Initial Relative Leaching Rate, R[a] $^{234}U/^{238}U$	$^{228}Th/^{230}Th$	Annealing Time τ_A (years)
1	1.11 ± 0.05	52.6 ± 6.3	550 ± 300
2	1.11 ± 0.03	10.2 ± 1.4	3,000 ± 1,100
3	1.12 ± 0.06	14.7 ± 2.3	2,300 ± 1,300
Average[b]			2,000 ± 1,300

*The data pertain to the three leaching series reported in Table 1.
[a]Initial relative leaching rates, R, given by the ratios of fractions of initial activities leached. The data pertain to leaching treatments L_1 to L_4 reported in Table 1.
[b]Average annealing time from experiments 1, 2 and 3.

also show a significant reduction in the leaching rate of ^{230}Th, and perhaps a certain increase in the dissolution rate of uranium, at the beginning of the dissolution process, as compared to the leaching rates obtained with the unheated and annealed (200°C) betafite.

CONCLUSIONS

The leaching experiments on metamict betafite demonstrate that individual alpha-recoil tracks are preserved for some time as disordered regions of higher chemical reactivity in already fully-damaged, aperiodic structures which result from alpha-decay events. This damage is most evident for the short lived nuclide ^{228}Th and to a much lesser extent for the long-lived nuclide ^{234}U because the alpha-recoil tracks fade with time. We have calculated an annealing time of 2,000 ± 1,300 yr in the metastable aperiodic structure. In crystalline thorianite, uraninite and monazite, localized alpha-recoil tracks have calculated mean lives of approximately 15,000 to 18,000 yr [4,5]. The present research demonstrates that alpha-decay products may be selectively leached from amorphous materials, but the damage is subject to low-temperature annealing in relatively short periods of time. Similar alpha-recoil tracks should be formed and annealed in aperiodic nuclear waste forms, such as borosilicate glass.

TABLE 3. LEACHING DATA FOR HEAT-TREATED BETAFITE*

Leaching Treatment	Cumulative Leaching Time (days)	Activity Leached $(dis/min \cdot g^{-1})^{a}$ ^{238}U	^{230}Th	Relative Leaching Rate, R^{b} $^{234}U/^{238}U$	$^{228}Th/^{230}Th$
Annealed at 200 °C (24 hours, nitrogen atmosphere)					
L_1	0.012	24.0 ± 2.8	5.4 ± 1.0	1.29 ± 0.06	10.0 ± 4.3
L_2	0.033	8.5 ± 1.2	2.7 ± 0.6	0.97 ± 0.18	12.2 ± 9.8
L_3	0.044	3.6 ± 0.5	2.5 ± 0.5	1.00 ± 0.18	11.1 ± 8.9
L_4	0.97	18.3 ± 1.1	11.8 ± 1.2	1.09 ± 0.04	13.0 ± 2.8
L_5	3.97	13.9 ± 0.8	14.9 ± 2.6	1.09 ± 0.07	5.4 ± 2.3
L_6	10.0	12.4 ± 0.8	8.3 ± 2.4	1.21 ± 0.09	1.4 ± 1.1
L_7	21.1	13.7 ± 1.0	13.8 ± 2.0	1.15 ± 0.09	8.5 ± 5.4
L_8	49.0	18.1 ± 1.0	13.2 ± 1.0	0.98 ± 0.06	1.3 ± 1.1
Annealed at 800 °C (2 hours, nitrogen atmosphere)					
L_1	0.010	34.4 ± 1.8	1.0 ± 0.3	1.03 ± 0.03	48.8 ± 22.5
L_2	0.021	16.1 ± 0.8	0.5 ± 0.4	1.05 ± 0.06	15.3 ± 12.5
L_3	0.032	8.2 ± 0.6	0.4 ± 0.4	0.99 ± 0.09	56.3 ± 38.0
L_4	0.91	32.0 ± 2.8	5.6 ± 0.9	1.00 ± 0.04	31.9 ± 9.8
L_5	3.91	26.4 ± 1.4	6.6 ± 0.7	1.02 ± 0.06	4.5 ± 2.8
L_6	10.0	27.7 ± 1.6	10.0 ± 1.0	0.93 ± 0.06	3.6 ± 2.7
L_7	21.1	24.8 ± 1.3	13.1 ± 2.0	1.06 ± 0.06	3.8 ± 2.8
L_8	48.9	28.8 ± 1.6	10.7 ± 1.4	1.01 ± 0.06	7.7 ± 2.6

*The results of two independent experiments are reported separately. Each experiment involved the leaching of 0.44 g of heat-treated mineral powder in 10 ml of 0.1 M $NaHCO_3$ and 0.1 M Na_2CO_3 solution at ∿ 22 °C.
[a]Measured radioactivities in units of disintegrations per minute per gram heat-treated betafite.
[b]Relative leaching rates, R, given by the ratios of fractions of activities leached.

ACKNOWLEDGMENTS

This research was supported by grants from the United States - Israel Binational Science Foundation, Jerusalem, Israel (YE and RCE), and the USDOE, OBES, under contract DE-FG04-84ER45099 (RCE). The DTA was kindly provided by D. E. Peterson and L. Foltyn of Los Alamos National Laboratory.

REFERENCES

1. G. R. Lumpkin and R. C. Ewing, in Scientific Basis for Nuclear Waste Management VIII, edited by C. M. Jantzen, J. A. Stone and R. C. Ewing (Materials Research Society, Pittsburgh, 1985) pp. 647-654.
2. Y. Eyal and A. Kaufman, Nucl. Tech. 58, 77-83 (1982).
3. Y. Eyal, in Scientific Basis for Radioactive Waste Management V, edited by W. Lutze (Elsevier, 1982) pp. 399-408.
4. Y. Eyal and R. L. Fleischer, Nature 314, 518-520 (1985).
5. Y. Eyal and R. L. Fleischer, Geochimica Cosmochimica Acta 49, 1155-1164 (1985).

AN INVESTIGATION OF URANIUM L-EDGES OF METAMICT AND ANNEALED BETAFITE

R.B. GREEGOR*, F.W. LYTLE*, B.C. CHAKOUMAKOS**, G.R. LUMPKIN** AND R.C. EWING**
*The Boeing Company, Seattle, Washington 98124 USA
**Department of Geology, University of New Mexico, Albuquerque, New Mexico 87131 USA

ABSTRACT

The uranium site in naturally occurring metamict minerals of the pyrochlore group ($A_{1-2}B_2O_6Y_{0-1}$) has been investigated using x-ray absorption spectroscopy (XAS). Pyrochlore structures are common phases in proposed polycrystalline waste forms. Betafite, a member of the pyrochlore group (B = 2Ti ≥ Nb+Ta), exhibits U-O bond lengths of 1.94 and 2.37 A for the metamict state, and 2.03 and 2.51 A for the crystalline (annealed) state. The U-O bond lengths decrease (~0.1 A) and there is a disruption in the second nearest neighbor periodicidy as material is converted from the crystalline to the metamict state.

INTRODUCTION

Derivatives of the pyrochlore structure type, $A_{1-2}B_2O_6Y_{0-1}$, are common constituents of polycrystalline nuclear waste forms (e.g. Synroc or tailored ceramics) [1-4]. Because naturally occurring pyrochlores are often metamict, an important concern has been the effect of alpha-recoil events on the crystalline structure and on properties such as fracture toughness and leachability. X-ray absorption techniques are well-suited to examine radiation effects since they can be applied to aperiodic materials. Hence, the extended x-ray absorption fine structure (EXAFS) and x-ray absorption near edge structure (XANES) are used in this study to investigate the coordination geometry of U in metamict and annealed betafite, a member of the pyrochlore group.

We have previously reported EXAFS/XANES studies on the Ti-site in pyrochlore group minerals [5]. The principal effects found in the fully-damaged material compared to annealed samples included an increase in the asymmetry of the Ti-site and a slight reduction in bond length and coordination number with concurrent loss of long range periodicity. The EXAFS of the Ti-site in metamict material most closely resembled that of titanite, which has chains of corner-sharing Ti-O octahedra in which the Ti atom is offset from the geometric center of each octahedron. Other studies have used x-ray absorption spectroscopy to examine the U-site in sodium silicate [6] and borosilicate [7, 8, 9] glasses. In the sodium silicate glasses, evidence was found for the uranyl configuration with clustering of U in the glass. The borosilicate glasses have not shown the presence of the uranyl structure. In this study, we extend the investigation to the U-site in the pyrochlore group mineral, betafite.

SAMPLES AND EXPERIMENTAL DETAILS

The betafite (UNM #178, AMNH 28426) is from an unspecified locality in Mexico. Electron microprobe and TGA analyses give the formula $A_{1.04}B_2O_{6.0}F_{0.04} \cdot 2.3H_2O$, where A = 0.46 Ca, 0.42 U^{4+}, 0.11 Fe^{2+}, 0.03 Ba, 0.02 Pb and B = 1.10 Nb, 0.88 Ti, and 0.02 Ta atoms. The sample is x-ray diffraction amorphous, but recrystallizes when annealed at 1000°C in air for one hour. A unit cell dimension of a = 10.344 ± 0.005 A was calculated for the annealed betafite. Standard reference samples included uraninite (UO_2), U_3O_8, UO_3 and uranyl nitrate hexahydrate ($UO_2(NO_3)_2 \cdot 6H_2O$).

The EXAFS/XANES experiments were performed at the Stanford Synchrotron Radiation Laboratory (SSRL) [10]. At SSRL the side stations of Wiggler Beam Lines IV-1 and VII-3 were used, with the synchrotron beam at a current of approximately 30-50 mA and an energy of 3 GeV. A Si (220) double crystal monochromator was used, detuned by approximately 50% in order to reduce the harmonic content of the beam thereby reducing sample thickness effects. The samples were powdered to pass a -400 mesh sieve and were held in 1/4 mil polypropylene bags. The beam in the sample chamber was 1 x 20 mm and the monochromator was moved using the Berkeley software [11] to provide a step of 0.050 ± 0.015 A^{-1}. The data were gathered at room and liquid N_2 temperatures using a fluorescent x-ray detector [12], with argon in the ion chambers.

RESULTS AND DISCUSSION

The EXAFS data gathered at SSRL were analyzed using techniques that have been described in detail [13]. Briefly, the EXAFS equation

$$X(k) = \frac{1}{k} \sum_j \frac{N_j}{R_j^2} F_j(k) \exp(-2\sigma_j^2 k^2) \sin[2k R_j + \phi_j(k)] \qquad (1)$$

is Fourier transformed

$$\Phi_n(R) = \left[\frac{1}{2\pi}\right]^{1/2} \int_{kmin}^{kmax} k^n X(k) \exp i [2kR + \phi(k)] \, dk \qquad (2)$$

to produce a radial structure function centered on the absorbing atom. The inverse transform of the first neighbor peak is taken and then the electron backscattering F(k) and phase shift $\phi(k)$ are extracted using a known reference compound, e.g. in this study uraninite for which the coordination number, N, and U-O bond length, R, are known to be 8 and 2.38 A, respectively. The F(k) and $\phi(k)$ from the reference can then be used to analyze unknown samples. Theoretical values for F(k) and $\phi(k)$ have also been calculated, and are available in tabular [14] and parameterized [15] form. We used the parameterized values to study the differences that might be expected between UO_2 and UO_3, and the range of data that would be required to resolve the short and long U-O bonds of UO_3 in the Fourier transform peaks. The parameterized form for F(k) and $\phi(k)$ are given by

$$F(k) = \frac{A}{1 + B^2 (k-C)^2} \quad (3)$$

and

$$\phi(k) = P_0 + P_1 k + P_2 k^2 + P_3/k^3 \quad (4)$$

The values for the coefficients used for UO_2 and UO_3 are given in Table I, and the calculated EXAFS spectra are shown in Figure 1.

Table I. Values of parameters used in calculating EXAFS spectra (a)

PHASE	N	R (A)	σ^2 (A^2)	A (A)	B (A)	C (A^{-1})	P_0	P_1	P_2	P_3
UO_2 (b)	8	2.37	0.001	1.08	0.41	1.90	14.0	-3.4	0.18	-30.0
UO_3 (c)	2	1.79	0.0005	1.08	0.41	1.90	14.0	-3.4	0.18	-30.0
	2	2.28	0.0010	1.08	0.41	1.90	14.0	-3.4	0.18	-30.0
	2	2.32	0.0010	1.08	0.41	1.90	14.0	-3.4	0.18	-30.0

(a) Values for A, B, C are from [15]; values for P's determined empirically from uraninite
(b) Values of N and R are from [16]
(c) Values of N and R are from [17]

Figure 1 shows that for UO_3 a beat occurs in the EXAFS spectra at ~ 9 A^{-1}, which is followed by an increase of signal amplitude in the range of 10 to 14 A^{-1}. However, for UO_2 the envelope of the amplitude asymptotically approaches zero and has a small value above k = 11 A^{-1}. Figure 2 shows the Fourier transforms of the calculated UO_3 spectra for different ranges of k. In the transforms the phase shift, ϕ, has been included in the transform kernel so the peak appears at the correct R position. For short ranges of k (3 to 8 A^{-1}) the different U-O bond distances are not resolved, and the oxygen peak at 2.08 A in the transform represents a weighted average of the bond lengths present which is shorter than the arithmetic average of 2.13 A. For transform ranges of 3 to 12 A^{-1}, the distances begin to be resolved; for the 3 to 16 A^{-1} transform, two peaks (1.83 and 2.30 A) are clearly resolved, with the 2.28 and 2.32 A distances included in the second peak. Although the amplitude envelopes of the spectra depend upon the actual value of σ^2,

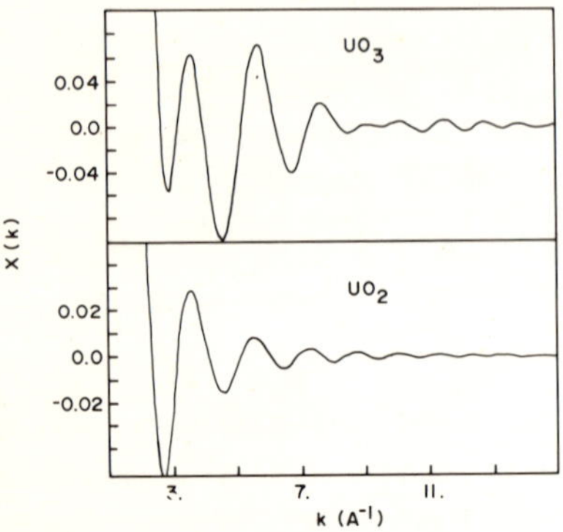

Figure 1. Calculated EXAFS spectra for UO_3 and UO_2 using equation (1).

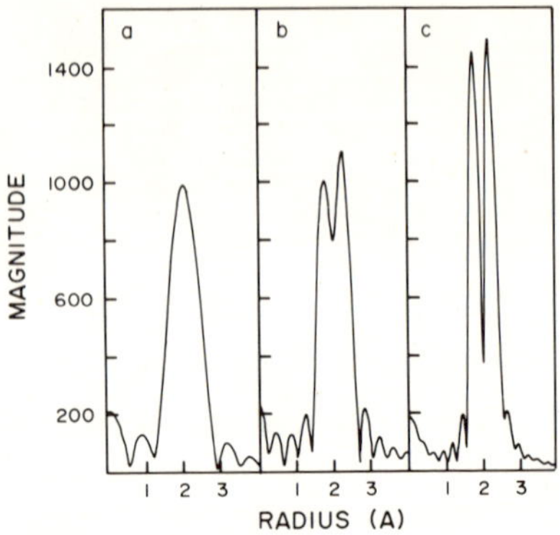

Figure 2. Fourier transforms of calculated UO_3 EXAFS spectra. All are k^3 weighted and use the U-O phase shift in the transform kernel, but have different ranges of k, a) 3-8 A^{-1}, b) 3-12 A^{-1}, c) 3-16 A^{-1}.

this analysis indicates that resolution of a short, UO_3, uranyl-like bond is enhanced using data beyond a k value 8 A^{-1}. In the absence of data over a long range of k, the presence of a UO_3 uranyl bond is probably indicated by average bond lengths less than 2.10 A. For a mix of short and long U-O bond lengths which differ by more than those in UO_3, the peak positions are more easily resolved over a short k range (3 to 8 A^{-1}) as shown in Figure 3, for $UO_2(NO_3)_2\ 6H_2O$. In this transform the 1.80 A oxygen peak is resolved but the 2.4 and 2.5 A peaks are combined in the second peak.

The XANES of the annealed and metamict betafite are compared in Fig. 4. The XANES for the annealed and metamict samples are indistinguishable.

To verify the theoretical analysis, Fourier transforms of the EXAFS data, using the same background removal technique, k weighting (k^3) and incorporation of the U-O phase shift in the transform kernel, were calculated for two ranges of data, k $\sim$3 to 8 A^{-1} and k $\sim$3 to 12 A^{-1}. These are shown in Fig. 5 for the metamict sample and Fig. 6 for the annealed sample. For both, the larger range of k improves the resolution. Broad peaks of 2.07 A (metamict) and 2.18 A (annealed) are revealed to be doublets, 1.94/2.37 A (metamict) and 2.03/2.51 A (annealed). For these bond

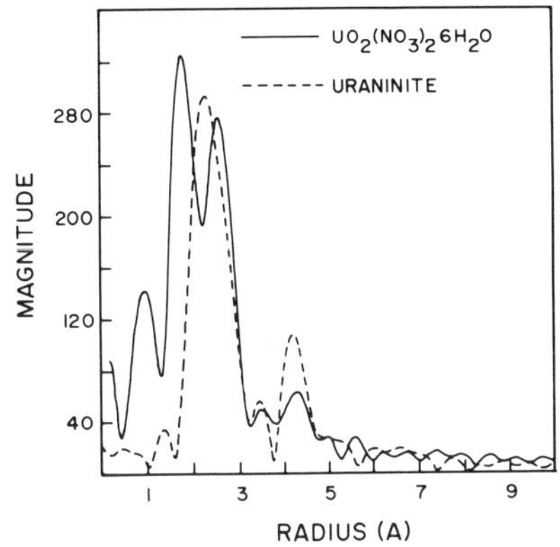

Figure 3. Fourier transforms of experimental EXAFS spectra for standard reference materials. The transforms were made using k^3 weighting, U-O phase shift in the transform kernel and k range of 3-8 A^{-1}. The uraninite has been scaled up for ease of peak position comparison.

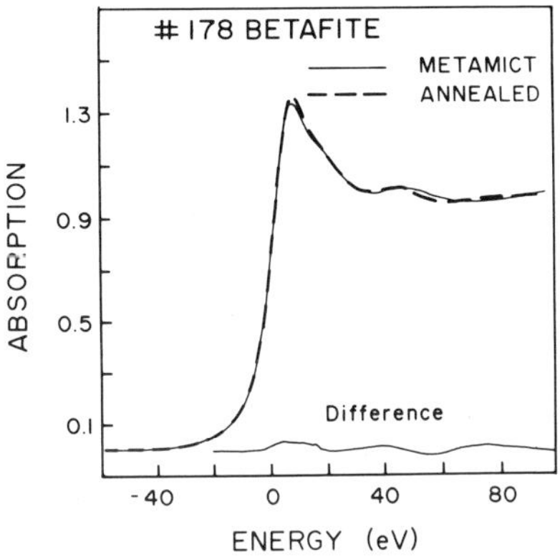

Figure 4. U L_{III}-edge XANES of betafite in natural metamict and annealed (1000°C) condition. The edges have been aligned in the region of -20 to 0 eV with respect to the U L_{III} edge.

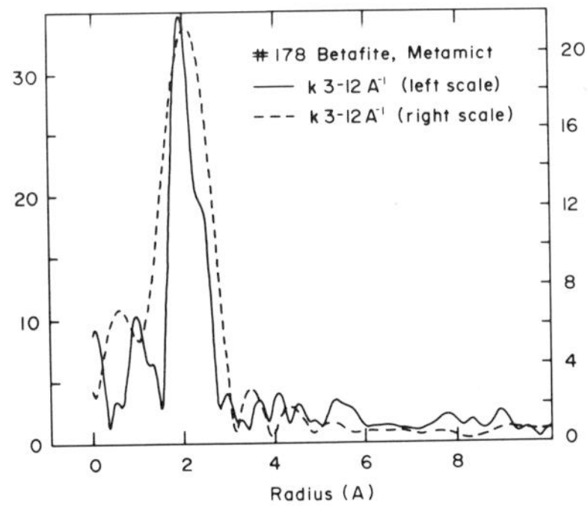

Figure 5. Fourier transforms of experimental U L_{III}-edge EXAFS spectra for metamict betafite.

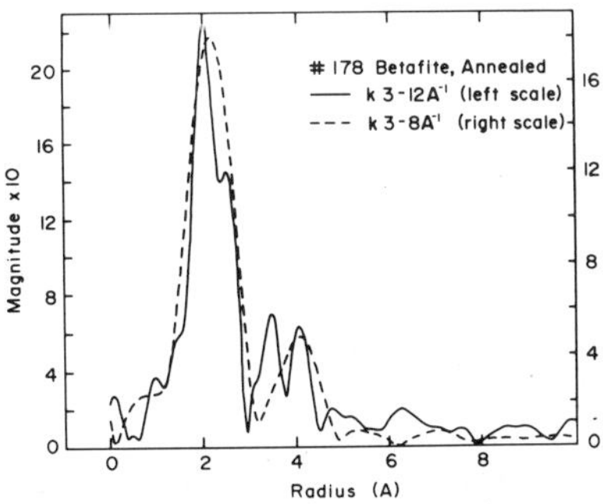

Figure 6. Fourier transforms of U L_{III}-edge EXAFS spectra for metamict betafite, annealed (1000°C).

lengths, a uranyl-like structure is plausible, however, further resolution is needed to make a definitive confirmation. Moreover, the U-O bond lengths do significantly shorten (0.1 A) in going from the crystalline (annealed) to the metamict states. Also, the second nearest neighbors of the crystalline (annealed) sample are more ordered than those of the metamict sample, as indicated by the metal-metal peaks in the Fourier transforms.

ACKNOWLEDGMENTS

This work was supported by the U.S. Department of Energy, Division of Materials Sciences, under Grant numbers DE-FG06-84ER45121 (Boeing) and DE-FG04-84ER45099 (UNM). Samples were kindly provided by Harvard University, and Gary Eller (Los Alamos National Laboratory). We are grateful to SSRL for beam time and help from their excellent staff. SSRL is supported by DOE, NSF and NIH.

REFERENCES

1. R.G. Dosch, T.J. Headley and P. Hlava, J. Am. Ceramic Soc. 67, 35 (1984).
2. P.E.D. Morgan, T.M. Shaw and E.A. Pugar, in Advances in Ceramics, Vol. 8, edited by G.C. Wicks and W.A. Ross, (American Ceramic Society, Columbus, Ohio 1984) p. 209-221.
3. A.E. Ringwood, S.E. Kesson, N.G. Ware, W. Hibberson and A. Major, Nature 278, 278 (1979).
4. S.E. Haggerty, in Ann. Rev. Earth Planet. Sci., Vol. 11, edited by G.W. Wetherill, A.L. Albee, and F.G. Stehli, (Annual Reviews Inc., Palo Alto, CA 1983) p. 133.
5. R.B. Greegor, F.W. Lytle, B.C. Chakoumakos, G.R. Lumpkin and R.C. Ewing, in Scientific Basis for Nuclear Waste Management - VIII, edited by C.M. Jantzen, J.A. Stone and R.C. Ewing (Materials Research Society, Pittsburgh, 1985) p. 655.
6. P.G. Eller, G. D. Jarvinen, J.D. Purson, R.A. Penneman, R. R. Ryan, F.W. Lytle and R.B. Greegor, submitted Radiochim. Acta. (1985).
7. G.S. Knapp, B.W. Veal, D.J. Lam, A.P. Paulikas and H.K. Pan, Mat. Letters 2, 253 (1984).
8. G.S. Knapp, B.W. Veal, A.P. Paulikas, A.W. Mitchell, D.J. Lam and T.E. Klippert, in EXAFS and Near Edge Structure III, edited by K.O. Hodgson, B. Hedman and J.E. Penner-Hahn (Springer-Verlag, Berlin, 1984) p. 305.
9. M. Antonini, A. Merlini and F.R. Thornley, in EXAFS and Near Edge Structure III, edited by K.O. Hodgson, B. Hedman and J.E. Penner-Hahn (Springer-Verlag, Berlin, 1984) p. 349.
10. H. Winick and A. Bienenstock, Ann. Rev. Nucl. Part. Sci. 28, 33 (1978).
11. J.A. Kirby, "EXAFS Data Collection Program," Univ. Calif., Berkeley (1978).
12. F.W. Lytle, R.B. Greegor, D.R. Sandstrom, E.C. Marques, J. Wong, C. Spiro, G.P. Huffman and F.E. Huggins, Nucl. Inst. and Meth. Phys. Res. 226, 542 (1984).
13. R.B. Greegor, F.W. Lytle, R.C. Ewing and R.F. Haaker, Nucl. Inst. and Methods B1, 587 (1984).
14. B.K. Teo and P.A. Lee, J. Am. Chem. Soc. 101, 2815 (1979).
15. B.K. Teo, P.A. Lee, A.L. Simmons, P. Eisenberger and B.M. Kincaid, J. Am. Chem. Soc. 99, 3854 (1977). P.A. Lee, B.K. Teo and A.L. Simmons, J. Am. Chem. Soc. 99, 3856 (1977).
16. B.T.M. Willis, J. Phys. 25, 431 (1964).
17. S. Siegel and H.R. Hoekstra, Inorg. Nucl. Chem. Lett. 7, 455 (1971).

SOURCE TERM FOR THE ACTIVITY RELEASE FROM A REPOSITORY FOR SPENT LWR FUEL

RUEDIGER WUERTZ AND M. ELLINGER
Kraftwerk Union Aktiengesellschaft, D-8757 Karlstein, Federal Republic of Germany

ABSTRACT

In order to provide basic data for the calculation of the dose burden from a repository for spent LWR fuel in a salt dome, a formula is given for determining activity release into the surrounding salt brine under accident conditions. The main characteristic of this source term is a time-independent, instantaneous release of part of the activity and mobilization of another part at a continuous and temperature dependant rate, using nuclide-specific leach rates. Changes in leach rates resulting from reduction of the waste surface during mobilization and precipitation are taken into account. The model also takes into account axial and radial activity distribution in the fuel rods.

INTRODUCTION

Within the framework of a comparison of different waste management systems in West Germany, several studies have been carried out for a salt dome repository for different waste forms. This included establishing a model for the activity released by leaching of spent fuel in a flooded repository regardless of the likelihood of such an event. This source term for the activity release is based on the results of three years of leaching experiments with spent fuel and various brines and on experimental data on activity distribution, specific surface and chemical condition of the waste [1] and [2].

SOURCE TERM MODELLING

It is known from leach tests and from spent fuel characteristics that radioactive elements are inhomogeneously distributed within the fuel rods. An evaluation of previous data [1] enabled the estimation of the fractional distribution of all of the radiologically relevant nuclides given in Table I. Three areas of nuclide residence are distinguished:

(1) gap between cladding and fuel stack including plenum of the fuel rod,
(2) UO_2 grain boundaries and
(3) UO_2 matrix.

Those nuclides which were not analyzed after the leach tests were treated by considering a chemically similar element for which experimental data are available and are included in Table I.

Table I: Fractional Distribution of Radionuclides in Spent Fuel

Nuclides analyzed	Nuclides of similar chem. characteristics	Fractional distributions in:		
		gap	grain boundaries	fuel matrix
Cs 137	Rb 87	0.05	0	0.95
	Cs 135	0.05	0	0.95
Eu 155	Sm 147	0.01	0	0.99
	Sm 151	0.01	0	0.99
Ru 106	Zr 93	0.03	0.2	0.77
	Tc 99	0.015	0.485	0.5
	Pd 107	0.015	0.485	0.5
	Nb 94	0.9	0	0.1
I 129	Se 79	0.015	0	0.985
Sb 125	Sn 126	0.015	0.485	0.5
U	U-Isotopes	0.008	0	0.992
Pu	Pu-Isotopes	0.008	0	0.992
Am 241	Am-Isotopes	0.008	0	0.992
Np 237	Np-Isotopes	0.008	0	0.992

The total activity inventory is taken from ORIGEN and CO-RIGEN codes.

Activity release for the purpose of safety calculations for a spent fuel repository is generally described by the following equations, which are compatible to the formalism used to carry out similar calculations for the vitrified waste from reprocessing [3]. The areas of the fuel mentioned above are initially taken as separate sources with individual release rates regardless of the real behaviour, which will be considered later. For the activity release of nuclides (i) from the three areas, the following expressions are used:

$$S_{1,i}(t) = R_{1,i} \cdot a_{1,i} \cdot I_i\ (t)$$

$$S_{2,i}(t) = R_{2,i} \cdot (t, T(t)) \cdot I_i\ (t)$$

$$S_{3,i}(t) = R_{3,i} \cdot (t.\ T(t)) \cdot a_{3,i} \cdot I_i\ (t)$$

where S are the activity release rates (Bq/a), a the fractional distributions of activity from Table I, I the activity inventories (Bq), t the time (a) and T the temperature (°C).

While a nearly instantaneous gap inventory release was observed and taken to be independent of time, leach tests described in [2] suggested a linear release of activity from both the grain boundaries and the fuel matrix for the long term leaching. This strongly supports the assumption of a congruent matrix dissolution of the fuel although present data do not yet fully prove this. Therefore, for these areas, a common linear leach rate R_2 was established from the leach tests as a function of temperature T, brine type, and fuel type for all relevant nuclides. Experimental data are available for rock-salt brine and quinary brine (1.45 wt % NaCl, 4.73 wt % KCl, 26.82 wt % $MgCl_2$, 65.60 % H_2O) at temperatures of 100 °C and 200 °C. All leaching experiments were carried out under a system pressure of 100 bar using commercial (standard) spent fuel of 37 GWd/t-U burn-up and highly-rated test fuel (HL).

As the waste surface was not significantly reduced during the leaching period, leach rates, R_2, could be related to the specific surface and weight of the fuel creating a "linear corrosion rate", k_2, as follows:

$$k_2 = R_2 / \rho \cdot O_B \qquad (1)$$

where k_2 = corrosion rate (mm/d)
R_2 = leach rate (d^{-1})
ρ = fuel density (g/cm^3)
O_B = fuel macroscopic specific surface (mm^2/g)

O_B was taken from [1] to be 277 mm^2/g for commercial fuel and 281 mm^2/g for the test fuel examined.

An Arrhenius-type dependence for temperature was assumed:

$$k_2 = C_2 \cdot e^{-E_2/R_o T} \qquad (2)$$

and coefficients C_2 and E_2 were calculated from the leach test results. The coefficients C_2 and E_2 representing the experimental results are summarized in Table II. Because experimental data are only available for 100 °C and 200 °C values for k should be restricted to this range of temperature. The coefficients C_2 and E_2 include all chemical and physical reactions involved in mobilization of radionuclides and therefore should rather be used to describe the activity release in a phenomenological way than to use it as a thermodynamic parameter for the interpretation of reaction mechanisms.

Table II: Activation Energies, E_2, and Coefficients, C_2, of the Corrosion Rates of Spent Fuel

Nuclides	E_2 (J/mol)	$C_{2,K,St}$ (mm/d)	$C_{2,Q,St}$ (mm/d)	$C_{2,K,HL}$ (mm/d)	$C_{2,Q,HL}$ (mm/d)
Sr 90	$1.4 \cdot 10^4$	$5.3 \cdot 10^{-5}$	$1.1 \cdot 10^{-4}$	$2.5 \cdot 10^{-4}$	$8.5 \cdot 10^{-5}$
Rb 87	$1.4 \cdot 10^4$	$5.3 \cdot 10^{-5}$	$1.1 \cdot 10^{-4}$	$2.5 \cdot 10^{-4}$	$8.5 \cdot 10^{-5}$
I 129	$1.4 \cdot 10^4$	$5.3 \cdot 10^{-5}$	$1.1 \cdot 10^{-4}$	$2.5 \cdot 10^{-4}$	$8.5 \cdot 10^{-5}$
Cs 135	$1.4 \cdot 10^4$	$5.3 \cdot 10^{-5}$	$1.1 \cdot 10^{-4}$	$2.5 \cdot 10^{-4}$	$8.5 \cdot 10^{-5}$
Cs 137	$1.4 \cdot 10^4$	$5.3 \cdot 10^{-5}$	$1.1 \cdot 10^{-4}$	$2.5 \cdot 10^{-4}$	$8.5 \cdot 10^{-5}$
Zr 93	$2.1 \cdot 10^4$	$5.0 \cdot 10^{-4}$	$3.6 \cdot 10^{-4}$	$2.1 \cdot 10^{-3}$	$4.6 \cdot 10^{-4}$
Nb 94	$2.1 \cdot 10^4$	$5.0 \cdot 10^{-4}$	$3.6 \cdot 10^{-4}$	$2.1 \cdot 10^{-3}$	$4.6 \cdot 10^{-4}$
Tc 99	$2.1 \cdot 10^4$	$5.0 \cdot 10^{-4}$	$3.6 \cdot 10^{-4}$	$2.1 \cdot 10^{-3}$	$4.6 \cdot 10^{-4}$
Pd 107	$2.1 \cdot 10^4$	$5.0 \cdot 10^{-4}$	$3.6 \cdot 10^{-4}$	$2.1 \cdot 10^{-3}$	$4.6 \cdot 10^{-4}$
C 14	$1.7 \cdot 10^4$	$9.1 \cdot 10^{-5}$	$1.4 \cdot 10^{-4}$	$4.1 \cdot 10^{-4}$	$3.0 \cdot 10^{-4}$
Sm 147	$1.7 \cdot 10^4$	$9.1 \cdot 10^{-5}$	$1.4 \cdot 10^{-4}$	$4.1 \cdot 10^{-4}$	$3.0 \cdot 10^{-4}$
Sm 151	$1.7 \cdot 10^4$	$9.1 \cdot 10^{-5}$	$1.4 \cdot 10^{-4}$	$4.1 \cdot 10^{-4}$	$3.0 \cdot 10^{-4}$
U	$1.7 \cdot 10^4$	$9.1 \cdot 10^{-5}$	$1.4 \cdot 10^{-4}$	$4.1 \cdot 10^{-4}$	$3.0 \cdot 10^{-4}$
Sn 126	$3.6 \cdot 10^3$	$8.0 \cdot 10^{-6}$	$3.0 \cdot 10^{-6}$	$4.5 \cdot 10^{-5}$	$5.7 \cdot 10^{-6}$
Am 241	$1.0 \cdot 10^4$	$7.6 \cdot 10^{-5}$	$7.5 \cdot 10^{-5}$	$1.0 \cdot 10^{-4}$	$7.0 \cdot 10^{-5}$
Ra	$1.0 \cdot 10^4$	$7.6 \cdot 10^{-5}$	$7.5 \cdot 10^{-5}$	$1.0 \cdot 10^{-4}$	$7.0 \cdot 10^{-5}$
Actinides	$1.0 \cdot 10^4$	$7.6 \cdot 10^{-5}$	$7.5 \cdot 10^{-5}$	$1.0 \cdot 10^{-4}$	$7.0 \cdot 10^{-5}$

K : rock-salt brine
St: standard fuel
Q : quinary brine
HL: highly-rated fuel

For long-term leach rates, the decrease in the fuel surface has to be taken into consideration. For this purpose, an idealized fuel surface with radial and horizontal cracks has been assumed. The number of cracks has been evaluated on a statistical basis from post-irradiation examinations [1] and the reduction of the fuel surface as a function of time has been taken into account using the idealized conditions of the fuel shown in Fig. 1. Congruent dissolution of the fuel matrix was assumed as already mentioned above.

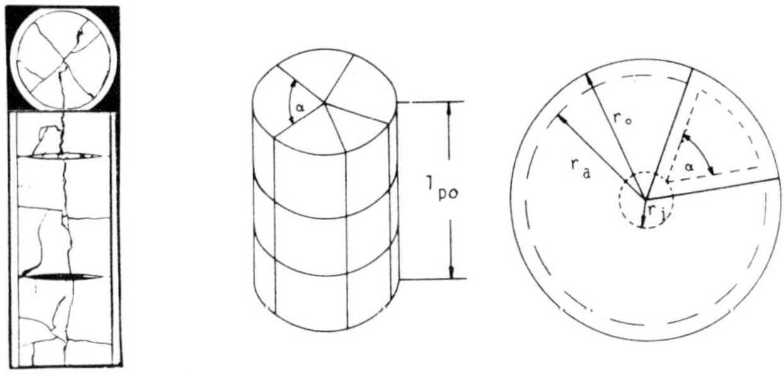

Fig. 1: Idealized Fuel Cross Section

With A(t), the activity content of fuel, at the time t and A_0, the initial average activity, the leach rate which is time dependent due to surface reduction can be set to

$$R_2 = \frac{\partial}{\partial t} \left(\frac{A_0 - A(t)}{A_0} \right) \qquad (3)$$

where the time dependence of the activity content is given by the congruent dissolution of the fuel sectors according to Fig. 1 resulting in the following:

$$A(t) = \int_{s_1}^{s_2} \int_{r_i(t)}^{r_a(t)} 2\pi \cdot L(t) \cdot A_M(r - r_i)\, F_R(r)\, F_A(s)\, dr\, ds \qquad (4)$$

where

$$r_i(t) = k_2 \cdot t \cdot \frac{1}{\sin \alpha/2}$$

$$r_a(t) = r_o - k_2 \cdot t$$

$$L(t) = 1 - \frac{1}{l_{po}} \cdot (k_2 \cdot t \cdot 2\,(n_H + 1))$$

with l_{po} = fuel pellet length (11 mm)

n_H = number of horizontal pellet cracks (≈ 2)

α = $\frac{360}{n_R}$; n_R = number of radial cracks (≈ 9)

A_m = average specific activity of fuel

$F_R(r)$ = radial activity distribution coefficient

$F_{A(s)}$ = axial activity distribution coefficient (according to 1)

In order to allow a common treatment of the leaching behaviour of chemically similar radionuclides, the half-life of the nuclides is not yet included in the formula. This is considered when the source term is fit into the formula for activity transport, which is not the aim of this contribution.

Although factors F_R and F_A concerning the inhomogeneity of the activity distribution in the fuel are available from [1], they are not included in further considerations because of their minor contribution to the results.

RESULTS

On the basis of the above formulae, the ratio $A(t)/A_0$ representing the relative time dependent activity inventory of the fuel was calculated for all nuclides not taking into account the instantaneously released part containing nuclides like Cs and I accounting for 5 % and 1.5 % of the initial inventory respectively.

Activity release of uranium to rock-salt brine and Q-brine at 100 °C and 200 °C is given in Fig. 2 showing how temperature and Q-brine enhance corrosion as compared to rock-salt brine. According to the parameters given in Table II, the same curves are received for C 14, Sm 147 and Sm 151 which are also radionuclides to be considered in safety calculations. As already mentioned, the half-life of the nuclides is not yet included in these curves in order to be able to give a common representation of the leaching behaviour of similar elements.

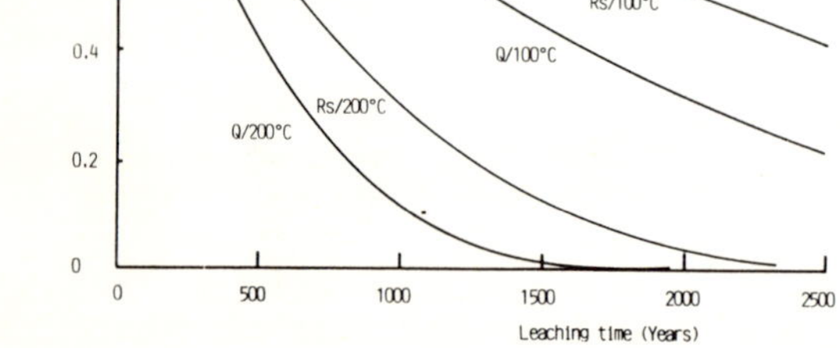

Q: quinary brine   Rs: rock-salt brine

Fig. 2: Leaching of Uranium from Spent LWR fuel

It should be mentioned here that spent fuel generally was attacked more heavily by Q-brine. The cladding material (zircaloy), however, was more strongly corroded or almost completely dissolved by rock-salt brine at 200 °C and exhibited no visible attack by Q-brine. All radioactive constituents of the cladding therefore were conservatively regarded as instantaneously released.

Increased leaching was observed for nuclides like Cs and I shown in Fig. 3, similarly enhanced by higher temperatures and by leaching in Q-brine. Again the curves do not yet include the half-lives of the nuclides and A/A_0 represents the relative activity inventory in the fuel at a given time.

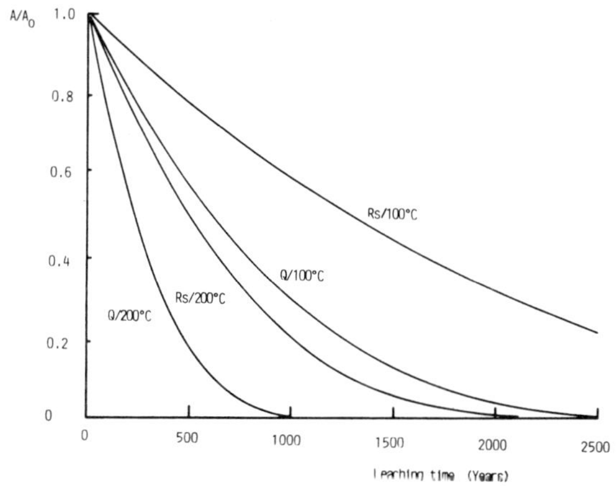

Q: quinary brine   Rs: rock-salt brine

Fig. 3: Leaching of Cs, I, Rb from Spent LWR fuel

Activity release for other radiologically relevant nuclides can be calculated using the data given in Table II.

LIMITATIONS OF THE MODEL

The present model is based on the leach-tests already mentioned and therefore restricted to the experimental parameters applied. Of particular importance in this connection is the ratio of fuel surface to brine volume which was 0.3 cm^{-1} for all tests. Specific surface was taken from [1]. The release rates to be calculated from Table II include as well the dissolved as the precipitated fraction of radionuclides and therefore are conservative representations of the release process. A more realistic approach is achieved by including the "mobilization factors", f, shown in Table III, which represent the ratios of dissolved and totally released activities for a surface to volume ratio of 0.3 cm^{-1} at the experimental conditions used.

Table III: Fraction, f_L, of Dissolved Nuclides Related to the Total Mobilization at 200 °C

Actinides	$f_{L,K,St}$	$f_{L,Q,St}$	$f_{L,K,HL}$	$f_{L,Q,HL}$
Ra Rb 87 Sr 90 I 129 Cs 135 Cs 137	1	1	0,6	0,8
Zr 93 Nb 94 Tc 99 Pd 107	0,4	0,6	0,1	0,1
C 14 Sn 126	0,5 0,15	0,5 0,1	0,5 0,1	0,5 0,1
Sm 147 Sm 151	0,6	0,7	0,2	0,5
U	0,5	0,9	0,1	0,2
Pu	0,4	1	0,02	0,04
Am 241 Actinides	0,7	0,8	0,5	0,6

Additional uncertainties exist because of the unknown pH- and redox conditions in the case of brine intrusion into the repository and the influence of the large amounts of iron from the container shielding. The effect of iron was examined and the results of these experiments will be published before long.

REFERENCES

1. R. Wuertz and M. Ellinger, BMFT Report No. BMFT-FB-K83-002, 1983

2. R. Wuertz and M. Ellinger, Systemstudie Andere Entsorgungstechniken, Technischer Anhang 9, KWA 5124/A7, 1984

3. F. K. Altenteim, W. Lutze, R. C. Ewing in Scientific Basis for Nuclear Waste Management V, W. Lutze ed., North Holland, N.Y., p. 45

ALPHA-RADIOLYSIS OF AQUEOUS SOLUTIONS

HILBERT CHRISTENSEN* AND ERLING BJERGBAKKE**
*Studsvik Energiteknik AB, S-611 82 Nyköping, Sweden
**Risö National Laboratory, DK-4000 Roskilde, Denmark

ABSTRACT

The effects of the radiolysis of water exposed to mixed alpha- and beta-radiation originating from spent fuel have been calculated. The water is assumed to have penetrated the copper canister and fuel cladding, and then to exist as a 30 µm thin surface film on the fuel pellets.

The combined effects of alpha and beta-radiation, and the presence of iron in the water have been found to be important parameters. The beta-radiation will lower the yield of hydrogen, as will the presence of low concentrations of iron ions. The most likely conditions for water exposed fuel give rise to a total production of 1 mol H_2 per m^2 fuel surface after 1 million years. A stoichiometric amount of oxygen is also formed.

Calculations have also been carried out in connection with experiments on alpha-radiolysis from Am-241 or from spent fuel pellets. In the case of alpha-radiolysis from Am-241 it was calculated that a 20 cm^2 thin layer of 4 mCi Am-241 produced 1.4×10^{-3} cm^3 hydrogen per day.

Calculations simulating the conditions of experiments on the leaching of spent fuel have shown that both alpha- and gamma-radiation accelerate the dissolution. The rate is strongly dependent on the dose rate and on the area of the water film covering the fuel surface and fuel fissures.

INTRODUCTION

According to a proposal by I. Neretnieks and co-workers [1] an excess of oxidizing radicals from the radiolysis of ground water may cause a migrating front of radioactive nuclides. Some nuclides become more soluble after oxidation, but will precipitate again after having diffused a distance due to reduction by Fe^{2+} ions. Molecular hydrogen is formed as a primary product in the radiolysis of water and - being rather inactive - may be responsible for the excess of the oxidizing radicals.

We have previously made calculations of the radiolysis of ground water outside the canister, assuming varying wall thicknesses [2] and have shown that radiolysis may have an effect only with wall thicknesses which are thinner than in the present design.

However, it is known that alpha-radiation decomposes water to a greater extent than gamma- and beta-radiation. We have, therefore, in the present investigation carried out calculations on the radiolysis of water which has penetrated the canister and fuel cladding, and thus has been exposed to the more densely ionizing alpha-radiation. These calculations represent an extreme accident case.

In the second part of this paper calculations simulating reactions which may take place in connection with the leaching of spent uranium fuel have been made. It is currently being discussed, whether radiolysis - and in particular alpha-radiolysis - can accelerate the dissolution of uranium oxides.

Mat. Res. Soc. Symp. Proc. Vol. 50. ©1985 Materials Research Society

CALCULATION METHODS

Computer program

A program developed by Lang-Rasmussen and Bjergbakke [3] was used. The program is based on DIFSUB [4]. In principle the program translates the chemical equations of the complete reaction system into a set of differential equations, which is solved by numerical integration after specifying rate constants, initial concentrations, G-values, irradiation dose and duration. The results are presented in tables giving the concentrations of the various species at various times during and after the irradiation. The applicable chemical reactions and rate constants are given in ref 2. The predictive power of the program has been tested by comparing calculated and measured results from gamma-radiolysis [2].

Yields of primary radiolysis products

Whereas the yields for beta- and gamma-radiation are well known [5] this is not the case for alpha-radiation. Therefore comparisons have been made between measurements of hydrogen and oxygen carried out by Bibler [6] and calculations using different sets of G-values: 1. Values given by Cohen [5], mainly based on data from Allen's monograph [7]. 2. Values calculated from data given by Bibler [6]. 3. Burn's values [8] at a mean LET of 200 eV/nm. 4. An evaluated mean of the values 1-3. 5-6. Slightly modified values, in an attempt to improve the agreement between the measured and calculated results.

The G-values used in the calculations are shown in Table I, and the results of the calculations can be compared with the measured yields in Table II. Set no 4 was chosen for the subsequent calculations. The primary G-values from irradiation with beta-radiation and mixed alpha- and beta-radiation are also given in Table I.

Table I. Primary G-values by radiolysis of water. The values are taken from various sources: 1) Allen-Cohen, 2) Bibler, 3) Burns, 4) "Best estimate", 5-6) modifications.

	α						β	α±β*
	1	2	3	4	5	6		
H_2	1.7	1.28	1.17	1.3	1.3	1.4	0.49	1.08
H	0.16	0.50	0.14	0.21	0.21	0.21	0.44	0.27
E^-	0.04	0.13	0.13	0.06	0.06	0.06	2.68	0.75
H_2O_2	1.3	0.98	0.92	0.985	1.00	1.10	0.74	0.92
HO_2	0.3	0.35	0.11	0.22	0.22	0.22	0.01	0.16
OH	0.1	0.18	0.44	0.24	0.21	0.21	2.59	0.86
H^+	0.04	0.13	0.13	0.06	0.06	0.06	2.77	0.77
$-H_2O$	3.34	2.97	2.63	2.71	2.71	2.91	6.86	3.79

*The contribution of beta-radiation is 26%

ALPHA-RADIOLYSIS FROM SPENT FUEL STORED IN COPPER CANISTERS

Irradiation dose

The dose rates used were based on calculations carried out by Klas Lundgren [9], who reported alpha, beta and gamma-dose rates in pores of spent fuel containing water, after storage times of 40 and 10^5 y, see

Table II. Yields of hydrogen and oxygen after irradiation of acid water solutions of Cm-244 with alpha-radiation using different sets of G-values. Comparison with measured yields.

Set of G-values*	Assumed gas volume, ml	$G(H_2)$	$G(O_2)$
1	1	1.68	0.093
	5	1.70	0.11
2	5	1.27	-0.024
3	5	1.15	0.13
	10	1.15	0.14
4	5	1.29	0.097
5	5	1.29	0.084
6	5	1.39	0.085
Measured yields		1.27±0.07	0.17±0.06

*See Table I for explanation

Table III. Dose rates on the surface of fuel pellets after various storage times

		Dose rate in rad/s after various storage times in y						
		40	100	300	1000	10^4	10^5	10^6
BWR33	α	28	23	15	7	1.5	7.5E-2	3E-2
	β	10	6.9	4.5	2.1	0.45	1.7-2	9E-3
PWR38	α	32	26	17	8	1.7	8.6E-2	3.4E-2
	β	12	8.3	5.4	2.5	0.54	1.4E-2	1.1E-2

Table III. Dose rates at intermediate times and after 10^6 y were estimated based on the decay curves of heavy nuclides [10], see Table III. The alpha-particle range was assumed to be 0.03 mm, and we have chosen to present production in a volume of 4.32 l, determined by the range and the cylindrical surfaces of all the fuel pellets in one canister, 144 m^2. This is also close to the gas void volume which is the maximum volume available for water intrusion.

Diffusion of hydrogen and oxygen

The diffusion constant of hydrogen was calculated on the basis of principles similar to those described previously [2]. In most cases a constant based on the diffusivity in bentonite (2×10^{-11} $m^2 \times s^{-1}$ [11]) was used. This is because it was assumed that the hydrogen must diffuse through the surrounding bentonite in order to escape from the system. In a few additional calculations the diffusivity in pure water (1.4×10^{-8} $m^2 \times s^{-1}$) was used. The diffusivity of oxygen in bentonite is expected to be close to that of hydrogen in bentonite, therefore the same value was used.

Results

Calculations have been carried out from 40 to 10^6 y on radiolysis from BWR fuel with a burnup of 33 Mwd/t, and PWR fuel with a burn up of 38 Mwd/t. In these calculations we have assumed an initial Fe^{2+} buffer of 2.0 mol to be present in 4.3 l of H_2O. We have used the diffusivities in bentonite. As long as Fe^{2+} is added to the system the radiolysis products are hydrogen and Fe^{3+}, and the G-values are high. The Fe^{2+} buffer is consumed in a short time (less than 1 year). When the iron buffer has been consumed, $G(H_2)$ decreases by three orders of magnitude. The reason for this is that Fe^{2+} and Fe^{3+} are present, and this redox couple acts as a recombiner. The results are presented for BWR fuel, as G-values, hydrogen production, and integrated hydrogen production, in Table IV. The results for PWR fuel are almost the same: After 1 million years PWR fuel gives 147 mol, compared with 143 mol for BWR fuel, corresponding to about 1 mol per m^2 surface area.

Table IV. Hydrogen production in water irradiated with mixed alpha- and beta-radiation from BWR fuel with a burn-up of 33 MWd/t

Storage time y	$C(H_2)$ at equilib µM	$G(H_2)$	Hydrogen* production mol/y	Integrated** production mol
41	120	1.8 x E-4	9.8E-4	0
100	105	1.9 x E-4	8.3E-4	5.4E-2
300	84	2.4 x E-4	6.7E-4	2.0E-1
1000	58	3.5 x E-4	4.6E-4	6.0E-1
10^4	31	8.9 x E-4	2.5E-4	3.84
10^5	22	1.3 x E-2	1.8E-4	23
10^6	11	1.6 x E-2	0.9E-4	143

* Production in 4.3 l H_2O corresponding to a fuel surface of 144 m^2.

** In addition 1 mol of hydrogen is produced initially during the period of consumption of the iron buffer.

Effect of diffusivity

Using the diffusivities in bentonite, the H_2-concentration in water irradiated with 40 y old BWR33 fuel increased to the saturation concentration, 3.9×10^{-2}M (at 5 MPa), within 49 d. After irradiation for an additional 250 d the Fe^{2+} present initially (2 mol) had been oxidized and an equivalent amount of hydrogen (1.0 mol) had been formed. During these 250 days hydrogen bubbles are formed. After that the hydrogen disappears by diffusion at a rate of about 10^{-3} mol/y (between 40 and 100 y). $G(H_2)$ decreases from 0.26 during the period when Fe^{2+} is added continuously, to a value of only 1.8×10^{-4} when the Fe^{2+} buffer is consumed.

Using the diffusivities in water ($D(H_2) = 1.4 \times 10^{-8}$ $m^2 \times s^{-1}$, $D(O_2) = 5 \times 10^{-9}$ $m^2 \times s^{-1}$) the Fe^{2+} present initially is consumed in 92 d, with the formation of 1.0 mol of hydrogen. The diffusion of hydrogen is fast enough to avoid the formation of hydrogen bubbles. After the consumption of Fe^{2+}, hydrogen diffuses out of the system at the rate of 0.6 mol/y. At the same time oxygen diffuses out at the rate of 0.3 mol/y.

Continuous addition of Fe^{2+}

In the bulk of the calculations a Fe^{2+} buffer of 2.0 mol in 4.3 l of water has been assumed to be present initially. This iron content is consumed in less than one year after the start of irradiation with 40 y old fuel. However, if it is assumed that Fe^{2+} is continuously transported from the surrounding bentonite to the water layer at the surface of the fuel, so that this water keeps its initial Fe^{2+} concentration, 70 mol of hydrogen per m^2 fuel surface are produced after 1 million years.

Irradiation of pure water

If pure water is irradiated the hydrogen production is so high for the first 10^5 y that it mostly disappears by bubble formation. The hydrogen production is 200 mol per m^2 after 1 million years.

Effect of beta-radiolysis

The effect of low LET radiation, such as gamma- and beta-radiations, on hydrogen and oxygen when they diffuse out of the alpha-irradiated layer has been calculated. The effect of gamma-radiation outside the canister is small - due to the low dose rates prevailing there. However, the relatively high betaradiation field existing inside the fuel cladding, but outside the alphairradiated layer (0.03 mm), may recombine the major part of the hydrogen and oxygen produced in this layer.

If pure water is irradiated, the results shown in Table V are obtained. It can be seen that the extent of recombination depends strongly on the diffusion time, i.e. whether diffusion takes place in pure water or in water saturated bentonite. If it is assumed that Fe^{2+} is supplied continuously to the water, the beta-radiation may either increase or decrease the hydrogen production, depending on the diffusion time, see Table VI.

Table V. Beta-radiolysis of pure water inside the cladding, including the effect of alpha-radiolysis of the thin water film

		Diffusion in water			Diffusion in bentonite			
Storage time y	Total $G(H_2)$	$C(H_2)$ at Equilib μM	$\frac{d(H_2)}{dt}$ mol/y	$IP(H_2)$* mol	$C(H_2)$ at Equilib μM	$\frac{d(H_2)}{dt}$ mmol/y	$IP(H_2)$* mol	$IP(H_2)$* from film mol
40	1.95	167	0.89	0	51	0.40	0	0
100	2.05	162	0.87	53	57	0.44	0.03	260
300	2.15	135	0.72	210	60	0.47	0.12	860
E3	2.15	58	0.31	570	48	0.37	0.40	2100
E4	2.16	18	0.097	2100	22	0.170	2.8	8600
E5	2.79	1.57	8.4E-3	7100	217	1.70	87	21000
E6	2.49	0.74	3.9E-3	12600	35	0.25	970	29000

* IP = Integrated production

Table VI. Beta-radiolysis of Fe^{2+} solutions inside the cladding, including the effect of alpha-radiolysis of the thin water film

		Diffusion in water			Diffusion in bentonite			
Storage time y	Total $G(H_2)$	$C(H_2)$ at Equilib μM	$\frac{d(H_2)}{dt}$ mol/y	$IP(H_2)$* mol	$C(H_2)$ at Equilib μM	$\frac{d(H_2)}{dt}$ mmol/y	$IP(H_2)$* mol	$IP(H_2)$* from film mol
40	0.77	330	1.80	0	1.33	10.4	0	0
100	0.85	280	1.53	100	1.51	11.8	0.7	58
300	0.93	220	1.19	370	1.72	13.5	3.2	220
E3	1.02	10.8	0.59	990	1.96	15.3	13.3	600
E4	1.10	3.4	0.185	4500	1.85	14.5	147	2800
E5	1.28	1.56	8.6E-3	13200	0.52	4.0	980	6900
E6	1.13	0.78	4.3E-3	19000	0.32	2.5	3900	9800

* IP = Integrated production

ALPHA-RADIOLYSIS FROM Am-241

In an experiment by Trygve Eriksen [12] water-saturated bentonite was irradiated using an Am-241 alpha-source, with a surface area of 20 cm^2 and an activity of 7.4 x 10^6 Bq cm^{-2}. The dose rate was calculated to be 2.5 kGy x h^{-1}. The hydrogen and oxygen produced in the thin alpha-irradiated layer containing 1.65 x 10^{-2} cm^3 of water diffused out through a 1 cm thick layer of water-saturated bentonite, and was measured outside this layer. Using the computer program a hydrogen production rate of 2.5 x 10^{-9} mol · h^{-1} was calculated. After less than 2 h an equilibrium with a hydrogen concentration of about 2 x 10^{-4} M was established, see Figure 1.

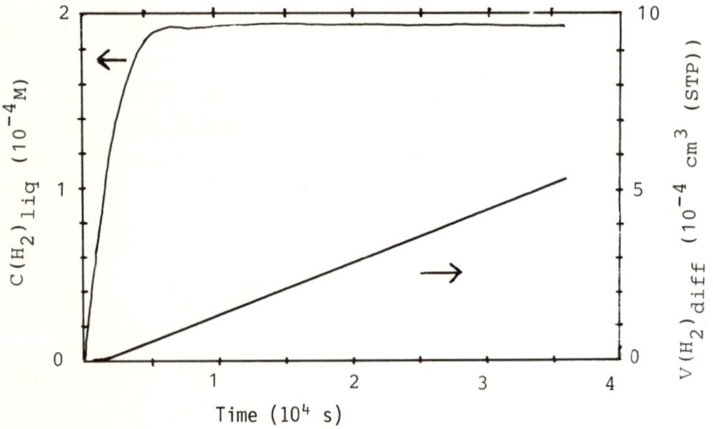

Figure 1. Production of hydrogen by alpha-radiolysis of water-saturated bentonite irradiated by an Am-241 source.

ALPHA-RADIOLYSIS FROM SPENT FUEL PELLETS

The leaching of spent UO_2 fuel in synthetic groundwater has been studied [13]. In connection with these and other experiments it has been discussed whether alpha-radiolysis of water could accelerate the dissolution of UO_2 fuel. It has been shown that gamma-radiolysis may accelerate the dissolution of UO_2 in water [14]. Computer calculations have, therefore, been carried out in order to simulate various experimental conditions.

Experimental conditions

A fuel sample containing 16 g UO_2 was leached in a closed glass container in 200 cm^3 Allard water (a synthetic groundwater [15], which was initially deaerated. Above the water was a 50 cm^3 air volume. The UO_2 fuel has attained either a low burnup (LBU) or a high burnup (HBU), giving the following calculated dose rates to the water on the fuel surface (Table VII).

Table VII. Calculated dose rates, rad/s

	α	β	γ	Total
HBU	25	110	25	160
LBU	0.04	150	60	210

The surface of the thin alpha-irradiated water layer was assumed to be 2 cm^2/g UO_2, corresponding to a water volume of 0.096 cm^3 (alpha-particle range 30 μm). The rate of reaction between UO_2 in the solid phase ($UO_2(s)$) and OH or O_2^- radicals in the liquid phase is not known. We have assumed that the OH radicals within a diffusion length (~14 μm) of the UO_2 wall may react with the wall with a high rate constant. The concentration of UO_2 "seen" by the OH radicals has been assumed to correspond to the dissolution of one molecular layer into the thin alpha-irradiated layer. The irradiation time is 22 d, as in the leaching experiment. This gives a total dose in the thin water layer of 3.0 x 10^5 and 4.0 x 10^5 krad for HBU and LBU fuel respectively. In the calculations a number of cases with different assumptions have been carried out, see Table VIII

Table VIII. Conditions in various calculation cases

Case	$C(UO_2)$, M	$C(O_2)$, M	$k(UO_2^{++} \rightarrow bulk)$
1a	2E-4	1.25E-4	2.08E3
1b	2E-4	0	2.08E3
1c	2E-6	1.25E-4	2.08E3
1d	2E-4	1.25E-4	2.08E-4
1e	2E-6	0	2.08E-4
2a*	2E-4	0	2.08E3
2c	2E-6	0	2.08E3
2d	2E-4	0	2.08E-4
2e	2E-6	0	2.08E-4

* in this and the following cases $C(H_2) = 8 \times 10^{-4}$ M (i.e. hydrogen-saturated)

Results

In Table IX the calculated results are given as mol uranium dissolved after 22 d. For comparison the experimental results are included.

Table IX. Calculated radiation-induced dissolution of UO_2. Results are given in mol uranium dissolved after 22 d

Air saturated water			Hydrogen saturated water		
Case/fuel type	HBU	LBU	Case/fuel type	HBU	LBU
1a	6E-5	7E-5	2a	1.5E-5	1.5E-5
1b	2E-5	2.2E-5			
1c	2E-6	2E-6	2c	1E-9	6E-10
1d	2.5E-5	3E-5	2d	1.4E-5	1.5E-5
1e	7E-7	7E-7	2e	1.4E-10	1E-10
Experiment	5E-7 - E-6	5E-6			

From the table it can be seen that the apparent concentration of UO_2 is an important parameter (case a 2 x 10^{-4} M, case c and e 2 x 10^{-6} M). The rate of UO_2^{++} transportation from the thin alpha-irradiated layer (0.096 cm^3) into the bulk phase (200 cm^3) is not an important parameter, compare cases a and d. The results indicate that the apparent UO_2 concentration is less than 2 x 10^{-4} M.

The calculated dissolution of uranium is slightly higher for the LBU fuel than for the HBU fuel. The difference in the experimental results is much larger, indicating that the difference is caused by other effects than radiolysis (one possible explanation could be difference in the specific surface area). The calculated dissolution of uranium is about four times lower in hydrogen-saturated water compared with air-saturated water (when $C(UO_2)$ = 2 x 10^{-4} M).

REFERENCES

1. I. Neretnieks, The Movement of a Redox Front Downstreams From a Repository for Nuclear Waste, KBS TR 82-16, 1982.

2. H. Christensen and E. Bjergbakke, Radiolysis of Ground Water From HLW Stored in Copper Canisters, KBS TR 82-02, 1982.

3. O.L. Rasmussen and E. Bjergbakke, Chemisimul-a Program Package for Numerical Simulation of Chemical Reaction Systems, Risö-R-395, 1984.

4. C.W. Gear, Algorithm 407, DIFSUB, Commun. ACM, 14, 185 (1971).

5. P. Cohen, Water Coolant Technology of Power Reactors, (Gordon and Breach, New York, 1969) p 93.

6. N. Bibler, J. Phys. Chem., 78, 211 (1974).

7. A.O. Allen, The Radiation Chemistry of Water and Aqueous Solutions, (D. van Nostrand, Princeton, 1961).

8. W.G. Burns and H.E. Sims, J. Chem. Soc., Farady Trans. 1, 77, 2803 (1981).

9. K. Lundgren, Radiation Levels and Absorbed Doses Around Copper Canisters Containing Spent LWR Fuel, KBS TR 82-11, 1982.

10. K. Lundgren, Strålningsnivå och till vatten deponerad strålningsenergi utanför kapslar i slutförvaret, (in Swedish), KBS TR-106, 1978.

11. I. Neretnieks and E. Skagius, Diffusivitetsmätningar av metan och väte i våt lera, (in Swedish), KBS TR-86, 1978.

12. T. Eriksen, (private communication), 1984-01-11.

13. R.S. Forsyth and L. Werme, The corrosion of spent UO_2 fuel in synthetic groundwater, this volume.

14. V. Gromov, Radiat. Phys. Chem., 18, 135 (1981).

15. B. Allard, B. Torstenfelt and K. Andersson in Scientific Basis for Nuclear Waste Management III, edited by J. Moore (Elsevier Science Publishing Co., New York, 1981) p 465.

PART V

Canisters—Design, Materials and Performance

SOME ASPECTS ON THE USE OF IRON CANISTERS FOR HLW

IVARS NERETNIEKS, Department of Chemical Engineering, Royal Institute of Technology, S-100 44 STOCKHOLM, Sweden

ABSTRACT

Iron canisters for high level nuclear waste embedded in compacted bentonite in deep geologic repositories will corrode forming hydrogen gas. The equilibrium pressure (when corrosion would stop) has been estimated to be between 500 and 1000 atm. under repository conditions. As this is much higher than the lithostatic pressure (weight of rock overburden) the gas must be allowed to escape before it disrupts the repository. Escape by diffusion alone is not sufficient but recent experiments have demonstrated that the larger pores in the bentonite are blown free of water and let the gas escape before excessive pressures build up.

The potential effect of a capillary breaking layer (CBL) has been explored. A fine layer nearest the canister (e.q. quartz sand) would have much lower capillary suction pressures than the bentonite clay and would keep the water out as long as there is sufficient overpressure. As long as the CBL is void of liquid water no radionuclides can escape, even if the canister is penetrated.

BACKGROUND

For high level waste (HLW) repositories, iron is considered in several countries as a candidate material for the canisters which will contain the HLW or spent fuel. For the HLW repositories, the iron canisters might be surrounded by a low permeability backfill such as compacted bentonite clay [1,2,3].

The iron will corrode by reaction with dissolved oxygen (if available - in deep locations the waters are reducing) and with water. Hydrogen gas will form and must escape. It can escape by diffusion in the water filled pores or by flow through the larger pores of the porous material. Water will simultaneously be drawn in by capillary forces and by flow in the initial stages where there is a pressure gradient towards the canister. If the hydrogen production is large, considerable pressures can build up as the hydraulic conductivity and diffusivity of bentonite is low.

The hydrostatic pressures are expected to be 50-130 bars at the depths considered. The rock stresses may be very much larger (150-250 bars in Sweden at 500 m depth). In Switzerland up to 350 bars rock stresses are expected at 1300 m. The rock may thus contain a body with much higher internal overpressure than the hydrostatic pressure. There are several questions to be approached:

1. Can the gas pressure reach such values that it will exceed the confining stress of the surrounding rock?

2. Can the hydrogen pressure reach such levels that corrosion no longer can take place due to thermodynamic reasons?

3. Can the porous structure of the clay materials decrease the activity of water in the two phase flow situation where so high capillary pressures develop that this will influence (lessen) the corrosion of the iron?

After an initial period where the entrapped oxygen reacts with the iron, the reaction will be between iron and water. Only these reactions will be considered below.

BENTONITE PROPERTIES

The bentonite clays considered consist of more than 80 % smectite minerals which exist in the form of very thin crystals. In the compacted state which will exist in the repositories (wet density ca. 2000-2100 kg/m^2) the average crystal distances are 15-20 Å. Because there are crystal agglomerates and foreign material with larger size, larger pores and voids are present. Water in the intercrystalline layer is very strongly bound by adsorption [4]. Up to 25-30 % by weight of water can be adsorbed strongly. Additional water will increasingly be looser bound in the material. At a compaction degree equivalent to a wet density of 2000 kg/m^3, the clays have a swelling pressure on the order of hundreds of atm. It varies somewhat with the source of the bentonite. Lower densities give lower swelling pressures [5,9]. Adsorption isotherms and energies as well as the activity of water in bentonite clays, have been measured [4]. Hydraulic conductivities of water saturated clay with different degrees of compaction have been measured by Pusch 1980 [5] as well as gas transport rates through compacted bentonite [6,7,8]. Gases dissolved in water may diffuse through a compacted bentonite layer and gas may also flow through the larger pores, provided sufficient pressure difference is imposed to overcome the capillary (and adsorption) forces in these pores. Figure 1 shows a conceptual picture of a bentonite barrier.

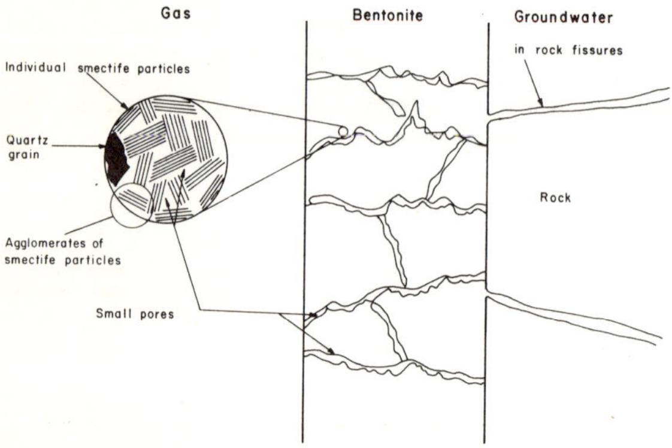

Figure 1: A compacted bentonite barrier may contain pores in addition to the voids between the individual smectite particles. Natural bentonites also contain a minor fraction of other minerals.

THERMODYNAMIC CONSIDERATIONS

In anoxic waters a probable iron corrosion product is magnetite. The reaction can be written

$$3Fe(s) + 4H_2O(l) \leftrightarrows Fe_3O_4(s) + 4H_2(g)$$

The equilibrium constant can be written

$$K_1^4 = \frac{f_{H_2}^4 \cdot a_{Fe_3O_4}}{a_{H_2O}^4 \cdot a_{Fe}^3} = \frac{\phi_{H_2}^4 \, P_{H_2}^4 \cdot a_{Fe_3O_4}}{a_{H_2O}^4 \cdot a_{Fe}^3} \tag{1}$$

f and ϕ denote fugacity and fugacity coefficient respectively, a denotes activity.

For 25° C, 1 atm and activity of water, iron and corrosion products equal to 1, the equilibrium fugacity of hydrogen is between 630 and 705 atm for the magnetite reaction. The fugacity coefficient at fugacities of 600-700 atm is around 1.3 - 1.4 (see below). If the water activity is 1 then about 400 atm equilibrium pressure would be expected.

This pressure is so high that the pressure influence on the equilibrium must be assessed. Furthermore the influence of elevated temperature must be investigated because the waste gives off heat causing the temperature in the repository to rise by 50-100° C depending on the layout of the repository and the age and concentration of the waste (or spent fuel).

Influence of temperature and pressure

The temperature influence can be assessed if the enthalpy of reaction ΔH_r is known

$$\left(\frac{\partial \ln K}{\partial T}\right)_P = - \frac{\Delta H_r}{RT} \tag{2}$$

The influence of pressure may be assessed by the relation

$$\left(\frac{\partial \ln a_i}{\partial P}\right)_T = V^o/RT \tag{3}$$

where V^o is the mole volume of the component. If the pressure were equal for all components equation 3 could be used directly to modify K. In the case where a mechanical stress exists between the porous clay and the iron canister and capillary or osmotic effects play a role this is not necessarily the case. The canister and corrosion products are subject to the mechanical swelling pressure P_M of the clay at the points of contact, whereas the water in the larger pores is at the hydrostatic pressure P_H. Figure 2a shows this case.

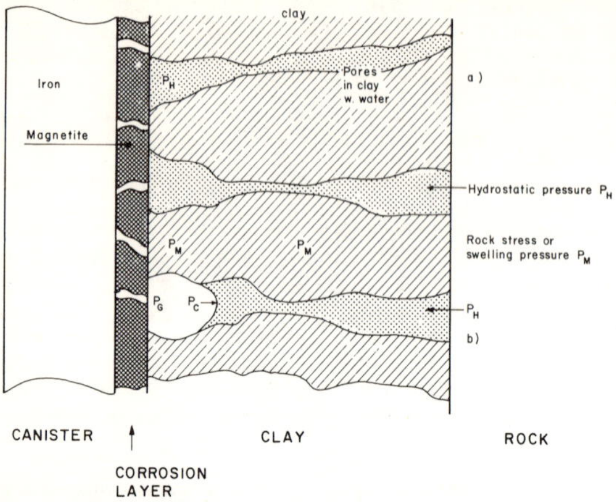

Figure 2: a) Showing the condition when the water in the pores is in direct contact with the canister $P_G < P_H$

b) Showing the conditions when a gas phase has formed with a pressure P_G. The capillary pressure is P_C.

When and if the hydrogen production rate becomes so high that hydrogen cannot escape at the same rate as it is produced, gas may form. (It will be shown later that the diffusion rate is not large enough at estimated corrosion rates to carry all gas away.) The gas pressure P_G will be larger than the hydrostatic by P_C - the capillary pressure in the pores.

$$P_G = P_H + P_C \qquad (4)$$

The corrosion products will (at least locally where further corrosion takes place) be subject to the gas pressure. This may vary between slightly above the hydrostatic pressure P_H and up to the lithostatic stress (pressure) P_L. At 1300 m depth (Nagra concept) P_H = 130 atm and P_L approx. 350 atm. If the gas cannot escape fast enough, P_M may increase from the swelling pressure to at most the lithostatic pressure P_L. The bentonite might then be compacted and leave a gas volume around the upper part of the canister.

If the larger pores open up before this happens and have a sufficient transport capacity to let the gas escape by flow, the pressure build up will reach a stationary value, lower than if gas must escape by diffusion only.

Fugacity can be seen as the "would be pressure" if the gas were ideal. As it is not at higher pressures, a correction is needed, ϕ_H.

For hydrogen at 80° C ϕ is estimated to [9, (Table 3-163)] about 1.24 at 400 atm.

By use of equation (1), (2), (3) we obtain the pressure P_G = 699 atm at t = 80.

The variability of the data is considerable and the predicted equilibrium pressures may well be in error by several tens of %. The equilibrium pressures will be well in excess of the lithostatic pressure and thus cannot be permitted.

ESCAPE OF HYDROGEN

The hydrogen which evolves by corrosion may escape by diffusion through the backfill and if this transport rate is not sufficient, flow will eventually take place. If the escape rate by these mechanisms is not sufficient to let the pressure stay below the lithostatic pressure, the rock will expand and the repository may be disrupted. Diffusivities in compacted fully water saturated bentonite have been measured [8,10]. An experiment by Pusch [6] may also be used to evaluate diffusivities of dissolved gas through bentonite. Experiments have recently been performed to investigate how gas may flow through a compacted initially water saturated bentonite layer [7].

The diffusivity for H_2 (aq) in bentonite was found to be $0.4\text{-}3 \cdot 10^{-11}$ m^2/s in the different investigations. The highest value was obtained in Pusch´s experiment where the pressure was kept at 90-100 atm. The other two experiments were performed at 1 atm and gave $0.4 \cdot 10^{-11}$ m^2/s [10] and $1.8 \cdot 10^{-11}$ m^2/s [8]. The differences may in part at least be due to differences in compaction of the bentonite.

Escape by diffusion

Fick´s first law may be used to estimate the rate of escape by diffusion of the dissolved hydrogen

$$N = - D_{H_2} A \frac{dC}{dz} \qquad (5)$$

The relation between the concentration C and the hydrogen pressure can approximately be expressed by a linear relation - Henry´s law.

$$P = H \cdot X = H \cdot C/C_0 \qquad (6)$$

within the pressure range of interest the errors involved are less than a factor 2. X is the mole fraction hydrogen in solution and C_0 is the molar concentration of the mixture (practically only water). $H = 6\text{-}10 \cdot 10^4$ atm (Perry) for $t < 100°$ C and $H/C_0 = 1.5 \cdot 10^3$. The hydrogen concentration at 130 atm pressure is 0.087 $kmol/m^3$.

The transport area A and barrier thickness Z can be estimated from the geometry of the tunnel. For the NAGRA concept they are: $A = 10$ m^2 and $\Delta Z = 1.3$ m. With $D = 10^{-11}$ m^2/s, $N = 2.1 \cdot 10^{-4}$ kmol/year. This is equivalent to $1.58 \cdot 10^{-4}$ kmol/year or 8.9 g/year iron corroded. Because of uncertainty in data it may be expected that the hydrogen that evolves by corrosion of between 2 and 50 g/year iron may escape by diffusion.

2 g/y is equivalent to an average corrosion depth of 0.032 μm and 50 g/y gives 0.8 μm/year iron corroded. This is far below the expected corro-

sion rate which has been estimated to be on the order of tens of μm/year. Even assuming that the hydrogen pressure would increase to the lithostatic pressure ca. 350 atm, this would still not be enough to let the hydrogen diffuse out.

The increasing hydrogen pressure will probably decrease the rate of corrosion because the reaction will have to take place nearer the equilibrium pressure, but assuming that the reaction rate is proportional to the difference between the equilibrium pressure and the existing pressure, this would give at most a factor 2 decrease in reaction rate.

With the data available at present the production rate of hydrogen by corrosion may be considerably larger than the escape rate by diffusion. If there were no other escape mechanisms the pressure would try to increase towards several hundreds of atmospheres above the lithostatic pressure.

Escape by flow

Pusch [11] has recently performed experiments to determine at which pressure differences the water in the pores of the bentonite will be blown out and thus when gas transport by flow can start. In the experiments the gas flow conductivity was also determined. The results can be summarized as follows: at an overpressure (differential pressures) of 20-90 % of the swelling pressure of the bentonite, flow starts to take place. The transport capacity rapidly increases from that due to diffusion by a factor of 100-1000.

The geometry in the repository differs from that of the experiments in at least one important respect. In the repository the gas will have to flow from the clay into the few thin fractures in the rock which will conduct the gas further out. This cross sectional area is very much smaller than the bulk of the clay and the gas flow must converge into a smaller area in order to reach the fracture. It is not immediately evident that this can be done. Data are lacking for a rigorous calculation of the two phase flow situation. To circumvent these problems an entirely different approach will be explored.

Corrosion rate

The corrosion rate of iron in anoxic water has been estimated to be on the order of 10-20 μm/year or less [12]. It is expected that the magnetite layers will form protective layers which will decrease the rate of corrosion [13,14] but as it has not been demonstrated to do so under repository conditions it seems prudent to assume that the protective layers do not decrease the average corrosion rate to well below 1 μm/year. It is known that for iron corrosion to take place there must be an electrolyte in contact with the iron [15,16]. The capillary forces in the small pores in the bentonite will pull in water to the canister to replenish that water which has reacted. If the liquid water could be kept away from the canister surface, the iron would not corrode. This idea will be explored below.

CAPILLARY BREAKING LAYER

If liquid water was not present to form an electrolyte in contact with the iron, corrosion would not take place [15,16]. Liquid water can be hindered from flowing into the iron if a Capillary Breaking Layer (CBL) is inserted between the canister and the bentonite. A capillary breaking layer may consist of a porous material which has a much larger pore size than the bentonite clay. If there then is a gas overpressure (formed by corrosion) $P_G = P_H + P_C$, the pores within the CBL which have a capillary suction pressure $P_{CBL} < P_C$ will be empty because they cannot draw the water in against the gas pressure P_G. Figures 3 and 4 illustrate this effect. The water-gas interface is in the larger pores of the bentonite which have the pore size just giving rise to a capillary pressure P_C. Thus as long as there is an overpressure there will be no water near the canister. The small pores of bentonite will not release the water to the CBL.

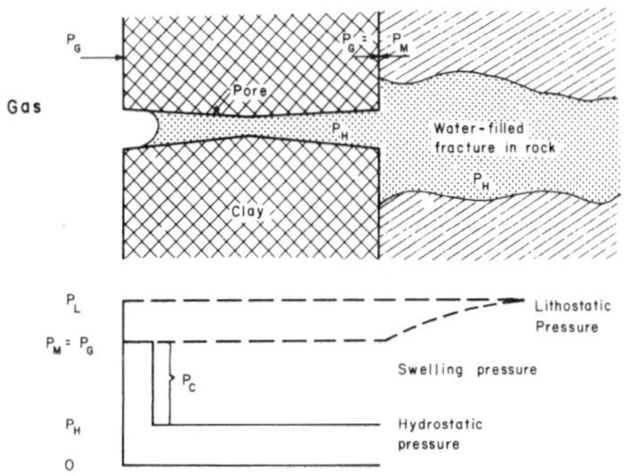

Figure 3: Pressure distribution in the bentonite when the gas pressure P_G exceeds the surrounding hydrostatic pressure P_H but is less than the lithostatic pressure P_L.

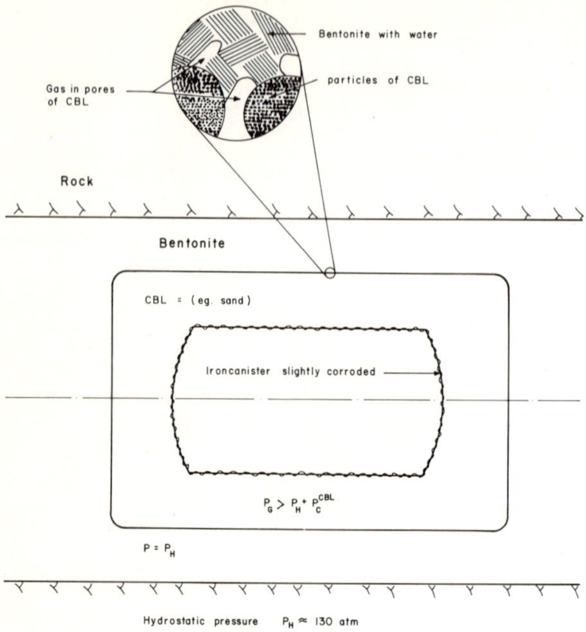

Figure 4: The capillary breaking layer (CBL) has no water in its pores when the gas pressure is higher than the capillary pressure and surrounding hydrostatic pressure.

Impact on corrosion

It must be expected that the corrosion products of the iron are porous and that they at least in some places form very fine pores. In the pores which are smaller than those in the bentonite which contain gas, a liquid water phase may form by capillary condensation. The activity of that water will be slightly less than that in the bentonite, because a driving force (activity difference) is needed to transport it from the bentonite to the canister wall. The transport can take place by diffusion in the gas phase and possibly by surface diffusion on the surfaces of the CBL particles. Furthermore the temperature at the canister surface is somewhat higher than the surrounding, because of heat produced by the radioactive decay of the waste in the canisters. This will also lead to a decrease in the size of the pores where liquid water might form due to capillary condensation. The relative humidity at the canister surface related to the water vapor pressure at the capillary pore opening in the bentonite will be less than 1 due to the above mentioned effects. At lower relative humidities corrosion slows down and at relative humidities less than about 60 % iron corrodes at extremely slow rates even in an oxygen-rich atmosphere. This effect is utilized to stop corrosion by having a low humidity atmosphere where iron objects are kept.

There are, however, several effects working in the same direction. The increased hydrogen gas pressure will tend to decrease the corrosion rate, the capillary breaking layer in the bentonite with the canister and the slight temperature gradient caused by radioactive heating will further decrease the driving force for the chemical reaction. The pore sizes of the magnetite will be considerably larger than those of the bentonite (on average).

At present these effects cannot be quantified due to lack of data but it seems quite conceivable that they combined will lead to a steady state pressure well below the lithostatic pressure.

If the pressure levels off at levels above the hydrostatic pressure but below the gas breakthrough pressure, the hydrogen will escape by diffusion only. In that case the corrosion rate can directly be assessed by the escape rate of the hydrogen. This was previously estimated to be equivalent to a corrosion rate between 0.03 and 0.8 μm/year (2-50 g/year). This low corrosion rate would make the iron in the canister last 0.2 - $5 \cdot 10^6$ years although penetration might occur sooner due to mechanical failure.

Impact on the escape of radionuclides

During the period when there is no liquid water in the capillary breaking layer the radionuclides which have no or very low vapor pressures will have no liquid water in which to migrate from the canister. This applies to practically all species of interest in nuclear waste. There will be some water adsorbed on the surfaces on the CBL but there are many candidate materials available that have very small sorption capacities of water i.e. quartz sand. In such materials the transport capacity for dissolved nuclides will be very small due to the low amount of water as well as due to the small surface of the CBL. This has been chosen with particle sizes so large (i.e. d_p = 0.5 mm) that the capillary forces are negligible compared to the capillary forces in the bentonite. The specific surface of a sand with particle diameter 0.5 mm will be about 10^6 times smaller than that of the bentonite. The amount of water would then be smaller by about the same amount or more, as the water sorbed on the grains is one or a few molecular layers thick only. The transport rate of nuclides in the absorbed layer of water would then become negligible if by no other reason then due to the lack of water.

As long as there is no liquid water present in the CBL, radionuclide transport will in practice be negligible.

DISCUSSION AND CONCLUSION

The thermodynamic data on iron corrosion by water reaction to form magnetite indicate that equilibrium pressures up to near 1000 atm might be expected. The small inaccuracies in the data have quite strong effects on the equilibrium pressure. It may very well be considerably lower.

The projected corrosion rates produce hydrogen at a rate much higher than what can escape by diffusion. Recent gas flow measurements in compacted bentonite indicate that hydrogen may escape by flow with over pressures less than the lithostatic pressure.

The introduction of a capillary breaking layer may considerably slow down the corrosion because it would not allow the water to come in direct contact with the iron. The water would have to be transported to the iron mainly by gas diffusion and must condensate in capillaries finer than those from where it evaporated in the clay. Hygroscopic effects may also cause the water to condensate at the iron surface.

The lack of experimental data does not allow these effects to be quantified at present. If it could be shown to work as projected, the canister life time might be prolonged by several orders of magnitudes and even after failure of the canister no radionuclide release could take place as long as a sufficient corrosion rate is kept up to keep the capillary breaking layer dry.

REFERENCES

1. KBS, "Final storage of spent nuclear fuel, KBS-3"; Swedish nuclear fuel supply co., Division KBS, May 1983.
2. Nagra, "Projekt Gewähr 1985"; Reports NGB 85-03, NGB 85-04, NGB 85-05, 1985.
3. Neretnieks I., "Some aspects of the use of iron canisters in deep lying repositories for nuclear waste." NAGRA technical report 85-35, Baden, Switzerland, Feb 1985.
4. Kahr G., Kraehenbuehl F., Mueller-Vonmoos M., Stoeckli H.F., "Wasseraufnahme und Wasserbewegung in hochverdichtetem Bentonit"; Personal Communication, and to be published as Nagra Technical Report.
5. Pusch R., "Swelling pressure of highly compacted bentonite"; KBS TR 80-13, August 1980.
6. Pusch R., "Gas migration through bentonite clay"; KBS TR 83-71, May 1983.
7. Pusch R., Personal Communication, November 1984.
8. Neretnieks I., Skagius K., "Diffusivity measurements of methane and hydrogen in wet clay (in Swedish)"; KBS TR 56, June 1978.
9. Perry R.H., Chilton C.H., "Chemical Engineers´ Handbook"; McGraw Hill, 1973.
10. Eriksen T., Jakobsson A., "Diffusion of hydrogen, hydrogen sulphide and large molecular weight anions in bentonite"; KBS TR 82-17, 1982.
11. Pusch R., Ranhagen L., Nilsson K., "Gas migration through MX-80 bentonite"; Nagra NTB 85-36, February 1985.
12. Grauer R., "Behältermaterialien für die Endlagerung hochradioaktiver Abfälle. Korrosionschemische Aspekte"; Nagra NTB 84-19, August 1984.
13. Heitz E., Graefen H., "Korrosion und Korrosionsschutz von Endlagerbehältern aus Eisenwerkstoffen unter besonderer Berücksichtigung des schweizerischen Endlagerkonzeptes"; Nagra NTB 84-04, October 1984.
14. Heusler K.E., "Ueberlegungen zur Thermodynamik und Kinetik der Reaktion zwischen Eisen und Wasser im Bentonit"; Nagra NTB 85-22, January 1985.
15. Mattsson E., Personal Communication, Stockholm, November 1984.
16. Grauer R., Personal Communication, November 1984.

EVALUATION OF THE LOCALISED CORROSION OF CARBON STEEL OVERPACKS FOR NUCLEAR WASTE DISPOSAL IN GRANITE ENVIRONMENTS

G. P. MARSH,* K. J. TAYLOR,* I. D. BLAND,* C. WESTCOTT,* P. W. TASKER,** and S. M. SHARLAND**
*Materials Development Division, **Theoretical Physics Division, AERE Harwell, Oxon, OX11 ORA, UK

ABSTRACT

The application of carbon steel corrosion allowance containers for the long term encapsulation of HLW or spent fuel requires data on the likely rates of corrosion attack so that the metal thickness needed to prevent penetration can be estimated. This paper describes a joint mathematical modelling and experimental approach to the evaluation of the rate of localised corrosion.

INTRODUCTION

Most feasibility studies for the disposal of heat generating nuclear waste include provision of a long endurance container system. One approach for achieving this objective is to fabricate overpacks from metals of comparatively low corrosion resistance, which can be used economically in large thicknesses to make allowance for corrosion penetration. This concept is being assessed in the UK with work to evaluate the feasibility of using carbon steel overpacks in granitic or argillaceous environments [1,2].

The UK programme has concentrated on evaluating the rates of both general and localised corrosion (ie pitting, crevice attack) so that the thickness of metal needed to prevent penetration can be estimated. In making these evaluations it has been recognised that long term predictions of corrosion rates can only be justified if they are based on a sound mechanistic understanding of the processes involved. Furthermore, it is considered that the best way to demonstrate this understanding is to apply it to develop mathematical models for the processes. Accordingly, a joint modelling and experimental approach has been adopted, with the experimental studies being aimed at testing the short-term predictions of the models, thereby supporting their application for making longer term predictions. The formulation of a model for the general corrosion of carbon steel overpacks has been described elsewhere [3]. This paper describes progress in a similar study aimed at evaluating the long term propagation of localised corrosion.

EXPERIMENTAL STUDIES

Likely Occurrence of Pitting Corrosion

The main ionic species dissolved in granitic and argillaceous waters are Cl^-, $SO_4^=$, $CO_3^=$, HCO_3^- and OH^-. Depending on the relative concentrations of these ions carbon steel may be subject to general corrosion or may be covered by a protective (ie passivating) oxide film. If the metal does attain a passive condition there is an attendant risk that the protective film may break down locally resulting in more rapid localised corrosion. To determine the chemical and electrochemical conditions under which these different modes of corrosion occur, cyclic polarisation tests were undertaken in electrolytes containing various concentrations of Cl^-,

$CO_3^{=}$ and HCO_3^- (all as sodium salts) at 50°C. The pH of these solutions was fixed by adjustment of the $CO_3^{=}$:HCO_3^- ratio.

The results of these tests, which were conducted with a plain 0.22% C steel, are shown in figure 1 in the form of empirical potential-pH diagrams. These diagrams depict the conditions under which carbon steel is thermodynamically stable, subject to general attack, passive or susceptible to pitting. In general it was found that carbon steel only passivated under comparatively alkaline conditions, and that the potential-pH conditions associated with passivity became increasingly limited as the $CO_3^{=}$ + HCO_3^- level was reduced. For example, in the 0.0001M $CO_3^{=}$/HCO_3^- solution which is representative of a low carbonate groundwater (ie 6.1 ppm HCO_3^-) passivation was not observed over the pH range under investigation. At high $CO_3^{=}$ + HCO_3^- concentrations passivation did occur, and local passive breakdown was induced by Cl^- ions. The electrode potential associated with this breakdown fell with increasing chloride concentration, and increasing the Cl^- content relative to the $CO_3^{=}$ + HCO_3^- level also tended to extend the conditions for general corrosion to higher pH's.

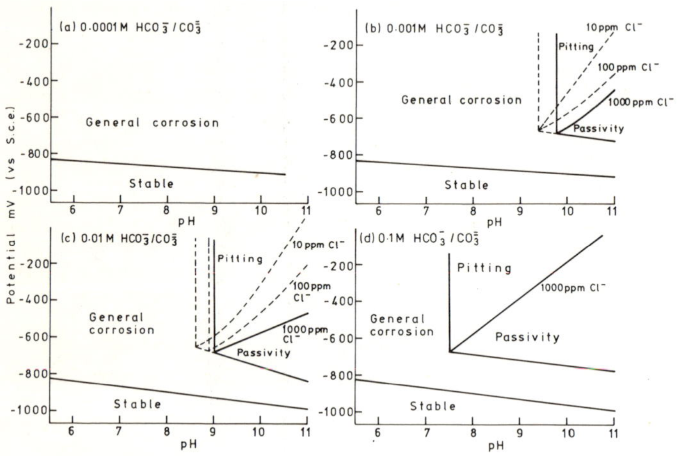

Fig 1 Experimentally determined potential-pH diagrams indicating the conditions under which carbon steel is subject to general corrosion and localised attack at 50°C.

Overall the results indicate that the conditions required to cause localised corrosion are quite specific and are unlikely to be attained in all groundwaters. It is possible, however, that some waters will contain the required balance of aggressive and passivating species. It was therefore concluded that carbon steel overpacks should be designed to make allowance for localised as well as general attack, and this raised the need for data on the propagation rate of localised corrosion sites.

Localised Corrosion Propagation Rates

The experimental study of the rates of localised corrosion has been based on an evaluation of the rate of pit propagation. It is considered that this approach is reasonable, because, while there are differences in the mechanisms for initiation of pitting and crevice attack, it is generally agreed that both propagate by the same acid occluded cell process. The experiments were undertaken in 0.1M $NaHCO_3$ + 1000 ppm Cl^- solution (pH 8.4) at potentials of -200 and 0 mV (NB all potentials are quoted relative to the saturated calomel electrode (SCE)) and 90°C (compare with figure 1(d)). Specimens were fabricated from a plain 0.17% C steel in the form of solid cylinders with the end faces perpendicular to the working direction. These cylinders were screwed into PTFE test cells so that only one end-face (area 8 cm^2) was exposed in an upward facing orientation. By this procedure it was hoped to investigate pit growth under 'worst case' conditions involving downward vertical growth along the preferred paths provided by elongated inclusions in the steel. Tests were carried out in groups of 5 with each group being run for a fixed exposure period of between 500 and 10,000h. The distribution of pit depths was determined by an incremental grinding and counting method.

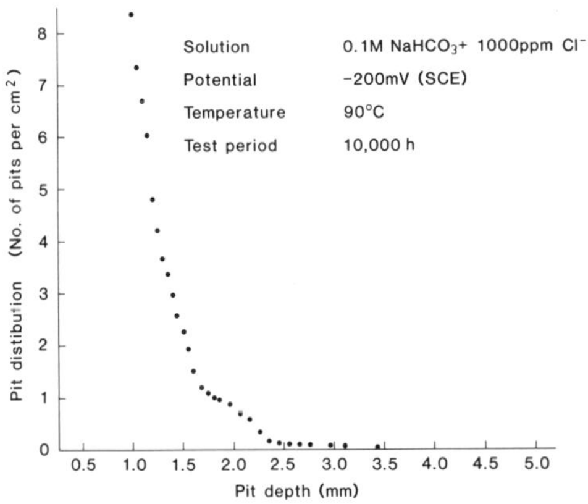

Fig 2 Pit size distribution obtained with carbon steel specimens in 0.1M $NaHCO_3$ + 1000 ppm Cl^- solution at 90°C.

Figure 2 shows the cumulative pit size distribution obtained from the group of 5 experiments conducted at -200 mV for 10,000h, and which is typical of the distributions obtained from all the tests. The figure does not show the distribution below 1.0 mm because below that level the density of pits was so great that it was not always possible to distinguish individual penetrations. The statistical nature of the pit size distribution (it is roughly exponential) indicates that any estimate of the maximum depth of penetration is likely to be dependent on the size of the sample used to make the estimate. Thus the maximum pit depth likely to develop over a set period in an overpack having a surface area of roughly 3 m^2 is likely to be greater than in a specimen of 8 cm^2 surface area. To take account of this area effect the pit depth data obtained in the experimental programme has been analysed using the 'Statistical Theory of Extreme Values' which can be used to extrapolate the small specimen data to estimate the maximum depth of penetration over the surface of an overpack. A full description of this method of data analysis is beyond the scope of this paper, but has been extensively reviewed by Gumbel [3], and descriptions of applications to pitting corrosion have been given by Aziz [4], Eldredge [5] and Marsh [2].

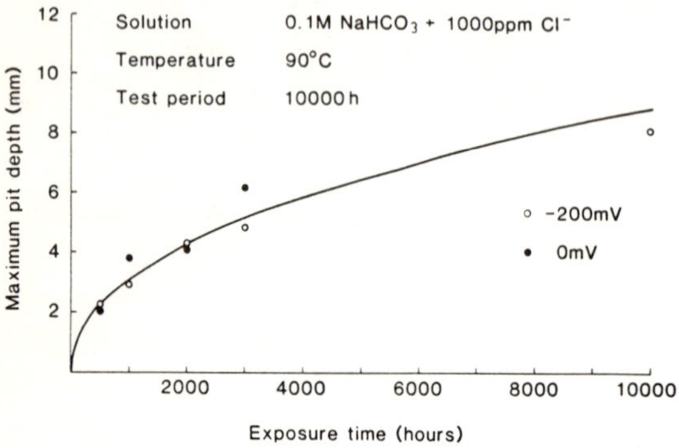

Fig 3 Variation of maximum pit depth with time for carbon steel in 0.1M $NaHCO_3$ + 1000 ppm Cl^- at 90°C.

Figure 3 shows a plot of the maximum pit depths in overpacks, estimated from the extreme value analysis, as a function of time. This plot includes the data from tests at -200 and 0 mV, and it should be noted that, within the scatter of the results, there was no apparent difference in the depths of penetration obtained at these potentials. A linear regression analysis of the logarithms of pit depth and time indicated that these factors were related by the expression

$$P_{max} = 8.35\ T^{0.46} \qquad (1)$$

where P_{max} is the maximum pit depth in mm and T is time in years. This expression had a correlation coefficient with the experimental data of 0.97, and is shown by the full line in figure 3.

Substitution of 1000 years in equation (1) yields a maximum pit penetration of 200 mm. While it is feasible to produce 200 mm thick carbon steel overpacks, this estimate is based on a large extrapolation (NB the maximum pit sizes in the experimental programme were 4-5 mm), and there are mechanistic reasons to believe that it may be over pessimistic. In particular it is by no means clear that the pitting process would favour the uninterrupted propagation of pits to such great depths. Thus the 'IR' drop within deep pits may become so large that, beyond some limit, the conditions may come to favour the initiation of new pits at the metal surface rather than the continued propagation of deep pits. Such phenomenon are not easily studied by real time experiments, and hence work has been initiated to develop a mathematical model of pit propagation in order to examine the growth of deep pits.

MATHEMATICAL MODEL FOR LOCALISED CORROSION

This section describes progress in the formulation of a mathematical model to describe the propagation of localised corrosion sites and gives some preliminary results.

The mathematical model aims to describe the propagation of pitting and/or crevice corrosion, by consideration of the electrochemical reaction kinetics, chemical equilibria and mass transport processes occurring in a localised corrosion site. The localised corrosion site is modelled as a parallel sided slot of variable dimensions.

The formulation of the model is based on the following assumptions:

(a) Changes in conditions within the corrosion site do not affect the external electrode potential of the metal.
(b) Transport in the through thickness direction can be neglected.
(c) The electrolyte within the corrosion site remains static.
(d) The rate of localised corrosion is slow compared to ionic migration rates such that moving boundary effects and induced electrolyte motion may be ignored.
(e) The external electrolyte resistance and the cathodic polarisation resistance of the external metal surface are both assumed to be negligible (ie the localised corrosion current will always be balanced by a cathodic current from the external surface without causing any change in the potential of that surface).
(f) Dilute solution theory can be applied.
(g) Electrochemical reactions are assumed to obey Tafel kinetics.
(h) Na^+ and Cl^- ions are assumed not to take part in any reactions but contribute to current transfer within the corrosion site.

At present the model considers six aqueous species (ie Fe^{2+}, $FeOH^+$, Na^+, Cl^-, H^+ and OH^-) which are produced or consumed by the following reactions:

$$Fe \rightarrow Fe^{2+} + 2e^- \quad (2)$$

$$2H^+ + 2e^- \rightarrow H_2 \quad (3)$$

$$2H_2O + 2e^- \rightarrow H_2 + 2OH^- \quad (4)$$

$$Fe^{2+} + H_2O \rightarrow FeOH^+ + H^+ \quad (5)$$

$$FeOH^+ + H_2O \rightarrow Fe(OH)_2 + H^+ \quad (6)$$

$$H_2O \rightarrow H^+ + OH^- \quad (7)$$

The chemical reactions (5)-(7) are assumed to be at equilibrium at all times, the equilibrium constants being given by the free energies of the reactions.

In dilute solution theory the transport of any aqueous species "i" is governed by the mass conservation equation

$$\frac{\partial C_i}{\partial t} = D_i \nabla^2 C_i + Z_i\, U_i\, \nabla\, (C_i \nabla \phi) + R_i \quad (8)$$

where R_i is the rate of production or consumption of species i, D_i is the diffusion coefficient and U_i is the mobility (ie $U_i = \frac{D_i}{RT}$).

The electrostatic potential of the system is governed by Poisson's equation.

$$\nabla^2\phi = \frac{\rho}{\varepsilon_0} \tag{9}$$

where ρ is the charge density. The large value of ε_o^{-1} ensures that any departure of the system from electroneutrality results in a very large electrical restoring force. This tends to remove charge gradients on a much faster timescale than those associated with the concentration gradient levelling processes. As a reasonable approximation, therefore Poisson's equation may be replaced by the equation of local charge neutrality

$$Z_iC_i = 0 \tag{10}$$

In addition to equation (10) the other boundary conditions of the problem are as follows;

(a) The concentrations of the species are fixed at the crack mouth and are equal to the values in the bulk solution outside the corrosion site.

(b) The flux of species involved in the electrode processes (ie Fe^{2+}, H^+ and OH^-) is proportional to the corresponding current at the base of the corrosion site. The flux of the other species at the base of the corrosion site is zero.

The seven coupled differential equations describing the corrosion process have been solved in the steady state (ie $\frac{dC_i}{dt} = 0$) for sites with corroding and non-corroding walls. An analytic solution of the equations would be extremely difficult since they are highly non-linear in nature. However by non-dimensionalising, rearranging and integrating once, the equations may be written in a form suitable for numerical integration. The method of solution involves making a reasonable estimate for the pH and potential profiles, calculating the various parameters dependent on these, solving the coupled equations and comparing the resultant pH and potential distributions with the estimates. If these do not coincide then the calculated profiles are used for the next iteration. The process is continued until convergence and hence a self-consistent solution is reached. The numerical solver subroutine chosen is a variable order backward differentiation formula method known as Gear's method. This method automatically chooses the stepsize and order of the integration formula. It is especially well suited to this problem since the variables show rapid variation over a very short distance near the pit mouth and a large number of evaluations are required here. A fuller description of the methods used to solve the differential equations has been published [7].

So far application of the model has been limited to an investigation of the steady state corrosion rate and concentrations of aqueous species within a 2 mm deep and 10 µm wide crevice existing in carbon steel immersed in 5×10^{-2}M Cl^- solution (ie 1750 ppm Cl^-) at 90°C and polarised externally to -200 mV (SCE). The diffusion coefficients for all the dissolved species were assumed to be 10^{-9} m^2 sec^{-1} except for H^+ and OH^- which were assigned values of 9.3×10^{-9} and 5.2×10^{-9} m^2 sec^{-1} respectively. The exchange current densities and charge transfer coefficients defining the kinetics of reactions (2) to (4) were those reported by Turnbull [8].

Figure 4 shows the predicted steady state concentrations of aqueous species for a crevice with corroding walls. The H^+, Fe^{2+}, $FeOH^+$ and Cl^- concentrations all increase markedly in the locality of the crevice mouth and then increase more slowly towards the base of the crevice. The model also predicts that the electrode potential at the base of the crevice should be 360 mV more negative than the potential of the external surface.

While the above predictions are generally in line with accepted theory for occluded cell corrosion the model's estimate of the rate of localised corrosion is less realistic. For example the anodic current density from the 2 mm deep and 10 µm wide crevice with corroding walls was 3×10^3 A m^{-2}, which is equivalent to a penetration rate of 3.5×10^3 mm/yr^{-1} (compare with figure 3).

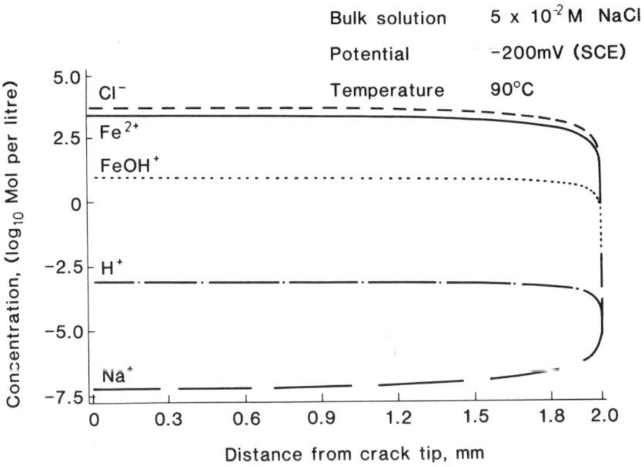

Fig 4 Predicted concentration of reactants and corrosion products within a 2 mm deep and 10 μm wide crevice in a bulk solution of 5×10^{-2}M NaCl.

One factor presently omitted from the model which could in reality limit the corrosion current is the presence of solid ferrous hydroxide within the corrosion site. This will restrict the diffusion and migration of both corrosion products and aggressive species. The significance of this factor has been investigated by scaling down the diffusion coefficients to take account of the proportion of crevice volume filled by solid. This has indicated that with a scaling factor of 10^{-4} the corrosion current density is reduced to 28 A m^{-2}, which is equivalent to a propagation rate of 32.8 mm yr^{-1}. For comparison differentiation of equation 1 gives a propagation rate for a 2 mm deep pit of 20.6 mm/yr.

In summary it is considered that while the comparison between the present model's predictions and experimental results is poor the work does show sufficient promise to merit its continuation. Future work will concentrate on developing a more mechanistically justifiable approach to adjusting diffusion coefficients to take account of the precipitation of $Fe(OH)_2$. Also the effect of some limitation on the amount of cathodic charge produced on the external surface of the metal will be considered since such a constraint may provide an important check on the corrosion current of the system. Electrochemical kinetic data for 90°C will also be obtained since part of the discrepancy between the theoretical and experimental results may be attributable to the use of Turnbull's data which were obtained at 25°C.

CONCLUSION

(a) The environmental conditions which will cause localised corrosion of carbon steel overpacks are quite specific and are unlikely to be attained in all granitic groundwaters. It is possible however that some groundwaters will contain the necessary balance of aggressive (eg Cl^-) and pasivating (eg $CO_3^=$, HCO_3^-, OH^-) ions, and for this reason it has been considered prudent to investigate the kinetics of localised corrosion penetration.

(b) The approach adopted to evaluate localised corrosion is based on a combination of experimental investigation and theoretical analysis.

The experimental results being used to test and hopefully validate a mathematical model for localised corrosion propagation.

(c) Experimental studies of pit propagation lasting for up to 10,000h, have indicated that the maximum penetration in a single overpack should increase with time according to the expression $P = 8.35T^{0.46}$. This indicates a maximum penetration of 200 mm over 1000 years.

(d) The formulation of a mathematical model for localised corrosion in carbon steel has been described and some preliminary results presented. The large difference between the model's predictions and experimental results is thought to arise because the model neglects the blocking effect of $Fe(OH)_2$ precipitated within the local corrosion site. Further work to adapt the model to take account of this factor is in progress.

ACKNOWLEDGEMENT

The financial support of the Department of the Environment and the Commission of the European Communities is gratefully acknowledged.

The work was commissioned by the Department of the Environment as part of its radioactive waste management research programme. The results will be used in the formulation of Government policy, but at this stage they do not necessarily represent Government policy.

REFERENCES

1. G P Marsh, Nuclear Energy, Vol 21, (No 4), 253-265, 1982.

2. G P Marsh, I D Bland, J A Desport, C Naish, C Westcott and K J Taylor, European Applied Res Reports - Nuclear Sci Technol, Vol 5, (No 2), 223-252, 1983.

3. G P Marsh, K J Taylor and A H Harker, Corrosion of Carbon Steel HLW Containers after Geological Disposal, 166th Meeting of the Electrochem Soc, Extended Abstracts, Vol 84-2, pp 403-404, 1984.

4. E J Gumbel, Statistical Theory of Extreme Values and Some Practical Applications, US Dept of Commerce, Applied Mathematics Series 33, 1954.

5. P M Aziz, Corrosion, Vol 12, 495-506t, October 1956.

6. G G Eldredge, Corrosion, Vol 13, 51-60t, January 1957.

7. S M Sharland and P W Tasker, AERE Report TP 1123, June 1985.

8. A Turnbull and M K Gardner, Corros Sci, 22, 661 (1982).

CORROSION RATE OF UNALLOYED STEELS AND CAST IRONS IN REDUCING GRANITIC GROUNDWATERS AND CHLORIDE SOLUTIONS

J.P. SIMPSON*, R. SCHENK* and B. KNECHT**
* Sulzer Bros Ltd., Winterthur, Switzerland
** National Cooperative for the Storage of Radioactive Waste (NAGRA), Baden, Switzerland

ABSTRACT

Measurements of the corrosion rate of unalloyed steel have been made under conditions representative for repositories in the granitic bedrock of Northern Switzerland using two independent methods: (1) Immersion tests in two representative groundwaters, with and without bentonite; (2) Hydrogen evolution measurements in these groundwaters. The immersion tests were carried out at 80°C and 140°C. In both cases the corrosion rate was higher than 50 µm/a as determined at the end of the first observation period of about 500 h. Corrosion rates of under 10 µm/a were estimated after the first 500 h. The corrosion rates were similar in both groundwaters with mineralisations of approx. 3 g/l and 14 g/l respectively, and were generally higher at 80°C than at 140°C. The hydrogen evolution measurements allow an hourly determination of the corrosion rate, with a sensitivity expressed as a corrosion rate of better than 0.1 µm/a. The results generally confirm the observation made in the immersion testing; high corrosion rates were observed over the first few days but then decreased to values well below 10 µm/a. The steady state corrosion rates measured were 1.1 µm/a, 6.5 µm/a, and 2.5 µm/a at 25°C, 50°C, and 80°C respectively in the water with the higher mineralisation. The inverse temperature effect on corrosion rate above 50°C is attributed to a change in the nature of the passive film at higher temperatures.

INTRODUCTION

Cast steel has been selected as reference container material for reprocessed high-level vitrified waste in the current feasibility projects in Switzerland [1]. The choice was justified on the basis of a literature evaluation and an experimental programme [2], [3], [4]. This experimental programme has been extended to improve the data base upon which a corrosion loss for 1'000 years of less than 20 mm was predicted [5]. The present paper presents the results obtained by the following methods:

1) Immersion testing in groundwaters
2) Corrosion of coupons embedded in bentonite saturated with groundwater
3) Instantaneous corrosion rates as derived from hydrogen evolution measurements

EXPERIMENTAL

Immersion Corrosion Testing in Groundwaters

Two groundwaters were used in the investigations. Säckingen water is a natural spring water with relatively high salinity. Böttstein water is a synthetic water based on the results of the analysis of samples obtained from the Swiss granitic bedrock at 1'200 m depth. The compositions of these waters are given in Table I. Deep groundwaters are expected to be oxygen-free, but in the early stages after closure oxygen may be present in a repository. Some tests in Säckingen water were therefore carried out with the oxygen level set at 0.1 µg/g, otherwise the solutions were purged with nitrogen (oxygen concentration less than 0.01 µg/g).

Table I. Composition of test solutions

	Säckingen water mg/l	Böttstein water mg/l
Na^+	900 - 1'070	4'800
K^+	75 - 86	54
Mg^+	12 - 15	3
Ca^{++}	124 - 158	1'100
Cl^-	1'480 - 1'820	8'100
F^-	2.3 - 2.8	3.8
SO_4^{--}	112 - 125	1'820

The cast steel proposed for the reference container is GS 40 [6]. A nodular cast iron, GGG 40 [7], was also included in most of the tests, the results for this material are also reported here for comparison.

Table II. Chemical compositions of the GS 40 and GGG 40 specimens

Material	C	P	S	Si	Mn	Al	Fe
GS 40	0.19	0.014	0.006	0.37	0.74	0.06	bal.
GGG 40	3.58	0.032	0.007	2.26	0.24	0.043	bal.

The corrosion rates measured in Säckingen water containing 0.1 µg/g O_2 decreased with time: at 140°C, from 11 to 2 µm/a for GS 40, and from 37 to 12 µm/a for GGG 40 after 700 and 6'000 hours exposure respectively.

The corrosion rates were higher at 80°C: 86 µm/a was measured for GS 40 after 500 hours reducing to 34 µm/a after 1'440 hours.

The corrosion rates in oxygen-free waters at 80°C, (Table III) show similar trends.

The surfaces of the corrosion coupons were examined after exposure. Macroscopically the coupons were evenly corroded. No pitting was observed. Microscopically the surfaces become roughened. Cross sections show a roughness (peak to valley height) which is typically comparable to the average corrosion rate after 500 - 700 hours, ndependent of the conditions. The roughness does not then increase appreciably with time [3].

Table III. Corrosion rates in oxygen-free waters at 80°C

Test period (h)	Medium	Material	Corrosion rate (μm/a)
0 - 700	Säckingen	GS 40	30
0 - 2'100	Säckingen	GS 40	16
0 - 6'300	Säckingen	GS 40	5
0 - 700	Böttstein	GS 40	42
0 - 2'100	Böttstein	GS 40	14
0 - 6'300	Böttstein	GS 40	10
0 - 700	Säckingen	GGG 40	45
0 - 2'100	Säckingen	GGG 40	8
0 - 6'300	Säckingen	GGG 40	6
0 - 700	Böttstein	GGG 40	25
0 - 2'100	Böttstein	GGG 40	24
0 - 6'300	Böttstein	GGG 40	5

Corrosion Tests in Compacted Bentonite

A bentonite swelling pressure apparatus was used for corrosion testing on specimens embedded in bentonite [3]. This test most closely simulates the conditions expected within the repository. The corrosion coupons were embedded in 110°C air-dried MX-80 bentonite [8]. Two compaction pressures were used to obtain water contents at saturation of approximately 21 % and 30 %.

The corrosion rate was determined by weight loss as for the immersion specimens. The measured corrosion rates are presented in Table IV. No localized attack was observed.

Table IV. Results of corrosion tests in bentonite

Test period (h)	Water content	Corrosion rate (μm/a) Medium			
		Säckingen		Böttstein	
		80°C	140°C	80°C	140°C
0- 816	30	66	40	68	40
	21	26	20	24	25
0-2'160	30	25	14	29	14
	21	13	8	14	12

Hydrogen Evolution Measurements

After an initial period in which oxygen from the air trapped in the repository is present, the conditions in the repository will be oxygen-free. It has been estimated that the oxygen trapped in the repository will under the most conservative assumptions corrode iron or cast steel to a maximum depth of 0.7 mm [5]. This corrosion is assumed to occur instantaneously as soon as water enters the repository. In the absence of oxygen, iron may be oxidised by the hydrogen ion, equation (1), or directly by water, equation (2).

$$Fe + 2H^+ \rightarrow Fe^{++} + H_2 \qquad (1)$$

$$Fe + 2H_2O \rightarrow Fe(OH)_2 + H_2 \qquad (2)$$

These reactions are thermodynamically equivalent, but follow different reaction paths and therefore differ in their kinetics. The reaction with the hydrogen ion is fast but at the pH values in the repository (pH 8-9) the concentration of H^+ is so low (10^{-8} to 10^{-9} m/l) that the direct reaction with water most probably dominates. In either case one gram atom of iron produces 1 gram molecule of hydrogen. This relationship forms the basis for the measurement of the corrosion rate via the hydrogen evolution rate.

In both cases iron is oxidised to its divalent oxidation state. Thermodynamically Fe_3O_4 is stable under repository conditions; Fe_3O_4 can form directly via the reaction:

$$3Fe + 4H_2O \rightarrow Fe_3O_4 + 4H_2 \qquad (3)$$

or by the conversion of $Fe(OH)_2$:

$$3Fe(OH)_2 \rightarrow Fe_3O_4 + 2H_2O + H_2 \qquad (4)$$

For the purposes of the conversion of the hydrogen evolution rate into a corrosion rate it was assumed that iron was only oxidised to Fe(II). Oxidation to Fe_3O_4 would give lower corrosion rates than those quoted for the same amount of hydrogen evolved.

The apparatus used is shown schematically in Figure 1: a nitrogen gas stream is passed over the corrosion cells where it picks up hydrogen evolved by the corrosion of the carbon steel sheet specimens (DIN Material no. 1.0333: 0.10 % C; 0.3 % Mn; 0.1 % Si; N, P, S less than 0.007 %). The hydrogen concentration is measured by gas chromatography (GC). The sensitivity is dependent upon the gas flow rate, the surface area of the specimens and the detection limit for hydrogen of the gas chromatograph. The apparatus in its present form has a detection limit corresponding to a corrosion rate of approximately 0.02 µm/a, corrosion rates down to 0.1 µm/a are thus readily measured. A more detailed description of the method is given in [9].

An advantage of the method over weight loss determinations is the in situ direct determination of the "instantaneous" corrosion rate; the 90 % response time to a step change in corrosion rate is about 2 hours. It should be noted that hydrogen evolution measurements could be used for calibration in electrochemical methods such as polarisation resistance measurements, thus obviating the need for weight loss determinations, which at the corrosion rates typically observed require long exposure times.

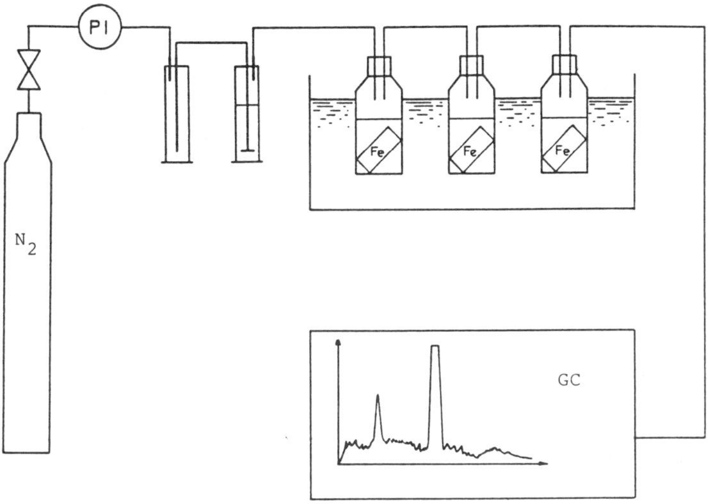

Figure 1. Determination of corrosion rate by the measurement of hydrogen concentration in a nitrogen gas stream

Figures 2 and 3 show typical hydrogen evolution vs. time plots. The spike in Figure 3 is due to local corrosion; in principle the method can detect this form of corrosion, but ideally should only be used for the measurement of general corrosion rates. The area under the curves can be integrated to give the total corrosion loss, which can be compared with the weight loss values. These are in good agreement in all the cases where local corrosion does not occur. In general, steady state corrosion rates were established after 2 - 3 days. The experiments were run for 7 - 10 days.

The results in Tables V and VI are the steady state corrosion rates established after the first 2 - 3 days.

Table V. Corrosion rate as derived from hydrogen evolution measurements on carbon steel sheet

Temperature (°C)	Böttstein water (μm/a)	Säckingen water (μm/a)
25	1.1	1.5
50	6.5	3.2
80	2.5	1.5

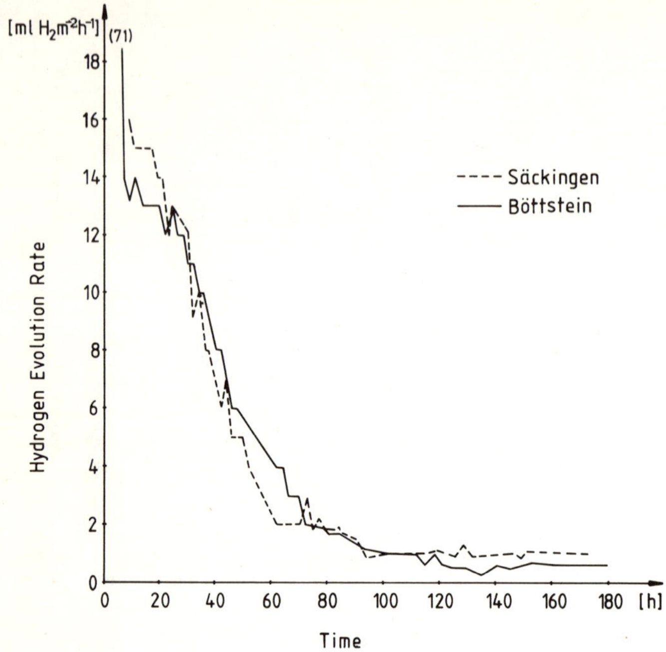

Figure 2. Hydrogen evolution rate as a function of time on plain carbon steel in Säckingen and Böttstein water at 80°C

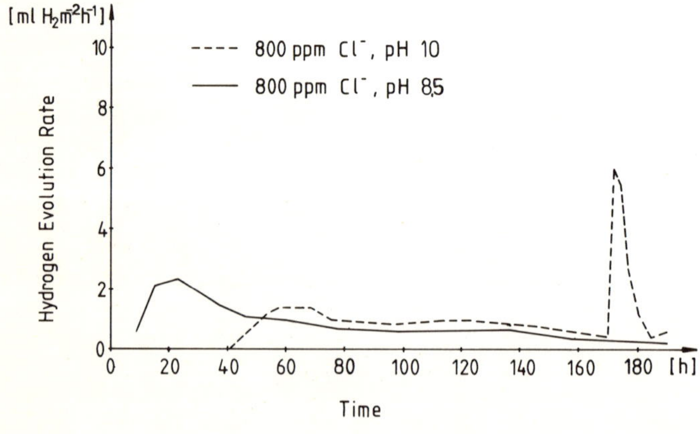

Figure 3. Hydrogen evolution rate as a function of time on plain carbon steel in 800 μg/g Cl^- solutions buffered with $HCO_3^-/CO_3^{=}$ at 80°C

Table VI. Corrosion rates, hydrogen evolution method for chloride media unbuffered pH 7, and buffered with 0.1 % $NaHCO_3/Na_2CO_3$ for pH 8.5 and 10 at 80°C (L) = local corrosion

Chloride (µg/g)	Corrosion rate µm/a pH = 7	pH = 8.5	pH = 10.0
0	0.9	0.3	0.5
80	3.8	1.5 (L)	2.5 (L)
800	5.0	0.5 (L)	1.0 (L)
8'000	0.8	2.5	0.5

DISCUSSION

There are two basic requirements which an unalloyed steel must fulfil with regard to corrosion before it can be used as a container material. These are that it should not suffer from local forms of corrosion such as pitting or environmental cracking and that the general corrosion rate should be acceptably low.

Environmental cracking has not yet been studied in detail under Swiss conditions; it is, however, not expected to occur. Examination of the corrosion coupons has revealed no pitting or excessive local attack on any of the specimens exposed to either of the reference groundwaters.

The observation of pitting in the chloride/carbonate solutions is in agreement with Marsh et al. [10], and demonstrates that this form of corrosion cannot be ignored in such media.

The experiments reported here were primarily intended to enable an upper limit to be set for the general corrosion rate.

Both the immersion corrosion tests and those in bentonite show a reduction in corrosion rate with increasing time. The hydrogen evolution method shows very high corrosion rates initially in both Säckingen and Böttstein water (Figure 2). The corrosion rate then decreases and reaches a steady value within a few days. Immersion test values are integral values and include this initial burst of corrosion; its contribution to the total weight loss decreases with time and given long enough exposure periods the corrosion rates from these tests should approach those obtained by the hydrogen evolution method.

More results are necessary for a detailed analysis of the corrosion behaviour but some general conclusions may be drawn:

1) The long-term corrosion rate to be expected is probably well below 10 µm/a.

2) The corrosion rate does not increase monotonically with increasing temperature. The corrosion rate was observed to be lower at 140°C than at 80°C in all the immersion and bentonite experiments. The hydrogen evolution method showed that the corrosion rate reaches a maximum below 80°C.

Magnetite formation is usually held responsible for the known reduction in corrosion rate above 100°C. A reduction in corrosion rate with temperature indicates a change in mechanism. Whether the change below 80°C is due to magnetite formation at a lower temperature than usually quoted remains to be investigated.

3) Bentonite has no large influence on the corrosion kinetics. The corrosion rates in bentonite are comparable for those in the waters at the same temperatures. The more highly compacted bentonite gave lower values, but this may be due to the fact that the amounts of air trapped in the bentonite decrease with increasing compaction.

The results reported in this paper confirm that unalloyed steels (cast or wrought) and cast iron are suitable waste container materials for final disposal in a crystalline bedrock with significant groundwater salinity. The corrosion rates measured are in agreement with those found by Marsh et al. [10] under comparable conditions. A waste container design based on the use of cast steel is presented in [11],[12]. Further work is planned, in particular electrochemical studies (with emphasis on local corrosion effects) and field testing in geological formations.

LIST OF REFERENCES

1. Nagra Gewähr Report NGB 85-04, Baden, Switzerland, 1985

2. Simpson J.P., Nagra Technical Report Series NTB 83-05, Baden, Switzerland, 1983

3. Simpson J.P., Nagra Technical Report Series NTB 84-01, Nagra, Switzerland, 1984

4. Grauer R., Nagra Technical Report Series NTB 84-19, Baden, Switzerland, 1984

5. Nagra Technical Report Series NTB 84-32, Baden, Switzerland, 1984

6. Swiss standard VSM 10697. Verband Schweizerischer Maschinenindustrieller, 1976

7. DIN Standard 1693. Deutsches Institut für Normung, 1973

8. Müller-von Moos M., Kahr G., Nagra Technical Report Series NTB 83-12, Baden, Switzerland, 1983

9. Schenk R., Nagra Technical Report Series NTB 83-16, Baden, Switzerland, 1983

10. Marsh G.P., Bland I.W., Desport J.A., Naish C., Westcott C., Taylor K.J., European Appl. Res. Rept.-Nucl. Sci. Technol. **5** (2), 223-252, 1983

11. STEAG Kernenergie GmbH, Motor-Columbus Ingenieurunternehmung AG, Nagra Technical Report Series NTB 84-31, Baden, Switzerland, 1984

12. Wick W.: Final storage container designs for highly radioactive waste. Poster paper presented at this conference and to be published.

CORROSION BEHAVIOR OF NODULAR CAST IRON CASKS FOR LOW AND INTERMEDIATE LEVEL WASTES

W. HAUSER AND R. KÖSTER
Kernforschungszentrum Karlsruhe GmbH, Postfach 3640, D-7500 Karlsruhe, Federal Republic of Germany

ABSTRACT

In applying new conditioning methods dehydrated wastes (e.g. concentrates, ion exchange resins) from operation of nuclear power plants and wastes from decommissioning (e.g. highly-activated core components) are conditioned using high integrity nodular cast iron casks without further solidification. Therefore more stringent requirements have to be made for the packaging material. The objective of the investigations was to evaluate the corrosion resistance of the cask material. For this purpose chemical corrosion tests were made applying corrosion media actually considered or postulated in accident scenarios during interim storage or after disposal in the abandoned iron ore mine Konrad or in a salt formation. The conclusion from the corrosion experiments, which have been performed up to now for the basic packaging material, indicates that nodular cast iron casks with wall thicknesses about 200 mm seem to ensure good protection for wastes especially from nuclear power plants with relatively short-lived radionuclides.

INTRODUCTION

At the present licensing status of repositories for radioactive wastes in the Federal Republic of Germany a period of 10 to 20 years must be anticipated for interim storage of low and intermediate level wastes. Therefore the use of volume effective conditioning and packaging techniques has a high priority. In the last few years several techniques have become available for conditioning these wastes while strictly reducing their volumes [1]. In these methods the dehydrated wastes (water content ~10-50 wt.%) are conditioned using high integrity casks without further solidification. Therefore more stringent requirements have to be made with respect to the corrosion resist ance of the packaging during interim storage above ground and for an adequate period in the repository.

Mainly for the wastes with the main activity carrier Co-60 from operaation and from decommissioning of nuclear power plants (e.g. concentrates, ion exchange resins, highly-activated core-components) high intergrity casks will be used. They are manufactured of nodular cast iron having a wall thickness between 150 and 215 mm to ensure radiation shielding. As an example of a typical cask design Fig. 1 gives an overview of a nodular cast iron cask for intermediate level wastes used at the Karlsruhe Nuclear Research Center.

In the scope of our activities to assess the barrier properties of waste packages, we previously investigated the corrosion behavior of 200 l steel drums [2]. The objective of the investigations presented here was to evaluate the corrosion resistance of the cask material. For this purpose chemical corrosion testes were made applying different corrosion media actually considered or postulated in accident scenarios during interim storage or after disposal in the abondoned iron ore mine Konrad or in a salt formation. In this work the effort was concentrated to investigate the corrosion behavior of the basic canister material. The investigations quatifying the corrosion resistance in the region of the canister lid are still running.

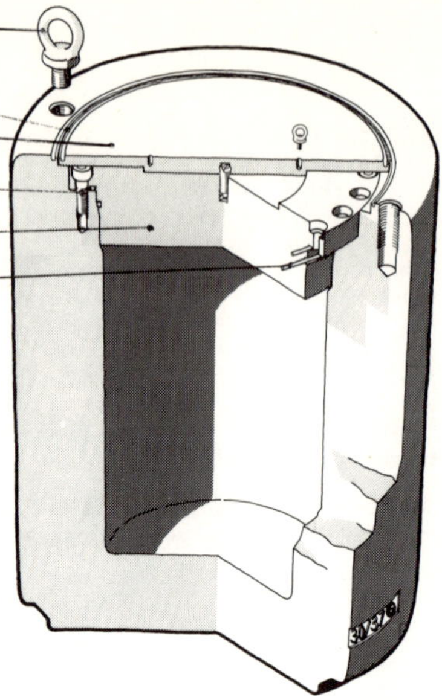

Fig. 1

Nodular cast iron cask for intermediate level wastes used at the Karlsruhe Nuclear Research Center.

Cask weight: 6420 Kg
Net volume : 0.27 m³

EXPERIMENTAL PROCEDURE

The canister material GGG 40 (DIN material No. 0.7040) was used in the investigations. It is a cast iron containing graphite in a spheroidal form to improve high mechanical stability. From cast iron plates with the composition (C: 3.47, Si: 1.98, Mn: 0.18, P: 0.013, Mg: 0.04 wt.%, base Fe) corrosion samples have been cut out with the dimensions 10x20x40 mm and with mechanically cleaned surface. Three samples were used for each corrosion medium and and each test duration.

The investigations have been carried out using test conditions which simulate internal and external canister corrosion. We used uncoated specimens. Samples for testing internal corrosion have been stored in closed glass-containers with simulated wastes. In the experiments the realistic ratio of metal surface to waste volume was applied.

In Table I the used simulated waste types and corresponding test conditions are summarized. Corrosion of cast iron was measured in contact with cemented waste forms (solid and liquid wastes) as well as in contact with unsolidified wastes like ion exchange resins and evaporator concentrates. These wastes were directly filled into the casks and were dehydrated by special techniques [1] to water contents about 10 to 50 wt.% .

For description of external canister corrosion the test conditions in Table II were applied corresponding to interim storage above ground and to disposal in geological formations characterized in detail here. In case of disposal a distinction was made between disposing of the canisters in ventilated or non-ventilated galleries and contact of the canisters with salt brines as postulated in accident scenarios. Nearly all of the tests were made under in-situ conditions. Only the contact of the cask material with salt brines was simulated in closed glass containers.

Tab. I Test Conditions for Simulation of Internal Canister Corrosion by the Waste Form

Waste Forms	Test Conditions	
	Temperatures	Corrosion Media
Cemented Waste Form (OPC 35F, W/C=0.4) for for non heat-producing wastes	22°C, 50°C	aquous solution of PZ 35F, pH ~12, O_2-content ~5 mg/l
Cemented Waste Form (PZ 35F, W/C=0.4) for heat-producing wastes	90°C	as above
Ion Exchange Resins Lewatit S100/M500, LiOH-/H_3BO_3-loaded, water content ~50 wt.%	22°C, 50°C	inactive waste simulate
Evaporator Concentrate from PWR, water content ~10 wt.%	22°C, 50°C	inactive waste simulate (1)

(1) Simulated PWR Evaporator Concentrate :
B 60000 ppm, Cl^- 800 mg/l, PO_4^{3-} 7 g/l, SO_4^{2-} 19.5 g/l,
NO_3^- 300 mg/l, SiO_3^{2-} 240 mg/l,
Na^+ 50 g/l, Fe^{2+} 1.8 g/l, Ca^{2+} 1.0 g/l, Mn^{7+} 500 mg/l,
Mg^{2+} 700 mg/l, NH_4^+ 10 g/l

Tab. II Test Conditions for Simulation of External Canister Corrosion by expected Storage Conditions

Storage Conditions	Test Conditions	
	Temperatures	Corrosion Media
Interim Storage (roof covered air space)	averaged over one year ~9°C	in situ-test at KfK-site rel. humidity ~80 %
Disposal in a Salt Dome :		
* ventilated gallery	30°C	in situ-test at -775 m level of Asse Salt Mine, samples covered with crushed rock salt rel. humidity ~30 %
* quinary salt brine	50°C, 90°C	saturated salt brine (1), pH ~5, O_2-content ~3 mg/l
Disposal at Iron Ore Mine Konrad :		
* ventilated gallery	25°C	in situ-test at -1250 m level of Konrad Mine, samples in contact with crushed iron ore, rel. humidity ~35 %
* non-ventilated gallery	42°C	as above, but rel. humidity ~80 %
* deep ground water from Konrad Mine	50°C, 90°C	deep ground water (2), O_2-content ~5 mg/l, pH ~6

(1) Quinary salt brine (Type Q) :
Mg^{2+} 87 g/l, Na^+ 7.6 g/l, K^+ 24.3 g/l,
Cl^- 270.2 g/l, SO_4^{2-} 23.7 g/l

(2) Main components of deep ground water from Konrad Mine :
Mg^{2+} 2.38 g/l, Na^+ 64.5 g/l, Ca^{2+} 13.5 g/l,
Cl^- 130 g/l, SO_4^{2-} 0.73 g/l

After 3, 6 and 12 months test duration, the samples were removed from the corrosion media and were evaluated using familiar corrosion test methods [3]. This was done by the following steps: mechanical and chemical cleaning of the samples, gravimetric determination of mass reduction and calculation of thickness reduction and general corrosion rate. In order to get additional data about non-uniform corrosion like pitting attack and shallow pit formation surface profiles, microscopic surface measurements, micrographs and scanning electron micrographs were recorded for evaluation of maximum corrosion dephts.

RESULTS AND DISCUSSION

Interpretation of Corrosion Experiments

In general, the investigations with this canister material revealed a pre-dominant uniform corrosion mechanism which in some cases was overlapped by pitting attack or shallow pit formation. As an example Fig. 2 illustrates the corrosion attack caused by ion exchange resins with a water content of ~50 wt.% after 95 days exposure time. Theses micrographs show an evident additional pitting attack. In the pits spheroidal graphite has been released from the cast material by dissolution of the surrounding iron. The same process occurs on the whole surface material in all media. In this example an additional crevice corrosion effect between ion exchange resin spheres and metal surface causes a significant amount of pits on the canister material.

(a)

(b)

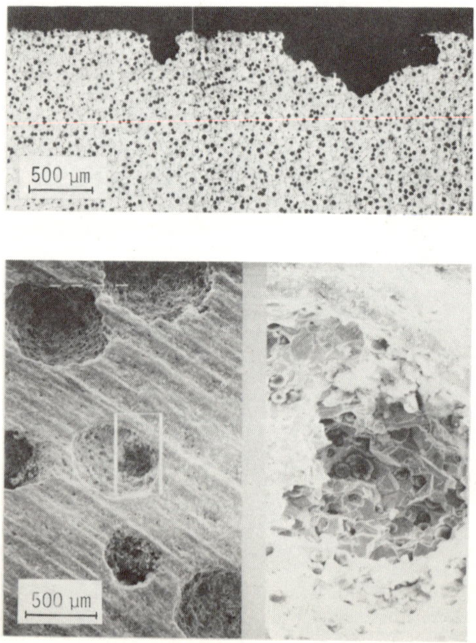

Fig. 2

Micrograph (a) and scanning electron micrograph (b) of corroded areas on a GGG 40 test specimen simulating internal canister corrosion by ion exchange resins (temp. 50°C, exposure time 95 days).

In order to get a complete estimation, corrosion depths by uniform corrosion and pitting corrosion were transferred into corrosion rates. Maximum pit depths were used to get a conservative extrapolation. Fig. 3 summarizes the data of internal corrosion evaluated for a totally filled canister. In

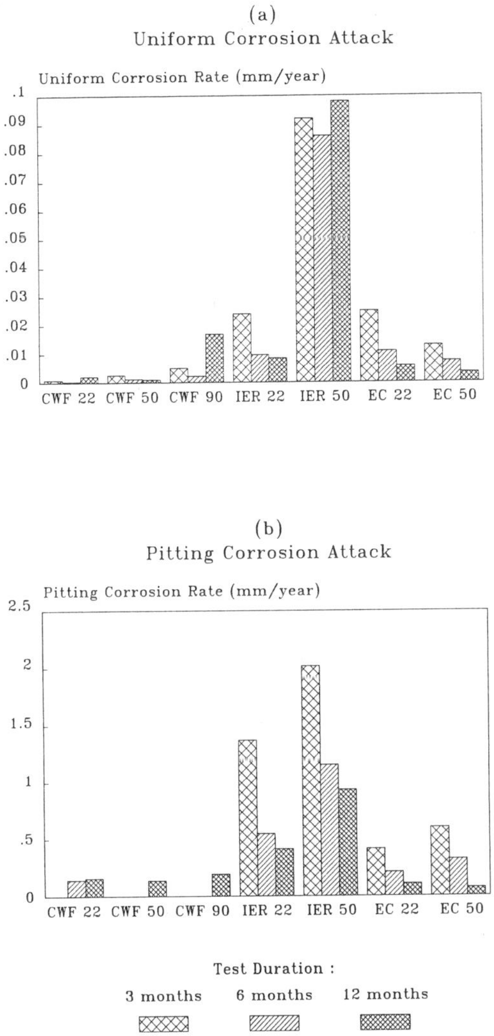

Fig. 3 Linear extrapolated corrosion rates for the internal side of a GGG 40 canister wall (canister totally filled). Simulated waste forms:

CWF 22: cemented waste form, temp. 22°C, CWF 50: cemented waste form, temp. 50°C, CWF 90: cemented waste form, temp. 90°C, IER 22: ion exchange resins, temp. 22°C, IER 50: ion exchange resins, temp. 50°C, EC 22: evaporator concentrate, temp. 22°C, EC 50: evaporator concentrate, temp. 50°C case of cemented wastes the passivation of the metal surface at high pH-values [4] causes a very small uniform corrosion rate of about 10^{-3} mm/y and only little pitting corrosion. For heat producing wastes which will cause a maximum temperature up to 90°C in the German disposal concept, the uniform corrosion rate raises to 10^{-2} mm/y. The most significant corrosion rates for the inner side of the canister were measured after exposure in ion exchange resins at 50°C because of the additional crevice effect. Provided there is sufficient supply of oxygen, a uniform corrosion rate of 0.1 mm/y will occur. This rate seems to be constant with time while the pitting corrosion rate is going down. A decreasing uniform corrosion rate with time was observed in evaporator concentrate (water content ~10 wt.%).

Fig. 4 is showing a uniform corrosion rate of about 0.01 mm/y for the external corrosion of a canister under the assumed interim storage conditions above ground. This rate decreases with increasing time as was observed for pitting corrosion. Under disposal conditions in salt or in ventilated galleries in an iron ore mine the observed corrosion rates were $<10^{-3}$ mm/y. The corresponding measurements were performed by contacting the canister material with crushed salt or crushed iron ore implying definite water contents. The same corrosion rates will occur after closing the disposal galleries of a repository in salt. For the selected iron ore mine conditions higher uniform corrosion rates of about 0.6 mm/y and pitting corrosion rates of 1.7 mm/y were measured. These tests were made in a closed gallery with a relative humidity of ~80 % at 42°C on the -1250 m level of the Konrad Mine. Nearly uniform corrosion rates of 0.01 mm/y and 0.02 mm/y were found after contacting the corrosion samples with saturated salt brine (Type Q) and deep ground water with low salt content from Konrad. By comparing these results with corrosion data measured in sea water [5], corrosion rates lower by a factor of 10 can be expected in salt brines. This seems to be due to the low oxygen content of 3-5 mg/l in these brines.

Extrapolation of Corrosion Behavior

A rough extrapolation of corrosion data beyond the test duration will only be possible for the canister wall material. Furtheron the simplifying assumption of constant conditions of corrosion has to be made. This means that a change with time in corrosion chemistry, caused for example by saturation effects or total consumption of oxygen, will not be considered.

After determining the type of corrosion dominant for longer periods of time, the lifetimes of the canister wall were estimated for a canister with a wall thickness of 200 mm corresponding to Fig. 3 and Fig. 4. All possible combinations of internal and external corrosion were considered. This estimation gave lifetimes limited by corrosion of about 10^3 years. In case of disposal in closed galleries in the Konrad Mine with a relatively high degree of humidity, shorter lifetimes of about 300 to 400 years can be concluded. This time span seems to be large enough especially for wastes from operation of nuclear power plants because the specific activity will have decayed by a factor of 10^3 to 10^4 at that time.

CONCLUSIONS

In order to describe the corrosion behavior of the canister material GGG 40, a common nodular cast iron material for high intergrity waste casks, chemical corrosion tests were performed using corrosion media which are expected to contact the canisters. Corrosion attack to a canister wall was simulated using cast iron samples for a test period of about one year. The investigations were based on different wastes (e.g. from the operation of nuclear power plants), on interim storage above ground, and on disposal accord-

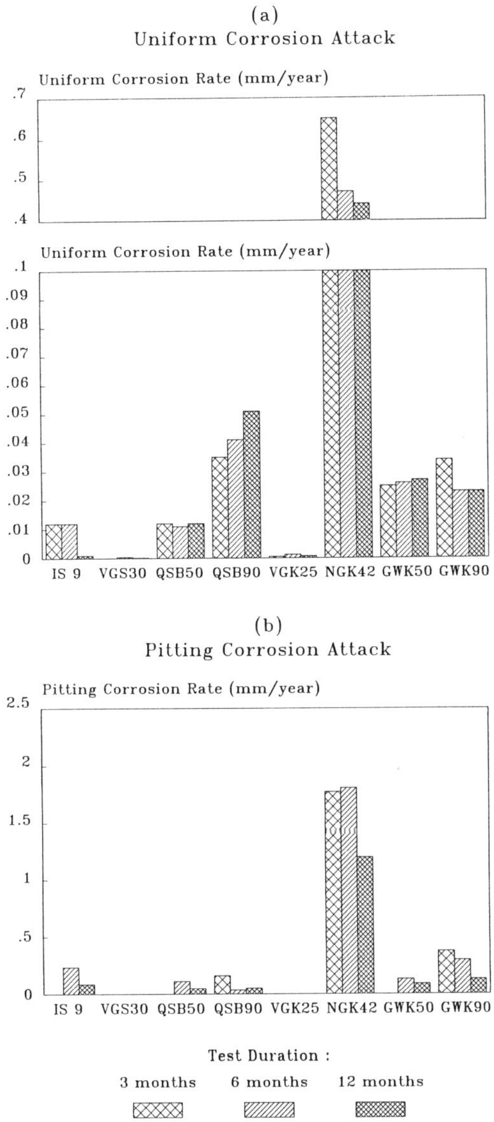

Fig. 4 Linear extrapolated corrosion rates for the external side of a GGG 40 canister wall. Storage conditions:

IS 9: interim storage above ground, temp. 9°C, VGS30: ventilated gallery in salt, temp. 30°C, QSB50: quinary salt brine, temp. 50°C, QSB90: quinary salt brine, temp. 90°C, VGK25: ventilated gallery in Konrad, temp. 25°C, NGK42: non-ventilated gallery in Konrad, temp. 42°C, GWK50: deep ground water from Konrad, temp. 50°C, GWK90: deep ground water from Konrad, temp. 90°C ing to the German disposal concept in a salt dome or in the abandoned iron ore mine Konrad.

With respect to internal corrosion attack, ion exchange resins with a water content of ~50 wt.% reveal the highest corrosion rate of 0.1 mm/y. In contrast cemented wastes cause a corrosion rate less by two orders of magnitude due to high pH-value.

In case of canister corrosion on the outside, dry storage conditions like roof covered interim storage above ground, disposal in a salt dome or in a ventilated gallery in Konrad Mine give a uniform corrosion rate of 0.01 mm/y. A corrosion rate of 0.03 mm/y was measured for contact of the canisters with salt brine. Test conditions prevailing in a closed gallery with a relative humidity of ~80 % at the -1250 m level of Konrad Mine caused a significant higher corrosion rate of 0.6 mm/y.

Based on these corrosion data for the canister material, an time extrapolation for a whole cask can only be made applying simple boundary conditions. The conclusion from the corrosion experiments which have been performed up to now, indicates that nodular cast iron casks with relatively high wall thicknesses seem to ensure good protection for wastes especially from the operation of nuclear power plants with relatively short-lived radionuclides.

REFERENCES

1. G. Peffrath, H.J. Schroeder
Water Chemistry 3. BNES, London 1984, pp. 131-136

2. W. Hauser, E. Smailos, R. Köster
Proc. of the Int. IAEA Symp., Utrecht, June 21-25, 1982, IAEA, Vienna 1983, pp. 283-292

3. DIN 50 905, Corrosion of Metals, Chemical Corrosion Tests, Beuth Verlag GmbH, Berlin (1975)

4. W. Stelzel
Dissertation, Technische Hochschule München (1963)

5. P. Drodten
Stahl und Eisen 90, Nr. 19 (1970), pp. 1038-1042

6. D. Rittscher et al.
Proc. of the Int. Symp. on Packaging and Transportation of Radioactive Materials, New Orleans, May 15-20, 1983
Oak Ridge (1983), pp. 215-223

HLW CONTAINER CORROSION IN GEOLOGICAL DISPOSAL CONDITIONS

MAX HELIE AND GEORGES PLANTE
Commissariat à l'Energie Atomique, IRDI-DERDCA-DCAEA-SCECF-SECA, Centre d'Etudes Nucléaires de Fontenay-aux-Roses, BP 6, 92265 Fontenay-aux-Roses Cedex, FRANCE.

INTRODUCTION

In order to prevent a possible return of activity to the biosphere, the current concept for High Level Waste disposal is to build a multi barrier system consisting of a slow leaching waste form, the waste container, backfilling materials and the geological formation. The main threat to the integrity of the container is aqueous corrosion and two concepts are currently accepted, one based on a thin "corrosion resistant" overpack, the other on a thick "corrosion allowance" canister.

In both cases however, further information is required about the corrosion behaviour of the candidate materials in order to fully demonstrate suitability.

This paper deals with the work carried out by the CEA and comprises two parts. One part concerns the "corrosion resistant" overpack concept and presents some results for the nickel-chromium alloys 625, C 276, C 4 and titanium 0.2 % palladium alloy whose localized forms of corrosion have been studied in solutions representative of three possible host formations, viz. granite, salt domes and clay. The other part, concerning the "corrosion allowance" overpack concept, is mainly focused on the corrosion kinetics of mild steel in deaerated water. However, some data on hydrogen embrittlement are also presented.

EXPERIMENTAL

All the samples were machined from sheets or rods in the as received conditions. In some cases, they were submitted to a thermal treatment prior to testing in order to simulate welding conditions. The chemical analyses of the tested alloys are given in table I.

Corrosion resistant overpack

The aim of the study was to determine for a given alloy the acceptable environmental conditions in order to compare the results with the worst forecast disposal conditions, i.e. immersion of the containers in flowing ground water. It should be noted that the composition of the water present in the geological formation before waste disposal does not correspond to the one of the water in contact with the containers. Besides the effect of the radiolysis and the presence of a backfilling material, the slow water renewal can locally lead to very important modifications of its composition. The solutions used in this study [1] are presented in table II. Moreover, the granitic solutions were in most cases acidified to simulate the local conditions of a crevice for instance.

The temperature of the containers being dependent on the duration of the intermediate storage period, the tests were carried out at both 90 and 170°C. The candidate alloys were studied with regard to pitting, crevice corrosion, stress corrosion cracking and, when relevant, hydrogen embrittlement.

Pitting and crevice corrosion were analyzed by means of polarization curves in granitic solutions whose pH at room temperature ranged from 1.3 to 8.5. Forward-backward curves were carried out starting from the

Table I . Chemical analysis of tested alloys (wt %, - means lower than 0.01)

	Fe	Cr	Ni	Mo	C	Si	Mn	Ti	Al	Nb	P	Co	W	V	Cu	Sn	Pd	O_2	N_2	Notes
625	3.95	22.1	60.8	8.88	0.02	0.26	0.12	0.21	0.18	3.43	0.01	-	-	-	-	-	-	-	-	a
C 276	6.2	15.7	56.7	14.8	0.01	0.15	0.39	-	0.3	-	-	2.2	4.0	-	-	-	-	-	-	b
C 276	6.0	14.9	57	16	-	-	0.45	0.1	0.3	-	-	2.3	3.6	0.15	0.2	-	-	-	-	c
C 4	0.50	15.6	68	15.4	-	0.03	0.22	0.11	-	-	-	0.1	-	-	-	-	-	-	-	a
Ti - 0.2 % Pd	0.06	-	-	-	0.01	-	-	bal.	0.01	-	-	-	-	0.03	-	0.07	0.16	0.14	0.01	a
Mild steel	bal.	0.04	0.03	-	0.11	0.27	0.66	-	0.04	-	0.02	-	-	-	0.01	-	-	-	0.01	a

Notes : a : all tests - b : all tests excepted S.C.C. - c : S.C.C. tests

Table II : Testing solutions

Granitic solutions		Argillaceous solutions		Q-brine
SG 2 ($mg.l^{-1}$)	SG 3 ($mg.l^{-1}$)	Oxidized ($g.l^{-1}$)	Natural ($mg.l^{-1}$)	(weight % at 55°C)
Cl^- : 17700	Cl^- : 1990	Na_2SO_4 : 13.916	Na_2SO_4 : 39.8	$MgCl_2$: 26.8
SO_4^{--} : 24	SO_4^{--} : 24		$MgSO_4$: 15.0	KCl : 4.7
HCO_3^- : 244	HCO_3^- : 244	K_2SO_4 : 0.870	NaF : 10.1	$MgSO_4$: 1.4
F^- : 1.9	F^- : 1.9		Na_2CO_3 : 74.2	NaCl : 1.4
SiO_3^- : 19	SiO_3^- : 19	$MgSO_4$: 19.190	$NaHCO_3$: 1344.5	H_2O : 65.7
Na^+ : 11600	Na^+ : 1200		KCl : 39.4	
Ca^{++} : 20	Ca^{++} : 20	$CaSO_4$: 3.606	NaCl : 19.9	(addition of NaCl above
Mg^{++} : 6.1	Mg^{++} : 6.1		Saturated with $CaCO_3$	55°C for saturation)

corrosion potential to assess pitting susceptibility while in the case of crevice corrosion the experiments were run starting with the alloy in the active state, which was achieved by a cathodic pre-polarization of the sample. In both cases the scan rate was 10 mV/min.

Slow strain rate experiments were carried out to study stress corrosion cracking and hydrogen embrittlement, using the method which has been developed by R.N. Parkins [2]. The tests were run in granitic solutions, Q.brine and argillaceous solutions at both 90 and 170°C. The potential of the electrode was controlled using a potentiostat, a platinum counter electrode and a calomel reference electrode.

Corrosion allowance overpack

When using a "corrosion resistant" alloy (i.e. Ti Pd, C 276) an attempt can be made to base a safety demonstration on intrinsic properties of the materials, regardless of the protection provided by the other engineered barriers. Considering the case of iron or copper, the oxygen content of ground waters leads to high corrosion rates which prevent the canister from being considered as an independent barrier.

The Swedish assessment for copper canisters [3] is based on the two following points :

- Oxidizing species transported by water or generated by the radiolysis are present in limited quantities.
- Water by itself cannot be considered as an oxidizing agent for copper.

The first point depends mainly on the qualities of the buffer material and on the thickness of the canister walls.

The second point, which derives from thermodynamic considerations, is not true in the case of mild steel. Therefore some experiments needed to be carried out on mild steel samples to determine the corrosion due to water regardless of other oxidizing species, the localized forms of corrosion such as pitting being studied by other workers [4].

General corrosion was studied by means of polarization resistance measurements in deaerated 3 g/l NaCl solution at 90°C, the oxygen being eliminated by nitrogen bubbling at the beginning of the experiment. Taking into account the fact that a perfectly oxygen free solution cannot be obtained, comparison was made between the mild steel samples and a copper control sample which can only be attacked by the remaining oxygen in that case. The corrosion rate de/dt is linked to the polarization resistance Rp by the following formula :

$$de/dt = A/Rp \qquad (1)$$

Where A is determined experimentally using Tafel extrapolation and Faraday's law [5].

The susceptibility of mild steel to hydrogen embrittlement was also studied using the slow strain rate technique.

RESULTS

Corrosion resistant overpack

• Crevice corrosion in granitic solutions

Susceptibility to that type of corrosion was assumed to be directly linked to the current peak observed on polarization curves starting with the alloy in the active state. The results are presented on figures 1.1 to 1.8.

As regards Ni-Cr alloys, we have plotted the current peak as a function of the pH of the solution. Furthermore, some 90°C tests were carried out on aged (TT) samples to simulate welding conditions.

In the case of Ti-Pd we have only represented the polarization curves for two values of the solution pH.

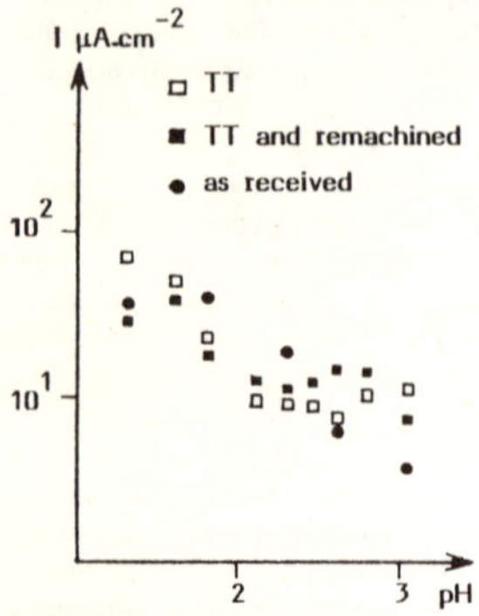

Fig. 1.1 . 625, SG2 90°C

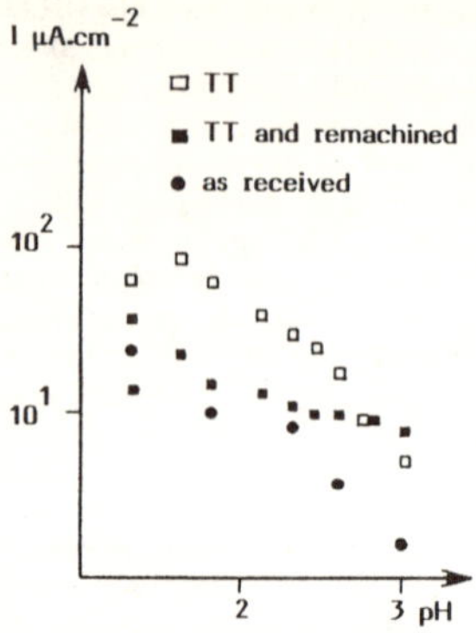

Fig. 1.2 . C276, SG2 90°C

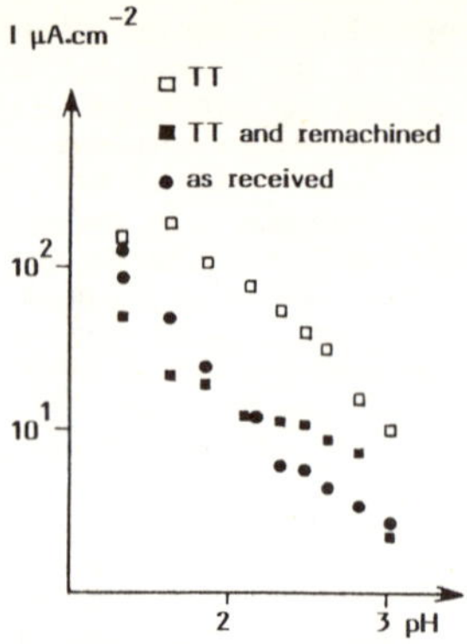

Fig. 1.3 . C4, SG2 90°C

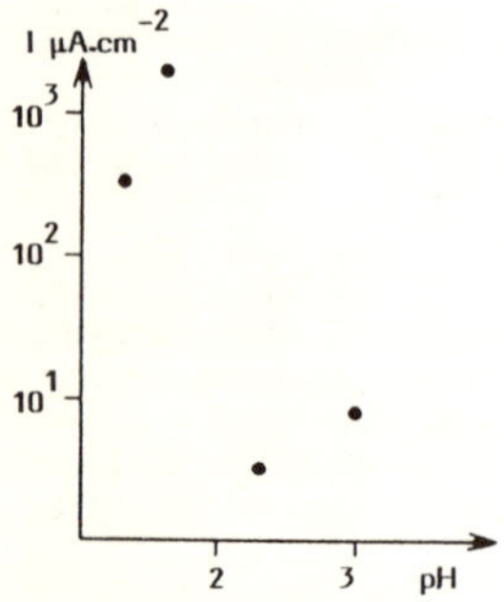

Fig. 1.4 . 625, SG3 170°C

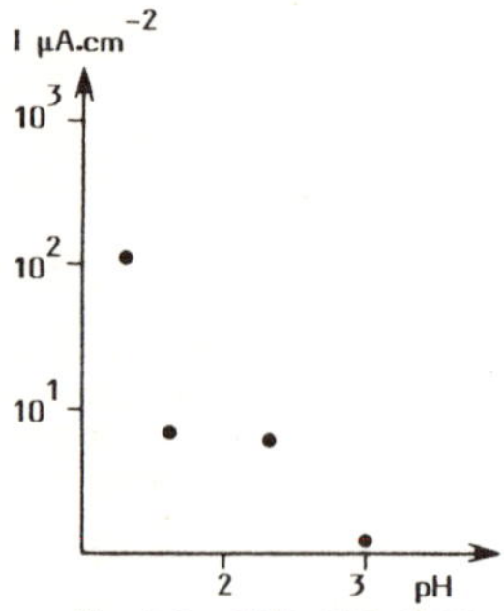

Fig. 1.5 . C276, SG3 170°C

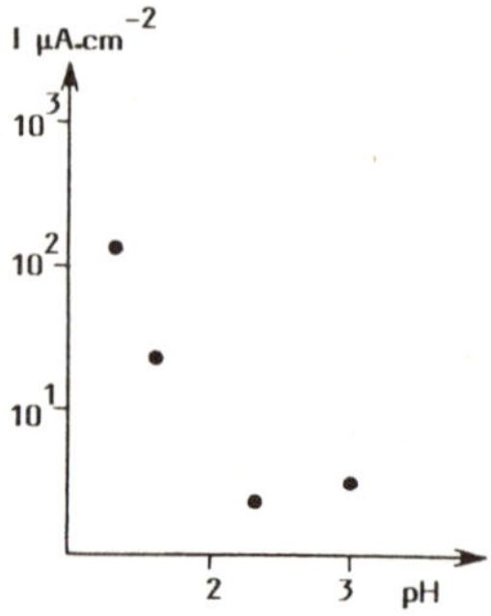

Fig. 1.6 . C4, SG3 170°C

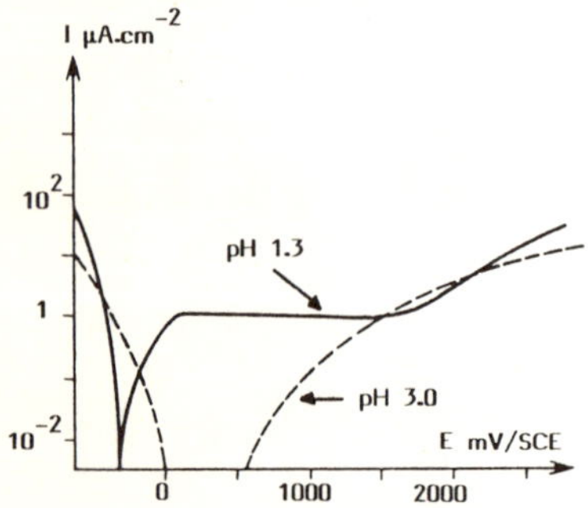

Fig. 1.7 . Ti-Pd, SG2 90°C

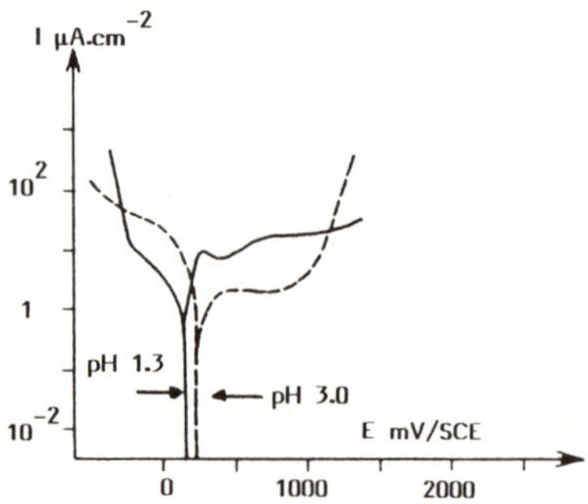

Fig. 1.8 . Ti-Pd, SG2 170°C

The results of the 90°C tests carried out on Ni-Cr alloys indicate that the thermal treatment had no effect on 625. Its influence was more significant on C 276 and C 4 and corresponds to an increased crevice corrosion susceptibility. However, the current peak values remain comparable to those of as received 625. Furthermore this phenomenon was confined to a thin superficial layer of the samples, since "TT and remachined" ones exhibited behaviours comparable to those of as received material.

170°C tests showed that the crevice corrosion susceptibility of the four alloys tested was greatly increased by the temperature. At both temperatures however, the results clearly indicated a great superiority of Ti-Pd over Ni-Cr alloys.

• Pitting corrosion in granitic solutions

The polarization curves are shown on figures 2.1 to 2.4. 90°C curves indicate no evidence of pitting susceptibility for the four alloys tested since when reversing the potential scan the current decrease was instantaneous.

As regard 170°C tests, although the solution contained ten times less chloride than in the case of Ni-Cr alloys, the reverse current remained relatively high. This non repassivation of Ni-Cr alloys is indicative of pitting. Titanium-palladium exhibits a good pitting resistance at the same temperature, even in a ten times as much concentrated solution.

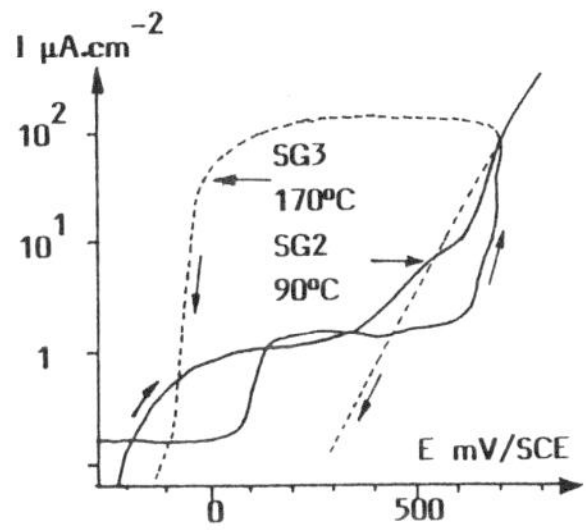

Fig. 2.1 . Pitting of 625

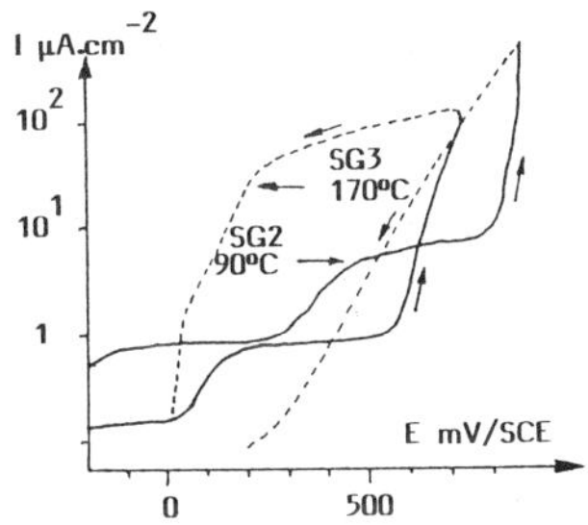

Fig. 2.2 . Pitting of C276

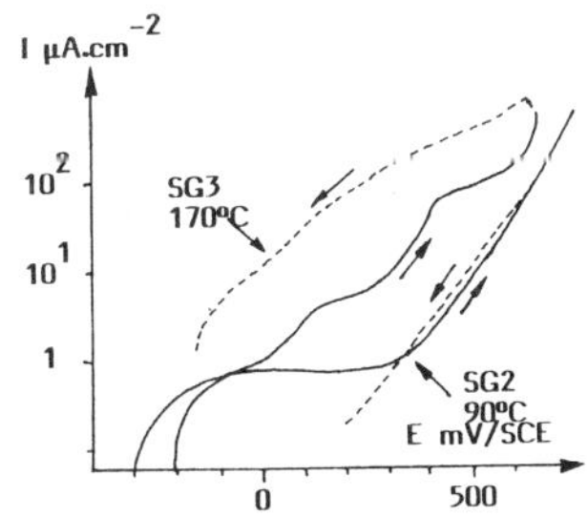

Fig. 2.3 . Pitting of C4

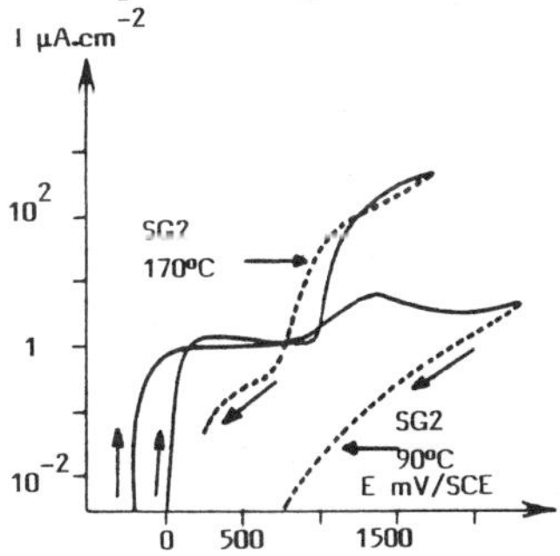

Fig. 2.4 . Pitting of Ti-Pd

• Stress corrosion cracking (SCC) and hydrogen embrittlement

The results of the SCC tests indicated that the Ni-Cr alloys could suffer from cracking at potentials close to the corrosion potential except in argillaceous solutions. The figure 3.1 presents the results obtained with C 276 in acidified solution SG 2 at 90°C.

It should be noted that the free corrosion potential of C 276 in neutral medium (- 290 mV/SCE) corresponds to the cracking domain in acidified solution, a fact which induces a risk of cracking in neutral medium if a local acidification occurs, such as inside a crevice.

Cracking of the Ni-Cr alloys was particularly marked when using Q.brine as the testing medium. This result, and the fact that no cracking was observed in argillaceous solutions, confirms the predominent effect of chloride.

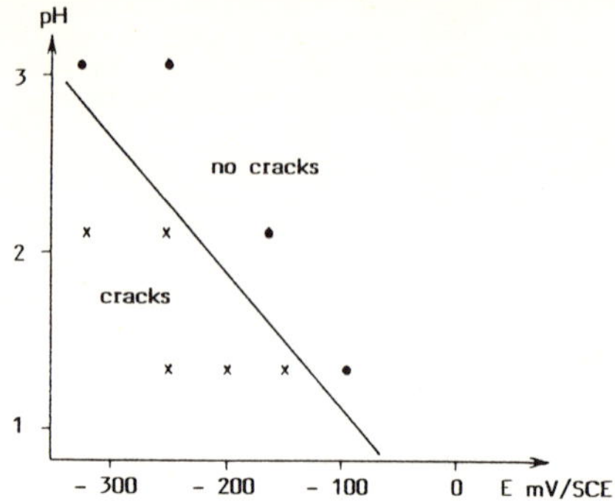

Fig. 3.1 . SCC of alloy C 276 as a function of solution pH and electrode potential in SG 2 at 90°C

As regards the Ti-Pd alloy, the SCC test indicated a good resistance to cracking in all the testing solutions. However, this alloy was found prone to hydrogen embrittlement provided a sufficient amount of hydrogen was produced. In our tests, the hydrogen was generated by a cathodic polarization of the samples at a rate of approximately 2.10-4 mole/cm² of atomic hydrogen per hour. It should be noted however that the greater part of the hydrogen produced escaped through the solution and thus was not involved in the embrittlement process.

Corrosion allowance overpack

- Generalized corrosion in deaerated 3 g/l NaCl solution at 90°C

Preliminary tests were run at 90°C and a record of the corrosion potentials of mild steel and copper (control sample) was made in order to assess the deaeration of the solution and the possible attack of mild steel by water. The evolution of those potentials are shown on figures 4.1 and 4.2.

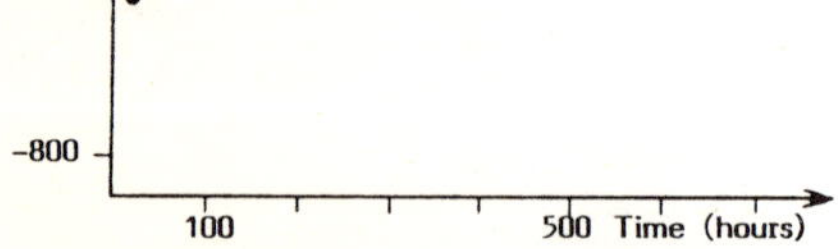

Fig. 4.1 . Corrosion potential of mild steel

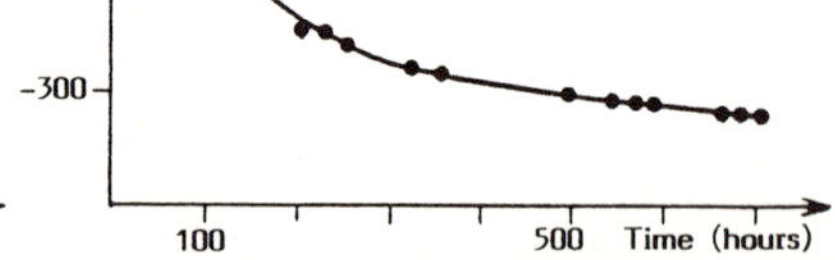

Fig. 4.2 . Corrosion potential of copper

While the potential of copper keeps on decreasing, the one of mild steel reaches a steady value, which points out that water reduction plays a part in its corrosion. As a matter o fact, the decreasing potential of copper is due to oxygen depletion. The steady value for mild steel shows that the attack due to the remaining oxygen has become negligible compared to water reduction.

Polarization resistance measurements were then carried out to study the time dependence of the corrosion rate of mild steel and the results are plotted on figure 4.3.

Fig. 4.3 . Time dependence of the polarization resistance of mild steel in deaerated 3 g/l NaCl solution at 90°C

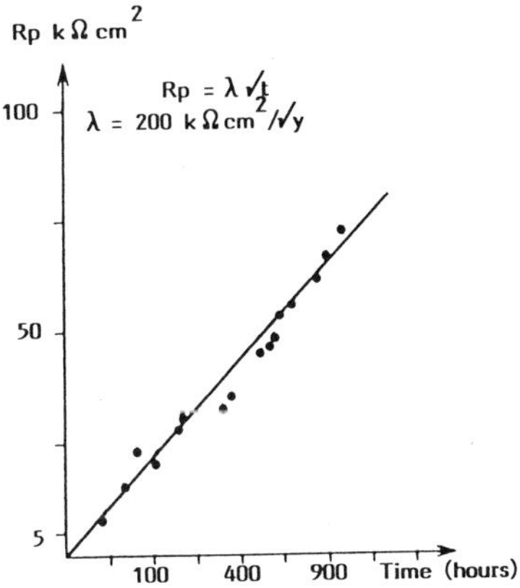

Using equation (1), the corroded thickness e can be derived from these results as a function of time :

$$e = k\sqrt{t} \qquad (2)$$

Calculations give a k value of 5 µm/$\sqrt{y}$ which would correspond to a corroded thickness of 150 µm after 1000 years, with an A value of 500 in equation (1). Nevertheless, assuming a complete breakdown of the oxide layer after each month, the corroded thickness due to a 1000-year attack would be equal to 20 mm.

• Hydrogen embrittlement in natural argillaceous solution at 90°C

The results obtained from slow strain rate experiments are presented in table III.

Table III . Hydrogen embrittlement of mild steel in argillaceous water at 90°C

Potential mV/SCE	Control sample in "millipore" water	Free (- 500)	- 700	- 1700	- 1800
Elongation (%)	16.5	16.6	18.4	14.3	11.7
U.T.S (MPa)	525	531	538	513	504
Remarks	-	oxidized	-	cracked	cracked

The presence of cracks and the elongation decrease of the samples show that mild steel is prone to hydrogen embrittlement in argillaceous water provided a sufficient amount of atomic hydrogen is produced.

CONCLUSION

The work summarized in this paper led to a better understanding of the corrosion behaviour of various alloys selected for the design of HLW container for geological disposal, with regard to pitting, crevice and general corrosion as well as stress corrosion cracking and hydrogen embrittlement.

As regards the "corrosion resistant" overpack, polarization curves have shown that crevice corrosion of alloys C 276 and C 4 was increased by a thermal treatment simulating welding conditions. However, the influence of such a treatment was limited to a thin superficial layer of the samples.

Comparison of polarization curves carried out at 90 and 170°C has shown the great influence of temperature on the crevice corrosion of Ni-Cr alloys, thus pointing out the importance of an appropriate intermediate storage period to achieve a sufficient cooling of the containers. The same temperature dependence has been found when studying pitting corrosion of these alloys.

The polarization curves carried out on Ti-Pd samples have demonstrated its superiority over Ni-Cr alloys against pitting and crevice corrosion.

Slow strain rate experiments have shown that Ni-Cr alloys were susceptible to SCC in all the solutions excepted argillaceous ones. However, it should be noted that neither chloride enhancement nor acidification were made in those solutions.

The same tests carried out on Ti-Pd indicated its superiority over Ni-Cr alloys regarding SCC. However, when carried out under cathodic polarization, those tests showed that Ti-Pd was prone to hydrogen embrittlement, provided a sufficient amount of hydrogen was produced.

As regards the "corrosion allowance" overpack, the tests carried out on mild steel have shown that the general corrosion rate only due to water had to be taken into account, since it could correspond to a corroded thickness of 20 mm after 1000 years of geological disposal.

The results of slow strain rate tests carried out under cathodic polarization of the samples have shown that mild steel was susceptible to hydrogen embrittlement. As for Ti-Pd, that corrosion process requires further experimentation to assess the influence of an hydrogen pre-charging of the material on its mechanical properties and compare the hydrogen concentration needed to induce embrittlement with the one that could be expected to be produced over long times by ground water radiolysis.

ACKNOWLEDGEMENT

This work was performed under contract with the Commission of the European Communities as part of its R & D Programme on Management and Storage of Radioactive Waste.

REFERENCE

1. CEC, Corrosion behaviour of container materials for geological disposal of high level waste, Nuclear Science and Technology. EUR 9570 EN (1985).

2. R.N. Parkins, in The Slow Strain Rate Technique, edited by Ugiansky and Payer (ASTM STP 665, 1970), p. 5

3. Final Storage of Spent Nuclear Fuel, KBS 3, published by the Swedish Nuclear Fuel Supply Company. SKBF/KBS Box 5864. S-10248 Stockholm, Sweden.

4. G.P. Marsh et al, European Appl. Res. Rept. - Nucl. Sci. Technol. Vol 5, N° 2 (1983), pp. 223-252.

5. G. Plante, M. Hélie, O. Sanatine and A. Chenière, in Etudes de corrosion des matériaux de conteneurs pour le stockage des déchets radioactifs en sites granitiques, (CEC, Sciences et Techniques Nucléaires, EUR 8762 FR, 1984), pp. 10-15.

METALLOGRAPHICAL STUDY OF HOT ISOSTATICALLY PRESSED COPPER ENCAPSULATION OF NUCLEAR FUEL ELEMENTS

B. LOBERG AND K. E. EASTERLING
Department of Engineering Materials
University of Luleå, S-951 87 Luleå, Sweden

ABSTRACT

Light optical and electron microscopy have been used to study both laboratory-scale and full scale hot isostatically pressed (HIPed) copper as employed for encapsulation of spent nuclear fuel elements. Particular attention has been paid to the degree of densification achieved, as well as to the status of the joins between the sleeve, lid and the HIPed powder. The work includes the development of computing models for predicting the optimal conditions for HIPing copper and other possible encapsulation materials such as lead and alumina. Mechanical testing of HIPed samples has also been carried out.

The current method of using hydrogen to reduce oxides at powder surfaces prior to HIPing is questioned on the basis that metallography reveals definite signs of hydrogen embrittlement. An alternative method of reduction is to use CO/CO_2 mixtures instead of hydrogen, and it is suggested that this approach could be strongly advantageous in future.

FULL SCALE ENCAPSULATION

In this encapsulation, made by the ASEA Company, dummy fuel elements were embedded in copper powder inside a forged copper sleeve. Top and bottom pieces were machined and wire-brushed. The capsule was surrounded by an outer container of stainless steel. After reduction of the powder with hydrogen (24 h, 400 °C), the container was evacuated, sealed and HIPed at 500 °C, 150 MPa, for 8 hours.

Samples for transmission electron microscopy (TEM) were prepared from the sintered powder and from the joint between the powder and the solid lid.

Subsequent examination of the specimens revealed that a large portion of the grain boundaries in the HIPed powder were liberally decorated with inclusions or pores, examples of which are shown in figure 1. Electron diffraction showed that they were mainly non-crystalline, while chemical microanalyses indicated the absence of all heavy elements ($Z > 10$). They were therefore interpreted to be micro-pores, probably containing entrapped water originating from the hydrogen reduction process. This may explain why they did not disappear during the HIP cycle.

Micro-pores were most abundant at incoherent and semicoherent grain boundaries. However, numerous examples of inclusions (oxides, as well as sulfides) were also found, particularly inside the grains, see figure 2. Larger oxides are more easily confirmed by optical microscopy of etched cross sections, using polarized light.

The boundary between the lid and the HIPed powder was conspicuous through a considerable difference in grain size and dislocation density on its two sides. It did not contain more inclusions or pores than the powder boundaries inside the sintered body. Parts of the boundary showed signs of having migrated into the powder mass. It was in general concluded that the bonding was satisfactory.

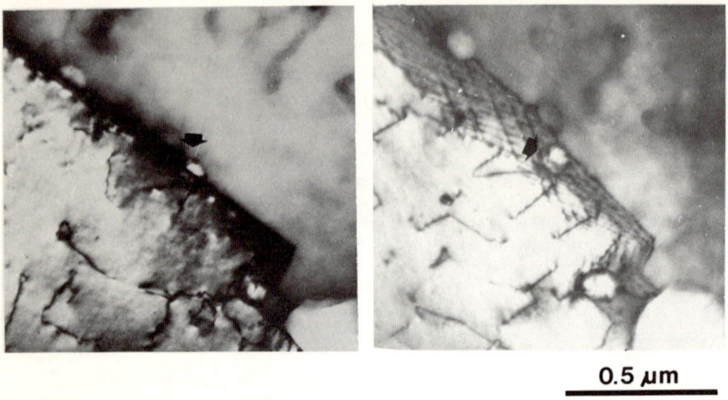

Fig 1 The same semicoherent grain boundary in the HIPed powder, imaged at two different orientations. The boundary is decorated with pores as indicated by arrows in both micrographs. TEM.

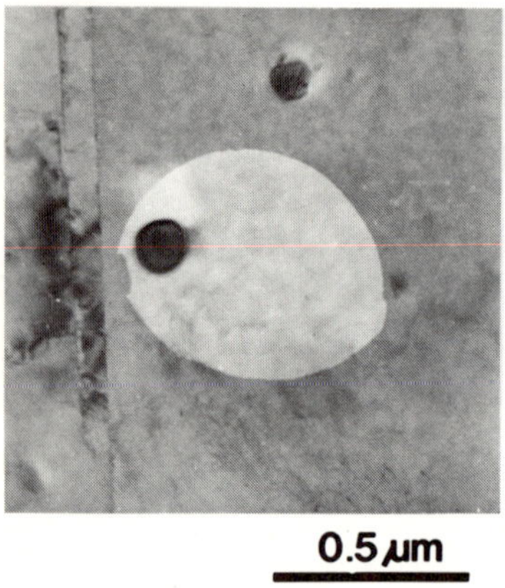

Fig 2 Examples of a fairly complex crystalline inclusion inside a grain in the HIPed material. Besides copper (and presumably oxygen), it also containes sulphur and a small amount of chromium and selenium. TEM.

LABORATORY-SCALE EXPERIMENTS

Three different powders were used. The first was fairly coarse and made from chopped copper wire (ALCAN AK91), the second was an argon atomized powder (ECKA AK91) and the third was a fine argon atomized powder (KOBELCO) which was shipped, handled and encapsulated in an inert atmosphere.

Three pairs of samples (length 300 mm, diameter 20 mm) were made. The first contained a mixture of ALCAN (60%) and ECKA (40%) powders and was reduced with hydrogen (16 h, 350 °C) before HIPing. The second pair contained the same mixture, but was HIPed without any reduction. The third was prepared from pure KOBELCO powder and also HIPed without reduction.

The parameters used were 500 °C, 150 MPa and 6 h. After HIPing, one sample from each pair was heat-treated at 400 °C for 1 hour in argon. Achieved densities and mechanical properties are given in table I.

Table I

Material	Heat treatment	Density (g/cm^3)	Yield strength $R_{0.2}$ (MPa)	UTS (MPa)	Elongation A_5 (%)
Reduced E/A	No	8.77	82	205	15
Reduced E/A	Yes	8.71	77	90	1
E/A	No	8.919	(80)	231	(10)
E/A	Yes	8.921	76	198	11
KOBELCO	No	8.928	135	239	40
KOBELCO	Yes	8.927	137	239	44

Full density for copper is around 8.935 g/cm^3.

The low densities achieved for the reduced samples indicates that the encapsulations may have developed small leaks during HIPing. Even so, the drastic decrease in tensile strength and elongation after heat treatment is evident. Hydrogen reduced material is clearly susceptible to hydrogen embrittlement, which might endanger its applicability as a material for long term storage.

The powder which has been handled in air does not develop proper bonding between the grains, as shown by the modest elongation for the E/A samples. The material is, however, unaffected by subsequent heat treatment.

The KOBELCO powder, handled under inert conditions, gave the best HIPed material with acceptable strength and ductility.

Metallographical studies of cross sections and fracture surfaces confirmed these conclusions. In addition, it was found that the KOBELCO powder was contaminated with a fairly coarse superalloy powder. The content of the contaminant was about 0.3 %. The mechanical properties in table I can therefore be considered as a lower limit for what can be achieved with a cleaner (uncontaminated) starting material.

Examples of fracture surfaces are given in figures 3 and 4.

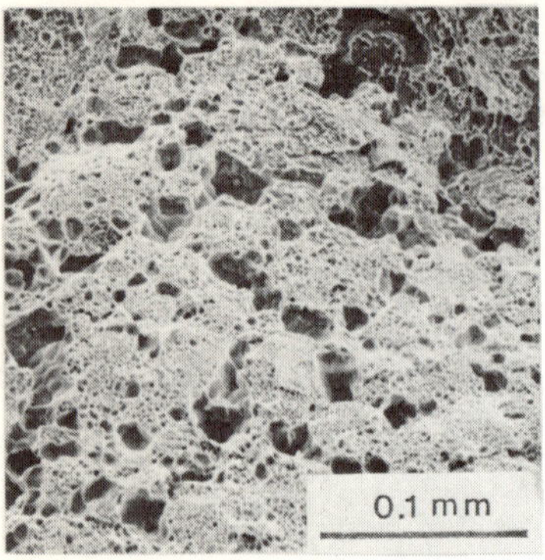

Fig 3

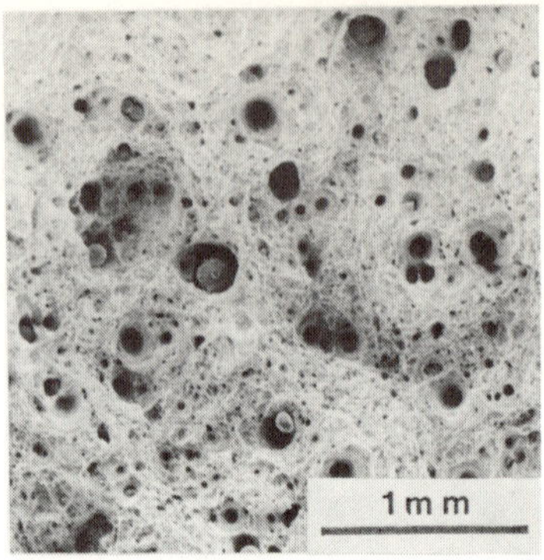

Fig 4

Fig 3. Fracture surface of the E/A powder mixture after reduction, HIPing and heat treatment. Note the large number of pores in the hydrogen embrittled material. SEM.

Fig 4. Fracture surface of the material made from KOBELCO powder. Note the round cavities, each caused by a contaminant superalloy particle, some of which are still remaining. SEM.

CARBON MONOXIDE REDUCTION

The mechanical testing discussed above showed that normal copper powder is susceptible to embrittlement after hydrogen reduction and HIPing. The standard method to avoid the problem would be to use a deoxidized starting material. However, small amounts of the deoxidizing agent (normally phosphorus) would be left in the powder. Its use as an encapsulation material has therefore been discouraged by corrosion experts.

An alternative method is to use carbon monoxide instead of hydrogen to reduce the powder surfaces before HIPing. Laboratory-scale experiments have been initiated to investigate the possibility, using fine powders (5 µm) of pure copper oxides as a model material. The results of reduction experiments are checked by weighing as well as X-ray diffractometry.

The results so far can be briefly summarized as follows:

a) CO has a tendency to dissociate into C and CO_2. To avoid contamination of the powder by carbon, the reduction agent must be a mixture of CO and CO_2 (typically less than 5% CO at 500°C).
b) The temperature must be above 500°C if fast reaction rates are to be achieved.
c) The reduction is exothermic. To avoid a spontaneous temperature increase, the flow of CO must be limited. (This applies to pure copper oxide, and is probably not important for lightly oxidized copper powder).
d) For laboratory-scale experiments (<10g), the reduction is complete in less than 4 hours at 550°C without any appreciable temperature increase.

The results so far are therefore encouraging, and trials on copper powder are now in progress. If these experiments prove successful, CO reduction would be greatly preferable to hydrogen, in that the CO essentially reduces only the oxides at the powder surface, whereas hydrogen penetrates into the copper with the resulting hydrogen embrittlement.

HIP DIAGRAMS

Computing models for predicting the optimal conditions for HIPing copper and other possible encapsulation materials such as lead and alumina have been developed. Densifying mechanisms taken into account are yielding, power law creep, and volume and boundary diffusion. In the case of polycrystalline powders, Nabarro-Herring and Coble creep are included. In the final stage, grain growth may play an important role in slowing down the densification rate. A full account of the model is given elsewhere [1].

Input data are mechanical properties, diffusion parameters and creep behaviour at the relevant temperature as well as HIPing parameters and particle and grain size.

The results are presented in diagrams, showing the achieved relative density as a function of HIPing temperature (or pressure) for different times and powder sizes. The different fields in the diagram indicates the mechanism that dominates the densification.

A diagram for copper is shown as an example in figure 5. R is the particle size and d the grain size. The density achieved for the full scale encapsulation discussed earlier is given as a point in the diagram.

The diagram predicts that the temperature should be increased to 550 ^{0}C if full density is to be reached within reasonable times.

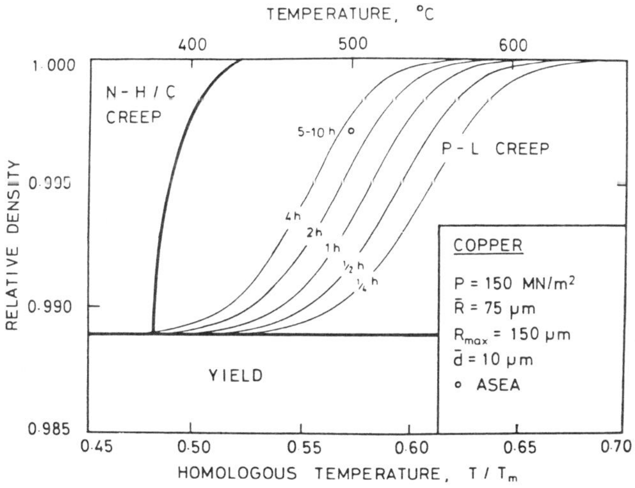

Fig 5. Density-temperature diagram for hot isostatic pressing of copper.

DENSIFICATION OF FULL-SCALE ENCAPSULATIONS

The above HIP maps calculate times of HIPing under the ideal conditions of uniform temperature throughout the compact. However, in practical HIPing work, particularly in the case of large powder masses as in the full scale encapsulations, this "ideal" condition is not met. It is thus important to be able to estimate the effect of sample dimensions - and even shape - on time for complete densification. Preliminary work [2] has shown that by taking into consideration the dependence of heat conductivity on density of the material, heat flow equations can be combined with the effect of pressure and temperature on densification mechanisms in HIPing, such as to predict the effect of sample dimensions on sintering. An example relevant to the present case is illustrated in Fig 6, in which the effect of increasing pressure from 100 to 150 MPa, assuming a sintering time of about 20 min. is given. The sintering temperature in both cases is 550°C. It is seen that densification of a large scale sample occurs from outside inwards, the densification front occurring initially over a very narrow distance. This reflects, of course, the problem of heat flow through the porous inner part of the sample. It is seen that in both cases, complete densification has not taken place, the sintering time being too short.

This work will now be continued to take into consideration actual HIP cycles in terms of realistic rates of increasing pressure and temperature, times at temperature, etc., and hence predict times necessary to obtain complete densification, as well as actual amounts of shrinkage expected in large scale encapsulation experiments.

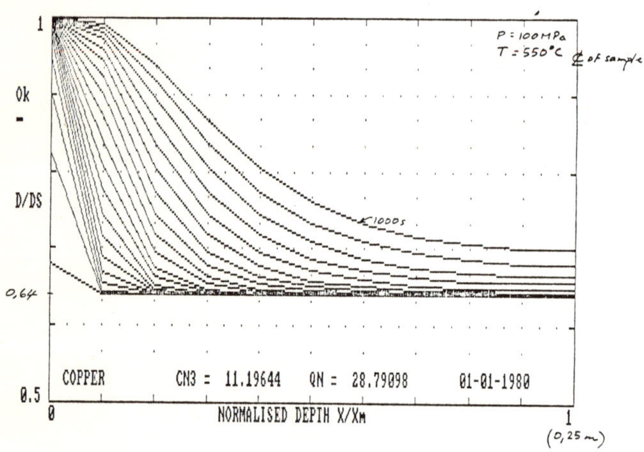

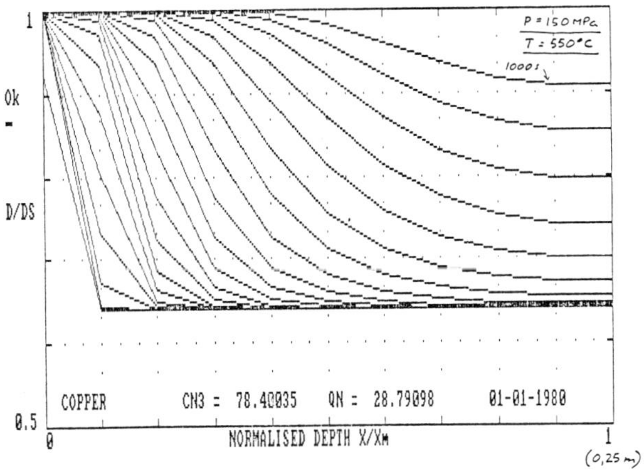

Fig 6 Computed densification diagrams of large scale copper samples during HIPing, showing the dependence of full densification on sample dimensions and applied pressure. The contours give equal increments in time up to 1 000 s.

ACKNOWLEDGEMENT

Financial support from Swedish Nuclear Fuel and Waste Management is gratefully acknowledged.

REFERENCES

1. A.S. Helle, K.E. Easterling and M.F. Ashby, Technical Report nr 1985:02T, Luleå University of Technology (1985).

2. M.F. Ashby and K.E. Easterling, "Non-Uniform densification during HIPing", to be published (1985).

THE DEVELOPMENT OF DURABLE CEMENTITIOUS MATERIALS FOR USE IN A NUCLEAR FUEL WASTE DISPOSAL FACILITY

N.C. BURNETT*, R.D. HOOTON*, R.B. HEIMANN**, AND M. ONOFREI**
* Ontario Hydro, Toronto, Ontario, Canada M8Z 5S4
** Atomic Energy of Canada Limited, Whiteshell Nuclear Research Establishment
Pinawa, Manitoba, Canada ROE 1L0

ABSTRACT

This paper describes the work on cement paste development and short-term leaching tests in Standard Canadian Shield Saline Solution (SCSSS) in the presence of bentonite at 150 °C. It has been found that:

- supplementary cementing materials such as silica fume or fly ash could significantly improve the properties of sulphate resistant portland cement (SRPC), in particular, permeability to water and pore size distribution.

- the addition of bentonite suppressed the normal tendency of the pH of groundwater to increase rapidly in the presence of cement.

- the presence of bentonite increased the release of potassium ions from the cements.

- SRPC blended with 20% silica fume resulted in a groundwater pH lower than that of SRPC, with and without bentonite. Moreover, its cumulative fraction of release of potassium was significantly lower than that of SRPC.

INTRODUCTION

The Canadian Nuclear Fuel Waste Management Program includes an evaluation of cements for vault grouting and final shaft sealing [1], and an assessment of concrete casks for the dry storage of used fuel and nuclear power generating stations [2]. A system that uses a dry storage cask for both transportation and disposal is also under consideration [3]. This so-called 'integrated' system would minimize the fuel handling required from storage through to disposal. The development of a suitable concrete and fabrication of a prototype integrated container have been described previously [4].

Concrete and cements could be used in a number of ways in a disposal vault, e.g., for the immobilization and containment of nuclear fuel waste and for vault backfilling, as well as for the more traditional construction uses such as bulkheads, shaft sealing and fracture grouting.

This report discusses the development and testing of cements for the containment of nuclear fuel wastes and their application to other uses in the disposal vault.

PERFORMANCE CRITERIA

The concrete cask, as part of a multi-barrier protection system in the disposal facility, would be expected to provide fuel waste containment for several hundred years. The following properties of a concrete containment barrier are therefore of particular concern: (1) durability, i.e., the concrete should be resistant to aggressive sulphate- and chloride-bearing groundwaters at elevated temperatures; (2) material interactions, i.e., deterioration of the concrete should not adversely affect the dissolution rate of the fuel waste or the effectiveness of other components of the multi-barrier system; and (3) radionuclide retention, i.e., once the groundwater reaches the fuel waste and dissolution commences, the concrete should provide some retention of the radionuclides. Until specific guidelines on container lifetime are produced, it is difficult to define the term 'durable' and to determine an acceptable limit to 'material interactions'. However, certain properties are required of a good quality cement, as the basis of the concrete, viz.,

(1) sulphate and chloride resistance,
(2) fine pore structure and low permeability,
(3) low free alkali and alkaline earth contents and, thus, low porewater pH, and
(4) high radiation stability.

To provide resistance to sulphate ions contained in the highly saline groundwaters expected in the Canadian Shield at depths of 800 to 1000 m [5], the most suitable choice of cement is a sulphate resisting portland cement (SRPC). The addition of supplementary cementing materials (mineral admixtures), has also been reported to increase the resistance of concrete to deterioration by aggressive ions such as chlorides [6] and sulphates [7], by reducing the permeability and the amount of easily leached calcium hydroxide in the hardened cement paste [8].

EXPERIMENTAL PROGRAM

A number of sequential tasks have been established to develop and evaluate the long-term durability of cementitious materials. This paper describes the work on cement paste development and short-term leaching tests. These tests will be followed by longer-term experiments to determine the extent of material interactions under simulated disposal-vault conditions, radionuclide migration in concrete, and radiation stability.

Development of a Highly Durable Cement Paste

The program involved a comparison of the properties of a sulphate resistant portland cement (SRPC), with the properties of cements made with various replacement levels of class F fly ash, blast furnace slag and condensed silica fume, i.e., blended cements [9]. The levels of replacement were based on a literature survey of high-durability cements. In addition, the effects of water-reducing agents and water/cement ratios (W/C) were investigated. Specimens were moist cured at room temperature. Future tests will include moist curing at 70°C and autoclaving at 185°C, to see if the desired properties can be obtained at shorter curing times.

Table I shows the results of tests on samples at W/C = 0.25 cured for 91 days. Of particular note is the reduction of porosity and portlandite ($Ca(OH)_2$) content with increasing amounts of silica fume. The $Ca(OH)_2$ is a by-product of the hydration of calcium silicates in cement and is primarily responsible for the high pH of the solution in contact with the material. The addition of silica fume appears to be more effective in reducing porosity than fly ash, but does not improve the compressive strength. These low strengths are anomalous and likely due to the difficulties in properly curing the silica fume in a paste. At the very low W/C ratio used in these tests, some self-desiccation may occur in the paste, causing the hydration reaction to cease. This type of paste must be cured under high-moisture conditions. The effect of replacement materials is reversed when a concrete is produced [4]. For concretes fabricated with magnetite and specularite aggregates the following compressive strengths were measured after 91 days curing:

with SRPC alone	43.0 MPa
with SRPC + Fly Ash	50.1 MPa
with SRPC + 20% Silica Fume	80.9 MPa .

Table I.

Properties of Optimum Blended Cement Pastes Cured 91 Days at 23°C

Properties	SRPC	SRPC + 35% Fly Ash	SRPC + 10% Silica Fume	SRPC + 20% Silica Fume
Compressive Strength (MPa)	109.0	122.8	87.6	97.9
Water Permeability (10^{-13} m/s)	1.3	0.5	0.6	0.3
Total Intruded Porosity (%)	11.1	19.1	11.7	8.1
Volume Portion of Pores > 25 mm Radius (%)	2.2	6.1	1.4	1.1
Most-common Size Pore Radius (nm)	16.5	25.0	18.5	12.5
Solid Volume Density (g/cm^3)	2.31	2.22	2.14	2.02
Porewater pH*	10.7	10.5	10.8	10.4
$Ca(OH)_2$ (by XRD)	19	6	8	0

* From analyses of solutions extracted from the water permeability experiments.

In general, the pore structure of blended cements reflects the permeability, i.e., a smaller volume of coarse pores (> 25 nm) results in a lower permeability.

As shown in Table I, the pore structure of the pastes became finer with increasing quantities of silica fume and with the 20% silica fume mix, the total intruded porosity decreased significantly. The volume portion of coarse pores > 25 nm radius, was 6.1% for the 35% fly ash paste but only 1.1% for the 20% silica fume paste. Primarily, because of its larger surface area (6000 m^2/kg), silica fume is much more reactive than fly ash (400 m^2/kg). However, it is anticipated that fly ash modified cement might eventually attain these superior properties, but after much longer curing periods.

It is most probable that the amorphous siliceous material has reacted with calcium hydroxide to form more calcium silicate hydrates, which results in reduced porosity and permeability [8,10]. While measured reductions in pore water pH were small, more sophisticated pore water expression tests [11] done elsewhere show more dramatic decreases with silica fume additions [12]. This gives an initial indication that an SRPC/silica fume mix would be more leach resistant and, affect groundwater chemistry less than normal portland cements.

In a related study, recent grouting experiments carried out at Ontario Hydro have also shown than a SRPC + 20% silica fume grout is less permeable than SRPC when filling narrow fractures (between 50 μm and 1.40 nm) at typically high-water-to-solids ratio (> 2.5). No water flow could be detected through this grout sheet and a permeability of less than 2.0 x 10^{-13} m/s was calculated. In comparison, at the same water to solids ratio, a plain SRPC grout did not completely seal the fine fracture and a permeability of only 7 x 10^{-5} m/s was measured.

Short-Term Leaching Experiments on Cements in a Saline Solution in the Presence of Clay

To examine the chemical interaction between cement and montmorillonite clay under hydrothermal conditions, a number of prisms of different pastes, described in section 1, were placed in an autoclave at 150°C for 14 days in contact with either sodium or calcium montmorillonite (exchanged with 1 mol/L NaCl and $CaCl_2$ solution, respectively) and either deionized water (DIW) or Standard Canadian Shield Saline Solution (SCSSS) [13]. The nominal concentrations of the principal ionic species in the SCSSS are (in mg/L): Ca, 15 000; Mg, 200; Na, 5050; K, 50; Cl, 34 260; SO_4, 790; with pH = 7.0

[14]. The leach solutions were sampled at various times to measure pH and to analyze for Si, K, Mg and Fe concentrations to determine the leaching rate of the cement.

Table II shows the pH after quenching to 25°C and the calculated pH, at 150°C, of the leach solution after 14 days in an autoclave. The majority of the pH values had stabilized after 14 days. The calculations were done with an algorithm provided by the SOLMNQ code [15].

Table II.

pH of Solutions After 14 Days Leaching at 150°C

System*	Quench pH at 25°C	Calculated pH at 150°C
SRPC/DIW	11.9	9.4
SRPC/SCSSS	9.3	6.1
SRPC/SCSSS/CaM	7.6	5.7
20SF/SCSSS/CaM	4.6	3.8
20SF/SCSSS/NaM	5.4	5.2

* SRPC/DIW - SRPC in deionized water
SRPC/SCSSS - SRPC in SCSSS
SRPC/SCSSS/CaM - SRPC in SCSSS + Calcium Montmorillonite
20SF/SCSSS/CaM - SRPC modified by 20% Silica Fume in SCSSS and Calcium Montmorillonite
20SF/SCSSS/NaM - SRPC modified by 20% Silica Fume in SCSSS and Sodium Montmorillonite

The addition of Ca-montmorillonite to the base cement-saline groundwater system (SRPC/SCSSS/CaM) led to a lower pH at both 25° and 150°C, owing to the acidic nature of this clay [16]. The addition of silica fume to the base cement resulted in a much lower pH value. This is attributable to the reduction and, finally, the complete disappearance of portlandite, $Ca(OH)_2$, as confirmed by X-ray diffraction measurements. Comparison of the pH of 20SF/SCSSS/CaM and 20SF/SCSSS/NaM showed that the presence of Na-montmorillonite led to a slightly more alkaline pH. The pH of deionized water heated to 150°C has been calculated to be 5.7; that of Standard Canadian Shield Saline Solution at 150°C is 4.9.

To compare the durability of the cement pastes under different experimental conditions, cumulative fractions of release (CFR) of potassium were calculated. Originally, the release rate of silicon was also considered. However, because the silicon concentration in the solution is subject to complex precipitation kinetics, leading to amorphous silica, magnesium silicates and, possibly, calcium (aluminium) silicate phases, the evaluation of cement durability based on the CFR of silicon could be misleading. In addition, potassium-containing compounds in the cements are much more soluble than the calcium silicate hydrates. Therefore, high potassium levels may not necessarily indicate cement deterioration.

Figure 1 shows that the potassium release was highest in SRPC/SCSSS/CaM. The CFR of potassium from 20SF/SCSSS/NaM was almost constant over the entire leach period, whereas from 20SF/SCSSS/CaM, the CFR increased with time to approach after 14 days the value obtained from 20SF/SCSSS/NaM. Finally, the CFR of potassium from the base cement in deionized water, without the addition of clay (SRPC/DIW), increased slightly with time.

The shape of the curves suggest for SRPC/SCSSS/CaM and 20SF/SCSSS/CaM a decrease of the rate of CFR of potassium with time caused by the formation of some kind of protective layer. Thin, apparently dense reaction layers of hydroxylapatite were found covering the surface of the cement [17].

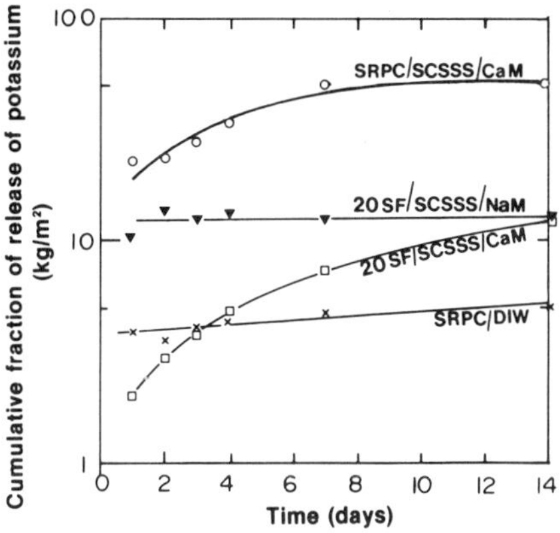

Figure 1. Cumulative Fraction of Release of Potassium

Phosphate is presumably supplied by the Ca-montmorillonite which contains around 0.15 weight% P_2O_5 [18].

It has been assumed that no significant ion exchange of potassium on the clay surface took place because montmorillonite is much more selective for Ca^{2+} than for K^+ [19,20].

The trends for changes in pH and potassium release were confirmed in leaching experiments on a heavy shielding concrete in SCSSS and DIW, with and without the addition of sodium montmorillonite [21]. The concrete was made with a sulphate resisting cement, and modified by blast furnace slag, silica flour and iron oxide aggregates and had been autoclaved at 185°C for 24 hours. The leach tests were performed at 100°C. The pH of the quenched leach solutions after 120 days at 100°C was 9.5 for concrete/DIW, 9.2 for concrete/SCSSS and 7.6 for concrete/SCSSS/NaM. The potassium levels in solution were increased markedly by the presence of the clay. This appears to be a synergistic effect, i.e., the concentration of potassium in the concrete/SCSSS/NaM leach solution was greater than that in a concrete/SCSSS solution plus that in a control system NaM/SCSSS solution. This effect is possibly indirectly due to calcium depletion of the SCSSS by ion exchange with sodium in the NaM. The SCSSS would then be more aggressive to the cement thus leaching calcium along with other elements including potassium.

To study the influence of leaching on the pore structure of the cements selected for this study, cumulative pore size distributions for the cement pastes after curing and leaching were determined by the mercury intrusion method. The results are shown in Figure 2.

The leached cement samples always showed a reacted surface layer to a depth of 1 to 2 mm. This layer was mechanically removed and analyzed separately. For the cement pastes, the 150°C leaching environment resulted in higher total intruded porosities compared with the as-cured condition (182 days at 23°C).

Comparing all the SRPC experiments (Figures 2 (a)-(c)) it can be seen that the most severe leaching was observed in deionized water, in terms of the formation of large-radius pores in the leached surface layer. This likely resulted from the severe leaching of calcium hydroxide from the paste (Figure 2(a)). However, the total porosity of the surface layer was not increased. In all three experiments with SRPC, the leached surface layer contained a significantly larger volume of coarse pores than the interior of the sample. Sample 20SF/SCSSS/CaM (Figure (2d)) exhibited the finest pore structure and the lowest intruded porosity of the pastes. For sample

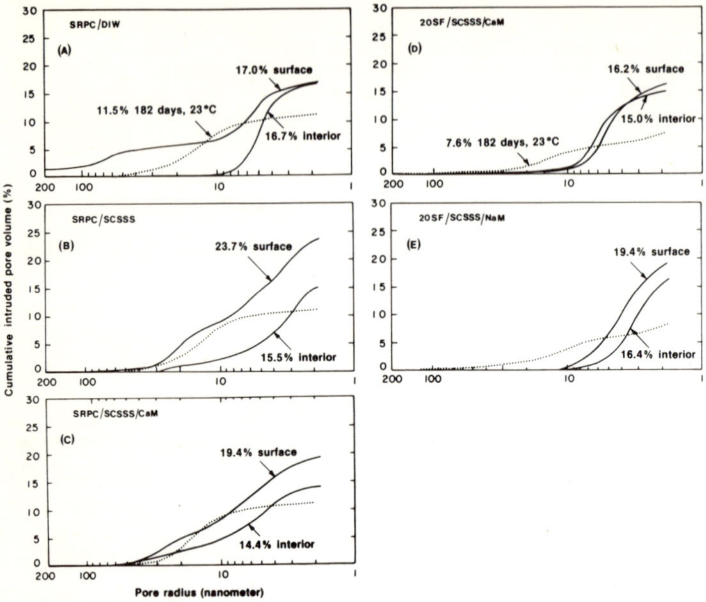

Figure 2. Cumulative Pore Size Distributions of As-Cured Cements (dotted lines) and Leached Cements (solid lines).

20SF/SCSSS/NaM (Figure 2(e)), the sodium montmorillonite was more severe in terms of increasing total porosity of the cement paste. A slight coarsening of the pore structure in the very fine end of the spectrum was noticed.

CONCLUSIONS

1. Cement pastes of various compositions were evaluated for their chemical durability. It was found that supplementary cementing materials, such as silica fume, could significantly improve the properties of sulphate resistant portland cement, in particular permeability to water and pore size distribution, and reduce the calcium hydroxide contents of the pastes. Addition of 20 percent by volume silica fume eliminated calcium hydroxide after moist curing for 91 days.

2. Due to the complexity of the system, no element could be found whose measured concentration in solution could accurately predict the durability of the cement. It was concluded, however, that monitoring the release of potassium would give some indication of durability. The presence of clay increased the release of potassium ions from the cements. The CFR of potassium steadily increased with time to reach a steady-state value in the presence of Ca-montmorillonite, whereas in the presence of Na-montmorillonite, or in leach solutions without clay, the cumulative fraction of release remained constant over the time investigated. The addition of 20 percent silica fume to the SRPC reduced the CFR of potassium giving some insight into the increased durability of blended cements in a disposal-vault environment.

3. The addition of clay suppressed the normal tendency of the pH of groundwater to rise rapidly in the presence of cement. The pH of the silica fume-modified cement was lower than that of SRPC, with and without the presence of clay.

4. SRPC modified by 20 percent silica fume was chosen for subsequent long-term material interaction experiments. A 'family' of cementitious materials based on this composition has the potential for use in a disposal vault, particularly for grouting of rock fractures, construction of concrete bulkheads, and containment of nuclear fuel waste.

REFERENCES

1. T.E. Rummery and E.L.J. Rosinger, The Canadian Nuclear Fuel Waste Management Program, AECL 7787, May 1983.

2. J. Freire-Canosa, J. Cipolla and N.C. Burnett, Dry Storage Systems Evaluation at Ontario Hydro. Proceedings of an International Workshop on Irradiated Fuel Storage, Toronto, Canada, November 1984.

3. R.D. Hooton and N.C. Burnett, Evaluation of Concrete Casks for the Management of Irradiated Fuel in Canada. Second International Symposium in Ceramics in Nuclear Waste Management. Advances in Ceramics 8, p.456, (1984).

4. R.D. Hooton and N.C. Burnett, Development of a Highly Durable Concrete Container for the Storage, Transportation and Disposal of Irradiated Fuel, to be published as American Concrete Institute, Special Publication.

5. S.K. Frape, P. Fritz and R.H. McNutt, Water-Rock Interaction and Chemistry of Groundwaters from the Canadian Shield, Geochim. Cosmochim. Acta 48, 1617 (1984).

6. R.F. Feldman, Proceedings of the Fifth International Symposium on Concrete Technology, Monterrey, California, 263 (1981).

7. P.K. Mehta, ibid p.35.

8. R.F. Feldman, Effects of Fly Ash Incorporation in Cement and Concrete, Materials Research Society Symp. Proceedings, 124 (1981).

9. R.D. Hooton, Permeability and Pore Structure of Cement Pastes Containing Fly Ash, Slag and Silica Fume. ASTM C-1, Symposium on Blended Cements, Denver, Colorado, June 1984.

10. Huang Cheng-yi and R.F. Feldman, Influence of Silica Fume on the Microstructural Development in Cement Mortars, Cement and Concrete Research, 15, 285 (1985),

11. R.S. Barneyback Jr. and S. Diamond, Expression and Analysis of Pore Fluids from Hardened Cement Pastes and Mortars, Cement and Concrete Research, 11, 279 (1981).

12. C.L. Page and O. Vennesland, Pore Solution Composition and Chloride Binding Capacity of Silica Fume-Cement Paste, Materials and Structures, 16, No. 91, 19 (1983).

13. R.B. Heimann and R.D. Hooton, Short-Term Dissolution Experiments of Various Cement Formulations in Standard Canadian Shield Saline Solution in the Presence of Clay, AECL Report (to be published).

14. F.P. Sargent, Applied Chemistry and Geochemistry Research for the Canadian Fuel Waste Management Program, Atomic Energy of Canada Limited Technical Record, TR-149*, July 1982.

15. B.W. Goodwin and M. Munday, A Reference Guide to SOLMNQ - Interactive Solution - Mineral Equilibrium Program, AECL 7800, December 1983.

16. R.B. Heimann and M.A.T. Stanchell, Reaction of Ca-Saturated Montmorillonite with Groundwater of Different Ionic Strengths at Temperatures Between 25 and 200°C, Geolog. Soc. America Abstracts 16 (6), 536 (1984).

17. R.B. Heimann and R.D. Hooton, Mineralogical Changes of Various Cement Formulations During Reaction with Groundwater in the Presence of Ca- and Na-Bentonite at 150°C, Can. Min., in preparation.

18. R.F. Hamon, Atomic Energy of Canada Limited, personal communication, 1985.

19. R.W. Lahann and H.E. Roberson, Dissolution of Silica from Montmorillonite: Effect of Solution Chemistry, Geochim. Cosmochim. Acta 44, 1937 (1980).

20. L.V. Benson, A Tabulation and Evaluation of Ion Exchange Data on Smectites, Environ. Geol. 4, 23 (1982).

21. M. Onofrei, Leaching Studies of Heavy Concrete Materials for Nuclear Fuel Immobilization Containers, AECL Report (to be published); also presented at the 87th Annual Meeting of the Amer. Ceram. Soc., Cincinnati, May 5-9, 1985.

---

* Unrestricted, unpublished report available from SDDO, Atomic Energy of Canada Limited Research Company, Chalk River, Ontario K0J 1J0.

OPTIMIZATION OF CONCRETE CONTAINER COMPOSITION IN RADIOACTIVE WASTE MANAGEMENT

I. B. PLECAS, Lj. L. MIHAJLOVIC and A. M. KOSTADINOVIC
Institute of Nuclear Science "Boris Kidrič", Vinca P.O.B.522, 11001 Belgrade, Yugoslavia

ABSTRACT

In this paper an optimization of concrete container composition, used for storing low and intermediate level radioactive waste from nuclear power plants in Yugoslavia, is presented.

Mechanical properties 37-52 MPa, permeability $1.07 \cdot 10^{-13}$ - $1.50 \cdot 10^{-11} cm^2$ and leakage rate $3.66 \cdot 10^{-6}$ - $1.77 \cdot 10^{-4}$ cm/d for concrete made of commercial materials, were tested.

INTRODUCTION

Low and intermediate waste represents 90% of volume of the total radioactive waste. After volume reduction and valuable components recovery, waste materials have to be conditioned for transport, storage and disposal. Storage of waste is made in concrete containers of 950 l volume, which are disposed in shallow ground trenches.

Engineering trench systems provide three biological protection barriers (Fig. 1).

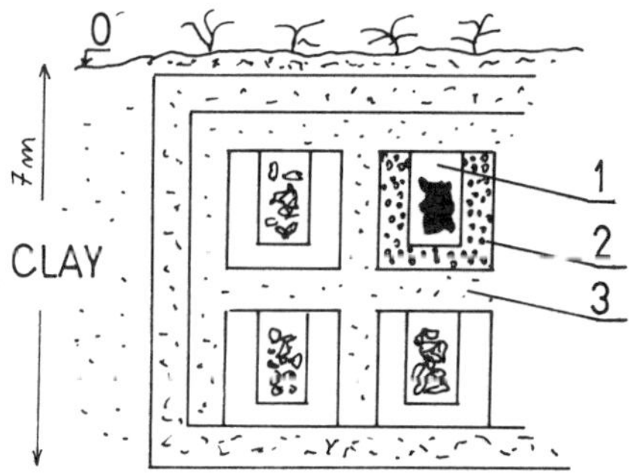

1. Mortar for immobilizing the waste and filling the concrete container
2. Concrete container
3. Concrete for filling trenches

Figure 1. Scheme of engineering trenches system.

The Granulometric composition calculation that rigorously controls main physico-mechanical characteristics, such as:

1. compressive strength
2. permeability
3. leakage test,

are especially important.

CONCRETE COMPOSITION FOR CONTAINERS

Formulations of concrete are represented in Table I.

Table I. Formulation of concrete

Formula number	1	2	3	4	5	6	7	8	9	10
Cement gm	400	400	400	400	400	400	400	400	400	400
Sand 0-2 mm	858	626	625	525	692	822	662	525	750	590
Granulate 2-4 mm	95	69	69	50	75	91	73	50	83	65
Granulate 4-8 mm	536	576	437	475	423	595	317	450	417	476
Granulate 8-15 mm	496	715	853	935	794	476	933	960	734	853
Water ml	140	140	140	140	140	140	140	140	140	140
Additive ml	8	8	8	8	8	8	8	8	8	8

Used materials

- Cement CPA-45 MPa(C_1) and cement CPA-55 (C_2), produced in Yugoslavia (Beočin Cement Factory).
- Sand "Moravac" from Serbia, fraction 0-2 mm, Gravel "Moravac" was used in the three following fractions: 2-4 mm, 4-8 mm and 8-15 mm.
- The water used for preparation of concrete was in accordance with JUS standard.
- Plastificator type additives, produced in Yugoslavia ("Chromos" - factory, Zagreb): Fluidal VX-OC (A_1) and Super fluidal (A_2).

CONCRETE QUALITY TESTING METHODS USED IN THE RADIOACTIVE WASTE SOLIDIFICATION PROCESS

Mechanical Characteristics

Testing of concrete compressive strength is classical method [1] which is practised in civil engineering. Cube shaped concrete samples 10 x 10 x 10 cm were used. Compressive strength (M) is expressed in MPa. Data are presented in Table II.

Permeability

Each concrete sample was subjected to nitrogen permeability measurement at an absolute pressure of 1 MPa and temperature of 20°C. Samples were held for 28 days at 20°C and 65% humidity [2]. The apparatus for permeability measurement is shown in Fig. 2. A gas stream through concrete follows Darcy's Law. By integrating Darcy's Law over the height and cross-section (for cylinder shaped samples). The permeability coefficient, K (cm^2), is calculated by:

$$K = \frac{V \cdot dP_i \cdot H \cdot \mu}{Pm \cdot t \cdot S(P-P_0)}$$, where

V - volume of sample (cm^3)
Q - gas flow (cm^3/s)
K - permeability coefficient (cm^2)
S - sample cross section (cm^2)
P_0 - absolute pressure (Pa)
μ - gas viscosity (Pa.s)

H - sample height (cm)
t - time (s)
P - starting pressure (Pa)
dP_i - actual pressure (Pa)
Pm - mean pressure (Pa).

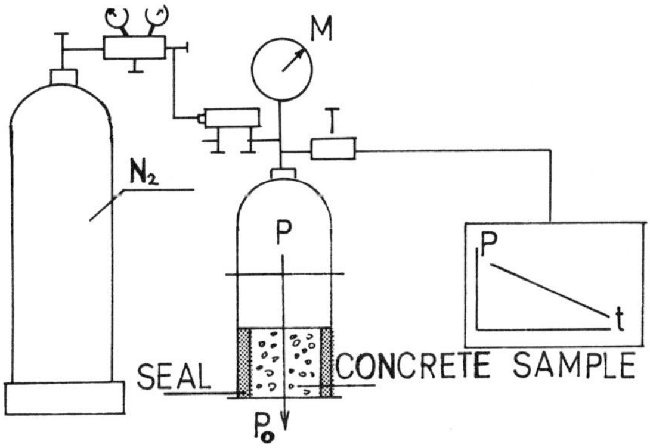

Figure 2. Scheme of the experimental equipment for permeability measurement.

Concrete samples were cylindrical in shape (diameter 10 cm, height 10 cm)

Leakage test

An apparatus for concrete leakage testing has been constructed at the Institute of Nuclear Science "Boris Kidrič" Vinča [3] . This original method provides a virtual image of the leakage rate of radionuclides in solidified waste materials by underground flows. This is not the standard procedure suggested by IAEA, worked out by F. Hespe [4] .

The apparatus made simulation of real leakage process in concrete container, possible.

Concrete discs with diameter 10 cm and height 1 cm were used for testing (Fig. 3).

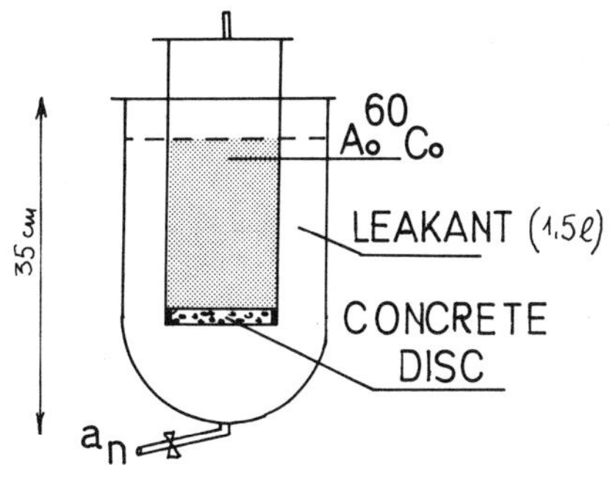

Figure 3. Scheme of the experimental equipment for leakage.

Every 15 days samples of leakant were taken and examined during the period of one year. By using following equation the incremental leakage rate, R (cm/d), which has the dimension of velocity, can be estimated as:

$$R\ (cm/d) = \frac{a_n}{A_o} \cdot \frac{V}{S} \cdot \frac{1}{t_n}\ ,\ \text{whereas}$$

a_n - radioactivity leaked during the leakage renewal period (Bq)
A_o - radioactivity initially present above concrete disc (Bq)
V - volume of concrete disc (cm^3)
S - exposed surface area of concrete disc (cm^2)
t_n - duration of leachant renewal period of 15 days.

RESULTS

Experimental results for 40 concrete mixture formulas for containers are shown in Table II.

Table II. Results of experimental measurement of mechanical property, M (MPa), permeability, K (cm^2), leakage rate, R (cm/d), of 40 concrete mixture formulas

Cement	Aditive	N^o	M (MPa)	K (cm^2)	R (cm/d)
		1	42.1	4.69×10^{-12}	
		2	47.6	6.50×10^{-12}	
		3	48.1	6.60×10^{-12}	1.68×10^{-4}
		4	43.4	1.48×10^{-12}	
	A_1	5	43.2	8.16×10^{-13}	
		6	40.6	1.50×10^{-11}	
		7	37.0	8.94×10^{-13}	
		8	40.0	8.16×10^{-13}	
		9	47.0	6.05×10^{-13}	
C_1		10	42.0	9.38×10^{-13}	
		1	44.4	2.68×10^{-13}	
		2	36.8	1.04×10^{-12}	
		3	36.8	7.61×10^{-13}	
		4	46.0	8.20×10^{-13}	
	A_2	5	44.6	9.71×10^{-13}	1.77×10^{-4}
		6	42.3	1.08×10^{-12}	
		7	37.0	1.94×10^{-12}	
		8	40.0	4.02×10^{-13}	
		9	47.0	2.34×10^{-13}	
		10	42.0	5.21×10^{-13}	
		1	44.2	2.80×10^{-13}	
		2	44.2	1.15×10^{-13}	
		3	49.8	1.17×10^{-13}	
		4	44.0	1.28×10^{-13}	
	A_1	5	51.2	2.66×10^{-13}	8.60×10^{-5}
		6	43.0	5.30×10^{-13}	
		7	38.0	8.12×10^{-13}	
		8	43.4	4.17×10^{-13}	
		9	47.5	6.62×10^{-13}	
C_2		10	43.0	9.38×10^{-13}	

Cement	Aditive	N°	M (MPa)	K (cm^2)	R (cm/d)
		1	43.2	4.30×10^{-13}	
		2	47.2	1.07×10^{-13}	
		3	52.0	1.12×10^{-13}	3.66×10^{-6}
		4	46.2	1.56×10^{-13}	
	A_2	5	47.4	3.41×10^{-13}	
		6	43.0	5.20×10^{-13}	
		7	45.0	2.81×10^{-13}	
		8	44.0	2.62×10^{-13}	
		9	50.0	4.08×10^{-13}	1.68×10^{-4}
		10	44.1	2.88×10^{-13}	

CONCLUSION

Results of permeability measurement and mechanical strength are very satisfactory. Leakage-rate of radionuclides through 5 tested concrete samples do not exceed $1{,}70 \times 10^{-4}$ (cm/d), which secure radionuclides preservation in solidified medium much longer than 300 years.

Functional relations between mechanical property M, permeability K and leakage rate R, have been analysed by polynominal least square approximation of experimental data.

Results presented in this paper are examples of results obatined in a 3-year concrete testing project which will influence the design of the engineered trench systems for a future central Yugoslav radioactive waste storing centre.

REFERENCES

1. JUS U.M1.012.
2. I. B. Plećaš, Waste management -83 Tucson, Arizona, USA, vol. 2, pp. 185 (1983)
3. I. B. Plećaš, XI Regional congress of IRPA, Vienna, pp. 206, (1983).
4. E. D. Hespe, At. Energy Rev. 9, pp. 195, (1971).

THE CORROSION OF STEEL AND ZIRCONIUM IN ANAEROBIC CONCRETE

CAROLYN M. HANSSON
The Danish Corrosion Centre
DK-2600 Glostrup, Denmark

ABSTRACT

An experimental investigation has been made of the corrosion behaviour of steel and Zircalloy 2 in deaerated synthetic pore solution and of steel in deaerated mortar. Zircalloy 2 was found to be passivated at all values of pH and electrochemical potentials investigated. Its passive corrosion rate was of the order of <0.01 μm/year. The steel exhibited active corrosion rates of a few μm/year with hydrogen evolution as the corresponding cathodic reaction. The measured free potentials indicated, however, that the hydrogen partial pressures were less than 1 atm.

INTRODUCTION

It is proposed that nuclear waste, encapsulated in steel or zirconium containers, be deposited in reinforced concrete silos buried underground. The concrete will be water saturated and, after a period of time, will probably be exposed to saline water. Initially, oxygen will be present both in the water phase and in air bubbles trapped in the concrete. The container material and the reinforcement will, thus, be exposed to the cement pore solution and, thereby, subject to corrosion.

In the highly alkaline conditions prevailing in concrete, both steel and zirconium are passivated, i.e. a thin, highly protective oxide or hydroxide film is formed on the surface. This film does not stop the corrosion but reduces it to a level which, for most normal constructions, can be considered as insignificant - typically 0.1 μm/yr and <0.01 μm/year for steel and zirconium, respectively. The film is maintained by a slow dissolution of the oxide or hydroxide into the cement pore solution and an equally slow oxidation of the metal. A continuous supply of oxygen is required in order to maintain this passive state.

In an underground system which is not exposed to air, the slow corrosion of the steel will eventually exhaust the available supply of oxygen and the passive film will not be stable. When this occurs, the metal will no longer be protected and will actively corrode with the evolution of hydrogen as the corresponding cathodic reaction.

In practice, the actual rate of corrosion is likely to be limited by the access of pore solution to the metal surface, by the removal of the corrosion products away from the surface and by the partial pressure of hydrogen which can exist in thermodynamic equilibrium under the prevailing electrochemical conditions. To the author´s knowledge, neither gaseous H_2 evolution from steel in concrete nor electrochemical potentials low enough to permit it have been observed in practice. In theory, however, it is possible that hydrogen pressures sufficient to crack the concrete could build up in the silos and the cracks could lead to seepage of radioactive material.

In order to gain a better understanding of these possibilities, an experimental investigation has been made of the corrosion of steel and zirconium in anaerobic synthetic cement pore solution[1] and of steel in anaerobic, water saturated mortar[2]. Concrete samples were not used because their large size would have necessitated a reduction in the number of samples to be studied and, in fact, the larger stones in concrete relative to those in mortar do not affect the electrochemistry of the system. Electrochemical techniques [3] were employed to monitor the behaviour of the samples and to measure the actual corrosion rates.

EXPERIMENTAL PROCEDURE

Steel Samples

For the study of steel for both reinforcement and containers, a plain carbon steel was used in the as-received condition. The "environments" investigated were synthetic cement pore solutions (mixtures of $Ca(OH)_2$, NaOH and KOH) of 4 different pH values and three different mortar mixes.

For the tests in synthetic pore solution, 8 mm ø steel rods were degreased and immersed in the appropriate solutions in closed cells together with a reference electrode and a second steel rod which was used as a counter electrode, as illustrated in Fig. 1.

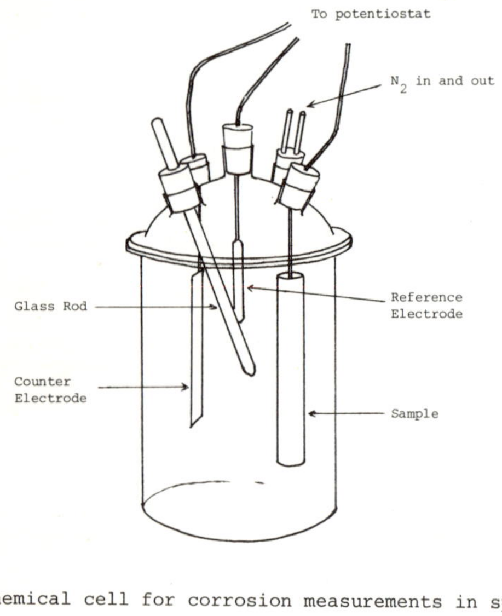

Fig. 1. Electrochemical cell for corrosion measurements in synthetic cement pore solution.

Oxygen-free nitrogen was bubbled through the solution to remove any dissolved oxygen and cathodic potentials were applied to the embedded steel to electrochemically reduce both the oxygen in the pore solution and any surface film on the steel.

After 3 days at a potential of -1400 mV SCE and a further few hours at -1600 mV SCE, hydrogen could be seen to bubble from the steel surfaces. Polarization curves were then obtained for the samples by repeatedly stepping the potential and measuring the current flowing between the sample and counter electrode at each potential step.

For the mortar samples, the mix proportions of cement:sand:water were 1:3:0.5 and the cements used were Swedish Standard cement (ordinary Portland cement), Standard cement + 0.8% CaS and Massiv cement (a blast furnace slag cement containing approx. 65% slag and 35% Standard cement). Prism samples of the mortars, 40 x 40 x 160 mm^3, each containing an 8 mm ø steel rod were cast and stored under water for approx. 18 months. Six samples of each mortar type were then immersed in a $Ca(OH)_2$ solution in closed tanks. They were deaerated in a similar manner to the bare steel samples described above but for a longer period because the oxygen dissolved in the pore solution of the mortars takes a longer time to be reduced. It should be noted that bubbles of hydrogen could not be observed even for samples which had been cathodically polarized to -1800 mV SCE for several weeks.

Polarization curves were obtained for each of the samples and then the applied potential was removed and the change in free potential, E_R, monitored. The values of E_R and of the corrosion current, i_{corr}, of each sample determined by the polarization resistance method[3,4] were measured periodically over the next 18 months.

Zircaloy 2 Samples

Zircaloy 2 tubes of the nominal composition given in Table 1 and with an outer diameter of 12 mm and a wall thickness of 0.1 mm were cut into samples 100 mm long. Single strand copper wire was spot welded to the inner surface of each as an electrical lead. The tubes were then filled with epoxy resin so that only the outer surface would be exposed to the solution and the weld joint was covered.

Table I. Composition of Zircaloy 2 (supplied by ASEA-ATOM)

Sn	1.50%	Ni	0.07%
Fe	0.10%	O	1400 ppm
Cr	0.10%	Remainder: Zr	

The samples were then immersed in synthetic pore solutions of pH 12.0, 12.6, 13.1 and 13.8 in closed cells as shown in Fig.1. Oxygen-free N_2 was bubbled through the cells for two days and then polarization resistance measurements were made on each of the samples. Two extra cells with the highest and lowest pH, resp., were also set up and the samples in these were held at -1850 mV SCE, which resulted in hydrogen evolution after two days. The applied potential was maintained for several days and then removed, and polarization resistance measurements were made the next day.

The passive film on zirconium is thermodynamically stable to potentials of -2565 mV SCE[5] and attempts to remove this film electrochemically were unsuccessful. Consequently, the film was removed mechanically by scratching the sample surface while it was immersed in the synthetic pore solution, and the repassivation was investigated by monitoring the corrosion current during and after scratching.

RESULTS AND DISCUSSION

Steel Samples

The polarization curves for two bare steel samples in deaerated "pore solution" of pH 13.0 are shown in Fig. 2. The potentials have been corrected for the "IR" drop across the measuring system. The "IR" drop is negligible for small currents (i.e. in the region of the free potential) but is significant at higher currents. It can be seen that the values of E_R mea-

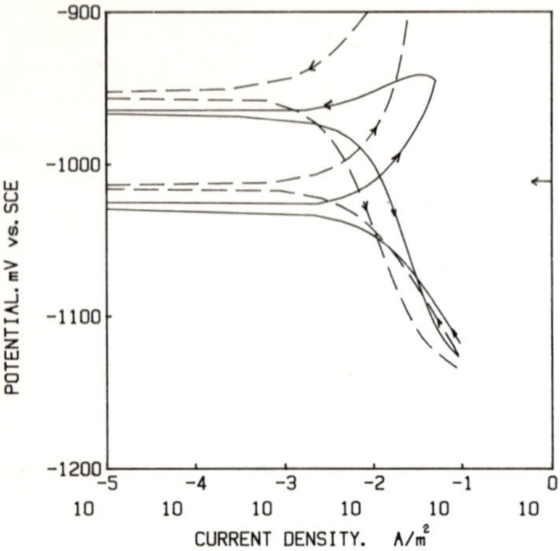

Fig. 2. Potentiostatic polarization curves for two bare steel samples in synthetic cement pore solution of pH 13.0. The curves were obtained by stepping the potential in the direction of the arrows.

sured during the first sweep of the potential were in the region of the equilibrium potential for hydrogen evolution at 1 atm. ($E_0(H_2)$) for this pH but that, after a slight positive polarization, E_R shifted to higher values. The change in slope of the cathodic curve on the negative sweep of the potential can be attributed to the onset of H_2 evolution. This experiment was repeated several times, both on the same and on different samples and the value of E_R was never observed to be lower than $E_0(H_2)$.

Figure 3 shows an example of the polarization curves for steel in each of the three mortar types. The curves were begun 24 hrs. after the applied potential of -1950 mV was removed. The value of E_R of all the samples was between -900 and -1000 mV SCE, i.e. around the value of $E_0(H_2)$ for 1 atm. The exact value of $E_0(H_2)$ is not known because the pH of the pore solution in the samples at the time of the measurement is not known.

The corrosion current for the samples represented in Fig. 3 can be estimated by extrapolating the anodic and cathodic curves to the free potential giving a value of approximately 10^{-3} A/m^2, or an average corrosion rate of 1.2 μm/year. Note that it is unreasonable to expect a more accurate estimate of the corrosion rate than within an order of magnitude for two reasons: firstly, the linear (Tafel) region of the curves, from which the extrapolation should be made, are not generally well defined; secondly, the data give only the average value of the corrosion rate over the whole steel sample whereas, in fact, corrosion is not normally uniform over the steel surface.

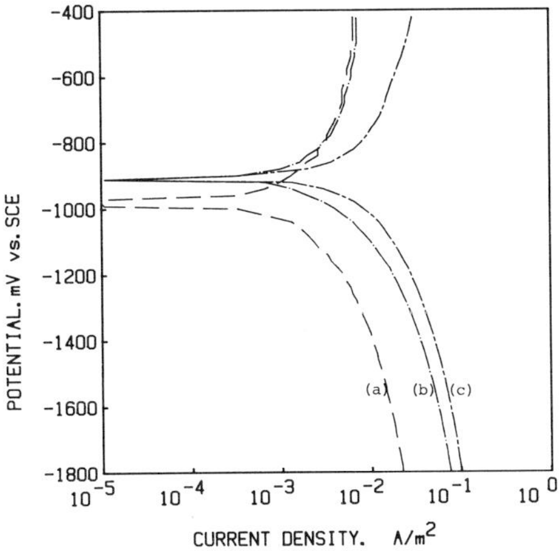

Fig. 3. Potentiostatic polarization curves for steel in deaerated mortar after cathodic polarization at -1950 mV SCE. (a)Massive mortar; (b) Standard mortar; (c) Standard mortar + 0.8% CaS.

The values of E_R and i_{corr} for the samples of steel in mortar measured over a period of 18 months were remarkable stable. Of the 18 samples, thirteen have values of E_R between -925 and -997, and these values have fluctuated within that range over the 18 months. The free potentials of the remaining five samples have risen slightly over this period to levels of -714, -883 and -870 (Massive), -660 and -824 (Standard + 0,8% CaS). The reason for this behaviour is not known but it is highly unlikely that it is due to oxygen diffusing into these samples as they have had the same treatment as those whose free potential has not risen. It is more likely that the change in potentials in the Massiv and Standard + CaS samples is a result of some internal change such as a Pozzolanic reaction but this cannot be confirmed without destructive testing.

The corrosion currents for all samples has been stable and of the order of 10^{-3} A/m^2 for the whole of this 18 month period. If this current is totally due to H_2 evolution, it is equivalent to ∿100 μl/day per sample.

Fick's first law for diffusion through the wall of a cylinder can be expressed as :

$$q = -D\ (2\pi l t)\, dC/d(\ln r)$$

where q is the quantity of diffusing species passing through the cylinder in time t, D is the diffusion coefficient, and dC/d(ln r) is the concentration gradient across the wall of the cylinder. Assuming the concentration of hydrogen at the steel/mortar interface to be equal to the maximum solubility in water at 1 atm. pressure and room temperature ($7.8.10^{-4}$ M [7]) and that the concentration at the surface of the mortar samples is zero, the quantity of hydrogen developed by a current of 10^{-3} A/m^2 can be continuously removed by diffusion if the diffusion coefficient is 9.10^{-11} m^2/s. This value may be compared with the value of $5.2.10^{-9}$ m^2/s for the diffusion of hydrogen in water[7] and a value of $1.6.10^{-12}$ m^2/s for diffusion of hydrogen through bentonite clay[8] .

These results indicate either that the hydrogen that is developed can escape through the samples without building up a gas pressure or, that within the period of the measurements, the Fe^{3+} ions have not all been reduced to Fe^{2+} ions.

Zircaloy 2

The values of free potential and corrosion current for Zircaloy 2 samples in "pore solution" are given in Table II. The currents are at the lowest limit of the detection range of the equipment used and should, therefore, be regarded as the maximum values. It is possible that the actual values are even lower.

Table II.

pH	E_R mV SCE	i_{corr} A/m^2
12.0	-401	2.10^{-5}
12.6	-533	3.10^{-5}
13.1	-562	3.10^{-5}
13.8	-473	3.10^{-5}
13.8*	-1045	9.10^{-5}

* Sample cathodically polarized prior to testing

In order to determine the stability of the passive film on the Zircaloy 2 samples, the free potential was first measured and the potentiostat set to maintain that potential. While held at this potential, the current was measured at 450 ms intervals and recorded by computer. At the free potential, the net current is zero but when the surface of the sample was scratched, the small area of bare metal had a significantly different free potential and resulted in a sharp rise in current. This experiment was conducted in a solution of 13.6 and with two samples, one virgin sample and one which had been held at a low potential for several days and had, thus, been exposed to violent hydrogen evolution and could, therefore, be expected to be saturated with hydrogen.

It was expected that if the samples were in a thermodynamically active region of the potential-pH diagram[5] , active corrosion would continue after the passive film was mechanically removed. On the other hand, if the region of passivity actually extended up to the high pH values, the samples should rapidly repassivate. In fact, the latter behaviour was observed for both samples as illustrated in Figs. 4 and 5. High corrosion rates were briefly observed but repassivation occurred within a few seconds and there was no observable difference in behaviour between the virgin sample and the one which had been cathodicaly polarized. The higher currents in Fig. 5 are most likely attributable to the fact this sample was scratched harder that the virgin sample.

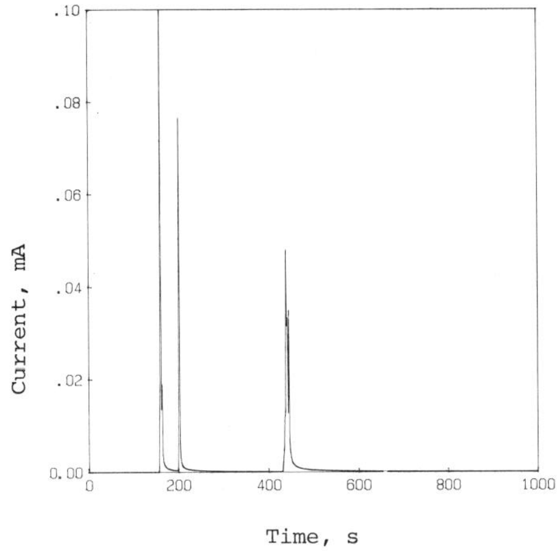

Fig. 4. Peaks in the anodic current resulting from scratching the surface of a virgin Zircalloy 2 sample potentiostatically held at its free potential (from Ref. 1).

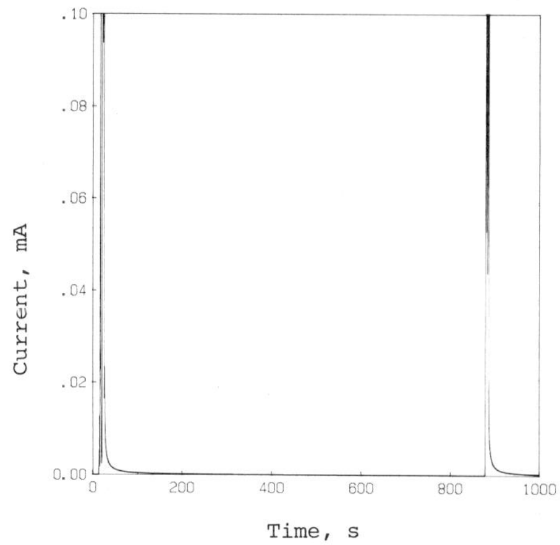

Fig. 5. Anodic current peaks resulting from scratching the surface of a Zircaloy 2 sample which had previously been cathodically polarized and, thereby, exposed to hydrogen (from Ref. 1).

SUMMARY AND CONCLUSIONS

Steel Samples

1. When the dissolved oxygen in the mortar and oxides on the steel have been reduced, the free potential of the steel returns to a value around the hydrogen equilibrium potential, independent of the value of the previous applied potential.

2. During the 18 month period of investigation, the corrosion current of the steel in deaerated mortar, determined by both potentiostatic polarization curves and polarization resistance measurements is of the order of 10^{-3} A/m^2 corresponding to a corrosion rate of a few μm/year. The measured currents were independent of the value of free potential and of the previous applied potential.

3. The values of corrosion current and the stability of the free potential indicate that it is unlikely that undissolved gaseous H_2 has been produced by the corrosion reaction during the 18 month period of investigation.

Zircaloy 2 Samples

1. The corrosion rate of Zircaloy 2 samples in synthetic "pore solution" is of the order of 2.10^{-5} A/m^2 and is independent of the pH within the measuring capability of the experiment.

2. It did not prove possible to remove the passive film electrochemically and, when the film was mechanically removed, the scratched area repassivated within a few seconds demonstrating that the film is stable even at pH = 13.6

3. It may be concluded that active corrosion of Zircaloy 2 in concrete will not occur. Moreover, by comparison with measurements on steel, it is probable that the corrosion rate of Zircaloy 2 in concrete will be even lower than that in the "pore solutions". Consequently, even if the potential were to fall to a negative level, the rate of hydrogen evolution would be correspondingly small.

REFERENCES

1. C.M.Hansson, SKBF KBS Technical Report, 84-13 (1984)

2. C.M.Preece, SKBF KBS Technical Report 82-19, (1982).

3. C.M.Hansson, Cement and Concrete Research, 14, 574 (1984)

4. M.Stearn and A.L.Geary, J.Electrochem.Soc. 104, 56 (1957)

5. M.Pourbaix, Atlas of Electrochemical Equilibria in Aqueous Solutions, (Pergamon Press, Oxford, 1966), p.312.

6. P.G.Shewmon, Diffusion in Solids, (McGraw Hill Book Company, New York, 1973) p.2.

7. R.H.Perry and C.H.Chilton, Chemical Engineer´s Handbook, 3rd ed. (McGraw Hill Book Company, New York, 1973) p.3-225.

8. T.E.Eriksen and A.Jacobsson, SKBF KBS Technical Report 82-17 (1982).

ESCA INVESTIGATION OF THE REACTION PRODUCTS FORMED ON TITANIUM EXPOSED TO WATER SATURATED BENTONITE CLAY

HÅKAN MATTSSON AND INGEMAR OLEFJORD
Department of Engineering Metals, Chalmers University of Technology,
S-412 96 Göteborg SWEDEN

ABSTRACT

Titanium is one of the materials proposed in the Swedish programme for the final containment of spent nuclear fuel. In the present investigation, the final repository environment was simulated on the laboratory scale by embedding titanium and a Ti-Pd alloy in dense, water-saturated bentonite clay. The temperature was 95°C and the exposures lasted between 4 months and 2 years. Analysis was performed using ESCA combined with ion etching.

The reaction products formed on the surface consists of TiO_2. Montmorillonite - the main constituent of bentonite - is incorporated in the oxide. Suboxides exist near the oxide/metal interface.

The oxide thickness is in the range 70-100 Å. The oxide growth between 4 months and 2 years is small. No significant influence of Pd could be noted. If it is assumed, that the oxide growth follows a logarithmic law, the expression giving the thickest oxide is $y = 5.5 \ln t$ (where y is the oxide thickness (Å) and t is the exposure time (s)).

INTRODUCTION

The storage of spent nuclear fuel calls for materials with exceptional resistance against corrosion. Titanium is one of the candidates in the Swedish programme for waste disposal. According to this programme, the storage will take place at a depth of 500 m in the Swedish bed-rock [1]. After deposition, the shafts will be back-filled with a sand-bentonite mixture which will be saturated with the ground-water present at this depth.

Investigations performed on titanium under similar conditions are rather few. In foreign studies, titanium and its alloys have primarily been considered for the storage of spent nuclear fuel in bedded salt or in sub-seabed sediments. Hence, a great deal of interest has been devoted to corrosion studies in brines and in sea-water [2, 3].

In this investigation, the composition and the thickness of the reaction products formed on commercially pure titanium and an alloy, Ti-0.2 weight% Pd in a simulated repository environment were investigated. The main departures from the real milieu were a higher oxygen content in the ground-water and the absence of a nuclear radioactive source. Analysis was performed with ESCA in combination with ion sputtering.

EXPERIMENTAL

The metals used were commercially pure titanium (Avesta ATi24) and Ti-0.2 weight% Pd alloy (Avesta ATi24Pd). The chemical analyses of the two materials are shown in Table I.

Prior to exposure the samples were ground on emery papers of successively finer grades down to grade 600.

The repository environment was simulated on a laboratory scale by embedding the titanium samples in water-saturated, dense bentonite at a high temperature (95°C). The equipment is described in detail elsewhere [4]. The titanium sample is placed in a block of bentonite which is compressed to give a density of 2.15 g/cm3. The block is then placed in a pressure cell, which prevents the clay from swelling. The bentonite is then exposed to the ground-

Table I. Chemical analyses of ATi24 and ATi24Pd.

Material	weight-%	C	O	Fe	N	Pd	Ti
ATi24		0.008	0.06	0.04	0.008	-	bal.
ATi24Pd		0.008	0.06	0.04	0.008	0.18	bal.

water. The clay absorbs 15 weight-% water under these conditions. The composition of the artificial ground-water is given in Table II. The difficulties of removing air trapped in the pores of the bentonite made any attempts to reduce the oxygen content futile. Therefore, the oxygen concentration in the artificial ground-water can be assumed to be the same as for water in equilibrium with air (1 ppm at 95°C).

Table II. Chemical composition of the artificial ground-water [5].

Species	HCO_3^-	SiO_2	SO_4^{2-}	Cl^-	Ca^{2+}	Mg^{2+}	K^+	Na^+
Concentration (ppm)	123	12	9.6	10	18	4.3	3.9	6.5

The exposure times were 4 months, 1 year and 2 years. Analyses were made using ESCA in combination with ion sputtering. Calibration of the ESCA binding energies was obtained by using the Cu $2p_{3/2}$ signal at 932.5 eV and the C 1s signal of graphite at 284.3 eV as internal standards. The sputtering rate - 7.5 Å/min - was obtained from titanium anodized at various potentials in 1M H_2SO_4, which has a well-known anodic growth (25 Å/V) [6].

RESULTS

Figure 1 shows the ESCA O 1s, Ti 2p, Mg Auger, Si 2s, and Al 2p spectra recorded from the Ti-0.2Pd alloy exposed for 2 years in bentonite clay. The positions of the peaks correspond to Ti^{4+}, Si^{4+}, Al^{3+} and Mg^{2+} in their simple oxides. The O 1s signal is composed of two peaks. From the binding energy, 530.2 eV, as well as from the intensity relative to Ti 2p it can be concluded that the low binding energy peak corresponds to TiO_2. The binding energy of the oxygen peak at 532.1 eV is different from the O 1s peaks representing oxygen in any of the simple oxides SiO_2 (533.5 eV), Al_2O_3 (532.8 eV) and MgO (531.0 eV). However, the intensity of the 532.1 eV peak matches the intensities of Si^{4+}, Al^{3+}, and Mg^{2+} if one assumes that these elements are present as montmorillonite, $Al_{5/3}Mg_{1/3}(OH)_2(Si_2O_5)_2^{1/3-}$, which is the main constituent of bentonite. The atomic ratios between the latter three elements at different etch depths are shown in Table IIIa. The corresponding values for pure bentonite are given in Table IIIb.

The atomic ratios correspond well with montmorillonite both before and after a slight ion etching. The same is true for pure bentonite. It can thus be concluded that Si, Al and Mg are elements from the bentonite clay which are absorbed in the oxide layer.

The other elements detected with ESCA were Ca, S, C and N. Of these, calcium is present in its divalent state between the negatively charged montmorillonite. The sulphur presumably originates from the sulphate present in the artificial ground-water (see Tab. II). Carbon and nitrogen are contaminants from the atmosphere and the rinsing procedure.

In the lower row of Fig. 1 are shown the spectra recorded after 50 Å ion etching. It is notable, that valence states lower than Ti^{4+} appear in the Ti 2p spectrum. These are created by decomposition of the surface oxide during

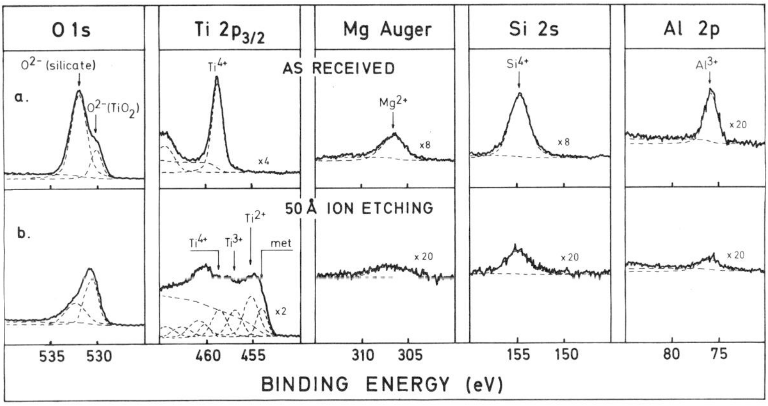

Fig. 1. ESCA spectra recorded from Ti-0.2Pd exposed 2 years to ground-water saturated bentonite clay at 95°C. Upper row: spectra recorded as received. Lower row: spectra recorded after 50 Å ion etching

Table III. Atomic ratios between Si, Al and Mg at different etch depths.

Element	Etch-depth (Å) 0	2	15	35	50	montmor-illonite
a) Ti and Ti-0.2Pd exposed to compacted bentonite clay saturated with ground-water at 95°C for times between 4 months and 2 years.						
Si	1	1	1	1	1	1
Al	.44	.49	.66	.78	.62	.42
Mg	.09	.10	.11	.13	.13	.08
b) pure bentonite						
Si	1	1	1	1	1	
Al	.44	.44	.54	.56	.52	
Mg	.06	.06	.06	.07	.09	

Ar^+ ion sputtering [7]. The presence of a signal from the metal substrate shows, that the thickness of the remaining oxide is of the same order of magnitude as the mean free path of the photoelectrons (i.e. a few tenths of Å). It is further noted, that the Si 2s, Al 2p and Mg Auger (KLL) peaks as well as the high binding energy component in the O 1s spectrum has decreased substantially. This is a further support for the theory that the high binding energy component in the O 1s spectrum originates from the montmorillonite species. Figure 1 and Table III show that the ratio between Al and Si has increased. This is probably due to a higher sputtering yield for Si than for Al [8]. The bentonite exhibits the same effect, although the Al enrichment is lower.

The in-depth distributions of Ti^{ox} (the sum of all Ti oxide states), Si^{4+} and Al^{3+} are shown in Fig. 2 for Ti samples exposed 4 months and 2 years. A high concentration of Al and Si is found at the surface which decreases towards the oxide/metal interface at 80 - 90 Å. The Ti-oxide concentration is low at the surface and increases to a maximum value of ca. 35 atomic % at 40 to 50 Å etch depth. Approaching the oxide/metal interface the Ti^{ox} concentration decreases considerably, but can still be found in significant amounts after 150 Å ion etching.

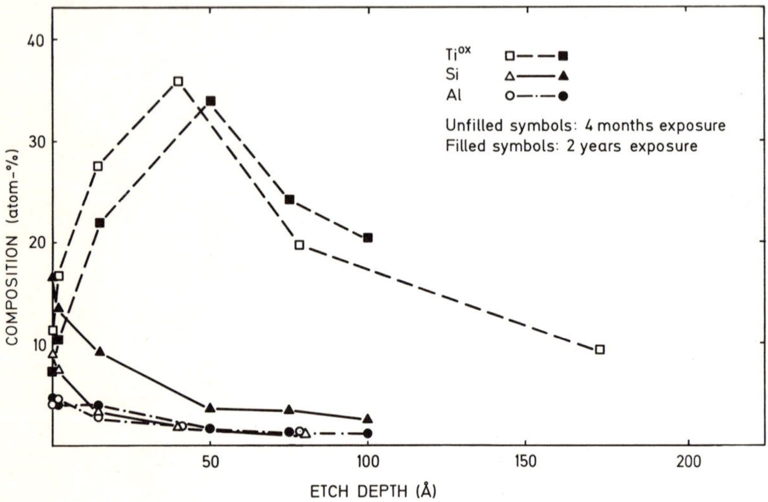

Fig. 2. Depth profiles of Ti exposed to water-saturated bentonite clay at 95°C for 4 months and 2 years respectively.

Essentially three models can be put forward for the interaction between bentonite and Ti; (a) incorporation, (b) agglomeration and (c) on-top adsorption. In Fig. 3 these models are illustrated together with the calculated ESCA depth profiles. The observed concentration profile (Fig. 2) is in best agreement with model a (incorporation of bentonite into the oxide): the concentration of the elements Si and Al is highest at the surface and is then monotonously decreasing throughout the oxide and the measured Ti^{ox} profile shows a maximum in the interior of the oxide The models b and c show less good agreement. In model b the bentonite is distributed as large agglomerates giving a constant intensity from the bentonite elements over several hundred ångströms. Model c comprises the adsorption of a few monolayers of bentonite on top of the titanium oxide. In this case the concentration of the bentonite elements would drop to zero over a depth of 2-10 Å while the Ti-oxide signal would increase to its full value over the same distance

The true in-depth composition of the oxide cannot be obtained by ion etching since the sputtering process reduces the oxide. However it is possible to make comparisons between the decomposition behaviour of the actual oxide with that of a homogeneous TiO_2 sample. In Fig. 4 the intensity fractions of the three Ti oxide states Ti^{2+}, Ti^{3+} and Ti^{4+} are given as a function of etch depth both for a homogeneous TiO_2 sample and a Ti-0.2Pd sample exposed for 2 years in bentonite. The figure shows that the two

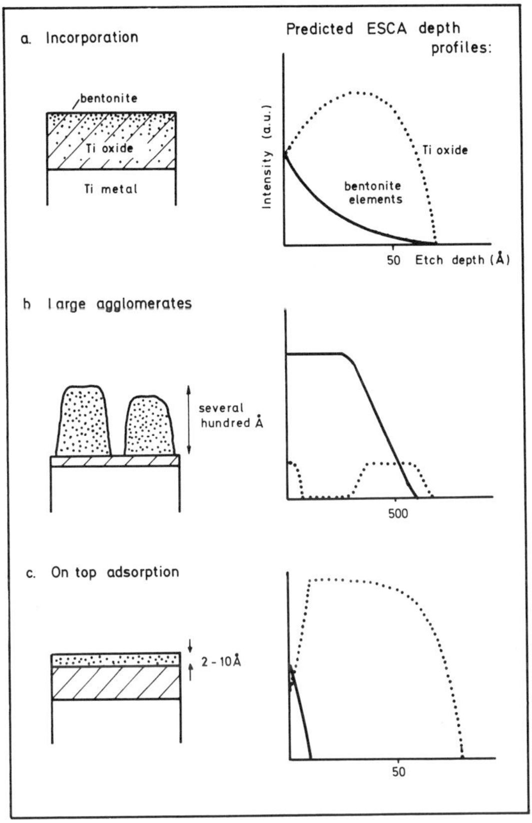

Fig. 3. Showing three proposed models and the corresponding ESCA depth profiles for the interaction between bentonite and titanium.

intensity fraction profiles are quite similar down to 35 Å. At etch depths greater than 50 Å contribution from the Ti^{2+} state is markedly higher for the Ti-0.2Pd alloy than for solid TiO_2. The conclusion is that lower valence oxides exist below this etch depth. In this respect the exposure time does not influence the Ti^{2+}/Ti^{4+} ratio. Thus, it is suggested that a quasi-thermodynamic equilibrium exists at the oxide/metal interface

The Pd 3d signal was recorded for the Ti-0.2Pd alloy. No Pd could be detected in the oxide or at the oxide/metal interface. At deeper etch depths, the concentration approached its bulk-value. Due to the finite extent of the oxide/metal interface it is difficult to draw any conclusions about Pd enrichment in the top metal layers. The function of Pd as an alloying element is to facilitate the repassivation of titanium in acid, non-oxidizing environments. This is accomplished by enrichment of the more noble Pd metal in the outermost metal layers. In this case the amount of titanium dissolved is low, about 50 Å, in this environment, so that the enrichment effect of Pd is low. This is further supported by the observation that no noticeable difference in composition and thickness exists between the alloyed and unalloyed material.

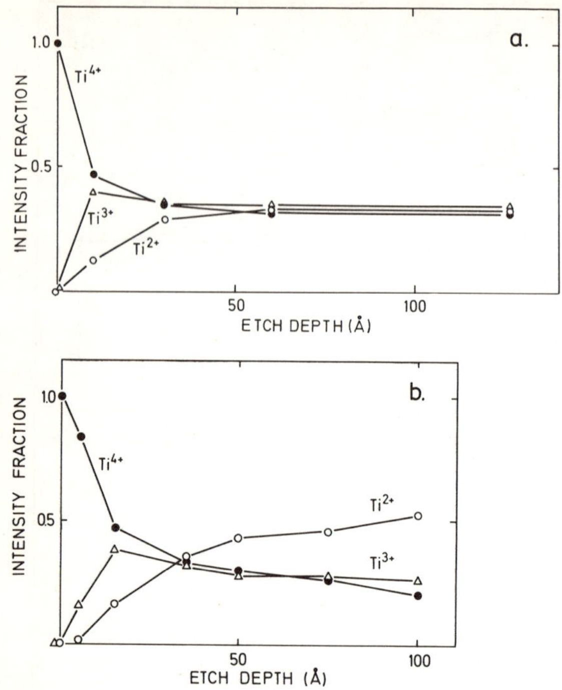

Fig. 4. Intensity fractions of the different Ti valence states (ESCA Ti 2p signal) vs. etch depth during etching with 1 keV Ar^+. a. TiO_2. b. Ti-0.2Pd exposed 2 years to water-saturated bentonite clay.

The oxide thickness was established by ion sputtering. In this case the oxide/metal interface was defined as the etch depth were the O 1s intensity had reached its half maximum value (this coincided with the depth where the Ti 2p metal signal had reached its half maximum value). The oxide thicknesses thus obtained are shown in Fig. 5.

The thickness of the reaction products formed on titanium and Ti-0.2Pd exposed to water-saturated bentonite clay for times ranging between 4 months and 2 years are in the range 70 - 100 Å with only a small tendency of thickening with exposure time. The Pd-alloyed samples exhibit only a few Å thicker oxide.

DISCUSSION

Montmorillonite is incorporated in the oxide formed on titanium exposed to bentonite clay saturated with artificial ground-water. The in-depth distribution shows that the concentration of the montmorillonite elements decreases monotonously through the oxide from a maximum value at the surface. It is suggested that montmorillonite is adsorbed on the surface at a constant rate while the oxide is growing outward with a rate that decreases with increasing oxide thickness.

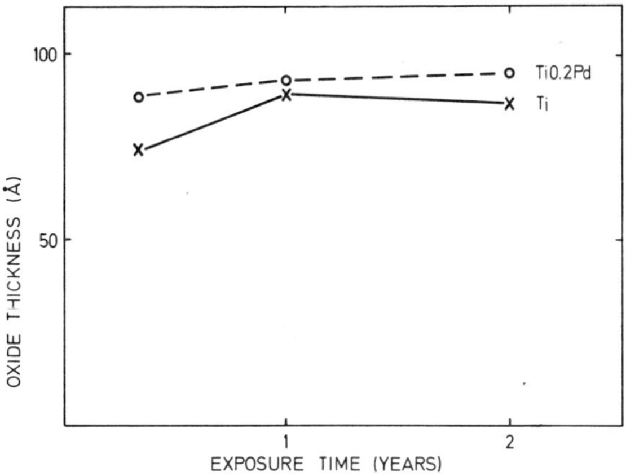

Fig. 5. Oxide thickness vs. exposure time for titanium and Ti-0.2Pd exposed to water-saturated bentonite clay at 95°C.

The oxide consists of an outer, thick layer of TiO_2. Suboxides are present near the oxide/metal interface. According to the phase diagram for Ti-O, several suboxides coexist with TiO_2 at 95°C [9]. Of these, the oxide with the lowest dissolution pressure is situated at the oxide/metal interface. The presence of TiO_2 on top of a layer of suboxides shows, that this phase diagram also is applicable for the oxidation of titanium in water-saturated bentonite clay.The thicknesses of the surface products formed on the samples are in the range 70 - 100 Å and shows only a small tendency of thickening with the exposure time. The results are in good agreement with those found by Henriksson and Pourbaix [10]. They obtained corrosion product thicknesses in the range 100 - 400 Å for titanium exposed 300 days at 100°C in bentonite clay. However, their results were of the same order of magnitude as the accuracy of their measurements.

A logarithmic growth law has been observed for titanium exposed between 1 minute and 3 months in aqueous solutions at 95°C [11]. This law yields only slightly thinner oxides than what was observed on the bentonite exposed samples. An explanation of the difference could be that the absorption of montmorillonite influences the oxide growth. It could also be explained as an apparent topographic effect. The bentonite-exposed samples were ground on emery paper and had therefore a rougher surface than the water-exposed samples. The latter were electropolished.

The oxygen content of the simulated ground-water was higher than what could be expected in a final repository environment. In the investigation quoted above of titanium exposed in aqueous solutions, the concentration of dissolved oxygen varied between 10^{-4} and 20 ppm. The results showed, that the thickness of the oxide is independent of the oxygen content. Thus, the oxidation kinetics of titanium is controlled by H_2O reduction. For this reason, the oxygen content of the simulated ground-water is assumed to have no influence on the oxide growth. Hence, the results obtained in our study are representative even for a repository environment with a very low concentration of dissolved oxygen. Further, the content of chloride ions in the artificial ground-water is rather low. It is considered unlikely that the NaCl content will exceed 0.5 weight-% in the Swedish bed-rock [12]. Even so, the exposures in aqueous solutions showed, that the oxide growth is not influenced by the NaCl content. The maximum concentration in the study was 1 weight-%.

If a logarithmic growth law is assumed to be valid, the highest growth obtained from the data of Fig. 5 is y = 5.5 ln t, where y is the oxide thickness (Å) and t is the exposure time (s). Extrapolation of this expression to 100,000 years, which is the desired lifetime for the canister material, yields an oxide thickness of about 160 Å which is equivalent to the dissolution of 90 Å of metal. Thus the limitation of titanium as a canister material seems not to be its general corrosion behaviour. Instead other phenomena such as localized corrosion and hydrogen embrittlement may limit the use of titanium as a canister material.

ACKNOWLEDGEMENTS

The Swedish Nuclear Fuel and Waste Management Co is gratefully acknowledged for the financial support.

REFERENCES

1. KBS Technical Report 31, Stockholm, 1977.

2. M.A. Molecke et al., Sandia Report SAND82-0429, Sandia National Laboratories, N.M., December 1982.

3 J.A. Ruppen et al., in Titanium for Energy and Industrial Applications, edited by D. Eylon (The Metallurgical Society of American Institute of Mining, Metallurgical, and Petroleum Engineers, 1981) pp. 355-369.

4. I. Olefjord and H. Mattsson, in Scientific Basis for Radioactive Waste Management - V, edited by W. Lutze (Elsevier Science Publishing Co, 1982) p. 669

5. B. Allard and G.W. Ball, J. Environ. Sci. Health, 14, 511 (1979).

6. R. Olier et al., J. Less-Common Met. 69, 73 (1980).

7. S. Hofmann and J.M. Sanz, J. Trace and Microprobe Techn. 1, 213 (1982/3).

8. G.K. Wehner, in Methods of Surface Analysis, edited by A.W. Czanderna (Elsevier, Amsterdam, 1975) p. 5.

9. E.M. Levin et al., Phase Diagrams for Ceramists, edited by M.K. Reser (The American Ceramic Society, Columbus, Ohio, 1964) p. 41.

10. S. Henriksson and M. de Pourbaix, KBS Technical Report 79-14, (Stockholm, 1979).

11. H. Mattsson and I. Olefjord, KBS Technical Report 84-19, (Stockholm, 1984).

12. KBS Annual Report 1980, KBS Technical Report 80-26, Stockholm, 1981.

PART VI

Near-Field Chemistry

THE "SOURCE TERM" FOR SAFETY ASSESSMENTS OF THE DISPOSAL OF VITRIFIED HIGH-LEVEL RADIOACTIVE WASTE

A. HOUGH,* J.A.C. MARPLES* AND I. HANDYSIDE**

*AERE Harwell, Didcot, Oxon, OX11 ORA, UK.
** Dept. of the Environment, Romney House, 43 Marsham Street, London, SW1 3PY, UK.

ABSTRACT

Two series of experiments have been carried out in an attempt to derive a source term for the release of some radioisotopes from vitrified high-level waste after disposal in hard rock. In the first type of experiment, the repository was modelled in miniature and the release of the isotopes was measured after they had leached from the glass and diffused through the backfill. In the second type, the "solubilities" of the isotopes were measured in the presence of the constituents of the repository.

INTRODUCTION

The work reported here is part of a wider study of the release of potentially important radioisotopes from vitrified high-level waste. The compositions of the borosilicate glasses examined M5 (189), M22 (209) and MW are given in Table I.

TABLE I
Glass compositions (wt%)

Oxide	189(M5)	MW	209(M22)
SiO_2	41.5	46.3	50.9
B_2O_3	21.9	16.4	11.1
Na_2O	7.7	8.3	8.3
Li_2O	3.7	4.0	4.0

Waste : (in each glass)
MgO: 6.2, Al_2O_3: 5.0, Fe_2O_3: 2.7, Cr_2O_3: 0.6,
NiO: 0.4, ZnO: 0.4, U_3O_8: 0.1, Fission product oxides: 9.9

Experiments have been carried out to obtain an estimate of the source term for safety assessments of the disposal of the vitrified waste in repositories constructed in hard-rock. For these investigations we have assumed that the vitrified waste will have been cast into steel canisters and covered before disposal with a thick iron (mild steel) overpack. (In an alternative scheme, a thin Ti alloy overpack might be used). These units could be emplaced in mined tunnels or shafts deep underground and the surrounding space "backfilled" with cement-concrete or bentonite for example.

It is assumed that the canister and overpack may last for 500 years or more before a significant amount of the glass is exposed to water: by this time the Sr-90 and Cs-137 will have decayed to insignificant levels but they were included in these experiments to provide information for examining possible accident scenarios. When water finally penetrates to a large fraction of the glass, it is assumed there will still be a considerable amount of metallic iron present to maintain reducing conditions. Radiologically, the most significant isotopes at this stage will include Tc-99, Np-237, Pu-239 and -240 and Am-241 and -243.

We have carried out two types of experiment in order to derive suitable source terms. In the first, we have modelled a repository on a very small scale and measured the concentrations of the isotopes at the outer edge of the backfill. In the second, we have measured the solubilities of the isotopes in the presence of the components of the repository. The maximum possible rate of release of each isotope from the repository - the 'near-field' - will then be this solubility multiplied by the rate of water flow through the repository: if saturation is not achieved, the actual release rate will be lower. The results of this type of experiment are of value since it is not necessary to make any assumptions about, for example, long term changes in the leach rate of the glass or the amount of cracking that has occurred therein.

SCALE-MODEL REPOSITORY EXPERIMENTS

It is assumed for these experiments that in a repository, the encapsulated cylinders of glass will eventually be surrounded by some form of backfill whose purpose is both to reduce the access of water to the glass and also to adsorb some of the active isotopes leached from the glass. Two backfills that have been suggested are bentonite and concrete and we have been investigating the effects of these on the leaching of the glass.

For both sets of experiments, batches of glass containing a full spectrum of the non-active isotopes of the fission products were doped separately with Sr-90, Tc-99, Cs-137, Np-237, Pu and Am-241. For Sr and Cs, the level was arbitrarily set at 0.05 mCi/g so that they could readily be detected in the leachates. The levels of the other isotopes were Tc-99: 0.05 mCi/g, Np-237: 0.003 mCi/g, Pu: 0.05 mCi/g and Am-241: 2mCi/g. These levels are approximately those which will occur in the real vitrified Magnox waste except for Np-237 where the level is ca. 10 x higher so that it could be more easily detected in the leachates. The glasses were cast into 12mm diameter rods and, after annealing, these were cut into 2mm thick discs. A separate disc was used in each experiment, i.e. six parallel sets of apparatus were used on each occasion.

Bentonite Backfill

Bentonite could be emplaced dry, so that, when water seeps into the repository, it will swell and help to seal the canisters in place. The apparatus used is shown schematically in Figure 1.

The sample chamber A was a tube ca. 30mm diameter and ca. 75mm long with a volume of ca. 50ml. Both ends of the tube were joined to standard cone fittings, the male ends being partly closed with porous silica filter discs. The lower cone fitting was sealed in place and the tube half filled with powdered bentonite. The doped glass disc (surface area 3cm^2) was placed on the powder and the filling completed. The total weight of bentonite was about 40g. The upper cone was then sealed in place. Water that had been equilibrated with crushed granite was then allowed to flow into the bottom of the column of bentonite. After the initial few mm, this took place very slowly as the bentonite swelled and partially sealed the tube: with a head of 120cms, the bentonite was saturated after 10-20 days. After this had occurred, the top cone was filled with water and the whole system transferred to an oven at 60°C. Both the water in the top cone and the reservoir of water above the granite were continuously purged with a mixture of 95% argon, 5% hydrogen to keep the conditions reducing.

At intervals, the water contained in the upper cone (10ml) was removed and replaced with fresh water. The pH of the water after about one month above the

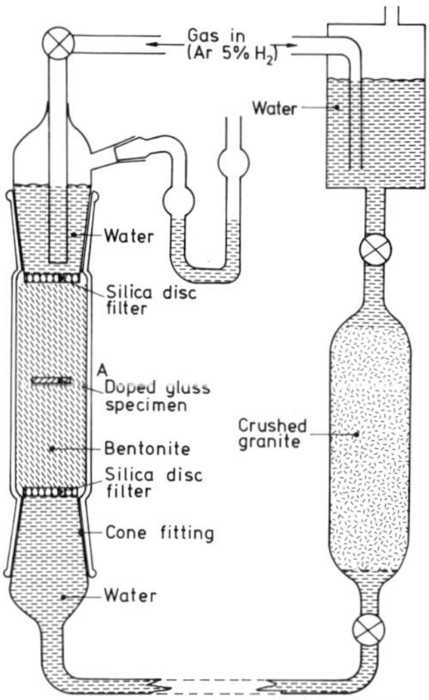

FIG.1.LEACH TESTING IN WATER - SATURATED BENTONITE

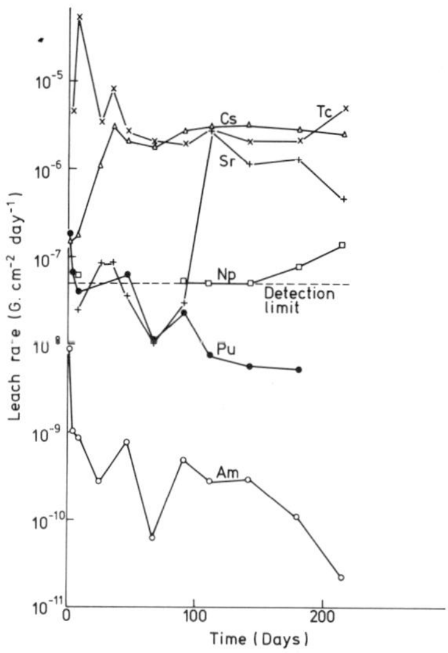

FIG.2.LEACH RATE OF UK GLASS 189 AT 60°C IN WET BENTONITE

bentonite filled tube is in the range 8.0 to 8.5. An aliquot was taken from this sample, evaporated to dryness and counted. A "leach rate" was calculated from the value obtained by dividing by the initial concentration of the isotope in the specimen and by the time since the last replenishment. This neglects any net adsorption on the apparatus or any activity diffusing down to the water below the column of bentonite. The system may be said to represent a repository where there is a flow of fresh water (10ml per month) past the surface of the bentonite backfill but only diffusion through the bentonite and past the specimen.

The results from the experiment are shown in Figure 2 for glass 189.

For glass 189, after 20–40 days the leach rate of Cs increased by about 10x and similarly for Sr after ca. 100 days the rate increased by about 15x. It is possible that this represented a breakthrough of the diffusion front. The other isotopes did not show this effect, the rates decreasing slowly with time except for Np which slightly increased from below the detection limit to only just above.

For glass 209, the leach rates were similar or lower than those for 189 except for Tc which was higher for 209 by about a factor of three. Some of the levels for Np, Pu and Am were below the detection limit, i.e. the number of counts recorded was less than the statistical variation in the background.

At the end of the experiment with glass 189, the glass tubes were removed from the oven and cut away from the bentonite columns. The latter were then cut into slices each about 2.5mm thickness which were separately soaked in water for 2 days after which acid was added and the soaking continued for one further day. Aliquots were taken, dried and counted to measure the amount of the radioisotope on each slice. After about 10 days a second aliquot was taken in some cases but showed no increase in activity – i.e. no further activity was being extracted from the bentonite. The activity levels found were converted to notional weights of glass by dividing by the specific activity of the latter in the usual way to enable the results from the different doped samples to be readily compared.

The glass specimens were dried and weighed: the gel layer was scraped off and the specimens reweighed. Leach rates were calculated from the weight losses both with and without the gel layer. These are given in Table II.

The gel layer was dissolved in acid, the solution diluted and counted and the result again converted to a notional weight of glass. The weights of glass equivalent to the amounts of activity adsorbed on the glassware of the apparatus and to the activity found in the static water reservoir at the bottom of the apparatus were calculated. The amounts found in the leachates during the whole run were totalled and a mass balance arrived at. (Table II).

The agreement with the actual weights of glass lost was remarkably good considering the complexity of the system: the amounts found by counting ranged from 67 to 113% of those found by weighing. The gel-layers formed on the Sr and Tc doped samples were scraped off in a glove box and were contaminated so that a mass balance could not be found.

TABLE II
Leach rates and mass balance for glass 189 in bentonite at 60°C

Sample	Sr	Tc	Cs	Np	Pu	Am
By weighing:						
Wt. loss (g)	.0192	.0073	.0180	.0256	.0084	.0101
Layer wt. (g)	.0498	.0249	.0533	.0493	.0205	.0163
Total (g)	.0690	.0322	.0713	.0749	.0289	.0264
From counting: "weight" loss (g)			.0569	.0504	.0327	.0245
% in bentonite			80	38	5	6
leachate			3	0.1	0.03	0.0008
layer			14	62	95	94
on apparatus			3	0.4	0.01	0.01
Leach rates ($g.cm^{-2}$ day^{-1})						
From weight loss	$2.6x10^{-5}$	$1.1x10^{-5}$	$2.7x10^{-5}$	$3.3x10^{-5}$	$1.2x10^{-5}$	$1.3x10^{-5}$
From weight loss including layer	$9.4x10^{-5}$	$4.7x10^{-5}$	$1.0x10^{-4}$	$9.7x10^{-5}$	$4.1x10^{-5}$	$3.9x10^{-5}$
From counting	$1.4x10^{-6}$	$2.6x10^{-6}$	$2.5x10^{-6}$	$6.2x10^{-8}$	$1.0x10^{-8}$	$2.0x10^{-10}$

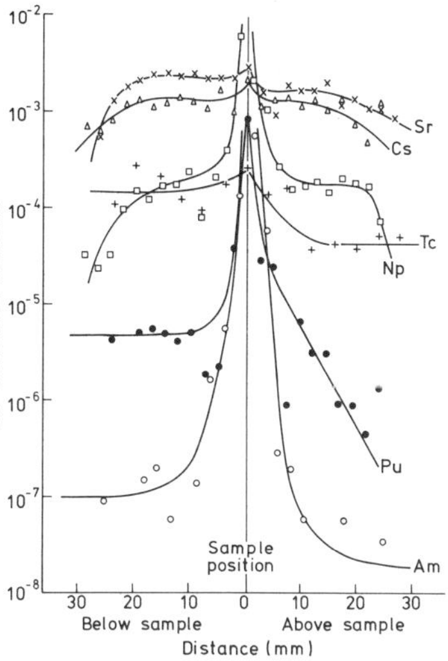

FIG.3.DISTRIBUTION OF ACTIVITIES FROM GLASS 189 IN A BENTONITE COLUMN AT 60°C

The quantities of the various isotopes (converted to notionally equivalent weights of glass for comparison) found in the slices of bentonite are plotted against position in the column in Fig. 3. The water samples were always taken from the top of the column and replaced by fresh whereas the water at the bottom of the column remained static. This explains why the activity levels were always higher in the bottom half of the column.

The activity levels in the column depend on a combination of two factors. Isotopes that are readily leached from the specimen tend to give high values but if the isotope diffuses rapidly through the column then the amount retained in the column is lower. Mathematical analysis of the curves in Fig. 3 was difficult because the radio-activity distribution was shown to be very dependent on any small amounts of radio-activity carried up the column from the specimen by the water during the initial non-equilibrium phase of the experiment. Experiments with longer columns (30 cm) are now in progress where the water diffused in from the top so as to avoid this problem. It has however taken longer than expected for the water to reach the specimen.

Source term leach rates have been obtained by multiplying the weight-loss leach rates by the fraction of each isotope released from the glass, i.e. not including that retained in the gel-layer. These are given in Table III, and represent the rate of leaching of these isotopes from the glass, discounting any adsorption on the backfill.

TABLE III

Leach rates ($g.cm^{-2}$ day^{-1}) in repository simulations

Glass		Cement backfill	Bentonite backfill	
		Including adsorption	Including adsorption	Leaching only
189	Sr	3×10^{-9}	10^{-6}	
(M5)	Tc	10^{-8}	3×10^{-6}	
	Cs	$< 5 \times 10^{-8}$	3×10^{-6}	7×10^{-5}
	Np	$< 5 \times 10^{-9}$	10^{-7}	3×10^{-5}
	Pu	$< 10^{-9}$	10^{-8}	3×10^{-6}
	Am	$< 5 \times 10^{-11}$	5×10^{-11}	2×10^{-6}
209	Sr	$< 2 \times 10^{-9}$	10^{-8}	7×10^{-6}
(M22)	Tc	10^{-7}	3×10^{-6}	9×10^{-6}
	Cs	10^{-7}	5×10^{-8}	10^{-5}
	Np	$< 5 \times 10^{-9}$	2×10^{-7}	10^{-5}
	Pu	$< 10^{-9}$	$< 10^{-9}$	7×10^{-7}
	Am	$< 2 \times 10^{-11}$	$< 2 \times 10^{-11}$	4×10^{-6}

Cement backfill

The system used to investigate the effect of cement was much simpler because it was thought that the flow or diffusion of water would be much slower and therefore that the preconditioning of the water by granite would have only a minimal effect. If concrete is used in the real case then it will probably be emplaced wet in the usual way, using granite chips as the aggregate portion. Because of size problems we decided to omit these and merely used an ordinary Portland cement/sand mixture, in the proportions (by weight): cement 1, sand 2.5, water 0.5.

A 2.5cm layer of cement was poured into a 3 cm. diameter closed ended tube and vibrated to remove air bubbles. The normal disc of doped glass was placed in position and a further 2.5cm of cement added and vibrated. The cement was then cured for 48 hours at 60°C. The volume of cement was about 40ml and the weight about 90g.

10ml of water was then poured into the tube on top of the cement: about half of this soaked into the cement. The tube was sealed and held in an oven at 60°C. At about monthly intervals the liquid on top of the cement was withdrawn and counted and replaced with 10ml of fresh water. No further liquid soaked into the cement. At the end of each month the pH of the "leachate" was 11.9.

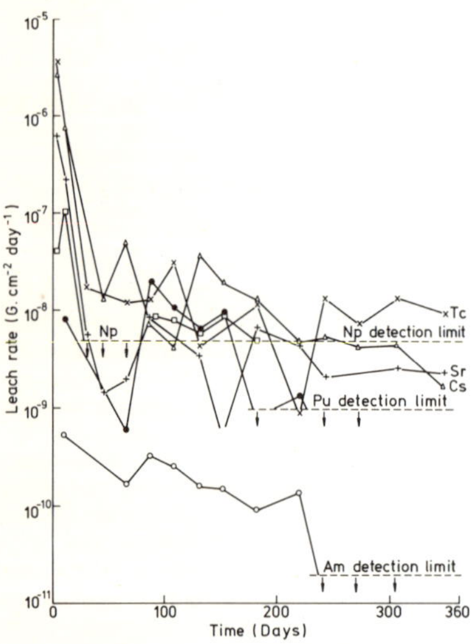

FIG.4. LEACH RATES OF UK GLASS 189 AT 60°C IN CONCRETE

The results obtained are shown in Fig. 4 for glass 189. Activity was detected in the overlying water from the start of the experiment – there was no incubation time before the diffusion front reached the water above the cement. Although the specimens were thoroughly cleaned before use it is possible that some loose activity was carried to the surface of the cement by air bubbles before the concrete set. Subsequently the activity release dropped to very low levels, near or below the analytical detection limits. Similar experiments were also carried out with glass 209.

At the end of the experiment, the system was dismantled but it was not possible to make a mass-balance as had been done with the bentonite experiment because the cement had stuck firmly to the glass. This also precluded the calculation of "source-term" leach rates as was done for the bentonite backfill experiments. The leach rates obtained for both glasses are given in Table III.

A new set of experiments have been started where the glass samples are located in a chamber in pre-cast cement.

EQUILIBRIUM SOLUBILITY EXPERIMENTS

In these experiments, restricted to Tc, Np, Pu and Am, samples of MW glass similarly doped to those described earlier, were ground to a powder and mixed with iron filings, to represent the corroding canister, and crushed pre- set cement mortar to represent the backfill.

The mixture was placed in a container and water added. The proportions used were: glass: 3, iron: 10, cement/sand: 15, water: 20. The container was sealed but small tubes allowed the gas space above the liquid to be purged with argon - 4% hydrogen and for small samples of the water to be removed at intervals for counting. Two of the containers were fitted with pH electrodes and two with Eh electrodes. The solids rapidly settled out of the slurry and in fact 'set' like cement: one of each type of electrode was in the solid and one in the liquid. The containers were held in an oven at 60°C. The pH in both liquid and solid was 11.0 whilst the Eh was -50mV in the solid and +220mV in the water. This difference strongly suggests that air is diffusing into the system through the teflon end caps. However, the measured solubilities should be the minimum values for the system, probably, for the isotopes used here, those corresponding to the most reduced state.

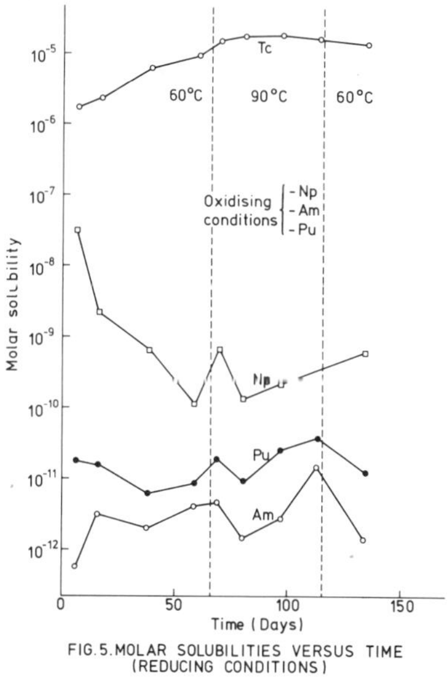

FIG.5.MOLAR SOLUBILITIES VERSUS TIME (REDUCING CONDITIONS)

The molar solubilities calculated from radiochemical analysis of the water are given in Figure 5, plotted against time. After about two months, the temperature was raised to 90°C but this did not seem to affect the solubilities significantly- as shown in figure 5.

In a similar but separate series of experiments, we also measured the solubilities under oxidising conditions. The mixture used in this case contained iron oxide (Fe_2O_3) instead of the metallic iron and the container was not purged with argon/hydrogen giving an Eh in the solid of + 150mV. The measured solubilities for the actinides at 90°C were much higher than before and are also marked in Figure 5. The solubility of Tc under these conditions was 10^{-5}M, about the same as under reducing conditions.

The solubilities under reducing conditions are compared with annual limits of intake in Table IV. A recent recommendation by the International Commission on Radiological Protection ICRP[1] is that the annual effective dose limit for members of the public shall not exceed 1mSv. Data derived from ICRP Publication 30[2-4] for the dose per unit intake is given in Table IV for the four isotopes studied and was used to derive an annual limit of intake equivalent to the dose limit of 1mSv. The corresponding level in drinking water (Bq/ml) was then derived by assuming a daily consumption of 2 litres. The factors by which the solubilities exceed this level are given in the last line of Table IV and range from x2 for Pu-239 up to x150 for Tc-99.

TABLE IV

Isotope	Tc-99	Np-237	Pu-239	Am-241
Dose per unit intake (Sv. Bq^{-1})	$3.3x10^{-10}$	$1.1x10^{-6}$	$5.8x10^{-7}$	$6.0x10^{-7}$
Annual intake (Bq) to give 1mSv	$3.0x10^{6}$	910	1700	1700
Corresponding* level in drinking water (Bq/ml)	4.0	$1.2x10^{-3}$	$2.4x10^{-3}$	$2.3x10^{-3}$
Solubility under reducing conditions (present work) (Bq/ml)	600	$7x10^{-3}$	$5x10^{-3}$	0.1
Multiple of drinking water level	x150	x6	x2	x43

*assuming a water intake of 2 litres per day.

Obviously in safety assessments, the possibility of concentration by plants (for example) must be allowed for, but on the other hand, the effect of dilution and absorption in the 'far-field' will considerable decrease the concentrations before the isotopes reach the biosphere.

CONCLUSIONS

1 The apparent leach rates for Sr, Tc, Cs, Np, Pu and Am from solid glass samples in a 'repository simulation' are very low when cement is used as the backfill, largely due to adsorption on the cement.

2 With a bentonite backfill, Sr, Tc and Cs are released much more readily and the rate of attack on the glass is fairly high.

3 Under reducing conditions, in the presence of cement and iron, the solubilities of Tc, Np, Pu and Am range from 2x(Pu) to 140x (Tc) drinking water levels. Under more oxidising conditions the actinide solubilities are much higher.

ACKNOWLEDGEMENTS

This work was sponsored jointly by the UK Department of the Environment and the Commission of the European Community as part of their radioactive waste management research programmes. The results may be used in the formulation of UK Government Policy but at this stage they do not necessarily represent such policy.

REFERENCES

1 Statement from the 1983 Washington Meeting of the International Commission on Radiological Protection. Ann. ICPR 14 (1) 1984
2 International Commission on Radiological Protection. Ann ICPR 2(3/4), 3, 4 (3/4), 5, 6(2/3), 7, 8.
3 National Radiological Protection Board Report NRPB-GS3 (1984)
4 DP Hodgkinson. Private Communication (1985)

AMERICIUM SPECIATION AND DISTRIBUTION COEFFICIENTS IN A GRANITIC GROUND WATER

GARY W. BEALL*, WILLIAM W.-L. LEE** and ABRAHAM E. VAN LUIK***
*United Catalysts, Inc., P.O. Box 32370, Louisville, KY 40232, USA
**R.F. Weston, Inc., 2301 Research Blvd, Rockville, MD 20850, USA
***Battelle, Pacific Northwest Laboratories, Richland, WA 99352, USA

ABSTRACT

We call attention to the possible importance of pH-dependent speciation in determining the distribution coefficient of americium in a granitic ground water. We obtained experimental distribution coefficients for americium. We used the geochemical code PHREEQE to calculate the amount present of americium complexes with hydroxides and carbonates over a range of pH, using formation constants taken from the literature and one estimated for $AmOHCO_3$, a neutral complex suspected of playing an important role in the aqueous chemistry of americium, although not experimentally evaluated. Combining the above inputs, we calculated individual distribution coefficients for americium complexes. We are able to show that the batch distribution coefficient for americium can be seen as the sum of individual distribution coefficients of various complexes.

INTRODUCTION

Radionuclides released to the subsurface environment will be likely transported by ground water. Various sorption processes retard the migration of such released radionuclides. Tracing the transport by ground water alone is a complex problem, and to superimpose chemical considerations compounds a difficult problem. Traditionally, a very simplified approach is used - a single variable called the retardation factor for each chemical element. The retardation factor, in turn, is based upon the familiar partition or distribution coefficient, K_d. The K_d is defined as the ratio of the mass of the sorbing species on the solid phase per unit mass (M/M) to the concentration of the sorbing species in solution (M/L^3).

Batch determined distribution coefficients for a particular nuclide on a specific rock or soil typically vary over many orders of magnitude [1]. Indeed, even with soil particles of the same size range, the K_d can vary over several orders of magnitude as a function of pH alone. Figure 1 gives data from a set of controlled laboratory experiments wherein the only variable was pH and the resultant K_d's varied over two orders of magnitude. It is this effect that we seek to investigate in this paper.

This study attempts to illustrate this speciation effect on the distribution coefficient by comparing the calculated partitioning of various americium species in an synthetic ground-water solution with observed batch distribution coefficients for a range of pH values. The predicted speciation was manipulated so as to produce a range of composite americium distribution coefficients which compared well with the observed K_d's. This type of treatment does not prove that speciation calculations will help better predict K_d's, but it suggests the possibility that the aqueous speciation of an actinide may partly explain the observed distribution behavior of that actinide.

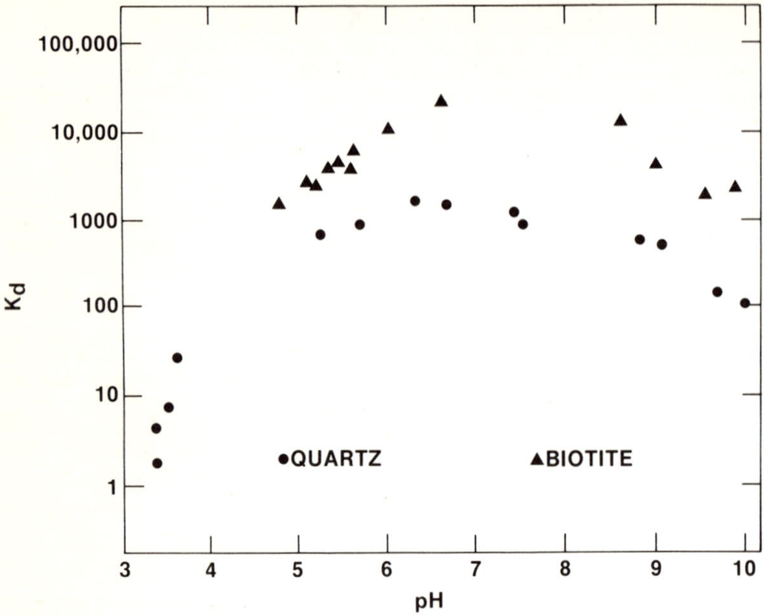

Figure 1. Batch distribution coefficients of americium on biotite and quartz.

EXPERIMENTAL

The experimental distribution coefficients utilized in the paper were taken from [2]. That study made distribution coefficient determinations of americium on thirty-four pure minerals as a function of pH in a synthetic granitic ground water (Table I).

Table I. Composition of Artificial Ground Water

Species	Concentration in Granitic Ground Water, mg/l[1]	Standard Water, ORNL, mg/l
HCO_3^-	60 - 400	123
SiO_2 (total)	5 - 60	12
$SO_4^=$	3 - 40	9.6
Cl^-	5 - 50	70
Ca^{2+}	10 - 60	18
Mg^{2+}	2 - 25	4.3
K^+	1 - 10	3.9
Na^+	10 - 100	65
F^-	0.01 - 5	3.8[2]
HPO_4^{2-}	0.01 - 0.5	0.89[2]
Fe^{2+}	0.5 - 20	Note 2
pH	7.2 - 8.5	8.2

Note 1 Probable concentration range for undisturbed deep ground water in contact with igneous rocks.
Note 2 Not added in laboratory artificial ground water.

Two minerals, quartz and biotite, were chosen for study in this paper. This choice was made based on the cation exchange capacities of the two minerals: quartz has a low cation exchange capacity and biotite a reasonably high cation exchange capacity. Calculations of aqueous speciation were conducted using the PHREEQE computer code [3]. The formation constants utilized in these calculations were selected from the open literature where available (Table II). In the case of the $AmOHCO_3$

Table II. Formation Constants for Americium with Hydroxides and Carbonates

Species	Reaction	log K	Source
$Am(OH)^{2+}$	$Am^{3+} + H_2O - H^+ = Am(OH)^{2+}$	-5.8	Ref. 5
$Am(OH)_2^+$	$Am^{3+} + 2H_2O - 2H^+ = Am(OH)_2^+$	-13.0	Ref. 5
$Am(OH)_3$	$Am^{3+} + 3H_2O - 3H^+ = Am(OH)_3$	-21.0	Ref. 5
$Am(OH)_4^-$	$Am^{3+} + 4H_2O - 4H^+ = Am(OH)_4^-$	-30.0	Ref. 5
$Am(CO_3)^+$	$Am^{3+} + CO_3^{-2} = Am(CO_3)^+$	6.0	Ref. 6
$Am(CO_3)_2^-$	$Am^{3+} + 2CO_3^{-2} = Am(CO_3)_2^-$	10.0	Ref. 6

complex, its existence in the solid phase is known but since no formation constant has been reported for the aqueous species, an estimated value was used. The PHREEQE code was used to calculate the amount of each aqueous species of americium present for the specified pH. The various species of the bicarbonate buffering system varied with pH. Chemical equilibrium was assumed, as were constant temperature and redox state. PHREEQE kept track of the saturation state of various natural minerals such as calcite, dolomite and gypsum, as well as solid americium species, to ensure that precipitation was not confused with sorption.

To calculate individual distribution coefficients for americium species, simultaneous equations were set up utilizing (1) the batch distribution coefficients for americium (Figure 1) and (2) the output of the PHREEQE code which gives the amount present of the various species at several pH values (Table III). These equations are of the form;

Table III. Relative Distribution of Aqueous Species of Americium in a Granitic Ground-water

Species / pH	4	5	6	7	8	9	10
Am^{3+}	97.97%	9.08%	0.015%				
$Am(OH)^{2+}$	1.01%	0.934%	0.015%				
$Am(OH)_3$						0.0025%	0.03%
$Am(OH)_4^-$						0.0274%	5.07%
$AmOHCO_3$	1.02%	89.999%	99.97%	100%	100%	99.97%	94.90%

$$1/(K^{*}+1) = \sum_{i=1}^{n} X_i[1/(K_i+1)] \qquad (1)$$

where K^{*}= the composite distribution coefficient at that pH

K_i= the individual distribution coefficients for species $i = 1,\ldots,n$

X_i= the mole fraction of each solution species present at a given pH.

RESULTS AND DISCUSSION

Americium has been chosen as an example radionuclide to illustrate the possible effect of speciation on sorption for two reasons. The first reason is that americium exists primarily as Am(III) in ground water, and the second reason is that americium sorption characteristics change radically as a function of pH [4].

Biotite and quartz were chosen as sorbents because they represent relatively high and low cation exchange capacity minerals, respectively, and also because experimental distribution coefficients for americium on both of these minerals are strongly pH dependent. Figure 1 gives experimental (batch) distribution coefficients for americium on the two minerals as a function of pH. It can be seen for quartz that at a low pH of 4, the K_d is in the range of 50; at pH=6, it has increased dramatically to about 1,300 and at the higher pH of 10 it has decreased substantially, to 100. The curve for biotite is higher, but has the same shape, offset by about one order of magnitude. This difference appears to correlate fairly well with the respective cation exchange capacities of these two minerals. The similarities in the shapes of the adsorption curves for these two minerals may indicate that the system is being dominated by the solution chemistry of americium.

In order to determine the number and type of species that could be causing these radical changes in the americium K_ds, a literature search was performed for formation constants for complexes between americium, hydroxide, and carbonate [4, 5, 6, 7, 8, 9, 10, 11, 12]. The values chosen for this study are shown in Table II.

When these values were utilized to calculate the species present in the granitic ground water, it was found that the unhydrolysed cation or aqueous americium hydroxide complexes dominated the speciation with very little of the americium carbonate ion-pair being formed at any pH. For example, at a pH of 5, the calculated distribution of species was approximately 96% Am^{3+}, 2% $Am(OH)^{2+}$, and 2% $Am(OH)_2^{+}$, and at a pH of 10, the predominant species were calculated to be 90% $Am(OH)_4^{-}$ and 8% $Am(OH)_3$. The smooth calculated gradation from a system dominated by cationic species to the system dominated by the anionic and neutral hydroxy complexes did not seem to either correlate with the observed behavior of americium K_d over the same pH range or fit the evidence for the importance of carbonate complexes reported in the literature.

In the work of Beall et al. [13], for example, which involved preparing and isolating pure rare earth carbonates, it was found that the predominant solid phase obtained was always a rare earth hydroxy-carbonate.

More recent work by Kim, Silva and Nitsche [14, 15, 16, 17] also indicated that for americium the controlling phase from a solubility standpoint is $AmOHCO_3$. Silva has speculated that there is probably an aqueous americium basic carbonate complex, with unknown properties. No formation constant has been experimentally determined or reported for $AmOHCO_3$, however. Based on a reported value for the formation constant of the analogous $AmCO_3^+$ ion, we initially estimated a log K value of 6, for the reaction

$$Am^{3+} + H_2O - H^+ + CO_3^{-2} = AmOHCO_3\ (aq) \quad (2)$$

and finally chose a value of 5.7 after several iterations. Using this hypothetical neutral complex, the observed data of Figure 1, and Equation (1), we obtained the comparison of Table IV and the K_d values for $AmOHCO_3$ in Table V.

Table IV. Comparison of Experimental and Calculated K_d's for Americium on Quartz and Biotite in a Granitic Groundwater

	pH	4	5	6	7	8	9	10
Quartz	K_d Obs	80	490	1,300	1,400	1,000	490	100
	K_d Cal	80	485	1,386	1,400	1,394	985	19
Biotite	K_d Obs	250	1778	10,000	21,000	16,000	668	158
	K_d Cal	250	1760	19,620	21,000	19,991	2875	19

The distribution of species when $AmOHCO_3$ is included at this assumed log K value is shown in Table III. At a pH of 7, $AmOHCO_3$ may be the only aqueous species present. This inference is supported by the results of electrochromatographic work by [18] which suggested that americium is essentially neutral to slightly anionic in aqueous solutions near pH 7. $Am(OH)_3$ and $Am(OH)_4^-$ were assumed to have K_d's of zero, but since they do not enter into the speciation significantly until pH = 10, this assumption is harmless.

Table V. Distribution Coefficients for Individual Americium Species in a Granitic Ground-water

Species	Quartz K_d (ml/g)	Biotite K_d (ml/g)
Am^{3+}	80	257
$Am(OH)^{2+}$	31	53
$Am(OH)_3$	0	0
$Am(OH)_4^-$	0	0
$AmOHCO_3$	1400	21000

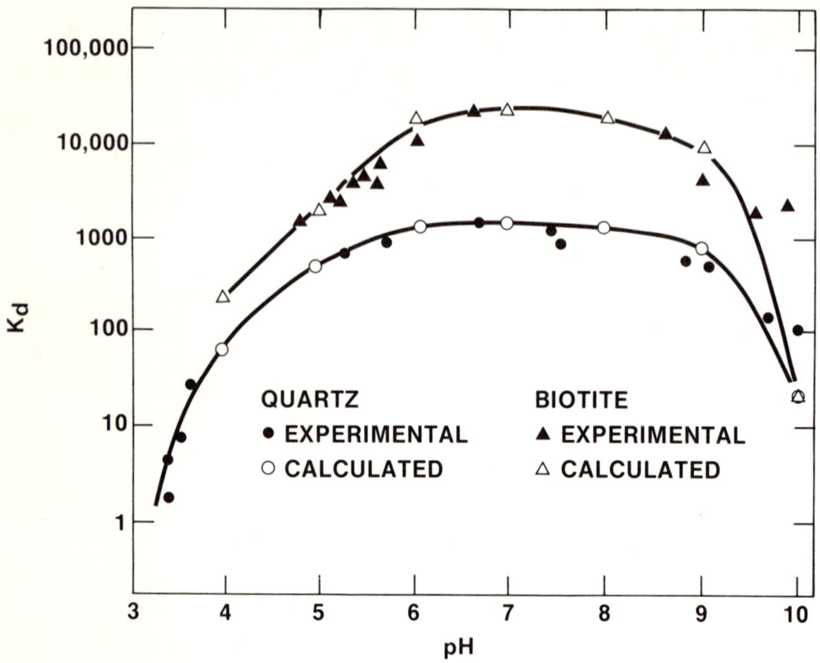

Figure 2. Comparison of batch distribution coefficients and calculated composite distribution coefficients of americium on biotite and quartz.

The calculated composite K_ds fit well with the batch, experimental values (Table IV) and were used to plot a curve for americium sorption on quartz and biotite (Figure 2). There is good agreement between the calculated and the batch distribution coefficients. The agreement shown in Figure 2 is partly due to the choice of the value for the formation constant of the dominant complex. However, the significance of this work does not depend on the good agreement between the calculated and observed K_ds, but lies rather in having demonstrated the idea that variation in aqueous speciation may be strongly correlated with variation in observed K_d value. The assignment of the significant sorbing attribute of aqueous americium to an as yet hypothetical neutral complex is problematic, but others have experimentally observed both americium's strongly sorbing behavior and neutral to slightly ionic character in near neutral aqueous solutions [18]. The results of individual distribution coefficients obtained in this study may not be unique, since they are dependent upon thermodynamic values for complexes obtained from the literature, and these thermodynamic data show large variations.

SUMMARY AND CONCLUSIONS

It is well known that changes in pH can have a profound influence on the sorption characteristics of americium. We have illustrated that these large variations may be explained in terms of speciation. If further work shows that this is so, then given proper thermodynamic data, such variations may be predicted via a combination of speciation calculations and experimentally derived sorption data.

ACKNOWLEDGEMENTS

This work was supported, in part, by the U.S. Department of Energy. Dr. Ben Tencer of R.F. Weston, Inc. provided additional computing time. Dr. I. Neretnieks and F.T. Ewart made many valuable suggestions on early drafts.

REFERENCES

1. M.I. Sheppard, D.I. Beals, D.H. Thibault and P. O'Connor, Soil Nuclide Distribution Coefficients and their Statistical Distributions. AECL-8364 (1984).

2. G.W. Beall and B. Allard, 1981. in Adsorption from Aqueous Solutions, edited by P.H. Tewari (Plenum, New York, 1981) pp. 193-212.

3. D.L. Parkhurst, D.C. Thorstenson and L.N. Plummer, USGS Water Resources Investigations 80-96, revised 1982.

4. D.G. Brookin, Geochemical Aspects of Radioactive Waste Disposal, (Springer-Verlag, New York, 1984).

5. B. Allard and G.W. Beall, in Proceedings of a Workshop on the Environmental Chemistry of the Actinide Elements, Warrenton, VA, 1978.

6. B. Allard, U. Olofsson and B. Torstenfelt, Inorg. Chemica Acta, 94, 205 (1984).

7. N. Edelstein, J. Bucher, R. Silva and H. Nitsche, Lawrence Berkeley Laboratory Report LBL-14325, 1983.

8. J.F. Kerrisk, Los Alamos National Laboratory Report LA-10040-MS, 1984.

9. D. Rai, R.G. Strickert, D.A. Moore and R.J. Serne, Geochem Cosmochim Acta, 45, 2257 (1981).

10. G. Bidoglio, A. Avogardo, A. de Plano and G.P. Lazzari, Radioactive Waste Management and the Nuclear Fuel Cycle, 5, 311 (1984).

11. A.B. Muller, Radioactive Waste Management & the Nuclear Fuel Cycle, 6, 19 (1985).

12. S.L. Phillips, C.A. Phillips and J. Skeen, Lawrence Berkeley Laboratory Report LBL-14996, 1985.

13. G.W. Beall, W.O. Mulligan and S. Mroczkowski, Acta Cryst., B32, 11 (1976).

14. R.J. Silva and H. Nitsche, in NRC Nuclear Waste Geochemistry '83, edited by D.H. Alexander and G.F. Birchard, NUREG/CP-0052. p.70.

15. R.J. Silva, in Scientific Basis for Nuclear Waste Management VII, edited by G.L. McVay (Elsevier Science Publishers, New York, 1984) pp. 875-881.

16. R.J. Silva, Lawrence Livermore National Laboratory Report UCRL-91821, 1985.

17. M.F. Bernkopf and J.I. Kim, Inst. fur Radiochimie der Tech. Univ. Munchen RCM-02884, 1984.

18. S. Fried, A.M. Friedman, J.J. Hines, R.W. Atcher, L.A. Quarterman and A. Volesky, in Actinides in the Environment, edited by A.M. Friedman (American Chemical Society, Washington, DC, 1976).

MODELLING RADIONUCLIDE SPECIATION AND SOLUBILITY LIMITS IN THE NEAR-FIELD OF A DEEP REPOSITORY

HANS WANNER
Swiss Federal Institute for Reactor Research, 5303 Wuerenlingen, Switzerland

ABSTRACT

In the safety analysis recently reported for a potential Swiss high-level waste repository, radionuclide speciation and solubility limits are calculated for expected granitic groundwater conditions. With the objective of deriving a more realistic description of radionuclide release from the near-field, an investigation has been initiated to quantitatively specify the chemistry of the near-field. In the Swiss case, the main components of the near-field are the glass waste-matrix, a thick cast steel canister horizontally stored in a drift, and a backfill of highly compacted bentonite.

Based on available experimental data, an ion-exchange model for sodium, potassium, magnesium, and calcium has been developed, in order to simulate the reaction of sodium bentonite backfill with groundwater. The model assumes equilibrium with calcite as long as sufficient carbonates remain in the bentonite, as well as quartz saturation. The application of this model to the reference groundwater used in 'Project Gewaehr 85' results in a significant rise in pH (by up to 3 units) as well as a marked increase in the carbonate concentration.

Neptunium and plutonium speciation and solubility limits are calculated for the reference groundwater chemistry gradually altered to that of saturated bentonite water and back again by a water exchange cycle model. The solubility limits estimated in this way generally turn out to be higher for the bentonite water than for the reference groundwater, mainly due to carbonate complexation of the actinide components AnO_2^+ and AnO_2^{2+}. Uncertainties are particularly large for neptunium solubility due to its strong Eh dependence in bentonite water.

INTRODUCTION

The extent of radionuclide release from a high-level waste repository to the geosphere depends strongly on the solubility of these nuclides in the available groundwater. The chemical properties of the groundwaters considered hitherto for the Swiss case, have been derived from chemical analyses of specific reference groundwaters [1]. The solubility limits of radionuclides in such groundwaters have been considered representative for their release into the far-field. In terms of a more realistic description of the release characteristics, the chemical properties of the backfill and corrosion materials should be taken into account for the safety analysis of a radioactive waste repository.

Based on experimental data for the effect of sodium bentonite on standard Swedish groundwater [2] and of the ion-exchange properties of sodium bentonite [3], a simple ion-exchange model is developed to simulate the changes in the chemical composition of any type of groundwater when brought into contact with bentonite. It should be noted that all calculations which model the situation at a repository site (including those which deal with Swedish groundwater) refer to the structure of the Swiss reference near-field.

THE BENTONITE MODEL

The model developed to simulate the effect of sodium bentonite on groundwater [4] is based on ion-exchange reactions as well as calcite and silica saturation. In this model, bentonite is assumed to consist of a number of equivalent sites (Z^-) which are capable of coordinating each of the considered cations (sodium, potassium, magnesium, and calcium) assuming dynamic equilibrium as shown in the picture below.

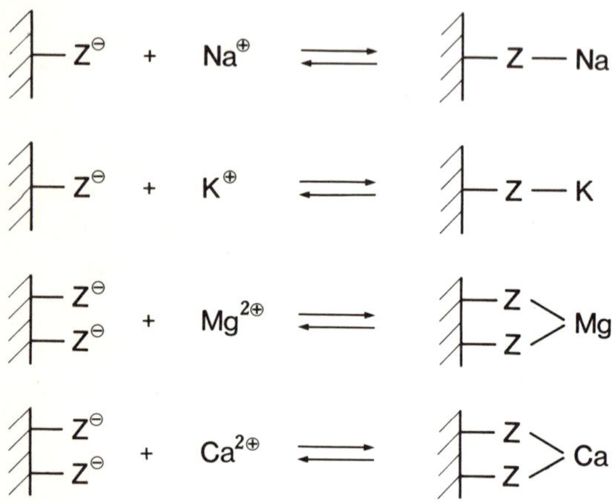

(1)

(2)

(3)

(4)

These sites are treated as simple ligands carrying a single negative charge. Their coordination to each of the mentioned cations is characterised by an equilibrium constant which is calculated from the available experimental data [2,3] as discussed in ref. [4]. By help of these analytical results and the ion-exchange capacities of sodium bentonite [3], the ion-exchange constants are calculated to be:

$$K(1) = \frac{\{ZNa\}}{\{Z\}\cdot\{Na\}} \;;\; \log K(1) = 20.0(\text{def}) \qquad K(2) = \frac{\{ZK\}}{\{Z\}\cdot\{K\}} \;;\; \log K(2) = 20.7$$

$$K(3) = \frac{\{Z_2Mg\}}{\{Z\}^2\cdot\{Mg\}} \;;\; \log K(3) = 42.0 \qquad K(4) = \frac{\{Z_2Ca\}}{\{Z\}^2\cdot\{Ca\}} \;;\; \log K(4) = 42.7$$

From the data absolute values of formation constants cannot be extracted, but they are chosen so high (log K(1) = 20.0, by definition) that virtually no uncoordinated sites exist. These values are used to model the effect of bentonite on groundwater. The process is envisaged as a titration experiment: granitic groundwater is assumed to react with an increasing amount of bentonite which is represented by Z^-. In this way an effort is made to account for the fact that in compacted bentonite not all the ion-exchange sites might be available for the reactions with groundwater.

The bentonite model is applied to the Swedish groundwater used in the experiment by M. Snellman [2] as well as to the reference deep groundwater in the Swiss case. Both calculations are carried out for a temperature of 55°C in order to be directly comparable. The calculated changes in the concentrations of the exchangeable cations, as well as in the alkalinity and pH, are presented in Figure 1a. Although the model is quite simple,

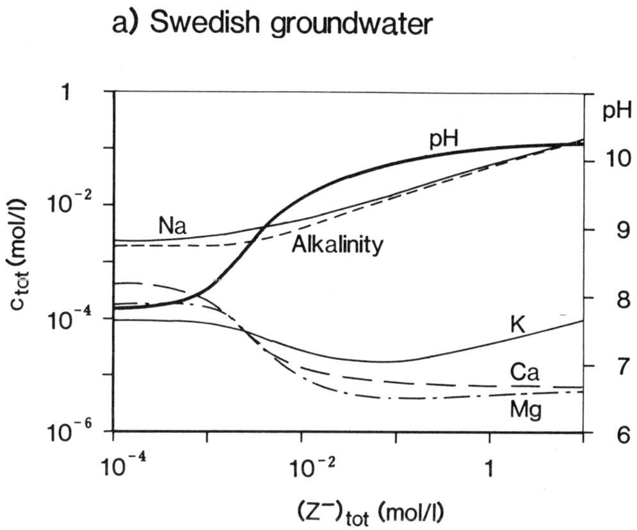

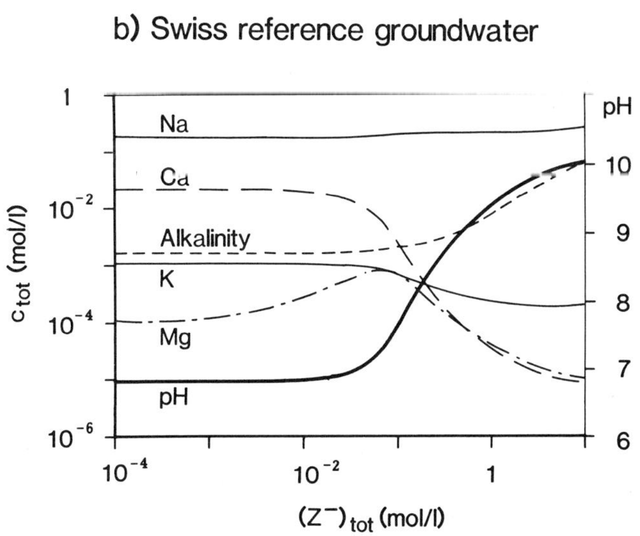

Figure 1. Modelled change in the composition of (a) Swedish groundwater [2] and (b) Swiss reference groundwater [1] when brought into contact with an increasing quantity of sodium bentonite (represented by the total concentration of ion-exchange sites Z^-). Alkalinity is expressed in eq/l.

reasonable consistency is found with the measurements performed to date with the Swedish groundwater (allowing for some possible experimental limitations). Figure 1b shows the results of the bentonite model application to the Swiss reference groundwater. The trends are roughly the same as for the Swedish groundwater. However, there are great differences in the absolute values of the concentrations, even on the right hand side of the plots (where exchange sites are present in large amounts and bentonite could be expected to strongly dominate the composition of the water). Obviously, the dissimilarities in the composition of the groundwaters are responsible for these differences.

Once the bentonite backfill in a waste repository is saturated and equilibrates with the pore water, slow diffusion of ions between the near-field and the far-field will take place. Consequently, the composition of the bentonite pore water will not remain constant, but will slowly change with time. As it is important to know more about these long-time alterations, a very simple model is applied in order to yield information on trends of variation in the composition of the bentonite pore water in the very far future. The following calculations are based on the assumption that, within a certain time period, the whole of the equilibrated water present in the backfill will be exchanged by the same amount of 'fresh' groundwater. This procedure is repeated a number of times (N). The parameter 'time', however, is not used as an explicit variable. The total number of ion-exchange sites in the bentonite is assumed to remain constant. The only parameters of bentonite which vary are the proportions of the single exchangeable cations available and the total content of carbonates.

Figure 2 shows results of the water exchange model. In the case of the Swedish groundwater (dashed curves) the total available calcite will be completely dissolved after 25 water-exchange cycles (Figure 2a). Thus, the groundwater pH (Figure 2b) will not be affected further by the bentonite. In the case of the Swiss reference groundwater, as many as about 70 water exchange cycles are required to reach a bentonite composition which remains more or less unaffected by the available groundwater. The sodium bentonite will then be converted to calcium bentonite. Contrary to the case of Swedish groundwater, the available carbonates present in the bentonite will not dissolve completely (only to about 20% at most). With the model assumptions for the Swiss case one might estimate the minimum time period required for one water exchange cycle to be roughly 30,000 years.

MODELLING ACTINIDE SPECIATION AND SOLUBILITY LIMITS IN THE BENTONITE MODEL WATER FOR THE SWISS CASE

The mathematical model used for the calculation of actinide speciation and maximum solubility in aqueous solution assumes dynamic equilibrium among all species considered. Consequently, actinide solubilities will be controlled by the least soluble solid which is assumed to be a pure oxide or hydroxide solid.

The database used for these evaluations was adopted from M. Schweingruber [5,6].

For each step in the water exchange model described above speciation and maximum solubilities of neptunium and plutonium are calculated for given Eh values ranging from -600 mV to 0 mV with intervals of 50 mV, using the code MINEQL/EIR [7]. The results for Np and Pu are shown is Figures 3a and 4a. As the redox conditions will most probably be controlled by the steel canister corrosion products, some pH dependent redox equilibria are chosen in order to get some idea of most probable Eh/pH combinations. The $H^+/H_2(g)$ equilibrium sets a lower Eh limit corresponding to the dissociation of water by reduction. The Figures 3b and 4b show the intersections along the Eh/pH functions for the following redox equilibria (reduction potentials calculated for 55^oC):

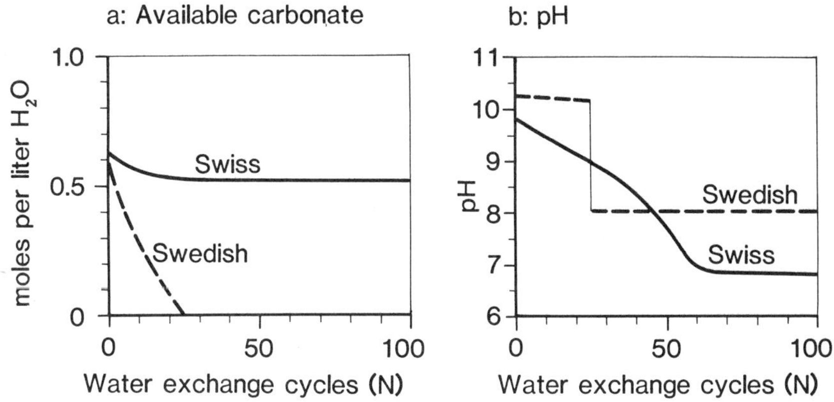

Figure 2. Change in available carbonate (calculated as calcite available for 1 liter of water) present in bentonite (a) as well as pH (b), calculated by application of the water exchange model for the Swiss reference groundwater and Swedish groundwater (dashed).

$Fe(OH)_3$(amorphous) + $3H^+ + e^-$	$\rightleftharpoons Fe^{2+} + 3H_2O$	E^o = +880 mV	[1]
FeOOH (goethite) + $3H^+ + e^-$	$\rightleftharpoons Fe^{2+} + 3H_2O$	E^o = +775 mV	[1]
Fe_2O_3(hematite) + $3H^+ + e^-$	$\rightleftharpoons Fe^{2+} + 3H_2O$	E^o = +680 mV	[1]
$3Fe_2O_3$(hematite) + $2H^+ + 2e^-$	$\rightleftharpoons 2Fe_3O_4$(magnetite) + H_2O	E^o = +215 mV	[1]
$2H^+ + 2e^-$	$\rightleftharpoons H_2$ (g)	E^o = 0 mV	(def)

The resulting curves yield probable solubility limits of neptunium and plutonium, respectively, as a function of N, assuming the redox pairs listed above to control Eh conditions.

Results for Neptunium and Plutonium

Under strongly reducing conditions, the soluble complex $Np(OH)_4$ will dominate in solutions with low carbonate content. In fully saturated bentonite water (N=0), $NpO_2(CO_3)_3^{5-}$ will predominate in solution due to the high concentration of carbonate which stabilises the +V oxidation state of neptunium by complex formation. As soon as the concentration of $Np(OH)_4$ becomes significant, the solubility limit decreases considerably. The solubility determining solid phase is NpO_2 (s) over the whole Eh/pH range modelled. Figure 3b shows that neptunium solubility is strongly dependent on Eh variations at high pH (N$\rightarrow$0). Under the assumption that the redox couple hematite/magnetite controls Eh in the saturated bentonite pore water (N=0), the solubility limit of neptunium turns out to be as high as $10^{-4.5}$ M.

Contrary to neptunium, plutonium may be reduced to the +III oxidation state in the Eh/pH range considered. PuO_2 (s) turns out to be the only solid phase. There are quite a number of plutonium complexes with comparable stabilities for virtually all Eh/pH combinations. For saturated bentonite water (high pH), $Pu(OH)_5^-$ will dominate in solution. However, its existence has recently been questioned [8]. As can be seen from Figures 4a and 4b, plutonium behaves inversely to neptunium. In saturated bentonite water, the solubility turns out to be low ($10^{-8.7}$ M) for the Eh range

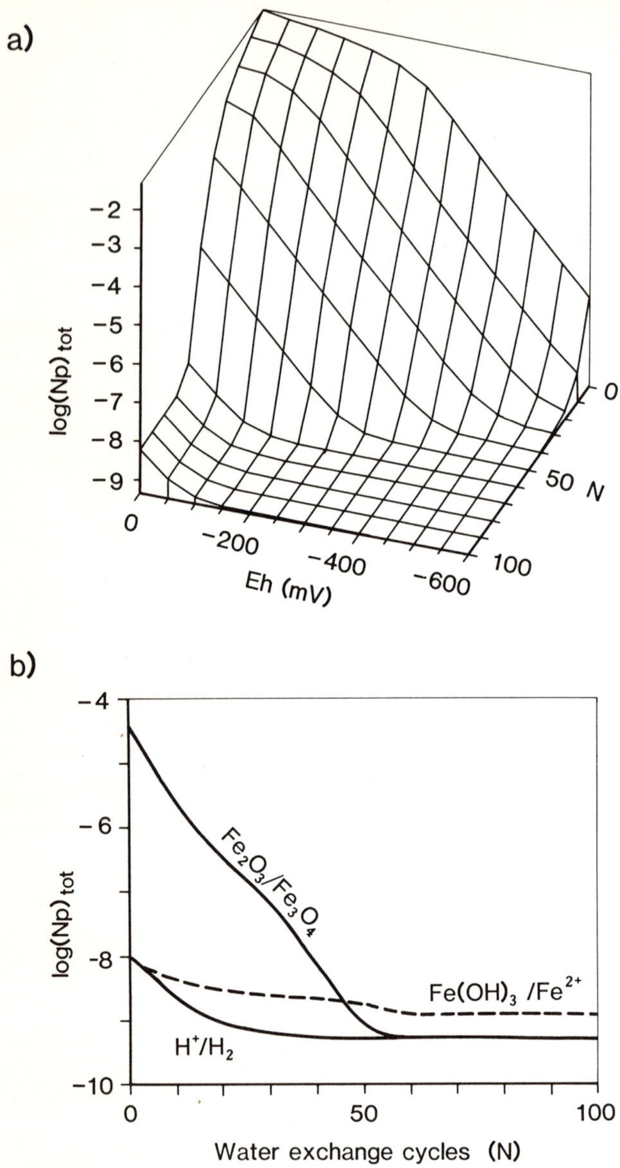

Figure 3. Representation of neptunium solubility limits (a) for the entire Eh range considered and (b) along Eh/pH functions of selected iron redox couples as mentioned in the text.

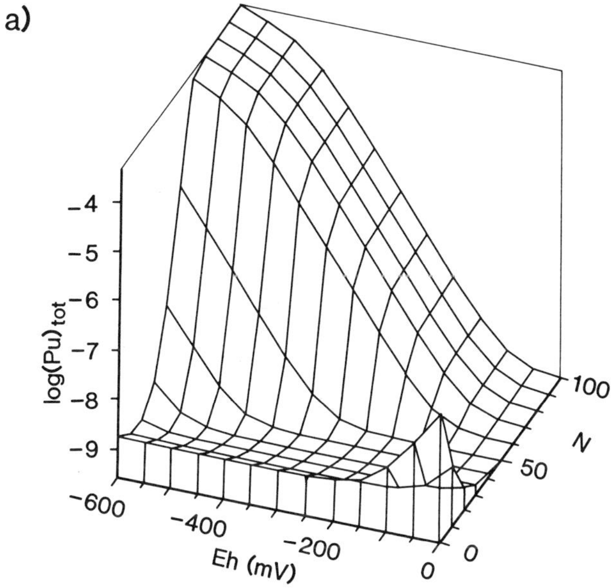

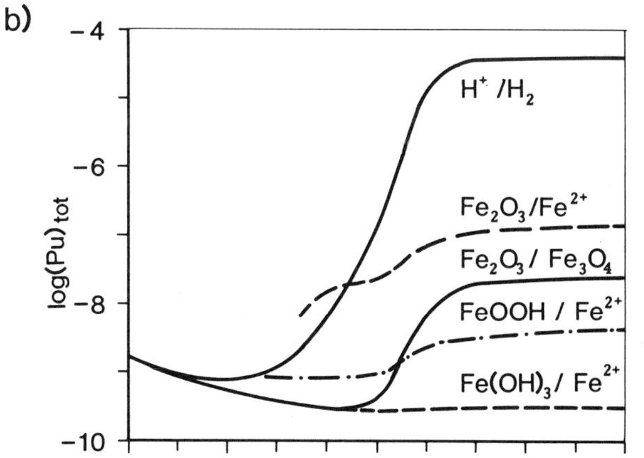

Figure 4. Representation of plutonium solubility limits (a) for the entire Eh range considered and (b) along Eh/pH functions of selected iron couples as mentioned in the text.

expected (-600 mV to -300 mV). After 70 to 100 water exchange cycles (where, corresponding to the bentonite model, the pore water composition will have roughly returned to that of the reference groundwater) plutonium solubility depends strongly on the redox potential of the solution.

CONCLUSIONS

From the estimations of actinide speciation and solubility limits in the bentonite water modelled, it can be seen that there are large uncertainties which are mainly due to the absence of a reliable Eh model. Neptunium solubility generally turns out to be higher in the bentonite model water than in the reference groundwater (see Ref.[5]). The reason for this is the very high carbonate content (and pH) of the bentonite water which, according to the model assumptions, is assumed to be saturated with respect to calcite. Clearly, the results obtained are strongly dependent upon the thermodynamic database used in this model. An updated database for selected actinides is currently being compiled at OECD [9].

Similar estimations for other nuclides such as uranium, americium, thorium, and technetium have also been done [10,11]. A more detailed redox model is about to be developed which will allow to more authentically predict redox conditions in the near-field of a high-level waste repository.

REFERENCES

1. M. Schweingruber, On Probable In-Situ pH, pCO_2, and Redox Conditions for Waters Sampled at 400, 600, and 1300 m Depth in Borehole KRISTAL I, Boettstein, EIR Internal Report TM-45-84-20 (Wuerenlingen, 1984)
2. M. Snellman, Chemical Conditions in a Repository for Spent Fuel, YJT 84-08 (Espoo, 1984)
3. H. Mueller-Vonmoos, G. Kahr, Mineralogische Untersuchungen von Wyoming Bentonit MX-80 und Montigel, Nagra NTB 83-12 (Baden, 1983)
4. H. Wanner, The Effect Sodium Bentonite on the Chemical Composition of Groundwater, EIR Internal Report TM-45-85-03 (Wuerenlingen, 1985)
5. M. Schweingruber, Actinide and Tc Solubilities in Deep-Crystalline Reference Groundwater Defined for NAGRA Project 'Gewaehr', EIR Internal Report TM-45-84-30 (Wuerenlingen, 1984)
6. M. Schweingruber, Actinide Solubility in Deep Groundwaters - Estimates for Upper Limits Based on Chemical Equilibrium Calculations, EIR Report No. 507 (Wuerenlingen, 1983)
7. M. Schweingruber, User's Guide for Extended MINEQL (EIR-Version) - Standard Subroutine/Data Library Package, EIR Internal Report TM-45-82-38 (Wuerenlingen, 1982)
8. D. Rai, J.L. Ryan, Solubility Constraint: An Important Consideration in Safety Assessment of Nuclear Waste Disposal, PNL-SA-11489 (Richland, 1984)
9. A.B. Muller, International Chemical Thermodynamic Data Base For Nuclear Applications, Rad. Waste Man. and Nucl. Fuel Cycle 6(2), 131, (1985)
10. H. Wanner, Modelling Uranium, Neptunium, and Plutonium Speciation and Solubility Limits in the Near-Field of a Deep Repository, EIR Internal Report TM-45-85-17 (Wuerenlingen, 1985)
11. H. Wanner, Modelling Americium, Thorium, and Technetium Speciation and Solubility Limits in the Near-Field of a Deep Repository, EIR Internal Report, in preparation

GEOCHEMICAL MODELLING OF THE ALTERATION OF A GRANITIC ROCK AROUND A REPOSITORY FOR SPENT NUCLEAR FUEL

BERTRAND FRITZ*, MARIE KAM* AND YVES TARDY**
* Centre de Sédimentologie et de Géochimie de la Surface (CNRS),
1, rue Blessig - F 67084 STRASBOURG Cedex - France
** Institut de Géologie, 1, rue Blessig - F 67084 STRASBOURG Cedex - France

ABSTRACT

The alteration of a granitic rock around a repository for spent nuclear fuel has been simulated considering the geothermal effect through an increase of temperature. The results of the simulation were considered in terms of mass transfers and their volumic consequences. The dissolution of primary minerals by alteration tends to increase the porous volume due to fissures and cracks, while the precipitation of secondary minerals and particularly here calcite and clay minerals will tend to decrease the porosity by sealing the cracks. The results of the simulations show that an increase of temperature from 10 to 100°C will favour the alteration of the granitic rock by the solutions circulating today in the rock. This rock is characterized by a very low fissure porosity and the water velocity is consequently particularly low. These conditions favour intense water-rock interactions and the water-rock system will probably reach the stability field of secondary clay minerals predicted by the calculations : the porosity will decrease. A combination of these thermodynamic mass balance calculations and hydrodynamic considerations, with a kinetic approach of mineral dissolutions gives a first attempt to calibrate the modelling in the time scale.

INTRODUCTION

The storage of spent nuclear fuel in a deep geological location will increase the temperature up to a maximum value in the surrounding rock. The temperature will then decrease progressively with time. During this period of tens of thousands years the rock-solution system around the repository will act as a small induced geothermal system : the water-rock interactions will be influenced by these temperature variations and some transformations of the mineral phases can be expected. These geochemical consequences must be studied in any safety analysis of the storage, in addition to the study of the hydrological consequences of the storage. This has been done within the extended survey of the swedish site of Stripa (SKB project). The temperature in the close vicinity of the storage is expected to increase from 10°C now to about 100°C. The geochemical modelling tools [1],[2], have been applied in order to try to predict if the mass balances of the induced hydrothermal alteration will lead to a sealing or opening tendancy for the fractures and fissures net in the rock surrounding the site of storage. The modelling approach itself, using only chemical potentials as energy functions, only gives informations on the tendencies of evolution of a given system. The geochemical path calculation describes then the mass transfer between the solid and aqueous phases, but the lack of data for activation energies for all minerals involved in these calculations makes impossible the direct thermodynamic and kinetic calculations. However, if one considers the predicted thermodynamic sequences of alterations together with data on the hydrology of the granitic massif, it is possible to find a rough calibration in the time scale. For this approach, some recent results on calculation of mass transfer as a function of time and surface area [3],[4], are very helpful as they concern the alteration sequence of felspars, major components of the granites.

THE MODELLING APPROACH

The geochemical modelling tools have been built during the last fifteen years on a thermodynamic base. One considers that any alteration process is due to the fact that, at least one mineral from a mineral assemblage is unstable with respect to the solution in contact with that mineral phase. The instability is then considered as the driving force for the geochemical evolution of a water-rock system : the instability corresponds to a potential energy for an irreversible dissolution. This reaction may occur spontaneously and will modify the composition of the system. This is a very general thermodynamic rule : natural systems always tend to come back to equilibrium conditions by minimizing the Gibbs free energy. The basic idea for the thermodynamic modelling is to suppose that the irreversible process occurs and to calculate the consequent evolution of the aqueous phase. In this reaction paths the formation of secondary minerals is taken into account as a reversible process at equilibrium [5],[6],[7]. In a non-kinetic model this corresponds to the simplest kinetic assumption one can make : secondary minerals in alteration processes are formed more rapidly than primary minerals are dissolved.

We have developed this approach combining irreversible dissolution processes and secondary reversible reactions (program DISSOL) for the study of:

- weathering processes at low temperature on granitic rocks [1],[8];
- hydrothermal alterations in a granitic body [2];
- diagenetic transformations in sedimentary basins [9],[10].

For the last applications a new computer code has been built using the irreversible temperature variations as a driving force for the geochemical evolution in the water-rock system (program THERMAL [2]).

The application of these tools to the problem of heat storage, due or not to the storage of nuclear waste, has been made [11] and is still in progress.A summary of the results obtained until now is presented here.

THE REACTING PHASES FOR THE SIMULATION

Mineralogical composition of the rock

The mineralogical assemblage chosen for the calculation fits the chemical composition of a swedish granodiorite from borehole 1 at Finnsjö Lake [12]. The mineralogical and chemical composition of the rock is given in Table I. Only major chemical elements are considered in the stoechiometric composition of the minerals used for the simulation.

Table I - Mineralogical and chemical composition of the granite used in the simulation (granodiorite type from Finnsjö Lake, Sweden).

Mineral	Formula	Mole %	Oxide	Weight %
Quartz	SiO_2	64.4	SiO_2	66.5
Microcline	$KAlSi_3O_8$	7.9	Al_2O_3	14.8
Albite	$NaAlSi_3O_8$	12.5	FeO	5.2
Anorthite	$CaAl_2Si_2O_8$	10.0	CaO	4.8
Biotite	$Si_3AlFe(2)_3O_{10}(OH)_2K$	3.7	K_2O	3.6
Tremolite	$Si_8Mg_5Ca_2O_{22}(OH)_2$	1.5	Na_2O	2.5
			MgO	2.0
			H_2O	0.6

Chemical composition of the reacting solution

The selected chemical composition of the reacting solution was taken from a so called "mean drilled well" from the Finnsjö area, corresponding to a temperature around 10°C. The initial conditions for the simulations at 25,60 and 100°C have been obtained by simulating the heating of that solution from 10 to 100°C using the computer code THERMAL. The major differences between these three initial solutions are related to the calcite deposit (Ca,alkalinity) and Eh-pH conditions.This appears in table II. The initial amounts of aluminium and iron are limited by the saturation with respect to kaolinite and goethite or pyrite depending on redox conditions. Those are consistent with two values of oxygen fugacity (log pO_2 = -45 for mean reducing conditions and -60 for reducing conditions). The corresponding Eh values may represent the range of Eh values in the swedish deep granites as shown by careful measurements [13]. The partial pressure of CO_2 was not fixed in the simulation but only calculated from the alkalinity and pH data.

Table II - Chemical composition of the solutions used for the simulations at 25, 60 and 100°C (initial conditions).

Temperature	25°C	60°C	100°C
pH	7.400	7.065	6.790
Eh (mV) -option(I)	+120.	+2.	-155.
-option(II)	-95.	-260.	-440.
log pO_2 (I)	-45.41	-44.27	-44.55
(II)	-59.94	-60.13	-59.92
log pCO_2	-2.113	-1.56	-1.135
Calcite deposit from 10°C to t °C (millimoles/l)	0.000	0.205	0 502
Ca (millimoles/l)	1.525	1.320	1.022
Al (micromoles/l)	0.001	0.100	1.000
Fe (nanomoles/l) (I)	0.001	0.100	5.000
" " (II)	10.00	1.000	0.050
Alkalinity (meq/l)	3.344	2.936	2.340
at all temperatures :	Na - 0.631; K - 0.19; Mg - 0.189 (millimoles/l)		
	Si = 0.185; SO_4 = 0.175; Cl = 0.556 (" " ")		

Simulation conditions

The programs DISSOL and THERMAL have been used with the closed system option. This means that the minerals produced in early stages of the alteration may be redissolved later and transformed into another secondary phase like gibbsite into kaolinite and kaolinite and goethite into clay minerals. This seems reasonable for simulating alterations occuring in deep aquifers with low porosity and low water flux through the rock matrix.

The six rock forming minerals (table I) are dissolved in the solution by small increments as long as their saturation state in the solution in not obtained. They have not been allowed to precipitate after, even if this was thermodynamically possible for some of theim (quartz, microcline and tremolite) considering that their rate of precipitation can be neglected at these temperatures.

Secondary minerals are pure minerals like goethite, pyrite, kaolinite and calcite, a multicomponent solid solution for 2:1 clays [14],[15], and a binary solid solution for chlorites (clinochlore + daphnite).

It is important to recall here that the modelling itself does not include kinetic constraints : reversible mass transfers are calculated versus irreversible supposed mass transfers.

SIMULATED STABILITY SEQUENCES FOR MINERALS

Primary minerals

The calculations show an increase of the mass transfer from 25 to 100°C. This is due essentially to an increase of solubility of quartz and microcline. The albite (sodic end-member for plagioclases) remains very soluble, compared to these minerals and only tremolite shows a decrease of solubility with increasing temperature.These trends (fig. 1) correspond to an increase of the alteration potential due to the heat released from the storage. The amount of rock dissolved before saturation of the different primary minerals is respectively at 25, 60 and 100°C :

- 0, 1.06 and 5.02 microunits/l for quartz
- 0.54, 6.84 and 17.56 microunits/l for microcline
- 85.9, 118.5 and 105.0 microunits/l for albite
- 52.9, 13.0 and 12.62 microunits/l for tremolite

One microunit of rock corresponds to the assemblage of minerals given in table I (100 moles of different minerals). The two remaining primary minerals, anortite and biotite (annite) never reach the saturation state in these calculations, even for biotite with the reduced conditions considered here. This is consistent with the fact that these minerals, calcic plagioclases and biotite are generally highly altered in weathering conditions and also under hydrothermal conditions (below 150°C). The increase of stability of the Mg-amphibole may contribute to the low Mg content of the corresponding solutions, if combined with the increase of secondary chlorite stability described later. These effects may explain the very low Mg concentration often described in dilute geothermal waters from crystalline rocks [2],[16].

Secondary minerals

For the whole temperature range calcite, goethite and kaolinite are produced from the beginning stages of the reaction.Later in the sequence as shown by figure 1, smectites and or chlorite are produced between microcline and albite saturations:

- at 25°C the 2:1 clay begins to precipitate for an alteration of 1.33 microunits of granite;
- at 60°C the same clay begins to precipitate for an alteration of 18.1 microunits of granite;
- at 100°C the Mg-chlorite precipitates first (5.43 microunits) and will redissolve (81.76 microunits) after the precipitation of the 2:1 clay (58.31 microunits).

These numbers show clearly the increase of alteration potential between 25 and 100°C: primary minerals are stabilized later and secondary minerals appear later too. Consequently, if the saturation state with respect to albite is reached (end of the calculations) the mass transfer due to the alteration is much more important at higher temperatures. At this point the composition of the smectite produced is respectively at 25, 60 and 100°C :

- $Si_{3.352}Al_{0.648}\ Al_{1.248}Fe_{0.582}Mg_{0.255}O_{10}(OH)_2K_{0.027}Na_{0.000}Ca_{0.210}Mg_{0.101}$
- $Si_{3.308}Al_{0.692}\ Al_{1.238}Fe_{0.695}Mg_{0.101}O_{10}(OH)_2K_{0.171}Na_{0.049}Ca_{0.197}Mg_{0.041}$
- $Si_{3.480}Al_{0.520}\ Al_{1.189}Fe_{0.628}Mg_{0.275}O_{10}(OH)_2K_{0.452}Na_{0.024}Ca_{0.022}Mg_{0.000}$

These results have been obtained using mean reducing conditions (option I,in table II). The calculations were also made with a lower oxygen fugacity (log fO_2 = -60. instead of -45.,see option II,table II). The corresponding range of Eh values (around -100 mV at 25°C, -270 mV at 60°C, -440 mV at 100°C) along the simulated sequences, has been suggested by field measurements and Eh recalculations for different wells in the swedish deep granites [13]. With these

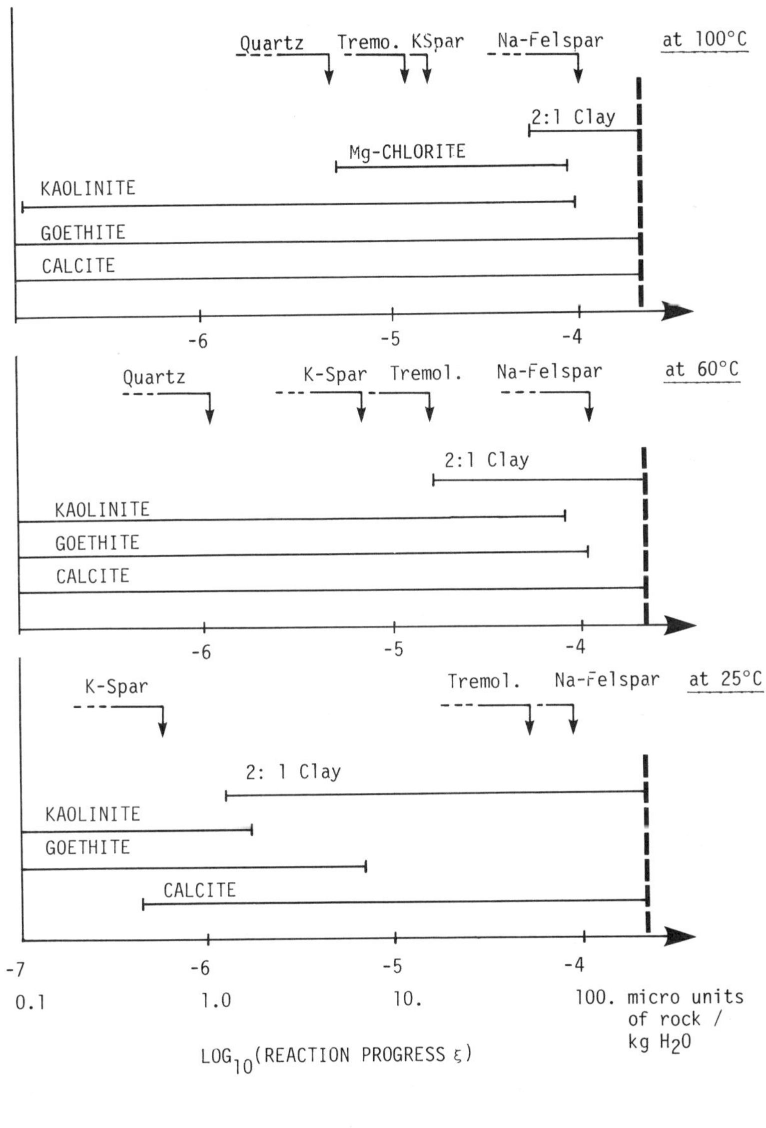

Fig. 1 - Calculated mineralogical sequences for the alteration of the granitic rock at 25, 60 and 100°C.

reducing conditions, the changes in the simulated sequences concern the iron bearing minerals :

- at 25°C, the simulated sequence is only lightly altered. The goethite is formed later in the sequence but still in the early alteration stage and the ferrous biotite is not stabilized in the total sequence.
- at 60°C, the pyrite is precipitated instead of the goethite and the solution is able to stabilize the ferrous biotite (dissolution of 5.9 microunits of rock).
- at 100°C, the pyrite also precipitates instead of goethite and the saturation state with respect to biotite is also reached (dissolution of 7.7 microunits of rock).

The clay phase produced in these sequences show a similar evolution : a 2:1 clay at 25°C, smectite + chlorite at 60°C, chlorite only at 100°C. The reducing conditions favour here the ferrous chlorite phase with respect to the ferric smectite.

MODIFICATION OF THE POROUS VOLUME

Alteration sequences and volumic balances

The aqueous phase reacting with the rock-forming minerals in the simulation (1 kg of water) is located in a very large volume of granitic rock if the porosity due to small fissures and cracks is very low :

- 0.2 m^3 of granite with a porosity of 0.5% contains 1 liter of solution.

The porous volume containing the reacting solution is decreased by any precipitation of secondary mineral, but increased by the dissolution of primary minerals. The calculations allow to follow this total volume change along the alteration sequences and to point out the opening or cealing tendency of the alteration process. In the sequences described previously this calculation underlines again the two major stages of the alteration :

- In the initial stage of the dissolution, the total volume of dissolved minerals Vd is superior to the total volume of precipitated minerals Vp but this difference Vd-Vp decreases progressively. However, in this initial step of alteration,the mass transfers are very low (less than 1mm^3) compared to the solution or porous volume (1l).
- Later in the sequence, when both calcite and clay minerals precipitate, the volumic balance shows an excess of volume for the solid phase (Vd-Vp is negative). The corresponding decrease of porosity has then to be taken into account.

This result shows that the same alteration sequence may lead to an opening or a sealing tendency for the porosity of the rock and the thermodynamic calculation will not give the answer to the question : where do the solution "goes" through the possible stability sequence in a given system ? However the two major stages mentioned above correspond to different types of alteration in the nature. The first type corresponds to largely open systems, where the drainage is high, the residence time of the solution in the reacting areas is short and the mass transfer per liter of solution is not important.This is typical from intense tropical alterations and active geothermal systems for example. At the contrary the second type of alteration will correspond to more closed systems where the porosity is low, the water velocity is very low and the rock/water ratio very high. In the swedish granites for example at a few hundred meters depth, porosities are very low (less than 1%), water velocities are very low (0.1 to 1 m/year) and the rock/water ratio can be as high as 500 (in weight). For these reasons it seems reasonable to assume that the solutions will reach in the predicted sequences the saturation state of clay minerals and calcite. consequently the alteration process will tend to diminish the porosity. With this hypothesis it is not necessary to know exactly where the equilibrium point will be (fig.1) within the (clay + calcite) stage. Indeed, the calculation of volumetric effect shows that the excess of volume of the precipitated minerals with respect to the volume of altered minerals is stabilized in the second stage of the reaction:

- at 25°C, the excess of mineral volume (Vp-Vd)/Vd is stabilized around 1.12 to 1.15 all along the calcite+clay production;
- at 60°C, the same ratio is stabilized between 1.12 and 1.20 when the clay is formed;
- at 100°C, the excess of volume is more important and may be as high as 35%.

These results concern sequences calculated at constant temperatures. The cooling effect on these solutions will increase all these effects on the volume because of the lower solubilities at lower temperature as shown by the mass transfer calculations. As a conclusion of these calculations, one may remember that the second stage of alteration of the granitic rock will lead to a decrease of the porous volume when producing the calcite+clay secondary phase. However, the volumetric effect remains very small compared to the volume of the reacting solution (1 cm^3 versus 1,000 cm^3).

Cumulative effects

In a completely closed water-rock system, the effect of the hydrothermal alteration induced by the heat released from the nuclear waste repository could probably be neglected : the calculation has shown a tendency to decrease the porosity by producing clays and calcite, but the total mass transfer per liter of reacting solution remains negligible compared to the mass of rock containing this solution in a very low porosity. The alteration will modify a lot the chemistry of the solution but the mineral transformations will be quite undetectable in the total rock :

- one liter of solution contained in 540 kg of granite (porosity 0.5%; density 2.7) will induce a total mass transfer of about 1g of matter !

This is true for a theoretical closed system and can be used to say that only repeated alteration in slightly open system can produce important mass transfers in a rock [1],[10]. The system studied in Sweden is slightly open,(the water velocity is low, but not zero) and it is important to know if a repeated alteration of a given rock unit will always show a decrease of porosity as obtained for one interaction.

The calculation shows that the solution migrating in the granite from one rock unit to another can only involve a greater decrease of the porous volume (if it is in the clay formation step). It is particularly true if the migration goes together with a temperature decrease, away from the vicinity of the storage.

KINETIC CONSIDERATIONS

The discussion above has shown the major difficulty for using mass transfer calculations made on a thermodynamic base only : the mineralogical sequences are not time dependent and one has to assume how complete the sequence will be in a given system for a given time. We have used the results of a recent publication [3], that shows by thermodynamic and kinetic modelling of quartz and felspars alteration(at pH = 4) that :

- the early stages of alteration need only a few years to be reached;
- the kaolinite-clay boundary (kaolinite-muscovite for pure microcline) will need about one to two centuries to be reached.

These calculations assumed an estimated mineral surface area of $10^{4.6} cm^2/kgH_2O$ which can be compared to an estimated values in the swedish granite of 10^5 cm^2/kgH_2O for a low porosity (0.5%) and a mean fissure width of 100 microns. From the work quoted above we may consider that one century is enough time for the solution to reach the clay saturation at 60 to 100°C. In the swedish granites, the average water velocities range between 0.2 and 0.5 m/year [17].If these velocities are not too much disturbed around the repository, the solution migration will be 20 to 50 meters per century : these distances are smaller than the distances of thermal influences around the repository [18] for the same period of time and the solutions will have enough time to complete the predicted reactions. More, assuming a decrease of porous volume per unit of time (DV/V = 2.10^{-5} to 2.10^{-4} or 20 to 200mm^3 of porosity lost per liter per century) we calculated the cumulative effect of volume decrease on the porosity of the rock. After 100.000years the loss of porosity is between 2 and 20 % of the initial porosity. This is a first approximation in this attempt to combine thermodynamic mass balances calculations, hydrological data and kinetic considerations.

This work was performed in the frame of a contract between SKB-Stockholm and Université Louis Pasteur-Strasbourg (1982-1985).

REFERENCES

1. B.Fritz, Sci. Géol.,Mém.,Strasbourg, 41, 152 (1975).
2. B.Fritz, Sci. Géol.,Mém.;Strasbourg, 65, 197 (1981).
3. H.C. Helgeson and W.M. Murphy, Math. Geol., 15, 109-130 (1983).
4. H.C. Helgeson, W.M. Murphy and P. Aaagaard, Geochim. Cosmochim. Acta, 48, 2406-2432 (1984).
5. H.C. Helgeson,T.H. Brown,A. Nigrini and T.A. Jones, Geochim. cosmochim. Acta, 34, 455-481 (1970).
6. H.C. Helgeson,T.A. Jones, T. Mundt, T.H. Brown, A. Nigrini,R.H. Leeper and D.H. Kirkham, Comp. Library H.C. Helgeson,U.C. Berkeley,USA (1971).
7. H.C. Helgeson,J.M. Delany, H.W. Nesbitt and D.K. Bird,Amer. J. of Sci., 278,A, 229 (1978).
8. B.Fritz and Y.Tardy, Bull.Soc.Géol. France, 26,4,339-367 (1976).
9. J.P. Cassan, M.C. Garcia Palacios, B.Fritz and Y.Tardy, Bull.Centre Rech. Exploit.-Prod.,Elf Aquitaine,5,1, 113-135 (1981).
10. B.Fritz, J.P. Cassan, N. Clauer and Y.Tardy, in Thermal Phenomena in Sedimentary Basins, edited by B.Durand (Editions Technip,Paris,1984),p.293.
11. B.Fritz, M.Kam and Y.Tardy, SKB Technical Report, 84-10, 27 (1984).
12. KBS Annual Report,Stockholm, II,Geology, 81 (1978).
13. P. Wickberg, I. Grenthe and K. Axelsen, KBS technical Report, 83-40,(1983).
14. Y. Tardy and B. Fritz, Clay Minerals, 16, 361-373 (1981).
15. Y. Tardy, J. Duplay and B. Fritz, in International Clay Conference 1981, H. van Olphen and F. Veniale editors,Developments in sedimentology,35, Elsevier,Amsterdam, 441-450 (1981).
16. G. Michard, Bull.Bur.Rech.géol.min.,sér.II,sect.III,2, 183-189 (1979).
17. L. Carlsson and A. Winberg, KBS Technical Report, TR 83-45, 208 (1983).
18. T. Tarandi, KBS Technical Report, TR 83-22, 30 (1983).

HUMIC SUBSTANCES IN GROUND WATERS

N. PAXÉUS*, B. ALLARD**, U. OLOFSSON*** AND M. BENGTSSON***
*Department of Analytical and Marine Chemistry, Chalmers University of Technology and University of Göteborg, S-412 96 Göteborg, Sweden
**Department of Water in Environment and Society, University of Linköping, S-581 83 Linköping, Sweden
***Department of Nuclear Chemistry, Chalmers University of Technology, S-412 96 Göteborg, Sweden

ABSTRACT

The humic materials isolated from ground waters of a borehole in Fjällveden (Sweden) were characterized by elemental and functional group analyses. Spectroscopic properties, molecular weight distributions as well as acid-base properties of the fulvic and humic fractions were also studied. The ground water humic substances were found to be quite similar in many respects (but not identical) to the Swedish surface water humics concentrated from the Göta River but appeared to be quite different from the American ground water humics from Biscayne Florida Aquifer or Laramie Fox-Hills in Colorado. The physico-chemical properties of the isolated humic materials are discussed.

INTRODUCTION

A major part of the organic matter in natural waters, soils and sediments consists of humic substances. Humic substances can be considered as a generic name referring to a great variety of different biochemically stable polymeric substances, which have been produced during the alteration/degradation of organic matter in natural environments.

One important property of humic substances is their ability to form complexes with a variety of polyvalent metals. Considerable attention has been focused on the formation of complexes between soil and surface water humic substances and transition metals, e.g. copper, zinc, cadmium, lead, mercury and vanadium [1-6] . The ability of humic substances to form complexes with aluminium, iron(II)/iron(III), europium(III), americium(III) and thorium(IV), has also been reported [7-10] . Humic substances were found to cause substantial reduction of iron (III) to iron (II) and subsequent dissolution and depolymerisation of colloidal and particulate iron [11] . Humic substances are able to react with clays and minerals either through adsorption on clay/mineral surfaces or by degradation processes. Thus, a gradual dissolution of certain minerals in the presence of humic substances has been reported [12]. This process has been associated with leaching of metals from the minerals.

Problems associated with the potential disposal of radioactive waste in deep bedrock attract attention to the presence of synthetic or naturally occurring complexing agents that may enhance the migration of the disposed radionuclides and thus facilitate their uptake by plants. Therefore, knowledge of the chemistry of ground waters and water-rock interactions is a requirement. Humic substances may, through their ablility to form complexes and to participate in redox reactions, play an important role in the transport of radionuclides in connection with underground storage of radioactive waste [10].However, these aspects of ground water humic substances are poorly understood.

In order to elucidate the role of humic substances in ground waters it is necessary to know their physico-chemical properties. Therefore, the main purpose of the present work was to characterize humic substances from ground waters and to compare these with humic substances from surface waters.

OCCURRENCE OF HUMIC SUBSTANCES IN GROUND WATERS

Occurrence of humic substances in ground water is poorly documented. Ground water humic substances originate presumably from a water-soluble fraction of the soil humus that is mobile enough to be transported downwards by surface waters. Since the vertical transport of dissolved humic substances in soils is restricted downward mainly to the B horizon [13], concentrations of these substances in shallow ground waters (reported range 0.1-1.3 mg/l C, mean 0.8 mg/l C; [14]) might be expected to be lower than the concentrations found in river waters (range 3-20 mg/l C; [15]). According to Thurman and Malcolm[16]humic substances in ground waters account for 30-50% of the DOC or TOC. This is consistent with the data obtained in this work (see Table I). In this respect it should be pointed out that the non-selectivity of TOC-based analytical methods bring about the uncertainty associated with contamination problems. Thus, there is a strong suspicion that the samples Fj 8:562 and Km 3:376 (Table I) might have been contaminated.

Table I. Variation in the concentration of humic substances with depth in the boreholes at Fjällveden and Kamlunge, Sweden.

Borehole	Depth m	TOC[10] mg/l	Humic* organic carbon, mg/l FL	SP	[10]	Humic carbon, % of TOC
Fj 4:349	349	6.7	2.20	2.36	2.16	35
Fj 8:402	402	1.6	0.40	0.42	0.55	25
Fj 2:409	409	3.3	1.74	1.74	2.15	53
Fj 4:420	420	7.9	3.20	2.91	3.40	37
Fj 8:562	562	15(?)	0.40	0.50	0.80	3(?)
Km 3:376	376	4.0(?)	0.12	0.14	0.20	3(?)

*determined fluorometrically (FL) and spectrophotometically (SP).

Tracing of a water movement in bedrocks in studies of groundwater is still a problem. Deep groundwaters usually have increased salinities compared with surface waters [17]. Saline groundwaters in crystalline bedrocks in Sweden are originated from either relict Baltic seawater (3,500-8000 year old water) or residual igneous/metamorphic fluids or mixtures of these [17]. In any case, these waters are characterized by absence or extremely low concentration of humic matter. Surface waters, on the other hand, are relatively rich in humic matter. Therefore, humic substances may in fact serve as natural tracers for studing the processes of mixing fresh surface waters with deep ground waters.

Thus, the concentration of humics is expected to decrease with depth unless direct intrusions of surface waters take place. Results presented in Table I indicate a possibility of such intrusion at about 420 m depth.

ISOLATION OF GROUND WATER HUMIC SUBSTANCES

Concentrations of humic substances in natural waters (surface waters, ground waters) are on the order of several mg/l, which in most cases is too low to investigate them directly in the aquatic environment. Therefore, the examination of physical and chemical properties of aquatic humic matter requires a preconcentration step. Fulvic and humic acids studied in this work were isolated from deep ground waters (depth 409 m) of the borehole at Fjällveden, Sweden. DEAE anion-exchange resin (DE52) was used to concentrate the humic materials from about 2 m water. The scheme of the concentration procedure is shown in Fig. 1.

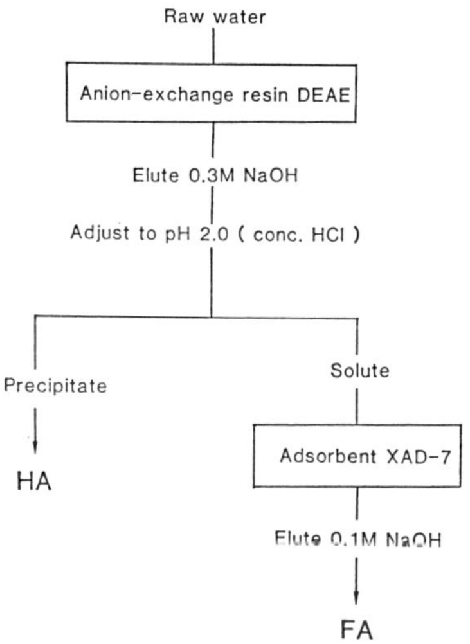

Fig. 1 - Concentration procedure used for isolation of humic substances from ground waters

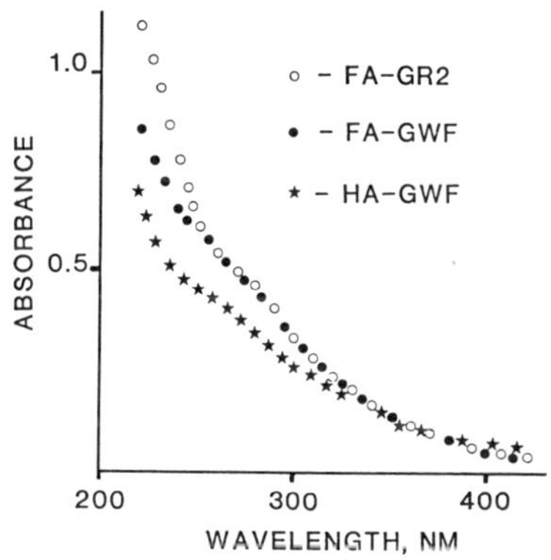

Fig. 2 - UV-VIS spectra of the humic substances (Cary 210 Varian spectrophotometer; cell length 1 cm; solvent, distilled water; concentration 30 mg/1).

The humic substances were eluted with 0.3 M NaOH. The eluate solution (5 1) was acidified to pH 1.5 and the humic acid fraction was centrifuged off. The fulvic acid fraction was concentrated by adsorption on XAD-7 and re-eluted with 0.1 M NaOH. The solution was then decationized by passage through a strong cation-exchange resin in H^+-form. The fulvic acid was recovered by freeze-drying. The fulvic and humic acid fractions were then purified from low-molecular weight organic impurities. Details of the procedures applied for purification of humic substances can be found elsewhere [18]. The yield of the fulvic acid (FA-GWF) was about 2.4 g and of the humic acid (HA-GWF) about 0.4 g.

CHARACTERIZATION OF AQUATIC HUMIC SUBSTANCES.

Fulvic acid represented the major fraction of the isolated humic substances (>70%). Therefore this work was mainly focused on characterization of the fulvic acid FA-GWF.

Results of the elemental analysis for humic and fulvic substances, obtained both in the present work and taken from the literature, are shown in Table II. The data from the analysis of methoxyl groups are also included in this table.

Table II. Elemental analysis, methoxyl content and average molecular weights of humic (HA) and fulvic (FA) acids.

Sample	%C	%H	%N	%O	%S	%Ash	OCH_3, meqv/g	Mol. weight* $\overline{M}_n$	$\overline{M}_w$	
				Surface waters						
FA-GR1	51.42	4.48	0.97	40.21	2.27	0.02	0.9	1040	2790	a
HA-GR1	55.10	4.83	0.58	38.10	1.02	0.06	1.0	1170	3100	a
FA-GR2	49.68	5.25	0.40	40.51	3.27	0.09	2.3	1100	2650	a
HA-GR2	55.15	4.60	0.33	37.95	1.88	0.05	1.9	1200	3140	a
				Ground waters						
FA-GWF	50.64	4.24	1.58	41.62	1.22	<0.1	0.7	1070	2690	b
HA-GWF	51.30	5.27	5.87	36.42	0.95	<0.1	0.3	1250	3090	b
FA-B	55.44	4.17	1.77	35.39	1.06	0.43	n.d	n.d		c
HA-B	58.28	3.39	5.84	30.14	1.43	0.1	n.d.	n.d.		c
FA-LFH	62.67	6.61	0.42	29.14	0.44	1.09	n.d	n.d.		c
HA-LFH	62.05	4.92	3.21	23.45	0.96	5.12	n.d.	n.d		c

GR-(1-2) - Göta River (Sweden); GWF - Fjällveden (Sweden); B -Biscayne Florida Aquifer (USA); LFH -Laramie-Fox Hills (Colorado,USA); n.d.=not determined; a - ref. [18]; b - this work, c - ref.[16]; * - number- ($\overline{M}_n$) and weight- ($\overline{M}_w$) average molecular weight.

As can be seen from the table, the humic materials isolated from different environments show many similarities in elemental composition, especially, with regard to the major elements - carbon, hydrogen and oxygen. There are also some variations. Thus, the nitrogen content of surface water humics is lower than that of the ground water humics. The content of C and H in the Swedish ground water humic substances is rather close to that of the Swedish surface water fulvic acids, but differs from that of the American. At present it is not possible to judge whether this difference is caused by different concentration procedures or reflects environmental influences. Compared with the Swedish surface water humics the ground water humic substances have much lower methoxyl content which may indicate their higher age.

Molecular weights for the aquatic humic and fulvic acids, determined by gel permeation chromatography, are shown in Table II. As can be seen from this table the samples have rather low molecular weights. It is of interest to note that the number-average molecular weight for the Swedish surface water and ground water fulvic acids is about 1000, which is very close to these reported for North American aquatic and soil fulvic acids, ranging between 800 and 1000[19,20].

Analysis of humic substances by spectroscopic methods.

In this work, UV-VIS, fluorescence, IR and NMR spectroscopy was used for analysis of humic substances.

a) UV - VIS and fluorescence spectroscopy

The spectra of the ground water humic and fulvic acids studied in this work are shown in Fig. 2. The spectra exhibit a broad shoulder at about 260-280 nm. Phenolic moities, present in the humic and fulvic acids contribute, at least partly, to the absorption in this region. It is also possible that some of the phenolic units are the components derived from lignin. The UV-VIS spectra of the ground water humic and fulvic acids are very similar to those of surface water humic materials. Thus, the spectrum of the ground water fulvic acid is almost identical with that of FA-GR2 from the Göta River (Fig. 2).

Fluorescence spectra of the ground water fulvic and humic acids exhibited one broad band in the excitation spectrum (maximum$\approx$ 350 nm) as well as in the emission spectrum (maximum $\approx$450 nm) and in this respect were similar to the corresponding spectra of the surface water humics from the Göta River. However, while the intensity of fluorescence for HA-GWF was about the same as that of the surface water humics, the ground water fulvic acid FA-GWF showed about three-times higher intensity.

b) Infrared spectroscopy.

The IR spectra of ground water humic and fulvic acids studied in this work together with a typical spectrum of surface water fulvic acid (FA-GR2, KBr pellet) are shown in Fig. 3. The broad band at about 3400 cm^{-1} and the inflection at about 2600 cm^{-1} arise from O-H stretching in carboxylic, phenolic and alcoholic groups of the sample. The bands at 2990 cm^{-1} and 2970 cm^{-1} are due to C-H stretching. Carbonyl stretching gives rise to a group of bands with a maximum at 1725 cm^{-1} (unconjugated C=O) and 1630 cm^{-1} (conjugated C=O). Signals from aromatic nuclei (skeletal vibrations) probably both contribute to the latter band and are also responsible, at least partly, for the band at 1510 cm^{-1}. The complex band at about 1200 cm^{-1} can be due to C-O stretching of phenolic and alcoholic moities and O-H deformation in carboxyls. It should be noted that the band at about 1725 cm^{-1}, caused mainly by carboxyl groups is dominant in the spectrum. This, however, does not necessarily mean that their content is high in the sample. Actually, the observed dominance of carboxyls is a result of their very high molar absorptivity in IR and consequently, a very high sensitivity of IR for their presence.

The IR spectra of the ground water humic and fulvic acids differ somewhat from that of the surface water humics. The bands in the IR spectrum of FA-GWF, especially in the region of skeletal frequences, are much broader and less resolved than those exhibited by FA-GR2. In this respect, the spectrum of the ground water fulvic acid is similar to the IR spectra of paleosol fulvics [21]. The uncertainty in the interpretation of the skeletal mode region in IR, however, does not permit to make any definite conclusions.

c) nuclear magnetic resonance (NMR) spectroscopy

The ^{13}C spectrum of the ground water fulvic acid is shown in Fig. 4.

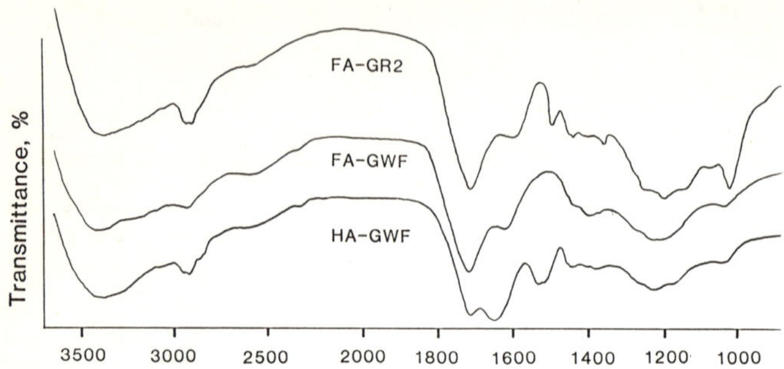

Fig. 3 - 3IR spectra of the humic substances.

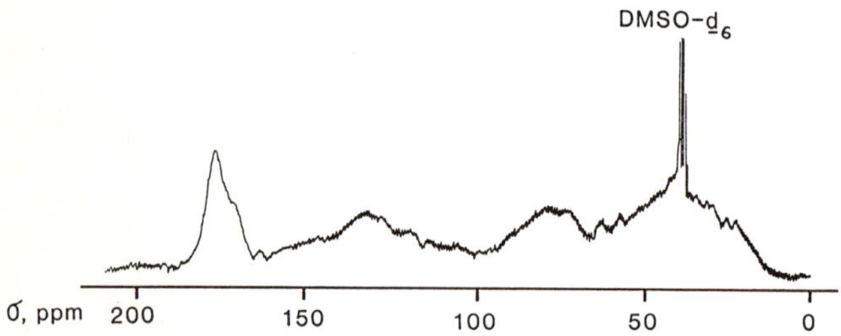

Fig. 4 - 4 ^{13}c NMR spectrum of FA-GWF (solvent, D_2O; DSMO-$\underline{d}_6$ added as a standard).

This spectrum was found to be very similar to the one of surface water fulvic acid [18]. Assignment of signals [18]in the spectrum of FA-GWF in terms of structural grouping contributing to the observed spectrum is as follows: 15 ppm (CH_3- in alkyl groups); 35.5 ppm (-CH_2-CO- or -CH_2- in hydrocarbon structures); 45.9 ppm (aliphatic >CH-, >CH-CO-); 55.6 ppm (Ar-OCH_3; 65 ppm (-CH_2-O-); 75 (>CH-O-); 114.5, 122.2, 128.0 ppm (aromatic >CH-); 150 ppm (quarternary, unsaturated carbon); 167 ppm (Ar-COOH); 172 ppm (R-COOH). The ^{1}H spectrum of the ground water fulvic acid was consitent with the interpretation given above.

Acid - base properties of humic substances

Potentiometric titrations was used in this work to study acid-base properties of the ground water humic and fulvic acids. Considering the rather low molecular weights for aquatic humic and fulvic acids (see Table II) it is clear that these should be treated as low molecular weight polyelectrolytes. Heterogeneity of the particular sample, i.e. presence of a large number of functional groups with different stability constants, is therefore the main factor affecting an appearence of the titration curve. Titration of a humic or fulvic acid sample in terms of constants and concentrations can be approximated as titration of a mixture of N components, regarding each component as n-basic acid (N,n = 1,2,3...). The model to describe acid-base properties of aquatic humic and fulvic acids based on a non-linear curve-fitting procedure with a varying number of parameters was published earlier [22]. The results obtained for the ground water fulvic and humic acids are shown in Table III. For comparison, results obtained for the surface water humic and fulvic acids are also included in this table.

Table III. Comparison of constants (Ki) and concentrations (Ci, in meqv per g of sample) of acidic groups for different humic and fulvic acids. Results obtained from the six-group model in the pH range about 2.4 - 10.4. Ionic strength 0.1 M, temperature 25 ^{o}C.

Sample	K1	C1	K2	C2	K3	C3	K4	C4	K5	C5	K6	C6	-COOH
				Surface water									
FA-GR1	2.64	2.49	4.20	1.70	5.37	1.15	6.69	0.40	8.19	0.28	9.69	0.44	5.34
HA-GR1	2.75	1.81	4.21	1.61	5.38	1.13	6.69	0.44	8.30	0.40	9.75	0.53	4.55
FA-GR2	2.59	1.94	4.17	1.28	5.30	0.88	6.59	0.33	8.11	0.35	9.70	0.64	4.10
HA-GR2	2.71	1.47	4.14	1.25	5.25	0.88	6.51	0.40	8.11	0.51	9.75	0.65	3.60
				Ground water									
FA-GWF	2.64	2.58	4.14	1.88	5.29	1.21	6.63	0.49	8.19	0.32	9.57	0.47	5.58
HA-GWF	2.62	1.20	4.25	1.74	5.31	0.94	6.46	0.65	8.01	0.52	9.86	0.61	3.88

FA-GR1,FA-GR2/HA-GR1,HA-GR2 -fulvic/humic acids from the Göta River[18]

Despite the fact that the source and concentration procedure vary, the agreement for the values of the stability constants is very good. The three most acidic groups were shown to be carboxylic[22]. For the ground water fulvic and humic acids this assumption was confirmed by the IR analysis of the acid- sodium salt forms. Details of the procedure can be found elsewhere [22]. As can be seen from the table, humic acid fractions are characterized by a lower content of carboxyls, especially the most acidic ones. This is, probably, the main reason for their lower solubility and thus lower content in the ground waters. Also, because of lower carboxyl content HA-GWF might be expected to have lower complexation capacity than FA-GWF.

The ground water fulvic acid obtained in this work is being used for further studies of the complex formation with several actinides, primarily americium and thorium, and, subsequently, of effects on their sorption and mobility.

ACKNOWLEDGEMENT

Financial support from Swedish Nuclear Fuel Supply Co.,Division KBS, is greatly appreciated.

REFERENCES

1. F.J. Stevenson, Soil Sci. 123, 10 (1977).
2. M.S. Shuman and J.L. Cromer, Environ. Sci. Technol. 13, 543 (1979).
3. J. Buffle, P. Deladoey, F.L. Greter and W. Haerdi, Anal. Chim. Acta 116, 255 (1980).
4. R.A. Saar and J.H. Weber, Geochim. Cosmochim. Acta 44, 1381 (1980).
5. G.D. Templeton and N.D. Chasteen, Geochim. Cosmochim. Acta 44, 741 (1980).
6. F.H. Frimmel and J. Geywitz, Fresenius Z. Anal. Chem. 316, 582 (1983).
7. E.L. Bertha and G.R. Choppin, J. inorg. nucl. chem. 40, 655 (1978).
8. P. Ruggiero, M. Attimonelli, F.S. Intresse and O. Sciacovelli, Agrochim. 23 37 (1979).
9. S. Arai and K. Kumada, Geoderma 26, 1 (1981).
10. G.R. Choppin and B. Allard, in Handbook of the physics and chemistry of the actinides, edited by A. Freeman, J. Landers and C. Keller (North Holland Publ. Comp., Amsterdam, 1984).
11. C.H. Langford, R. Kay, G.W. Quance and T.R. Khan, Anal. Lett. 10, 1249 (1977).
12. M. Schnitzer, in Soil organic matter, edited by M. Schnitzer and S.U. Khan (Elsevier Scientific Publishing Company, Amsterdam, 1978), p. 52.
13. F.J. Stevenson, Humus Chemistry, (Wiley-Interscience, New York, 1982), p. 427.
14. J.A. Leenheer, R.L. Malcolm, P.W. McKinley and L.A. Eccles, J. Res. US Geol. Surv. 2, 361 (1974).
15. J.H. Reuter and E.M. Perdue, Geochim. Cosmochim. Acta, 41 325 (1977).
16. E.M. Thurman and R.L. Malcolm, Envir. Sci. Technol. 1981, 463.
17. B. Allard, S.Å. Larson, E.-L. Tullborg and P. Wikner, KBS Technical Report 83-59, Swedish Nuclear Fuel Supply Co./Division KBS, Stockholm, 1983.
18. N. Paxéus, PhD. Thesis, Chalmers University of Technology and University of Göteborg, Sweden, 1985.
19. K. Ghosh and M. Schnitzer, Soil Sci., 129, 226 (1980).
20. J.H. Reuter and E.M. Perdue, Geochim. Cosmochim. Acta 45, 2017 (1981).
21. G. Calderoni and M. Schnitzer, Geochim. Cosmochim. Acta 48, 2045 (1984).
22. N. Paxéus and M. Wedborg, Anal. Chim. Acta, 169, 88 (1985).

LABORATORY AND MODELLING STUDIES OF MICROBIAL ACTIVITY IN THE NEAR FIELD OF A HLW REPOSITORY

Julia M West*, Ian G McKinley+, Helen A Grogan+, Susan C Arme*

Fluid Processes Research Group, British Geological Survey, Keyworth, Nottingham NG12 5GG, U.K.
+ Eidg. Institut fur Reaktorforschung, 5303 Wurenlingen, Switzerland.

INTRODUCTION

Microbial effects are one of the possible perturbations to the expected performance of a deep geological repository which must be examined as part of a comprehensive safety analysis. Recent literature surveys (eg [1, 2]) and reconnaisance sampling studies [3, 4] have concluded that:

a) microbial contamination of a repository is inevitable
b) even for high level nuclear waste, conditions in the near-field are insufficiently extreme to ensure complete sterilisation.

Work is now focussed on examining the constraints on microbial populations and their influence on repository performance. In this paper, laboratory studies are described which examine the tolerance to extreme conditions of important groups of micro-organisms which have been identified in relevant geological formations. This work is put into context by using simple models of near field processes. These allow the extent of possible microbial processes to be compared with purely inorganic effects.

LABORATORY STUDIES

The object of the current laboratory studies has been to ascertain the capacities of microbial groups isolated from relevant formations to tolerate and thrive under repository conditions of temperature, pressure and radiation. A series of experiments have been carried out using two microbial isolates of sulphate reducing bacteria (SRB) from Konrad, FRG and Matlock, Derbyshire, U.K., Thiobacillus spp from South Crofty, Cornwall, U.K. and Stripa, Sweden, and Sulfolobus spp, a thermophilic organism related to the thiobacilli [3, 4]. In each case growth curves were defined to establish basal activity and experiments were conducted during the same growth phase to allow their comparison. Optimum conditions for growth were provided in each experiment, the only constraint being the specific tolerance tested. Epifluorescence microscopy and adenosine triphosphate (ATP) assay were used to estimate population numbers and subsequent culture on established viability. For a more detailed description of methods used and results, see West and Arme [5].

The results showed in general that microbial populations studied will tolerate increased temperature, pressure and radiation but the extent of resistance varies between groups and species. Consequently in all the tests the SRB were very tolerant to the environmental extremes whereas the sulphur oxidisers were more sensitive. Fig. 1a shows that SRB from Matlock were thermophilic, growing more rapidly at 50°C than at 30°C. However, when both temperature and pressure are increased (Fig. 1b) at 80°C and 30 MPa population numbers are drastically reduced but exhibit remarkable recovery when subsequently cultured-on under ambient conditions. The ATP assay confirmed these trends. SRB appear to be unaffected by gamma radiation from 10^2 up to 10^5 rad, single dose (Fig. 1c) but the sulphur oxidisers showed a declining trend with increasing gamma dose.

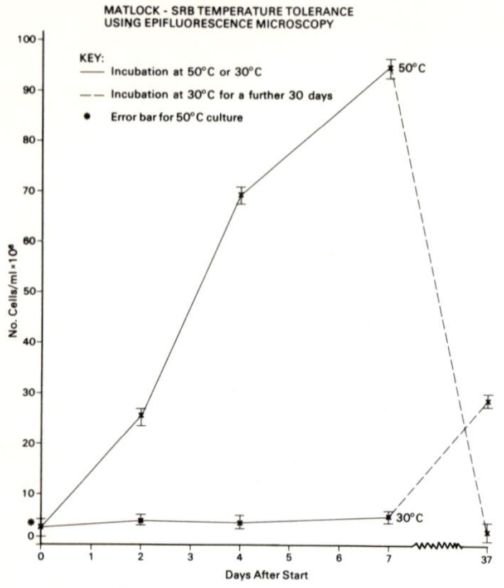

Figure 1a. Matlock SRB temperature tolerance using epifluorescence microscopy.

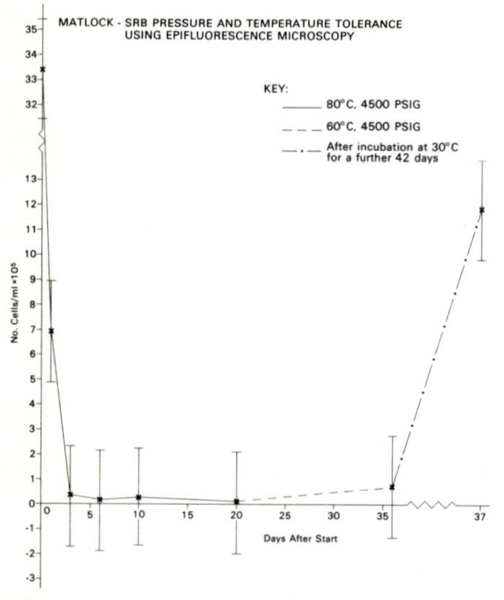

Figure 1b. Matlock SRB pressure and temperature tolerance using epi-fluorescence microscopy

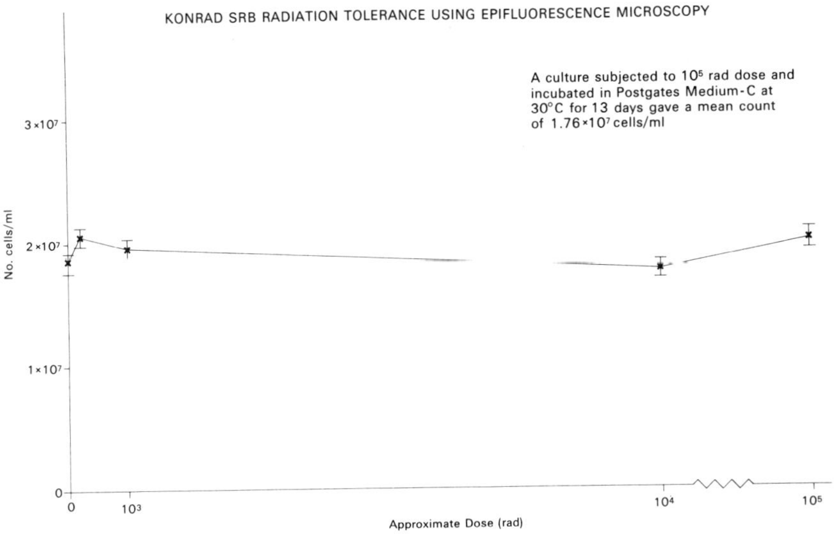

Figure 1c. Konrad SRB radiation tolerance using epifluorescence microscopy.

In conclusion, although the sulphur oxidisers appear intolerant to extreme conditions this situation may be altered by gradually increasing stresses e.g. temperature, pressure, in order to allow adaptation to take place. Such a situation is probably more realistic when considering a repository under construction and during operation where temperature and radiation would increase as waste emplacment occurs. In addition to this, after repository closure, a temperature/radiation gradient will be set up in the backfill around a canister. This environment could allow microorganisms to gradually adapt to this novel ecological niche.

MODELLING OF THE NEAR FIELD

Current swiss disposal concepts envisage vitrified waste disposal in cast steel canisters emplaced horizontally into bentonite infilled drifts (Fig. 2). As described in detail elsewhere [6] , the corrosion of the steel canister will provide a large amount of energy ($\sim$3 x 10^9J/canister) which could be utilised by chemolithotrophic organisms. * Essential nutrients could be supplied by groundwater or from the bentonite and the absolute constraint on possible population size is set by available carbon.

* Free energy is assumed to be 'available' from specific reactions which may be used to produce ATP in a fully coupled biochemical reaction chain (see e.g. [7]).

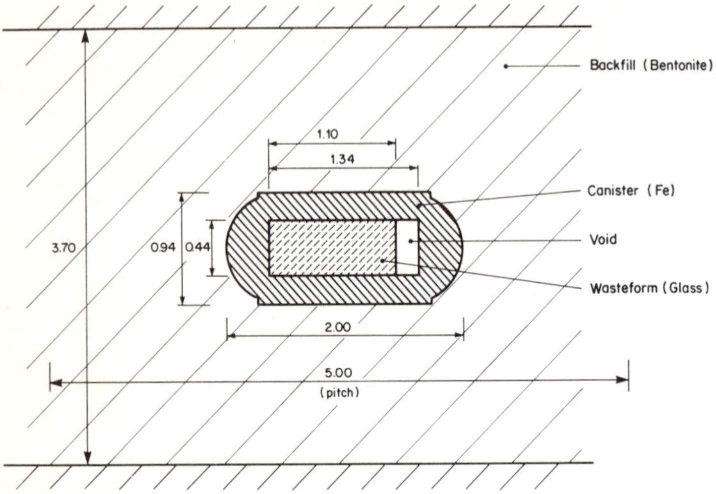

Figure 2. Reference Swiss disposal geometry (dimensions in metres)

Such general observations can be made more specific by considering the laboratory data presented above. Although extremely heat tolerant organisms are known [1], the experiments indicate that the aerobic sulphur bacteria are less heat tolerant than the anaerobic SRB. Thus they would not be expected to be present at the canister surface during the first 50y after emplacement when temperatures are above 100°C. It would therefore seem unlikely that the potentially very large amounts of energy available from aerobic canister corrosion ($\sim 2 \times 10^8$ kJ/canister) would be utilised by the bacteria. The radiation dose rate used in the experiments (2.5×10^5 Rad/hr) is very much higher than that at the canister surface ($\sim$100 m Rad/hr declining to $\sim$2m Rad/hr after 10^3y) and this factor would not seem to constrain colonisation - especially by the more resistant SRB.

In contrast to aerobic corrosion, unaerobic corrosion is likely to be very slow and may well be limited by the rate of supply of electron acceptors, as no case of microbial utilisation of the direct unaerobic Fe/H_2O reaction is known. For example taking the main electron acceptor as SO_4^{2-} in the groundwater (1.6×10^{-2} mol/l) and the reference case water flux of 0.7 l/canister/year (neglecting limitations of solute transport within the near field) and simplistically representing the overall reaction as

$$3\ Fe + SO_4^{2-} \longrightarrow S^{2-} + Fe_3O_4$$

the resultant free energy produced would be 2 kJ/canister/year. Previous calculations [6] have indicated a required energy input of 64 kJ to synthesize and maintain 1g (dry) of microbial cells. This is significantly greater than the estimated 2 kJ/canister/year calculated above and would thus limit maximum population size and activity level.

For illustrative purposes a steady state is assumed in which the annual rate of production of microbial cells ($\sim$0.03 g) is balanced by equivalent losses from the near field. The standing population would thus correspond to $\sim 2 \times 10^{13}$ organisms in a bentonite volume of 50 m^3 (4×10^{11} organisms/m^3) which compares with a typical soil content of $\sim 10^{14}$ organisms/m^3. To put these data into context some potentially significant microbial nearfield processes should be examined, i.e.

1 Physical disruption of the bentonite backfill
2 Enhanced canister corrosion
3 Enhanced glass degradation
4 Organic by-product complexation of important nuclides
5 Alteration of Eh/pH conditions
6 Microbial gas generation
7 Direct microbial uptake of radionuclides.

Since the total volume of the microbial cells above would only be 6 ml it is reasonable to discount immediately any physical effect on 50 m^3 of bentonite. The corrosion rate used for the energetics calculation corresponds to oxidation of only 2.5 g of Fe/canister/year out of an original mass of 6.5×10^3 kg.This would not affect the 10^3y canister life assumed for safety analysis. Glass degradation is not expected to provide a useful energy source or key nutrients for microorganisms and hence no direct microbial effect is to be expected. In this organic carbon poor system, the carbon is likely to be recycled with reasonably high efficiency but losses could occur as either metabolic byproducts (eg organic acids) or gases (CO_2, CH_4). If the entire steady state organic carbon loss (0.03g/canister/y) occured as CH_4 this would correspond to only 2×10^{-3} moles or 0.04 litres (at STP). This is much less than expected production rates of H_2 from inorganic anaerobic corrosion (30 moles/canister/year) and hence the effects of microbial gas production would be negligible. If, as an alternative, the organic carbon loss is assumed to be in the form of organic acids, it could be of greater significance. Assuming an average molecular weight of 10^4 daltons, the loss corresponds to $\sim 3 \times 10^{-6}$ moles/canister/year which could be important if it complexed and increased the solubility of particular actinides (eg U, Np). A more reasonable assumption is that all waste components compete for available complexant which does not significantly affect release rates to the far-field. Although micro-environments with greatly altered Eh/pH conditions can be expected in and around microbes, their small quantity relative to the large pH (bentonite) and Eh (canister corrosion products) buffers negates such effects on a larger scale. Finally, direct uptake of radionuclides into or onto microbes would have a detrimental effect only if it was irreversible and such organisms were mobile within the nearfield. If, as expected, uptake is even partially reversible, the large exchange capacity of 8.8×10^7g of bentonite relative to 0.06g of organisms would minimise this effect. In addition, strong evidence exists to suggest that the average pore size in compacted bentonite ($\sim 50\mu m$) is sufficiently small to ensure extremely limited microbial mobility.

CONCLUSIONS AND FUTURE RESEARCH

For the particular case of the Swiss concept for deep geological disposal of HLW, there is increasing evidence that although microorganisms will be present throughout the repository their effects on near-field performance can be demonstrated to be negligible. The main area of uncertainty is in the energy availability from microbial corrosion processes since, if it were much larger than calculated above, organic complexants might affect the conclusion of current safety analyses. Some further work on microbial degradation energetics and on the tolerance of key microbial groups should allow resolution of this problem. Subsequent microbial research should then focus on low/intermediate level waste disposal options where the significance of microbial processes is likely to be very much greater.

ACKNOWLEDGEMENTS

The U.K. component of this work was funded by the U.K. Department of the Environment and the Commission of the European Communities and is published by permission of the Director of the British Geological Survey (NERC) and the Department of the Environment. The Swiss component was financially supported by NAGRA. We would also like to thank the Radiochemistry Section at the University of Loughborough for the use of their ^{60}Co cell, and D. Savage for the use of hydrothermal equipment.

REFERENCES

1. West, J. M., McKinley, I. G. 1984. The geomicrobiology of nuclear waste disposal. Proc. Boston MRS Meeting, 1983. Scientific basis for nuclear waste management VII, G. McVay (ed.) North Holland, 487-494.
2. West, J. M., Christofi, N., McKinley, I. G. 1985. An overview of recent microbiological research relevant to the geological disposal of nuclear waste. Radioactive Waste Management and the Nuclear Fuel Cycle, 6, 79-95.
3. Christofi, N., West, J. M., Philp, J. C., Robbins, J. E. 1984. The geomicrobiology of used and disused mines in Britain. Rep. Brit. Geol. Surv., FLPU 84-5.
4. Christofi, N., West, J. M., Philp, J. C. 1985. The geomicrobiology of European Mines relevant to nuclear waste disposal. Rep. Brit. Geol. Surv., FLPU 85-1.
5. West, J. M., Arme, S. C. 1985. Tolerances of micro-organisms to extreme environmental conditions. Prog. Rep. August 1984 - February 1985. Rep. Brit. Geol. Surv. FLPU 85-14.
6. McKinley, I. G., West, J. M., Grogan, H. A. 1985. An analytical overview of the consequences of microbial activity in a Swiss HLW repository. NAGRA NTB 85-43.
7. Thauer, R. K., Morris, J. G. 1984. Metabolism of chemotrophic unaerobes: old views and new aspects. The Microbe, 1984 (D. P. Kelly and N. G. Carr, Eds.) Cambridge University Press, 123-168.

The Chemistry of the Near Field Environment.

F.T. Ewart[1], S.M. Sharland[2] and P.W.Tasker[2]
[1]Chemistry Division, [2]Theoretical Physics Division,
U.K.A.E.A., Harwell, Oxford, OX11 0RA , U.K.

Introduction.

The chemistry of near field, i.e. the pH, Eh and potential complexing ions, determines the solubility of radionuclides from a waste source. It is often assumed that outside the repository this chemistry is that of the appropriate groundwater. However, the chemical environment around the repository itself is perturbed both by the near field components, e.g. wasteform, backfill etc., and by the act of building the repository. For example, a shallow land site in clay might be created by excavating a trench in which would be constructed a concrete structure, later to be surrounded with returned oxygenated clay from the original excavation. The repository structure would be filled with cemented waste in metal containers and a concrete backfill.

Each of these components of the repository will establish chemical equilibrium with the aqueous phase in the interstices. The soluble constituents of the cement and concrete will dissolve producing an alkaline water saturated with lime and calcite. This water will react with the metallic containers, reducing the redox potential, initially to that of the corroding metal (probably iron) and subsequently to the Fe(II)/Fe(III) equilibrium. This water will be available to migrate within the repository and then into the surrounding clay. This clay may be subjected to several 'waves' of differing chemistries. The first will be the diffusion of oxygen while the trench is open, followed by oxygenated water from the returned clay, the reducing alkaline water from the repository structure and finally any possible released radionuclides and associated radiolysis products. These chemical waves will modify the sorption properties and to some extent the rheological properties of the clay

The purpose of the programme is to establish to what extent the geological and chemical environment may be perturbed from its undisturbed state and at what distance from the repository the water chemistry is again controlled only by the geology of the far field. Our approach to the problem of predicting these overall chemical perturbations is divided into two areas. These are
(1) experimental studies of clay to investigate the chemical composition and the buffering capacity against changes in pH and redox potential, and
(2) theoretical models which aim to study the evolution of the chemistry over longer time scales and assess the importance of individual chemical reactions to the overall behaviour of the system The behaviour of the potential repository site, over extended periods of time, can only satisfactorily be predicted by a thoroughly validated mathematical model. This present exercise is one stage in the construction of such a model. The comparison of experimental and theoretical results gives confidence in the model predictions and in the extrapolation over longer time scales.
The first problem we shall be concerned with is the oxidation of clay which generates acidity and the intrusion of alkaline water from the concrete backfill. Thus, of particular importance is the response of the clay to changes in pH.

Experimental Studies.

The pH buffering capacity of the clay was determined by carrying out acid and alkali titrations, in steps, and monitoring the pH change. Core samples were taken from the Oxford clay bed at Elstow in Bedfordshire, U.K. The cores were cut into 30cm lengths and wrapped in plastic film then aluminium foil and wax sealed immediately after the core was extracted. This was to minimise atmospheric oxidation in the sample. A section from a 25mm diameter core taken at 6.97 to 7.21 metres depth was used for these experiments. This core sample contained a significant number of fossil residues. The samples used were full cross sections from the core.

The clay was analysed in some detail by chemical analysis for elemental composition and by sedimentation and X–ray diffraction to determine the mineralogical composition. The results of the analysis of an adjacent core to the one used for the titration experiment are given in table I below.

The clay was titrated against acid by suspending about 4g of clay in 20 ml of degassed

Table I. Mineralogical analysis of Elstow clay at 6.8 to 7.2m depth

Mineral	Content (% by weight)
Kaolinite	12.4
Chlorite	2.9
Illitic clays	31.4
Gypsum	2.7
Smectite	1.1
Quartz	26.4
Calcite	20.3
Pyrite	2.8

demineralised water. This suspension was transferred to the sealed vessel of an automatic metering burette. The acid titration was made by adding incremental volumes of 1M hydrochloric acid where each increment was equivalent to 10 milliequivalents of acid per 100g of clay. Separate experiments were mounted to determine the equilibration time needed between additions: this was found to be 30 minutes. It was not possible to perform the alkali titrations by the same method because of the interference from atmospheric carbon dioxide. The method adopted was the batch titration method. A number of separate samples of 2g of clay were suspended in 10ml of degassed demineralised water in separate sealed vessels. To each of the vessels was added a quantity of .021M calcium hydroxide solution to produce a series of vessels with ratios of 2 to 16 milliequivalents of alkali to 100g of clay. The vessels were agitated for 48h then opened and the pH measured.

The aqueous phase from each of the above experiments was recovered by centrifuging and saved for analysis. The residue of the clay was analysed for exchangeable ions by mixing with 10 ml of 0.5M ammonium acetate, centrifuging and recovering the aqueous phase. The aqueous solution were analysed by emission spectroscopy for all elements present at concentration greater than 0.1 ppm. The as-received clay sample was also suspended in water to extract any water soluble species, and contacted with ammonium acetate solution to obtain any exchangeable species. These solutions were also analysed by emission spectroscopy. The results of the analyses of the aqueous phases are given in Table II.

Table II. Analysis of aqueous phases (in ppm.)

Element	1	2	3	4	5	6	7	8
Al	.22	27	1.8	.55	<.06	.06	.09	.07
Ca	42	.29%	69	97	360	300	370	435
Fe	.065	51	.56	.025	<.002	<.002	.015	<.002
Mg	15	125	21	.055	54	13	53	47
Si	1.0	11	1.5	1.0	5.3	1.2	4.9	4.3
Na	45	555	63	38	8.5	42	12	6
K	20	600	25	14	30	81	27	30
S	80	50	155	360	7	14	13	10

Key
1. water extract of clay
2. endpoint of acid titration (vol ca. 30ml water/3g clay)
3. alkali titration after 2meq/100g step
4. alkali titration after 16meq/100g step
5. soln.1. after ion exchange
6. soln.2. after ion exchange
7. soln.3. after ion exchange
8. soln.4. after ion exchange

The acid titration of the clay is shown in figure 1 as a plot of pH against amount of acid added, expressed in units of milliequivalents per 100g of clay . The reaction that is represented by this curve is primarily the neutralisation of the calcite in the clay. The acid required for the reaction is equivalent to 12% by weight of calcium carbonate. This corresponds reasonably well with the mineralogical analysis of this clay, considering the heterogeneous distribution of the fossil remains in the sample. The calcite neutralisation completely masked any buffering reaction which may have been due to the clay minerals[1] but the high concentration of the polyvalent ions iron, aluminium and magnesium in the aqueous phase after the reaction was complete suggests that some proton substitution of these ions on the clay minerals may have occurred.

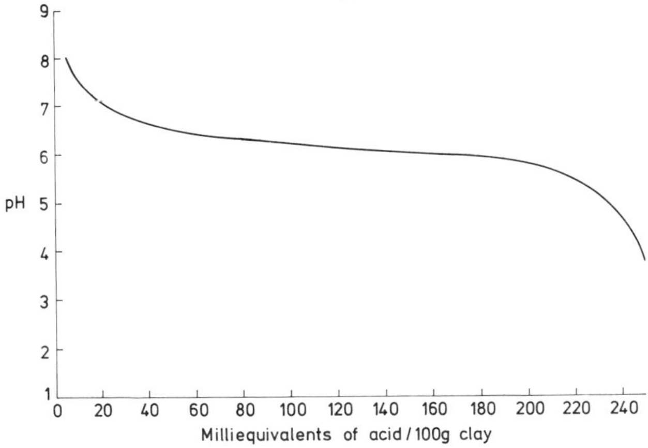

Fig.1. pH buffering of Elstow clay-acid titration.

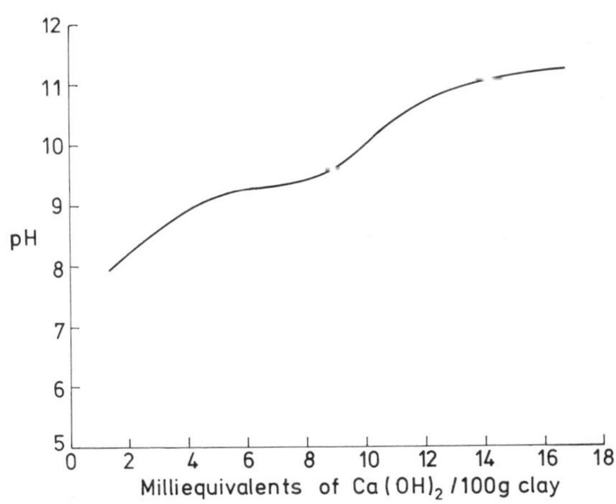

Fig.2. pH buffering of Elstow clay-alkali titration

When the clay was titrated against calcium hydroxide, there were two clear steps in the curve of pH against added alkali. These are shown in figure 2 which plots the reaction on similar axes to the previous figure. These buffering reactions are most probably due to the substitution of calcium for some polyvalent ions in the clay minerals. Such ions could be iron aluminium or magnesium. All of these form hydroxides which have smaller dissociation constants than calcium hydroxide and thus lower the pH of the solution. This view is supported by the analyses of the aqueous phase which shows an initial increase in the concentration of these ions. This increase does not continue during the titration since the solubility of the hydroxides decreases with increasing pH. The concentration of calcium in solution does confirm the existence of some substitution reaction since the concentration does not increase with each addition of calcium hydroxide solution. These effects are illustrated in Figure 3 where the concentration of calcium, iron, magnesium and aluminium in the aqueous phase are plotted against the alkali addition. The two steps in the alkali titration reaction do not appear to have any corresponding perturbation on the curves of polyvalent ion concentrations, although there is a change in slope of the concentration curve which corresponds to the plateau in the titration curve. The two steps can probably be attributed to the existence of two mineral phases which will accept calcium substitution but with different equilibrium constants.

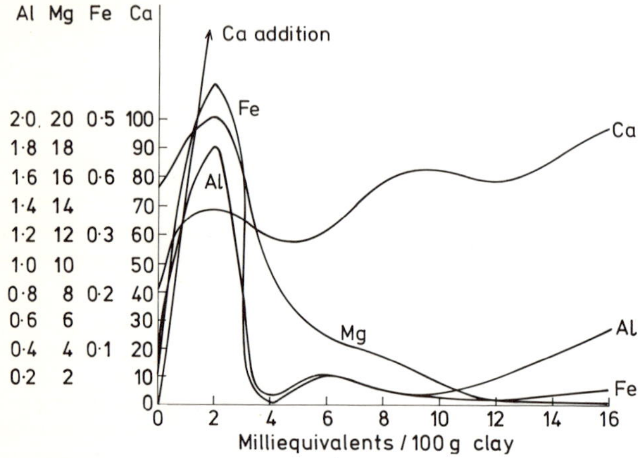

Fig.3. Concentrations of cations in aqueous phase-alkali titration

The ions that were involved in the substitution reactions did not appear to be readily replaced by ion exchange, nor was there any strong variation in the concentrations of the exchangeable ions during the titration. The exception to this was strontium which showed some small increase in exchangeable fraction at the plateau in the titration curve. Calcium alone was present in significant concentrations after ion exchange. The amount was substantially constant and independant of the amount added during the titration. A possible conclusion from this is that the exchangeable calcium is present in a totally different mineral phase from that involved in the substution reaction, and that the ions that are substituted are not available for ion exchange. These observations are at variance with some previous studies on soil liming[2] where a good correlation was obtained between the exchangeable ion concentration and the pH buffering. In that work, however, the ion exchange was made apparently using a calcium salt.

Theoretical Modelling.

The theoretical model is concerned, in this first study, with the oxidation of the pyrite component. This oxidation of one molecule of pyrite produces two molecules of sulphuric acid which will tend to modify the pH of the clay. The first objective then is to model the pH and the oxygen profile in the clay. We know from the above experiments the proportion of pyrite present and also that the clay should show a substantial buffering capacity. As a starting point we consider an empty trench and the effect of diffusing oxygen on the clay. We consider a narrow horizontal section extending from the trench wall into the clay and model the progression of oxygen through static pore water. The model may be extended later to include the effect of some groundwater flow. Oxygen is consumed in a complex reaction which converts pyrite (FeS_2) into ferric hydroxide ($Fe(OH)_3$).The aim of the model is to predict the concentrations of a number of aqueous species as functions of time and distance from the trench wall. The equation governing one dimensional mass transport of aqueous species i may be written

$$\frac{\partial C_i}{\partial t} = D_i \frac{\partial^2 C_i}{\partial x^2} + R_i$$

where R_i represents the rate of production or depletion of species i by chemical reaction.

Initially we model the oxidation of pyrite using the following stoichiometric reactions ,

$$2FeS_2 + 7O_2 + 2H_2O \rightarrow 2Fe^{2+} + 4SO_4^{2-} + 4H^+ \quad (1)$$
$$4Fe^{2+} + O_2 + 4H^+ \rightarrow 4Fe^{3+} + 2H_2O \quad (2)$$
$$Fe^{3+} + 3H_2O \rightleftharpoons Fe(OH)_3 + 3H^+ \quad (3)$$
$$FeS_2 + 14Fe^{3+} + 8H_2O \rightarrow 15Fe^{2+} + 2SO_4^{2-} + 16H^+ \quad (4)$$

However there is little available data on these individual reaction rates and a preliminary sensitivity analysis of the results shows a strong dependence on the rates for (2) and (4). Therefore this system is replaced by a single reaction

$$FeS_2 + 7/2O_2 + 15/4H_2O \rightleftharpoons Fe(OH)_3 + 2SO_4^{2-} + 4H^+ \quad (5)$$

The equilibrium constant of this reaction is sufficiently large that it may be regarded as a single direction reaction. The rate constant is chosen large so that the results are unaffected by changes in this parameter.

We have used a numerical model to solve the coupled differential equations. The variation of solution chemistry with distance from the trench wall is treated by dividing the section into cells. Within each cell the solution chemistry proceeds according to the concentrations within the cell, the diffusion is inserted by using finite difference representations of the diffusion equations. The concentration of oxygen is kept constant in the first cell (made much smaller than the others) and no chemical reactions occur here. In this way the supply of oxygen to the grid can be kept constant. At the other end of the grid there is free diffusion of all species.

The model uses rate equations throughout rather than a mixture of rate equations and equilibrium equations. This means that there is little algebraic manipulation required before the model is run. In the cases in which the rate constants are not well known they are chosen to reproduce the known equilibrium constants with the faster direction going at a rate set so these processes occur on a much faster timescale than the diffusion i.e. the rates are made sufficiently fast to simulate the driving force for the equilibrium. This gives rise to stiff components in the coupled differential equation system. The integration is done using the FACSIMILE computer program[3], which is based on Gear's method[4].

We consider events over a period of 1 year although it is envisaged that the trench will be lined with concrete soon after excavation. The diffusion coefficients of the species are sufficiently small that the associated migration length scales are of the order of a few centimetres. The model section is 3 cm in length with 10 grid points (the first cell is a thousandth the size of the section length). The results from an initial run which postulated pyrite in an inert matrix showed that the amount of oxygen present beyond the first few cells is minimal due to its consumption in the pyrite oxidation reaction. The consequent acid generation produced a pH near the trench wall of about 5. While such pH values are found in some soils it is unrealistic in these calcite bearing clays. Subsequent calculations included the reaction of the calcite component with the sulphuric acid oxidation product. The resultant calculated pH near the trench wall was 8.15 after 1 year. The chemical equilibria involved in buffering a solution in equilibrium with calcite include a reaction which releases gaseous carbon dioxide. There is therefore a difference between a system in equilibrium with the atmosphere with a partial pressure of carbon dioxide of $10^{-3.5}$ atm. and one that is closed. The pH of an open system is calculated as 8.2 while a closed system has a pH of 9.9. We start our calculation with a water chemistry equilibrated with air (the initial concentrations of the species are given in table III). However, the subsequent reactions in the clay are equilibrated with atmospheric CO_2 at the trench wall. Equilibration deeper in the clay can only occur by diffusion of dissolved CO_2 from the open boundary. Further buffering reactions may occur due to various mineral reactions; one such reaction has been included in this model – the anorthite/ kaolinite reaction[5],

$$CaAl_2Si_2O_8(s) + 2H^+ + H_2O \rightleftharpoons Ca^{2+} + Al_2Si_2O_5(OH)_4(s) \quad (7)$$

We note that when this reaction is included the equilibrium pH in the pore water is 9.0, close to that observed and the buffering of the system is more effective. However, we have not included other important equilibria involving clay minerals and so our description of this system is not complete.

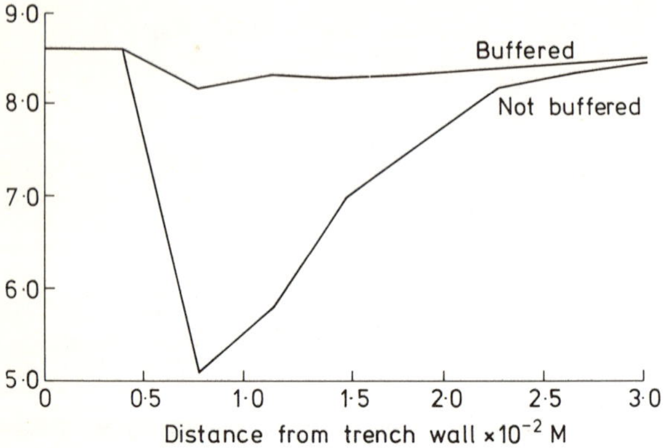

Fig.4. Calculated pH profiles in model section

Figure 4 shows a comparison of the pH profiles in the model section for the cases of no buffering and buffering by the calcite equilibria after 1 year from exposure to the atmosphere. The decrease in pH at about 0.8 cm into the grid marks the cell in which oxidation of pyrite is occuring. At this time the pyrite in the first cell has been totally consumed and the pH has recovered to its initial value by diffusion from the boundary cell. The amount of oxygen present in the clay is negligible for the duration in which the pyrite is converted into ferric hydroxide. As soon as the supply of pyrite is exhausted in a given cell the oxygen is free to diffuse in without depletion. This is illustrated in figure 5 which shows the variation in oxygen concentration in the first full size cell with time for a fully buffered system. After 100 days the pyrite is totally removed and there is a sharp increase in oxygen level. The oxygen profile after another 265 days (i.e. one year from exposure) is illustrated in figure 6 with the pyrite distribution at this time also shown. These indicate that oxygen penetration of clay is very limited over the likely time scales of repository construction.

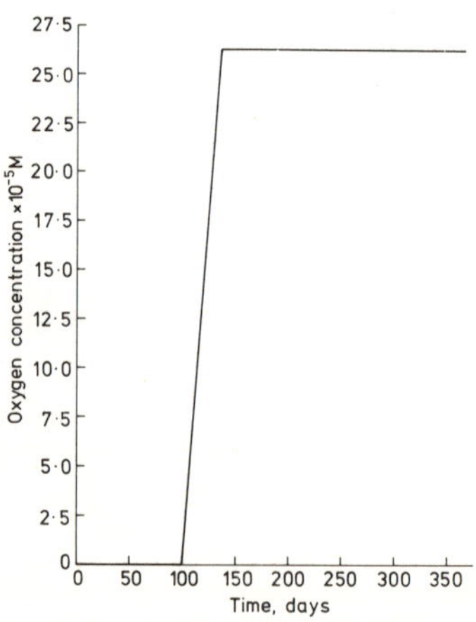

Fig.5. Calculated oxygen concentration in first cell with time for a calcite buffered run

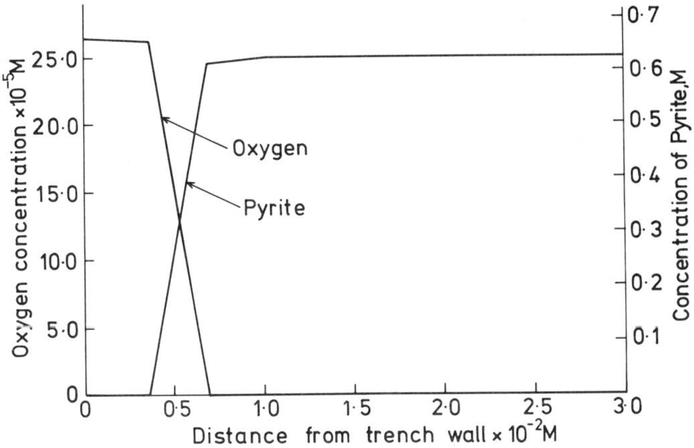

Fig.6. Calculated oxygen and pyrite profiles after 1 year for a calcite buffered run

Conclusions.

The experimental measurements on the clay have demonstrated the substantial capacity of the clay to buffer the pH against the addition of both acid and alkali. The acid buffering is due largely to the calcite content but the alkali buffering reaction is more complex, probably involving two different substitution reactions.

The diffusion of oxygen into the clay body will oxidise the pyrite component, producing significant quantities of sulphuric acid which, without the clay buffering reactions would result in a dramatic shift in the pH of the system. However, the modelling has shown how the calcite reaction and the equilibration of kaolinite and anorthite makes this change in pH almost negligible.

We have shown that the initial effects of oxygen intrusion and consequent acid generation in clay are neutralised by the natural buffering capacity. However the composition of the aqueous phase will also have been changed by the release of the substituted elements. This may effect other clay mineral equilibria. Of particular importance are in the reactions that will occur when the alkaline front from the cement moves into the clay. Further developments of the model will include more and competing mineral reactions by incorporating a geochemical code to resolve the chemical equilibria at each stage. A test of such a model will be to reproduce the alkali titration experiment.

Table III. Physical Parameters used in model runs with starting chemistry for the case of buffering by calcite.

Species	Chemical Energy of Formation kJ mol^{-1}	Effective Diffusion Coefficient $\times 10^{-10} m^2 s^{-1}$	Initial Concentration mol dm^{-3}
$Fe(OH)_3(s)$	–39.79	–	0.0
H_2O	–14.03	–	1.0
H^+	0.0	9.3	2.487×10^{-9}
OH^-	– 8.985	5.3	3.933×10^{-6}
O_2	0.0	1.0	2.648×10^{-4}
SO_4^{2-}	–42.53	1.0	0.0
$FeS_2(s)$	– 9.54	–	0.625
HCO_3^-	–33.53	1.0	3.421×10^{-5}
CO_3^-	–30.16	1.0	1.854×10^{-13}
$H_2CO_3(s)$	–35.596	–	1.077×10^{-5}
Ca^{2+}	–31.62	1.0	1.418×10^{-4}
CO_2	–22.77	–	3.162×10^{-4}
$CaCO_3(s)$	–64.48	–	4.06

Acknowledgements.

This work was funded by the Nuclear Industry Radioactive Waste Executive (NIREX) and permission to publish is gratefully acknowledged. The authors would like to thank Mr.A.J.Smith of Harwell for performing some of the experiments.

References.

1. E.J.Russell,in 'Soil Conditions and Plant Growth' Longmans,Green & Co. London 1950
2. R.C.Turner et al., Soil Science 93 pp374–382 1961 Soil Science 94 pp58–63 1961 Soil Science 95 pp186–191 1962 Soil Science 99 pp88–92 1965
3. E.M.Chance,A.R.Curtis,I.P.Jones and C.R.Kirby, AERE-R8775 1977
4. C.W.Gear,Commun. ACM 14,176–179 1971
5. W.Stumm and J.J.Morgan, in Aquatic Chemistry J.Wiley,1981

THE CHEMICAL ENVIRONMENT IN CEMENT MATRICES

M.J. ANGUS AND F.P. GLASSER
Dept. of Chemistry, University of Aberdeen, Meston Walk, Old Aberdeen, AB9 2UE, Scotland

ABSTRACT

General progress in modelling the behaviour of cement matrices and their interactions with wastes is reviewed in terms of appropriate modelling concepts and their formulation in terms of parameters which are susceptible to evaluation. The redox behaviour, particularly of slag-cement blends is shown to differ sharply from that of Portland cement matrices. Experimental methods of measuring E_h and poising capacities are described and the composition of pore fluids determined. Slag cements contain appreciable soluble S^{n-} species as well as $S_2O_3^{2-}$ and SO_4^{2-}

INTRODUCTION

The leach behaviour of cemented radioactive waste forms is controlled to a large extent by the intrinsic matrix properties. For cements, these properties are both physical and chemical. The interaction of these factors control the observed leach rates. However, leach tests rarely provide data which can be extrapolated with confidence to longer ages because the underlying contributory factors have not been identified with certainty and isolated for separate study.

Table I.

REQUIREMENTS OF A CHEMICAL MODEL FOR CEMENT SYSTEMS

1. Model must rest on a foundation of physiochemical data, e.g. solubilities, pH, E_h, thermodynamic functions, phase equilibria.
2. Models must be 'open ended' to allow for input of data concerning such factors as
   - (i) matrix behaviour, including ageing.
   - (ii) effect of chemically reactive matrix modifiers, e.g. slag, PFA.
   - (iii) waste-matrix interactions.
   - (iv) matrix-waste-near field and environmental interactions.
3. Models must be capable of interfacing with kinetic data, e.g. leach rates.
4. Time must be included as an essential variable.
5. Modelling exercise must have as its aim maximum validation of predictive exercises.
6, Treat colloid formation

It has also proven to be difficult to develop satisfactory models for cement systems. These difficulties arise partly from the reasons given above and partly because of the complexity of the system, which includes not only matrix itself but also its interactions with radwaste materials as well as with the near field environment. In devising appropriate models, it seemed to us important to ask three questions (i) what expectation should one have of the model? (ii) how can the model be formulated to make it compatible with our knowledge of cements, yet amenable to solution using either available data or readily-obtainable data inputs? (iii) how useful will the model be for a variety of purposes, and what credence can be placed on its predictions?

We have taken preliminary steps towards the formulation of a model by attempting to define the internal environment of cements in terms of three components - CaO,

SiO_2 and water - and of its known phase composition [1]; in the present study we extend some of these concepts. We begin by listing, in Table 1, some of the principal requirements of a model to which this as well as future contributions will be addressed. Listing these requirements is, however, very different from formulating them in terms of measurable parameters. Table II suggests in more detail, how the "open end" concept might be applied to cement systems. Beginning with a simple model, in which the cement is viewed as a diphasic system of solid and liquid, they increase in complexity gradually encompassing not only the known phase inhomogeneity of the solid cement hydration products as well as the variable composition of the hydrogel, but expanding to allow for the presence of additional chemically-reactive cementitious components e.g. pozzolanic additives of reactive materials containing radwaste entities, as well as environmental reactions. Progress has thus been made in developing a conceptual framework and in assessing what relevant cement properties need to be determined. We envisage that models should rest mainly on an equilibrium or quasi-equilibrium approach. Cement systems are however inherently dynamic and ageing effects occur. Moreover, practical measures of performance such as leach testing reflects both kinetics and equilibria. But we are confident that this approach is the best way to proceed; it permits us to utilize thermodynamic and related data, e.g. solubilities and subsequently to develop interfaces with kinetic data.

RESULTS

The Aqueous Phase

Cements are a microporous matrix whose pores are normally filled with an aqueous phase which is in intimate contact with the high surface area matrix phases. Accordingly, its composition assumes considerable importance in determining the properties of the composite. Studies of synthetic cement pastes made, for example, by hydrating pure calcium silicates disclose that the aqueous phase has a high pH arising mainly as a consequence of Ca solubility. Real cements, which contain substantial alkali, are also strongly alkaline but the source of the alkalinity is as a consequence of high Na, K solubility; that of Ca is greatly reduced. These pore fluids can be extracted using a powerful hydraulic press and Table III, modified from [2,3], gives chemical compositions of these fluids in cements of two different alkali contents. Material balance calculations disclose that K and Na especially are partitioned between the solid and aqueous phases, so that the solid contains appreciable amounts especially of Na. Experi-

Table II.
SPECIFIC CHEMICAL FEATURES INHERENT IN MODELLING CEMENT SYSTEMS

CEMENT-ONLY MODELS, VARYING IN SOPHISTICATION	- SIMPLE TWO-PHASE MODEL: HOMOGENEOUS CEMENT + AQUEOUS PHASE - INHOMOGENEOUS SOLID MODELS INCLUDING CaO, SiO_2 and H_2O COMPONENTS - MORE COMPLEX MODELS TO INCLUDE Na^+, K^+, Fe^{3+}, $A\ell^{3+}$, SO_4^{2-}, etc.
CEMENT-MATRIX WASTE MODELS	- INTERACTIONS BETWEEN SPECIFIC WASTE FORMS OR CONSTITUENTS AND CEMENT ALLOWED USING ONE OF THE MODELS ABOVE.
CEMENT/ENVIRONMENTAL INTERACTIONS	PERMITS INTERACTIONS BETWEEN A CEMENT MODEL AND ENVIRONMENTALLY-SENSITIVE EXCHANGES AND UPTAKES
HYBRID MODELS	ALLOWS CEMENT/WASTE/ENVIRONMENTAL INTERACTIONS TO BE ASSESSED, BUT ARE COMPLEX

Table III.

COMPOSITION OF TYPICAL CEMENT PORE FLUID

Aqueous Phase Properties	Curing Time (days)	Ordinary Portland Cement	High Alkali Portland Cement
pH	90	13.4	13.8
Na (moles/ℓ)	30	0.058	0.066
K (moles/ℓ)	30	0.222	0.551
Ca (moles/ℓ)	90	2.7×10^{-3}	1.2×10^{-3}

ments of several years duration disclose that for OPC, the alkali partition reaches steady state in 30-90d.

However, other considerations are important with respect to radwaste immobilization. Some radwaste species, notably Tc, Mo, I, U and some transuranics readily undergo redox reactions, so the E_h or redox potential, of the pore fluid is also of importance. Table IV lists some of the parameters for the characterization of the aqueous phase.

In normal cements, a water/cement ratio exceeding 0.22-0.24 is necessary to ensure complete hydration: water, normally in excess of that ratio fills the micropores, the size distribution of which has been extensively studied on dried material by mercury intrusion porosimetry and helium inflow pycnometry. However, the presence of blending agents such as slag modifies substantially the microstructure and chemical nature of the pore fluid. In order to measure these changes the pH-E_h functions in slag cements have been determined.

Basically, the measurement of E_h in cements is an electrochemical function. At first, cells were devised in which a Pt electrode was cast directly into the cement, but these proved unsatisfactory owing to polarization effects and the variable, but rather high resistivity of the matrix. A satisfactory technique which was eventually used as standard, consisted of three stages. (i) production, formulation and maturation of ca 200g, test cylinders, (ii) pore fluid squeezing to obtain a sample of the aqueous phase free from solids and (iii) measurement of its E_h in a small polythene cell with a Pt electrode. The arrangements for collecting the aqueous phase and subsequent measurement of its E_h enabled all stages to be carried out with the exclusion of air. Table V shows the E_h values thus obtained. The pore fluid from an OPC typically has an E_h of ∿ 80mV. As Portland cement is normally fired under slightly oxidizing conditions at ∿ 1450°C this value seems reasonable, although of course reducing conditions in the kiln may sometimes occur; wider experience of other cements will be required to determine the range of E_h values encountered in commercial cements.

Table IV.

CHARACTERIZATION PARAMETERS FOR THE AQUEOUS PHASE

Significant Parameters Include:
Volume, distribution
Chemical composition
- inorganic species
- organic species
pH
E_h
Colloid content

Addition of Mg and Zn do not lead to a permanent decrease of the E_h of matured cements. This is perhaps not surprising: these metal powders react rapidly and completely

Table V.

REDOX POTENTIAL IN CEMENT AND SLAG SYSTEMS

SYSTEM(1)	QUANTITY OF ADDITIVE (moles/g OPC)	E_h (mV)	pH
OPC	none	+82	13.5
OPC-Mg	$3x10^{-3}$, Mg powder	+99	13.4
OPC-Zn	$3x10^{-3}$, Zn powder	+63	13.6
OPC-Mn^{2+}	$4.5x10^{-4}$ $MnC\ell_2$	+158	12.7
OPC-S^{2-}	$4.5x10^{-4}Na_2S$	-335	13.9
OPC-Fe^{2+}	$4.5x10^{-4}Fe(NH_4)_2(SO_4)_2$	+34	13.2
BFS(2)	No OPC	-315	12.7

1. Aged 25d at 20°C prior to expressing pore fluid. OPC to British Standard specification.
2. Frodingham Blast Furnace slag, mixed with $Ca(OH)_2$ in the ratio 10:1.

before setting is completed and the reduced product of reaction, H_2, is swiftly eliminated. Self-potential measurements made on less electropositive metals in cement, e.g. mild steel, suggest that they will lower the internal E_h provided their surfaces are not passivated [4]. The presence of Mn^{2+} raises the E_h value slightly: much of the Mn is probably precipitated in the course of hydration; these reactions consume OH^- leaving $C\ell^-$ in solution to reduce somewhat its pH. Ferrous iron behaves similarly, slightly depressing the E_h but leaving its pH largely unaffected since much of the SO_4^{2-} is removed by precipitation in the AF_m and AF_t phases. However, the presence of S^{2-}, either added as Na_2S or liberated from blast furnace slag, very significantly lowers the E_h of the composite by ca 400 mV relative to OPC.

The special role of slags in controlling E_h deserves further exploration. In practice, slag is normally blended with activators to promote hydration. The most commonly used activator is OPC; Table VI shows E_h values obtained from slag-cement blends.

Table VI.

REDOX POTENTIAL OF SLAG CEMENTS AFTER CURING AT 20°C. SOLUBLE S^{2-} CONTENTS MEASURED BY ION CHROMATOGRAPHY

OPC	BFS (Frodingham)	E_h (mV)	$[S^{2-}]$ (ppm)
100	0	+82	0
75	25	+79	-
50	50	+68	12
25	75	+35	6
15	85	-240	120
10	90	-269	110
5	95*	-330	1000
2.5	97.5*	-259	1100

*0.5% $Ca(OH)_2$ added to ensure setting.

It can be seen that at slag contents up to 75% the E_h remains comparatively high, and S^{2-} in the aqueous phase remains low but at higher slag contents, the E_h drops sharply while S^{2-} rises. These data were obtained after 25d hydration, at which time approximately 25% of the slag had reacted. Therefore much of the total S present in the slag, which contained 1.31% S, remains locked in the unhydrated portion. It is suggested that the existence of a 'break' in E_h values at > 75% slag may result from the comparative short ageing, and that the existence of a sharp break in E_h values, and its occurrence at a particular slag content, should await test data obtained after longer ageing. However, leach testing is frequently commenced after 25-30d ageing, and in this context the break at ∿ 75% is significant.

The absolute values of E_h have been complemented by measurements of the ability of the system to resist changes in its redox potential. This measure is termed its poising capacity. The concept is analogous to the buffering capacity of a system whose pH function is being measured; the two types of measurement complement each other and together more completely characterize the properties of the system. The practical measurement of poising capacity is largely operational, and we have chosen to define the total oxidizable content of the system in terms of the half-reaction: $Ce^{4+} \rightarrow Ce^{3+} + e^-$. The method consists of placing ∿ 1g of solid, dried at 105°C, in a 250 mℓ conical flask to which 50 mℓ of 0.05M ceric ammonium sulphate dissolved in 2M H_2SO_4 are added. After stirring for 1h, the solid is mostly dissolved and the Ce^{3+} formed determined by an electrometric back titration, using 0.1M Fe^{2+} solution made from ferrous ammonium sulphate, and a Pt electrode with calomel reference. The end point of the reaction

$$Fe^{2+} + Ce^{4+} \rightarrow Fe^{3+} + Ce^{3+}$$

at which

$$\frac{[Fe^{2+}]}{[Fe^{3+}]} = \frac{[Ce^{4+}]}{[Ce^{3+}]}$$

is found by averaging the formal potential of the two half cells:

$$E_{cell} = \tfrac{1}{2}(E^{o/}{}_{Fe^{III}/Fe^{II}} + E^{o/}{}_{Ce^{IV}/Ce^{III}}) = 1.057V$$

and the results are expressed as moles electrons g^{-1} cement, which is effectively moles Ce^{4+} reduced per gram.

The oxidizable content of a series of cement and cement blends thus determined is shown in Table VII.

Table VII.

POISING CAPACITY OF VARIOUS UNHYDRATED CEMENTS AND CEMENT BLENDING AGENTS

Material	Oxidizable Content, moles electrons g^{-1} solid
OPC	9.59×10^{-5}
High alumina Cement	2.30×10^{-4}
PFA*	3.43×10^{-5}
Natural Pozzolan+	1.73×10^{-4}
Slag (Ravenscraig)	1.14×10^{-3}
Slag (Frodingham)	1.10×10^{-3}

*Eggborough, U.K.

+Pozzolanico Grade 325, Italy.

These measurements reflect the oxidizable content available upon acid dissolution, in the course of which attack was only partially complete on natural pozzolan and less so on PFA, but rather complete on slag, OPC and HAC. The trend of values is much as expected and agrees well with theoretical expectations: that slags should have the highest poising capacity, the numerical value of which is commensurate with their content of Mn^{2+}, Fe^{2+} and S^{2-}, as found by chemical analysis. The poising capacity of HAC, made under more reducing conditions than OPC, is the highest of the cements tested. The acidic solvent system attacks OPC and HAC cement hydration products, but attack on unhydrated slag and PFA is less complete and the values of poising capacity obtained for these materials probably represent minimum values. The method is however, suited to measuring the poising capacity of hydrated products such as are developed in blended cement.

Table VIII records selected data for two 85% slag blends, matured 25d at either 20°C or 40°C. In both cases, selective dissolution using an EDTA-based solvent disclosed that the slag was ~ 25% hydrated. Even at this stage, however, the slag has clearly altered the aqueous phase composition relative to that of OPC. The pore fluid is strongly reducing, and the concentrations of S^{2-} are comparable with those reported by Tuutti [5] whose measurement of $[S^{2-}]$ is, to our knowledge, the only other value reported in the literature. However, the presence of thiosulphate appears not to have been reported previously. Thiosulphate was determined using a Dionex ion chromatograph; in this recently-developed technique, trial injections of known thiosulphate solutions were used to establish retention times for this anion, and quantitative determination was carried out by comparison with the peak area of a series of thiosulphate solutions of known concentrations. Sulphide was determined by an S^{2-}-selective AgS electrode against a series of standards adjusted to the same pH with KOH solution. The pH value is important, since this affects the distribution of S between HS^- and S^{2-}. $S_n{}^{2-}$ species may also be present at high concentrations, but are thought less likely to occur at the lower concentrations encountered here. The presence of thiosulphate, a potentially powerful complexing agent, is noteworthy; moreover, the presence of reduced or partially reduced S species also seems to enhance the solubility of $SO_4{}^{2-}$ relative to the levels encountered in OPC. The pore fluid composition remains essentially identical with temperatures, as does the proportion of slag hydrated over the range 20°C - 40°C.

Table VIII.

PHYSIOCHEMICAL PROPERTIES OF THE AQUEOUS PHASE IN SLAG-CEMENT

Function		25d, 20°C	25d, 40°C
pH		12.6	12.5
E_h		-227 mV	-202 mV
$[S^{2-}]$	ppm	120	145
$[S_2O_3{}^{2-}]$	"	310	320
$[SO_4{}^{2-}]$	"	297	198

DISCUSSION

The nature of slag cements, particularly at high slag contents, has been shown to differ substantially from other Portland cement systems, including those OPC systems modified with PFA, silica fume, etc. It is true that all these systems share certain similarities: they are, for example, strongly alkaline. But on account of their generally higher Na and K contents, systems rich in OPC, or OPC and PFA, may have pH's up to 1 unit more alkaline than slag-rich systems. In circumstances where ions form anionic hydroxy complexes, e.g. Cr^{3+}, these differences could be important. Much more significant, however, are the differences in redox state, as measured on the E_h scale, and poising capacities of the two types of system. Slags maintain much the most reducing conditions and have sufficient poising capacity to buffer significant quantities of potentially reducible components which may be added in the form of radwastes.

The magnitude of the E_h developed in slag-rich formulations is comparable with expectations that the E_h is dominated by the presence of S-containing species. Fig. 1 shows a portion of the $S-H_2O$ system modified after Pourbaix [6] showing the pH-E_h relationships. The stippled area superimposed on the diagram shows the pH-E_h values obtaining in the pore fluids of slag-rich cements. As would be expected, the pore fluids contain $SO_4{}^{2-}$, $S_2O_3{}^{2-}$ and S^{2-}. While such treatments can only be approximate,

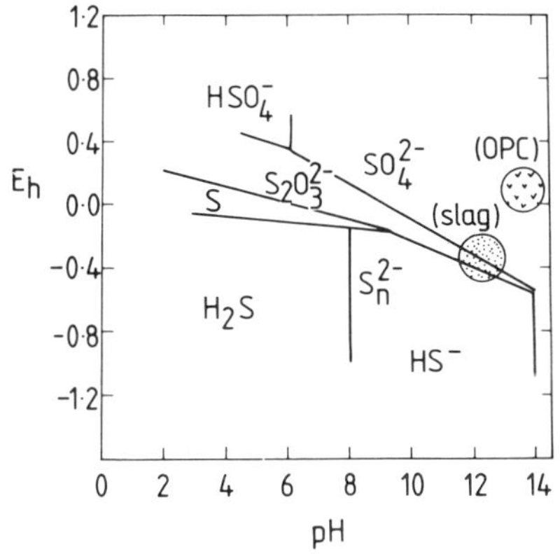

Fig. 1.

pH-E_h diagram for the S-H_2O system. OPC matrices contain S as SO_4^{2-} but measured relations in slag rich systems straddle the SO_4^{2}, $S_2O_3^{2-}$ and S^{2-} (or HS^-) fields.

owing to the chemical complexity of the aqueous phase in real cements the good general agreement between simple theory and experiment is encouraging.

The interactions between the internal environment may now be considered. Table IX shows some of the interactions which may be expected as a consequence, in whole or in part, of the reducing conditions and internal chemistry. If we consider first changes in oxidation state, it is apparent that some ions, e.g., Cs, Sr which have only one stable oxidation state in aqueous solution will remain unaffected as a consequence of the lowered E_h. However, other ions are likely to be reduced, iodine being an example where in alkaline reducing conditions, as occur in slag cements, I^- is likely to be the principal oxidation state: Iodine, if present in other oxidation states, is likely to undergo reduction. Other ions are borderline with respect to reduction. Uranium is an example. Fig. 2 shows a portion of the U-H_2O diagram on which the pH-E_h conditions typical of OPC and slag-rich cement systems have been superimposed. It appears that while U^{6+} is stable in OPC systems, it may well be reduced in whole or in part to U^{4+} in slag cements. This has practical consequences because, while neither U^{4+} or U^{6+} are expected to complex strongly with S-containing species, U^{4+} is in general much less soluble in alkaline solutions than U^{6+}.

In other cases, the sophistication of predictions concerning the effect of the cement environment would have to be increased, so as to embrace interactions between more components. Consider the state of Ru. Fig. 3 shows the Pourbaix diagram for Ru-H_2O. The

Table IX.

POTENTIAL INTERACTIONS OF RADWASTE ENTITIES WITH SLAG CEMENT MATRICES

1. Reduction of radwaste ions to lower valence states
2. Precipitation of radwaste as
   - hydroxides, hydrates, sulphates, silicates
   - sulphides, thiosulphates
   - mixed anion salts
3. Complexation reactions leading to solubility
   - sulphate, thiosulphate and polysulphide complex formulation.

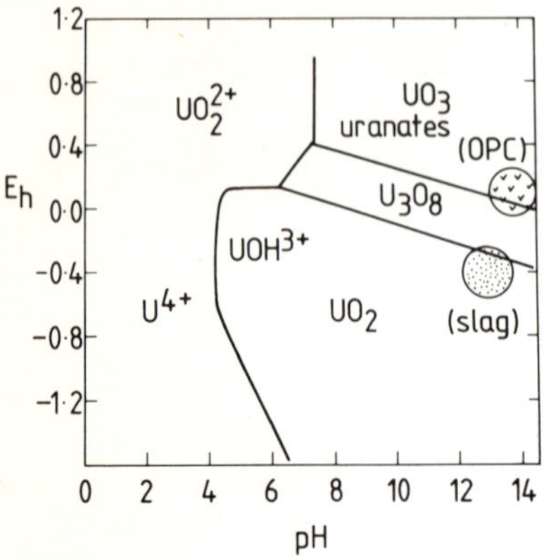

Fig. 2.

pH-E_h diagram for the U-H_2O system showing the approximate conditions superimposed by OPC and slag-rich matrices

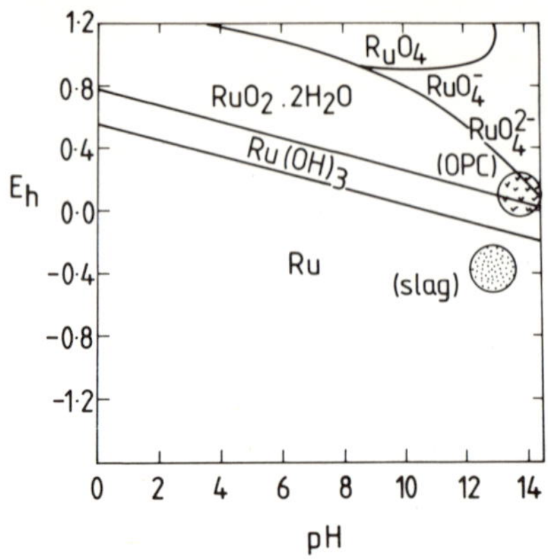

Fig.3.

pH-E_h diagram for the Ru-H_2O system showing the approximate conditions superimposed by OPC and slag-rich matrices

slag cement conditions, superimposed on this diagram, suggest that Ru may be reduced and precipitated as Ru metal. However, the low E_h of slag cements is achieved at the expense of creating a pore fluid rich in S^{2-} and, as the sulphides of Ru are stable in aqueous solutions but very insoluble, it may well be that precipitation would occur but as a sulphide rather than as Ru. Much of the poising capacity of the slag system is associated with the insoluble portions, and it is believed that the presence of insoluble elemental S may be responsible. Certainly, the solid hydrated phases contain substantial reserves of low-valent S which are potentially available for sulphide formation.

The discovery that the aqueous phase of slag cements has a complex sulphur chemistry raises some additional problems. Materials such as stainless steel, normally passive in OPC formulations, may be depassivated by S species; on the other hand, thiosulphate may have corrosion inhibitor properties. The presence of these S species, especially $S_2O_3^{2-}$, introduces yet another complexity owing to its ability to act as a strong complexing agent. We are unable to find data on the stability constants of thiosulphate complexes of many common radwaste species, but thiosulphate is known to be a strong complex former for Cu, Hg and Ag [7]; so strong that even metallic silver is slowly soluble in dilute thiosulphate solutions. It appears likely that thiosulphate will act as a strong complex former for many elements adjacent in the periodic table to the above examples. Other possibilities remain to be explored. Thus, although the oxidation state of Cs is unaffected by the presence of S^{2-}, etc., Cs may form relatively insoluble sulphide polysulphide complexes in alkaline solution [8], the nature of which have not been explored in recent decades.

We have thus far considered only the internal environment developed within slag cements. In radioactive waste immobilization, this environment will inevitably become coupled with the near field environment. It is therefore a matter of interest as to (i) how long reducing conditions are likely to persist in slag cements and (ii) if radwaste ions, perhaps complexed by some of the unique anions present in slag cement formulation, e.g. $S_2O_3^{2-}$, escape what will be their fate in the near-field environment? At present, we can only indicate some qualitative answers to these questions which, it is hoped, will stimulate more detailed examination. The reducing conditions obtaining in slag cements can only be mitigated by dissolution of the matrix or, if it retains its integrity, by uptake of oxygen. This will lead to gradual oxidation of S^{2-}, $S_2O_3^{2-}$, etc. to SO_4^{2-}. We are therefore concerned with the balance between diffusion of oxygen into the matrix and the matrix capacity to buffer this oxygen, as well as the availability of oxygen in the near field environment. If, as has been suggested, many repository environments are nearly depleted in O_2, it may be possible to envision reducing conditions persisting indefinitely. Similar considerations apply to the outward diffusion of complexed ions, although the cement matrix itself will also act to modify the local groundwater composition by contributing alkalis, calcium etc. Further analysis of this part of the problem will be deferred until work on the chemical solubility model of the dissolution of cements is more advanced.

ACKNOWLEDGEMENT

This work has been commissioned by the United Kingdom Department of the Environment as part of its radioactive waste management programme. The results may be used in the formulation of government policy but at this stage they do not necessarily represent Government policy.

REFERENCES

1. F.P. Glasser, M.J. Angus, C.E. McCulloch and A.A. Rahman. Scientific Basis for Nuclear Waste Management VIII (Ed., C.M. Jantzen, et al). Materials Res. Soc., Pittsburgh Pa, Vol. 44 (1985). ISBN 0-931837-09-X. pp 849-858.
2. J. Marr and F.P. Glasser. Proc 6th Intl. Conf. on Alkalis in Concrete. (Ed. G.M. Idorn and S. Rostam), DBF, Copenhagen (1983). pp 239-242.
3. J. Marr and F.P. Glasser. Il Cemento (in press).
4. Corrosion of Steel in Concrete. Ed., K. Tuutti) Swedish Cement and Concrete Res. Stockholm Inst., Research Report Fo4 (1982) ISSN 0346-6906: Chapter 5 pp. 145-216.
5. op.cit., Chapter 3 pp 123-136; also Chapter 5, pp 287-289.
6. M. Pourbaix. Atlas of Electrochemical Equilibria in Aqueous Solutions. Pergamon Press (Oxford) 1966).

7. S. Kotrlý and L. Šucha. Handbook of Chemical Equilibria in Analytical Chemistry. Elis Horwood, Chichester 1985. ISBN 0.85312-107-9.

8. Gemelin. Handbuch der Anorganische Chemie. System Nummer 56, Verlag Chemie, Berlin (1936). pp 216-227.

PART VIIA

Thermal and Stress Effects on Fluids in Rocks

NEAR-FIELD THERMAL ANALYSIS OF THE NUCLEAR WASTE DISPOSAL VAULT

H.S. RADHAKRISHNA*, K-C. LAU** AND A.M. CRAWFORD***
*Ontario Hydro, Research Div., 800 Kipling Ave., Toronto, Canada
**Strata Engineering Corp., Toronto, Canada
***University of Toronto, Dept. of Civil Eng., Toronto, Canada

ABSTRACT

This paper describes an experimental and theoretical analysis of the near-field thermal performance of a backfilled disposal vault in which the waste containers are emplaced in boreholes and surrounded by a bentonite-sand buffer. Effects of heat and moisture diffusion on the thermal performance of clay-based buffer and backfills were studied by means of laboratory scale experiments. The bentonite-sand buffer exhibited cracking due to desiccation. The impact of such cracking on the perforamnce of buffer as an engineered barrier is being investigated.

An integrated finite difference computer code 'TRUCHAM' was written to analyze the transient heat and moisture flow regime in a disposal vault and was validated by a scale model heater experiment. It is hoped that with proper characterization of the material properties 'TRUCHAM' can be employed as a useful tool in assessing the thermal response of the prototype disposal vault. Further work is needed to integrate the effects of heat and moisture on the shrinkage and expansion of clay-based buffer and backfills.

INTRODUCTION

The Canadian Nuclear Fuel Waste Management program currently focuses on assessing the concept of a disposal vault in plutonic rock at a reference depth of 1 km below the surface in a suitable pluton in the Canadian Shield [1]. A conceptual design for such an underground disposal facility has been developed under the auspices of Atomic Energy of Canada Limited [2]. The reference design consists of a series of long, parallel and horizontal drifts called waste rooms.

Figure 1 shows the two emplacement schemes that are being studied. In TEC-1 the waste containers are placed in the rooms whereas in TEC-2 the containers are placed in the vertical boreholes drilled in the floor of the waste room. In both these schemes the waste container will be surrounded by a bentonite-sand buffer and the rest of the room will be backfilled with a clay-crushed rock mixture.

The performance of buffer and backfills as engineered barriers depends upon the physical and chemical properties of the materials and the vault environment to which they are exposed. Since the radiogenic heat from the emplaced waste sets up a thermal field subjecting the buffer and backfills to high temperatures and temperature gradients, the ability of buffer and backfill materials to dissipate the heat and to withstand the resultant temperatures in the near field deserve a careful consideration. For this purpose a laboratory scale experiment was performed in which the thermal drying of buffer and backfill materials under simulated vault conditions was studied.

The purpose of the work described in this paper was to:

1. assess thermal performance of the buffer and backfill materials

2. validate the numerical model TRUCHAM developed for the analysis of transient heat and moisture diffusion in the near field of the disposal vault.

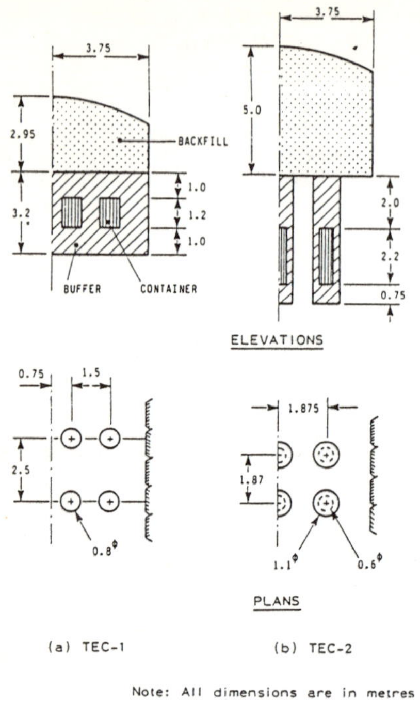

Figure 1. Waste Container Emplacement Methods

HEATER EXPERIMENT

A one-eighth scale experimental model was built simulating the near-field thermal environment of a backfilled disposal vault in which the waste is emplaced in boreholes (TEC-2 design, Figure 1). The model consisted of a concrete block to simulate the thermal mass of rock and the electric heaters to simulate the heat production from the fuel waste containers. (Figure 2).

Since the waste containers are placed in long rows of boreholes (3 per row in TEC-2 design) and the disposal vault has several disposal rooms parallel to each other, only one half of a single row of waste containers in a disposal room was studied in this experimental model. This was achieved by emplacing one full round heater and one half round heater in one and a half boreholes as shown in Figure 2. Four adiabatic boundaries were simulated by insulating the vertical faces of the concrete block with 20 cm thick polystyrene insulation. Infinite extent of rock mass above the disposal room and beneath the disposal boreholes could not be simulated very well in this model due to the limitations on its size. These boundaries were also insulated to simulate the long heat paths in the vertical direction.

The experimental model was cast in-place with a 30 MPa superplastized concrete. The boreholes and the disposal room were hollowed while casting the concrete by using strippable moulds and forms. The heaters were made out of double-walled stainless steel tubing with an electrical heating element embedded between the two walls. The dimensions of the heater, boreholes and concrete block were proportioned to model the disposal room (TEC-2 design) to one-eighth scale.

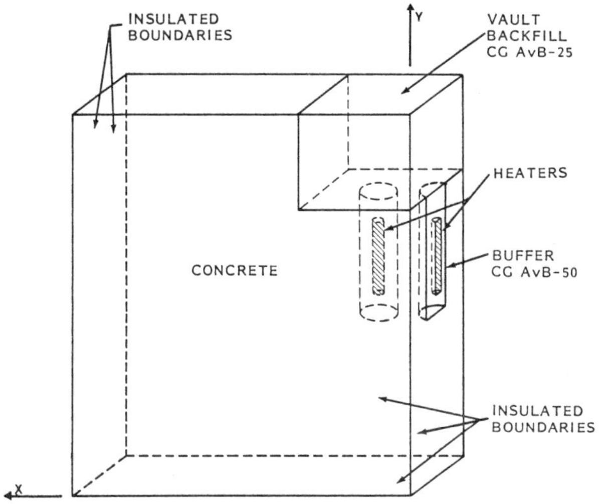

Figure 2. Experimental model simulating the Nearfield Thermal Environment in a backfilled disposal vault.

A total of 80 temperature sensors (thermocouples) were installed at various strategic positions in the heater, buffer, backfill and concrete block. About 60 of these were placed in a vertical mid-plane through the heaters. About 4 moisture sensors (psycrometers) and 4 thermal conductivity probes were also placed in the buffer, backfill and concrete to monitor the changes in moisture content.

The heat input was controlled by a programmable power supply and the thermocouples were monitored by a digital data acquisition system controlled by a desk top computer. The data was collected, stored and displayed by this computer.

BUFFER AND BACKFILL MATERIALS

An extensive program of laboratory evaluation of the clay-based materials for buffer and backfill applications has been carried out in the Canadian program [3]. Various sources of bentonite and illite rich clays were tested in combination with silica sand and crushed granite. While some of these studies are still in progress the data available suggests that the potential buffer material will be a mixture of equal proportions of silica sand or crushed granite and Avonlea bentonite (sodium enriched). The backfill composition is likely to be a mixture of crushed granite and Avonlea bentonite or an illite rich clay.

The buffer material used in this experimental program is a mixture of crushed granite (maximum 2 mm size) and Avonlea bentonite in equal proportion by dry weight. This mix material was compacted around the heaters to a dry density of 1500 kg/m^3 at a moisture content of 22.7 per cent. The backfill material was made up of 75 per cent of crushed granite (maximum 19 mm size) and 25 per cent of Avonlea bentonite. This material was compacted to a dry density of 1800 kg/m^3 at a moisture content of 14.0 per cent. Both buffer and backfill materials were placed in thin lifts and compacted by hand.

The thermal conductivity of the compacted buffer and backfill materials were determined on test specimens prepared at the same moisture and density conditions as those in the experimental model. Figure 3 shows the moisture dependency of the thermal conductivity of the buffer and backfill compositions. Also included in this

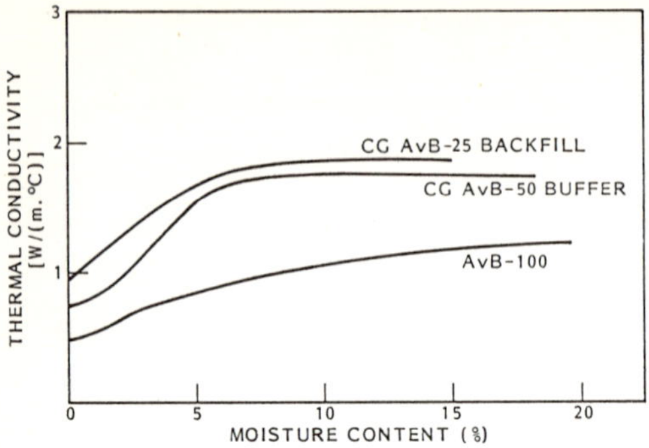

Figure 3. Thermal Conductivity vs Moisture Content Relationship for Crushed Granite and Avonlea Bentonite Buffer and Backfill

figure for comparison purposes are the data for 100 per cent Bentonite. These measurements were made at a constant dry density by subjecting the specimens to drying in stages.

Earlier studies have shown that the thermal conductivity of clay-based materials is dependent on its composition, dry density and moisture content [4]. Once the material compositions are selected and placed at a controlled density the thermal conductivity of buffer and backfill are only governed by their moisture content.

COUPLED MOISTURE AND HEAT DIFFUSION THROUGH BUFFER AND BACKFILL

In the process of its radioactive decay the emplaced nuclear waste produces heat. For the reference fuel waste container with 72 fuel bundles the initial heat production rate is 269 watts per container which decays with the radioactive decay of the waste. This radiogenic heat is dissipated through the buffer and backfill zones to the surrounding host rock. In the absence of large volumes of water circulation through the disposal vault, conduction is the predominant mode of heat transfer.

The buffer and backfills which are essentially 3 phase materials exhibit a coupled phenomena of heat and moisture diffusion. This is especially true in the low and medium saturation ranges. For accurate prediction of the near field thermal regime in the disposal vault it is necessary to model the coupled heat and moisture diffusion.

Considering a scenario that the disposal vault will be operated for several years and maintained in a dry or a semi-dry condition during this phase, the emplaced buffer and backfill in the disposal rooms will undergo an initial phase of thermal drying. This is followed by a wetting phase as the disposal vault is sealed and the ground water regime is allowed to re-establish itself. The drying phase will be critical to the overall performance of the engineered barriers. This is so because drying will degrade the thermal properties of the buffer and may also cause the formation of shrinkage cracks in the buffer. The presence of such cracks could alter the integrity of the buffer as a barrier to contaminant migration. The experimental and theoretical analysis presented in this paper deals with this situation.

Many researchers have attempted to model the heat and moisture transport through porous media. The major differences between the various theories proposed are in the assumptions of how the moisture transport takes place through an unsaturated media and the effect of thermal gradient on the vapour and liquid phase transfers [5]. The model proposed by Philip and DeVries [6] includes the interaction between the vapour and liquid phases. Also it showed good correlation with the laboratory results obtained by Cassel, et al [7].

The governing equation for moisture movement in unsaturated soils under combined thermal and moisture gradients is [6]

$$\frac{\partial \Theta}{\partial t} = \nabla \cdot (D_T \nabla T) + \nabla \cdot (D_\Theta \nabla \Theta) + \frac{\partial K_\Theta}{\partial z} \qquad (1)$$

in which Θ = volumetric moisture content,
∇T = temperature gradient,
$\nabla \Theta$ = moisture content gradient,
D_T = thermal moisture diffusivity,
D_Θ = isothermal moisture diffusivity,
and K_Θ = unsaturated hydraulic conductivity.

The governing equation for heat transport through a porous media is [6]

$$C \frac{\partial T}{\partial t} = \nabla \cdot (\lambda \nabla T) - L\nabla \cdot (D_{\Theta V} \nabla \Theta) \qquad (2)$$

in which C = volumetric heat capacity,
λ = thermal conductivity,
L = latent heat of vapourization,
and $D_{\Theta V}$ = isothermal vapour diffusivity.

Equations 1 and 2 are the simultaneous equations describing coupled heat and moisture transfer in unsaturated soils. In view of the coupling effect between Equations 1 and 2 and non-linear material properties involved, no exact solution can be obtained.

Computer Code - TRUCHAM

The TRUCHAM computer code, developed at Ontario Hydro [8] which utilizes an integrated finite difference technique, can accommodate a variety of linear and non-linear boundary conditions and material properties.

The special features of this program are that it can successfully deal with the coupled moisture-heat induced flow; accommodate a wide variation of material properties; its formulation in multi-dimensional space is relatively simple and there is an automatic choice of time step to ensure the stability and convergence of the numerical solution [8].

This computer code is ideally suited for the near-field thermal analysis of the disposal vault with very dissimilar materials, complex boundaries and time dependent heat sources.

RESULT OF STUDIES

In this experimental study only the initial thermal drying phase of the disposal vault is examined. A constant heat input of 33 watts/per heater, one eighth of the heat production by a waste container, was maintained for a period of 40 days. The heater temperature reached saturation at about 18 days after which the near field thermal regime stabilized.

Figure 4 shows the temperature profile along the mid plane through the heaters after 30 days of heating. The heater surface temperatures reached a maximum of about 105°C after 30 days of heating and rather steep thermal gradients were obtained in the buffer zone.

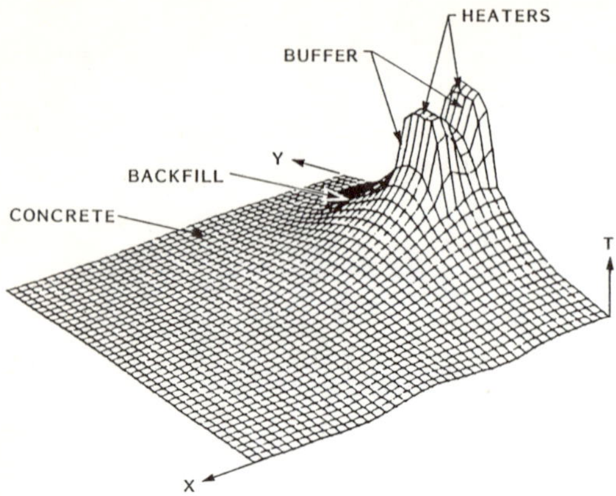

Figure 4. Temperature contours along the mid-plane of Heaters.

The transient heat conduction process through this experimental set-up was modelled in TRUCHAM computer code by defining the boundary conditions, heat sources and material properties. A 3-dimensional analysis was performed to predict the thermal regime in the buffer, backfill and concrete.

The following three cases were analyzed:

(1) constant thermal properties with no moisture migration

(2) thermal properties allowed to change with moisture migration (effect of thermal drying)

(3) thermal properties allowed to change with moisture migration and including the effects of desiccation cracking of the buffer.

After the 40 days of heating and then cooling down, the condition of buffer and backfill was examined. The desiccation cracking of buffer was carefully mapped and it was input into the computer analysis (Case 3). The desiccation cracking was predominantly horizontal in direction. The buffer mass also had shrunk away from the walls of the borehole. The horizontal cracks were about 1 to 5 mm wide and were spaced at 5 to 8 cm apart. The vertical gap between buffer and concrete in the boreholes was about 0.5 mm in width.

Figure 5 shows the comparison between the predicted and measured temperature profiles along the centreline of the heaters and in the mid-plane through the model. A good agreement exists between the predicted and measured data for the case in which the thermal properties are allowed to change with moisture loss and the thermal cracking of buffer is considered. The predictions made for the other two cases underestimate the near-field temperatures. The results of this analysis indicates the importance of coupling the moisture migration effect with the heat transfer process in the near-field analysis.

The implications of thermal drying and desiccation cracking of the buffer on its performance as an engineered barrier need to be investigated. For this purpose a new experimental model is being constructed which will have provisions for introducing

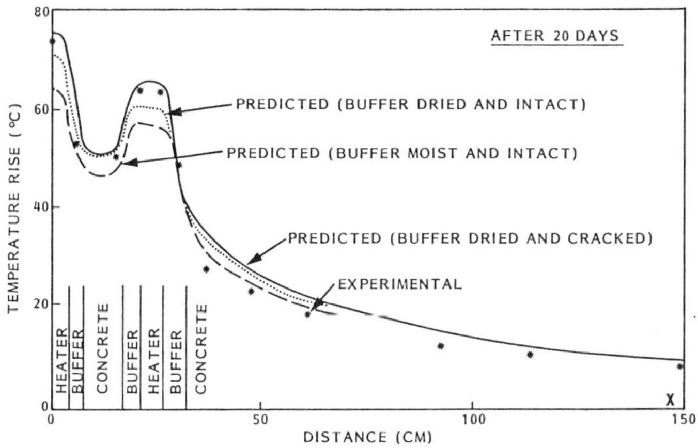

Figure 5. Temperature profile through the mid-plane of Heaters.

water under pressure after the initial thermal drying phase. Such an experiment aided by the computer code TRUCHAM can provide a clear insight into the complex process of thermal drying and rewetting of the buffer and backfill.

CONCLUSIONS

The experimental program has demonstrated the tendency for clay-based buffer material to desiccate and crack under the radiogenic heat from the emplaced waste. The performance of the buffer as an engineered barrier under the anticipated thermal and hydraulic gradients is being evaluated. The results of these model experiments will assist in optimizing the placement parameters to enhance the performance of the buffer.

The computer code 'TRUCHAM' can model the coupled heat and moisture diffusion through the clay-based buffer and backfills under realistic field conditions in the disposal vault. The code is found useful for parametric study of the factors affecting the thermal field in the disposal vault. It can also be used in evaluating alternative schemes of waste emplacement and buffer-backfill geometry in the design phase. Potential exists to enhance the application of TRUCHAM to the disposal vault analysis by coupling the process of heat and moisture flow with the volume change process (swelling or shrinkage) in the clay-based engineered barriers.

ACKNOWLEDGEMENTS

The work of K-C. Lau was supported by an NSERC Industrial Research Fellowship held at Strata Engineering Corp. and that of A.M. Crawford by NSERC, Canada and the University of Toronto. Mr. J. Westland was responsible for producing Figure 4.

REFERENCES

1. J. Boulton (Editor), "Management of Radioactive Fuel Wastes": The Canadian Disposal Program, Atomic Energy of Canada Limited Report, AECL-6314 (1978).

2. Acres Consulting Services Ltd. and Associates, "A Disposal Centre for Irradiated Nuclear Fuel: Conceptual Design Study", Atomic Energy of Canada Limited Report, AECL-6415 (1980).

3. R.S. Lopez, S.C.H. Cheung and D.A. Dixon, "Geotechnical Aspects of the Sealing of Underground Nuclear Fuel Waste Vaults", Technical Note, Canadian Geotechnical Journal, No. 3, Vol. 21 (1984).

4. H.S. Radhakrishna, "Thermal Properites of Clay-based Buffer Materials for Nuclear Fuel Waste Disposal Vault", Atomic Energy of Canada Limited, AECL-7805 (1984).

5. O.N. Abdel-Hadi, "Flow of Heat and Water Around Underground Power Cables", Ph.D. Thesis presented to the University of California, Berkeley, California (1978).

6. J.R. Phillip and D.A. DeVries, "Moisture Movement in Porous Materials Under Temperature Gradients", Trans American, Geophysical Union 38, (The American Geophysical Union of the National Academy of Sciences, Washington, DC, 1957) pp. 222-232.

7. D.S. Cassel, D.R. Nielson and J.W. Biggar, "Soil Water Movement in Response to Imposed Temperature Gradients", Soil Science Society of America, Vol 33 (1969).

8. H.S. Radhakrishna, K.C. Lau and A.M. Crawford, "Coupled Heat and Moisture Flow through Soils", Journal of Geotechnical Eng., American Society of Civil Engineers, Vol. 110, No. 12 (1984).

DETERMINATION OF NEAR FIELD EXCAVATION DISTURBANCE IN CRYSTALLINE ROCK

ROBERT KOOPMANS and ROBIN W. HUGHES, Ontario Hydro, 800 Kipling Avenue, Toronto, Ontario, Canada, M8Z 5S4

INTRODUCTION

A procedure has been established to assess the effectiveness of various excavation techniques using a computerized borehole dilatometer system developed by the authors[1]. This system, which incorporates data acquisition and computer capabilities, allows data to be collected and analyzed quickly in the field. The zone of disturbance is determined by conducting borehole tests at relatively close intervals and profiling the results with depth from the excavation wall. A similar concept of using deformation moduli measurements from plate bearing tests to detect changes in bedrock conditions is discussed by Koopmans[2]. In general, lower moduli are obtained in disturbed zones which have become stress relieved due to rock relaxation and/or the presence of rock discontinuities while higher values are obtained in relatively undisturbed zones which have not become stress relieved due to an absence of rock relaxation and/or rock discontinuities.

The determination of near field excavation disturbance was conducted at two sites in Canada. The first program was undertaken within the shaft collar of an underground research laboratory (URL) located in crystalline rock near Pinawa in south-eastern Manitoba, while the second was conducted within the sandstone rock mass surrounding two parallel circular tunnels in the Donkin Mines near Sydney, Nova Scotia. Although the majority of our studies are concentrated in crystalline rock, the opportunity to test two completely different excavation techniques in the same rock type arose. Regardless of the fact that these two tunnels were excavated in low modulus sandstone, it was decided that valuable information relating to excavation disturbance could be obtained from this study.

COMPUTERIZED BOREHOLE DILATOMETER SYSTEM

The computerized borehole dilatometer is shown schematically in Figure 1 and discussed in detail by the authors[1]. Only a brief description is provided here for completeness. The system is highlighted by an intensifier which is operated by an electrically-driven hydraulic pump to allow controlled pressure loading and unloading of the dilatometer. The intensifier has a 6:1 piston differential ratio between the oil and water chambers, respectively. This allows lower oil pressures to be administered from the hydraulic pump through flexible hoses to the intensifier where the pressure is increased six times to allow water inflation of the dilatometer.

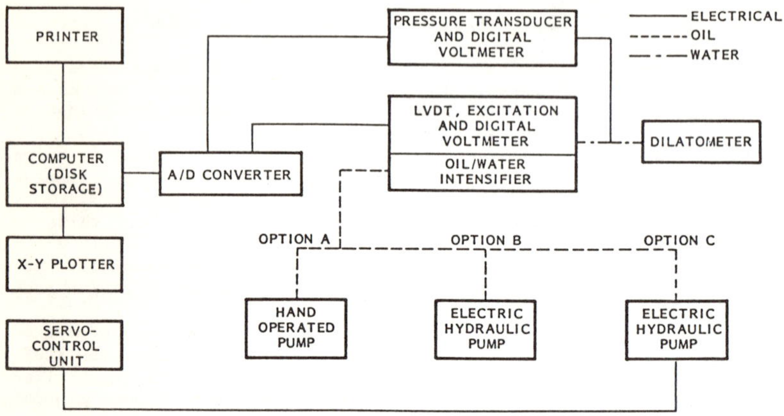

Fig. 1. Schematic of automated computerized dilatometer system

A linear variance differential transformer (LVDT) has been fastened to the top of the intensifier and measures linear movement of the high pressure water piston during loading and unloading. Linear movement represents volume change and is determined by calibration. A pressure transducer has also been connected to the water side just outside of the intensifier to measure changes in pressure. Analog signals from both the LVDT and pressure transducer are sent to an A/D converter for conversion to digital signals and then into a computer. The dilatometer itself consists of a high-strength, stainless steel heat-treated shaft and an adiprene membrane capable of applying pressures in excess of 70 MPa. Lead beads or other fillers may be placed in the membrane cavity to reduce water volume and thereby increase the system stiffness. High-pressure, thick-walled tubing is used to connect the dilatometer to the intensifier.

CALCULATIONS

The complete derivations of equations required for the calculation of deformation modulus of rock is provided by Hustrulid and Hustrulid[3]. The basic equations are presented here. The only change in these equations is that calculations are now expressed as pressure/volume instead of pressure/turn. The intensifier moves in a linear direction and eliminates the need to read rotational turns inherent in the original CSM system.

Step 1 - The pressure-volume relationship (Mc) for the cylinder alone is calculated:

$$Mc = \frac{\gamma Gc}{\pi L\, r_{ic}^2 \left[\frac{1 + \beta c}{1 - \beta c} - \frac{2\mu c\ \beta c}{1 - \beta c}\right]} \tag{1}$$

where: γ = volume of fluid displaced
Gc = modulus of rigidity of the calibration cylinder
L = length of membrane
$\beta c = (r_{ic}/r_{oc})^2$
r_{ic} = inner radius of calibration cylinder
r_{oc} = outer radius of calibration cylinder
μ - Poisson's ratio

Step 2 - The slope of the pressure-volume curve (Mm) for the system plus the calibration cylinder is determined.

Step 3 - The stiffness of the pressurizing system (Ms) alone is calculated:

$$Ms = \frac{Mc\ Mm}{Mc - Mm} \tag{2}$$

Step 4 - The slope of the pressure-volume curve (Mt) for the pressurized system plus rock is determined.

Step 5 - The pressure-volume relationship (Mr) for the rock alone is calculated:

$$Mr = \frac{Ms\ Mt}{Ms - Mt} \tag{3}$$

Step 6 - The modulus of rigidity (Gr) is calculated:

$$Gr = \frac{Mr\ \pi L r_{ir}^2}{\gamma} \tag{4}$$

where r_{ir} = inner radius of borehole

Step 7 - The deformation modulus (E) can be calculated if Poisson's ratio (μ) is known (or can be assumed):

$$E = 2(1 + \mu)\ Gr \tag{5}$$

Both deformation modulus and Poisson's ratio are affected by excavation disturbance. A stress decrease results in lower values while a stress increase results in higher values. The effect of estimating Poisson's ratio in determining deformation modulus is generally minimal provided a realistic value is assumed.

TESTING PROCEDURE

At the URL site in Manitoba, four horizontal Ex (3.81 cm diameter) boreholes were tested in the shaft collar to depths of approximately 15 metres into the crystalline rock mass surrounding the shaft. The shaft was excavated to a size of 3 m by 5 m using drill and blast techniques. The first five metres were tested at 0.25 metre intervals, while the next ten metres were tested at one metre intervals due to time constraints.

At the Donkin Mines in Nova Scotia, testing was conducted in the sandstone rock mass surrounding two parallel 7.5 m diameter circular tunnels. One tunnel (T2) was excavated using a tunnel boring machine while the second tunnel (T3) was mined using drill and blast techniques. An Ex (3.81 cm diameter) borehole was drilled from tunnel T3 to T2, a distance of approximately 42 metres. Testing was conducted to a depth of about 8.5 metres in T3 and 15 metres in T2. Wherever possible, tests were undertaken at 0.25 metre intervals to a depth of 5 metres and 0.5 metre intervals at depths in excess of 5 metres.

The following test procedure was used at both test sites. Prior to testing, the dilatometer was calibrated in both an aluminum and steel thick-walled cylinder. Six cycles were run for each calibration after which the dilatometer was placed at the bottom of the borehole using placement rods. Once again, six cycles were run. Due to the very small volume changes which occur within the dilatometer, especially in high modulus rocks, it was determined that more accuracy could be obtained by averaging the slopes of the curves together. Pressures in the order of 2 MPa were used to seat the dilatometer while pressures of approximately 25 MPa were used at the top of each cycle. Borehole testing progressed by moving the dilatometer from the bottom of the borehole towards the collar. Upon completion of testing the borehole, calibrations were taken once again to verify the original calibrations. With the aid of a computer, both pressure and volume were recorded every two seconds and stored on floppy disk. At the same time, the points were both plotted and printed out to enable immediate analysis as desired. The straight line portion between 15 and 25 MPa was selected as having the most elastic behaviour (Figure 2). The slope was determined by using the least squares method for each of the runs and averaging them together with the aid of a computer. Another computer program was then used to determine the deformation modulus. The results of all test locations were then plotted to show the profile of deformation modulus with depth for each borehole. A typical example from the URL shaft collar is shown in Figure 3.

DISCUSSION OF RESULTS

URL-Manitoba

The calculated deformation moduli for each borehole are shown in Table I and the borehole profiles are shown in plan view around the shaft in Figure 4. In general, each borehole shows a tendency for the deformation

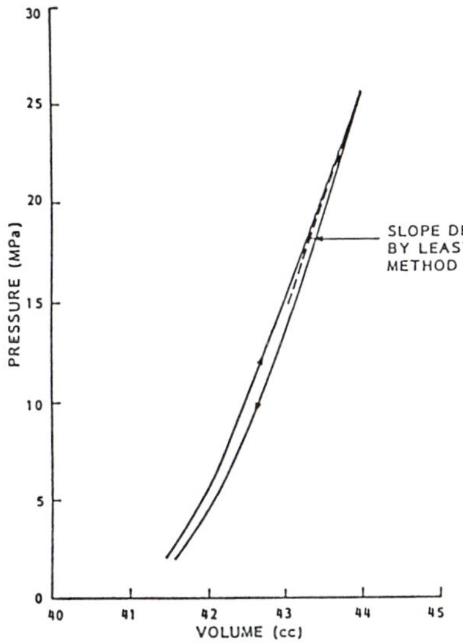

Fig. 2. Typical run for borehole test

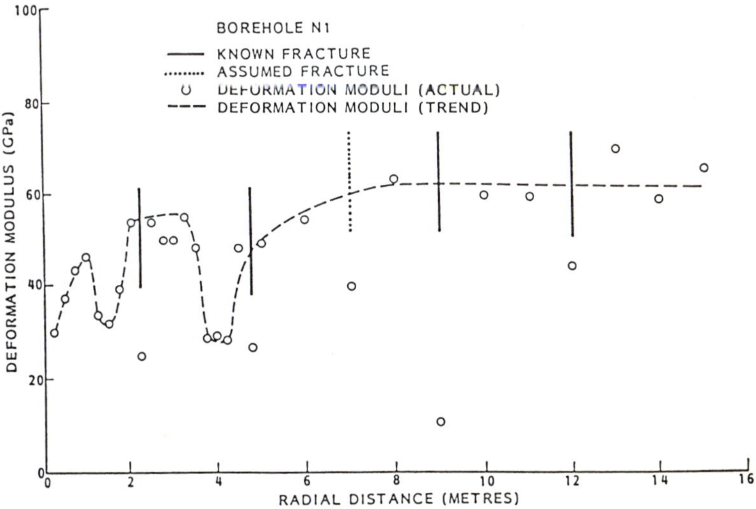

Fig. 3. Deformation modulus with depth from shaft wall

modulus to increase with depth from the shaft wall and level out at a constant value between one and two shaft diameters. The extent of excavation disturbance therefore varied from 4 to 8 metres in this portion of the shaft. Where individual test results have been plotted, it can be noted that low deformation moduli represent fractures intersecting the borehole (Figure 3).

TABLE I

DEFORMATION MODULUS WITH DEPTH - URL SHAFT COLLAR

	Deformation Modulus (GPa)			
Depth (m)	N1	N2	S3	S4
0.25	29	59	35	25
0.50	37	48	41	44
0.75	43	56	59	36
1.00	47	43	42	28
1.25	34	38	36	28
1.50	32	46	49	41
1.75	39	47	51	50
2.00	54	38	44	51
2.25	25	46	51	47
2.50	50	46	42	46
2.75	47	55	37	42
3.00	50	55	47	44
3.25	52	59	57	54
3.50	44	68	55	45
3.75	28	68	52	60
4.00	29	57	56	61
4.25	28	56	63	
4.50	49	49	62	
4.75	26	46	64	
5.00	50	44	60	49
6.00	50	48	65	44
7.00	40	61	59	49
8.00	74	81	64	60
9.00	11	21	62	40
10.00	56	55	64	63
11.00	56	68	60	69
12.00	44	62	58	76
12.50			71	
13.00	77	57		72
14.00	59	43		63
15.00	72	56		

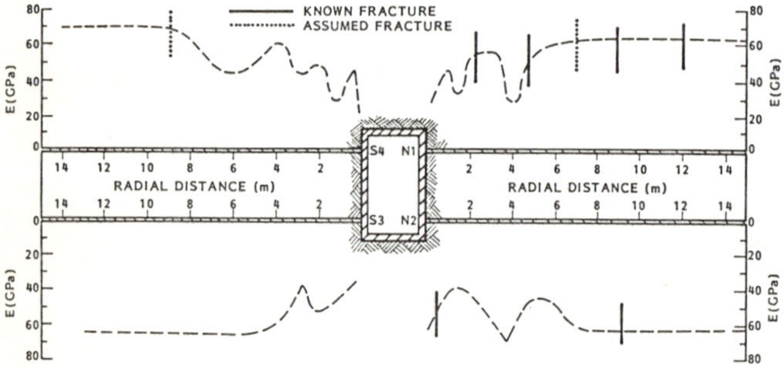

Fig. 4. Deformation modulus with depth from shaft wall

Fractures designated as known fractures were verified by vacuum permeability measurements[4] or by using the borehole dilatometer as an impression Packer (ie wrapped with tape). The cause for the undulating appearance of the profiles close to the shaft wall is probably due to variations resulting from blast-induced rock block movement. A comparison of deformation moduli with other techniques is shown in Table II. Average results for the dilatometer testing in all four boreholes is 54 to 63 GPa in the elastic zone of the rock mass including fractures and 63 to 67 GPa in the elastic rock mass excluding fractures. Cross-hole ultrasonics provided values in the order of 65-70 GPa for the rock mass surrounding the shaft[5]. Average values for both biaxial cell testing[6] and uniaxial compression testing[7] were determined for samples obtained from borehole URL-1 which is located near the URL shaft. These results were 54 GPa and 73 GPa, respectively. The comparison indicates that the dilatometer results obtained were very good especially when one considers that dynamic testing and compression testing of good quality core samples typically provide higher moduli than static in situ testing.

TABLE II

COMPARISON OF DILATOMETER RESULTS WITH OTHER TECHNIQUES

Technique	Borehole	Average Deformation Modulus (GPa) Including Rock Mass Fractures	Average Deformation Modulus (GPa) Excluding Rock Mass Fractures	Reference
Borehole Dilatometer	N1	54	64	This Report
"	N2	56	63	"
"	S3	63	63	"
"	S4	63	67	"
Cross-Hole Ultrasonics	N1-N2 S3-S4	65-70		[5]
Biaxial Cell	URL-1		54	[6]
Uniaxial Compression	URL-1		73	[7]

Note: URL-1 is located near the URL shaft

Donkin Mine, Nova Scotia

The calculated deformation moduli for each borehole are shown in Table III and the borehole profiles surrounding tunnels 2 and 3 are shown in Figure 5. Results indicate that high moduli (up to 27.39 GPa) are encountered in the rock mass close to the opening of the TBM excavation while lower moduli are obtained in the rock mass close to the drill and blast excavation. In the TBM excavation, reasonably constant values averaging 9.92 GPa are found at a depth greater than 7 metres. In the drill and blast excavation, moduli gradually increase from 7.44 GPa near the excavation wall to 11.60 GPa at a depth of 5.5 m. Beyond this point, the values increase to a high of 36.68 GPa and then drop to 9.92 GPa at about 9 m.

Based on these results, it appears evident that the drill and blast procedures cause greater excavation disturbance than the TBM technique as a result of both blast damage and stress relaxation. Figure 5 shows the

TABLE III

DEFORMATION MODULUS WITH DEPTH - DONKIN TUNNELS

	Deformation Modulus (GPa)	
Depth	2T	3T
0.25	27.39	
0.50	25.76	7.44
0.75	25.95	9.45
1.00	24.67	10.22
1.25	20.86	8.98
1.50	14.11	10.26
1.75	21.13	10.83
2.00	24.16	9.70
2.25	28.19	10.26
2.50	19.72	9.70
2.75	19.60	9.95
3.00	15.18	10.00
3.25	11.99	10.50
3.50	11.85	11.60
3.75	10.98	11.39
4.00	11.36	11.65
4.25	11.50	10.93
4.50	11.09	11.87
4.75	10.53	11.80
5.0	10.92	11.64
5.5	11.11	11.60
6.0	10.82	36.68
6.5	9.90	26.30
7.0	10.18	23.76
7.5	11.04	22.89
8.0	10.61	18.76
8.5	10.42	9.92
9.0	9.08	
9.5	10.57	
10.0	10.61	
10.5	9.79	
11.0	8.89	
11.5	9.87	
12.0	9.95	
12.5	9.64	
13.0	8.78	
13.5	9.78	
14.0	9.80	
14.5	10.00	
15.0	9.70	

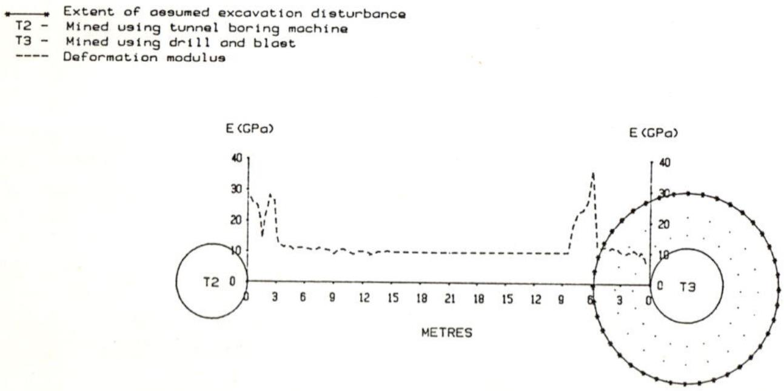

Fig 5. Deformation modulus with depth from tunnels 2 and 3

extent of the assumed excavation disturbance surrounding Tunnel 3.

The results obtained for both tunnels are stress related. This means that high stresses are indicative of high moduli and vice versa. Typically, an excavation which has been carefully mined will show high tangential stresses surrounding the opening. The stress values will then drop off with depth and level out at a distance of approximately one tunnel diameter. Tunnel 2, which was excavated using a tunnel boring machine, shows this type of behavior based on the deformation moduli obtained. On the contrary, if an excavation is mined using less cautious procedures, the tangential stresses will be low surrounding the opening due to destressing as a result of rock damage. This disturbed zone will extend with depth to the point of no further damage. At this boundary, the tangential stresses increase where the rock is more elastic and then drop with depth before levelling out. Tunnel 3, which was mined using drill and blast procedures, is indicative of this type of behavior.

CONCLUSION

The computerized dilatometer system has rapidly and economically provided deformation moduli of low and high modulus rock, determined the extent of excavation disturbance surrounding an underground opening and located open fractures within a rock mass. Results from both test sites indicate that the moduli obtained were influenced by the in situ tangential stress field. It has been shown that the near field excavation disturbance is kept to a minimum through the use of careful excavation techniques such as the tunnel boring machine. In turn, the in situ tangential stress levels and deformation moduli are maximized while the corresponding permeability is minimized.

ACKNOWLEDGEMENTS

The authors wish to thank Mr. G.R. Simmons and the rock mechanics staff of Atomic Energy of Canada Limited and Dr. C. Yuen and Mr. J. Gilby of Golder Associates for their co-operation during our testing programs.

REFERENCES

1. Koopmans, R. and Hughes, R.W. (1984). An Automated and Computerized Dilatometer System to Measure Deformation Modulus of Rock, Western States Mining Expo-84, Reno, Nevada.

2. Koopmans, R. (1983). Bedrock Degradation as Determined from Plate Bearing Tests, International Symposium on Soil and Rock Investigations

by In-Situ Testing, Paris, vol 2, pp 77-81.

3. Hustrulid, W. and Hustrulid, A. (1975). The CSM Cell-A Borehole Device for Determining the Modulus of Ridigity of Rock, 15th Symposium on Rock Mechanics, South Dakota.

4. Jakubick, A.T. (1984). Personal Communication.

5. Wong, J., Hurley, P.A. and West, G.F. (1984). Small Scale Cross-Borehole Seismology for Mapping of Dynamic Elastic Moduli and Detection of Cracks Within Rock Masses, prepared for Atomic Energy of Canada.

6. Koopmans, R. (1982). Near Surface In-Situ Stress Measurements - Lac du Bonnet Batholith, Manitoba, Ontario Hydro Research Report No 82-73-H.

7. McKay, D.A. (1982). Lac du Bonnet Geomechanical Test Results, Ontario Hydro Research Report No 82-441-K.

SIMULATION OF WATER TRANSPORT IN HEATED ROCK SALT

M. SCHLICH* and N. JOCKWER**
*RWTH Aachen, Institut für Elektrische Anlagen und Energiewirtschaft, Lehrauftrag Leistungsreaktoren (LL), Schinkelstraße 6, D-5100 Aachen, Federal Republic of Germany
**Gesellschaft für Strahlen- und Umweltforschung mbH, Institut für Tieflagerung (GSF-IfT), Theodor-Heuss-Allee 4, D-3300 Braunschweig, Federal Republic of Germany

ABSTRACT

This paper summarizes computer simulation studies on water transport in German rock salt. Based on JOCKWERS's experimental investigations on water content and water liberation, the object of these studies was to select a water transport model, that matches the water inflow which was measured in some heater experiments in the Asse Salt Mine. The main result is, that an evaporation front model,with Knudsen-type vapor transport combined with fluid transport by thermal expansion of the adsorbed water layers in the non evaporated zone, showed the best agreement with experimental evidence.

INTRODUCTION

Basic investigations, done at the GSF showed that German domal salt contains water mainly as adsorbed water at crystal grain boundaries and as hydrated water of the minerals kieserite and polyhalite /1/. The study of the liberation and the transport of this water content due to temperature fields as they occur by the disposal of High Level Waste (HLW) is at present the subject of intensive laboratory and in situ experiments and of theoretical computer simulations. The latter are carried out at the LL in commission of and in close cooperation with the GSF /2/.This paper presents the progress in this computer modeling of water transport in heated rock salt.

WATER IN ROCK SALT

JOCKWER's laboratory experiments proved that kieserite shows significant water liberation rates at temperatures higher than 270 ^{0}C and polyhalite at a temperature level beyond 230 ^{0}C; the latter fact was confirmed by the heater experiment 5 (TV5) : post test analysis of rock samples revealed dehydrated polyhalite crystals only in a small area around the heater, where 230 ^{0}C had been exceeded /3/. Thus, water adsorbed at crystal grain boundaries will probably be the main source of water which migrates in a borehole of a disposal field in a German salt dome (if carnallitic domains are avoided by appropriate dislocation of the disposal fields). Therefore the emphasis of the simulation work was to study the release of adsorbed water by investigating the measured water collection data of the TV5 and the US-FRG Brine Migration Test (BMT i, i: number of the specified test site - totally 4) /4/. The TV5, which was finished in February 1983, provided the main basis of the model developments. It was a single heater experiment in which the rock salt was heated up to a final temperature of c. 250 ^{0}C in 5 successive heater power stages of c. 60 d duration. The BMT test sites consist of a central borehole containing an electric heater (BMT 2,3) or radioactive heat sources together with electric heaters (BMT 1,4). Each borehole is surrounded by 8 peripheral heaters to achieve a high temperature of c. 210 ^{0}C. The BMT

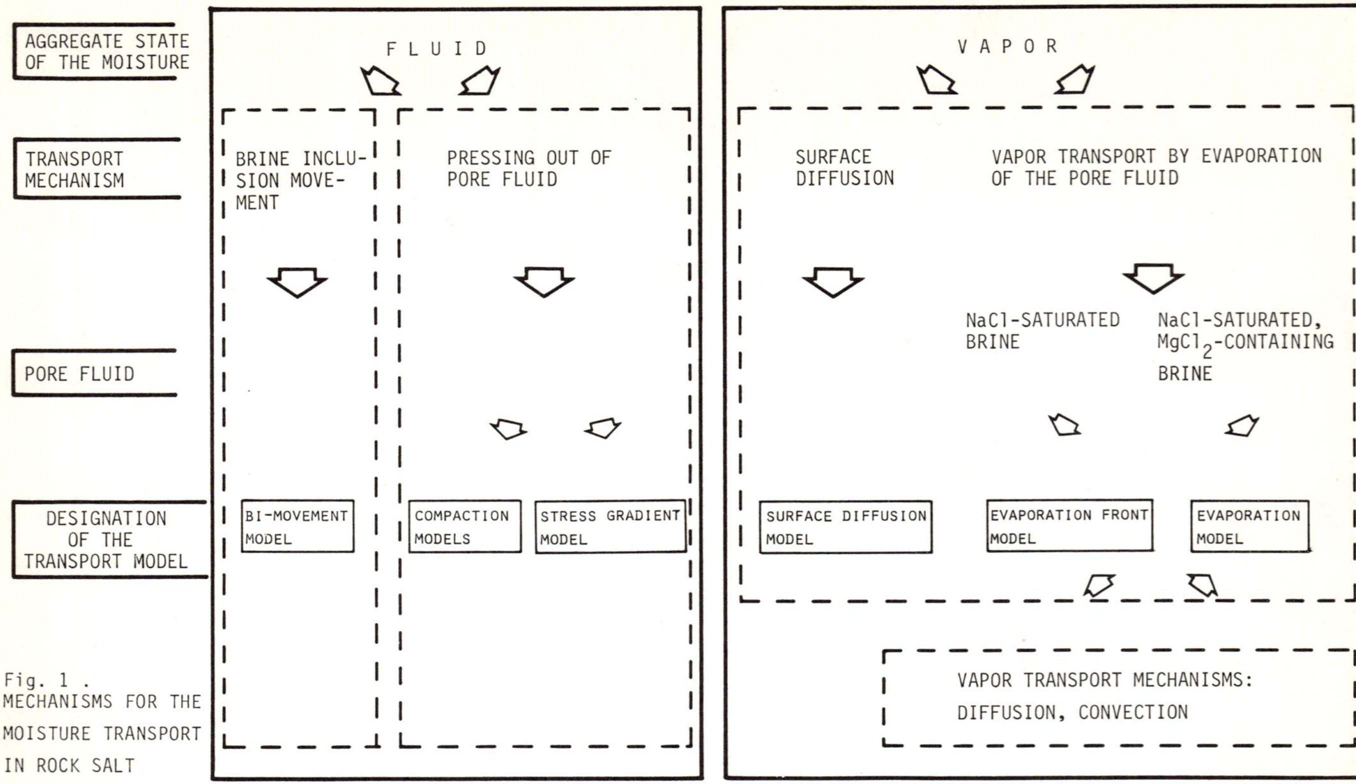

Fig. 1 . MECHANISMS FOR THE MOISTURE TRANSPORT IN ROCK SALT

will be completed in the end of 1985.

WATER TRANSPORT MODELS

Basing on similar preceding American treatises /5/ several different transport models (fig. 1) have been taken into account. They can be subdivided into two major groups : the first comprises models to explain a transport of fluid moisture, the second consists of models for a vapor state moisture transport (surface diffusion is difficult to classify and is arbitrarily put into the second group). Both model types were investigated theoretically because it could not be decided in advance in which aggregate state adsorbed water is set free : on the one hand adsorbed water films can be squeezed out by compaction of the intergranular pore space or by stress gradients or move partially the same way as liquid brine inclusions; on the other hand drying models can be applied. For the latter special German investigations exist for the water liberation of heated concrete in hypothetical accidents of sodium cooled reactors.

SIMULATION RESULTS

TV5

These physical models were transformed into suitable mathematical descriptions and numerical calculation codes. They were applied to the TV5 in order to select relevant transport mechanisms /6/. The simulation strategy was to adapt free parameters in the transport models so that the amount of the water inflow measured 150 days after heater power switch-on was reproduced. Then the calculated flow rates were compared with the measured ones looking for similarities. The simulation results showed that a compaction and a stress-gradient model cannot be applied to explain the measured water collection. The fluid inclusion model with the empirical formula of JENKS /7/ for the migration velocity of brine inclusions yielded the result in fig. 3 for the water inflow rate into the borehole. In this calculation it was assumed that the adsorbed water (0.04 wt %) moves as liquid inclusions. The calculation result doesn't agree with the measurements : the kinetics of the measured water liberation in fig. 3 and also the quantity of the freed water could not be reproduced. Better results were obtained by the experiment-simulation with water vapor transport models : the pertinent mass flux and filtration velocity equations are listed in fig. 2. Fig. 4 shows the calculated water flux for an evaporation front model with a Knudsen-type vapor flow; it reproduces well the initial peak of the water liberation and shows also (but fainter) peaks after heater power elevations at 63 days and 122 days after heater power switch-on. Fig. 4 also presents similar results of an evaporation front model if combined Knudsen- and Darcy-flow is considered; the main outcome is that Darcy-dominated transport did not occur, because it cannot reproduce the initial peak and the high water liberation within the first two heater power stages when the total pressure gradient between the evaporation front and the borehole pressure is not very high. The results of the evaporation model and the surface diffusion model simulations of the TV5 showed no better if not a weaker correspondence to the measurements. Better harmony with the measured data was obtained with an evaporation front model comprising the thermal expansion of the adsorbed water layers : fig. 5 shows the results for several parameter values of the space-dependent part of the pore fluid which can thermally expand into the borehole. This thermal expansion of the adsorbed water layers in the non evaporated zone causes a dislocation of the evaporation front towards the borehole; it can be regarded as a special kind of fluid

FIG. 2 . SET OF EQUATIONS FOR THE CALCULATION OF THE MASS FLUX DENSITIES AND FILTRATION VELOCITIES IN A TWO - COMPONENT GAS MIXTURE IN A POROUS MEDIUM

VAPOR TRANSPORT MECHANISM	MASS FLUX DENSITY EQUATIONS	FILTRATION VELOCITY EQUATIONS
CONVECTION — DARCY-FLOW	$\dot{m}_1=-\frac{p_1}{R_1T}\frac{k}{\eta_m}\frac{\partial p}{\partial r}$ $\dot{m}_2=-\frac{p_2}{R_2T}\frac{k}{\eta_m}\frac{\partial p}{\partial r}$	$v_1=-\frac{k}{\eta_m}\frac{\partial p}{\partial r}$ $v_2=-\frac{k}{\eta_m}\frac{\partial p}{\partial r}$
DIFFUSION — KNUDSEN-DIFFUSION	$\dot{m}_1=-\phi\frac{D_{1K}}{R_1T}\frac{\partial p_1}{\partial r}$ $\dot{m}_2=-\phi\frac{D_{2K}}{R_2T}\frac{\partial p_2}{\partial r}$	$v_1=-\phi\frac{D_{1K}}{p_1}\frac{\partial p_1}{\partial r}$ $v_2=-\phi\frac{D_{2K}}{p_2}\frac{\partial p_2}{\partial r}$
BINARY DIFFUSION	$\dot{m}_1=-\frac{D_{12}}{R_1T}\frac{p}{p_2}\frac{\partial p_1}{\partial r}+\frac{R_2}{R_1}\frac{p_1}{p_2}\dot{m}_2$ $\dot{m}_2=-\frac{D_{12}}{R_2T}\frac{p}{p_1}\frac{\partial p_2}{\partial r}+\frac{R_1}{R_2}\frac{p_2}{p_1}\dot{m}_1$	$v_1=-D_{12}\frac{p}{p_1p_2}\frac{\partial p_1}{\partial r}+v_2$ $v_2=-D_{12}\frac{p}{p_1p_2}\frac{\partial p_2}{\partial r}+v_1$
DIFFUSION AND CONVECTION	$\dot{m}_1=-\frac{p_1}{R_1T}\frac{k}{\eta_m}\frac{\partial p}{\partial r}-\phi\frac{D_{1K}}{R_1T}\frac{\partial p_1}{\partial r}+\frac{D_{1K}}{D_{12}}(\frac{R_2}{R_1}\frac{p_1}{p}\dot{m}_2-\frac{p_2}{p}\dot{m}_1)$ $\dot{m}_2=-\frac{p_2}{R_2T}\frac{k}{\eta_m}\frac{\partial p}{\partial r}-\phi\frac{D_{2K}}{R_2T}\frac{\partial p_2}{\partial r}+\frac{D_{2K}}{D_{12}}(\frac{R_1}{R_2}\frac{p_2}{p}\dot{m}_1-\frac{p_1}{p}\dot{m}_2)$	$v_1=-\frac{k}{\eta_m}\frac{\partial p}{\partial r}-\phi\frac{D_{1K}}{p_1}\frac{\partial p_1}{\partial r}+\frac{D_{1K}}{D_{12}}\frac{p_2}{p}(v_2-v_1)$ $v_2=-\frac{k}{\eta_m}\frac{\partial p}{\partial r}-\phi\frac{D_{2K}}{p_2}\frac{\partial p_2}{\partial r}+\frac{D_{2K}}{D_{12}}\frac{p_1}{p}(v_1-v_2)$
DIFFUSION — SURFACE-DIFF.	$\dot{m}_1=-K\frac{\partial y_F}{\partial r}$; $K=C_0\exp(-E_a/(R_mT))$	$v_1=-K\frac{1}{y_F}\frac{\partial y_F}{\partial r}$

E_a :ACTIVATION ENERGY

r :SPACE COORDINATE

p_1, p_2 :PARTIAL PRESSURE

p :TOTAL PRESSURE

ϕ :POROSITY

T :TEMPERATURE

k :PERMEABILITY

η_1, η_2 :DYNAMIC VISCOSITY

η_m :MIXTURE VISCOSITY

C_o :PROPORTIONALITY CONSTANT

κ :MOISTURE CONDUCTIVITY

y_F :MOISTURE CONCENTRATION IN ROCK SALT

D_{1K}, D_{2K} :KNUDSEN-DIFFUSION COEFFICIENT

v_1, v_2 :FILTRATION VELOCITIES

$\dot{m}_1, \dot{m}_2$:MASS FLUX DENSITIES

D_{12} :DIFFUSION COEFFICIENT

R_1, R_2 :SPECIFIC GAS CONSTANTS

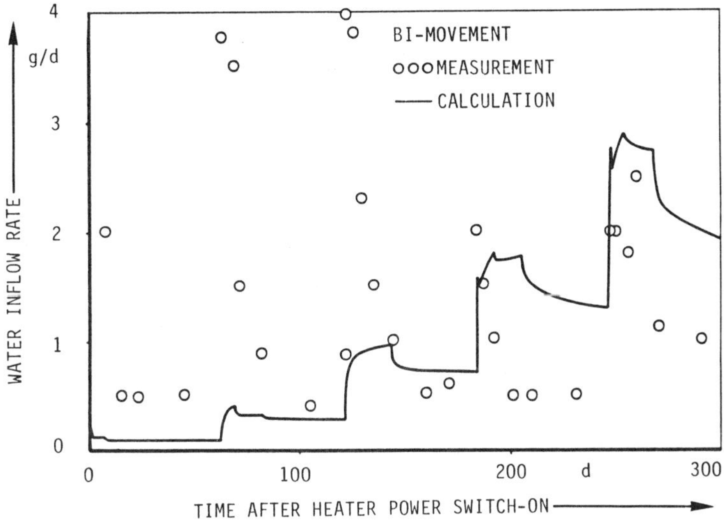

FIG. 3. WATER INFLOW RATE IN THE TV5 HEATER BOREHOLE FOR ASSUMED BRINE INCLUSION MOVEMENT

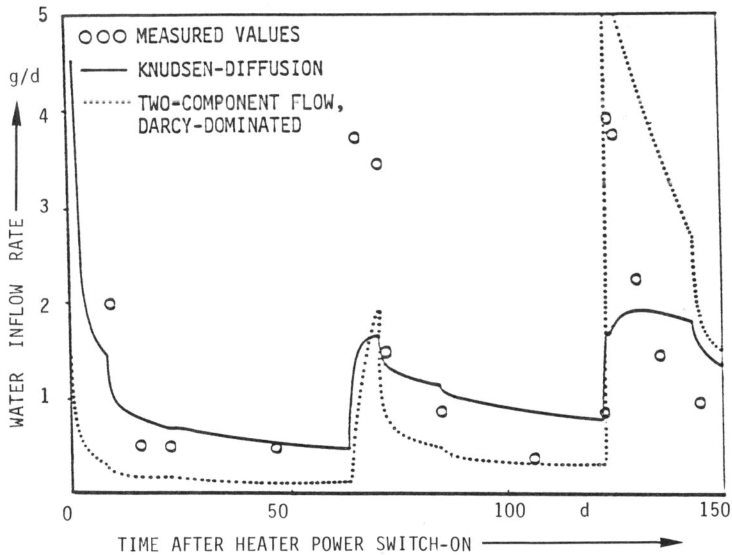

FIG. 4. WATER INFLOW RATE IN THE TV5 HEATER BOREHOLE FOR SEVERAL VAPOR-TRANSPORT MECHANISMS

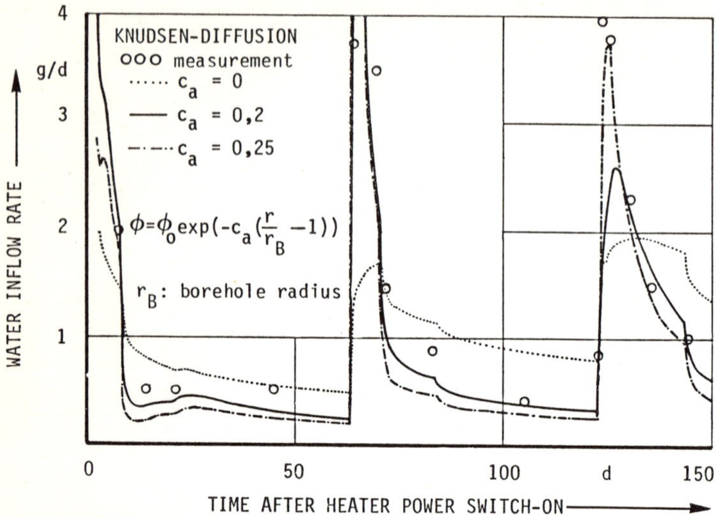

FIG. 5. WATER INFLOW RATE IN THE TV5 HEATER BOREHOLE -CONSIDERATION OF THE THERMAL EXPANSION OF THE PORE FLUID

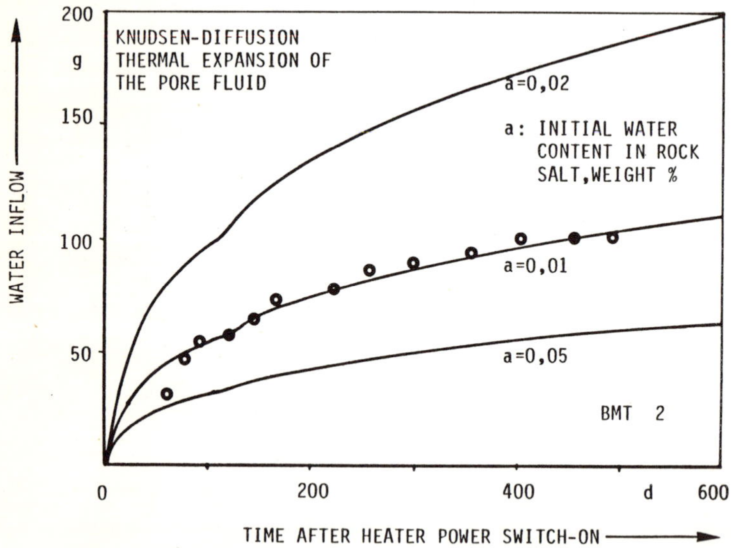

FIG. 6. COMPARISON OF CALCULATED AND MEASURED WATER INFLOW IN THE BMT 2 CENTRAL HEATER BOREHOLE

transport. Depending on the properties of the salt investigated, it may even be possible that this mechanism leads to a penetration of fluid moisture into the borehole leading to efflorences. The simulation results for German salt didn't show this effect - this is confirmed by the TV5 post test investigation where no hints at efflorences could be found.

BMT

The parameters of the best-fit model of the TV5 simulation were taken to calculate the water inflow into the boreholes of BMT 2 and BMT 4. Fig. 6 shows that the measurements in the BMT 2 experiment are very well reproduced for a water content of 0.01 wt.%; an analogous result was obtained from the BMT 4 simulation. Salt core investigations of the BMT salt region yielded an adsorbate water content range of 0.01 wt.% - 0.02 wt.%. This conformity with the BMT measurements is a strong hint, that the postulated transport model (evaporation front motion, Knudsen-flow, thermal expansion of the pore fluid) represents the real water transport process.

HLW repository

Fig. 7 - fig. 8 contain the results of calculations for a 300 m borehole in a HLW-disposal field. The temperature history at the wall of a borehole in the midplane of the disposal region was obtained from a so called unit cell calculation /8/. Fig. 7 presents the time dependent convergence of an initial gap between a HLW-canister and the borehole wall for different initial gap widths. These curves were calculated with EDSPA, a new one-dimensional thermo-mechanical computer code on the basis of the coarse-mesh-method /9/. The time for the closure of a 30 mm initial gap is 170 days. During this time water vapor transport into the borehole will appear. After closure this transport process will be stopped by the sealing of pore space channels due to high mean (pressure) stresses in the surrounding zone of the borehole : this was the main result of the calculations with the compaction models mentioned above.

Fig. 8 presents the result of an extrapolatory calculation for the water inflow in a borehole of a HLW repository for the first 250 days after emplacement; as in the BMT calculations the best-fit moisture transport model of the TV5 and its parameters were applied : the total amount of water inflow is lined out as a function of time for several initial adsorbate water contents of the rock salt. It is very likely that the repository relevant range lies between 0.01 wt.% and 0.04 wt.% : the total amount of water inflow per meter will not exceed 200 g if a convergence of the gap occurs within the first year after emplacement. It has to be emphasized, that this water inflow quantity is only the first estimation for the constraints of German rock salt; it is necessary to undertake further experimental efforts to confirm the underlying water transport model.

SUMMARY

This paper presents the essential results of the water transport model development with regard to the constraints in German rock salt. The measurements of the TV5 showed that the hydrate water of polyhalite is set free above 230 °C. In the repository relevant range below 230 °C only adsorbed water was freed as water vapor. Extensive numerical simulations by testing several transport mechanisms of adsorbed water yielded the following most important results :

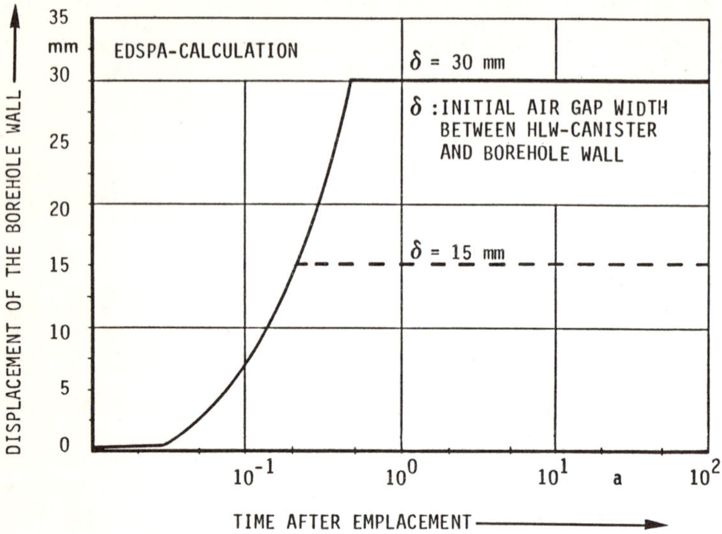

FIG. 7 . DISPLACEMENT OF A BOREHOLE WALL IN A HLW-DISPOSAL FIELD

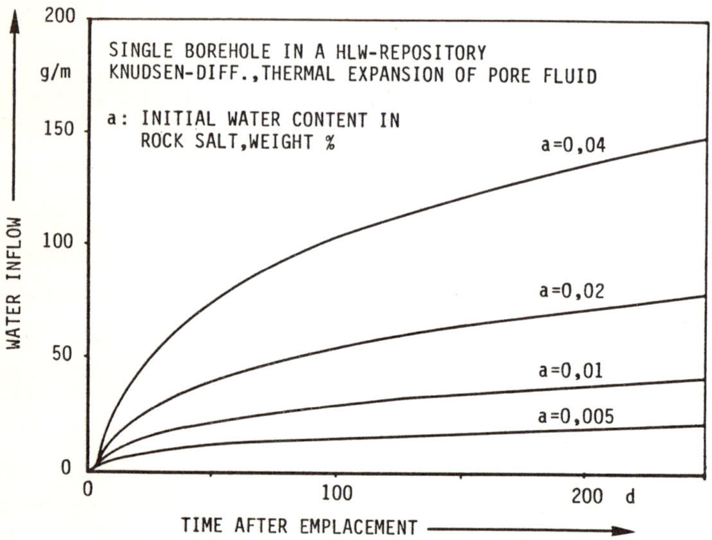

FIG. 8. WATER INFLOW PER METER IN A HLW-BOREHOLE OF A REPOSITORY FOR DIFFERENT INITIAL WATER CONTENTS OF ROCK SALT

- o the kinetics of the water inflow into the heater borehole could be well described by an evaporation front model;
- o the water vapor flow in this model is Knudsen-type dominated;
- o the main influence parameter on the kinetics of the calculated water inflow is the fluid transport by thermal expansion of the adsorbed water film causing a dislocation of the evaporation front towards the borehole wall.

This model development was mainly based on the simulation of the TV5. The application on the Brine Migration Test in the Asse salt mine yielded a good agreement with the measured water inflow. An extrapolatory calculation for a borehole in a repository showed that the water vapor inflow during the time of convergence of the initial gap between canister and borehole wall will not exceed 200 g per meter.

The main objective of present and future experimental and simulation work is to validate further the proposed transport model and to reduce the ranges of critical model parameters.

REFERENCES

1. Jockwer, N.: Untersuchungen zu Art und Menge des im Steinsalz des Zechsteins enthaltenen Wassers sowie dessen Freisetzung und Migration im Temperaturfeld endgelagerter radioaktiver Abfälle, Thesis, TU Clausthal 1981

2. Schlich, M.: Entwicklung von physikalischen Modellen und ihre Umsetzung in zugehörige Rechenprogramme zur Beschreibung der Bewegung von Feuchte im Salzgestein infolgde des Wärmeeintrags durch ein Endlager für hochradioaktive Abfälle, Interim reports, GSF-IfT and LL, 1983, 1984

3. Rotfuchs, I.: Untersuchung der thermisch induzierten-Wasserfreisetzung aus polyhalitischem Steinsalz unter in situ Bedingungen - V6 im Salzbergwerk Asse, Final report, GSF-IfT, 1985

4. Asse Salt Mine - BMT Test Program, Office of Nuclear Waste Isolation. ONWI-245, GSF-T118, April 1981

5. Shefelbine, H.C.: Brine Migration: A Summary Report, SAND-82-0152, Sandia National Laboratories, Albuquerque, 1982

6. Schlich, M.: Simulation der Bewegung von im natürlichen Steinsalz enthaltener Feuchte im Temperaturfeld, Thesis, TH Aachen, 1985

7. Jenks, G.H.: Effects of Temperature, Temperature Gradients, Stress an-Irrigation on Migration of Brine Inclusions in a Salt Repository, Oak Ridge National Laboratory, ORNL-5526, Tenessee, 1979

8. Hahne, K., Schlich, M., Korthaus, K.: Numerical Calculation of the Time and Space Dependent Temperature Distribution in a HLW Repository of a Finite Extension for a Temporal Step-by-Step Disposal Procedure, Poster Session at this Symposium

9. Schlich, M., Elsen, R.: Beschreibung und Anwendungshandbuch für das Rechenprogram EDSPA, program developed at LL, 1985, program code available at NEA Data Bank, Gif-sur-Yvette

NATURAL, THERMAL AND RADIOLYTICAL GAS LIBERATION IN ROCK SALT AS A RESULT OF DISPOSED HIGH-LEVEL RADIOACTIVE WASTE

NORBERT JOCKWER AND SABINE GROSS
Gesellschaft für Strahlen- und Umweltforschung mbH München, Institut für Tieflagerung, Theodor-Heuss-Strasse 4, D-3300 Braunschweig, Federal Republic of Germany

ABSTRACT

Rock salt contains crystal water of the hydrated minerals, water in the form of brine inclusions or adsorbed to the crystal boundaries, as well as primary gases. Further gases are produced or liberated due to the disposed high-level, heat producing wastes resulting from the thermal and radiolytic decomposition of the main and secondary constituents.

All volatile components which are present in a final repository have an influence on the disposal horizon, promote the spreading of possibly liberated radionuclides, increase corrosion of the containment and may lead to a pressure increase in a disposal borehole. Therefore a qualitative and quantitative determination of the components which could be liberated from the rock salt as a result of the disposal of high-level wastes is of significance for a final repository.

INTRODUCTION

In the Federal Republic of Germany it is planned to dispose of high-level radioactive waste in rock salt layers of the North German salt domes. These rock salt layers consist of the main mineral halite, of minor and trace minerals and of gases and fluids trapped within small inclusions or adsorbed to the crystal boundaries. During disposal of high-level radioactive waste the rock salt in the near vicinity of the emplacement borehole will be heated up to 200 °C and will be irradiated with a dose rate of 10^5 rad/h and with a dose of 10^{11} rad within 50 years.

This contribution is concerned with the mineralogical and geochemical composition of rock salt and the influence of the disposed high-level radioactive waste on the host rock, especially the liberation and generation of volatile components.

INVESTIGATIONAL PROCEDURE

In order to determine the mineralogical composition of the rock salt, samples from the Asse salt mine from different depths and varying stratigraphic layers have been investigated. The methods used were X-ray diffraction, microscopic investigation of thin sections, and chemical analyses. The results are shown in Table 1. Due to the presence of hydrated minerals, polyhalitic and kieseritic rock salt contains small amounts of water. Furthermore, water may be adsorbed to the crystal boundaries or

Mat. Res. Soc. Symp. Proc. Vol. 50. 1985 Materials Research Society trapped in small fluid inclusions. The water contents of 500 salt samples were determined by Karl Fischer titration, Du Pont moisture analyser, and by thermogravimetry. This investigation indicated that 55 % of the samples have a water content less than 0.1 weight % and 75 % of the samples less than 0.2 weight %. The distribution maximum is at about 0.04 weight %.

Table I. Mineralogical composition of 365 rock salt samples from the Asse mine

	mineral	present in ... samples	weight %
main minerals	halite	365	80 - 100
	polyhalite	335	0 - 18
	anhydrite	209	0 - 18
	kieserite	39	0 - 5
	sylvine	10	0 - 5
minor and trace minerals	langbeinite	5	5
	bloedite	2	5
	clay carbonates bitumen	more or less in all samples	1

The gaseous components which are initially present in the rock salt will already be liberated at natural mine temperatures when galleries are built or boreholes drilled. In order to determine these liberated components, horizontal and vertical boreholes have been dry-drilled in different areas of the Asse salt mine. Immediately after drilling these boreholes were sealed gastight and rinsed with pure nitrogen. During the following months gas samples were taken to determine the gas components besides nitrogen by gas chromatography and mass spectrometry both qualitatively and quantitatively. Table II illustrates the main results of this investigation.

Table II. Maximum concentration of the different gas components liberated into gastight boreholes at natural mine temperature

CO_2	up to	100	vpm (cm^3/m^3)
CO	up to	2	vpm
H_2	up to	2	vpm
HCl	up to	280	vpm
Cl_2	up to	1	vpm
H_2S	up to	6000	vpm (= 0,6 vol. %)
SO_2	up to	25	vpm
CH_4	up to	100	vpm
further hydrocarbons	up to	5	vpm

In order to determine the thermally generated and liberated gases from the natural rock salt, about 100 samples from the Asse mine have been investigated. The samples were heated up to 300 °C and the liberated components were determined by a gas-chromatograph or a mass spectrometer. The main results are shown in Table III.

Table III. Gases liberated thermally at 300 °C from rock salt. Maximum values from 100 samples

1.	CO_2	up to 400 ppm = 900 g/m^3 salt = 450 Nl/m^3 salt
2.	H_2S	up to 5 ppm = 12 g/m^3 salt = 6,2 Nl/m^3 salt
3.	HCl	up to 150 ppm = 350 g/m^3 salt = 195 Nl/m^3 salt
4.	gaseous hydrocarbons, especially CH_4	up to 60 ppm = 140 g/m^3 salt = 156 Nl/m^3 salt

To determine the effects of radiolysis by γ-radiation, salt of different mineralogical composition with a water content between 0.01 and 38 weight % has been irradiated in the reactor core. 500 g of ground salt were filled into a glass ampoule with a diameter of 55 mm and a length of 650 mm. Only half of the volume is required for this amount. The ampoules were rinsed with very pure nitrogen and sealed gastight.

Subsequently, they were placed in the reactor core for about 65 hours while the reactor was not in operation, therefore only γ-radiation took place. The dose adsorbed varied between 10^8 and 10^9 rad. The temperature during irradiation was 60 °C.

After irradiation the gas components besides nitrogen were determined in the residual volume of the ampoule and, moreover, the irradiated samples were heated up in order to liberate the adsorbed components. Figures 1 to 3 show the gas content (H_2; O_2; HCl) of some irradiated and unirradiated samples of different mineralogical composition which may be found in the disposal area.

This investigation indicated that there is no relationship between the water content and the generated gas components. In some cases the irradiated samples contain less oxygen than those unirradiated, which points to an oxygen-consuming chemical reaction. There is no relationship between the chloride content of the samples and the generation of chlorine (Cl_2) or hydrochloric acid (HCl). In the irradiated samples of pure halite (NaCl), for example, neither HCl nor Cl_2 could be detected, whereas the other samples with minor and trace minerals contain a significant amount of these components. This indicates that the mineral composition has a great influence on the gas generation. Further investigations of this point are necessary.

Table IV. Radiolytical gas generation as a function of the absorbed dose for a rock salt sample with 85 wt % halite, 13 wt % polyhalite, 2 wt % traces

	H_2	O_2	CO_2	HCl	CH_4
unirradiated	/	$2{,}5 \cdot 10^3$ ppm 9,2 Nl/m^3 salt	50 ppm 0,2 Nl/m^3 salt	240 ppm 0,9 Nl/m^3 salt	5 ppm 0,02 Nl/m^3 salt
irradiated with 10^8 rad	$3{,}2 \cdot 10^3$ ppm 11,7 Nl/m^3 salt	$2{,}4 \cdot 10^3$ ppm 8,8 Nl/m^3 salt	$1{,}1 \cdot 10^3$ ppm 4,2 Nl/m^3 salt	458 ppm 1,7 Nl/m^3 salt	124 ppm 0,44 Nl/m^3 salt
irradiated with 10^9 rad	$24{,}7 \cdot 10^3$ ppm 92 Nl/m^3 salt	$4{,}1 \cdot 10^3$ ppm 15,0 Nl/m^3 salt	$12{,}6 \cdot 10^3$ ppm 46 Nl/m^3 salt	554 ppm 2,0 Nl/m^3 salt	84 ppm 0,3 Nl/m^3 salt

Table IV shows the influence of the absorbed dose on the generation of the components H_2; O_2; CO_2; HCl and CH_4. It increases with increasing dose. It is not yet known, however, whether there is a saturation level at a certain absorbed dose.

The G-value can be calculated with the results of the generated gas components . It is the amount of generated or consumed molecules per 100 eV absorbed energy. With the G-value it is possible to calculate the amount of the different gas components which are generated in an emplacement borehole. The G-factors for the sample shown in Table IV and the dose of 10^8 rad are:

$$\left.\begin{array}{ll} O_2 & : -4.3 \times 10^{-5} \\ H_2 & : \ 1.8 \times 10^{-3} \\ CO_2 & : \ 6.2 \times 10^{-4} \\ HCl & : \ 1.3 \times 10^{-4} \\ CH_4 & : \ 6.7 \times 10^{-5} \end{array}\right\} \frac{\text{molecules}}{100\ \text{eV}}$$

Table V shows the results for components O_2, H_2, CO_2, HCl and CH_4 for five samples with different mineralogical composition.

Table V. Gas components generated by radiolysis within an emplacement borehole in liter per year and per meter borehole

Mineralogical composition of the salt	O_2	H_2	CO_2	HCl	CH_4
85 wt % halite 12 wt % polyhalite 1 wt % anhydrite 2 wt % kieserite	$-85\cdot10^{-3}$	4,2	0,88	0,12	$29\cdot10^{-3}$
100 wt % halite	1,7	0,23	$13\cdot10^{-3}$	$1\cdot10^{-3}$	-
90 wt % halite 5 wt. % anhydrite 2 wt % kieserite 3 wt % bloedite	0,55	6,7	0,11	$-63\cdot10^{-3}$	$33\cdot10^{-3}$
97 wt % halite 2 wt % anhydrite 1 wt % clay	0,85	$91\cdot10^{-3}$	0,66	$0,43\cdot10^{-3}$	$0,37\cdot10^{-3}$
85 wt % halite 13 wt % polyhalite 2 wt % traces	$-35\cdot10^{-3}$	1,5	0,51	0,1	$55\cdot10^{-3}$

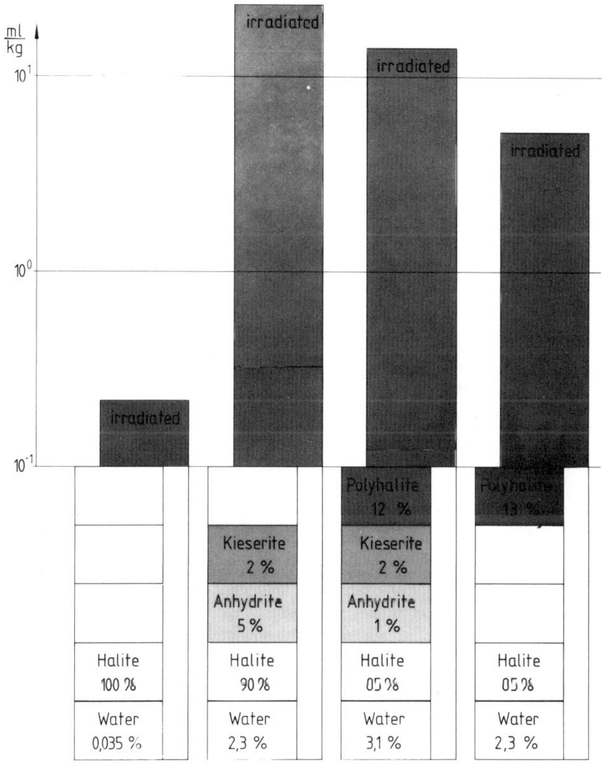

Figure 1. Content of hydrogen in rock salt of different mineralogical composition after irradiation with 10^8 rad

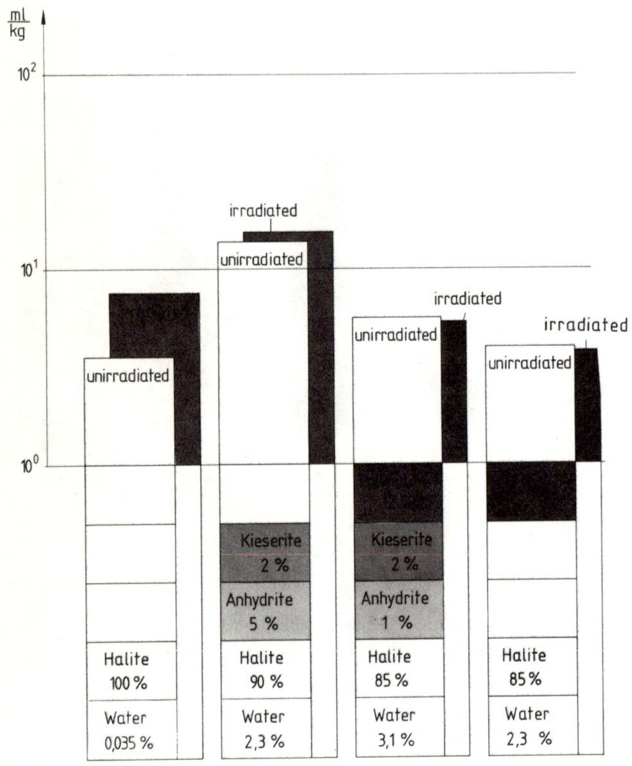

Figure 2. Content of oxygen in unirradiated rock salt of different mineralogical composition and after irradiation with 10^8 rad

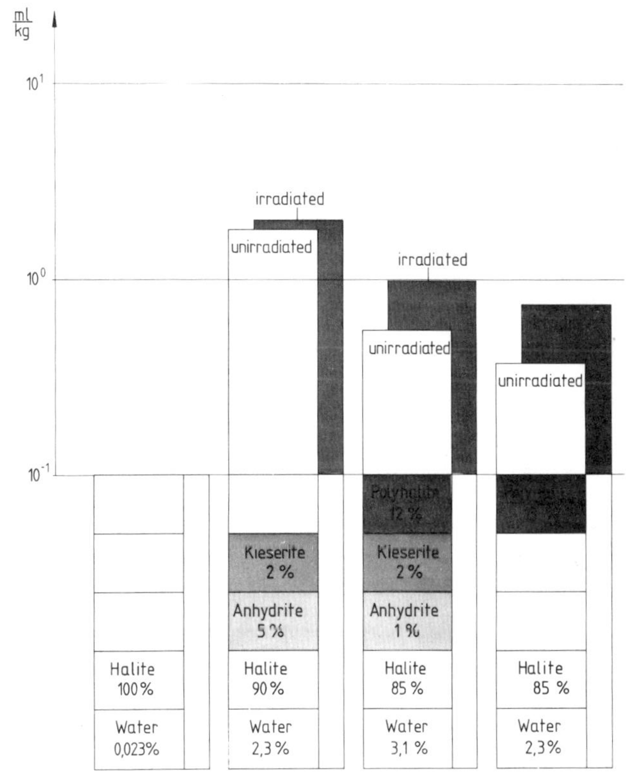

Figure 3. Contents of hydrochloric acid in unirradiated rock salt of different mineralogical composition and after irradiation with 10^8 rad

SUMMARY AND CONCLUSION

Rock salt formations which are foreseen for disposal of high-level radioactive wastes contain water derived from the hydrated associated minerals or from brine inclusions. Primary gases as well as further water are adsorbed to the crystal boundaries. Mining activities and elevated temperatures liberate these components. The disposed waste generates further gases by thermal and radiolytical decomposition. Components such as HCl and H_2S may enforce the corrosion of the containment and the liberation of radionuclides. Components such as CO_2, H_2, O_2 and hydrocarbons may pressurize the emplacement borehole or influence the integrity of the disposal horizon and sealing of the borehole. Therefore, all of these volatile components may promote the spreading of radionuclides.

In order to develop a long term safety concept for the disposal of high-level radioactive waste the liberated and generated gases are of importance both qualitatively and quantitatively. The liberation and generation itself depends on the

- mineralogical and chemical composition of the salt
- temperature
- absorbed dose
- absorbed dose rate

As the influence of these parameters is not yet well known, further investigations on this subject are essential.

REFERENCES

1. W.R. Aufricht and K.C. Howard, Salt characteristics as they affect storage of hydrocarbons, J. Petrol Technol. 1961, 733 - 739.

2. O. Braitsch, Entstehung und Stoffbestand der Salzlagerstätten, (Springer Verlag Berlin 1962).

3. A.G. Herrmann, Geowissenschaftliche Probleme bei der Endlagerung radioaktiver Substanzen in Salzdiapiren Norddeutschlands, Geol. Rundschau 68, 1076 - 1106 (1979).

4. E. Hofrichter, Zur Frage der Porosität und Permabilität von Salzgesteinen, Erdöl-Erdgas Zeitschrift 92, 3, 77 - 80 (1976).

5. N. Jockwer, Laboratory investigation on the water content within the rock salt and its behaviour in a temperature field of disposed high-level waste, Proc. of Scientific Basis for Nuclear Waste Management, Boston (1980).

6. N. Jockwer, Untersuchungen zu Art und Menge des im Steinsalz des Zechsteins enthaltenen Wassers sowie dessen Freisetzung und Migration im Temperaturfeld engelagerter radioaktiver Abfälle, D. rer nat. thesis, TU Clausthal, 1981.

7. N. Jockwer, Laboratory investigations on radiolysis effects on rock salt with regard to the disposal of hig-level radioactive wastes, Proc. of Scientific Basis for Nuclear Waste Management, Boston (1983).

HARD CLAY AS HOST MEDIUM
THERMO-MECHANICAL EXPERIMENTS AND MODEL

G. BALDI*, M. BORSETTO*, T. HUECKEL*, A. PEANO*, E. TASSONI**,
*ISMES SpA, V.le G. Cesare, 29, Bergamo, Italy
** ENEA, Casaccia, Rome, Italy

ABSTRACT

Clay's impermeability is an essential feature for the quality of a possible repository. However it results in a substantial pore water pressure rise induced by heating whose magnitude and dissipation rate depend on the clay's deformational response to heating. Experiments on thermal strains and effective stress are reported and the mathematical model implemented in a computer code is described. It allows study of a repository performance due to waste heating in both continental and subseabed clay formation.

INTRODUCTION

Clay has been chosen as a candidate host rock for nuclear waste disposal for its good quality as water migration barrier and for its high capacity of ion sorption as well as excellent plastic properties obviating the requirement for an engineered backfill. The latter property implies also a self healing capacity in the case of thermal or mechanical cracking [1].

The above properties of clay are connected with its very limited permeability, which reaches values as low as of the order 10^{-8} to 10^{-11} m/sec and with a particular character of water-mineral micromechanical structure.

However, the high impermeability and the particular water-mineral interaction may on the other hand give rise to a potentially adverse effect in the clay deposit caused by heating. This effect consists in high pore water pressure caused primarily by the thermal expansion of water. This hypothesis is supported by experimental evidence as seen in Fig. 1. Considerable values of the pore pressure excess were obtained during heating of a specimen of Orte Clay in undrained conditions under constant total isotropic stress. Due to the low permeability the excess needs a long time to dissipate. The pore pressure is enhanced by the unusual tendency of clays to compact under heating below 120°C. Such tendency which was found experimentally for all tested clays was previously reported for oceanic sediments by Campanella and Mitchell [2] and by Heremans et al. [3]. It has been found for overconsolidation states and even more pronounced for normal consolidation states. In the latter circumstance it leads to significant irreversible (plastic) thermal strains.

HYPOTHESES OF CLAY REPOSITORY PERFORMANCE UNDER HEATING

In principle the pore water pressure increase could lead to two undesirable scenarios. Both apply to continental as well as subseabed clay repositories.

The first one consists in a generation of a hydraulic

gradient in the field around buried container. In fact, the pore pressure is expected to increase in relation to local temperature of clay. The heat diffusion curves for clays indicate a high space gradients of the temperature field, diminishing with time, Fig. 2. In consequence one may predict an enhanced flow of the water off the deposit.

Although it is not any adverse effect by itself it may result in undesired changes in hydrologic equilibrium in large scale and in concomitance with some ill function of insulation system may lead to a migration of radioactive substances toward the human environment.

The second critical scenario is provoked by a loss of shear resistence of clay due to a reduction of the effective isotropic stress in the clay skeleton. The latter one may be in fact expected by virtue of the principle of effective stress due the thermal pore pressure rise. Therefore as heating proceeds the skeleton compression reduces till the failure state is reached even at constant shear stress corresponding to the in situ one.

A similar scenario has been discussed by Davis and Banerjee [4], in the context of a buoyancy study for an oceanic sediment repository, and has been found plausible for hard clays in the experiments presented in Section 5.

It is important to prove that the above scenarios are not realistic. This can be achieved by developing a capability to predict correctly the interaction between the clay skeleton and fluid deformation and/or flow caused by heating.

In fact, the interstitial water pressure build up results roughly speaking from the difference between deformational response to heating of the water entrapped in the pores and that of the clay structure. The water-skeleton interaction may be simplified to the idealized scheme, Fig. 3, considering as known the local temperature increase.

The central rule of the interaction under the hypothesis of full saturation is the well known water flow continuity condition. This condition requires the equality of the rate of outflow of the water to the pore volume change referred to an elementary volume. The former one depends on the hydraulic gradient, the latter one is a function of deformation of the skeleton, mineral solid and water due to the pore pressure, temperature and effective stress, u, T, σ'.

$$\frac{\partial}{\partial \underline{x}}(\underline{v}) = \frac{\partial \epsilon_v}{\partial t}(\delta T, \delta\sigma') - (1-n)\frac{\partial \epsilon_v^*}{\partial t}(\delta T, \delta\sigma', \delta u) - \frac{\partial \epsilon_w}{\partial t}(\delta T, \delta u) \quad (1)$$

where ϵ_w , ϵ_v^* are the deformation of water and of mineral solid contained in the soil, respectively, while ϵ_v is that of the soil medium as a whole and $\underline{v}$ is the water flow velocity.

When the drainage is not allowed the left hand side of (1) is equal to zero.

In order to quantify the respective volume changes in the r.h.s. of (1) mechanical and thermal compliances of all three constituents must be established. The novel point in this context arises from the complexity of functions which may be generically referred to as coefficient of thermal expansion and thermally sensitive mechanical compliance of clay skeleton.

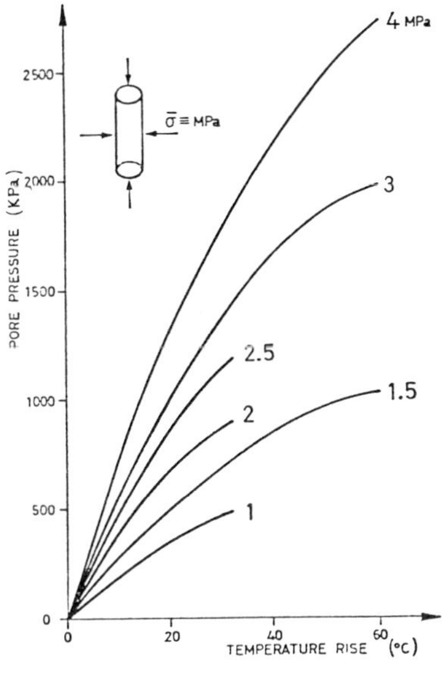

Fig. 1 - Pore water pressure rise under heating in undrained Orte clay sample under various constant total stress.

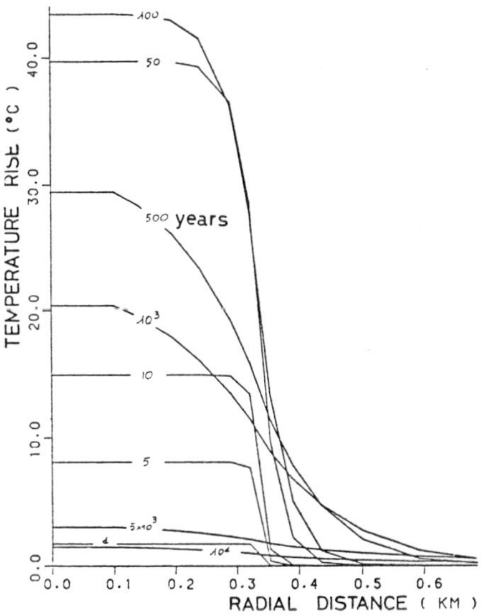

Fig. 2 - Temperature horizontal distribution around a deposit. Emplacement delay 25 years.

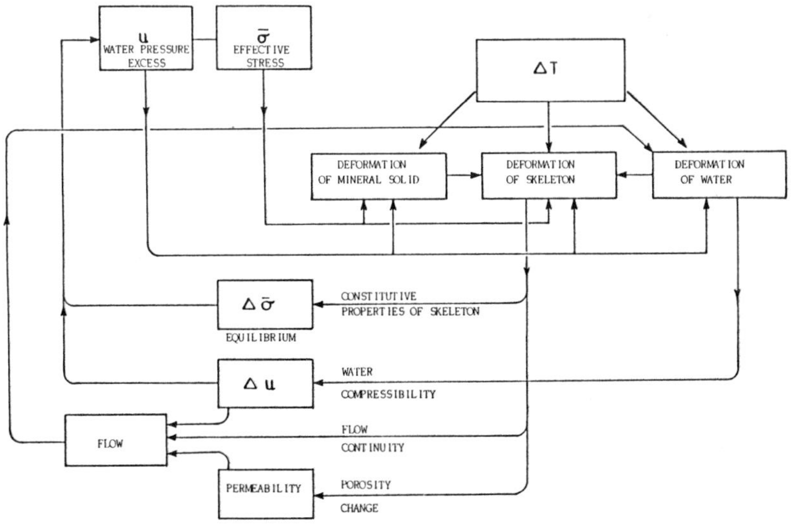

Fig. 3 Scheme of thermo-mechanical-hydraulic coupling in deformation of heated clay

Fig. 4a Temperature-pressure test programme

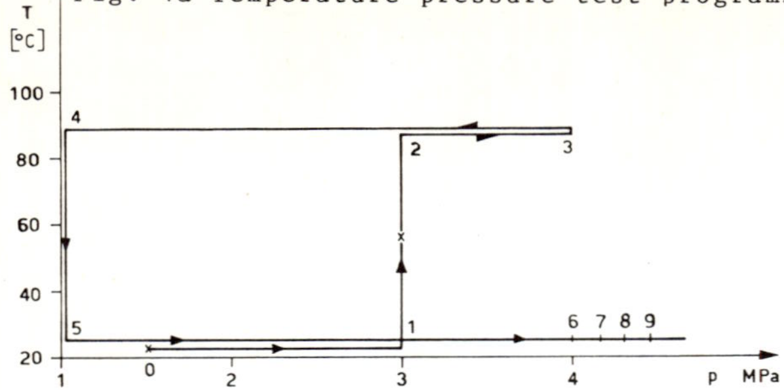

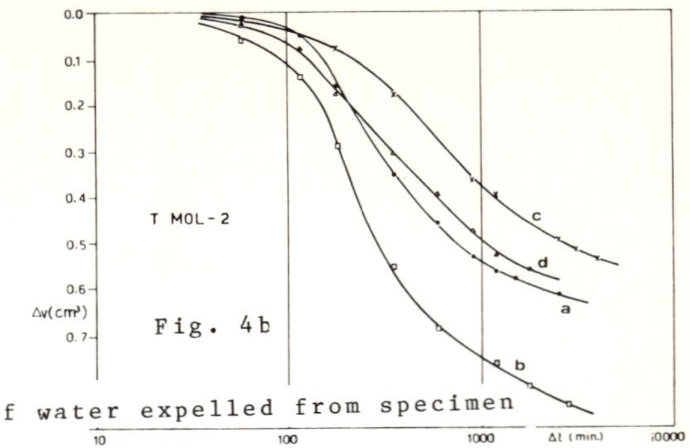

Fig. 4b

Volume of water expelled from specimen

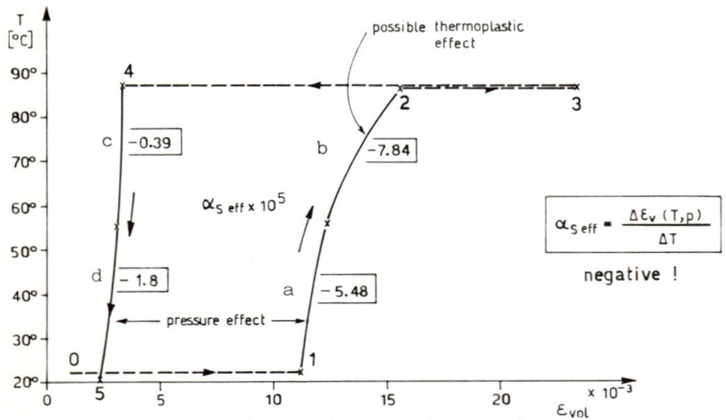

Fig. 5a Thermal strains in Boom Clay

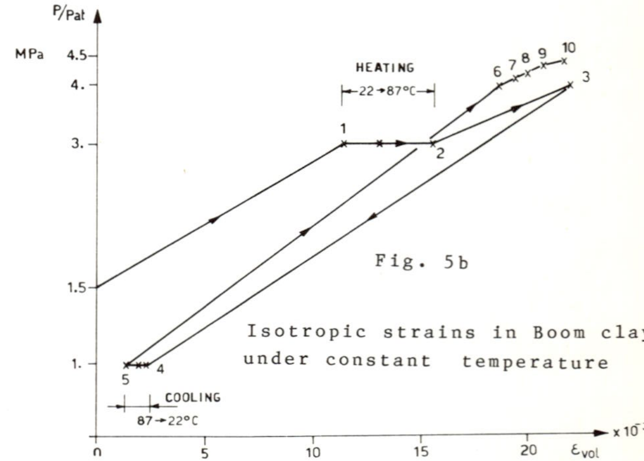

Fig. 5b

Isotropic strains in Boom clay under constant temperature

LABORATORY RESULTS

Series of experiments have been undertaken at ISMES in order to establish variations of the above functions in overconsolidated and normally consolidated range in isotropic stress, drained conditions.

The materials tested were a remoulded kaolin, remoulded Pontida, silty clay and intact Boom clay from Mol 260 m deep site.

Tests were performed using the HITEP Triaxial System developed at ISMES for high pressures and high temperatures operating in a condition to avoid uncontrolled temperature variations. Loads are applied by means of high pressure air system. Heat is applied via a system of resistances attached to the cell and embedded in the apparatus base. For technical details of the apparatus and tests see [5].

In the first test program on Boom clay a cycle has been performed, Fig. 4a, consisted of loading (0-1) at room temperature, heating under constant isotropic pressure (1-2) at OCR = 1.8, further loading (2-3) and unloading (3-4) at constant elevated temperature and finally cooling at constant, low pressure (4-5) followed by an isothermal reloading involving normal consolidation.

Typical experimental curves of volume of water expelled from the specimen as a function of time are presented in Fig. 4b. Because of their similarity to the mechanical consolidation curves the classical technique has bee adopted for defining termination of the non-viscous part of the process. Further elaboration of the curves consisted in correcting the obtained volumes for the expansion of water contained in the testing system outside the specimen (porous stone, tubes etc.) appropriately calibrated. Furthermore the expansion of pore water contained in the specimen itself was predicted on the basis of the standard formula for nonlinear thermal expansion coefficient of pure water.

As a result, clay solid phase volumetric strain has been obtained as a function of temperature and then presented in Fig. 5a in which logarithmic isotropic stress-strain curves relating to the mechanical loading part of the programme are also given.

It should be noted first of all, that the thermally induced volume changes are compactive.

Therefore the ratio of the volume change to the temperature difference, referred to as effective coefficient of thermal expansion of clay solid is negative in sign.

Such a result has been obtained previously for ocean sediments by Campanella and Mitchell [2] and for plastic clays by Heremans et al [3].

The former authors attributed it to clay structure changes in the course of heating. In fact adsorbed water films bounding clay solid particles are known to reduce with temperature giving rise to the global volume reduction, see also [6].

Finally, note a remarkable influence on thermal strain of the effective isotropic stress. Moreover, a reduction of bulk compliance modulus of clay upon heating is seen producing, under loading, additional strains of the order of the heating strains, Fig. 5b.

Doubts may arise concerning the reversibility of such complex processes. A test was undertaken to check it on a remoulded material (Pontida Silty Clay-1) in which first a preconsolidation state has been produced and then a closed effective pressure-temperature cycle has been performed obtaining full reversibility of strain within the cycle. The quantitative results have been published previously [7]. Moreover, no influence on thermal strains of the overconsolidation has been verified repeating the heating at the same effective state at two different OCR. Effective pressure confirms to have an essential role in thermal strains. Some hysteresis of thermal strains is seen upon heating and cooling.

In the normal consolidation range a dramatic increase of thermal compaction is observed. Fig. 6 presents the result of heating till 87°C of Boom clay at the constant effective pressure equal to 6 MPa. The resulting thermal strain is about 4 times greater than this in the overconsolidated range. Also with respect to cooling strain, (which are affected by incidental measurement uncertainty), the difference is remarkable. Note the non-linear character of the curves.

Similar results have been seen for the remoulded material (Pontida silty clay-1). A comparative result is given in Fig. 7a, b. Note first that heating in the same range (20°-87°C) gives three times greater strains at normal consolidation. A 60% percent of the final strain is irreversible upon cooling. A further mechanical loading has been undertaken after completion of cooling in order to investigate the normal consolidation range sensitivity to cooling. In fact, a localized modulus increase is seen at the first reloading step followed by a decrease to the value of hardening modulus measured in the plastic loading before heating.

The above experimental results lead to the following conclusions: thermally induced strains in clay are compactive; there is strong effective stress dependence of thermal expansion coefficient (negative), a temperature dependence is also observed; there is essential enhancement of thermal volumetric strains in the normal consolidation range with respect to those in the overconsolidation range; the normal consolidation range limit is sensitive to heating-cooling cycle in the drained conditions.

MATHEMATICAL MODELLING OF CLAY THERMO-MECHANICAL BEHAVIOUR

The most complex task in this context is determination of the total deformability of clay medium. The main lines of the present modelling has been stated elsewhere [7]. In this paper a particular form of the constitutive model will be given and some details will be discussed.

The modelling is performed within the framework of the theory of thermo-plasticity, fundamentals of which are given in [8]. On this basis a generalization is proposed for the modified Cam-Clay model broadly accepted for geotechnical applications. The original Cam Clay model has been described in [9].

Due to dependence of clay strain-stress behaviour on irreversible strain history (e.g. on plastic preconsolidation) the material deformability may be defined explicitly for incremental loading only. Thus the incremental deformation may be written as

$$d\underline{\varepsilon} = \underline{\underline{A}}\ d\underline{\sigma}' + (\hat{\alpha}\, dT - \hat{\beta}\, du)\underline{m} \tag{2}$$

where $d\underline{\sigma}'$, dT, and du are effective stress vector, temperature and interstitial water pressure increments in an incremental elapse of time, while $\underline{\underline{A}}$, $\hat{\alpha}$ and $\hat{\beta}$ are mechanical compliance matrix, thermal expansion incremental coefficient and pore pressure induced incremental deformability modulus. The vector $\underline{m}$ is composed of units for normal strain components and of zeros for shear components.

The behaviour of clay may be either thermo-elastic or thermo-plastic if the clay is in overconsolidated (O.C.) or normally consolidated (N.C.) state respectively, i.e. if

$$\text{OC: } f<0\ ;\ f=0,\ df<0 \qquad\qquad \text{NC: } f=0,\ df=0 \tag{3}$$

respectively. The yield function $f = 0$ describes the thermo-elastic domain limit, which undergoes modifications due to plastic strain-hardening and/or heating.

In particular, clay behaviour in the overconsolidation range may be considered as elastic as supported by experimental evidence in section 3. Therefore thermo-elastic strain may be given in non-incremental form. Then using secant moduli, account taken of dependences established in experiments the thermo-elastic strain may be given as

$$\varepsilon_v^e = \frac{-K_i}{1+e_0}(1+\beta\Delta T+\gamma\Delta T^2)\ln\frac{tr\,\underline{\sigma}'}{tr\,\underline{\sigma}_g'} + [\alpha^* + \alpha_0 + (\alpha_1+\alpha_3\Delta T)\ln\frac{tr\,\underline{\sigma}'}{tr\,\underline{\sigma}_g} + \alpha_2\Delta T]\Delta T - C^*u - \varepsilon_g\ ; \qquad \underline{e}^e = \frac{1}{2G}(\underline{S} - \underline{S}_g) \tag{4}$$

$$\underline{e}^e = \underline{\varepsilon}^e - \tfrac{1}{3}\varepsilon_v^e\,\underline{m} \qquad\qquad \underline{S} = \underline{\sigma}' - \tfrac{1}{3}tr\,\underline{\sigma}'\,\underline{m}$$

where ε_v^e, e^e denote elastic volumetric and deviatoric strain components, K_i and G are the isothermal bulk and shear moduli, α^* and C^* are the thermal expansion coefficient and the bulk compliance of mineral solid, ε_g is a fictious strain corresponding to the geostatic stress $\underline{\sigma}_g$, e_0 is the initial void ratio, β, γ, $\alpha_0 \ldots \alpha_3$ are constants.

In this sign convention expansion is taken as positive.

The normal consolidation behaviour is described by a thermo-plastic law. Plastic yielding or consolidation refers to the soil skeleton only. The limit of the overconsolidation range is defined by the yield function, which makes it expand as plastic compaction occurs and shrink with temperature increase. At the presence of plastic dilative strain rate, $\dot{\varepsilon}_v^p > 0$, the yield surface shrinks, while it expands with cooling.

The elliptic yield surface is taken as a generalization of the modified Cam Clay

$$f = \left(\frac{p'}{a(\Delta T, \varepsilon_v^p)} - 1\right)^2 + \left(\frac{q}{M\,a(\Delta T, \varepsilon_v^p)}\right)^2 - 1 = 0 \tag{5}$$

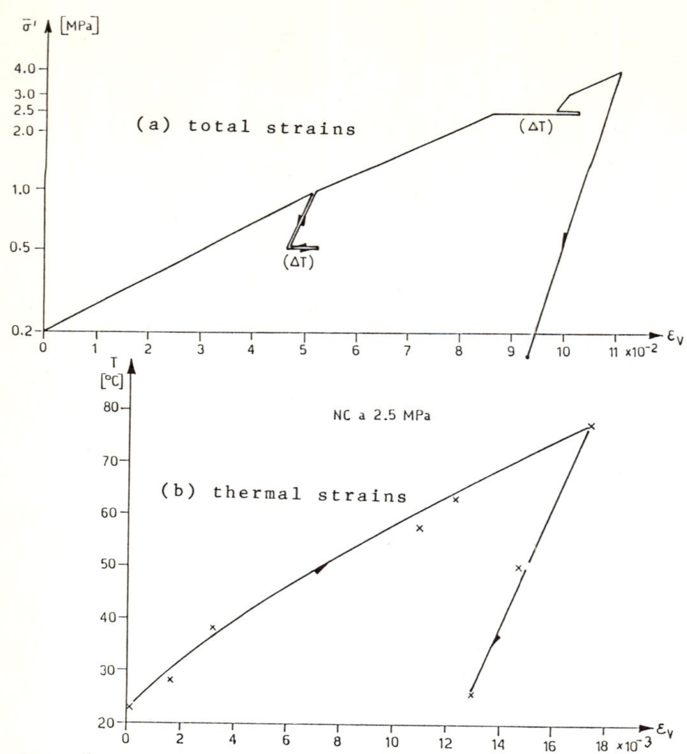

Fig. 7

Plastic thermal strain under constant isotropic loading of N.C. Pontida silty clay

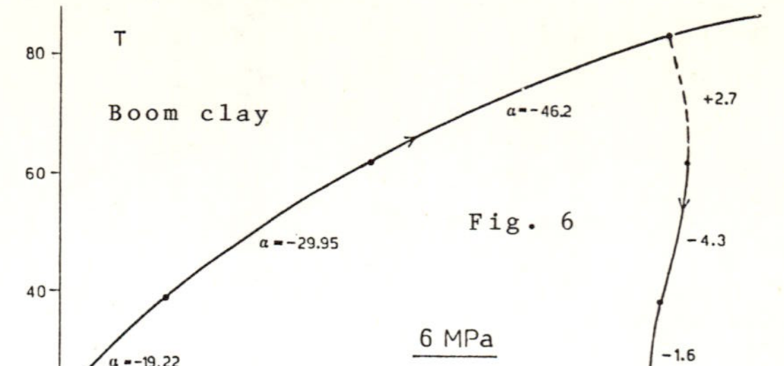

Fig. 6

Plastic thermal strains under constant isotropic loading

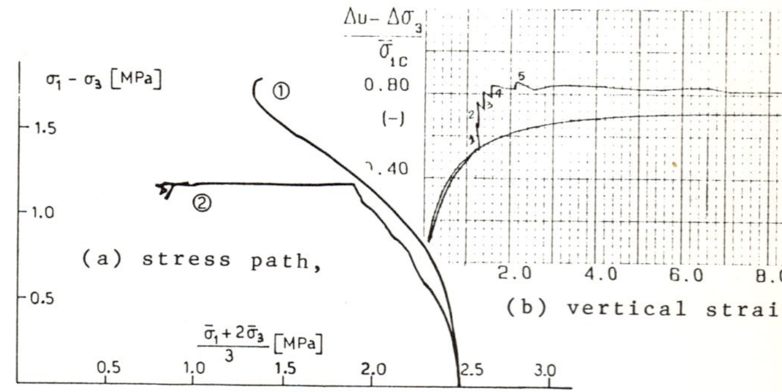

Fig. 8

Thermally induced failure in the sample of Pontida silty clay where: $p' = -\frac{1}{3} \, tr \, \underline{\sigma}_o'$; $q = \sqrt{\frac{3}{2}} (\underline{s}^T s)^{1/2}$; $M = const.$

The dimension of the overconsolidation domain is controlled by the ellipse semi-axis a which depends on ε_v^p and ΔT in the following manner:

$$a = \frac{1}{2} \bar{a} \, \exp\left\{ \frac{1}{\lambda \, K_i} \left[e_1 - [e_g + (1+e_o)\varepsilon_v^p][1 - a_o \Delta T] \right] \right\} + a_1 \Delta T + a_2 \Delta T^2 \quad (6)$$

where: a_o, a_1, a_2 are constants, $\bar{a} = -1$ kG/cm^2 ; λ is the plastic hardening modulus, e_1 is the void ratio at $p = \bar{a}$, e_g is a void ratio connected with the geostatic stress state.

The irreversible strain increment due to heating and/or mechanical loading is defined in the form of the associated flow rule.

$$d\underline{\varepsilon}^P = d\hat{\lambda} \, \frac{\partial f}{\partial \underline{\sigma}'} \quad (7)$$

The deformation mode defined by the yield surface gradient is thus not alterned by heating. In particular it predicts thermally induced shear irreversible strain. The plastic multiplier $d\hat{\lambda}$ which determines the entity of flow is defined from the consistency equation, df = 0, and thus depends both on effective stress and temperature increments

$$d\hat{\lambda} = \frac{1}{H} \left(\frac{\partial f}{\partial \underline{\sigma}'}^T d\underline{\sigma}' + \frac{\partial f}{\partial T} dT \right); \qquad H = -\frac{\partial f}{\partial \varepsilon_v^p} \frac{\partial f}{\partial p'} \quad (8)$$

where H is the isothermal plastic hardening modulus.

In the plastic flow the elastic and plastic strain increments are summed up and after classical plasticity theory manipulations an inverse equation to (2) is found in the form

$$d\underline{\sigma} = \underline{\underline{D}}^{ep} \, d\underline{\varepsilon} + \frac{1}{3} \left(\underline{\underline{D}}^{ep} \underline{m} \, C^* - \underline{m} \right) du + \frac{1}{3} \left(\alpha_T \underline{\underline{D}}^{ep} \underline{m} - \underline{q}^{pl} \right) dT \quad (9)$$

$$\underline{\underline{D}}^{ep} = \underline{\underline{D}}^{-1} - \frac{1}{H - H_c} \underline{M} \, \underline{M}^T ; \quad \underline{M} = \underline{\underline{D}}^{-1} \frac{\partial f}{\partial \underline{\sigma}'} ; \quad H_c = -\frac{\partial f}{\partial \underline{\sigma}'}^T \underline{\underline{D}}^{-1} \frac{\partial f}{\partial \sigma'}$$

and $\underline{\underline{D}}^{-1}$ is the inverse elasticity incremental matrix.

$$\underline{g}^{pl} = \frac{1}{H - H_c} \frac{\partial f}{\partial T} \underline{\underline{D}}^{-1} \frac{\partial f}{\partial \sigma'} \quad (10)$$

α_T is a combination of thermal expansion coefficients of skeleton, water and mineral solid.

According to the scheme presented in Fig. 3 the isotropic part of the strain rate in eq (7) is substituted into continuity equation (1). Introducing moreover in (1) water and mineral solid compressibility and thermal expansion coefficients, C_w , C^* and α_w, α^* respectively, and combining it with classical Darcy equation the following system is obtained

$$\begin{bmatrix} 0 & 0 \\ 0 & \frac{k_s}{\mu}\frac{\partial^2}{\partial \underline{X}} \end{bmatrix} \begin{Bmatrix} \underline{\varepsilon} \\ u \end{Bmatrix} + \overline{\begin{bmatrix} \underline{\underline{D}}^{ep} & \frac{1}{3}\underline{\underline{D}}^{ep}\underline{m}\,C^{*} - \underline{m} \\ -\underline{m}^T + \frac{1}{3}C^{*}\underline{m}^T\underline{\underline{D}}^{ep} & C^{**} + \frac{1}{9}C^{*2}\underline{m}^T\underline{\underline{D}}^{ep}\underline{m} \end{bmatrix} \begin{Bmatrix} \underline{\varepsilon} \\ u \end{Bmatrix}}^{\bullet} +$$

$$\begin{Bmatrix} -\frac{1}{3}\alpha_T\,\underline{\underline{D}}^{ep}\,\underline{m} + \underline{g}p^{l} \\ \frac{1}{3}C^{*}\underline{m}^T\underline{g}p^{l} + \alpha_v - \frac{1}{3}\alpha_T C^{*}\,\underline{m}^T\underline{\underline{D}}^{ep}\underline{m} \end{Bmatrix} \dot{T} = \begin{Bmatrix} \dot{\underline{\sigma}} \\ 0 \end{Bmatrix}; \quad \begin{aligned} C^{**} &= (1-n)C^{*} + nC_w \\ \alpha_v &= (1-n)\overset{*}{\alpha} + n\alpha_v \end{aligned} \tag{11}$$

Note that the constitutive matrix obtained is symmetric.

The above system describes coupled hydraulic-mechanical response of the heated clay-water medium at a given temperature increase.

CONCLUSIONS

The model has been implemented into a computer Finite Element code in two versions: for transient flow and for undrained conditions corresponding to long term and short term analyses respectively. It is applicable both to continental and subseabed clay waste disposal analyses and design.

By means of the mentioned code it is possibile to simulate both scenario for a particular geomechanical situation. For instance, the second scenario concerning the thermally induced failure as simulated in laboratory on a triaxial specimen can be numerically reproduced. The test has been performed on Pontida silty clay in undrained conditions corrected for thermal expansion of water in the loading system. The stress path is presented in Fig. 8a indicating a failure reached at the critical line manifested experimentally by a dramatic increase of vertical strain (Fig. 8b). Clearly, a realistic boundary value problem should be solved numerically in order to verify plausibility of the phenomenon in a full scale case.

ACKNOWLEDGEMENT

The work described has been carried out at ISMES as a part of a research programs supported by ENEA (National Board for Alternative Energies of Italy) and by the Commission of European Communities.

REFERENCES

1. L.H. Baestle and M. Mittempergher, in Radiactive Waste Management and Disposal. Proc. 1st European Comm. Conference, Luxembourg 1980, edited by R. Simon and S. Orlowski, (Harwood Academic Publishers, Brussels and Luxembourg 1980) p. 442

2. R.G. Campanella and J.K. Mitchell, J. SMFE Proc.ASCE, 94 NO SM3, 709 (1968)

3. R. Heremans, A. Barbreau, P. Bourke, H. Gries, in Radioactive Waste Management and Disposal, Proc. 1st Eur.

Comm. Conference, Luxembourge 1980, edited by R. Simon and S. Orlowski (Harwood Academic Publishers, Brussels and Luxembourg, 1980) p. 468

4. T.G. Davies and P. L. Banerjee, Constitutive relationships for ocean sediments subjected to stress and temperature gradients, AERE Harwell report NO HL 80/2604 (C22), August 1980

5. G. Baldi, T. Hueckel, E. Tassoni, presented for Int, Symp. on Environmental Geotechnology, Allentown, PA, 1985 (to be published)

6. D. Resendiz, in Proc. 6th ICOSDMEF Montreal, 1965, Vol 1 p. 97 (1965)

7. M. Borsetto, D. Cricchi, T. Hueckel, A. Peano, in Numerical Methods for Transient and Coupled Problems, edited by R. W Lewis et al. (Pineridge Press. Swansea U.K., 1984) p. 608

8. N. Prager, Proc. Konickl Nederl. Akad. van Wetenschappen series B, 61, 3, p. 176, 1958

9. K.H. Roscoe and J.B. Burland, in Engineering Plasticity edited by J. Hayman and F.A.A Leckie, (Cambridge University Press, Cambridge 1968) p. 535

PART VIIB

Diffusion and Transport in Clay Buffers

NMR-STUDIES OF PORE WATER IN BENTONITE/WATER/ELECTROLYTE

TORBJÖRN CARLSSON
University of Luleå, Division of Soil Mechanics
S-951 87 LULEÅ, Sweden

ABSTRACT

The properties of water saturated bentonite systems are highly dependent on the pore water chemistry, water content and temperature. In this study relative changes in the physical state of pore water are studied by means of the nuclear magnetic resonance technique (NMR). It is found that the relative water mobility increases with increasing water content and/or increasing temperature, while an increase in the electrolyte content lowers the water mobility.

INTRODUCTION

The use of clay minerals as buffer and sealing materials requires a detailed knowledge of the physico-chemical interaction between clay minerals, water and electrolyte. However, at present the clay chemistry, which is a rather specialized branch of physical chemistry, does not fully describe the relationship between various physico-chemical parameters and the properties of a clay/water/electrolyte system. This means, that it is not yet possible to predict accurately from a purely theoretical point of view how e.g. the properties of a clay barrier may change due to changes in the pore water composition.

The main object of this paper is to illustrate the influence of dissolved ions, temperature and water content on the physical state of the pore water in some water saturated bentonite/water/electrolyte systems. Using nuclear magnetic resonance technique (NMR), the state of the pore water can be described in terms of the observed proton T_2-relaxation time (T_2). The value of T_2 may in this case be regarded as a measure of the mobility or viscosity of the water in a clay/water system. In short, T_2 increases with increasing water mobility. Thus, "bound" water close to clay mineral surfaces is associated with relatively small T_2-values, while the comparatively "free" water at large distances from mineral surfaces is associated with relatively high T_2-values.

The NMR-studies suggest that the water mobility in water saturated MX-80 systems is a function of 1) type and 2) concentration of electrolyte in the pore water, 3) water content and 4) temperature.

MATERIALS

The bentonite used in this work is the main KBS reference material; commercially available Na-bentonite, type MX-80 (trade name) from the American Colloid Company. According to Müller-Vonmoos and Kahr [1], the dominating constituent in MX-80 is Wyoming montmorillonite (75% by weight), while the remaining contents is quartz (15.2%), feldspar (5-8%), carbonate (1.4%), pyrite (0.3%), mica (<1%), kaolinite (<1%), organic carbon (0.4%) and some not specified minerals (2%).

The chemical formula of the montmorillonite component is [1]

$$(Si_{3.96}Al_{0.04})(Al_{1.55}Fe^{3+}_{0.20}Fe^{2+}_{0.01}Mg_{0.24})O_{10}(OH)_2\ M^{+}_{0.29}$$

where $M^{+}_{0.29}$ represents the amount of exchangeable cations compensating the charge deficit in the interior of the montmorillonite crystal lattice. The exchangeable cations are Na^+, Ca^{2+}, Mg^{2+} and K^+, which balance 85.5%, 10.1, 4.1 and 0.3 per cent of the charge deficit, respectively [1].

The montmorillonite mineral consists of negatively charged unit layers with a thickness of 9.6 Å and lateral dimensions of several hundreds of Å. In dry montmorillonite the exchangeable cations are situated between the unit layers and on the outer surfaces of the layers. During wetting, water penetrates the interlayer spacings and the mineral expands. At high water contents the exchangeable cations may no longer be adsorbed on the surfaces but form diffuse ion atmospheres in the vicinity of the unit layers.

It is well known, that the crystal structure of the montmorillonite unit layer consists of a layer of aluminium-octahedrons sandwiched between two layers of silica-tetrahedrons. However, there is still an uncertainty as to whether all or just a fraction of the silica-tetrahedrons point towards the octahedral layer. Figure 1 shows two possible models of the montmorillonite crystal structure.

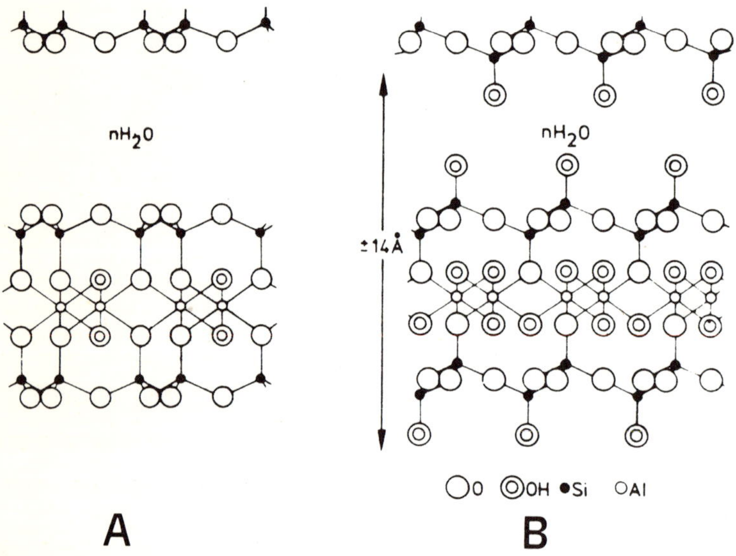

Figure 1. The structure of montmorillonite according to
A) Hofmann, Endell and Wilm, 1933
B) Edelman and Favejee, 1940

SAMPLE PREPARATION

Water saturated samples of MX-80/water/electrolyte were made by mixing untreated MX-80 bentonite (i.e. MX-80 of commercial quality) with distilled water or an electrolyte solution. The latter consisted of distilled water containing one of the following electrolytes; NaCl, KCl, $CaCl_2$ and $SrCl_2$. Water saturated samples with densities less than 1.5 g/cm^3 were made by mixing the constituents by hand, while samples with higher densities were made by means of a hydraulic press. In the latter case, samples with high water contents were inserted into a steel cylinder and compressed to volumes corresponding to the desired density. The drainage of supernatant liquid was performed by means of a filter which did not let through the MX-80 bentonite.

EXPERIMENTAL

For a review of the application of NMR to the study of clay minerals, the reader is referred to a recent work by Fripiat [2]. In a short and somewhat over-simplified way, the measurement of the proton T_2-relaxation time was carried out in following way. A glass tube containing a small amount (<1 cm^3) of a water saturated MX-80 sample was placed in a magnetic field in a NMR-spectrometer. As a result of the influence from the magnetic field the hydrogen protons in the sample becomes distributed on two different energy levels. At equilibrium the distribution is given by the Boltzmann equation

$$N_\alpha = N_\beta \cdot \exp(-\Delta E/kT) \qquad (1)$$

where N_α and N_β are the number of protons on energy level E_α and E_β, respectively, $\Delta E = E_\alpha - E_\beta$, k is Boltzmann's constant and T is the absolute temperature.

The equilibrium distribution can be disturbed by sending a short radiofrequency signal with a suitable frequency towards the sample. After the pulse the protons tend to obtain the original equilibrium population distribution between the two energy levels. In short, the higher the mobility of the water, the longer is the time needed in order for the system to obtain equilibrium. Since this time is directly proportional to the proton T_2-relaxation time, the value of T_2 may be used as a measure of the water mobility thus, an increase in T_2 is associated with a corresponding increase in the water mobility. In principle, the above reasoning is correct, at least when only one type of water is present in the clay/water system.

The measurements were performed by means of a Bruker CXP-100, NMR-spectrometer operating at 90 MHz. All T_2-values presented here were measured by means of the line width at half peak height in the adsorption spectrum.

MX-80 bentonite contains a rather high amount of paramagnetic Fe^{3+} (cf. the chemical formula of the montmorillonite component). Therefore, the observed T_2-values are considerably smaller as compared to if no paramagnetic ions were present in the sample. No T_2-valuses presented in this report have been corrected for the paramagnetic effects. However, as long as only relative values in one type of system is discussed, such corrections are not needed.

RESULTS AND DISCUSSION

This NMR-study indicate that the physical state of the pore water in water saturated MX-80 systems is influenced by 1) the pore water composition 2) the water content and 3) the temperature. The results refer to rather dense MX-80 systems with water contents below 150%.

Influence of temperature

The relationship between temperature and T_2 seemed to be associated only with the thermal motions of the water molecules; there was no evidence of any other thermal influence on the MX-80 systems. It was found that T_2 increases linearly with temperature within the experimental limits, i.e. from +15 °C to +80 °C. Figure 2 shows the increase of the T_2 caused by increasing the temperature from +33 °C to +80 °C.

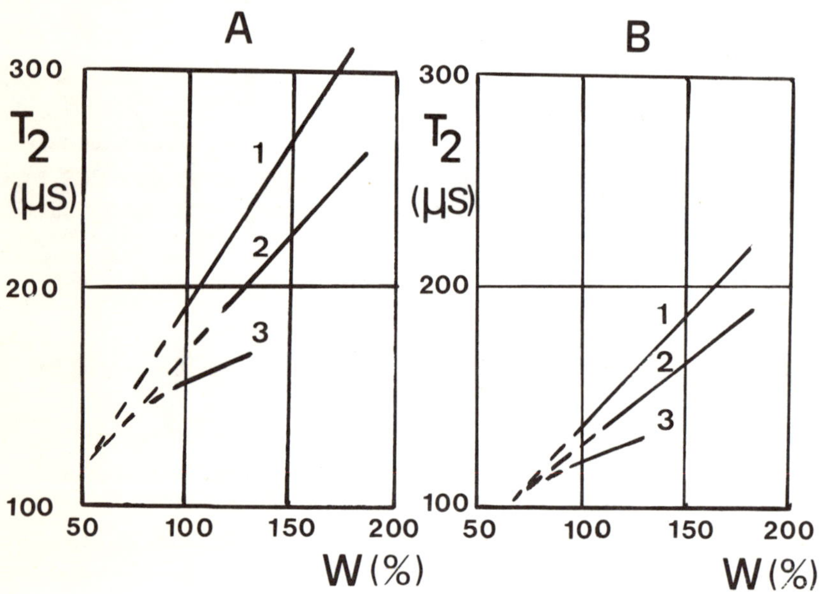

Figure 2. Influence of different electrolytes on the proton T_2-relaxation time (μs) in some water saturated MX-80 systems. Electrolyte concentration in the solutions added to the MX-80 bentonite: 0.175 moles/litre. Type of electrolyte added: 1) NaCl, 2) KCl, 3) $CaCl_2$ or $SrCl_2$. Temperature: A) +81 °C, B) +33 °C

Influence of water content

It was found that within the experimental limits, T_2 increases almost linearly with increasing water content (cf. Figure 2). Only homogeneous MX-80 samples were studied in this work, which means the water content in all cases was low enough to avoid flocculation. According to the graphs in Figure 2 corresponding to system containing $CaCl_2$ and $SrCl_2$ are not extended beyond water content of roughly 140%, since this value is close to where flocculation starts. In the case when 0.175 M NaCl- or KCl-solutions are added to MX-80, flocculation takes place when the water content exceeds roughly 300% and 150%, respectively.

It is clearly seen, that the graphs in Figure 2 depend on the pore water composition. The dotted extrapolations of the experimentally obtained results suggest that there is a point of intersection corresponding to a water content of approximately 60%. This water content is tentatively interpreted as a limit below which the electrolyte content in the MX-80 systems does not significantly influence the T_2-value. The relation between water content w(%), specific surface area S (m^2/g) and interparticle spacing, d (Å) is given by

$$d = 200 \cdot w/s \qquad (2)$$

According to this equation, a water content equal to 60% corresponds to an average interparticle spacing of 20 Å, provided that the specific surface area of MX-80 is 600 m^2/g. In a series of swelling pressure measurements on water saturated MX-80 systems Pusch [3] found that the electrolyte composition only influence the swelling pressure when the water content exceeds approximately 25%. This value corresponds to a bulk density of 2 t/m^3 provided that the specific gravity is 2.7 t/m^3. In principle the two methods seem to lead to the same conclusion despite the fact that their results are somewhat different: the electrolyte composition becomes decisive to the properties of the water phase only when the water content is high enough to allow the formation of double layers.

Influence of pore water composition

The influence of electrolyte on the water mobility in the investigated MX-80 systems is indicated by the series $SrCl_2 \approx CaCl_2 < KCl < NaCl < H_2O$, where H_2O represents a water saturated system obtained by mixing MX-80 with distilled water. Figure 3 indicates that T_2 decreases with increasing electrolyte content other things being equal. It is suggested, that the observed effect on the T_2-value, dT_2, caused by the addition of small amounts of one of the above electrolytes to a water saturated MX-80 system can be written in the differential form

$$dT_2 = \Sigma\, a_i \cdot dc_i \qquad (3)$$

where a_i is a constant characteristic of ion "i" and dc_i is a differential change in the concentration of ion "i" in the pore water. Equation (3), which has been experimental supported in some pilot-tests, is probably applicable to most water saturated MX-80 systems, provided that the ions added do not take part in any chemical reactions. According to literature [4], Cl^--ions in water solutions tend to increase the water mobility. Thus, provided that this effect is the same in water saturated MX-80 systems, the presence of Cl^- in the pore water results in a small positive contribution to the observed T_2-value which implies that the influence of the pore water composition on the water mobility is mainly caused by the cations.

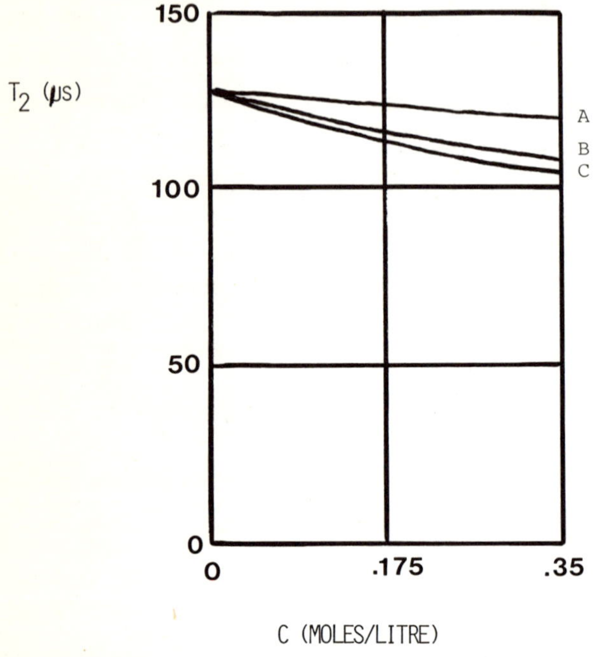

Figure 3. Proton T_2-relaxation time vs electrolyte concentration, c in solutions used in the preparation of MX-80/electrolyte water samples. C=0 refers to the system MX-80/water. The sample temperature was in all cases +24 °C and the water content 100%. Type of electrolyte added: A) NaCl, B) KCl, C) $CaCl_2$ or $SrCl_2$

REFERENCES

1. M. Müller-Vonmoos and G. Kahr, Technischer Bericht 83-12, ETH, Zürich, 1983

2. J.J. Fripiat in Advanced Chemical Methods for Soil and Clay Minerals Research, edited by J.W. Stucki and W.L. Banwart (Reidel Publ Co., 1980), pp. 245-315

3. R. Pusch, SKBF/KBS Technical Report 80-13, 1980

4. G.E.Walrafen, J. Chem. Phys. 44, 4176 (1979)

DIFFUSION MECHANISMS OF STRONTIUM, CESIUM AND COBALT IN COMPACTED SODIUM BENTONITE

ARTO MUURINEN, JUHA RANTANEN, PIRKKO PENTTILÄ-HILTUNEN
Technical Research Centre of Finland, Reactor Laboratory,
Otakaari 3 A, SF-02150 Espoo, Finland

ABSTRACT

For a porous water-saturated material where diffusion in the pore-water, sorption on the solid material and diffusion of the sorbed ions (surface diffusion) occur, a diffusion equation can be derived where the apparent diffusivity includes two terms. One represents diffusion in the pore-water, the other surface diffusion.

In this research diffusion mechanisms were studied. The apparent diffusivities of strontium, cesium and cobalt in compacted sodium bentonite were measured by a non-steady state method. The sorption factors were adjusted using different sodium chloride solutions, groundwater and addition of EDTA for saturation of the bentonite samples. The corresponding sorption factors were measured by a batch method.

The results suggest that cations diffuse also while being sorbed. A combined pore diffusion-surface diffusion model has been used to explain the transport and the corresponding diffusivities have been evaluated. The surface diffusivities (D_s) of Sr and Cs were $8\text{-}9 \cdot 10^{-12}$ m^2/s and $4\text{-}7 \cdot 10^{-13}$ m^2/s respectively. The pore diffusivity εD_p of Cs was $3.5 \cdot 10^{-11}$ m^2/s which has been used also for Sr.

The sorption mechanism of Co seems to be different from that of Sr or Cs and the results allow no specific conclusions of the diffusion mechanism of Co. The apparent diffusivity of Co ranged from $2 \cdot 10^{-14}$ to $7 \cdot 10^{-14}$ m^2/s.

The anionic Co-EDTA seems to follow some other diffusion mechanism than the cations.

INTRODUCTION

The concept for the disposal of high level wastes receiving most attention in Finland involves emplacement of the wastes in hard crystalline rock. In this concept engineered barriers have been considered as a redundant element against the release of radionuclides to the biosphere. The buffer material planned for the spent fuel is sodium bentonite. In the repository bentonite is saturated by groundwater and corrosive substances and radionuclides may be transported by diffusion through the bentonite. For safety analysis the mechanisms by which the substances migrate and the corresponding parameters must be known.

The diffusivities of Sr and Cs in compacted bentonite and the corresponding sorption factors have been measured by many researchers [1,2,3,4]. In these studies higher diffusivities were found than could be expected considering the high sorption factors. Surface diffusion has been suggested by way of an explanation.

The evaluation of diffusion rates is based on Fick's diffusion law. The measurement of the rates described by diffusivity (D) can be carried out by different methods, where the diffusion equation is solved using appropriate initial conditions.

Fick's I and II law in one dimension are expressed as follows [5]:

$$J = - D \frac{\partial c}{\partial x} \quad (1)$$

$$\frac{\partial c}{\partial t} = D \frac{\partial^2 c}{\partial x^2} \quad (2)$$

In Eq's 1 and 2,
J is the diffusion flow density in the positive x-direction (Bq/m²s)
c(x,t) is the concentration of the diffusing substance (Bq/m³)
t is the time
D is a proportionality constant called diffusivity or diffusion constant (m²/s)
Eq. 2 is generally called the diffusion equation.

Assuming the following boundary and initial conditions:

$$c(0,t) = c_o \text{ and} \quad (3)$$

$$c(x,0) = 0,$$

Eq.2 can be solved, the solution being

$$c(x,t) = c_o \mathrm{erfc}\left(\frac{x}{2\sqrt{Dt}}\right) \qquad (x \geq 0) \quad (4)$$

where

$$\mathrm{erfc}(y) = \frac{2}{\pi} \int_y^{\infty} e^{-t^2} dt.$$

Assume that the substance diffusing in the pore water of the medium is sorbed on the surfaces of the porous medium, and that the sorption reaction is instantaneous and reversible and follows a linear sorption law:

$$c_s = \frac{K_D \rho}{\varepsilon} c_p, \text{ where} \quad (5)$$

c_s is the bulk concentration of sorbed ions (Bq/m³)
c_p is the bulk concentration of free ions in pore water (Bq/m³)
ρ is the dry bulk density of the medium (kg/m³)
ε is the porosity of the medium ($0<\varepsilon<1$) and
K_D is the sorption factor (m³/kg).
The diffusion equation (Eq.2) has now the form

$$\frac{\partial c_p}{\partial t} = D_p \frac{\partial^2 c_p}{\partial x^2} - \frac{\partial c_s}{\partial t}, \text{ which can be further developed:}$$

$$\frac{\partial c}{\partial t} = D_a \frac{\partial^2 c}{\partial x^2}, \text{ where} \quad (6)$$

$$D_a \equiv \frac{D_p}{1+K_D\rho/\varepsilon} \text{ and} \quad (7)$$

c is the bulk concentration of migrating substance, $c \equiv c_p + c_s$,
D_p is the diffusivity of ions in pore water excluding sorption.

Thus, with the sorption present, the situation is reduced to another Fick's law equation with different parameters; the constant D_a defined by Eq. 7 is called the apparent diffusivity.

Furthermore, it is possible that the sorbed ions migrate in their own phase [6]. The theory must again be extended to include the new phenomenon. Assume that the migration obeys Fick's laws, i.e.

$$J_s = -D_s \frac{\partial c_s}{\partial x}, \text{ where} \tag{8}$$

J_s is the net diffusion flow density of the sorbed ions (Bq/m^2s) and D_s is the diffusivity of the sorbed ions (m^2/s).

The diffusion equation (6) can now be modified:

$$\frac{\partial c}{\partial t} = D_{as} \frac{\partial^2 c}{\partial x^2}, \text{ where} \tag{9}$$

$$D_{as} \equiv \frac{D_p + \frac{K_D \rho}{\varepsilon} D_s}{1 + K_D \rho/\varepsilon}. \tag{10}$$

If the surface diffusivity D_s is constant in the system, Eq.9 can be solved as Eq's 2 and 6. It can be seen that the new apparent diffusivity D_{as} reduces to D_a (Eq.7) if $D_s = 0$.

It can be seen from Eq.10 that the apparent diffusivity is a function of K_D; multiplying both sides by the denominator of the right hand side and ε, we obtain for the effective diffusivity

$$D_e \equiv (\varepsilon + K_D \rho) D_{as} = \varepsilon D_p + K_D \rho D_s. \tag{11}$$

From Eq. 7 can be obtained for the effective diffusivity without surface diffusion:

$$D_e \equiv D_a(\varepsilon + K_D \rho) = \varepsilon D_p \tag{12}$$

Equations (11) and (12) can be used in checking the validity of the surface diffusion hypothesis. If there is no surface diffusion, $D_a(\varepsilon + K_D \rho)$ should be constant. If the surface diffusion exists, $D_a(\varepsilon + K_D \rho)$ should be linear in K_D and Eq. 11 can be used in determining εD_p and D_s.

MEASUREMENTS

In this research diffusion mechanisms were studied. The apparent diffusivities of strontium, cesium and cobalt in compacted sodium bentonite were measured in varying porewater conditions using a non-steady state method described earlier [7]. The corresponding sorption factors were measured by batch method using 10 g bentonite in 100 ml of solution. The mixture was shaked for four weeks. The solution was separated by centrifugation. The following measurements were carried out:

a) The apparent diffusivities of Sr and Cs were measured in compacted bentonite saturated with various sodium chloride solutions in order to adjust K_D.

b) The apparent diffusivities of Sr, Cs and Co were measured in compacted bentonite saturated with groundwater where the salt concentration was appr. 10 g/l [7].

c) The apparent diffusivities of Sr, Cs and Co were measured in compacted bentonite saturated with groundwater where 0.5 g/l of EDTA was added.

RESULTS AND DISCUSSION

The results of the sorption measurements in different sodium chloride solutions are presented in Fig. 1. The sorption factors of Cs and Sr are clearly dependent on the salt concentration of the solution. The sorption of cobalt shows some scattering but seems, however, not to be dependent on the salt concentration.

With low salt content the sorption factors of different ions are about equal.

For the ion-exchangers Eq. 13 [8] is valid.

$$K_D \propto [B]^{-[a/b]} \qquad (13)$$

where

$[B]$ is the concentration of macro ion in solution (in this case Na^+)

b is the charge of the macro ion

a is the charge of the tracer ion

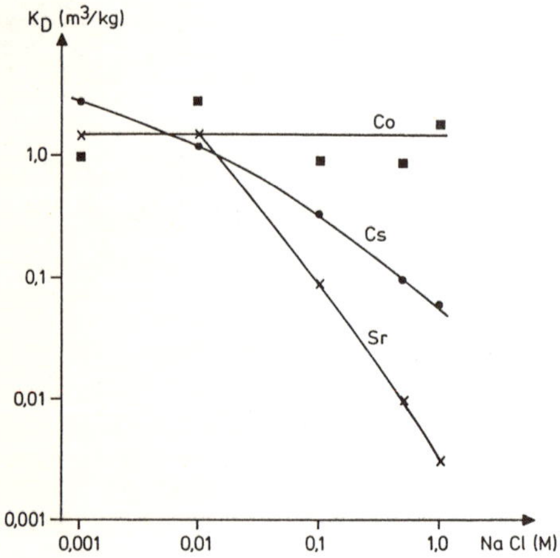

Fig. 1. Sorption factors (K_D) of strontium, cesium and cobalt for sodium bentonite in different sodium chloride solutions.

The theory predicts that the slopes of the curves in Fig. 1 would be -2 for Sr^{2+} and -1 for Cs^+. The absolute values of the slopes are somewhat less than predicted by the theory; the order is, however, correct. Because the K_D of cobalt does not seem to be dependent on the salt concentration, it probably follows a different sorption mechanism than strontium and cesium.

Table 1 gives the diffusivities of Sr-85 and Cs-134 in compacted bentonite, which was saturated with sodium chloride solutions. The concentration of sodium varied from 0.001 M to 1.2 M. The sorption factors in Table 1 are based on Fig. 1. The diffusivities D_{as} signify the measured apparent diffusivity and D_e signifies the effective diffusivity, which was calculated from Eq. 11.

Figure 2 presents the apparent diffusivities D_{as} and the effective diffusivities D_e versus K_D. On the basis of the D_e- and $K_D\rho$-values, εD_p and D_s have been solved by fitting Eq. 11 on the points. The fitting for cesium has been done by weighting the least squares sum terms with a factor $1/(K_D\rho)_i$ so that the terms representing small K_D and D_e values are also taken into account.

For strontium only two measurements are available which allow D_s (the slope) to be solved. A negative value is, however, obtained for εD_p because of inaccuracies in the measurement. For εD_p of strontium, however, the corresponding value of cesium is used until more accurate measurements are available.

Only part of the K_D-values could be covered by adjusting the salt concentration as can be seen in Fig. 1. Dotted lines are used in Figures 2 and 3 for the fitting curves where measured values are not available.

Table I. Sorption factors and apparent diffusivities of strontium and cesium in different porewater conditions.

Nuclide	Density [g/cm³]	Porewater	K_D [m³/kg]	D_{as} [m²/s]
Sr-85	1.74	0.6 M NaCl	0.007	$6.9 \cdot 10^{-12}$
Sr-85	1.74	0.001 M NaCl	1.5	$9.1 \cdot 10^{-12}$
Cs-134	2.03	1.2 M NaCl	0.05	$6.5 \cdot 10^{-13}$
Cs-134	1.98	0.6 M NaCl	0.09	$6.9 \cdot 10^{-13}$
Cs-134	1.87	0.2 M NaCl	0.21	$8.5 \cdot 10^{-13}$
Cs-134	1.91	0.02 M NaCl	0.85	$3.5 \cdot 10^{-13}$
Cs-134	1.94	0.003 M NaCl	1.80	$2.5 \cdot 10^{-13}$

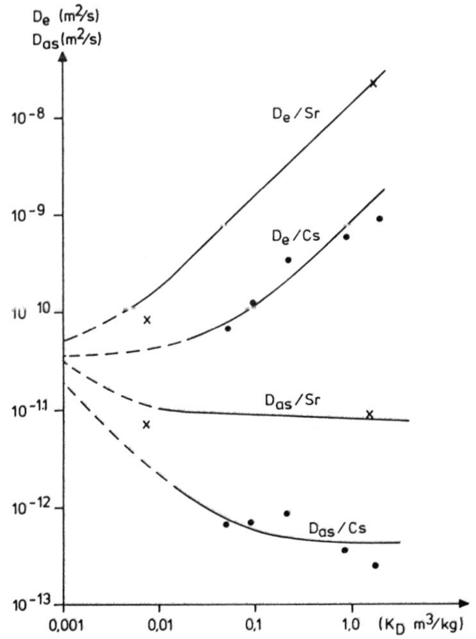

Figure 2. Apparent diffusivities (D_{as}) and effective diffusivities (D_e) of strontium and cesium versus K_D in compacted bentonite saturated with different sodium chloride solutions.
Sr: $\varepsilon D_p = 3.5 \cdot 10^{-11}$ m²/s
$D_s = 8.8 \cdot 10^{-12}$ m²/s
Cs: $\varepsilon D_p = 3.5 \cdot 10^{-11}$ m²/s
$D_s = 4.3 \cdot 10^{-13}$ m²/s

The measured points in Figure 2 fit qualitatively fairly well with the curves calculated from Eq. 10 and 11, with those K_D-values which are available by adjusting salt concentration. Some scattering is, however, evident, which is probably caused by the differences in the densities of the samples as can be seen in Table I.

The results of the measurements using groundwater and groundwater + EDTA are presented in Table II. Figure 3 gives the diffusivities D_{as} and D_e versus K_D for groundwater and the fitted curves based on Eq. 10 and 11. For εD_p the value of cesium from the experiments with sodium chloride solutions has been used. The results of the groundwater experiments correspond fairly well with those of the sodium chloride experiments with the same salt concentration.

While considering the results of Co one has to keep in mind that the sorption mechanism of Co probably is different from that of Sr and Cs and the K_d of Co could not be adjusted by changing the salt concentration. In Figure 3 the surface diffusion theory according to Eq. 10 and 11 has been used also for Co. The results for Co do not, however, allow any specific conclusions of the diffusion mechanism. If Eq. 12 would be used for Co, εD_p would range from $1 \cdot 10^{-10}$ to $5 \cdot 10^{-10}$ m^2/s.

Table II. Sorption factors and apparent diffusivities of strontium, cesium and cobalt in compacted sodium bentonite.
gw = groundwater, total salt appr. 10 g/l
gw + EDTA = groundwater + EDTA 0.5 g/l

Nuclide	Porewater	K_D [m^3/kg]	D_{as} [m^2/s]
Sr-85	gw	0.050	$9.5 \cdot 10^{-12}$
Sr-85	gw	0.050	$7.8 \cdot 10^{-12}$
Sr-85	gw+EDTA	0.042	$8.4 \cdot 10^{-12}$
Sr-85	gw+EDTA	0.042	$1.5 \cdot 10^{-11}$
Cs-137	gw	0.393	$6.1 \cdot 10^{-13}$
Cs-137	gw	0.393	$1.0 \cdot 10^{-12}$
Cs-137	gw+EDTA	0.184	$6.4 \cdot 10^{-13}$
Cs-137	gw+EDTA	0.184	$7.3 \cdot 10^{-13}$
Co-60	gw	3.74	$7.0 \cdot 10^{-14}$
Co-60	gw	3.74	$2.0 \cdot 10^{-14}$
Co-60	gw+EDTA	0.0013	$5.7 \cdot 10^{-14}$
Co-60	gw+EDTA	0.0013	$4.4 \cdot 10^{-14}$

The diffusivities D_{as} of Sr, Cs and Co in Figure 3 when groundwater was used for the saturation, seems to be inversely proportional to K_D. However, comparing Figures 2 and 3 it can be seen that D_{as} versus K_D with different ions depends on the salt content. With low salt concentration the sorption factors of Sr, Cs and Co are about constant and the D_{as} versus K_D would be vertical.

The sorption factors and diffusivities of strontium and cesium with groundwater + EDTA do not differ essentially from the values received without EDTA, whereas K_D of cobalt is clearly less with EDTA. The diffusivities (D_{as}) of cobalt are, however, about equal with and without EDTA. The anions, as the Co-EDTA-complex, seem thus to follow some other diffusion mechanism than cations.

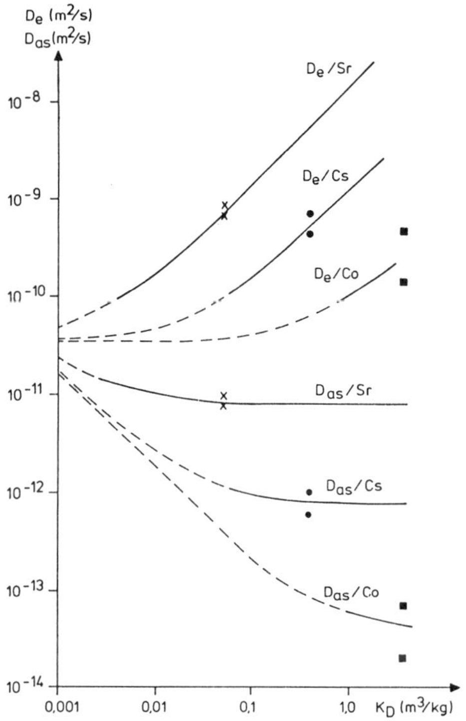

Figure 3. Apparent diffusivities (D_{as}) and effective diffusivities (D_e) of strontium, cesium and cobalt versus K_D in compacted bentonite saturated with groundwater

Sr: $\varepsilon D_p = 3.5 \cdot 10^{-11}$ m²/s
$D_s = 8.2 \cdot 10^{-12}$ m²/s
Cs: $\varepsilon D_p = 3.5 \cdot 10^{-11}$ m²/s
$D_s = 7.6 \cdot 10^{-13}$ m²/s
Co: $\varepsilon D_p = 3.5 \cdot 10^{-11}$ m²/s
$D_s = 3.9 \cdot 10^{-14}$ m²/s

CONCLUSION

The sorption factors of strontium, cesium and cobalt with sodium bentonite in low salt concentrations are almost equal. The effect of the salt concentration on K_D of Sr and Cs follows the ion-exchange theory with some small modifications. This means that the sorption factor of strontium with the charge 2+ can be adjusted more by changing the salt concentration than that of cesium with the charge 1+. The sorption factors of cobalt do not seem to be dependent on the salt concentration.

The apparent diffusivities of strontium and cesium do not change much when the salt concentration used for the saturation of the samples is changed and the sorption factors change. The surface diffusion model agrees fairly well with the observed diffusion-sorption behaviour. The surface diffusion might noticeably increase the diffusion through a barrier because the concentration difference related to the sorbed phase is often much higher than the concentration gradient in the porewater.

The results for cobalt do not allow any specific conclusions of the diffusion mechanism.

The anionic Co-EDTA complex seems to follow some other diffusion mechanism than the cations.

REFERENCES

1. I. Neretnieks, "Diffusivities of some dissloved constituents in compacted wet bentonite clay -MX80 and the impact on radionuclide migration in the buffer", KBS-TR-82-27, Stockholm 1982.
2. B. Torstenfelt, B. Allard, K. Andersson, H. Kipatsi, L. Eliasson, U. Olofsson, H. Persson, "Radionuclide diffusion and mobilities in compacted bentonite", KBS-TR-83-34, Stockholm 1983.
3. T.E. Eriksen, A. Jacobsson, "Diffusion in clay - Experimental techniques and theoretical models", KBS-TR-84-05, Stockholm 1984.
4. A. Soudek, F.M. Jahnke, C.J. Radke, "Ion-Exchange Equilibration and Diffusion in Engineered Backfill", NUREG/CP-0052, CONF-8308126---, 1984, pp. 172-203.
5. J. Crank, The mathematics of diffusion. (Oxford University Press, London 1956.)
6. A. Rasmuson, I. Neretnieks, "Surface migration in sorption processes", KBS-TR-83-87, Stockholm 1983.
7. A. Muurinen, J. Rantanen. R. Ovaskainen, O.J. Heinonen, "Diffusion measurements in concrete and compacted bentonite", in Scientific Basis for Nuclear Waste Management VI, (Elsevier Science Publishing Co., Inc., 1983) pp. 777-784.
8. J.A. Marinsky, Ion Exchange. (Marcel Dekker, Inc., New York, 1966) p. 290.

PART VIII

Transport in Fractured Rocks

FLOW AND TRACER EXPERIMENTS IN CRYSTALLINE ROCKS: RESULTS FROM SEVERAL SWEDISH IN SITU EXPERIMENTS

H. ABELIN, L. BIRGERSSON, J. GIDLUND, L. MORENO, I. NERETNIEKS, S. TUNBRANT
Department of Chemical Engineering, Royal Institute of Technology, S-100 44 Stockholm, Sweden

BACKGROUND

In a final repository for radioactive waste in crystalline rock, water flowing in the fractures may transport the radionuclides eventually leached from the waste. To be able to predict the migration of the radionuclides the processes involved must be understood. To quantify the processes, data from flow and transport in real fractures under realistic conditions are needed. Models used for prediction must include descriptions of the important processes and mechanisms.

Most previous studies concerning water flow in rocks are based on the assumption that the water flow can be described as porous media flow. This might be true for very large distances where the flow would encounter a multitude of fractures, and some averaging may be conceivable at the scale considered.

Transport over short distances, i.e. in the vicinity of the canister, most probably occurs in individual fractures. On an intermediate scale where more than a few fractures conduct the flow, well-type tracer tests alone cannot give the detailed information needed to understand dispersion and sorption phenomena in fissured rock. Actual tracer tests are needed to obtain data on pathways and flow porosities. Tests with sorbing tracers are needed to compare predictions based on models and laboratory data, i.e. to validate the transport models.

The migration modelling in the safety analysis for a repository in granitic rock is based on the assumption that if and when any radionuclides are leached from the waste, practically all of the important radionuclides will interact chemically or physically with the bedrock and will thereby be considerably retarded. The magnitude of this retardation depends upon the flow rate of the water, the uptake rates and equilibria of the reactions as well as the surface area in contact with the flowing water.

PURPOSE

The main purpose of the two investigations performed in the Stripa mine is to understand and describe water movement in sparesly fractured crystalline rock. The first experiment, now finished, concerned water flow within a single fracture and also dealt with sorbing tracers interaction with the rock matrix, such as sorption and diffusion into the matrix. The second experiment, now underway, concerns water flow paths within a larger block of rock.

Flow and migration within a single fracture (2-D)

The main purposes of this experiment were to investigate if it is possible to extend results on sorption and retardation of radionuclides in granitic rock, obtained from laboratory experiments, to a real environment with migration distances up to 5 m and to try to determine the extent of channelling within single fractures.

Flow and migration within a large rock volume (3-D)

The main purpose of this experiment is to investigate the spatial distribution of water flow paths in a larger block of rock. Migration distances of up to 50 m are utilized. Flow parameters like mean residence time, dispersion, longitudinal as well as transverse, will also be studied.

EXPERIMENTAL DESIGN

The experiments are performed at the 360 m level. Water flows constantly into drifts at this level since it is located well below the water table.

The 2-D experiment

The design of the experiment was based on the idea that reasonably well defined individual fractures can be located and that tracers, non-sorbing and sorbing, can be introduced into the natural water flow within a single fracture without a large disturbance of the flow field.

Almost all water coming out from the fracture at the face of the drift was collected. Figure 1 shows the layout of the test site with two fractures and five injection holes intersecting the fracture planes. Each of the fractures had about 15 sampling holes at the face of the drift. Tracers, sorbing and non-sorbing, were only injected into fracture number 2.

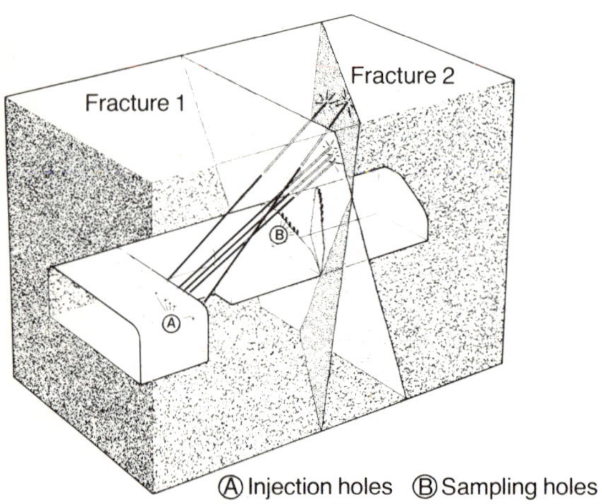

Figure 1. Schematic view of the 2-D test site.

Migration distances of 5 and 10 m have been used, which is more than one order of magnitude larger than distances that have been used in the laboratory experiments, Moreno et al. 1984 [1]. Conservative (non-sorbing) tracers were used to characterize the flow within the fracture.

The following sorbing tracers, analogs for radionuclides, were used: Cs (I), Sr (II), Eu (III), Nd (III), Th (IV), U (IV and VI). The sorbing tracers were continuously injected at one injection point during seven months (Sr 5 1/2 months, U 3 months).

The results from the runs with the non-sorbing tracers and laboratory data on sorption and porosity have been used to predict the breakthrough curve for a sorbing tracer. The predicted curve would later be compared with the experimentally obtained breakthrough curve.

Of the sorbing tracers only Sr was predicted to reach the sampling holes within the time of the experiment, the rest of the tracers would be sorbed in the rock surrounding the injection point. To make it possible to analyse surface concentration and penetration depth of the sorbing tracers, part of the fracture around the injection point was excavated.

The 3-D experiment

From the test site, which has a total length of 100 m, three vertical holes (length 70 m) have been drilled.

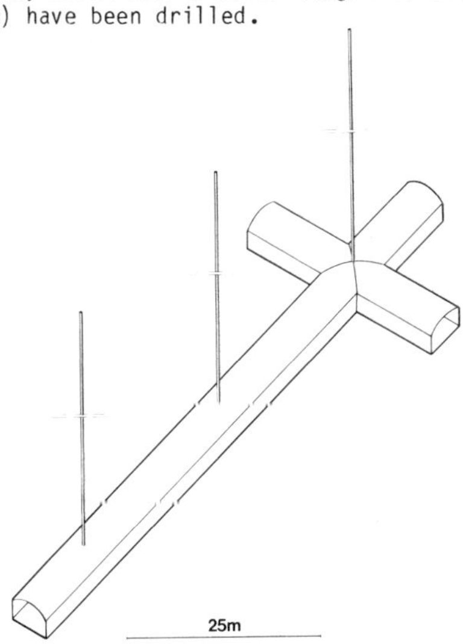

Figure 2. Layout of the 3-D test site with 3 vertical injection holes.

Injection of conservative (non-sorbing) tracers are carried out from a total number of 9 separate high permeability zones within the three vertical holes, each zone about 2.5 m in length. The injection zones are located between 10 and 55 m above the test site. The tracers will be injected continuously for 1 year. The injections are carried out with a "constant" over pressure, approximately 10-15% above the natural pressure. The concentrations of the injected tracers are between 1000 and 2000 ppm, and the different flow rates vary from 1 to 20 ml/h. The following tracers are injected: Uranine, Eosin blueish, Eosin yellowish, Phloxine B, Rose Bengal, Elbenyl Brilliant Flavine, Duasyn Acid Green, Bromide and Iodide.

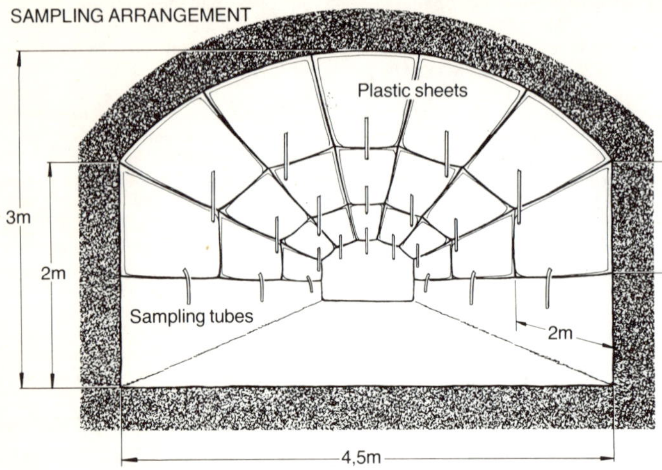

Figure 3. Sampling arrangement at the 3-D test site.

The upper part of the test site has been covered with plastic sheets, each sheet with an area of about 2 m^2. A total number of about 350 sheets serves as sampling areas (points) for water emerging into the upper part of the test site. The sampling arrangement completely covers a surface area of 700 m^2. When sampling over an area the spatial distribution of water flow path ways can be obtained. The presence of the cross increases the resolution of the 3-dimensional distribution of tracers in that part of the test site.

MODELLING OF TRACER TESTS

Tracer tests are analysed using different models. Hydraulic data such as permeabilities, fissure widths and flow porosities are evaluated by means of tracer test with the non-sorbing species. Tracer interaction mechanisms with the matrix for sorbing species are evaluated and compared with laboratory data when available.

The following models have been considered:

- Advection-dispersion-surface sorption model (A-D). This model includes hydrodynamic dispersion in the fissures and sorption onto the surface of the fissures, but no diffusion into the matrix. Flow through several individual fissures may be considered.

- Advection-dispersion-matrix diffusion model (A-D-D). The model describes flow through an individual fissure including, moreover, diffusion into the rock matrix and sorption onto the surface of the microfissures within the rock matrix.

- Advection-channeling-matrix diffusion model. In this case the dispersion is taken in account by channeling. Channeling occurs when water flows through non-interconnected or seldom connected channels with different water velocities. Dissolved species will be transported different distances over a given time.

The concentration at the outlet, for the advection-dispersion model (A-D), can be written in a condensed form as:

$$C(t) = C_o\ P_f\ f(Pe,t_w,R_a,t)$$

C_o is the injection concentration, P_f accounts for dilution effects, Pe is the Peclet number, t_w the water residence time, R_a is the surface retardation factor. When fitting the data from the runs with the conservative tracers (R_a=1), values for the parameters P_f, t_w and Pe will be obtained.

The corresponding outlet concentration for the advection-dispersion-matrix diffusion model (A-D-D) is:

$$C(t) = C_o\ P_f\ f(Pe,t_w,A,t)$$

The A parameter accounts for the interaction with the solid matrix and includes data on diffusion into the matrix and sorption within the matrix. The fitting of experimental data to this model was done in two ways. It was either a three parameter fit giving values of Pe, t_w and P_f where A is calculated from laboratory data or a four parameter fit where also the value of A was determined by the fitting procedure.

The corresponding outlet concentration for the advection-channeling-matrix diffusion model is:

$$C(t) = C_o\ P_f\ f(\sigma_1',t_w,D_D,t)$$

The D_D parameter accounts for the interaction with the solid matrix and includes data on diffusion into the matrix and sorption within the matrix. σ_1 is the standard deviation in the log-normal distribution of channel widths. The fitting of experimental data to this model was done in two ways. It was either a three parameter fit giving values of σ_1, t_w and P_f where D_D is calculated from laboratory data or a four parameter fit where also the value of D_D was determined by the fitting procedure.

In general, the models were solved for a constant concentration in the injection hole. This boundary condition gives large errors when the concentration in the matrix near the injection point is calculated using A-D-D model. In this case a constant flux is used as boundary condition at the injection point.

RESULTS

Water flow monitoring (2-D)

Figure 4 shows the results from the water flow monitoring in the sampling holes of the two fractures. Also the tracer flux from the different sampling holes is presented.

As can be seen in Figure 4, the water flow is unevenly distributed within single fractures. This type of flow pattern was also seen in the fracture used in a preparatory investigation, Abelin et al. (1982) [2].

Water flow monitoring (3-D)

The results from the water flow monitoring in the 3-D site shows that water does not flow uniformly in the rock over the scale considered (700 m^2), but seems to be localized to wet areas with large dry areas in between.

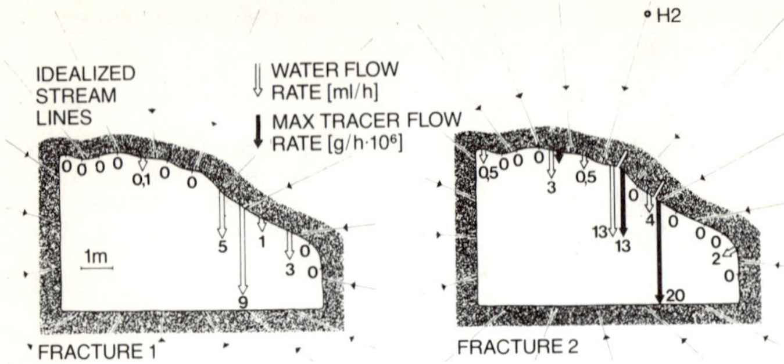

Figure 4. Water flow rates in fractures 1 and 2 (2-D).

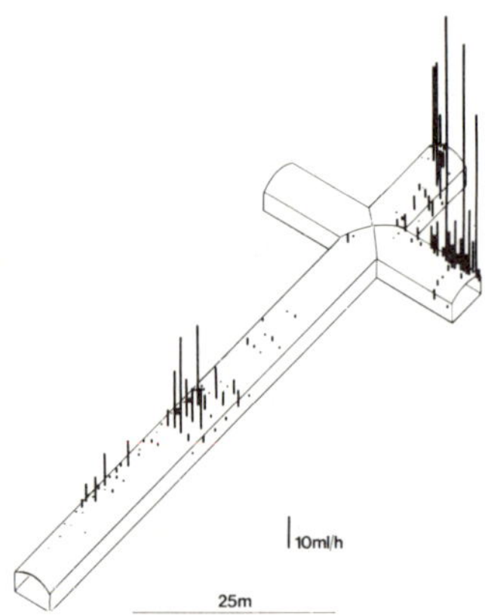

Figure 5. 3-D site. Water inflow rates into the sampling areas (before drilling the injection holes).

Measurable amounts of water emerge into just 1/3 of the 350 sampling areas. 10% of these "wet" sampling areas give more than 50% of the total water inflow.

Non-sorbing tracer injection and modelling (2-D)

Figure 6 shows the breakthrough curve for Uranine at one of the sampling holes in fracture 2. It also shows the fit between different models and experimental data.

As can be seen in Figure 6 it is possible to obtain a good fit with any of the choosen model concepts. It is not possible to select between the models with the results from this experiment only. Some of the mechanisms and their parameters have to be determined independently.

Table I and II shows the flow parameters obtained when fitting the models to experimental data, obtained from two different sampling holes.

The A-parameter in the three and four parameter fits of the A-D-D' model for sampling hole 2-6, Table I, only differs with a factor 1.4. On

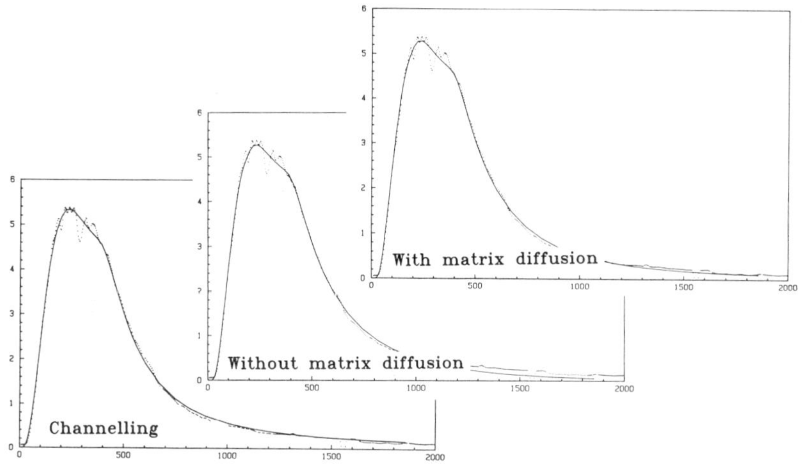

Figure 6. Breakthrough curve for fracture 2 and the fit of different models (2-D).

the other hand the A-parameter for sampling hole 2-8, Table II, in the four parameter fit, was found to be 133 compared with a value of 30 000 when independent laboratory data were used to determine A. This indicates a high interaction with the rock matrix. This seemingly large interaction may be due to presence of stagnant water zones into which the tracer can diffuse, Moreno et al. (1983) [3].

Calculation of fracture "widths" (2-D)

Any calculations of the fracture openings will include some assumptions on the properties within the fracture.

The average volume of the fracture could be determined from measuring the flow rate (Q) over a breadth (B) of the fracture and the residence time (t_w) of the water. We call this the mass balance fracture width.

$$d_f = f(Q, t_w, L, B)$$

L is the path length. When calculating d_f one assumes that all water collected over the breadth B passes the injection point and that the injected tracer directly enters this flow.

Another property of interest is that equivalent fracture width which would permit a certain flowrate at a given pressure drop (Δh). This we call the cubic law fracture width.

$$d_c = f(Q, \Delta h, L, B)$$

A third equivalent fracture width is that which would give a certain water velocity for a given pressure drop. This we call the frictional loss fracture width.

$$d_l = f(\Delta h, t_w, L)$$

Table I. Flow parameters from model fitting, sampling hole 2-6 (2-D).

Flow par.	A-D	A-D-D 3 par.	A-D-D 4 par.	Channelling 3 par.	Channelling 4 par.
P_f	3.0	2.8	2.7	2.8	2.5
t_w, h	366	293	279	348	291
Pe	1.96	2.33	2.46	-	-
A	-	4926	3548	-	-
σ_1	-	-	-	.40	.36
$D_D \cdot 10^7$	-	-	-	2.65	7.88

Table II. Flow parameters from model fitting, sampling hole 2-8 (2-D).

Flow par.	A-D	A-D-D 3 par.	A-D-D 4 par.	Channelling 3 par.	Channelling 4 par.
P_f	10.5	10.5	8.5	11.5	8.4
t_w, h	710	548	44.5	530	90
Pe	0.86	1.16	36.9	-	-
A	-	30000	133	-	-
σ_1	-	-	-	.47	.17
$D_D \cdot 10^7$	-	-	-	2.65	48.49

Table III. Calculated equivalent fracture widths (2-D).

Sampling hole	$K_{pf} \cdot 10^6$ [m/s]	linear d_f [μm]	linear d_1 [μm]	linear d_c [μm]	radial d_f [μm]	radial d_1 [μm]	radial d_c [μm]
2-6	1.1	240	1.1	6.7	130	1.2	5.5
2-8	3.4	280	2.0	10	150	2.0	8.5
4*	1.5	2200	1.3	16	1200	1.4	13
5*	0.75	1500	0.9	11	820	0.96	9.1

* Fracture in preliminary investigation. Abelin et al. (1982) [2].

There is a large difference between the different calculated equivalent fracture widths, Table III. It should be noted that d_c underestimates the mean residence time for the water and d_1 corresponds to a fracture that can not carry all the water actually measured, there would have to be several of these fracture within the fracture "plane".

Comparison with other field data

The results from the 2-D experiment, Table I, II and III, are compared with results from another tracer test in crystalline rock over a larger distance (30 m), carried at Finnsjön area. Iodide and strontium were here simultaneously injected, during 350 h, at a depth of about 100 m. The detection hole was located at 30 m distance where water was continuously pumped up to get a drawdown zone. The results (Moreno et al., 1983) [3] are shown in Table IV.

Table IV. Flow parameters from model fitting for experiments at the Finnsjön area.

Flow par.	A-D-D 3 par.	A-D-D 4 par.	Channelling 3 par.	Channelling 4 par.
P_f	370	334	383	336
t_w, h	38.5	18.3	41.4	18.8
Pe	5.24	86.9	-	-
A	30000	320	-	-
σ_1	-	-	0.285	0.072
$D_D \cdot 10^7$	-	-	0.155	7.24
Fracture width				
d_f, µm	981	466	1060	479
d_1, µm	61	89	59	88
d_c, µm	155	155	155	155

In the Finnsjön tracer test the water residence time was shorter (about 20-40 hrs) than in the Stripa 2-D runs.

A dispersion length of about 1.9 m was obtained for sampling hole 2-6. For sampling hole 2-8 these lengths were 0.12 and 3.9 m when 4 or 3 parameters are included in the fitting process. In tracer test at Finnsjön these values were 0.34 and 5.7 m respectively. The low dispersion obtained when 4 parameters are fitted may be explained by the low value of A. Part of the spreading is accounted by the model as diffusion into the matrix and thus not included in the dispersion.

These comparisons show that the Stripa test 2-8 has a shorter dispersion length than the test in Finnsjön. On the other hand σ_1 is higher in the Stripa run. The principal difference between the two tests is the different travel lengths used (4.5 m in Stripa and 30 m at Finnsjön). A compilation of dispersion lengths from experiments in crystalline rocks, Neretnieks 1985 [4], seems to indicate that the dispersion length increases with distance.

Non-sorbing tracer injection (3-D)

After 6 months of injection (April-85), tracers from at least 5 injection zones are seen in about 35 sampling areas (see Figure 7 and 8).

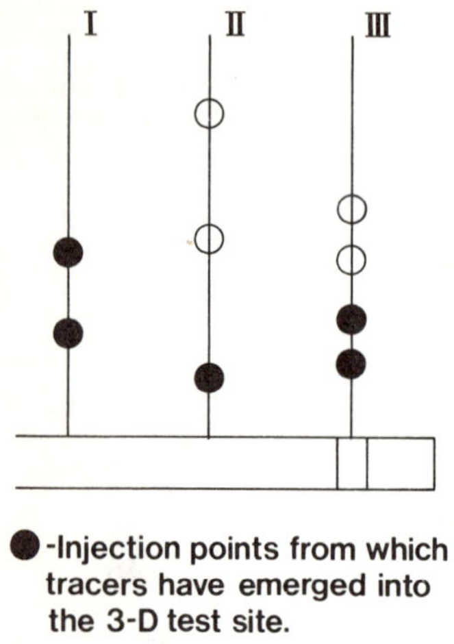

Figure 7. Tracers emerged into the test site (3-D).

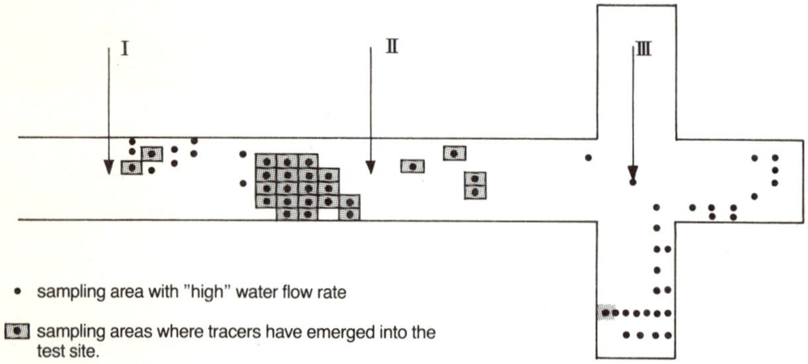

Figure 8. Tracer distribution (3-D).

Looking at Figures 7 and 8 it is remarkable that the tracer injected closest to the drift in hole III (Br^-), has not been found in any of the sampling areas close to hole III, but in many of the sampling areas around hole II.

Figure 8 shows those sampling areas that have "high" water flow rates and in which of these tracers have been found.

Results from the injection of sorbing tracers

The flow parameters obtained in the runs with the conservative tracers and independent laboratory data on sorption and porosity have been used to predict the breakthrough curve for Sr. Sr was predicted to reach the sampling holes well within the time of the experiment. However, Sr did not arrive in detectable concentrations at the sampling holes. As mentioned earlier, part of the fracture has been excavated. Figure 9 shows the concentration of Cs on some points of the fracture surface. It also shows the concentrations in the matrix.

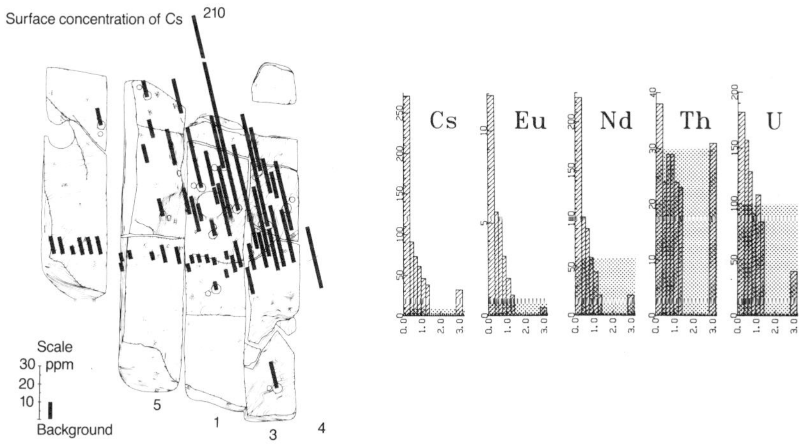

Figure 9. Excavated fracture with surface and matrix concentrations of Cs (2-D).

It can be seen in Figure 9a that there are elevated concentrations of Cs around the injection point, but there seems to be an area to the right of the injection hole which has the highest concentrations. This could indicate that the injected water was not evenly distributed around the injection point but had a prefered direction of flow.

Figure 9b shows concentration profiles into the rock obtained in one of the samples taken close to the injection hole.

It can be seen in Figure 9b that elevated concentrations of Cs, Eu and Nd, at this sampling point, can be found to a depth of 1.4 mm into the matrix.

In the Finnsjön test, a very high concentration solution of Sr was injected. This high concentration of Sr gives a low volume sorption coefficient, Sr behaved almost as a non-sorbing tracer. Probably precipitation of Sr also occurred within the fracture due to the high concentration.

DISCUSSION AND CONCLUSIONS

There are some inherent difficulties to interpret tests with sorbing tracers. Among these one could mention the variation in the natural content of the tracer over the fracture surface, variation in mineral composition giving different K_d values at different sites in the fracture and variations of the thickness of the fracture filling and coating materials.

As can be seen in Figure 6 it is possible to obtain a good fit with all of the models. The fitting thus cannot differentiate between the mechanisms. The correct mechanism(s) must be selected by some other independent process.

The calculated surface concentrations and the concentration profiles within the rock matrix could partly be confirmed by experimental data. They show sorbing tracers concentration profiles into the rock matrix, which can be explained by matrix diffusion.

Supporting in-situ experiments with the aim to ascertain a connected pore system in "undisturbed" granitic rock have been performed in the Stripa mine at the 360 m level (Birgersson et al., 1984) [5]. A mixture of conservative tracers (Uranine, Cr-EDTA and I^-) were injected for a time period of months. The subsequent analysis of tracer content close to the injection hole showed that all three tracers had migrated cm´s to tens of cm´s into the rock matrix. The penetration depths were in all cases beyond the zone disturbed by the presence of the injection hole. Therefore, these experiments indicate the existence of a connected pore system in "undisturbed" granitic rock at levels relevant for a nuclear waste repository.

Available observations show that it is not realistic to treat a fracture as two planar parallel surfaces with a constant width. Blocks of rock can not hover over each other but have to have points where they are in contact. If this is the only effect that causes channelling the direction and magnitude of the channels would be randomly distributed and there would not be any preferred direction for the water flow. The flowing water in the channels might, however, in some way affect the channels, so that preferred flow directions might develop with time. If there is a preferred direction of the channels, the results of a tracer test must be different depending on if the induced pressure drop is perpendicular to or parallel with the direction of the channels. The present experiment has given no information which could be used to determine preferential flow directions.

Our present concept of flow in a fracture is that there are different channels which may mix their waters at irregular distances. The channels may have zones of stagnant or near stagnant water which may be reached mainly by diffusion. This is illustrated in Figure 10.

With this concept of the fracture it is easy to see the difficulties to model the response of tracer tests in natural fractures in detail.

In the 2-D experiment, diffusion into the matrix was observed by direct observation of penetration depths, channeling was observed by using a large amount of collecting points. The presence of stagnant "lakes" was

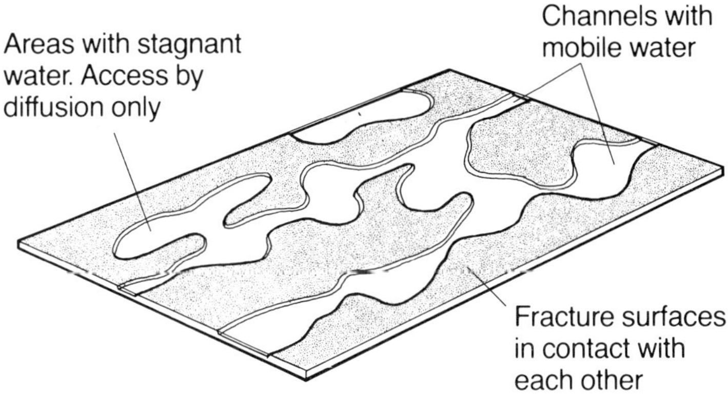

Figure 10. Concept of water flow within a fracture.

not observed directly but inferred from general observations on the necessity of points of contact between rock blocks. The water monitoring in the 3-D test site shows that water does not flow uniformly in the rock but seems to be localized to wet areas with large dry areas in between.

The observed fracture openings determined from the water volume that resides in the fractures is considerably larger than what is needed to induce the observed pressure drop. This means that hydraulic testing may not be used to obtain the actual flow porosity of a fissured natural rock using the presently available interpretation methods.

REFERENCES

1. Moreno, L., Neretnieks, I., Eriksen, T.: "Analysis of some laboratory tracer runs in natural fissures". KBS TR 84-04, Nuclear Fuel Safety Project, Stockholm (1984). To be published in Water Resources Res. 1985.

2. Abelin, H., Gidlund, J., and Neretnieks, I.: "Migration in a single fissure". Scientific Basis for Nuclear Waste Management V, p 529-538, North-Holland 1982.

3. Moreno, L., Neretnieks, I. and Klockars, K.E.: "Evaluation of some tracer tests in the granitic rock at Finnsjön", KBS TR 83-38, Nuclear Fuel Safety Project, Stockholm (1983).

4. Neretnieks, I.: "Transport in fractured rocks". Paper at the IAH 17th International Congress on the Hydrology of Rocks of Low Permeability. Tucson, Arizona, Jan 1985. Proceedings in print.

5. Birgersson, L., Neretnieks, I.: "Diffusion in the matrix of granitic rock. Field test in the Stripa mine". Scientific Basis for Nuclear Waste Management VII, p 247-254, North-Holland 1984.

RADIONUCLIDE TRANSPORT MODELLING
IN FISSURED ZONES AND CHANNELS

ANDERS RASMUSON AND IVARS NERETNIEKS
Department of Chemical Engineering, Royal Institute of Technology, S-100 44 Stockholm, Sweden

ABSTRACT

Radionuclides escaping from a repository for high level nuclear waste in crystalline rock may eventually be carried by the flowing water in fissure zones. In such zones the rock is broken in blocks of varying sizes and shapes. Also, the water velocity may vary very much along the flow path. The integrated finite difference method (IFDM) is proposed for numerical calculations of radionuclide transport in such zones. A method for lumping blocks of various sizes and shapes into a single PSEUDOBODY is tested by comparing it with an exact analytical solution which can account for the diffusion into blocks of any size distribution. The errors obtained in using the PSEUDOBODY-approach are found to be small. Furthermore, a method for determining an "average" Peclet number in a strongly varying velocity field is tested and found to give comparatively small errors.

Recent investigations give strong indications that water flows in largely isolated channels in fissured rock. The second part of the paper sets out to investigate the radionuclide transport in channels including matrix diffusion. It is shown that a cylindrical hole in a matrix is much more effective, per unit contact area, than a flat surface. Comparison of diffusion from cylindrical and slit-formed channels shows that the differences in interfacial flux are minor. A slit-formed channel may, therefore, be approximated by a cylindrical one. Using the integrated finite difference method a number of radionuclide transport calculations are performed for the case of flow and dispersion in a cylindrical channel coupled to diffusion and sorption in the matrix.

BACKGROUND

The Swedish concept for a repository for spent nuclear fuel is to emplace the fuel in copper canisters and to store these at large depths in crystalline rock. A typical depth considered is 500 m below the ground surface. The bedrock is fissured even at these depths, and water seeps through the fissures. At larger distances (of the order of 1 kilometer), large long zones have been observed, where the rock is more fissured than in the rest of the rock. These zones may be very long and extend to large depths. They are usually vertical or near vertical. These zones are called lineaments.

The repository will be situated in less fissured ("good") rock. If the copper canisters degrade, and water contacts the spent fuel, the radionuclides may be leached by the seeping groundwater. The radionuclides will first move in the "good" rock and then eventually emerge into the lineament. The lineament is more porous and more fissured, and there are also parts which are crushed to small particles. The surface area available for sorption in the lineament may be considerably larger per water volume than in the "good" rock. This may enhance sorption and nuclide retardation. On the other hand, there may be parts of the lineament where the water movement is upwards, and as the seepage rate is larger, the residence time of the water may be shorter than if a continuum were assumed. This gives less time for nuclide interaction with the rock matrix.

Mat. Res. Soc. Symp. Proc. Vol. 50. © 1985 Materials Research Society

This paper sets out to investigate the impact of strongly varying velocities along the flow path. In the paper also the impact of variations in block sizes along the flow path is investigated. It is further attempted to devise a simple but for practical purposes sufficiently accurate numerical scheme to calculate the nuclide movement along the whole flow path.

Water flow in fractured crystalline rock takes place in the fractures or in those parts of the fractures which are open to flow. Several recent investigations indicate that only a small portion of the fractures carry water. Most fractures carry little or no water. Comparisons of the hydraulically conducting zones (from two-packer tests) and the fracture maps (from core logs), have shown that only 20-50% of all fractures carry water [1]. Investigations in the Stripa mine in mid Sweden have shown that the flow in individual fractures takes place in channels [2]. In the program in Switzerland several deep boreholes have been drilled into granitic rock [3]. It was found that at intervals of less than 100 m there were shear zones (kakeritic) which were strongly altered. In these zones, quartz-lined holes were found with openings up to a cm in diameter. Their shape varies from round to slit-like openings. These tubes are deemed to be the major pathways for water flow in the deep lying rock. These observations support a concept of flow in fissured media which is more based on channeling than on the more often used idea of flow in a (more or less) homogeneous medium. In this paper the effect of such channels is also treated.

CONCEPTUAL MODELS

Crystalline bedrock has been observed to be fissured on several scales. There are the large scale lineaments at distances of kilometers apart or more with widths of up to several tens of meters. Their vertical extension can probably be many kilometers and the lineaments can often be followed for many kilometers or tens of kilometers on the ground surface. In the lineaments the rock is broken and crushed and consists of blocks of various sizes. There may be a considerable amount of "fine" material. This has been observed in all the areas investigated in Sweden for the Nuclear Fuel Safety project [1]. The bodies of rock between the lineaments are also fractured, but the fractures are usually at larger spacings and the rock is on the average less permeable [1]. The hydraulic conductivity of the lineaments as well as of the "good" rock decreases with depth. Also the driving force for the water movement - the hydraulic gradient - decreases with depth [4]. Figure 1 gives a view of an idealized lineament.

In principle, it is possible to do rigorous calculations of the three dimensional flow field and to superimpose the transport equations on the flowlines in a full three dimensional calculation, including the effects of a velocity dependent dispersion tensor and the effects of interaction by the radionuclides with rock blocks of different sizes along the flow path. In fact, rigorous three dimensional flow path calculations have been performed for such cases [4] and there are model formulations which account for the above effects for nuclide transport [5,6]. There also seem to be no difficulties in principle to fully incorporate them into existing codes. The major drawbacks at present seem to be that larger computer memories and computing times are needed and that the results may not be quite as transparent as when only a few effects must be accounted for. In this study we attempt to investigate the impact of some simplifying assumptions which lead to considerably simpler computing schemes than the rigorous approach.

The observations of highly localized flow together with other visual observations in Stripa and other tunnels in crystalline rock has led us

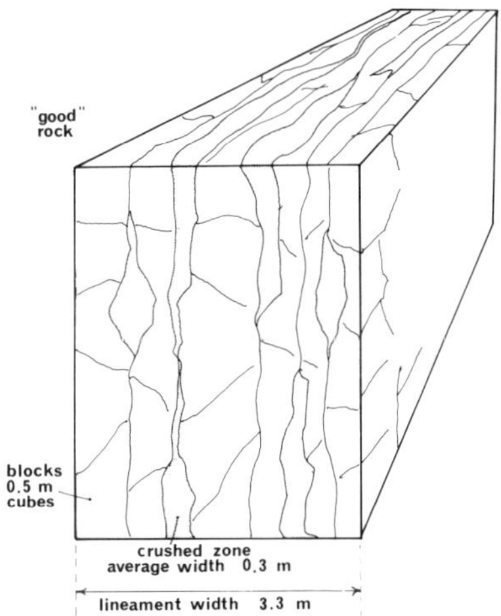

Figure 1. A lineament consisting of larger blocks interspersed with crushed zones.

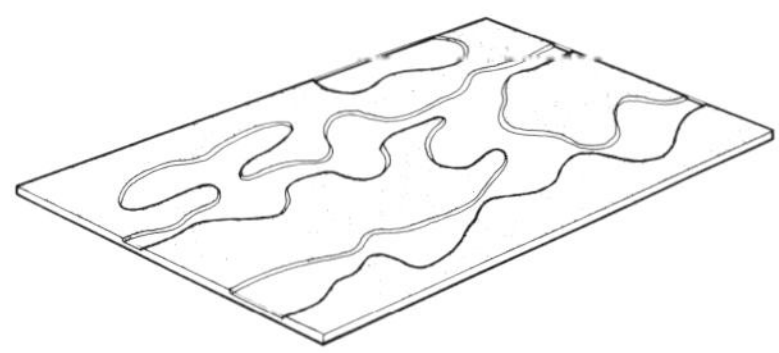

Figure 2. Channels with mobile water within a fracture plane are formed between the areas of contact of the surfaces.

to form an idea of flow in fissured media such as those described which is more based on channeling than on the more often used idea of flow in a (more or less) homogeneous porous medium. The idea is that the water flows in channels which are quite widely separated and which may not for quite considerable distances intersect other channels. The channels are located in fracture planes and may or may not connect to channels in other fracture planes at fracture intersections. The connected channels may extend for considerable distances before they loose their identity by mixing with other channels. The individual channels can vary in shape and size; they may be circular tubes at some location and then widen into planar slits with constantly changing widths and openings. The pressure drop is caused by friction in the narrower parts of the conduit and also by fracture filling materials and surface roughness. The flow porosity (channel volume) is not directly related to the pressure drop. Figure 2 shows the main notions of the concept.

Transport of tracers or radionuclides in such conduits may then be described by the conventional advection-dispersion equation with matrix diffusion effects added. The surface area where flowing water is in contact with the rock surface is much less than if it is assumed that all the surfaces of the rock blocks are actively participating in the exchange. The geometry for the diffusion from a conduit is also quite different from that for diffusion into a block from all sides. In the first case the original contact area is small but the further the migrating species is from the surface the larger the cross section for transport becomes. For the diffusion into blocks, the situation is the opposite.

MATHEMATICAL MODELS

For flow and sorption from the flowing water we have:

$$R_a \frac{\partial C_f}{\partial t} + V \frac{\partial C_f}{\partial z} - D_L \frac{\partial^2 C_f}{\partial z^2} = - S_k - R_a \lambda C_f \quad (1)$$

The terms in this equation represent accumulation in the water in the fissure and on the interface to the solid rock, advective transport, transport by axial dispersion, exchange with the rock matrix, and radioactive decay.

The term S_k accounts for the depletion of the nuclide in the water due to diffusion into the porous rock. For a discrete block-size distribution S_k becomes:

$$S_k = - \frac{1}{m} \sum_i \frac{\alpha_f^i + 1}{b^i} F(b^i) D_p^i \varepsilon_p^i \left(\frac{\partial C_p}{\partial r}\right)_{r=b^i} \quad (2a)$$

The quantity $(\alpha_f^i + 1) F(b^i)/mb^i$ is the total outer particle surface of size group "i" per unit bed-void volume. α_f is 0, 1 or 2 for slabs, cylinders or spheres respectively. For the individual size fraction "i" the transport equation in the matrix becomes:

$$K^i \frac{\partial C_p^i}{\partial t} = \frac{1}{r^{\alpha_f^i}} \frac{\partial}{\partial r} \left(r^{\alpha_f^i} D_p^i \varepsilon_p^i \frac{\partial C_p^i}{\partial r}\right) - K^i \lambda C_p^i \quad (3)$$

In this equation the terms give sorption on the interior surfaces and accumulation in the pore fluid, diffusion in the pore fluid, and radioactive decay.

For a cylindrical channel S_k is given by:

$$S_k = - \frac{2}{a} D_p \varepsilon_p \left(\frac{\partial C_p}{\partial r}\right)_{r=a} \quad (2b)$$

and equation (3) holds with $\alpha_f = 1$.

Equation (3) is coupled to equation (1) by the relation:

$$C_p^i (b, z, t) = C_f (z,t) \quad (4)$$

The surface and volume retardation coefficients are as defined by Rasmuson and Neretnieks [7].

To evaluate S_k in equation (2a) it is in principle necessary to evaluate equation (3) the same number of times, N, as there are size fractions if conventional numeric methods are used to solve the equations. To avoid this we have introduced the PSEUDOBODY-concept which lumps all block sizes and shapes into one single block. Only one equation for transport in the blocks is then needed.

The basic assumptions for lumping the various block sizes into one shape seem to have first been proposed by Pruess and Narasimhan [8]. They called it the MINC (Multiple Interacting Continua)-approach. In the following we denote this approach the PSEUDOBODY approach. In this method the blocks in an element of space are divided into a number of shells and it

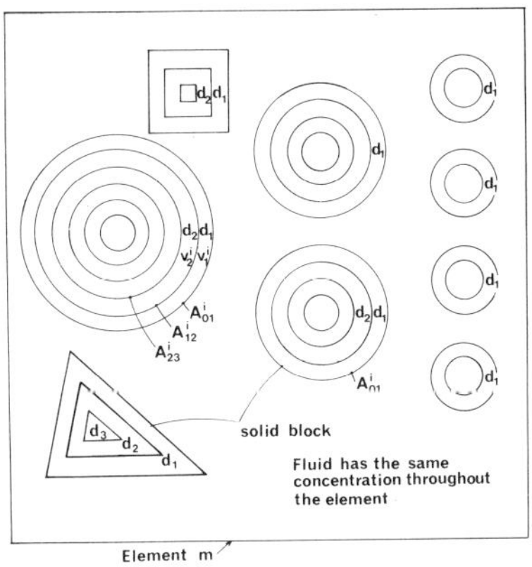

Figure 3a. The principle of gathering all shells of equal distance from the surface into one volume.

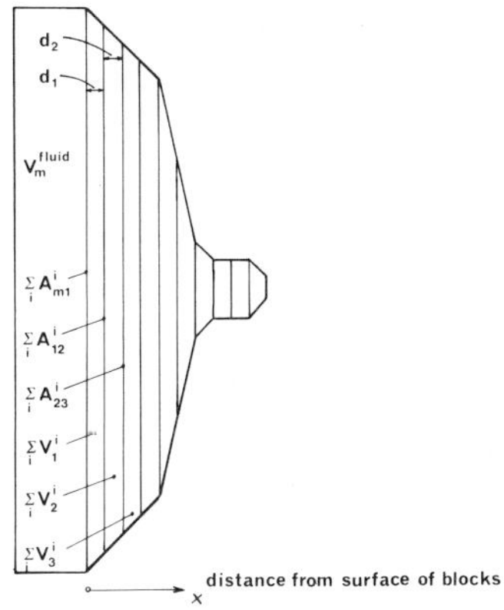

Figure 3b. The sum of the areas and volumes of the blocks for various distances from the surface for the pseudobody.

is assumed that all the shells, regardless of blocksize and shape, at the same distance from the surface behave alike (Figure 3). This is of course an approximation because the shape of the blocks influences the transport in a shell at a given distance, and even for the same shape a shell at a given distance reacts differently if the area ratio of adjacent shells is different [6].

The PSEUDOBODY-assumption makes it possible to construct equally sized pseudoblocks having a given area A(x) for diffusion at a distance x from their surface. The pseudoblocks have the same A(x) and the same volume as the real block size distribution in the element. For these pseudoblocks the transport equation (3) can be written

$$K \frac{\partial C_p}{\partial t} = \frac{1}{A(x)} \frac{\partial}{\partial x} (D_p \varepsilon_p A(x) \frac{\partial C_p}{\partial x}) - K \lambda C_p \qquad (5)$$

A(x) can be interpreted as the crossectional area normal to the transport direction in a pseudoblock at a distance x from the block´s surface. For real regular bodies like infinite slabs, cylinders and spheres $A(x) \propto r^{\alpha_f}$ if the distance from the surface is exchanged for the radius. Equation (5) can now be used once instead of using equation (3) N times to evaluate S.

Finally, in this section, it is noted that if we follow a streamtube, where $V = V(z)$ and $D_L = \alpha V(z)$ [9], equation (1) still holds.

RESULTS AND DISCUSSION

To investigate the accuracy of the PSEUDOBODY-approach an exact analytical solution of (1), (2a) and (4), for constant velocity V, was derived [10]. This solution is not given here. The two methods are compared by investigating the differences in the effluent concentration of Np-237 for some sample cases. The data are summarized in Table 1.

The physical situation modelled is the flow in a lineament which contains two blocksizes and is also in contact with the surfaces of the

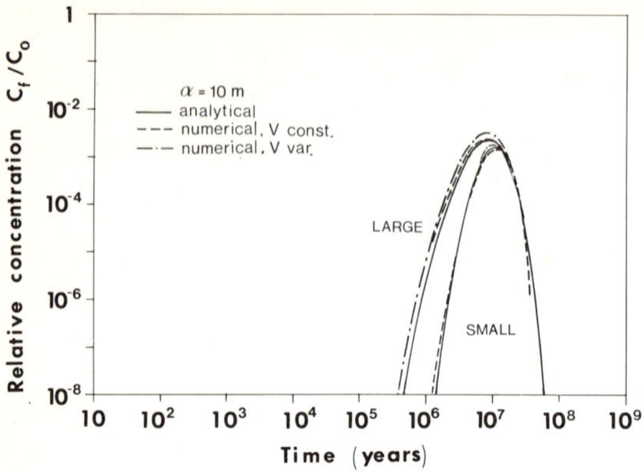

Figure 4. Comparison between numerical PSEUDOBODY calculations (V = const. and V = variable respectively) with analytical solution with averaged (equation (7)) Peclet number but V = const. α = 10 m.

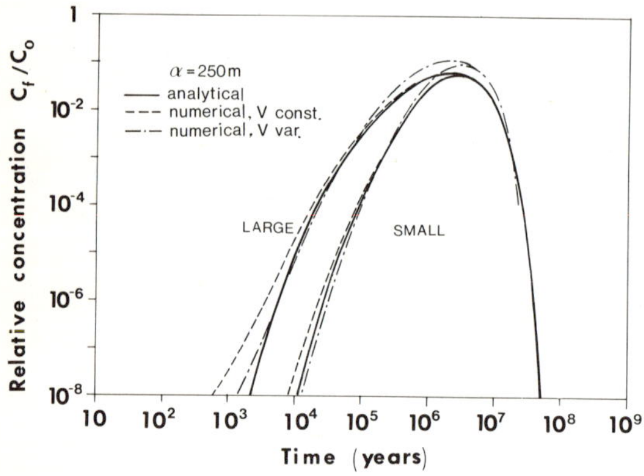

Figure 5. Same as Figure 4 but α = 250 m.

"good" rock on both sides of the lineament (Figure 1). Note that in the analytical solution, the blocks are approximated by spheres. The PSEUDOBODY approach does not distinguish between cubes and spheres. Two block size distributions are used. In the one called "Small" the fines are made up of 0.02 m blocks. In "Large" the fines consist of 0.1 m blocks. In all other respects the two distributions are alike.

In the example the water velocity and dispersivity are assumed to be constant in order to utilize the analytical solution exactly. The initial and boundary conditions used assume zero initial concentration of Np-237 and at times larger than zero the inlet condition is a decaying band release with infinite duration. $C_f(z = 0) = C_0 \cdot e^{-\lambda t}$ for $t \geq 0$.

The results are depicted in Figures 4 and 5. The agreement is surprisingly good. The PSEUDOBODY approach tends to give somewhat earlier breakthrough in this case compared to the exact solution, but the peak is

Table 1a. Block-size data used for comparison of exact solution and PSEUDOBODY approach.

Block size diameter m		Number of blocks in REV*	Volume fraction $F(b^1)$	Outer surface area $A(b^1_2)$ in REV m^2
Case "small"	0.02	9375	0.0909	22.5
	0.5	6	0.9091	9
	∞ (200 m)**	(2)	-	0.5
Case "large"	0.1	75	0.0909	4.5
	0.5	6	0.9091	9
	∞ (200 m)**	(2)	-	0.5

* The representative elementary volume (REV) is 3.3 m wide, 0.5 m high and 0.5 m long. It also includes two surfaces from which diffusional transport takes place into two very large blocks.

** In the exact solution the very large blocks are modelled by a part of a large sphere (diam = 200 m). Since the penetration depth is very small in comparison to the radius the flat walls adjacent to the lineament are accurately accounted for. The volume of these blocks is not included in the REV.

Table 1b. Data used comparing exact and PSEUDOBODY solutions.

Parameter	Dimension	Value
Travel distance, L	m	475
Water residence time, t_w	s	$7.8125 \cdot 10^7$
Peclet number, Pe	-	21.875 and 0.875
Surface retardation coefficient, R_a	-	1
Volume equilibrium coefficient, K	m^3/m^3	$1.35 \cdot 10^4$
Effective diffusivity in blocks, $D_p \varepsilon_p = D_e$	m^2/s	$5 \cdot 10^{-14}$
Half life of Np-237, $T_{1/2}$	years	$2.14 \cdot 10^6$
Flow porosity, ε_f		$5 \cdot 10^{-4}$

Table 1c. Additional data for the varying velocity case.

Parameter	Dimension	Value
Dispersion lengths, α	m	10 and 250
Hydraulic gradient, i	m/m	1/Z for Z>25
Water flux, Q	$m^3/m^2 \cdot s$	$0.1/Z^3$ for Z>25
Water residence time, t_w	s	$7.8125 \cdot 10^7$

very accurately predicted. It may be noted that there is a noticable difference between the two cases "Large" and "Small" as there is an impact of the size of the fines in this case. Fines mainly influence the early part of the breakthrough curve.

The analytical solution used in the previous analysis is derived for the case of constant velocity. This solution is also exact for the case with no dispersion but variable velocity. Neretnieks and Rasmuson [6] showed this by introducing

$$\zeta = \int_0^z \frac{dz}{V(z)} \tag{6}$$

to transform the velocity into a residence time.

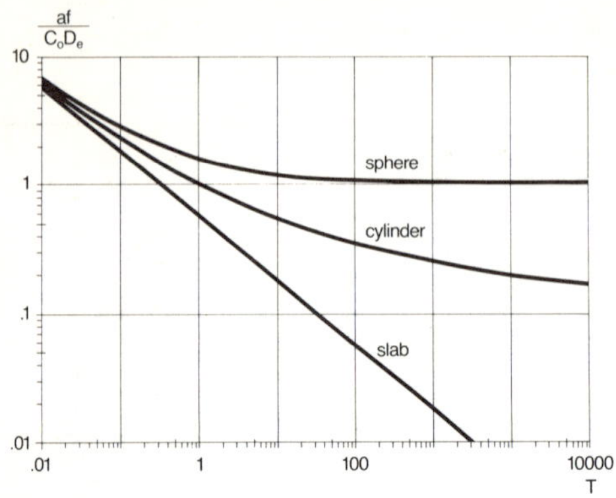

Figure 6. Flux, f, at the surface of regions bounded internally by a circular cylinder or sphere of radius a and semi-infinite slab, with zero initial concentration and constant surface concentration C_0.

When Fickian type dispersion cannot be neglected numeric methods must be used at present in cases where the velocity varies along the flow path. The longitudinal dispersion may be approximately accounted for in a varying flow field, however, by using the following averaging method [6]:

$$Pe = \frac{t_w^2}{\alpha \int_0^L \frac{dz}{[V(z)]^2}} \tag{7}$$

The point-dashed curves in Figures 4-5 are numerical calculations where the velocity varies along the flow path according to $V(z) = 200/z^3$. The Peclet number in the analytical calculation is chosen according to equation (7). Considering that the averaging has been done over a velocity ratio span of 8000, the differences are surprisingly small.

Two conclusions are drawn from the results. The PSEUDOBODY approach should be a powerful tool in numeric calculations with block size distributions, especially where chain decay must be accounted for, as compared to "exact" numeric calculations where every block size and shape must be accounted for separately. This will significantly reduce the computing effort. For single species the analytical solution should provide a powerful tool when very accurate calculations are needed.

In Figure 6 the diffusional fluxes from the surface of the region bounded internally by a cylinder with the region bounded internally by a sphere and the semi-infinite solid are compared. It is seen that for short times all the fluxes are identical. This is to be expected since only a thin layer adjacent to the source is affected by diffusion. For large values of T the cylinder flux tends to zero as 2/(a ln T) while the sphere flux tends to a constant 1/a.

However, both are much more effective than the slab (with the same interfacial area) where the flux tends to zero as $1/\sqrt{t}$. This is due to the fact that for divergent systems like the cylinder and the sphere a much larger volume is affected at longer times. A further illustration to the difference between diffusion into a semi-infinite solid from a flat

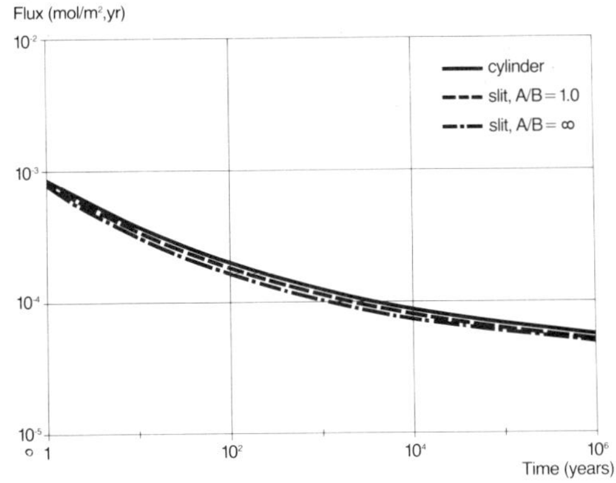

Figure 7. Flux, f, at the surface of the region bounded internally by a cylinder and by slits with various length/breadth ratios.

surface and a cylindrical region is obtained by noting that the latter case is analogous to the case of diffusion into a semi-infinite flat solid ($x \geq 0$) with linear increasing diffusivity $D = D_0(1+\beta x)$.

It is also of interest to compare diffusion from cylindrical and slit-formed channels. Choosing the same interfacial area we would expect that the uptake should be the same at short times since only the surface region is affected. For larger penetration the concentration profiles, from the slit, are governed by the length/breath ratio of the slit source. However, for large penetration, the uneven concentration distribution is smoothed and approximately resembles the one from a cylindrical source. Comparing fluxes from slits of various length/breath ratios with the cylindrical source we would therefore expect deviations at intermediate times only.

Calculations, for the slit, were made with the computer program TRUMP [11]. A mesh generator was written to define the subvolumes and their interconnections. Calculations were made with the interfacial area $\pi \cdot 10^{-2}$ m^2/m corresponding to a cylindrical tube with diameter 10^{-2} m. The diffusivity, $D_p \varepsilon_p$, is $5 \cdot 10^{-14}$ m^2/s and K = 1.0. The length/breadth ratios were 1.0, 2.0, 4.0, 16.0 and infinity. In Figure 7 the fluxes f in mol/m^2, yr are plotted versus time (in years). The concentration C_0 at the surface is set to 1 mol/m^3. It is seen that the differences are surprisingly small.

The data used in numerical calculations for combined channel flow and matrix diffusion are given in Table 2. The calculations are done for Np-237. The density of cylindrical tubes is assumed to be one tube/m^2 cross-sectional area. This is based on the experimental observations in the Swedish KBS-3 study [1]. In the bore-holes it was found that, on the average, the fracture spacing is 1 m. However, only every fifth fracture carried any water. The assumption taken here is that all fractures transport water but only in channels in the fracture plane. The experimental observation could then be explained by channels 0.2 m wide separated by impermeable zones 1 m wide.

As the geometrical shape of the channel seems to be of minor importance (compare with Figure 7) we approximate the flat channel with interfacial area 0.4 m^2/m with a cylindrical channel with the same contact area giving $a = 6.3662 \cdot 10^{-2}$ m. Diffusional transport in the matrix is then treated as 1-D instead of 2-D which gives a considerable numerical

Table 2. Data used in numerical calculations of flow in cylindrical tubes with matrix diffusion.

Parameter	Dimension	Value
Travel distance, L	m	100
Tube radius, a	m	$5.0 \cdot 10^{-3}$ and $6.3662 \cdot 10^{-2}$
Volumetric flow rate, Q	ℓ/m^2,yr (ℓ/tube,yr)	0.1 and 1.0
Peclet number, Pe	-	10
Surface retardation coefficient, R_a	-	1.0
Volume equilibrium constant, K	m^3/m^3	$1.35 \cdot 10^4$
Effective diffusivity in rock matrix, $D_p\varepsilon_p$	m^2/s	$5.0 \cdot 10^{-14}$
Half-life of Np-237, $T_{1/2}$	years	$2.14 \cdot 10^6$

simplification. The value $a = 5.0 \cdot 10^{-3}$ m is taken as an extreme case. In this case the interfacial area is $3.14 \cdot 10^{-2}$ m^2/m only.

The values of the flow rate are typical for undisturbed rock (Q = 0.1 ℓ/m^2,yr) and a crushed zone (Q = 1.0 ℓ/m^2,yr) respectively. The value for undisturbed rock is based on the Swedish KBS-3 study [1]. The flow rate in the crushed zone is assumed to be higher due to higher hydraulic conductivity [6]. Finally, the values of the volume equilibrium constant, K, and the effective diffusivity, $D_p\varepsilon_p$, are as given in the KBS-3 study [1].

The results for Q = 0.1 ℓ/m^2,yr ("undisturbed rock") and Pe = 10 are depicted in Figure 8 and for Q = 1.0 ℓ/m^2,yr ("crushed zone") and Pe = 10 in Figure 9. For comparison the breakthrough curves with surface retardation only are also shown as well as a calculation where the rock is assu-

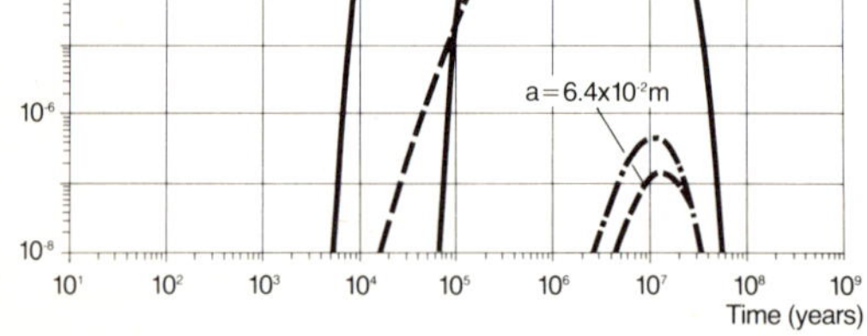

Figure 8. Flow in cylindrical fractures with radius $a = 5 \cdot 10^{-3}$ and $6.3662 \cdot 10^{-2}$ m and matrix diffusion (solid lines) and surface retardation (dashed lines). Q = 0.1 ℓ/m^2,yr and Pe = 10. Case of cubical system of orthogonal fractures (Pe=2) also shown (point-dashed lines).

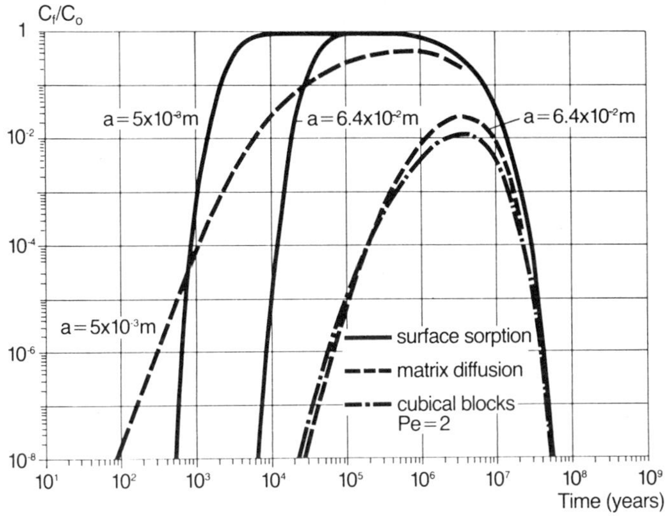

Figure 9. Same as Figure 8 but Q = 1.0 ℓ/m^2,yr.

med to consist of a cubic system of orthogonal fractures. The surface retardation coefficient is obtained from:

$$R_a = \left(\left(1 + \frac{\eta}{a}\right)^2 - 1\right) K + 1 \tag{8}$$

where η is the penetration depth. Under the assumption that η = 0.1 mm, that is 0.1 mm of the rock adjacent to the channel is instantaneously equilibrated, we get R_a = $5.4640 \cdot 10^2$ (a = $5.0 \cdot 10^{-3}$ m) and R_a = $4.3445 \cdot 10^1$ (a = $6.3662 \cdot 10^{-2}$ m). The cubical blocks are approximated by spheres having the same area-to-volume ratio. Assuming S = 5 m gives r_0 = 2.5 m. The interfacial area is 1.2 m^2/m. The Peclet number is 2 in these calculations indicating a system with large hydrodynamic dispersion.

It may be seen that matrix diffusion from the cylindrical channel has a large impact on the breakthrough curve. If surface retardation only is considered, in effect the breakthrough is delayed but no significant reduction in effluent concentration is obtained. In the case of matrix diffusion the negative effect of a small surface area is substantially compensated for by diffusion into the cylindrical matrix and the peak-heights are much lower. Note again that for the case of spherical blocks Pe = 2. For Pe = 10 the peak-heights are much lower, < 10^{-11} (Q = 0.1 ℓ/m^2,yr) and $6 \cdot 10^4$ (Q = 1.0 ℓ/m^2,yr). Comparing Figure 8 and Figure 9 it is seen that the flow rate has a large impact. Additional results are given in the paper by Rasmuson and Neretnieks [12].

NOTATION

$A(x)$	crossection for diffusion at distance x from block surface	L^2
a	radius of cylindrical channel	L
b	block radius	L
C_f	concentration in water in fissures	M/L^3
C_p	concentration in water in microfissures	M/L^3
C_o	inlet concentration in the water	M/L^3
D_a	$= (D_p \varepsilon_p)/K$, apparent diffusivity in microfissures	L^2/T
D_e	$= D_p \varepsilon_p$, effective diffusivity in microfissures	L^2/T
D_L	longitudinal dispersion coefficient	L^2/T
D_p	diffusivity in water in pores	L^2/T
$F(b^i)$	volume fraction of blocks with radius b^i	
f	molar flux from flowing water to rock matrix	$M/L^2,T$
i	hydraulic gradient	L/L
K	volume equilibrium constant	L
L	travel distance	L
m	$= \varepsilon_f/(1-\varepsilon_f)$	
Pe	$= (zV)/D_L$, Peclet number	
Q	water flow rate	$L^3/L^2,T$
R_a	surface retardation coefficient	
r	radial coordinate	L
r_o	effective spherical radius	L
S	fissure spacing	L
S_k	reaction rate due to diffusion into block	M/L^3 T
T	$= (D_a t)/a^2$	
$T_{1/2}$	half life	T
t	time	T
t_w	water residence time	T
V	average velocity of water in fissures	L/T
x	radial distance from surface of block	L
Z	depth below ground level	L
z	distance in flow direction	L

Greek letters

α	dispersion length	L
α_f	form factor; 0, 1 and 2 for slab, cylinder and sphere, respectively	
ε_f	porosity of fissures	
ε_p	porosity of rock matrix	
ζ	residence time, defined in equation (6)	T
η	penetration depth	L
λ	decay constant of radionuclide	T^{-1}

Superscript

i	block size group

REFERENCES

1. KBS - Nuclear Fuel Safety Project, Final storage of spent nuclear fuel, KBS-3, Technical Report, Stockholm (1983).
2. H. Abelin, J. Gidlund, L. Moreno and I. Neretnieks, Scientific Basis for Nuclear Waste Management VII, Boston 1983. Proceedings North-Holland 26, 239 (1984).
3. NAGRA - Nationale Genossenschaft für die Lagerung radioaktiver Abfälle, Projektbericht NGB 85-04 (1985).
4. L. Carlsson, A. Winberg and B. Grundfelt, KBS Technical Report 83-45 (1983).
5. R.J. Hopkirk and D.J. Gilby, Third International Conference on Numerical Methods in Thermal Problems, Seattle 1983. Proceedings (1984).
6. I. Neretnieks and A. Rasmuson, Water Resour. Res. 20, 1823 (1984).
7. A. Rasmuson and I. Neretnieks, Submitted to Water Resour. Res., Nov. 30 (1984).
8. K. Pruess and T.N. Narasimhan, Sixth Soc. Petr. Eng. of AIME Symp., New Orleans 1982. Proceedings 269 (1982).
9. J.J. Fried and M.A. Combarnous, Adv. Hydrosci. 7, 170 (1971).
10. A. Rasmuson, Chem. Engng. Sci. 40, 621 (1985).
11. A.L. Edwards, TRUMP: A computer program for transient and steady state temperature distributions in multidimensional systems. National Technical Information Service, National Bureau of Standards, Springfield Va., (1972).
12. A. Rasmuson and I. Neretnieks, Manuscript to be submitted to Water Resour. Res. (1985).

TRANSPORT OF RADIONUCLIDES IN INHOMOGENEOUS CRYSTALLINE ROCKS

JOERG HADERMANN*, FRITZ ROESEL*+ AND CHARLES MCCOMBIE**
* Swiss Federal Institute for Reactor Research, CH-5303 Wuerenlingen
** Nagra, National Cooperative for the Storage of Radioactive Waste, CH-5401 Baden, Switzerland
+ Present address: University Computer Center, CH-4056 Basel, Switzerland

ABSTRACT

In Switzerland high-level radioactive wastes are planned to be disposed of in deep-lying crystalline formations. The modelling approach for geospheric transport in the framework of a HLW safety assessment is described. The processes taken into account are advection and dispersion in water conducting veins or fractures. Out of these, nuclides may diffuse into a spatially limited zone of altered rock. Sorption of nuclides on the rock is taken into account. Diffusion out of fractures and sorption in the adjacent altered rock may often be represented by an effective surface sorption on the fracture walls. For transport of the 237-Np nuclide chain we present a sensitivity analysis. The performance of the geologic barrier depends most critically on the extent of the altered zones.

THE MODELS

The host rock under consideration in Switzerland for disposal of high-level wastes is crystalline overlayed by a series of sedimentary rocks. Deep drill hole measurements show [1,2] that the crystalline is inhomogeneous and three different zones can be distinguished.

First are the water conducting structures. These come essentially in two forms: (i) Veins in zones of crushed and altered granites (kakirites). The kakirites originate from the global stress field and form essentially planar structures of appreciable width with well defined dip and strike angles. The water carrying veins are modelled as an idealized network of tubes. (ii) Fractures in aplite/pegmatite dykes forming intrusions in the granite. The fractures are modelled as network of parallel plates. In these zones with two different geometries, we take into account advection of radionuclides, hydrodynamic dispersion and sorption on their surfaces.

The second type are altered zones adjacent to the water conducting zones. In these, advection of contaminant can be neglected. However, diffusion and sorption in the altered rock are important processes.

The third, and generally largest part of deep crystalline is intact rock. In view of its much lower hydraulic conductivity [2], its much lower porosity [2] and the lack of an unambiguous validation of matrix diffusion in undisturbed rock (see e.g. [3]), we conservatively neglect this part of the crystalline rock in transport calculations for safety purposes. The present modelling approach differs from previous ones (e.g. [4,5]) in that a differing geometry for the advective transport in veins embedded in an extended zone of altered crystalline (kakirites) and diffusion only into the limited zones of altered rock (limited matrix diffusion) is considered.

Considering the mass balance for a representative elementary volume the transport equations for a decay chain of radionuclides 1 --> 2 --> 3 ... can be written down (for details see ref. [6]):

$$R_f^i \frac{\partial}{\partial t} C_f^i = L_f (C_f^i) + S - \lambda_i R_f^i C_f^i + \lambda_{i-1} R_f^{i-1} C_f^{i-1} \qquad (1)$$

for transport in the water conducting zones and

$$R_p^i \frac{\partial}{\partial t} C_p^i = L_p (C_p^i) - \lambda_i R_p^i C_p^i + \lambda_{i-1} R_p^{i-1} C_p^{i-1} \tag{2}$$

for diffusive transport in the altered zones. Here, i is the index for nuclides in a linear decay chain, f and p the indices characterizing quantities in the water conducting systems and in the altered zones, respectively. C is the nuclide concentration in liquid phase and λ the decay constant. The following quantities have been defined:

$$R_f^i = \begin{cases} 1 + (2/R)\, K_a^i & \text{for the system of veins} \quad (3a) \\ 1 + (1/b)\, K_d^i & \text{for the system of fractures} \quad (3b) \end{cases}$$

$$R_p^i = 1 + (1+\varepsilon_p)\rho\, K_d^i / \varepsilon_p \tag{4}$$

are retardation constants from sorption (R being the vein radius, b the fracture half-width, ε the porosity and ρ the rock density of the altered zone, and K the usual sorption distribution constants).

The differential operator describing dispersive-advective transport in the water conducting systems is given by

$$L_f = a_L v_f \partial^2 / \partial z^2 - v_f \partial / \partial z \quad , \tag{5}$$

where a_L is the longitudinal dispersivity and v_f the water velocity along a water trajectory described by the coordinate z, and for diffusive transport in the altered zones by

$$L_p = \begin{cases} D_p \partial^2 / \partial r^2 + D_p (1/r)\, \partial / \partial r & \text{for the system of veins} \quad (6a) \\ D_p \partial^2 / \partial x^2 \, , & \text{for the system of fractures ,} \quad (6b) \end{cases}$$

where D_p is the diffusion constant in the altered zones, r the radius perpendicular to a vein and x the coordinate perpendicular to a fracture.

The term S ensures flux continuity at the interface altered zone / water conducting zone:

$$S = \begin{cases} (2/R)\, \varepsilon_p D_p \, \partial C_p^i / \partial r \Big|_{r=R} & \text{for transport in veins} \quad (7a) \\ (1/b)\, \varepsilon_p D_p \, \partial C_p^i / \partial x \Big|_{x=b} & \text{for transport in fractures.} \quad (7b) \end{cases}$$

The last two terms, finally, describe radioactive decay and build-up.

In deriving eqs. (1) and (2) a series of approximations has been made, especially (i) transport of each radionuclide can be described by transport of an effective single species, (ii) a linear sorption equilibrium isotherm, (iii) one-dimensional transport along a water trajectory neglecting transversal dispersion, (iv) piecewise constant parameters with no variability in time. The admissibility of these approximations - as well as the basic assumptions on relevant transport processes - has to be validated. This has been done partially, only, and constitutes a major task for future research.

Eqs. (1) and (2) have to be solved with appropriate boundary conditions (the initial condition being $C = 0$ for $t \leq 0$): (i) At the interface water conducting/altered zone we have continuity of concentration. (ii) At the geosphere inlet we take

$$C_f^i (0,t) = C_0^i(t) \quad \text{and} \quad C_p^i = 0 \text{ for } r > R \, / \, x > b \, , \tag{8}$$

where $C_0^i(t)$ is determined from the near-field model (e.g. a solubility limit). (iii) At the outlet we take C = 0 modelling a considerable dilution in an aquifer. (iv) Diffusion into the altered zone is limited by the extent of that zone and hence we have, depending on the geometry

$$\partial C_p^i / \partial r \Big|_{r=D} = 0 \quad \text{or} \quad \partial C_p^i / \partial x \Big|_{x=D} = 0 \ , \tag{9}$$

where D is half the width of the kakirite (assuming sufficiently spacing of the veins) or the width of the altered zone adjacent to the fractures, respectively.

Provided the following relation holds for transport in the fractures

$$\left| (1/6)\ (R_p / D_p)\ D^2\ \partial \{ \ln (C_f e^{\lambda t}) \} / \partial t \right| << 1 \tag{10}$$

the kinetic process of diffusion into and instantaneous sorption in the altered zone can be approximated by an effective surface sorption: The term S in eq. (1) can be dropped substituting the retardation factor, eq. (3b), by

$$R_f^i = 1 + (1+b) \{ K_a^i + (1 - \varepsilon_p)\, \rho\, D\, K_d^i \} \ . \tag{11}$$

This reduces the numerical expenditure appreciably.

NUMERICAL RESULTS AND DISCUSSION

Especially because of the influence of radioactive decay the effectiveness of a particular physico-chemical mechanism and the results of transport calculations are extremely dependent on parameter choices. A large body of calculations has been done for the safety analysis within "Project Gewaehr 1985" [7] for the Swiss high-level waste repository in deep crystalline. Where data were available site specific parameters have been chosen [1]. We present here some of the results [6] for transport of the important 237-Np chain. The nuclide independent parameters are given in table I, the nuclide dependent parameters are shown in table II. The numbers given are judged to be realistic but on the conservative side; parameters for variations are indicated in parentheses.

Table I. Nuclide independent parameters for the geosphere transport calculations. The numbers in parentheses define parameter variations of a base case.

Water flow through the repository $[m^3/yr]$	4.2
Hydraulic conductivity of kakirites/dykes [m/s]	1.25 E-9
Hydraulic gradient	0.012
Vein frequency $[1/m^2]$	1 (9)
Fracture frequency [1/m]	10 (1)
Vein radius R [m]	5 E-3 (5 E-4)
Fracture half-width b [m]	5 E-5
Extent of altered zone D [m] a) kakirites	0.5 (0.1 - 0.4)
b) dykes	1 E-3 (0.01,0.1)
Density of altered zone ρ $[kg/m^3]$	2160
Porosity of altered zone ε	0.033
Diffusion constant in altered zone D_p $[m^2/s]$	1.515 E-10 (1.515 E-11)
Dispersion length a_L [m]	50 (250)
Migration distance [m]	500

Table II. Nuclide dependent parameters for the geosphere transport calculations.

Nuclide	Half-life [yr]	Sorption constant [m^3/kg]
237-Np	2.14 E+6	1 (0.1)
233-U	1.592 E+5	1 (0.05)
229-Th	7.34 E+3	1 (0.01)

With respect to the effect of retardation by sorption in the altered zones, sorption on surfaces of water conducting structures can be neglected. The boundary concentration, eq. (8), is given in Fig. 1 taken from ref. [8]. The results are presented as concentration distribution (Fig. 2) or nuclide flow into an aquifer after 500 m migration normalized to the water flow through the repository (Figs. 3 to 7).

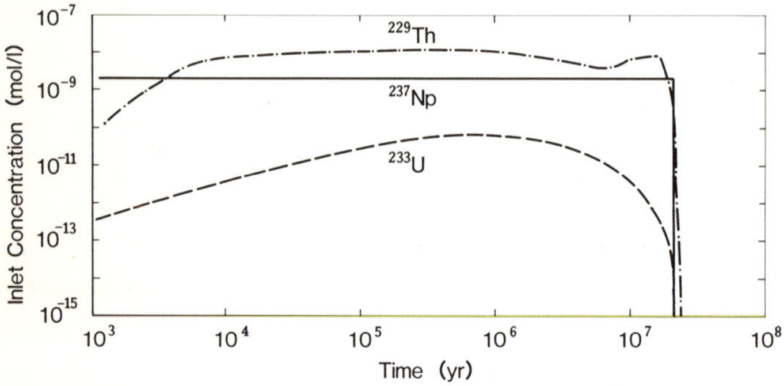

Figure 1. Inlet boundary concentrations for the 237-Np chain as a function of time after emplacement into the repository. Canister life-time is 1000 years.

Transport in kakirites

Fig. 2 shows the concentration of 237-Np in the water conducting veins at various times as a function of distance from the geosphere inlet for base case parameters. Only negligible amounts of 237-Np reach the aquifer. For these parameters the inhomogeneous crystalline represents an extremely efficient geologic barrier. The mechanism giving rise to this result is diffusion into and sorption in the altered zone. This is exemplified in Fig. 3 where the kakirite width has been varied. The parameters are those of the base case except for the sorption distribution constant where more conservative values have been chosen. For a model with limited matrix diffusion the spatial extent of the altered zone is an extremely important parameter when transport times are larger than half-lives.

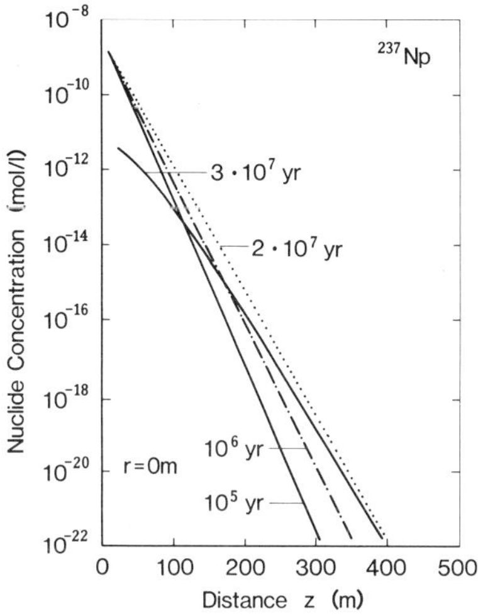

Figure 2. Concentration of 237-Np in the water conducting kakirite veins as a function of distance from the inlet for various times after canister failure. The parameters are those of the base case.

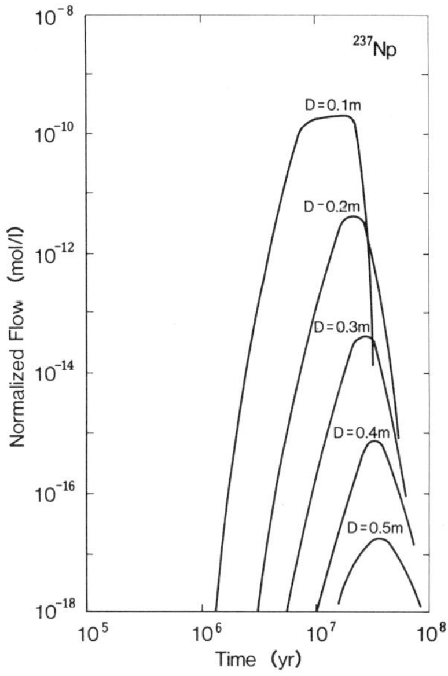

Figure 3. Effect of kakirite width on normalized 237-Np flow after a migration distance of 500 m. The parameters are those of the base case except for the sorption distribution constants (low values) and for the kakirite half-width which is indicated.

Taking the lower value for the diffusion constant in the altered zone yields the normalized nuclide flows presented in Fig. 4. The porous kakirite is less rapidly saturated with nuclides and, correspondingly, retardation is smaller than in the base case. Because 229-Th is released at considerable quantities into the geosphere, up to 100'000 years the inlet contribution can clearly be seen. For larger times all daughters are in radioactive equilibrium to their precursor.

Fig. 5 shows the results of a combined parameter variation: both for sorption distribution constants and for the diffusion constant in the altered zone the lower values have been taken. At times larger than 10 million years radioactive equilibrium is seen. At smaller times the source term contribution to outflowing 229-Th is pronounced as a result of low sorption for this element. In this context it is interesting to note that because of diffusion into the altered zone break-through time even for an inert tracer would be 2'400 yr whereas water transit time is only 83 years.

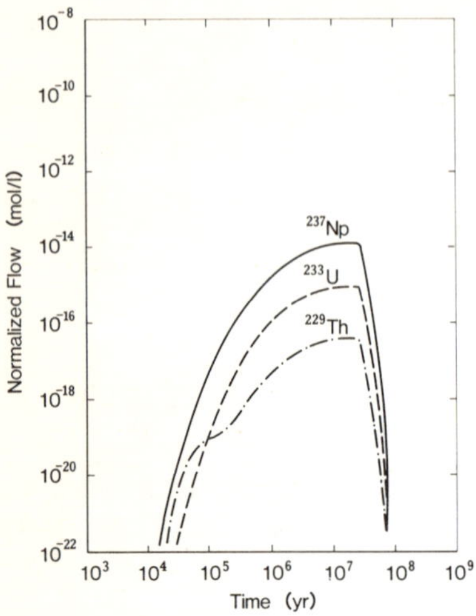

Figure 4. Normalized nuclide flow after a migration distance of 500 m in kakirites. The parameters are those of the base case except for the diffusion constant in the kakirite zone (low value).

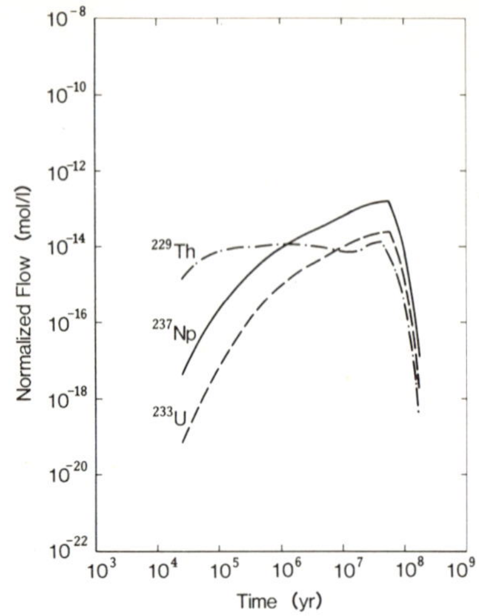

Figure 5. The same as figure 4. The parameters are those of the base case with exception of sorption distribution constants (set of low values) and the diffusion constant in the kakirite zone (low value).

Transport in fractured dykes

In this subsection we compare the results for the full solution of limited matrix diffusion, eqs. (1) and (2), to the effective surface sorption approximation, eqs. (1) and (11). In the rising part of the curves the results for effective surface sorption lie consistently lower since kinetic effects of diffusion in the altered zone are neglected. However, for the parameters chosen effective surface sorption constitutes an excellent approximation.

Fig. 6 shows normalized nuclide flows for the base case parameters. The geologic barrier performs much less effectively than for transport in the kakirite zones. This is caused by the much smaller extent of the altered zone adjacent to the fractures compared to the width of the kakirites. Hence, limited matrix diffusion is less effective. The daughters of 237-Np are in radioactive equilibrium and the calculations with an effective surface sorption are virtually indistinguishable from those with kinetic matrix diffusion.

Fig. 7 shows the results of a combined parameter variation: a decreased fracture frequency (increasing water velocity by a factor of 10), a low diffusion constant in the altered zone and the set of low sorption distribution constants. Retardation is small in this case and 237-Np flow is not lowered by geologic transport. For 229-Th the inflow structure can clearly be seen, decay leading to a lowering of roughly an order of magnitude. Here, again, effective surface sorption is an excellent approximation.

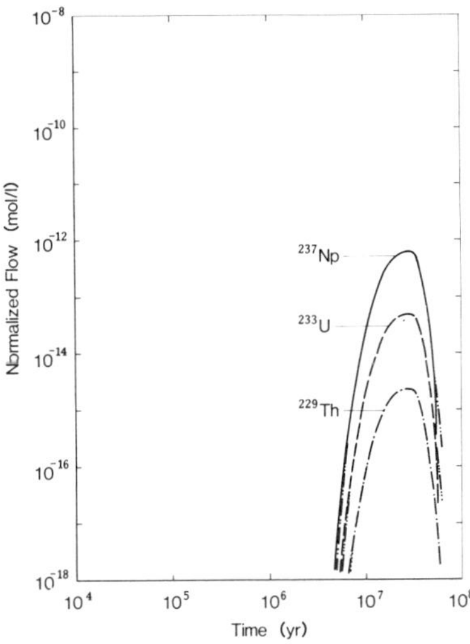

Figure 6. Normalized nuclide flows after a migration distance of 500 m in fractured dykes. The dash-two-dots, dash-three-dots and fully dotted lines are those for the effective surface sorption approximation, the parameters those of the base case.

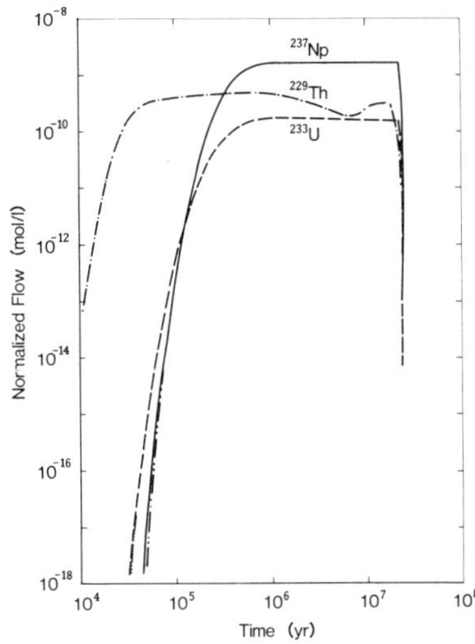

Figure 7. The same as figure 6. The parameters are those of the base case except for the fracture frequency (low value), diffusion constant (low value) and sorption constants (set of low values).

CONCLUSIONS

We have developed models for radionuclide chain transport in inhomogeneous crystalline rock. Advective and dispersive nuclide transport takes place in water conducting veins (represented as tubes) in kakirites or fractures in dykes, respectively. Out of these nuclides diffuse into a spatially limited zone of adjacent altered rock. For the case of transport in dykes we have investigated the conditions under which kinetic limited matrix diffusion can be represented by an effective surface sorption.

Starting from site-specific data and performing extensive parameter variations the following conclusions can be drawn:

a) The most important parameter is the spatial extent of altered crystalline rock adjacent to the water conducting structures. Depending on its width, radionuclide outflow from the geosphere may be totally insignificant or essentially equal the inflow.
b) Transport in kakirites and aplite/pegmatite dykes differs strongly. This is less a consequence of different geometry of water conducting zones but more a result of the point emphasized above (extent of altered zones).
c) A small value of the diffusivity in the altered zone resulting from tortuosity and constrictivity of the individual connected pores strongly reduces the retardation power of an extended altered zone.
d) On the other hand, diffusion into the intact rock could yield a further retardation.
e) Mechanical dispersion in the water conducting zones is relatively unimportant provided its value is not too large (less than $a_L = 100$ m in

the present calculations, say). Though this is largely a free parameter we feel that a_L = 250 m is unphysically large.

f) It is known for decades, now, and needs not to be stressed especially that sorption is an efficient mechanism for retardation. For the present models the sorption capacity of the altered zones is particularily important. However, also non-sorbing nuclides are retarded to some extent.
g) For transport in the fractures limited matrix diffusion may often be represented by an effective surface sorption. This reduces the numerical expenditure appreciably.

Finally, we also want to mention some open problems:

a) Though not the topic of the present work, a better knowledge of parameters is indispensable. This implies systematic experiments in the field, in the laboratory and analogue studies.
b) The same is needed for further validation of conceptual models. In this respect matrix diffusion is most crucial. Though diffusion into a limited zone of altered rock seems to be conservative it has to be validated in greater detail.
c) Transport of colloids needs to be considered. Little is known on their formation and stability, on their transport properties, especially the ability to diffuse into the pore space of the altered zones. Probably little can be gained in realism of description without taking into account kinetic chemical reactions between colloids and dissolved ionic species.
d) The accumulating data suggest that non-linear sorption is the rule rather than the exception. In order to gain realism and reduce conservatisms, non-linear isotherms have to be taken into account.

Work along these lines is in progress.

REFERENCES

1. Nagra, Projekt Gewaehr 1985, Endlager fuer hochaktive Abfaelle: Das System der Sicherheitsbarrieren, Projektbericht NGB 85-04 (Baden, 1985)
2. Nagra, Sondierbohrung Boettstein - Untersuchungsbericht, NTB 85-01 (Baden, 1985)
3. N.A. Chapman and I.G. McKinley, The Potential of Natural Analogues in Assessing Systems for Deep Disposal of High-Level Radioactive Waste, NTB 84-41 (Baden, 1985) and EIR-Bericht Nr. 545 (Wuerenlingen, 1985)
4. G.E. Grisak and J.F. Pickens, Water Res. Res. 16, 719 (1980)
5. I. Neretnieks, J. Geophys. Res. 85, 4379 (1980)
6. J. Hadermann and F. Roesel, Radionuclide Chain Transport in Inhomogeneous Crystalline Rocks - Limited Matrix Diffusion and Effective Surface Sorption, NTB 85-40 (Baden, 1985) and EIR-Bericht Nr. 551 (Wuerenlingen, 1985)
7. Nagra, Projekt Gewaehr 1985, Endlager fuer hochaktive Abfaelle: Sicherheitsbericht, Projektbericht NGB 85-05 (Baden, 1985)
8. R. Hartley, Release of Radionuclides to the Geosphere from a Repository for High-Level Waste - Mathematical Model Results, NTB 85-41 (Baden, 1985)

THE EFFECTS OF SIMULTANEOUS HEAT AND MASS TRANSPORT ON RADIONUCLIDE MIGRATION

FRANK B. WALTON, J.P.M. ROSS AND D.G. JUHNKE
Whiteshell Nuclear Research Establishment, Atomic Energy of Canada Ltd., Pinawa, Manitoba, ROE 1L0, Canada

ABSTRACT

Five thermal-convection loops, constructed of granite, were operated for 350 d with a hot-side temperature of 60°C and a cold-side temperature of 20°C, to examine the effects of a thermal gradient on the convective transport of seven radionuclides. Radionuclide concentrations on the surface of the loop flow channels indicated that ^{125}Sb, ^{60}Co and ^{54}Mn and precipitated iron oxyhydroxides were concentrated in the hot side of the loops. Concentrations of ^{144}Ce and ^{99}Tc were higher in the cold side of the loops. Both of these patterns were exhibited by ^{75}Se. The concentration of ^{137}Cs was not significantly affected by the thermal gradient. Radionuclide solubility and sorption were affected by variations in pH, Eh and ionic strength. Incorporation of the radionuclide into mineral lattices was found to play a key role in retarding the transport of all radionuclides tested.

INTRODUCTION

The thermal transient in an underground nuclear fuel waste disposal vault may induce convective heat and mass transfer for thousands to tens of thousands of years (depending on the form of waste) after vault closure. Experiments have been conducted to examine the potential effects of simultaneous heat and mass transport on radionuclide migration in granitic systems. These experiments were conducted in closed-loop systems constructed entirely of granite. Convective flow in the loops was induced by a 40°C applied thermal gradient across the systems. The design of the experiments ensured that any radionuclide concentration gradients around the granite surface of the convective flow channels at the end of the experiments were only the result of thermal effects.

EXPERIMENTAL

Five loops were fabricated by machining 12-mm-square flow channels into the faces of 40x20x6-cm granite blocks (see Figure 1). This configuration was chosen to eliminate the need for external pumps and metal fittings, which might distort the effect of the granite alone on the radionuclide deposition gradients. Granite coupons (50x14x3 mm) were positioned diagonally across the flow channels and the loops closed by bonding 2-cm-thick granite lids to the faces of the blocks containing the flow channels. A two-component silicon rubber sealant (Canadian General Electric, RTV 31) was used to bond the lids to the loop block. The outside of the blocks was sealed with a coating of a high-temperature epoxy resin (3-M, Scotch Cast 280). Thermocouples were placed into grooves cut into the back faces of the loops, to measure temperature distribution. The five loops, plus one solid block at each end, were clamped together to form a single unit. Electrical heaters were clamped around the flow channel on one side of the loops and copper cooling coils were clamped to the other side (see Figure 1). The centre three loops, containing the radioactive tracers, were connected via Teflon tubing to a common zirconium header. The header was connected to a zirconium tank containing degassed demineralized water with a nitrogen cover gas. An identical arrangement was provided for the other two loops, which were operated without radioactive tracers. Further details of loop

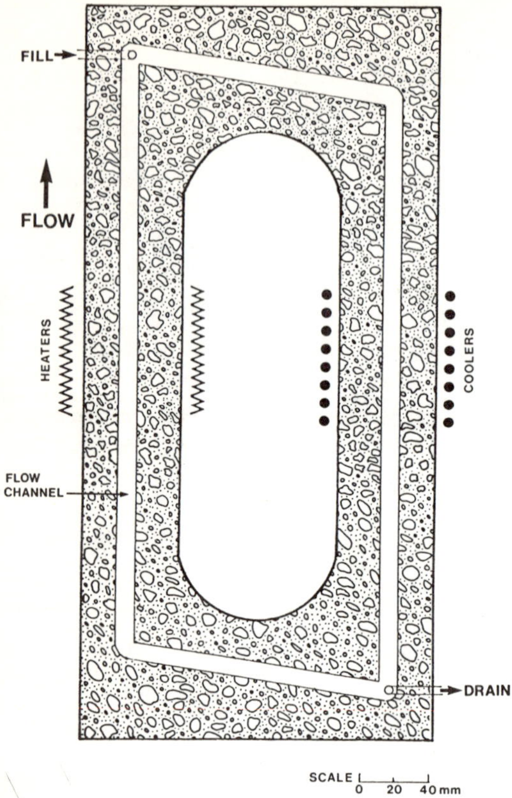

Figure 1. Schematic diagram showing 12-mm-square flow channel machined into the face of a 6-cm-thick granite block. A granite lid was bonded to the face of the block.

construction and petrographic analysis of the granite are described in reference 1.

At the start of the experiment, loops 2, 3 and 4 were filled with demineralized water (pH=7) doped with a mixture of seven radionuclides (^{99}Tc, ^{137}Cs, ^{75}Se, ^{125}Sb, ^{144}Ce, ^{60}Co, and ^{54}Mn). These tracers were chosen to give a wide spectrum of chemical behavior. With the exception of ^{99}Tc, all the radionuclides were gamma emitters. The other two loops were filled with demineralized water only, to provide inactive samples. The experiment lasted for 401 d with 350 d at full operating temperature. Outages due to equipment malfunctions accounted for the other 51 d. After the loops were shut down, the loop water was sampled for dissolved gases and ions, pH and conductivity. Water samples from loops 2, 3 and 4 were analyzed for the radioactive tracers. After sampling, the loops were drained and opened for inspection. Coupons removed from the radioactive loops were subjected to selective chemical extraction according to the following procedures, given in order of application. Demineralized water was used to rinse the coupons between procedures.

Exchangeable radionuclides were displaced by contacting the coupons at room temperature for 1 h with a solution of 1 mol/L $MgCl_2$ adjusted to pH=7 with $Mg(OH)_2$.

Amorphous iron, $Fe(OH)_3$, and associated radionuclides were dissolved by contacting the coupons for 30 min at 50°C with a solution of 0.25 mol/L $NH_2OH.HCl$ and 0.25 mol/L HCl.

Crystalline iron (i.e., Fe_2O_3) and associated radionuclides were dissolved by using dithionate/citrate buffered with bicarbonate (BCD). The coupons were contacted for 15 min at 80°C with 15 mL of 0.3 mol/L $Na_3C_6H_5O_7$, 2 mL of 1 mol/L $NaHCO_3$ and 0.38 g of $Na_2S_2O_4$. This procedure was repeated four times.

The final step was dissolution of the granite coupons. The coupons were placed in a Teflon container and contacted with 40 mL of nitric acid plus 20 mL of hydrofluoric acid at 80°C until dissolution was complete.

Gamma spectrometry was used to analyze for all tracers except ^{99}Tc. Technetium-99, a beta emitter, had to be extracted for liquid scintilation counting. A new extraction procedure was used to extract ^{99}Tc in the presence of ^{125}Sb. The method consisted of oxidation of Tc to Tc(VII) and extraction of Tc(VII) into cyclohexanone. Technetium was then stripped from the organic phase by adding carbon tetrachloride and extracting the Tc into water. The final phase involved extraction of Tc into tributyl phosphate. Details of this procedure are to be published [2].

RESULTS AND DISCUSSION

The measured temperature gradient around the loops is shown in Figure 2. Dye injection tests in a full-sized Plexiglass model of a loop indicated a thermal convection rate that recycled the loop water about once every 3 min. This produced about $1.7x10^5$ cycles during the 350-d operating period. This experiment was not intended to reproduce convective flow rates that may result from the thermal transient in an underground nuclear fuel disposal vault. The relatively high convective rate observed in these experiments was used to ensure that any observed tracer concentration gradients on the granite were the result of local thermal effects and not the product of limited convective transport.

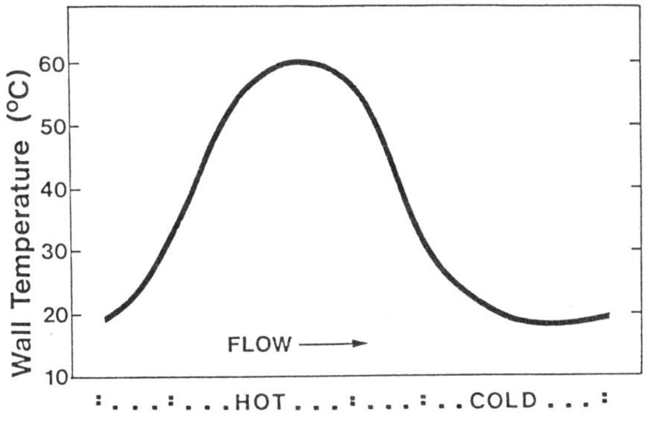

Figure 2. Temperature distribution in granite block as a function of position in flow channel.

Loop 5 developed an excessive leak early in the experiment and was isolated from loop 1. The measured leak rate for loop 1 was 0.21 mL/d. It was impossible to measure individual leak rates for the active loops because of the common header. However, the summary of loop chemistry for the four loops, given in Table I, allowed the relative leak rates to be estimated. Because the conductivity and total dissolved ion concentrations in loop 2 were higher than in loop 1, the leak rate for loop 2 was estimated to be less than that for loop 1 (i.e., 0.21 mL/d). The combined leak rate for loops 3 and 4 was estimated to be about 8 mL/d based on the measured total leak rate for loops 2, 3 and 4.

The calculated pH, based on dissolved CO_2, agreed with the measured values, with the exception of loop 2. In loop 2, the dissolved Fe^{2+} was rapidly oxidized to Fe^{3+} on exposure to air. The subsequent hydrolysis produced a lower pH than that calculated on the basis of dissolved CO_2. A visible precipitate of $Fe(OH)_3$ was produced by the hydrolysis reaction in samples removed from loop 2.

Two estimates of oxidation potential, Eh, were made. The first was calculated from the Fe^{2+} and Fe^{3+} concentrations and pH, and the second from dissolved O_2 concentrations. A direct comparison of dissolved ions in the loops was complicated by the fact that the active loops (i.e., 2, 3 and 4) had to be analyzed separately and their analysis was not as comprehensive as for loop 1. A comparison of the active loops indicated higher concentrations of dissolved ions in the loop with the lowest Eh and vice versa. Since the concentrations of the dissolved ions were an inverse function of the loop leak rate, and the supply of dissolved O_2 to the loops was proportional to the loop leak rate, then the relative oxidation potential of the loops (i.e., Eh for loop 4 > loop 3 > loop 2) was consistent with the other loop chemistry data shown in Table I.

A comparison of the relative importance of the various sorption mechanisms in loops 2, 3 and 4 is made in Figure 3. Figure 3 also shows the results of an analysis of the hydroxylamine hydrochloride and BCD extraction solutions for iron. The hydrofluoric acid solution was not analyzed because residual iron deposits could not have been differentiated from iron dissolved from the granite matrix.

The starting concentrations and a final mass balance for the tracers are given in Table II. Tracer concentrations on the surface of the loop walls were calculated by multiplying the coupon tracer inventory (determined from results of the selective chemical extraction procedures) by the ratio of the wall surface area to coupon surface area for a particular location in the loop. Tracer losses were estimated from the difference between the starting inventory and the final water concentrations plus the total granite surface inventory. Normalized tracer concentrations on the granite, as a function of position around the loop, are compared for loops 2, 3 and 4 in Figure 4. Normalized total iron deposits on the surface of the granite are also compared in Figure 4, as a function of position in the loops.

As can be seen from Figure 4, the thermal convective transport produced a variety of tracer distributions around the loops. Three distribution categories were observed. The first category consisted of the transition metals Co, Mn and Fe, and Sb for which the highest deposits were in the hottest part of the three loops. Total deposits for Co, Mn and Fe were highest in loop 2 and lowest in loop 3. The highest concentrations of iron coincided with visual observations of a reddish-brown stain (likely Fe_2O_3) on the loop walls and coupons in loop 2. Deposits of Sb were highest in loop 2 and lowest in loop 3. Figure 4 shows that exchangeable Co and Mn were only detected in loop 2, where the tracer concentration in the water was considerably higher than in loops 3 and 4. Small amounts (<1%) of exchangeable Sb were detected in loops 2 and 4. In the case of Co, Mn and Sb, more than 50% of the sorbed tracers was extracted by the BCD procedure, or was in the residual fraction. This indicated that large percentages of these elements were bound to the surface of the granite by a mineralization process (i.e., incorporation into mineral lattice structures). Separate tests [3] have shown that these elements can be incorporated into the

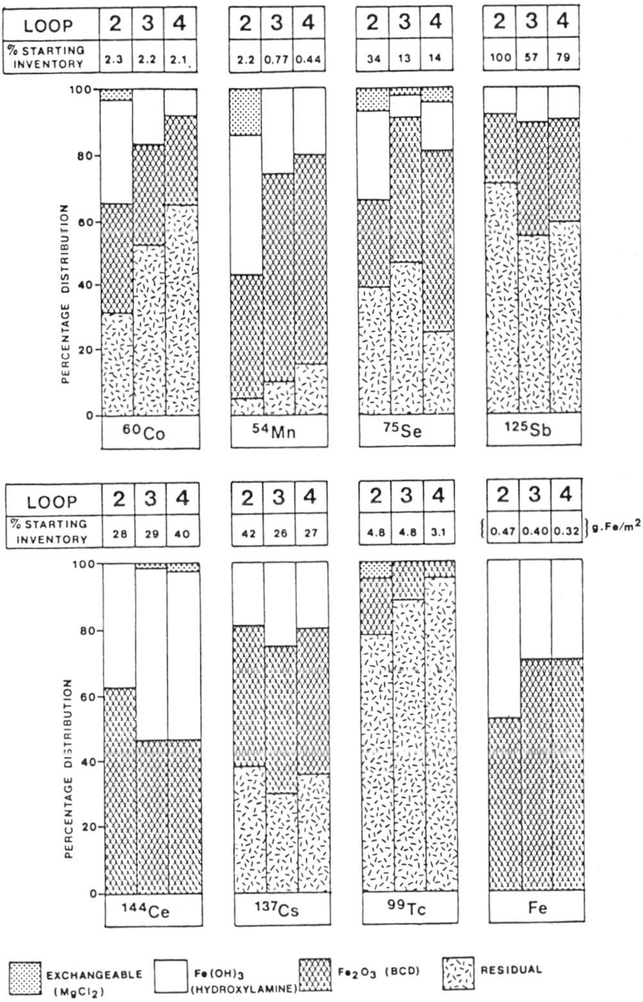

Figure 3. Percentage of starting tracer inventory on granite surfaces for loops 2, 3 and 4. The amount removed from the granite surfaces by each of selective extraction procedures is shown as a percentage of the total surface deposits. Also shown is the average Fe extracted from the granite surfaces expressed as g Fe/m^2. The iron extracted by the hydroxylamine and BCD procedures is shown as a percentage of the total iron extracted.

Table I. Summary of Loop Water Analysis

Measurement	Loop 1	Loop 2	Loop 3	Loop 4
pH	4.7	3.9*	5.5	5.5
pH(CO_2)	4.9	4.9	5.1	5.2
Eh(Fe^{2+}/Fe^{3+}) (V)	1.04	0.844	0.858	0.998
Eh(O_2) (V)	0.790	0.794	0.796	0.792
Conductivity (mS/m)	38.0	35.0	1.0	1.7
Dis. Gases (cm^3/kg)	28.9	29.8	23.9	17.3
N_2 (%)	55.0	52.0	74.0	83.0
O_2 (%)	3.4	7.3	10.0	5.0
Ar (%)	1.0	0.79	1.2	0.77
CO_2 (%)	41.0	40.0	15.0	11.0
Fe^{2+} (mg/L)	$<1.0 \times 10^{-8}$	1.1×10^{-4}	1.7×10^{-6}	$<1.0 \times 10^{-8}$
Fe^{3+} (mg/L)	2.4×10^{-6}	1.0×10^{-5}	4.5×10^{-7}	7.5×10^{-7}
Na (mg/L)	31.9	84.0	1.31	1.44
Ca (mg/L)	45.3	-	-	-
Mg (mg/L)	7.2	143.0	0.84	0.15
K (mg/L)	65.0	-	-	-
Sr (mg/L)	0.15	1.86	<0.03	<0.03
Si (mg/L)	380.0	29.0	4.1	2.6
Cl (mg/L)	30.9	1420.0	2.3	1.6
NO^{3-} (mg/L)	0.24	22.0	2.2	2.1
SO_4^{2-} (mg/L)	34.3	-	-	-
F^- (mg/L)	2.34	-	-	-

Notes:
- * measured pH is low due to hydrolysis of Fe^{2+} on exposure to air.
- pH(CO_2) was calculated from dissolved CO_2.
- Eh(O_2) was calculated from dissolved O_2.
- Eh(Fe^{2+}/Fe^{3+}) was calculated from dissolved iron concentrations and pH.

Table II. Tracer Starting Concentration and Final Mass Balance

Tracer	Measurement	Loop 2	Loop 3	Loop 4
^{99}Tc	Starting (mol/L)	6.9×10^{-4}	6.9×10^{-4}	6.9×10^{-4}
	Water (%)	8.6	0.33	0.33
	Granite (%)	4.8	4.3	3.1
	Losses (%)	87.0	95.0	96.0
^{137}Cs	Starting (mol/L)	1.8×10^{-9}	1.8×10^{-9}	1.8×10^{-9}
	Water (%)	18.0	0.10	0.20
	Granite (%)	42.0	26.0	27.0
	Losses (%)	40.0	74.0	73.0
^{144}Ce	Starting (mol/L)	5.0×10^{-12}	5.0×10^{-12}	5.0×10^{-12}
	Water (%)	56.0	0.0	0.0
	Granite (%)	28.0	29.0	40.0
	Losses (%)	16.0	71.0	60.0
^{75}Se	Starting (mol/L)	6.5×10^{-10}	6.5×10^{-10}	6.5×10^{-10}
	Water (%)	25.0	78.0	0.26
	Granite (%)	34.0	13.0	14.0
	Losses (%)	41.0	87.0	86.0
^{125}Sb	Starting (mol/L)	2.2×10^{-9}	2.2×10^{-9}	2.2×10^{-9}
	Water (%)	0.80	0.60	0.20
	Granite (%)	100.0	57.0	79.0
	Losses (%)	0.0	43.0	21.0
^{60}Co*	Starting (mol/L)	4.2×10^{-6}	4.2×10^{-6}	4.2×10^{-6}
	Water (%)	57.0	0.03	0.007
	Granite (%)	2.3	2.2	2.1
	Losses (%)	41.0	98.0	98.0
^{54}Mn	Starting (mol/L)	3.1×10^{-9}	3.1×10^{-9}	3.1×10^{-9}
	Water (%)	80.0	0.03	0.005
	Granite (%)	2.2	0.77	0.44
	Losses (%)	18.0	99.0	99.0

Note: * Cobalt starting concentration includes carrier

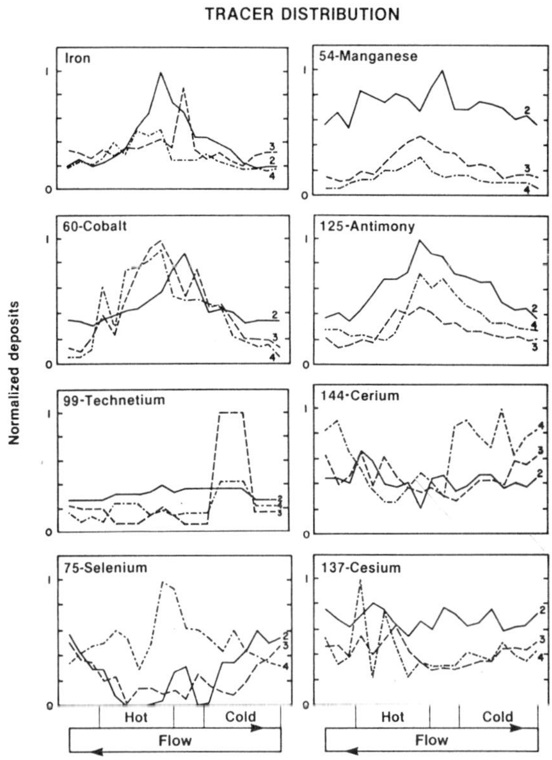

Figure 4. Total tracer deposits on granite surfaces in loops 2, 3 and 4 as a function of position in flow channel. Also shown is the total Fe removed from the granite surfaces by the hydroxylamine and BCD extraction.

lattice of Fe_2O_3 during the conversion of the thermodynamically unstable $Fe(OH)_3$ to Fe_2O_3 via a first-order reaction [4].

The second category includes Tc and Ce. These tracers produced the largest surface deposits in the cold side of the loop. The largest distribution gradients were produced in loops 3 and 4, with little or no gradient in loop 2. Some exchangeable Tc was observed in loop 2 where Tc concentrations in the water were highest. Exactly the opposite was observed for Ce (i.e., no exchangeable activity was found in loop 2 where the Ce concentration in the water was highest). In other experiments under oxic conditions, over 90% of the Ce was removed from solution by a 0.45 μm filter [5]. Similar filtration of the loop waters did not remove significant quantities of any of the tracers. In addition, Ce was the only tracer not extracted in the residual fraction, indicating that Ce was not incorporated to the same extent as the other tracers by mineralization processes. In contrast, the largest fraction of Tc was extracted in the residual fraction. Measurements of connected porosity in the Lac du Bonnet granite ranged from 1.5 to 1.8×10^{-3} [6]. Using the starting Tc concentration and the maximum porosity, the largest possible inventory associated with pore diffusion is more than a factor of 30 less than the Tc inventory recovered by the extraction procedures. Separate tests have shown that small amounts of Tc have been incorporated into the lattice of synthetic Fe_2O_3 under both oxic and anoxic conditions [7]. These same tests also indicated that sorption of small amounts of Tc onto $Fe(OH)_3$ retards the conversion of $Fe(OH)_3$ to Fe_2O_3. These combined observations suggest that Tc was incorporated by a slow mineralization process in the loop tests. Information on Tc or Ce speciation is not available from these tests.

The third category consists of Se. Figure 4 shows that the maximum surface concentration occurred in the hot side of loop 2 and the cold side of loops 3 and 4. Differences in surface concentration profiles in the three loops are believed to be primarily a function of differences in pH and/or Eh, which produced changes in the Se solubility gradient with temperature. The greater than 60% extraction of the Se in the BCD and residual fractions indicates that Se was primarily incorporated by mineralization processes on the granite surface. Separate tests have shown that Se can be incorporated into the lattice of iron oxyhydroxides [3].

Cesium does not fit into any of the preceding categories. Figure 4 shows that there is little correlation between surface concentrations in the three loops as a function of temperature. This is not surprising considering the high solubility of Cs.

In general, all tracers except Cs exhibited some form of concentration distribution on the granite surface as a function of loop temperature. The distribution differed from loop to loop for a given tracer, and differed between tracers for a given loop. This suggests that the independent transport of each tracer was controlled primarily by local tracer solubility gradients. Solubility gradients within a given loop were a function of temperature, while differences between loops are believed to have been a function of loop leak rate. Loop leak rates controlled overall tracer inventory and also determined solution pH, Eh, and the concentration of other ions in solution.

Tracer distribution coefficents, k_a, were calculated for each loop and have been compared to distribution coefficients for 28-d static tests [8] performed under oxic conditions with granite obtained from the same source as the loops in Table III. In loop 2, the k_a values for Cs, Ce, Se, and Co were less than those for the static tests with low-ionic-strength, granite groundwater (GGW). The values for Ce, Se and Co were even lower than those

Table III. Tracer Distribution Coefficients, k_a (mL/cm^2)

	Loop			Static Tests (8)	
Tracer	2	3	4	GGW	SCSSS
^{99}Tc	0.068	1.6	1.2	-	-
^{137}Cs	0.28	32.0	17.0	0.85	0.23
^{144}Ce	0.061	n.c.	n.c.	7.4	0.99
^{75}Se	0.17	2.0	6.6	1.3	0.23
^{125}Sb	15.0	12.0	48.0	d.l.	0.08
^{60}Co	0.0049	8.9	37.0	1.1	0.72
^{54}Mn	0.0034	3.1	11.0	-	-

Notes: GGH - Granite Groundwater 8
SCSSS - Standard Canadian Shield Saline Solution 8
n.c. - not calculated because tracer concentration in water below detection limits.
d.l. - concentration on granite surface below detection limits.

for high-ionic-strength, standard, Canadian Shield saline solution (SCSSS). The k_a value for Cs was slightly larger than that for SCSSS. This is consistent with the relatively high-ionic-strength in loop 2. The low k_a values for Ce, Co and Se are most likely a function of the higher ionic strength (i.e., competition for sorption sites by ions other than the tracers), and lower pH and Eh (i.e., higher solubility) in loop 2 than in the static tests. The behavior of Mn closely paralleled that of Co, as might be expected for a transition metal.

Previous dynamic tests with Sb showed little sensitivity of the k_a value to ionic strength, and from kinetic considerations, k_a values that were about an order of magnitude higher than those from static tests [8]. The higher k_a value for Sb in the loop experiments was, therefore, likely a function of slow mineralization reactions. This is consistent with the data for Sb, shown in Figure 4, where a large percentage of the surface inventory was associated with Fe_2O_3 and residual phases.

The k_a values for loops 3 and 4 were much higher than those for loop 2. They were also higher than those for the static tests with GGW. This was due to a combination of low tracer concentrations in the loop water and mineralization processes, which had considerably slowed, or prevented, desorption and removal of tracers from the loop by leakage. The extraction data in Figure 4 indicate that only small percentages of the sorbed tracers can be displaced from the surface by exchange with $MgCl_2$. With the exception of Ce, more than 50% of tracer surface inventories was incorporated into crystalline iron oxyhydroxides or other mineral structures. Mineralization processes appear to be the most significant mechanism in the retardation of tracer transport on longer time scales.

CONCLUSIONS

The primary objective of these experiments was to determine whether an imposed thermal gradient would induce a corresponding gradient in radionuclide sorption patterns in a closed-circuit granite convection loop. This objective was met in spite of problems with leakage from the loops during the 401-d experiments. The following conclusions were reached:

(1) With the exception of Cs, temperature-gradient effects were observed for all tracers and iron oxyhydroxides. In spite of the lack of direct measurements of solution concentration gradients, the wide variation in tracer and iron concentrations on the granite surface around the loops can only be interpreted in terms of temperature-related solubility gradients. Variations between loops are thought to be due to differences in pH and/or Eh.

(2) Little of the sorbed tracers was exchangeable. Most of the sorbed tracers were bound to the granite surface by mineralization processes. Selective extraction procedures indicated that iron oxyhydroxides played a major role in this.

(3) From 3 to 5% of the Tc starting inventory was bound to the granite surface by slow mineralization processes. The formation of Fe_2O_3 is thought to play a key role in this process, and its rate of formation has been shown to be slowed by the presence of Tc.

(4) Distribution coefficients for Ce, Se and Co in one loop were lower than those reported for static tests with the same granite and SCSSS under oxic conditions. The reduced sorption is consistent with the relatively high ionic strength of the water and the lower pH and Eh, produced as a result of the alteration of the fresh granite surfaces, in the loop.

REFERENCES

1. F.B. Walton, D.G. Juhnke and J.P.M. Ross, "Radionuclide Transport in Granite Thermal-Convective Loops", Atomic Energy of Canada Limited Report (in preparation).

2. P.A. Beeley, private communication.

3. F.B. Walton, unpublished data.

4. U. Schwertmann and E. Murad, "Effect of pH on the Formation of Goethite and Hematite from Ferrihydrite", Clays & Clay Min. 31, 4, 277 (1983).

5. F.B. Walton, T.W. Melnyk, J.P.M. Ross, and A.M.M. Skeet, "Determination of Radionuclide Sorption Mechanisms and Rates on Granitic Rock by Selective Chemical Extraction Techniques", A.C.S. Symposium Series, 246, 45 (1984).

6. T.W. Melnyk, private communication.

7. F.B. Walton, J. Paquette, J.P.M. Ross and W.E. Lawrence, "Tc(IV) and Tc(VII) Interactions with Iron Oxyhydroxides", submitted to Nuclear and Chemical Waste Management, (1985).

8. F.B. Walton, T.W. Melynyk, R.G. Abry and K. Fleury, "The Determination of Nuclide-Geologic Media Reaction Kinetics Using Mixing-Cell Contactors", Chem. Geo., 36, 155 (1982)

$^{234}U/^{230}Th$ RATIO AS AN INDICATOR OF REDOX STATE, AND U, TH, AND RA BEHAVIOR IN BRINEY AQUIFERS

J. C. LAUL, M. R. SMITH AND N. HUBBARD*
Pacific Northwest Laboratory, P.O. Box 999, Richland, WA 99352 USA
*Battelle Memorial Institute, Columbus, OH 43201 USA

ABSTRACT

The $^{234}U/^{230}Th$ ratio serves as an in-situ indicator of the redox state in groundwater aquifers. The higher this ratio, the more U there is in the +6 valance state and thus a less reducing environment. Radium sorption is retarded in the shallow aquifer and is dependent on the $CaSO_4$ content and the redox state. Relative to Ra, U and Th are highly sorbed. The total retardation factor for Th is $\sim$1400 and mean sorption time for ^{228}Th is $\sim$10 days in the shallow zone. The desorption rate of Ra is significantly slower in the shallow than in the deep aquifer. There is no effect of colloids in brines.

INTRODUCTION

Investigations of natural radionuclides of the uranium and thorium series in groundwaters can provide direct and analog information about the expected behavior of radwaste-radionuclides and the redox state of the environment in an aquifer. We have been studying deep and shallow briney aquifers in the Palo Duro Basin, Texas, for the natural radionuclides of ^{238}U and ^{232}Th series shown in Figure 1. Our findings [1,2] for the deep aquifers are: (1) $^{226}Ra/^{222}Rn$ ratio is $\gtrsim$1, indicating that Ra is highly soluble and thus, not retarded in brines; (2) relative to Ra (or Rn), U and Th are highly retarded by sorption; and (3) the $^{234}U/^{230}Th$ ratio is nearly unity, suggesting that U like Th is in the +4 state and indicating a reducing environment in these deep brines.

<u>^{238}U CHAIN</u>

^{238}U (4.4 x 10^9 yrs) → ^{234}Th (24.1 d) ------ ^{234}U (2.4 x 10^5 yr) → ^{230}Th (7.5 x 10^4 yr) →

^{226}Ra (1600 yr) → ^{222}Rn (3.82 d)

<u>^{232}Th CHAIN</u>

^{232}Th (1.4 x 10^{10} yr) → ^{228}Ra (5.76 yr) ------ ^{228}Th (1.91 yr) → ^{224}Ra (3.66 d)

Fig.1. Radionuclides Used IN ^{238}U AND ^{232}Th Chains

However, in the Wolfcamp deep aquifer sampled from the J. Friemel No. 1 (Zone 7) well in Deaf Smith County, Texas, results are substantially different. These are: (1) $^{226}Ra/^{222}Rn$ ratio is 0.30, suggesting that Ra is sorbed three times more than Rn; (2) $^{234}U/^{230}Th$ ratio is about 6, indicating that U is more in the +6 than in the +4 valence state which suggests a less reducing environment for this part of the

Wolfcamp aquifer[2]. Here we report the radionuclide data for the shallow (750 ft.) zone where the conditions are expected to be far less reducing and investigate any correlation with the in-situ redox state of groundwater aquifers. We also compare the U, Th, and Ra radionuclide data in unfiltered and filtered (0.8 μm) brine water to test any effect of colloids.

FIELD SAMPLING AND ANALYSIS

Groundwater (brine) was pumped from various depths using oil field drilling technology. Since drilling fluid was used in coring, contamination of the brine waters with drilling fluid was monitored with a NH_4CNS chemical tracer. Brine was pumped until the tracer concentration decreased to 0.5% or less of its original concentration in the drilling fluid.

The details of field sampling and analysis of brines for U, Th, Ra and Rn are described elsewhere by Laul et al.[3]. In brief, for Ra radionuclides several gallons of brine waters were passed through a Battelle Large Volume Water Sampler[4] containing several filters (0.8 μm) and two quarter-inch thick Al_2O_3-$BaSO_4$ adsorption beds. The adsorption beds were then dried in the laboratory and counted by gamma spectroscopy (γ.s) for ^{228}Ra, ^{226}Ra and ^{224}Ra radionuclides. A direct counting of an aliquot was made to measure ^{226}Ra and Ra collection yields. For Th and U nuclides, unfiltered and filtered (0.8 μm) waters were collected and acidified with HCl in the field. A few liters of each were subsequently analyzed in the laboratory using carrier-free radiochemical separations[3,5,6] and ^{233}U and ^{230}Th as tracers. Uranium-238, ^{235}U and ^{234}U were measured by mass spectrometry (m.s) , and ^{232}Th, ^{230}Th and ^{228}Th by alpha spectroscopy (α.s) and ^{234}Th by beta counting (β.s). These measurements were made in duplicate.

RESULTS AND DISCUSSION

The concentrations (dpm/L and ng/L) of ^{238}U and ^{232}Th chains (Figure 1) in unfiltered and filtered (0.8 μm) brines from the Seven Rivers Formation aquifer, Mansfield Well No. 2, Zone 1 (750 ft.) are shown in Table 1. The errors given are due to counting statistics or the precision of duplicate analyses, whichever is larger. The data for ^{222}Rn are from The Bendix Laboratory, Grand Junction, Colorado, USA. The radionuclide contents of ^{238}U, ^{234}U, ^{232}Th and ^{230}Th between the unfiltered and filtered brines, are the same at $\pm 2\sigma$ level, indicating that there is virtually no effect of colloids in brine (high-ionic strength) waters. This observation is consistent with our earlier observations in deep brines[2].

In comparison with deep aquifers (Wolfcamp carbonate, granite wash and Penn. granite wash) of the Palo Duro Basin, the U and Th nuclide concentrations are considerably higher in the shallow aquifer (Seven Rivers Formation), while the Ra content is considerably lower in the shallow aquifer. Following the approach of Hubbard et al.[1] and Laul et al.[2], the radionuclide concentrations are normalized to ^{226}Ra in the ^{238}U chain and to ^{228}Ra in the ^{232}Th chain. These activity ratios are tabulated in Table 1.

Table I. CONCENTRATION OF RADIONUCLIDES IN BRINE WATER FROM ZONE 1 (750 ft) OF MANSFIELD WELL NO. 2 (PALO DURO BASIN)

Isotope	$T_{1/2}$	Method Used	Zone 1* Seven Rivers Formation Unfiltered (dpm/L)	Filtered (dpm/L)	Filtered (ng/L)	Activity Relative to Ra(a)
U-238	4.47×10^9y	M.S	1.43±.02 (2)	1.11±.02 (2)	1500	0.062
Th-234	24.1 d	β.S	<3	<3	$<.6 \times 10^{-7}$	<.16
U-234	2.44×10^5y	M.S	2.96±.05 (2)	2.30±.05 (2)	0.17	0.13
Th-230	7.5×10^4y	α.S	0.16±.02	0.14±.02	3.0×10^{-3}	7.8×10^{-3}
Ra-226	1600 y	γ.S	--	18.0±1.0 (2)	8.2×10^{-3}	1.00
†Rn-222	3.82 d		--	250±30	7.3×10^{-7}	13.9
Th-232	1.40×10^{10}y	α.S	0.14±.015	0.15±.015	614	4.4×10^{-3}
Ra-228	5.76 y	γ.S	--	34.0±1.5 (2)	5.6×10^{-5}	1.00
Th-228	1.91 y	α.S	0.41±.04	0.36±.04	2.0×10^{-7}	0.010
Ra-224	3.66 d	γ.S	--	72.0±3 (2)	2.0×10^{-7}	2.12

Tracers used are: ^{233}U for M.S; ^{230}Th for ^{234}Th, ^{232}Th, ^{228}Th; ^{133}Ba for ^{228}Ra, ^{226}Ra, ^{224}Ra.

* Values in parentheses are the number of analyses. †Data from Bendix Lab.

(a) ^{226}Ra for the ^{238}U chain; ^{228}Ra for the ^{232}Th chain.

Radium Behavior and Retardation Factors

If all radionuclides in a decay series are in radiochemical equilibrium, then the activity ratio (AR) of each daughter/parent will be unity. This is not the case in brine waters, as expected from the different chemical properties of U, Th and Ra. The activity ratios (AR) of the various radionuclides for the Wolfcamp deep aquifer (5926-5825 ft.) of J. Friemel Well No. 1 are displayed as histograms in Figure 2 and that for the shallow zone (750 ft.) of Seven Rivers Formation in Figure 3. In nine out of ten deep briney aquifers, the ^{226}Ra/^{222}Rn ratio has been ≳1. Since ^{222}Rn is a noble gas and thus not sorbed to a significant degree, the ^{226}Ra/^{222}Rn ratio ≳1 indicates that ^{226}Ra is also not sorbed significantly. Thus, the retardation factor (RF) relative to Ra is just the inverse of activity ratio (RF = 1/AR). In the Wolfcamp deep brine (Figure 2), the ^{226}Ra/^{222}Rn ratio is 0.32, indicating that ^{226}Ra is sorbed about 3 times more than Rn. However, relative to Ra, U and Th nuclides are still highly sorbed. For example, the RF for ^{234}U and ^{230}Th are 5×10^3 and 3×10^4 respectively. These R.F.'s are about three times higher relative to Rn.

In the shallow zone, ^{226}Ra/^{222}Rn is 0.072, indicating that over 92% of the Ra is sorbed (Figure 3). Relative to Ra, R.F.'s for ^{234}U, ^{230}Th and ^{228}Th are 8, 128 and 100 respectively. However, the absolute RFs relative to Rn are 14 times higher for these radionuclides. Perhaps the Ra sorption is governed by the molality of NaCl and/or the dissolved content of $CaSO_4$. The NaCl content for both deep and shallow aquifers (Figures 2 and 3) is nearly the same (∼20%). However, the Ca content (0.16%) in the shallow aquifer is about 5.4 times less than the Ca content of the Wolfcamp aquifers. The SO_4^{--} content in the shallow zone is 0.57% and it is higher by 4.4 times than the deep aquifer. The SO_4^{--}/Ca^{++} ratio in the shallow zone is about 3.5 versus 0.15 in the deep aquifer. Thus, there is an excess of SO_4^{--} ions corresponding to the Ca content in the shallow zone, which can explain more sorption of Ra in the shallow versus deep aquifers. This correlates well with the

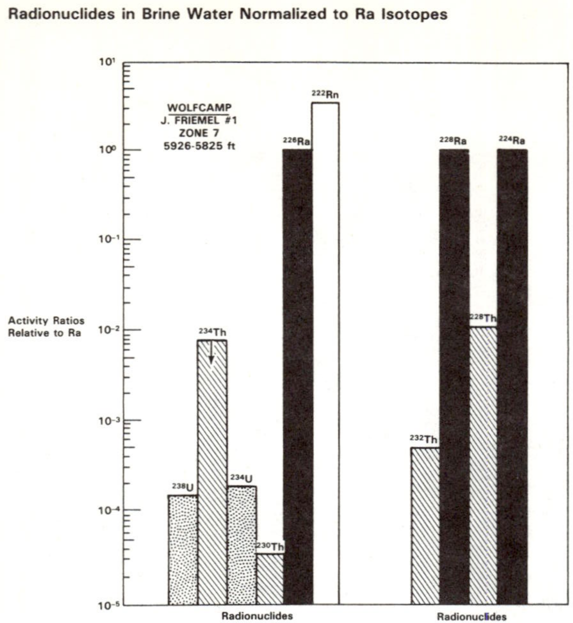

Fig. 2. Radionuclide data are normalized to ^{226}Ra in the ^{238}U chain; ^{228}Ra in the ^{232}Th chain, in deep aquifer

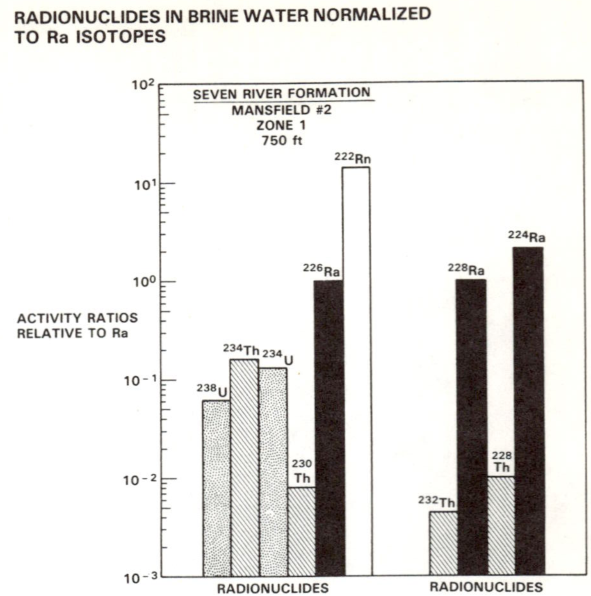

Fig. 3. Radionuclide data are normalized to ^{226}Ra in the ^{238}U chain; ^{228}Ra in the ^{232}Th chain, in shallow aquifer

SO_4^{--}/Sr ratio in the shallow versus deep aquifers. On the other hand, Ca and SO_4^{--} contents in the Wolfcamp aquifer (Zone 7) are about the same level as in other parts of the Wolfcamp aquifers, yet Ra is not retarded in other Wolfcamp aquifers. Perhaps, Ra sorption is dependent on both the SO_4^{--} content and redox state of the environment (discussed below).

Redox State of the Environment

As discussed earlier[2], the ^{234}U-^{230}Th pair can reveal information on the valence state of U and thus on the redox state of the environment. Quadravalent U has chemical properties very similar to that of Th^{+4} based on the same charge and ionic radii U^{+4} = Th^{+4} = 1.07 A°. The solubility products of the various species of U^{+4} and Th^{+4} are also very similar[5,6]. The chemical properties of U^{+6} (0.81 A°) are very different from Th^{+4}, and U^{+6} is far more soluble because it forms soluble complexes. Thus, in highly reducing environments U is expected to be in the +4 state and the activity ratio of $^{234}U/^{230}Th$ is expected to be about unity. However, in the Wolfcamp brine from J. Friemel Well (Figure 2), the $^{234}U/^{230}Th$ ratio is higher 6, indicating that U is significantly in the +6 instead of +4 valence state and thus suggests a relatively less reducing environment than for other deep brines. The $^{234}U/^{230}Th$ ratio in the shallow zone is even higher at about 14, suggesting U largely in the +6 state and thus indicating a far less reducing environment (possibly oxidizing). Thus, the magnitude of $^{234}U/^{230}Th$ ratio appears to reflect the degree of redox state of the environment.

Interestingly, we see a correlation in the ratios of $^{234}U/^{230}Th$ versus $^{226}Ra/^{222}Rn$ in the Wolfcamp and shallow aquifers (Table 2) which suggests that Ra sorption may be related to the redox condition. We analyzed two additional shallow aquifers Zeeck No. 1 Zone 4 (2970-2930 ft.) and Harman No. 1 dissolution zone (1187-1157 ft.). The $^{234}U/^{230}Th$ ratios and Ra contents for Zeeck and Harman brines are shown in Table 2. Radon-222 was not measured in these aquifers. However, based on the low ^{226}Ra content, it appears that Ra is considerably sorbed relative to radon, which appears common for shallow groundwaters [7,8]. Figure 4 shows a negative correlation of Ra activity (dpm/L) versus $^{234}U/^{230}Th$ ratio. The lower the Ra activity the higher is the $^{234}U/^{230}Th$ ratio. When the $^{234}U/^{230}Th$ ratio is near unity, as observed in most of the deep aquifers, the Ra/Rn ratio is also about unity, indicating no sorption for Ra. The declining Ra activity (dpm/L) with increasing $^{234}U/^{230}Th$ ratio suggests that Ra sorption in these groundwaters is related to the redox state of the environment.

Table II. URANIUM-234, ^{230}TH, ^{226}RA, ^{222}RN DATA IN PALO DURO BRINES

Aquifer	Depth ft.	$^{234}U/^{230}Th$	$^{226}Ra/^{222}Rn$	^{226}Ra (dpm/L)
Wolfcamp J. Friemel No. 1	5926-5825	6.0	0.32	220
Seven River Formation Mansfield Well No. 2	750	16	0.072	18
San Andres Zeeck Well No. 1 Zone 4	2970-2930	2.7	----	450
Yates and Salado Formation Harman Well No. 1 Dissolution Zone	1187-1157	12	----	30

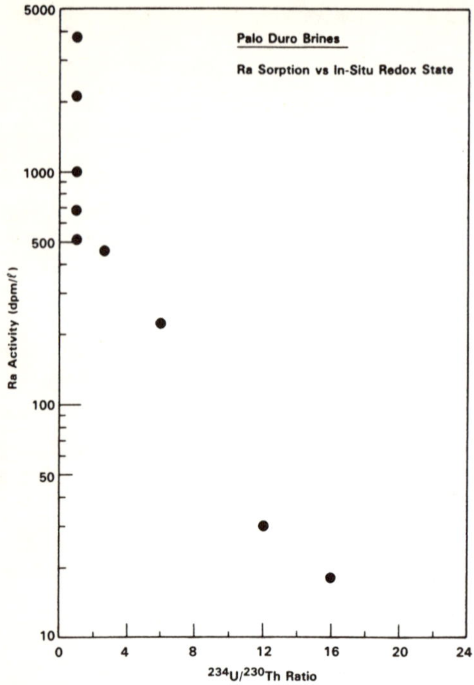

Fig. 4. Ra Activity (dpm/L) vs. $^{234}U/^{230}Th$ Ratio in Palo Duro Brines

Sorption and Desorption Rate of Th and Ra

In the ^{232}Th-^{228}Ra-^{228}Th-^{224}Ra chain, ^{232}Th decays to ^{228}Ra, which is soluble and goes into solution (Figure 5). Radium-228 atoms ($t_{1/2}$ = 5.76 y) decay by beta emission to ^{228}Th ($t_{1/2}$ = 1.91 y) atoms, which tend to sorb on the substrate. Thorium-228 atoms, in turn, decay by alpha emission to ^{224}Ra ($t_{1/2}$ = 3.66 d) atoms, which are soluble and go back into solution. The troika ^{228}Ra-^{228}Th-^{224}Ra gives us information on the sorption rate of Th and desorption rate of Ra.

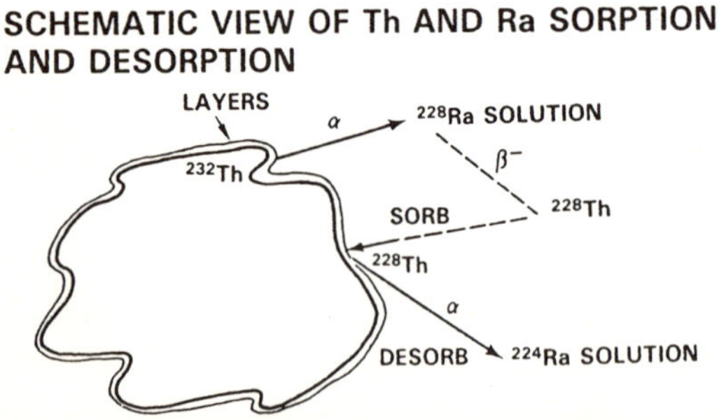

Fig.5 Schematic view of how ^{228}Th is sorbed and ^{228}Ra and ^{224}Ra are desorbed.

The mean activity ratio of $^{228}Th/^{228}Ra$ in both the deep and shallow aquifers is 0.010, indicating that 99% of ^{228}Th atoms born in solution from the decay of ^{228}Ra are sorbed on the substrate. Using the half-life of ^{228}Th (1.91 y) the mean sorption time for ^{228}Th is about 10 days. However, the desorption rate of Ra is different for deep versus shallow aquifers. The $^{228}Ra/^{224}Ra$ ratio for deep aquifers is about unity (Figure 2). This indicates the desorption time of ^{224}Ra to be very short relative to the 3.66 d half-life of ^{224}Ra (a few minutes), while the $^{228}Ra/^{224}Ra$ ratio in the shallow aquifer is 0.48, indicating a slow desorption rate for Ra. Using the $t_{1/2}$ of ^{224}Ra, the desorption rate of ^{224}Ra in the Seven Rivers groundwater is of the order of its half-life.

CONCLUSIONS

1. Radium is retarded in shallow brines and its sorption behavior is related to both the SO_4^{--}/Ca^{++} ratio in the groundwater and redox state of the environment.

2. Uranium-234/^{230}Th ratio in the shallow zone is higher at about 16, indicating that U is largely in the +6 state and thus a far less reducing environment exists in this brine than in the deep brine aquifers.

3. Relative to Ra, U and Th are highly sorbed and the retardation factors for Th are about 100 in both deep and shallow aquifers. The absolute RF for Th is 310 for deep (Wolfcamp aquifer, Zone 7) and 1400 for the shallow (Seven Rivers) aquifer.

4. The mean residence time for ^{228}Th sorption is about 10 days in both deep and shallow aquifers, while the desorption rate of Ra in the shallow Seven Rivers aquifer is considerably slower than observed for the deep aquifer systems.

5. There is no concentration gradient of U, Th and Ra between unfiltered and filtered brines and thus no colloids effect.

ACKNOWLEDGMENTS

We thank J. S. Schmitt and J. G. Pratt for their assistance in this experiment. This work was supported by the Salt Repository Office under the U. S. Department of Energy, Contract DE-AC06-76RLO 1830.

REFERENCES

1. N. Hubbard, J. C. Laul, and R. W. Perkins, "The Use of Natural Radionuclides to Predict the Behavior of Radwaste Radionuclides in Far-Field Aquifers", Scientific Basis for Radioactive Waste Management VII, Mat, Res. Soc. Smp. Proc., Vol. 26, 891-897, (1984).

2. J. C. Laul, M. R. Smith and N. Hubbard, "Behavior of Natural Uranium, Thorium and Radium Isotopes in the Wolfcamp Brine Aquifers, Palo Duro Basin, Texas", Scientific Basis for Radioactive Waste Management VII, Mat. Res. Soc. Smp. Proc., Vol. 44, 475-482, (1985).

3. J. C. Laul, M. R. Smith, C. W. Thomas, P. O. Jackson and N. Hubbard, "Analysis of Natural Radionuclides from Uranium and Thorium Series in Briney Groundwaters", J. of Radioanalytical Chemistry, in press, (1985).

4. W. B. Silker, "Collection and Analysis of Radionuclides in Seawater", in Advances in Chemistry Series, American Chemical Society, No. 147, 1633-1675, (1982).

5. J. E. Grindler, "The Radiochemistry of Uranium", Nuclear Science Series, NAS-NS, 3050, (1960).

6. E. K. Hyde, "The Radiochemistry of Thorium", Nuclear Science Series, NAS-NS, 3004, (1960).

7. M. Asikainen, "State of Disequilibrium Between ^{238}U, ^{234}U, ^{226}Ra, and ^{222}Rn in Groundwater from Bedrock," Geochimica et Cosmochimica Acta, 48, 201-206 (1981).

8. P. T. King, J. Michel and W. S. moore, "Groundwater Geochemistry of ^{228}Ra, ^{226}Ra and ^{222}Rn," Geochimica et Cosmochimica Acta, 46, 1173-1182 (1982).

PART IX

Radionuclide Speciation and Sorption

SORPTION OF SOME FISSION PRODUCTS AND ACTINIDES IN CONCRETE SYSTEMS

S. HÖGLUND*, L. ELIASSON*, B. ALLARD**, K. ANDERSSON*** AND B. TORSTENFELT*
* Department of Nuclear Chemistry, Chalmers University of Technology, S-412 96 Göteborg, Sweden
** Department of Water in Environment and Society, Linköping University, S-581 83 Linköping, Sweden
*** Studsvik Energiteknik AB, S-611 82 Nyköping, Sweden

ABSTRACT

The sorption of some actinides (Th, U, Np, Pu and Am) and fission products (I, Cs) was measured on two types of Standard Portland cements as well as on samples from old (70 years) hydro power dam constructions using a batch technique. Pore water compositions were analysed, and artificial pore water solutions were used as aqueous phases in the experiments. Measurements were also performed on five other concrete types (not reported in detail in this paper) to illustrate the effects of the cement matrix composition on the sorption behaviour of the radionuclides.

The sorption of actinides in the trivalent (americium), tetravalent (thorium) pentavalent (neptunium) and hexavalent (uranium) states was high in all the studied concrete systems.

Generally, the sorption of cesium was low due to the low exchange capacity of the cement and the high concentration of competing cations in the pore waters.

The sorption of iodine was much higher than in most silicate minerals of geologic origin.

The differences between the various concrete systems were generally minor in terms of their sorbing capacities.

INTRODUCTION

Concrete is used as solid matrix for long term storage of several categories of low and medium level radioactive waste. The radionuclide contents vary with the category of waste. Some important categories are:

- Operational waste from reactors, mainly ion exchange resins and some solids and concentrates with activation products (Co-60), fission products (Sr-90, Cs-137, etc) and minor amounts of actinides (Pu-239,240, Am-241,243)
- Reprocessing waste containing ion exchange resins, precipitates, solvents, etc from the solvent extraction but also solid metals; activation products (Co-60, C-14, etc), fission products (Sr-90, Tc-99, I-129, Cs-137, etc) and actinides (U-235,236,238, Np-237, Pu-238,239, 240,241,242, Am-241,243) are found in this category
- Decommissioning waste containing activation products from metallic construction materials (Zircalloy, etc) and other reactor meaterials; common nuclides are C-14, Co-60, Ni-59,63, Nb/Zr-95, etc.

A release of radionuclides incorporated in a cement matrix requires a dissolution and formation of a soluble fraction of the nuclide and its subsequent diffusion through the pore system of the cement. This transport will be highly effected by sorption phenomena.

A study of the sorption interactions of some fission products and actinides with various cement types in pore water environments is presented in this paper.

CONCRETE SYSTEMS

Solids

Standard Portland (SP) cement is the most common type of cement used for waste conditioning and storage. In this study one Swedish and one French type of SP were used, as well as three samples of about 70 years old, weathered concrete from hydro power dams. Five other cement types were studied for comparison (sulfate resistant, blast furnace slag, high alumina, fly ash and silica cement). Concrete samples were prepared with compositions according to Table I. A more detailed description of the concrete components and additives etc. is given in [1].

Pore waters

The pore water compositions of the cements were determined after more than 4 months of storage under water (see detailed description in ref. 2). Pore water compositions for the two Standard Portland cements as well as for the old concrete samples (from Porjus) are given in Table II. For the old concretes (P-1, P-2) the concentration ranges illustrate the difference in pore water composition within the samples (related to the distance from the water exposed surface). Sample P-1 represents a fairly intact concrete, while P-2 appears to be highly leached, as indicated from the low pH and cation concentrations.

EXPERIMENTAL

Radionuclides

The radionuclides studied in the experiments and their initial concentrations are given in Table III.

Artificial pore waters

Artificial pore waters, representative of the various cement pastes, were prepared from reagent grade chemicals (cf. Table II).

Sorption measurements

The cement samples which had been stored under water for more than one year were crushed and sieved in carbon dioxide free atmosphere. A known amount of the selected size fraction, 0.090-0.125 mm, was washed and pre-equilibrated with artificial pore water. A small volume (0.1 ml) of the radionuclide solution was added.

The contact time for most of the systems were up to 3 months (sampling after 6 h, 1 d, 6 d, 6 weeks and ca 3 months). Phase separation, prior to collecting samples from the aqueous phase, were made by centrifugation at 17000 rpm for 30 min. The radioactivity of the withdrawn samples was measured using either a NaI-well type scintillation counter or a liquid scintillation technique.

The experiments were performed in carbon dioxide free atmosphere at ambient temperature (ca 24°C). The solid/liquid ratio was usually 0.9 g per 45 ml.

Table I. Solid cement phases [1]

Type	Water/cement ratio	% Ballast[a]	Additives, etc.
Standard Portland	0.5	52	-
French Standard Portland	0.36	15[b]	≤0.5% thinning agent
Sulfate resistant	0.5	52	-
Blast furnace slag	0.5	52	-
High alumina	0.5	52	-
Fly ash	0.5	52	10% fly ash
Silica	0.32	49	8% silica fume + 1.2% naphtalene

[a] Quartz sand, 0.1-0.25 mm
[b] Fontaine Bleau sand, <0.3 mm

Table II. Pore water composition (mg/l) [2]

	SP	FP	P-1	P-2	PW
Na	1500	1800	1500-3650	100-260	1600
K	6300	5700	1530-5080	90-180	6300
Ca	90	20	33-92	480-570	40
Al	<5	30	n.d.	n.d.	
Si	<6	6	n.d.	n.d.	
Fe	0.5	0.5	n.d.	n.d.	
SO_4^{2-} [a]	(<300)	(<2000)	640-740	120-250	
pH	13.4	13.2	12.9-13	12.4-12.6	<13.4

[a] Estimated concentration ranges, assuming charge balances and an uncertainty of 0.05 in the pH-measurements, are given within parenthesis.

Legends: SP = Standard Portland, FP = French Standard Portland, P = old concrete (Porjus); PW = artificial pore water.

Table III. Studied radionuclides

Radionuclide	Initial conc. (mol/l)
I-125	$1.9x10^{-9}$
Cs-134	$4.3x10^{-10}$
Th-234 + Th-232 as carrier	$2.0x10^{-10}$
U-233	$2.2x10^{-7}$
Np-235 + Np-237 as carrier	$1.9x10^{-7}$
Pu-239	$1.2x10^{-9}$
Am-241	$2.3x10^{-9}$

RESULTS AND DISCUSSION

The distribution of the radionuclide between the solid phase and the solution was evaluated in terms of the distribution coefficient, K_d, defined as

$$K_d = q/C = [(C_o - C)V]/(Cm) \quad m^3/kg$$

where q = concentration in the solid phase [mol/kg], C = concentration in the aqueous phase [mol/m^3], C_o = initial concentration in the aqueous phase [mol/m^3], V = volume of the aqueous phase [m^3], and m = mass of the solid phase [kg]. The distribution coefficient versus time for the various cement systems is given in Figure 1.

The tri- and tetravalent actinides Am(III) and Th(IV) are both strongly sorbed on concretes. For americium, present in a concentration well below the solubility limit, anionic hydrolysis products are likely to dominate at the low carbonate concentrations of the pore water. Thorium is close to or above the solubility limit and is probably present as $Th(OH)_4$ [3]. Both americium and thorium usually also exhibit a very high sorption in geologic systems in the environmental pH-range [4].

Also the penta- and hexavalent actinides represented by Np(V) and U(VI) exhibit a high sorption in the concrete systems. Anionic hydrolysis products, possibly of low solubility in the presence of high concentrations of cations (Na^+), would dominate rather than highly charged anionic carbonate complexes which dominate in groundwaters [3]. The sorption is significantly higher than for natural water systems (lower pH and higher carbonate concentrations [4]).

Plutonium may be present in several oxidation states in natural waters. The tetra- or possibly the pentavalent state dominate under surface water conditions. Possibly oxic pore water conditions would favour the higher oxidation states (penta- or even hexavalent states) [1]. The observed sorption behaviour resembles more that of uranium-neptunium than of thorium.

The low sorption of cesium, which is a non-hydrolyzed cation under pore water conditions, is in agreement with previous studies [5,6]. The main sorption mechanism for a non-complexed cation is ion exchange, which is inhibited by the high ionic strength. The observed differences in sorption between various concrete types probably reflect the properties of the ballast rather than of the cement matrix.

The sorption of iodine is higher than would be expected for I^- at high ionic strength [7].

The range of distribution coefficients for all concrete systems are illustrated in Figure 2. Evidently the distribution coefficients are largely of the same order of magnitude regardless of the concrete type. Some deviations were however noted and can be summarized as follows (cf. ref. 1):

- Generally the sorption of actinides was highest in lower end of the observed pore water pH-range
- The sorption of penta- and hexavalent actinides was slightly lower on high alumina cement than on the others for unknown reasons
- The old concrete were better sorbents for cesium, probably due to the properties of the ballast materials
- The sorption of iodine was lower on slag cement than on the other cement types for unknown reasons.
- The sorption was generally increasing with time during the first weeks of contact (diffusion into the cement matrix ?)

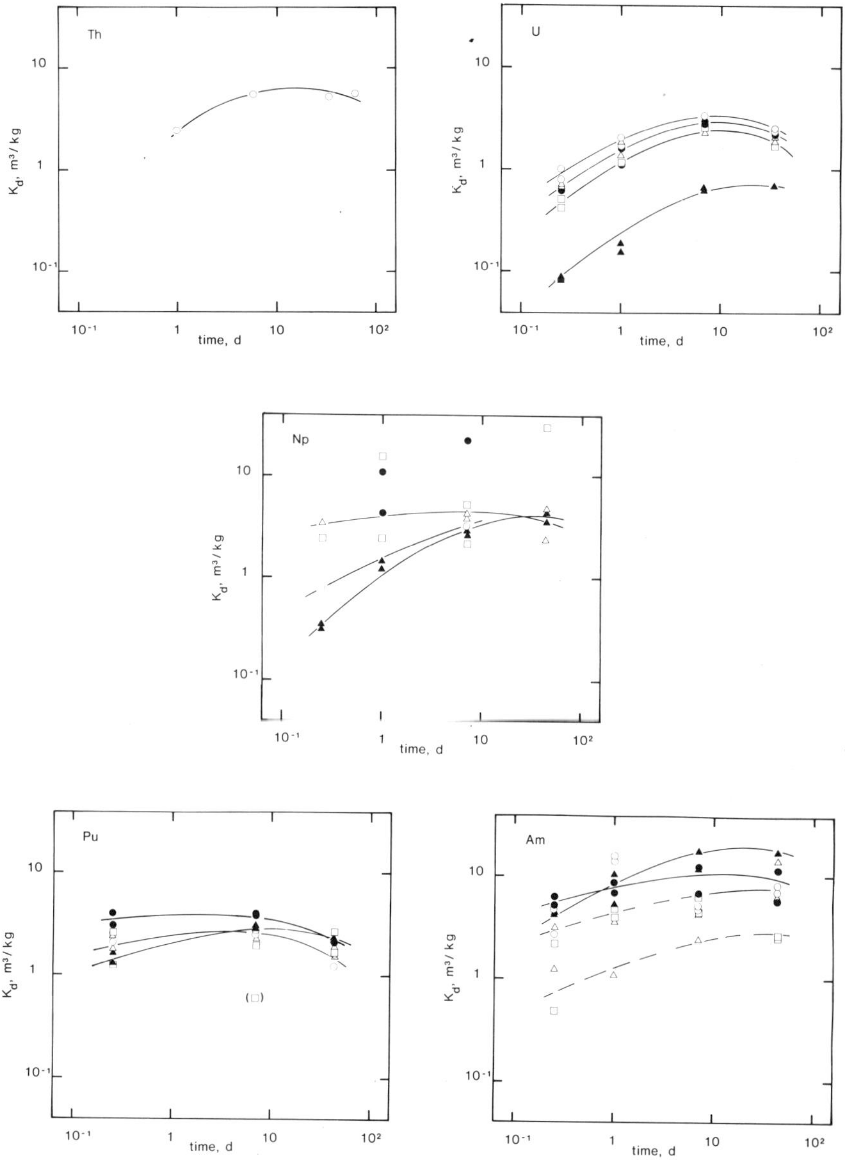

Figure 1. Distribution coefficients (K_d) vs time for Th, U, Np, Pu and Am
○ SP, △ FP, □ T, ● P-1, ▲ P-2 (T = old concrete from Trollhättan;
other legends as in Table I)

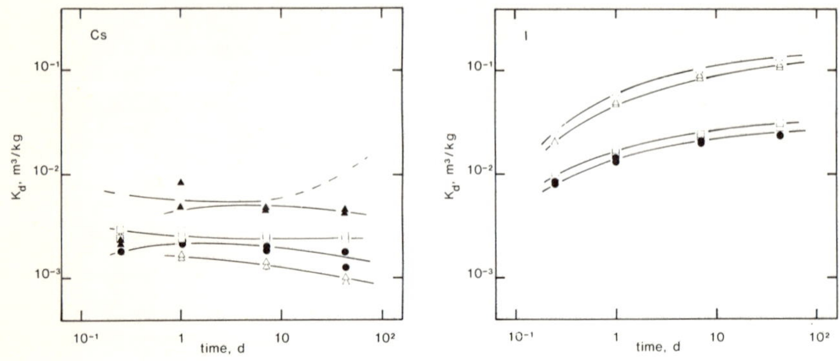

Figure 1 (cont.). Distribution coefficients (K_d) for Cs and I

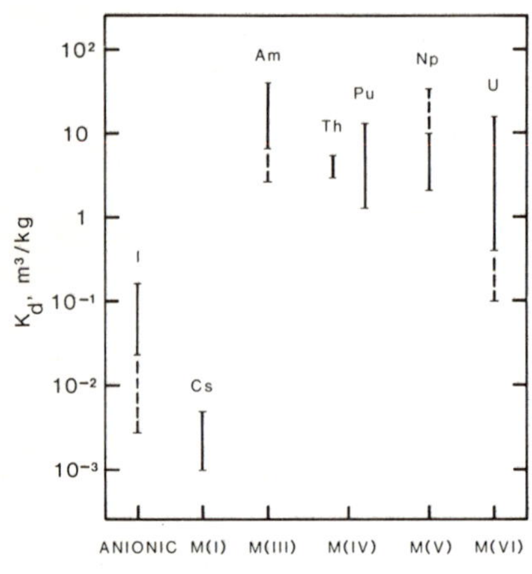

Figure 2. Ranges of distribution coefficients (K_d) for the various systems given in Table I. (The solid lines indicate the range for the majority of the systems)

CONCLUSIONS

The chemical conditions in the vicinity of a waste repository in a geologic system will be highly influenced by the presence of concrete, either as a waste form or in constructions or as engineered barriers. The conditions will vary between two extreme cases, as illustrated in Figure 3:

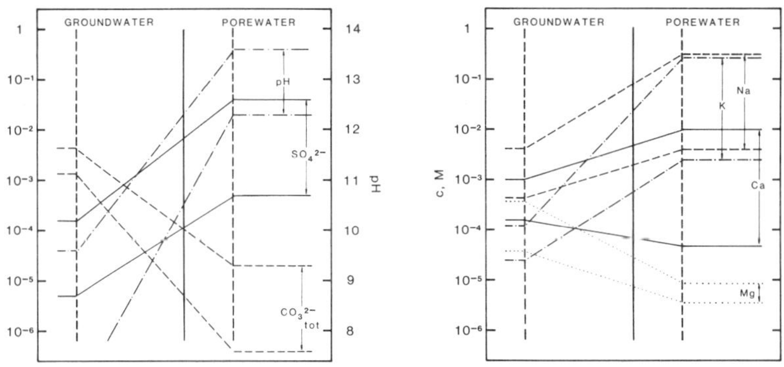

Figure 3. Concentration ranges for dominating an- and cations in cement pore waters and groundwaters from crystalline rock

- o The concrete environment; pore water with high concentrations of sodium, potassium and possibly sulfate as well as a very high pH, buffered by the presence of (Na,K)OH and solid $Ca(OH)_2$; low concentrations of carbonate
- o The undisturbed repository environment; groundwater with a fairly low total salt content, intermediate pH buffered by the presence of calcite and clay minerals and significant concentrations of carbonate as well as organics (for a granitic rock system)

Reactions in the cement/rock interface are likely to occur, notably a precipitation of calcite [1]. The interface layer may possibly lead to a general increase in the transport resistance for dissolved micro components released from the concrete system into the surrounding aqueous phase.

For tri- and tetravalent actinides (americium and thorium, respectively) the mobility can be expected to be very low both in concrete and in the undisturbed rock system.

For penta- and hexavalent actinides (neptunium and uranium, respectively) a low mobility can be expected in the concrete system, while an increase (possibly by orders of magnitude) would be expected in the rock/groundwater system due to the formation of low-sorbing carbonate complexes. Such complexes would not dominate in the pore water system.

Also plutonium would be highly immobile in a concrete environment, regardless of the oxidation state (III, IV, V or VI). An oxidation to the penta- or hexavalent state can not be excluded.

The mobility of cesium in the cement can be expected to be fairly high. However, the mobility in the surrounding rock will be much lower, especially in the presence of high-capacity minerals (e.g. a clay back-fill).

Iodine is likely to exhibit a lower mobility in concrete than in geologic media representative of the rock. The reason for the high sorption of iodine (probably as I^-) in comparison with silicates of geologic origin is not clear.

Diffusion measurements in various fresh concretes as well as in old and artificially degraded concretes are in progress to verify the qualitative conclusions from the batch-type experiments in the present project.

ACKNOWLEDGEMENTS

The assistance in selecting suitable concrete types by Å. Grudemo, L. Rombén and V. Lehtinen, Swedish Cement and Concrete Research Institute, Stockholm, assistance with pore water pressing and analysis by O. Gjørv, Cement and Concrete Research Institute, University of Trondheim, as well as experimental help by W. Johansson and M. Bengtsson are gratefully acknowledged. The French Standard Portland cement was provided by H. Forsström and the old concrete samples from Porjus and Trollhättan by A. Bergström, SKB/KBS, Stockholm. The project has been partly financed by SKB/KBS.

REFERENCES

1. B. Allard, L. Eliasson, S. Höglund and K. Andersson, KBS TR 84-15, SKB/KBS, Stockholm, 1984.
2. K. Andersson, B. Allard, M. Bengtsson and B. Magnusson, Cement and Concrete Research, in press.
3. B. Allard, U. Olofsson and B. Torstenfelt, Inorg. Chim. Acta 94, 205 (1984).
4. B. Allard, KBS TR 82-21, SKB/KBS, Stockholm, 1982.
5. K. Andersson, B. Torstenfelt and B. Allard, in Scientific Basis for Nuclear Waste Management. Vol. 3, edited by J.G. Moore, (Plenum Publ. Corp., New York, 1981), p. 235.
6. B. Torstenfelt, K. Andersson and B. Allard, Chem. Geol. 36, 123 (1982).
7. B. Allard, B. Torstenfelt, K. Andersson and J. Rydberg, in Scientific Basis for Nuclear Waste Management. Vol. 2, edited by C.J. Northrup, (Plenum Publ. Corp., New York, 1980), p. 673.

INVESTIGATIONS INTO THE MOBILIZATION OF ACTINIDES FROM ILW CONDITIONED IN CEMENT AND BITUMEN AND STUDIES OF THEIR SORPTIVE AND MIGRATION BEHAVIOUR IN BOTH THE NEAR- AND FAR-FIELDS FOR A REPOSITORY CONTAINED IN A DEEP SALT ENVIRONMENT

A. R. FLAMBARD, H.-U. FUSBAN, Ch. KEILING AND G. MARX
Dept. of Radiochemistry, Institute of Inorganic & Analytical Chemistry,
Free University of Berlin, Fabeckstraße 34-36, D-1000 Berlin 33,
Federal Republic of Germany.

ABSTRACT

Experiments have been carried out into the leaching of ILW conditioned in cement and bitumen in concentrated salt solutions. Although difficult to observe with real waste leachates, investigations into the leaching of a simulated waste in cement have indicated leached transuranic levels of ca. 10^{-9} M ; the amount of activity leached by a quinary salt solution being 10^2 - 10^3 times higher . This has been interpreted in terms of a pH effect. For the real waste, γ-spectroscopy has indicated a significantly larger release of Cs from cement than from bitumen. For all waste samples a notable absence of colloidal material was observed ; an observation which can be explained in terms of the high solution ionic strengths and the corresponding influence upon radionuclide solvation.

Transuranic mobility studies through salt and sand from a salt dome in Northern Germany have shown the presence of at least two types of species of wildly differing mobility ; one migrating with approximately the same velocity as that of the solvent front and the other strongly retarded. Actinide recoveries (i.e. that passing through the columns) could be strongly influenced by either changing the system pH or by the addition of a competitor such as Ce ; the latter effect pointing to a competitive sorption.

INTRODUCTION

For some time now, the possibilities of disposing radioactive material (either reprocessing wastes or spent reactor fuel elements) in deep salt environments has been considered by many countries |1,2,3| and since 1982 salt has also been included as an option by the Commission of the European Communities for study at community level |4|.

Salt formations posses a number of advantages for radioactive waste disposal over other types of geological media |5|, not the least of them being the absence of groundwater flow and their isolation from aquifiers under normal (undisturbed) conditions |6|. It is conceivable however, that water or brines may force entry via sealed access tunnels and back-fill during the first critical years after the closing of a repository where salt creep (as a result of lithostatic pressure) has not yet proceeded to completion |7|. In the event of such an occurrence it is therefore possible for the intruding brines to come into contact with the waste inventory so that at some later point in time, when these brines are expelled from the cavities as a result of the continuing convergence of the host evaporite, the transport of radionuclides to the overlying rock strata and aquifier region must be comtemplated. The complete szenario for the mobilization and transport of radionuclides out of a salt repository in the event of flooding has been hypothesized elsewhere |8|.

In order to be able to assess the radiological impact on the geosphere and ultimately on men's environment in the event of such a szenario it is necessary to determine accurate data concerning the mobilization and transport/ retardation of radionuclides for a salt environment |9|. The work presented here describes the results of some integral migration experiments where the processes described

Mat. Res. Soc. Symp. Proc. Vol. 50. © 1985 Materials Research Society

above have been simulated in the laboratory.

EXPERIMENTAL DETAILS

Detailed descriptions of the various experimental techniques employed may be found elsewhere |10-14|.

Materials

Salt solutions were prepared either with bidestilled water or with a saline groundwater from the aquifier above a diapir in Northern Germany. Either commercially available salt (ex. Merck) or a natural rock salt (Na3γ orange salt containing anhydrite inclusions) taken from the same formation (at ca. 729 m) were employed. The quinary Q solution was prepared from commercial materials according to D'Ans |15|. Samples of sand for column migration experiments were taken from the overlying strata of the salt dome at a depth of ca. 215 m.

The waste forms employed in this study (both real and simulated), were made available by the Commission of the European Communities under the auspicies of the MIRAGE programme |9, 14|. The first, an evaporator concentrate from the WAK conditioned in Portland 35 F cement (Reference Waste Form, RWF No. 8) was obtained from KfK and the second, consisting of sludges from chemical precipitation solidified in Shell Mexphalte R 90-40 (RWF No. 9) was obtained from CEA/CEN Cadarache. The specification of these RWF's have been published by the CEC |16|. A further cemented waste form, corresponding to the composition of RWF No. 8 but with the fission products inactively modelled and doped with extra LWR-plutonium (Pu-239 and Pu-240) and Am-241 was also obtained from KfK.

A standard americium (III) hydroxide as reference system was prepared by the dissolution of americium (IV) oxide in 8 M hydrochloric acid |17| followed by precipitation with ammonium hydroxide solution |18| and ageing in an autoclave |11|. A reference plutonium (IV) hydroxide was similarily prepared by precipitation at a pH of 4 from a 3 M hydrochloric acid solution by the addition of sodium hydroxide solution |19| followed by accelerated ageing |11|.

Physico-chemical characterization of salt brines

Solution pH-values were measured by means of a Ross electrode |10|, for which the usual reference electrode electrolyte of 3 M KCl had been replaced by a saturated NaCl solution in order to minimize the occurrence of a diffusion potential |20,21|. Previous experiments have indicated little evidence of extraneous electrode behaviour |22|. For the measurement of solution Eh, an Ingold redox electrode was similarily employed |10,22|. Carbon dioxide/carbonate contents were determined by classical precipitation with barium hydroxide and back-titration |23| after acidification and displacement by nitrogen. Solution densities were measured with the aid of a digital densitometer using the ultra-sonic vibration principle |24|.

Waste-leaching and actinide solubility determinations

Samples of crushed waste were leached in polyethene flasks at 298 K with a fixed solid to solution volume of 0.14 g cm^{-1}(approximately ten times that expected for an ILW storage chamber in a salt repository in the event of flooding |25,26|), and intermittant stirring. Previous experimentation had indicated radionuclide adsorption on container walls to be less than ca. 3-4 % |27|. Samples were periodically removed and filtered through membranes of varying pore-sizes (800 to 1.3 nm) in order to determine colloidal-size distributions |11,26|. Solutions derived from the pure actinide compounds were prepared in a similar fashion, although in this case ca. 20 mg of hydroxide were contacted with 20 cm^3 of solution.

Radionuclides were determined either by liquid scintillation counting (α-emitters), using either Packard Pico-Fluor 30 or Beckman Ready-Solv MP "cocktails" |11,26,28|, or by γ-spectoscopy (Ge-Li detector).

Sorption and Mobility investigations

Column migration experiments were carried out with the apparatus illustrated in Fig. 1 |12-14,22,26|. With the aid of a peristaltic pump, solutions were

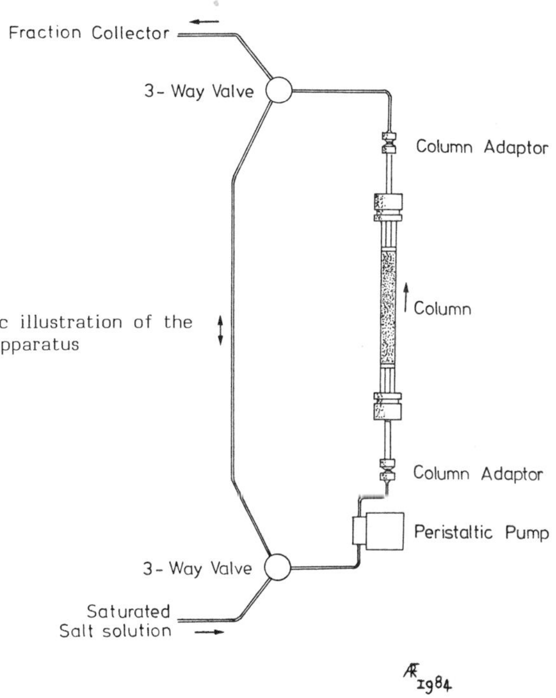

Fig. 1. Schematic illustration of the column apparatus

passed through columns filled either with rock salt or sand after which effluents were automatically collected by a fraction collector for subsequent analysis. A by-pass system assisted with the "spiking" of the columns and enabled experiments to be carried out in the recirculation mode when desired. The columns themselves were commercially available chromatographic columns consisting of precision borosilicate glass tubing of diameter 2.5 cm. The column packing was effectively held between two adaptors and although a standard column length of 10 cm was arbitrarily adopted, this simple arrangement allowed columns of various lengths to be easily facilitated. A LHSV of 1.5 cm^3 was employed and column porosities were determined in the usual fashion by measuring the breakthrough curves using tritiated water as tracer |29-31|.

RESULTS AND DISCUSSION

Table I indicates some physical characteristics of the salt solutions, agreeing well with the findings of other workers |22,23|. It can be seen that the NaCl solution and the Na3γ in bidestilled water exhibit very similar properties indicating that the latter's impurity levels (eg. Fe) are not sufficient to affect solution Eh or pH to any significant extent. Not surprisingly, the rock salt solution in groundwater contains appreciably more carbonate |33|, with the result of an increase in pH and decrease in Eh |10|. In the case of the Q solution, the relatively low pH of 4.2 can be shown to be responsible for the high positive redox potential by the application of the Nearnst equation |10|.

Table I Physical Properties of concentrated salt solutions at 298 K

Parameter	sat. NaCl	sat. Na3γ(bides)	sat. Na3γ(gw)	Q solution
pH	6.6	6.0	7.0	4.2
Eh/mV	89	96	49	243
Carbonate cont. /mole dm^{-3}	3×10^{-5}	7×10^{-6}	6×10^{-4}	2×10^{-5}
ρ / g cm^{-3}	1.2041	1.1954	1.2036	1.2930

Some representative results from waste leaching experiments in these concentrated brines are listed in Tables II a and b.

Table IIa The leaching of plutonium from simulated ILW conditioned in cement (corresponding to RWF No. 8) in concentrated salt solutions at 298 K (m/V = 0.14 g cm^{-3})

System/Time	Molar concentrations after filtering through filters of pore-size: 800 nm	450 nm	220 nm	100 nm	50 nm	2 nm
Na3γ(bides)						
54 d	9.5 E-9	-	-	-	-	-
156 d	6.5 E-9	5.2 E-9	5.2 E-9	5.2 E-9	-	4.9 E-9
280 d	15.3 E-9	7.2 E-9	5.6 E-9	-	-	-
285 d	13.4 E-9	8.6 E-9	8.4 E-9	6.9 E-9	6.0 E-9	-
Q soln.						
1 d	1.7 E-6	1.6 E-6	1.5 E-6	-	-	-
31 d	3.3 E-6	1.9 E-6	1.8 E-6	-	-	-
187 d	3.5 E-6	-	-	3.3 E-6	-	-
273 d	7.1 E-6	7.0 E-6	7.1 E-6	7.0 E-6	7.0 E-6	7.0E-6
460 d	3.7 E-6	-	3.3 E-6	-	-	-

For the simulated RWF No. 8 results are given in terms of leached plutonium concentrations only because the levels of leached americium were found to be in the same ratio as that in the waste (ca. 1 : 10) indicating the absence of preferential leaching. Furthermore, it can be seen from Table IIa that more radionuclides were leached by the Q solution than by Na3γ(bides), a result believed to be due to pH effects. Cements are of course, alkaline |32|, and it has been shown |34|, that the pH of brines in contact with cement change very rapidly to a final value of ca. 6.5 (in the case of the Q solution) or 12 (NaCl solution) |34-36|, thereby strongly affecting radionuclide solubility. This strong pH effect is made more apparent from the solubility data of Table III |13,35| for

Table IIb The leaching of radionuclides from real ILW conditioned in cement and bitumen (RWF No. 8 and RWF No. 9) in saturated Na3γ (bides) at 298 K (m/V = 0.14 g cm^{-3})

System/ RN	Molar concentrations after time: 1 d	3 d	10 d	27 d	34 d	45 d
RWF No.8						
Cs-137	2.0 E-7	2.2 E-7	2.2 E-7	2.3 E-7	1.9 E-7	2.0 E-7
Ru-106	-	4.7 E-11	-	4.9 E-11	4.3 E-11	4.2 E-11
Co-60	-	7.1 E-12	5.2 E-12	6.0 E-12	4.4 E-12	4.8 E-12
RWF No. 9						
Cs-137	0.6 E-12	1.2 E-12	2.5 E-12	2.0 E-12	2.1 E-12	2.0 E-12
Sb-125	0.8 E-13	2.1 E-13	3.8 E-13	-	2.4 E-13	2.5 E-13
Ru-106	2.4 E-12	3.9 E-12	4.8 E-12	5.0 E-12	5.8 E-12	5.2 E-12
Co-60	2.4 E-13	7.2 E-13	9.5 E-13	8.8 E-13	14.3 E-13	9.6 E-13

the standard plutonium and americium hydroxides where it may also be seen that the effects of groundwater on radionuclide solubilities are only slight.

Table IIb indicates some results for the leaching of real waste. Unfortunately, due to very high counting backgrounds resulting from the decay of fission products, it was not possible to follow the leaching of α-emitters with any great certainty, although traces were able to be descerned |37|. Care must be taken when comparing the leachabilities of both RWF's because of the different extents of surface exposed to the brines, but a very striking difference between the two waste forms concerns the leaching of Cs-137. The levels of Cs-137 are approximately the same for both waste forms, although the cement leachate contains a factor 10^5 more Cs-137 as that from bitumenized waste ; a significant result since the variation in surface areas of matrix present cannot differ by more than a factor 10^2–10^3. Indeed, it is well known that cement matricies have a poor retentivity for species such as Cs which are soluble in alkaline environments |32,38|, so that this high leachability difference is explainable in terms of differences in matrix corrosion mechanisms and sorption of radionuclides on corrosion products |32,38,39|. A further observation made with that Cs leached from both waste forms could not be made to precipitate with sodium hexanitrocobaltate |40| unless previously evaporated to dryness with concentrated nitric acid, even though tracer Cs added to the solutions could be made to precipitate quantitatively |27|. The reason for this is unknown at present but further investigations into Cs speciation are currently in progress.

Table III Equilibrium solubilities of plutonium and americium hydroxides in concentrated salt solutions

System/pH @ start	Solubilities (mole dm^{-3}) of: Pu	Am
Sat. NaCl		
pH 1.0	3.0 E - 4	> 1.4 E - 3
pH 5.5	4.5 E - 6	1.0 E - 6
pH 10.0	1.7 E - 8	3.6 E - 7
Sat. Na3γ(bides) pH 6.0	1.3 E - 6	2.5 E - 6
Sat. Na3γ(gw) pH 7.0	2.0 E - 7	9.0 E - 7
Q solution pH 4.2	2.9 E - 6	2.4 E - 5

Ultrafiltration experiments involving waste leachates and solutions derived from the standard hydroxides have indicated colloidal material within the size-range 800 to 1.3 nm (nominal filter porosities) to be largely absent with the exception of leachates derived from simulated RWF No. 8 in Na3γ solutions (cf. Table IIa). It is quite probable however that these colloids are due to abraision effects |41,42|. A decrease in colloidal particle size when HLW has been leached in brines as opposed to fresh water has already been reported |42,43| and is be-

Table IV Results of column migration experiments with leachates derived from simulated ILW in cement (RWF No. 8)

Column Material	Effective Porosity %	Solution in	Solution pH	Actinide Am	dose Pu	Actinide Recovery %	K_D ($cm^3 g^{-1}$)
NaCl(u)	36.6	NaCl	12.0	6.0 E-11 (t)		70(t)	0.04
NaCl(c)	36.6	NaCl	11.9	4.2 E-11 (t)		77(t)	0.005
Na3γ(u)	41.0	Na3γ(bides)	12.1	7.6 E-13	9.7 E-12	5(Am) 81(Pu)	0.09(Am+Pu)
Na3γ(c)	40.9	Na3γ(bides)	12.0	2.1 E-12	5.8 E-10	0.3(Am) 5.6(Pu)	0.03(Am+Pu)
Na3γ(u)	44.1	Na3γ(gw)	12.1	7.5 E-13	6.6 E-10	0.4(Am) 48(Pu)	0.09(Am+Pu)
Sand	40.0	Na3γ(bides)	12.0	8.0 E-11 (t)		0.3(t)	0.12
Sand	40.0	Na3γ(bides)	12.0	11.1 E-11 (t)		0	-

u: unconditioned column
c: column preconditioned to pH 12
t: Am, Pu not determined separately

lieved to be a consequence of the high ionic strengths of brine solutions reducing colloidal particle size through a reduction in the degree of solvation of the hydration shell.

Results of some column migration experiments are summerized in Table 4. In all cases breakthrough curves similar to Fig. 2 were obtained. The following points have been noted. (i) Preconditioning a column to pH 12 before the administering of radionuclide only had an influence upon the mobility characteristics of plutonium through Na3γ salt, where the actinide retardation (measured as the inverse of recovery) increased quite dramatically. Even though not directly measured, for the sodium chloride columns this does not appear to be the case. (ii) Americium appears to be preferentically retarded with respect to plutonium. (iii) In all cases, the mobilities of the species which are recovered from the columns are identical to that of the solvent front (ie. K_D ca. 0). Experiments involving in-

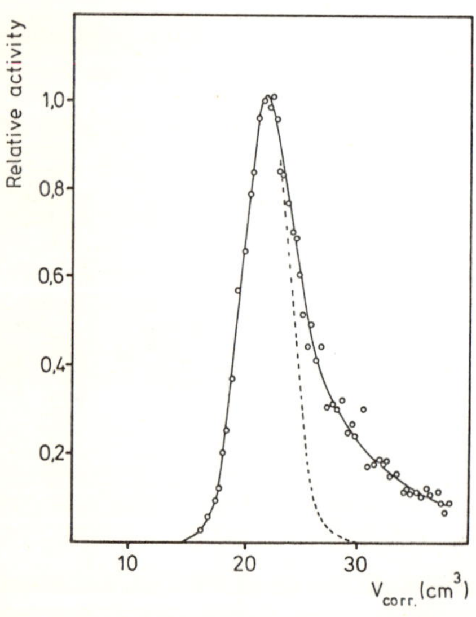

Fig. 2. Breakthrough curve for plutonium derived from simulated RWF No. 8 through Na3 rock salt

creasing the LHSV or **reversing** the direction of flow |13| were not successful in increasing the levels of actinide recovery. Only after accidification of the salt solutions to pH-values in the region of 1-3 was it possible to quantitatively remove the actinides retained by the columns. (iv) Actinide retention by sand appears to be stronger than by salt, a result also observed by other workers |42|.

Table V Results of column migration experiments with solutions derived from standard hydroxides in Na3γ(bides) through Na3γ salt - The effect of competitive sorption -

Column Porosity	Solution derived from	Solution pH	Presence of Cerium	Actinide Recovery %	K_D ($cm^3 g^{-1}$)
30.7	$Am(OH)_3$	5.8	-	0.2	0.07
38.9	$Am(OH)_3$	6.0	+	100	0.01
35.4	$Pu(OH)_x$	6.2	-	6	0.05
41.9	$Pu(OH)_x$	6.1	+	88	0.03

In view of the complexities of ILW |16|, it was decided to investigate the influence of a competing species such as Ce on the mobility characteristics of the actinides |13,44|; whereby the low α-activities leached from conditioned waste forms has necessitated investigations being carried out with solutions derived from the standard hydroxides. From the Results of Table V it can be seen that the incorporation of cerium into the spiking solutions dramatically increased the actinide recovery levels, although the K_D values for the migrating species remained unchanged. This is a strong indication that the actinides were normally retained upon special active sites within the columns; sites which have now been effectively blocked. Further experiments into competitive sorption effects coupled with preliminiary investigations into the influence of organic residues upon radionuclide migration are currently in progress, but the importance of such influences when considering safety analyses for radionuclide migration in geological systems need not be overemphesized.

ACKNOWLEDGEMENTS

This work was carried out under contract to the Commission of the European Communities (Nr. 415-83-7-WAS D (B)) as part of their Indirect Action Programme on the Management and Storage of Radioactive waste, and also to the Bundesministerium für Forschung und Technologie (Nr. 5691 KWA 5313) as part of the PSE Programme. The assistance of Mr. V. Eßer in determining the physical propeties of the salt solutions and of Mrs. E. Rohmann in the preparation of the manuscript is gratefully acknowledged.

REFERENCES

1. OECD-NEA/CEC Report, Geological Disposal of Radioactive Waste (Published by the OECD, Paris, 1984), p. 25
2. A.G. Herrmann, Radioaktive Abfälle, (Springer-Verlag, Berlin, 1983), pp. 159, 160.
3. D.W. Powers, S.J. Lambert, S.E. Sheffer, R.R. Hill and W.D. Weart, SNL-Report (SAND-78-1596 1 and 2, 1978).
4. JRC-Ispra/CEC Report, edited by N. Cadelli, G. Cottone, G. Bertozzi and F. Girardi EUR 9220, Luxembourg, 1984, pp. 42-50.
5. A.G. Milnes, M. Buser and W. Wildi, Z. dt. geol. Ges. 131, 359 (1980).
6. Z. Dlouhý, Disposal of Radioactive Wastes, Studies in Environmental science 15, (Transl. form Czech, Elsevier Scientific Publishers, New York,1982) p 220.

7. C.A. Baar, The in-situ Behavior of Salt Rocks, Applied Salt Rock Behavior 1, (Elsevier Scientific Publishers, New York, 1977).
8. Projekt Sicherheitsstudien Entsorgung (PSE), Zusammenfassender Abschlußbericht 1, edited by P. Faber and A. Huf (Published by the Project directorship via HMI, Berlin, 1985) p 29.
9. CEC Report, EUR 9304 EN, Luxembourg 1984.
10. V. H.-G. Eßer, Wissenschaftliche Hausarbeit i.R. d. 1. Staatsprüfung, Freie Universität Berlin, 1984.
11. Ch.D. Keiling, Diplom-Arbeit, Freie Universität Berlin, 1984.
12. A.R. Flambard, H.- U. Fusban and G. Marx, HMI-Berlin Contract ; 055138/531 /925, Abschlußbericht, Freie Universität Berlin, 1984.
13. A.R. Flambard, G. Marx, V. H.-G. Eßer, H.-U. Fusban and Ch.D. Keiling, presented at the 1st International Conference on Nuclear and Radiochemistry, Lindau, Federal Republic of Germany, 1984 (unpublished).
14. B. Côme, G. Bidoglio, N. Chapman, J. Delmas, F. Ewart, M. Jebrak, A.R.Flambard and G. Marx, presented at the 2nd EC Conference on Radioactive Waste Management and Disposal, Luxembourg 1985 (to be publ. by the CEC, 1985).
15. J. D'Ans, Die Lösungsgleichgewichte der Systeme der Salze ozeanischer Salzablagerungen, (Verlagsgesellsch.f. Ackerbau mbH, Berlin, 1933), p. 254.
16. CEC Reports, edited by R.A.J. Sambell, EUR 8663 EN, Luxembourg, 1982 and EUR 9423 EN, Luxembourg, 1984.
17. C. Keller, The Chemistry of the Transuranium Elements, (Verlag Chemie, Weinheim, 1971), 3, 497.
18. G. Koch, Transurane, publ. as "Gmelin Handbuch der Anorganischen Chemie", 8th ed. (Springer-Verlag, Berlin, 1975), 20 (D 1), p. 113,
19. O.T. Wick (ed.), Plutonium Handbook, (Gordon and Breach Science Publishers, New York, 1967), 1, 345.
20. K. Schwabe, pH-Messung, (Akademie-Verlag, Berlin DDR, 1980), p. 78.
21. R.G. Bates, Determination of pH (John Wiley & Sons Inc., New York, 1964), p. 41.
22. Ref. 8, ibid, Fachband Nr. 14, p. 1-15.
23. A.E. Markham and K.A. Kobe, J. Amer. Chem. Soc. 65, 2030 (1943).
24. O. Kratky, H. Leopold and H. Stanbinger, Z. angew. Phys. 27 (4), 273 (1973).
25. R. Brüggemann, TU Berlin (private communication).
26. G. Marx and A.R. Flambard, CEC Contract 415-83-7-WAS D (B), Abschlußbericht, Freie Universität Berlin, 1985 (in preparation).
27. Ch. D. Keiling, A.R. Flambard and H.-U. Fusban, unpublished results.
28. Ch. D. Keiling, A.R. Flambard and G. Marx, subm. to Nuc.Count. Exch. (1985)
29. U. Mühlenweg, Diplom-Arbeit, Freie Universität Berlin, 1982.
30. D. Klotz and H. Moser, Meyniana 34, 95 (1982).
31. D. Klotz and H. Hirth, Deut. Gewäss. Mitteil., 26 (5), 128 (1982).
32. F.P. Glasser, A.A. Rahman, R.W. Crawford, C.E. McCulloch and M.J. Angus in Testing Evaluation and Shallow Land Burial of Low and Medium Radioactive Waste Forms, Radioactive Waste Management Series 13 (Harwood Academic Publishers, Chur, 1984), p. 95.
33. B. Skytte Jensen, Migration Phenomena of Radionuclides into the Geosphere, Radioactive Waste Management Series 5 (Harwood Academic Publishers, Chur, 1982), p.B 13
34. A.R. Flambard and H.-U. Fusban, Sonderbericht zum PSE-Projektstab, Freie Universität Berlin, 1984
35. Ref. 8, ibid, Fachband Nr. 14, p. 3-15 to 3-18.
36. CEC Report, Research and Development on Radioactive Waste Management and Storage, Radioactive Waste Management Series 12 (Harwood Academic Publishers, Chur, 1984), p. 307.
37. ibid, pp. 31, 37.
38. P. Vejmelka, G. Rudolph and R. Köster, KfK Report 3413, 1982.
39. K. Brodersen, Ref. 34, ibid, p. 147.
40. H. Remy, Lehrbuch der Anorganischen Chemie, (Akademische Verlagsgesellschaft Geest & Portig KG, Leipzig 1959), II, p. 361.
41. D. Klotz and F. Oliv, Gwf-wasser/abwasser, 1983, 124.
42. G. Bidoglio, A. Avogadro, A. de Plano and G.P. Lazzari, Radioact. Waste

Management and Nuc. Fuel Cycle 5, 311 (1984).
43. A. Avogadro, C.N. Murray, A. de Plano and G. Bidoglio, Porc. Int. Symp. on Migration in the Terrest. Environment of Long Lived R'nuclides from the Nuc. Fuel Cycle, IAEA, Knoxville, 1981.
44. Ref. 8. ibid, Fachband Nr. 14, p. 4-15

THE SOLUBILITY OF ACTINIDES IN THE NEAR-FIELD

F.T. EWART*, R.M. HOWSE*, H.P. THOMASON*, S.J. WILLIAMS* and J.E. CROSS**
*Chemistry Division, Atomic Energy Research Establishment, Harwell, Oxford OX11 ORA, U.K.
**University of Wales Institute of Science and Technology, PO Box 13, Cardiff, CF1 3XF, Wales, U.K.

ABSTRACT

The solubility of some of the radiologically important actinide elements have been determined in a water whose chemistry was representative of that believed to exist in potential repositories in the U.K. The solubilities of the actinides have been determined as a function of the pH of the water. The results of these experiments have been compared with the results of predictions made using the PHREEQE geochemical modelling code in order to test and validate the data base used in the model. In the light of these comparisons, suggestions are made for alternative values for the thermodynamic data and for further studies.

INTRODUCTION

A nuclear power programme produces various categories of waste which are classified in terms of their heat emission and actinide inventory. A general disposal option under consideration for these wastes is their immobilisation in suitable matrices and burial at appropriate depths in carefully selected geological formations. The depth and construction of such repositories will depend on the category of the waste to be entombed. The most probable route for the return of radionuclides to the biosphere from the undisturbed repository is via migration in the aqueous phase. If the concentrations that can be attained by the radionuclides in the repository are known, the likely release rate of radionuclides from the near field can be derived from these values, known as the source term, and the water velocity through the repository. If the water flow is slow, the source term concentration may reach maximum equilibrium values limited either by nuclide availability or by solubility. If sorption from solution occurs the concentration will be reduced below this value. Thus, in order to model any future disposal option it is necessary to obtain data on the aqueous chemistry and sorption behaviour of many radionuclides. The information obtained from such studies will be applicable to a range of repositories regardless of the precise nature of the disposed waste. Before studies of sorption from solution onto solid components of the repository can be undertaken it is necessary to obtain solution concentrations for those nuclides which may be solubility limited; such nuclides will include the actinides, plutonium, americium and neptunium.

A potential disposal route for low and intermediate level wastes in the U.K. is encapsulation in a cement matrix followed by burial in a concrete lined repository at a suitable depth. In these repositories the aqueous phase will be essentially the pore water of the cement/concrete matrix with an alkaline solution chemistry dominated by the soluble components of the cement and a redox potential controlled by the corrosion of the metallic (iron) components of the waste canister and repository. The pH of the pore water will decrease as the cement ages [1] in a range between pH 13 and pH 11. Initially the redox potential will be held at about -800 mV vs standard hydrogen electrode (SHE) by the Fe/Fe(II) equilibrium and subsequently by the Fe(II)/Fe(III) equilibrium at approximately -500 mV in the presence of excess ferrous iron, at pH 11. Reliable experimental data for the solubilities of plutonium, americium and neptunium in this range of chemical conditions are sparse. Whilst some experimental results have been reported [2,3], discussions of the management of radioactive waste under UK conditions [4] have been based on chemical equilibrium models. Much of the data used in these models have been extrapolated from measurements at low pH or estimated by comparison with other elements [5]. In the work reported below, direct measurements have been made of the solubility limits of these actinides in concrete-equilibrated water. The pH of this water was varied over a limited range to simulate the ageing of the cement/concrete and the consequent changes in pH and to enable validation of the solubilities predicted by the currently available models and data bases.

EXPERIMENTAL

The water used for these experiments was representative of the pore water of a cement-based structure. In young cements the pH is about 12.5 due to the presence initially of sodium and potassium hydroxides and in older cements the alkalinity is controlled mainly by excess calcium hydroxide. In both cases sufficient carbonate impurity is present to saturate the water with respect to calcium carbonate. The reference water was prepared by equilibrating aged concrete with demineralised water until the pH and carbonate levels had reached equilibrium: this normally took several weeks. Analyses of pore water recovered from mature concretes [20] suggest that the concentrations of sodium and potassium may be lower in these experiments than in true pore water. After equilibration the water was decanted from the solids and filtered through a 50 nm Millipore cellulose filter. Samples were taken, after filtration, and analysed by photon correlation spectroscopy for colloids [6]; no significant amounts were found. The pH was varied in this water by the addition of sodium hydroxide or hydrochloric acid.

The results of the chemical analysis of the water is given in Table I below. A separate analysis by ion chromatography was made for carbonate. In the case of the experiments at low redox potential, the conditioned water was sparged with argon/5% hydrogen for some hours. This reduced the redox potential to ca. 50 mV (vs. SHE); at this redox potential the oxygen content of the water is ca. 10^{-37} molar. The water contains no redox couple which could be significant at this redox potential and further reduction could be achieved solely by the removal of oxygen from the solution.

Table I. Chemical Analysis of Water

	Component (millimolar)						
	Ca	Na	Mg	Cl	$SO_4^{=}$	$CO_3^{=}$	pH
Content	10	.05	.005	2	3	.03	12

This removal was accomplished, in a glove box with a nitrogen atmosphere containing less than 50 vpm of oxygen, by contacting the water with clean stainless steel mesh. This removed oxygen by forming a stable oxide film and contributed a negligible amount of iron or chromium to the solution. Because of the low concentration of redox species in these waters it was not possible to measure the redox potential with any accuracy, but this method has been shown in previous experiments to produce a redox potential reducing with respect to Tc(IV)/Tc(VII) [7]. In the absence of any redox couple in the water, the experiment in a closed vessel in an inert atmosphere glove box, is essentially a closed system. The redox potential will therefore be controlled by the valence state of the actinide in the source solution (since the concentration of the actinide (> 10^{-10} M) vastly exceeds that of the residual oxygen in the system) and the inadvertant oxygen ingress. The measured redox potential in the plutonium experiments was -300mV and in the americium and neptunium experiments +200mV.

The actinide source solutions were used as chlorides. Some difficulties had arisen from the use of nitrate solutions: trivalent plutonium was found to be readily oxidised in the presence of nitrate and the colour quenching produced by nitric acid in the scintillant for the liquid scintillation counter gave rise to some uncertainty in the observed count rates. Both of these problems were overcome by the use of chloride solutions (acidified with hydrochloric acid). The source solutions of plutonium and americium were prepared by precipitation of the hydroxide from the (usually) nitrate stock solution. The hydroxide was washed with water and then dissolved in hydrochloric acid. Plutonium was obtained in the tetravalent state and was reduced with hydroxylamine hydrochloride; the completion of the reduction was confirmed by measurement of the 560, 600 and 650 nm absorption peaks on a spectrophotometer. The Pu(IV)/Pu(III) ratio was less than 1:100. Solutions of neptunium were obtained as Np(V) chloride in hydrochloric acid solution, solvent extraction [8] and UV spectroscopy were used to confirm that Np(V) was the only oxidation state present. The source solutions were sequentially diluted before use to produce the desired concentration of nuclide and acidity. The acidity was generally such that the aliquot added to the water sample did not significantly change the water chemistry.

Small aliquots of the source solutions were added to 10 ml of the conditioned water, in a polypropylene vessel, and mixed well. For the experiments at pH values other than the equilibrium values, the pH was adjusted by the pre-addition of calculated amounts of sodium hydroxide or hydrochloric acid. Samples of this solution were taken and filtered, using the technique of Rai [9], through pre-conditioned Amicon CF25 filter cones into 0.1 or 0.25 ml of 1M hydrochloric acid. The acidified samples were transferred to a plastic counting vial, 10 ml of Ready Solv MP (Beckman Instruments) added and counted by liquid scintillation counting using a Beckman LS1801 liquid scintillation counter after the decay of the chemiluminescence. The pH of the residual solution was measured, and, when possible, also the redox potential. The equilibration time after mixing depended on the behaviour of the actinide and this was determined by monitoring the change in the filtered concentration with equilibration time. The time used was that after which no further significant change occurred; for plutonium and americium it was found to be 30 minutes, for neptunium 24 hours.

The amount of actinide added was such that pre-filter concentrations of approximately 10^{-5} molar were produced. A comparison of the activity added with the filtered activity showed the limiting solubility. The limit of detection of the actinides in this experimental procedure is restricted by the small volume of sample available. To enhance the sensitivity and lower the detection limit, some shorter lived isotopes were used, namely 238-Pu and 235-Np [19]. These isotopes were added to the 'normal' isotopes in the amounts necessary to give suitable counting statistics in the final solutions.

RESULTS

The solubility data for americium have been obtained only under oxidising conditions since only one valence state can exist in these experiments. Data have been obtained for neptunium under oxidising conditions as a precursor to later measurements under reducing conditions, in order to determine the possible interference from unreduced Np(V).

The experiments investigating the effect of pH on the solubility of the actinides have been simulated using the geochemical code PHREEQE [10], which computes the chemical equilibrium of a system. The chemical characteristics (e.g. component concentrations, Eh) of each experiment were defined and used as input to the code, together with a thermodynamic database comprising formation constants. Throughout a calculation, the mass balance is maintained and so it correctly represents an experiment with solid phases, hence enabling the maximum solubility of each actinide to be predicted. The thermodynamic database used for all the non-radioactive components was the original USGS [10] compilation. The actinide database was compiled at UWIST [11], based primarily on the earlier compilations of Lemire and Tremaine [12], Allard [13], and Benson [14], but with considerable review and updating of their data. In particular, estimated values have been revised in light of the recent experimental results of Rai et al [15], Kim [3,16] and Ryan and Rai [17]. The final formation constants for the predominant species are given in table II. The simulation results show the reference water to be oversaturated with respect to calcite. This is considered to reflect the real situation since the carbonate concentration used in the calculation was obtained by ion chromatography, which is more accurate than the titration method in alkaline conditons.

DISCUSSION

The comparison between the experimental data and the first series of calculations enabled us to refine the database. The experimental results and the final simulations are shown in figures 1-3. In the plutonium case the most thermodynamically stable solid was PuO_2, however, its use in the simulation gave rise to predicted solubilities that were several orders of magnitude too low. The use of amorphous $Pu(OH)_4$ as a solid phase gave good agreement with the experimental results (Fig.1). The increase in solubility at low pH was attributed to Pu(III) solution species.

The modelled results for americium reflect the general trends observed experimentally. It is predicted that the solid phase changes from $AmOHCO_3$ at pH < 9 to $Am(OH)_3$ at pH > 9.5 and this is supported by the point of inflection on the experimental curve. Modelling between pH 8.5 and pH 11.5 is very sensitive to the exact values of the formation constants taken for the different mixed hydroxycarbonate species. Differing values are found in the literature and initial modelling used the most recent data by Kim [3], corrected to the required ionic strength using the Davies equation. In the final model (fig. 2) the constant for the $Am(OH)_2CO_3^-$ species was reduced by ten, giving a considerably better fit.

Examination of initial modelled and experimental results for neptunium clearly indicate that the neptunium solubility was not limited by NpO_2, it was remodelled assuming the formation of NpO_2OH. Under the conditions of this experiment, as in the plutonium case above, the hydroxide is the more likely solid phase. This gave a reasonable fit at low pH but increased solubility was predicted at pH > 11 due to the formation of Np(V) hydrolysis products. The subsequent removal of these species from the model, leaving

Table II.

Selected Thermodynamic data used for final simulations
(correct at Ionic Strength I=0 and 25°C)

Solution Species	Formation Constant (log K)	Reference	Notes
$Pu(OH)^{3+}$	-1.19	[3]	
$Pu(OH)_2^{2+}$	-1.79	[3]	
$Pu(OH)_3^{+}$	-4.75	[3]	
$Pu(OH)_4^{0}$	-10.54	[3]	
$Pu(OH)_5^{-}$	-22.7	[17]	By analogy with uranium
$Am(OH)^{2+}$	-8.2	[15]	
$Am(OH)_2^{2+}$	-17.1	[15]	
$Am(OH)_3^{0}$	-27.0	[15]	
$Am(OH)_4^{-}$	-40.0	[15]	
$Am(OH)CO_3^{0}$	-0.59	[3]	
$Am(OH)_2CO_3^{-}$	-9.71	[3]	Adjusted (see text)
$AmOH(CO_3)_2^{2-}$	3.42	[3]	
$NpO_2(OH)^{0}$	-11.7	[3]	Excluded from final simulation
$NpO_2(OH)_2^{-}$	-23.35	[3]	" " " "
<u>Solid Phases</u>	<u>Solubility Product (log K_s)</u>		
PuO_2	-7.0	[12]	
$Pu(OH)_4$	0.4	[3,12,13,14]	Interpolated
$Am(OH)_3$	17.5	[15]	
$Am(OH)CO_3$	-7.0	[3]	
NpO_2	-2.7	[14]	
$NpO_2(OH)$	4.9	[13]	

NOTE All hydrolysis reactions are expressed in terms of the general reaction:

$$pAn^{n+} + qH_2O + CO_3 \longrightarrow An_p(OH)_qCO_{3r}^{(pn-q-2r)} + qH^+$$

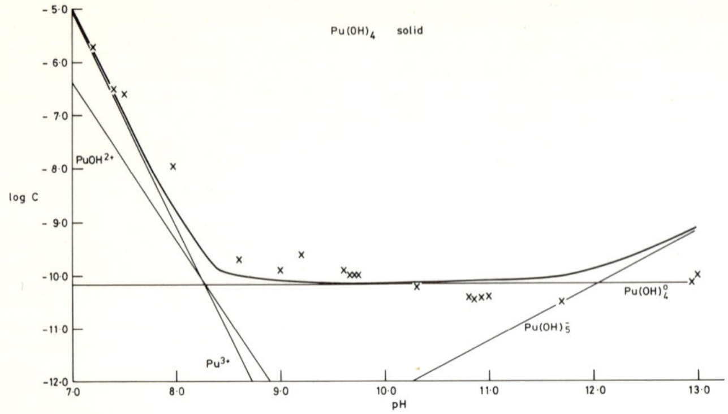

Figure 1 Plutonium

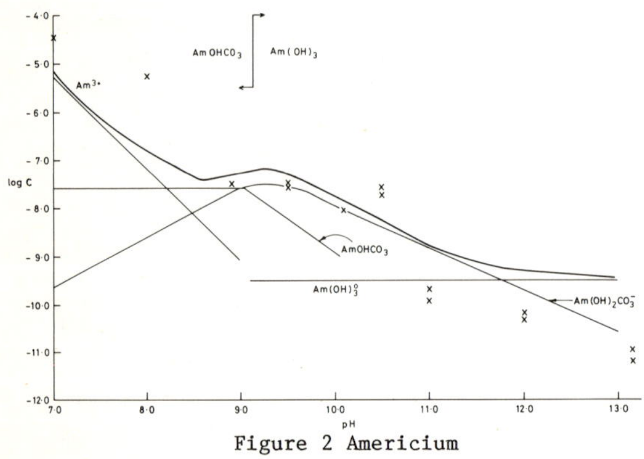

Figure 2 Americium

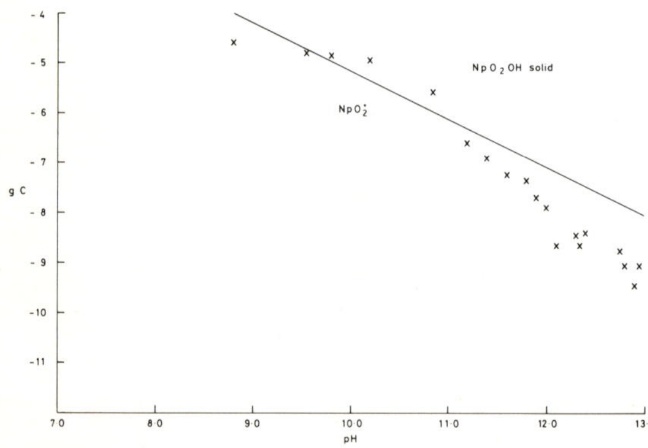

Figure 3 Neptunium

Figures 1-3. Experimental and calculated values for the solubilities of the actinides as a function of pH, with major contributing solution species. Points experimental data, solid lines calculated values Abscissa - Logarithm of actinide molarity essentially only NpO_2^+ in equilibrium with NpO_2OH solid, gave good agreement with the experimental results (fig. 3). Hence, under the conditions studied, neptunium solubility was controlled by the formation of NpO_2OH and the formation of the thermodynamically stable NpO_2 and Np(V) hydrolysis products was probably kinetically inhibited.

CONCLUSIONS

This experiment was designed to address two major aspects of the source term definition for a radioactive waste repository. These were the determination of the maximum possible solubility for the actinides under the conditions of water chemistry, which will exist in the repository, and to determine the extent to which these conditions can be represented by a computer code, modelling thermochemical equilibrium based on existing thermodynamic data.

The source term concentrations have been determined; their significance lies in the fact that these are the maximum values that can be attained in the near field. The pH will remain above pH 11 for about a million years, within the concrete structure [1], and thus the solubility will be maintained at these low levels for this period of time. The solution concentrations will be reduced in the repository by sorption on the surfaces of the cement and concrete in the repository structure. Preliminary results [18] suggest that sorption coefficients for actinides may be substantial, i.e. greater than 1000. Thus if the sorption sites on the near field components are sufficient to accommodate the actinides present, then the actual concentrations of actinides in the water passing from the near field to the far field will be several orders of magnitude below the measured solubility limit.

The simulation of these experiments using the PHREEQE code has shown that a satisfactory description of the experiments can be achieved after revision of the database. There may be a number of combinations of thermodynamic data which would enable the theoretical solubility to fit the experimental data; those given here appear to be the most reasonable. There is a clear need for the development of a speciation method which would enable the solution species to be determined at the low concentrations found in these experiments, and thus enable a better refinement of the data to be achieved.

ACKNOWLEDGEMENTS

This work was supported by the UK Nuclear Industries Radioactive Waste Management Executive, NIREX, and their permission to publish these results is gratefully acknowledged.

REFERENCES

1. A.Atkinson UKAEA Report AERE-R 11777 (1985).

2. D.Rai in Scientific Basis for Nuclear Waste Management VII ed. G.L.Mcvay (Elsevier, New York, 1984) pp.805-815.

3. A.Avogadro, A.Billon, A.Cremers, P.Henrion, J.I.Kim, B.Skytte Jensen, P.Venet and P.J. Hooker 2nd European Conf. on Radioactive Waste Management, Luxembourg, May 1985 (to be published).

4. R.H.Flowers Nucl. Energy 2 24 105-110 (1985).

5. J.H.Rees J. Nucl. Mater. 130 336-345 (1985).

6. J.D.F.Ramsay Harwell, Personal Communication (1985).

7. F.T.Ewart, R.M.Howes, A.J.Smith, H.P.Thomason, S.J.Williams, M.Young, and B.E.Chidley UKAEA Report AERE G-3411 (1985).

8. F.L.Moore Anal. Chem. 29 941 (1957).

9. D.Rai Radiochim. Acta 35 97 (1984).

10. D.L.Parkhurst, D.C.Thorstenson and L.N.Plummer Report USGS/WRI 80-96 (1985).

11. J.E. Cross, G.L. Smith and D.R. Williams UK DoE Report DoE SYVAC Tech. Note TN-UWIST-3 (1984).

12. R.J.Lemire and P.R.Tremaine J. Chem. Eng. Data 25 361-370 (1980).

13. B.Allard SKBF/KBS Technical Report 83-35 (1983).

14. L.V.Benson and L.S.Teague LBL-11448 (1980).

15. D.Rai, R.G. Strickert, D.A. Moore and J.L. Ryan Radiochim. Acta 33 201-206 (1983).

16. J.I.Kim in EUR 9543 13-44 (1984).

17. J.L.Ryan and D.Rai Polyhedron 2 947-952 (1983).

18. D.C.Pryke, J.H.Rees and R.J.W.Streeton UKAEA Report AERE-R 11391 (1984).

19. Supplied by Chemistry Division UKAEA Harwell

20. B. Allard, L. Eliasson, S. Hoglund, K. Andersson, SKBF/KBS Technical Report 84-15 (1984).

INFLUENCE OF REDOX ENVIRONMENTS ON THE GEOCHEMICAL BEHAVIOUR OF RADIONUCLIDES

G. BIDOGLIO, A. AVOGADRO AND A. DE PLANO
Commission of the European Communities, Joint Research Centre,
Ispra Establishment, Radiochemistry and Nuclear Chemistry Division, 21020
ISPRA (VA) Italy

ABSTRACT

Studies were carried out on the effect of redox conditions on the release from simulated vitrified HLW of Np-237, Pu-238 and Tc-99 and their migration through columns packed with site-specific porous material representative of clay and salt aquifers. Results obtained under oxic and anoxic conditions are compared. The release rates of Np, Pu and Tc were found to be lower in the absence of air than with oxygenated groundwaters. Speciation studies showed that neptunium and technetium were released in the oxidized forms Np(V) and Tc(VII) even under the investigated anoxic conditions where reduced species are thermodynamically more stable. In spite of that, a stronger interaction was observed when the leachate was passed through the column in the absence of air. A surface reduction process accounts for the observed behaviour. Under more reducing conditions, plutonium appears to be present in the tetravalent state in colloidal form.

INTRODUCTION

After the initial perturbation of the geochemical environment associated with the construction of a radioactive waste repository, re-establishment of low Eh conditions will slowly occur. Since radionuclides in nuclear wastes exist in several oxidation states, their release and migration through the geochemical barrier are influenced by the encountered redox environments. The high degree of hydrolysis of reduced species of actinides should result in a low solubility and a strong interaction with surrounding geomedia. A better understanding of processes taking place in undisturbed repository horizons is therefore needed.

In this paper the results of experimental studies investigating the effect of redox conditions on the dissolution of the waste form, the equilibrium concentration of Tc-99, Np-237, Pu-238 and their migration behaviour through soil columns are reported.

MATERIALS AND METHODS

Site specific porous material representative of geological systems surrounding potential nuclear waste repositories (either the clay disposal option in Mol-Belgium or the rock salt option in Gorleben-West Germany) was used for laboratory elution studies. The first geological material was a sand containing about 25% of glauconite, the second a quartz sand with small amounts of pyrite and magnetite. Each soil sample was placed inside a glass column and a constant water flow rate was maintained with a liquid displacement pump. The inflow solution was contaminated by leaching of radionuclide doped glass beads. Nuclepore membranes of different pore sizes were used for the ultrafiltration experiments.

The borosilicate glasses were separately traced with 0.58 wt% of Tc-99, 2 wt% of Np-237 and 0.05 wt% of Pu-238. The preparation of the Tc-99 glass deserves a special attention. Because of its great volatility, most of technetium initially added to the inactive glass matrix as a NH_4TcO_4 solution, was lost during the melting in a graphite crucible and under an argon atmosphere. Less than 1% of the introduced Tc-99 was found to remain in the glass. Slightly higher amounts were however measured when technetium was added to the vitrified matrix in the reduced form TcO_2.

Experiments under anoxic conditions involved the use of two different anaerobic chambers containing a nitrogen atmosphere. In one of these boxes where the percolation tests of neptunium and technetium were carried out, only slightly reducing environments could be simulated. In this case, Eh values measured with a platinum electrode were around -15mV (argillaceous material) and 0 mV (salt system). These Eh readings gives only a qualitative evaluation of the redox state of the solutions used. Typical uncertainties of Eh measurements in poorly poised solutions are estimated around 50 mV. More reducing conditions could be attained with the second anoxic installation where leaching measurements of Pu-238 reported in this paper were carried out. Constant Eh values of -140 mV were maintained.

RESULTS AND DISCUSSION

Preliminary considerations on the geochemical behaviour of radionuclides can be made referring to Eh/pH diagrams like that given in Figure 1 showing the field of predominance of technetium species. The same figure reports what is known as the Baas-Becking field, gathering combinations of pH and oxidation potentials typical of aqueous natural environments [1]. Since natural geological materials generally have cationic exchange capacity, two different regions with two different geochemical behaviours are identified in Figure 1. Under conditions lying in the stability field of the oxidized species, interactions with the minerals are limited by the negative charge carried by the migrating species TcO_4^-. At more reducing Eh, cationic species react with the solid phases more easily. This stresses the importance that chemical forms rather than total concentrations have in affecting the migration rate of radionuclides. The redox environment is likely to affect also the release properties of radionuclides from the waste form. It has been pointed out that leaching of radionuclides from borosilicate glasses is a process which is also governed by solubility limits [2]. Hence, lower leach-rates are expected in reducing environments, where less soluble reduced forms of radionuclides are formed. Flow-leach experiments of borosilicate glasses doped with single radionuclides : Tc-99, Np-237, Pu-238 were carried out under oxic and anoxic conditions. Two different leaching solutions, either a synthetic groundwater simulating that present in the aquifer overlying the Boom clay formation in Mol or a saturated NaCl brine typical of the saline area surrounding the Gorleben salt dome, were tested (see Table I for compositions). The average leach-rates, calculated from mass balance

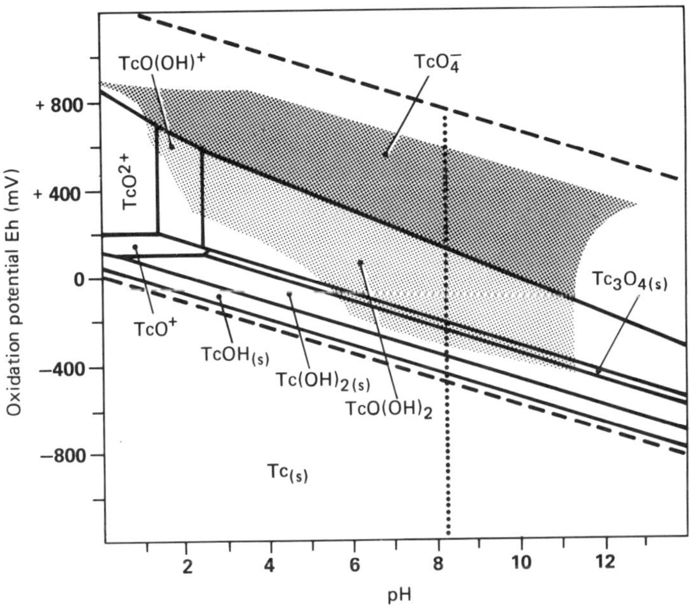

Figure 1 - Eh/pH equilibrium diagramme of the system Tc-H_2O displaying the Baas Becking field. The dotted line indicates the pH of the bicarbonate groundwater used in the reported migration tests.

considerations of the total leached activity, are reported in Table II. It can be observed that lower release rates of all redox sensitive radionuclides occur under anoxic conditions.

First migration tests were conducted with Tc-99 and the Mol groundwater under normal oxygen atmosphere. That only a few percent of leached Tc-99 was retained by the glauconitic sand (see Table II) supports the previously discussed presence of the non-sorbing anionic form TcO_4^- under oxic conditions. Nevertheless, EXAFS (Extended X-ray Absorption Fine Structure) analyses of the glass prepared under an argon atmosphere showed clusters of Tc metal and TcO_2 in the bulk of the vitrified matrix [3]. This implies that, when released by an oxidizing solution, the reduced forms of technetium are not stabilized by complexation with ligands in the leachant, but undergo a quick oxidation to pertechnetate. The successive set of experiments was carried out under an anoxic atmosphere. Despite the reducing environment (Eh about -15mV), determination of chemical species of technetium in model solutions (Mol groundwater, sea water and humic acid solution equilibrated with solid TcO_2 under the same anoxic conditions) using electromigration and radiochemical techniques, still indicated the presence of pertechnetate [4] . This result is contrary to thermodynamic previsions of the Eh/pH diagram in Figure 1. Nevertheless, Figure 2 shows that when steady state conditions are reached, the technetium concentration in the leachate is lower for the anoxic case, as also reported for static leaching tests [5]. Furthermore, a higher retention of Tc-99 in the glauconitic sand column was measured in the absence

Table I. Chemical Composition of the Groundwaters Used. Concentrations are given in Molarity.

Constituent	Clay groundwater	NaCl brine
Na^+	$2.4x10^{-3}$	5.48
K^+	$2.0x10^{-4}$	$2.0x10^{-3}$
Fe_{tot}	$1.8x10^{-6}$	$1.2x10^{-6}$
Ca^{2+}	$7.6x10^{-5}$	$4.1x10^{-3}$
Mg^{2+}	$1.3x10^{-4}$	$1.1x10^{-3}$
Cl^-	$1.8x10^{-4}$	5.48
SO_4^{2-}	$5.2x10^{-6}$	$5.6x10^{-3}$
Carb. Alkal.	$2.6x10^{-3}$	10^{-4}
pH	8.35	~ 6

Table II. Glass Leach Rates ($g\ cm^{-2}\ d^{-1}$) and Sorption Behaviour of Tc, Np and Pu under Oxic and Anoxic Conditions.

Bicarbonate Groundwater/ Glauconitic Sand		OXIC	ANOXIC
Tc-99	Leach-rate (*)	$9.5x10^{-8}$	$1.8x10^{-8}$
	% Activity sorbed	3.9	72
Np-237	Leach-rate	$1.5x10^{-6}$	$1.7x10^{-7}$
	Retardation factor	$\sim 10^3$	$> 5x10^3$
Pu-238	Leach-rate (**)	$2.4x10^{-7}$	$9x10^{-8}$
NaCl Brine/Quartz Sand		**OXIC**	**ANOXIC**
Np-237	Leach-rate	$7.4x10^{-7}$	$3.6x10^{-7}$
	Retardation factor	28	119

(*) Normalized to the amount of Tc-99 introduced before the melting, not to the actual total activity remaining in the glass.

(**) Test carried out under more reducing conditions (Eh-140 mV).

of oxygen (see Table II). This behaviour is typical of cationic species of Tc(IV), which, however, were not found in the glass leachate entering the column.

The migration behaviour of neptunium, another redox sensitive radionuclide , was also investigated both in the Mol groundwater and in the NaCl saturated brine. Ultrafiltration analyses of the leachates under oxic conditions showed the absence of colloids of neptunium which appeared to be released from the glass as a soluble species of Np(V). Nevertheless, similar results were also found (both Mol groundwater and saturated brine) under the anoxic conditions investigated, where the stronger colloid-forming tetravalent neptunium should prevail [6]. Retardation factors were then measured for the columns loaded with these Np(V) containing leachates in the reduced oxygen-atmosphere box. Results are given in Table II and in the Figures 3 and 4, where the compared contamination profiles were measured either by γ-scanning or by slicing the columns at the end of the percolation. In the NaCl brine system, the concentration of neptunium adsorbed by the soil is doubled when the percolation is performed under anoxic conditions. The shape of the retention profile of the glauconitic sand (Mol groundwater as a carrier) reveals a stronger interaction of neptunium in the first centimetres of the column.

Under anoxic conditions, the lower concentration of neptunium and technetium in the leachate might be explained with the lower solubility of the reduced species existing in the glass gel layer, where they can react with the oxygen present in the fluid. The importance of their dissolution

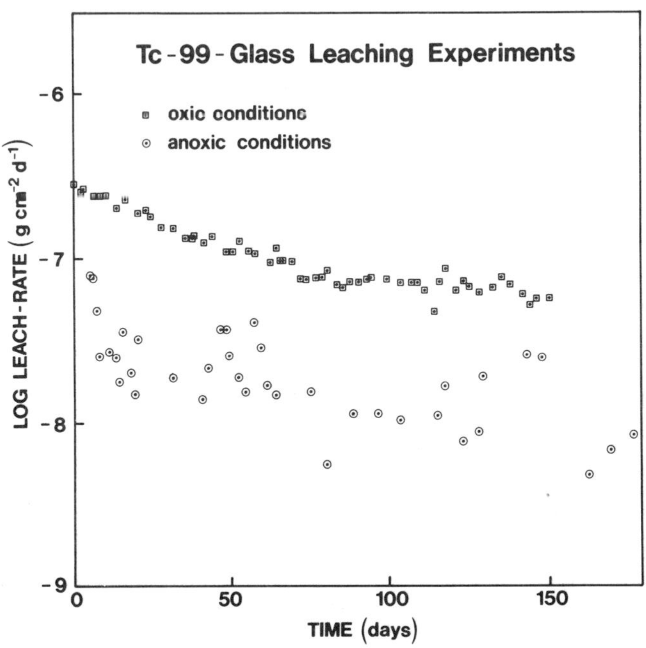

Figure 2 - Dynamic leach tests of Tc-99 doped glasses with a clay groundwater under oxic and anoxic conditions (see footnote Table II).

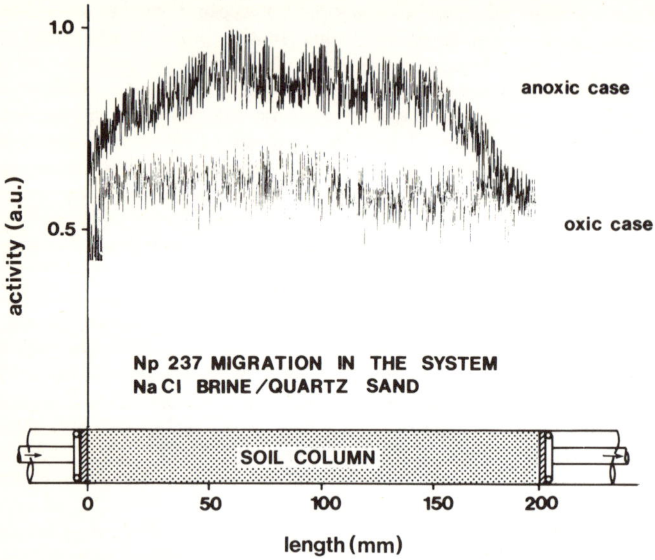

Figure 3 - Np-237 profiles of sand columns loaded under oxic and anoxic conditions with a brine leachate. The γ-activity given in arbitrary units is reported as a function of column length.

depends on the amount of oxygen accessible to the system. However, under the slightly reducing conditions studied, neptunium and technetium were found to be present in the bulk of the groundwater as oxidized species. The increase of adsorption on the soils may therefore be accounted for by inferring a surface process. When an available source of electrons is present, radionuclides will be reduced and sorbed simultaneously. Potential electron donor/acceptors are organic compounds coating soil particles and minerals containing sulfur or Fe(II) species, which are known to be reducing agents for Np(V). The kinetic of this reaction in acidic solutions was found to be described by the equation

$$-\frac{d[\mathrm{Np(V)}]}{dt} = 7.8\times10^{-2}\ [\mathrm{Fe(II)}][\mathrm{H^+}][\mathrm{Np(V)}] \quad \mathrm{mole\ l^{-1}\ sec^{-1}} \tag{1}$$

obtained by considering a rate limiting step involving hydrolyzed species of neptunium [7]. Although relative to acidic conditions, Eq(1) can be used to get an estimation of the time needed for 50% reduction ($t_{50\%}$) of Np(V) to Np(IV) in the bulk of the groundwater leachate. Considering constant the H^+ and Fe(II) concentrations, Eq (1) reduces to a pseudo-first order reaction which can be integrated yielding

$$t_{50\%} = \frac{2.82\ \times10^{-7}}{[\mathrm{Fe(II)}]\ [\mathrm{H^+}]} \quad \mathrm{years} \tag{2}$$

Even considering iron totally present in the Fe(II) state, we would expect for the experimental conditions in Table I a $t_{50\%}$ in the order of 10^5 years for the brine solution and 10^7 years for the fresh groundwater. A larger concentration of Fe(II) would be necessary to decrease $t_{50\%}$ to the order of magnitude of the groundwater residence time in the column (30 hours). This high concentration can be found at the interface of iron bearing minerals which can establish more reducing local potentials when the amount of oxygen accessible to the solution is limited. The reduction process may involve also an activated complex of neptunium with the surface of appropriate minerals. These findings are in agreement with static contact measurements reported in the literature, showing that technetium and neptunium sorbed on rock samples were mainly associated with iron minerals [8,9]. Similar considerations apply to the redox couple Tc(VII)/Tc(IV). As in the case of neptunium, a surface reduction process involving TcO_4^- and reducing minerals may occur. Coprecipitation of Tc_2S_7 can also take place when sulphur containing minerals are present.

Results from the dynamic leach tests in the clay groundwater of a glass traced with Pu-238 are given in Table II. Also the release of plutonium into the solution increases when oxygen is present in the system. However, peculiar to the plutonium experiments are the much more reducing conditions (Eh-140mV) as compared with neptunium and technetium. Ultracentrifugation and ultrafiltration analyses of colloidal fractions released from the glass surface indicated that suspended plutonium colloids are far more abundant in the reducing leachate. An explanation for this observation is the presence of

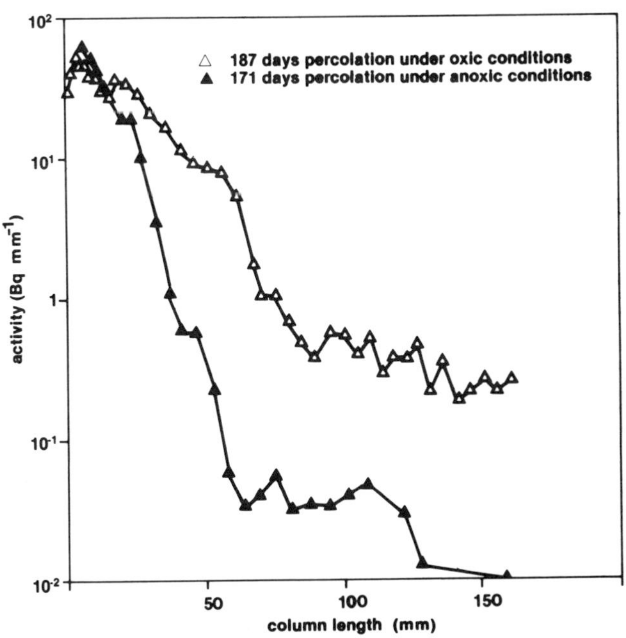

Figure 4 - Influence of anoxic environments on the dynamic uptake of Np-237 in the system glauconitic sand/Mol groundwater.

Pu(IV) which is the most readily hydrolyzed of the four oxidation states and exhibits great sorption properties.

CONCLUSIONS

These studies have shown that a considerable difference in leachability of borosilicate glasses exists between oxic and anoxic conditions, where lower release rates of Tc-99, Np-237 and Pu-238 were measured. The nature of the released species is however kinetically rather than thermodynamically determined. In fact, under the redox conditions investigated, reduction of neptunium and technetium in the groundwater phase appears to be a very slow process which is not induced by naturally occurring redox couples at environmental concentration, but rather seems to proceed through a heterogeneous reaction taking place at the surface of electron exchanging minerals in the soil. Strong reducing conditions do not appear to be needed for radionuclide removal from the water flow, as long as two requirements are met. Firstly, reducing atoms must not be firmly bound in the mineral lattice and then, oxygen has to be excluded from the system to allow redox sensitive minerals work as scavenging agents. Transport processes differ under the more reducing conditions found in deeper aquifers. As shown for plutonium, reduction to lower oxidation states can occur and colloidal transport become more important.

REFERENCES

1. L.G.M. Baas Becking, I.R. Kaplan, D. Moore, J. Geol. 68, 243 (1960)
2. F. Lanza, G. Bidoglio, E. Zamorani, presented at the International Seminar on Radioactive Waste Products, Jülich, June 1985 (to be published in the proceedings)
3. M. Antonini, A.E. Merlini, R.F. Thornley, J. non Cryst. Solids 71, 219 (1985)
4. G. Bidoglio, A. Chatt, A. De Plano, F. Zorn, J. Radioanal. Chem. 79, 153 (1983)
5. D.G. Coles, M.J. Apted, in Scientific Basis for Nuclear Waste Management VII, edited by G.L. McVay (Elsevier Science Publishers, New York, 1984), pp. 129-136; M.Y. Khalil, W.B. White, ibid., pp. 655-662
6. B. Allard, in Actinides in Perspective, edited by N.M. Edelstein (Pergamon Press, Oxford, 1982), p. 553
7. D. Gourisse, Report CEA-N-1441, 1971
8. T.T. Vandergraaf, K.V. Ticknor, I.M. George, in Geochemical Behaviour of Disposed Radioactive Waste, edited by G.S. Barney, J.D. Navratil, W.W. Schulz (ACS Symposium Series 246, 1984) p. 25
9. R.E. Meyer, W.D. Arnold, F.I. Case, ORNL Report No. 5978, 1984

STUDIES ON THE RADIONUCLIDE COPRECIPITATION-SOLID SOLUTION FORMATION. THE UO_2(S)-La$(OH)_3$(S) COPRECIPITATION AS AN ANALOGUE FOR THE UO_2(S)-Pu$(OH)_3$(S) SYSTEM

JORDI BRUNO, INGMAR GRENTHE AND MARIA MUÑOZ
Dept. of Inorganic Chemistry, Royal Institute of Technology, Stockholm, Sweden.

ABSTRACT

The UO_2(s)-La$(OH)_3$(s) coprecipitation equilibria have been studied in 0.5 M $NaClO_4$ medium, at 25^oC under anoxic conditions. The data obtained in the pH range between 2 and 5, and at different initial La(III)/U(IV) ratios (10%,5%,3%,2% and 1%), is explained by applying the Doerner-Hoskins[8] logarithmic law. The same law can also be used to describe previously collected UO_2(s)-Pu$(OH)_3$(s) coprecipitation data[2] . A method is proposed which uses the logarithmic distribution constants and the thermodynamic quantities for the pure phases to describe the solubility of radionuclides under repository conditions.

INTRODUCTION

A first estimate of the release and transport of radionuclides from spent nuclear fuel in contact with groundwater is often made by using thermodynamic models. Much work is devoted to the collection and critical evaluation of the data needed for these exercises. Most of the solubility information published relates to pure phases, even though it is well known that solid solutions of various types may form in the systems of interest for radioactive waste modelling. One important example is related to changes in the redox state in the near and far field region, caused e.g. by oxidation due to the radiolysis in the near field region followed by reduction in the far field. Another example is the release of the minor components from the UO_2(s) matrix.

The dissolution of uranium appears to be solubility controlled[1], but very little is known about the behaviour of the minor components. Some of these occur as separate phases while others form solid solutions with the matrix elements. In the later case the dissolution cannot be modelled by using thermodynamic data for the pure phases.

Reprecipitation of uranium either as $UO_2(OH)_2$(s) under oxidizing conditions or as UO_2(s), or some higher oxide phase UO_{2+x}(s) under reducing ones, should result in the coprecipitation of some of the minor components (particularly other actinides) in the nuclear waste system.

In both cases the migration of the minor radionuclides is affected. The behaviour of the system is no longer dependent only on the thermodynamic and kinetic properties of the various pure phases for the different elements, but also on the distribution properties of the components in the solid phase, e.g. the formation of solid solutions. As a consequence, any attempt to model the solubility and/or migration behaviour of the heavy radionuclides should include a quantification of their solubility related to that of uranium.

The aim of this work is to investigate to which extent coprecipitation will affect the solubility of minor radionuclide under repository conditions. We will also try to make a quantitative description of the coprecipitation phenomenon. In this way it should be possible to incorporate it into the various transport models.

Coprecipitation is a well studied chemical phenomenon. It has been widely used in analytical chemistry mostly for separation of trace elements[3]. As a consequence it has also been used in the nuclear industry in enrichment[4,5]. Coprecipitation phenomena have also been applied in the field of geochemistry, for example to explain the anomalous behaviour of the Sr-Ca carbonates[6].

Two different approaches can be used in the mathematical treatment of the coprecipitation phenomena. These can be considered as two limiting cases.

In the first case the initial distribution of the cations in the precipitates is random. Still, structural transformations might occur upon aging and the foreign ions can move by diffusion within the host lattice, to achieve an orderly structural arrangement at equilibrium. Under these circumstances true thermodynamic equilibrium can be assumed with the whole solid phase and the so called Berthelot-Nernst homogeneous law is followed. The equilibrium condition requires:

$$\left\{[Pu(III)]/[U(IV)]\right\}_{solid} = D\left\{[Pu(III)]/[U(IV)]\right\}_{solution} \qquad (1)$$

In the second case no structural arrangement has occurred and the distribution equilibrium is established just between the solid surface and the solution. Kroeker[7] made an elegant deduction of the equation valid for this case which was later applied by Doerner and Hoskins in a pioneer work on the Ba-Ra sulphate coprecipitation[8]. A logarithmic distribution law is obtained and the equilibrium condition between the solid surface and the solution can be written:

$$\log\left\{[Pu(III)]/[U(IV)]\right\}_{solid\ surface} = k\log\left\{[Pu(III)]/[U(IV)]\right\}_{solution} \qquad (2)$$

This study has been undertaken in order to

- decide which distribution law describes the coprecipitation of U(IV) and actinoid(III) ions

- get quantitative estimates of the solubility products for the minor components. The solubility product of the major component is also obtained and serves as a useful test of the experimental technique used.

We have previously made some work in this area. The first concerns the dissolution of spent nuclear fuel[1], where we can assume that uranium and plutonium are present as U(IV) and Pu(IV) in the $U_{1-x}Pu_xO_2(s)$ solid solution. Plutonium(IV) is reduced to Pu(III) in aqueous solution in the presence of U(IV). Hence, the dissolution of U(IV) and Pu(IV) under reducing conditions will give a solution of general composition:

$$1-x/2\ U(IV)\ ;\ x/2\ U(VI)\ ;\ x\ Pu(III)$$

The uranium(VI) concentrations may be much larger under oxidizing conditions in which plutonium will be present as Pu(IV). Groundwaters containing uranium(VI) can be reduced e.g. by contact with reducing minerals, a well-known mechanism in uranium mineralization.

We also performed some preliminary experimental work on the $Pu(OH)_3(s)$ coprecipitation in the presence of $UO_2(s)$ which showed Pu(III) solubilities much lower than the ones expected from the thermodynamic data for the pure Pu(III)-hydroxide precipitate.[2].

The Choice of La^{3+} as an Analogue for Pu^{3+}

To achieve the objectives already mentioned extensive experimental work under controlled conditions is required. The work with plutonium has some experimental limitations which can be solved by using a proper analogue. The two main factors controlling the coprecipitation are ion size and charge. The closer these two quantities are for the precipitating ions the less are the structural rearrangements required in the host lattice and consequently the higher is the probability that coprecipitation occurs. A comparison between the ionic radii ratio for $La^{3+}/U^{4+}=1.14$ and $Pu^{3+}/U^{4+}=1.00$[9], indicates that they are close enough for isomorphous substitution, at least at low mole fractions of the minor components

The aqueous chemistry of the ions will also affect the possible coprecipitation. Since hydroxo complexes will be the ones determining the solution behaviour of the trivalent ions at the acidities under study, it is of interest to note that Pu(III) and La(III) have very similar hydrolytic properties[10].

EXPERIMENTAL

We performed two different sets of experiments. In the first we studied the time dependence of the U(IV)-La(III) coprecipitation. In the second we studied how the La(III) and U(IV) concentrations varied in solution and in the solid formed with pH and the relative initial $[La(III)]^{\circ}/[U(IV)]^{\circ}$ ratios. We carried out these investigations both in batch and as titrations in a thermostated room at 25 0.5°C.

The Time Dependence of the Coprecipitation

In order to check the time for the attainment of an equilibrium state the following experiments were performed.

An initial volume V_0 of solution T_0 was introduced into a titration vessel. The composition of the solution T_0 was: B M UO_2^{2+} , H M H^+ and 0.5 M $(Na)ClO_4$. The initial U(VI) solution was reduced to U(IV) by bubbling $H_2(g)$ through a Pd-catalyst. Once the reduction was completed the hydrogen bubbling was replaced by $N_2(g)$ and the Pd-net was removed from the solution, at this point the precipitations started. A small volume of $UO_2(ClO_4)_2$ solution was then added.

We followed the changes in $[H^+]$ with time by means of a combined glass electrode (CGE) Metrohm EA-121. The internal saturated KCl solution of the electrode had been replaced by a 0.5 M $NaClO_4$; 0.01 M $AgClO_4$ solution in order to avoid large diffusion potentials. The E_0 of the CGE was determined in 0.5 M $NaClO_4$ prior to the introduction of the uranyl perchlorate solution into the vessel. We followed the time dependence of the redox potential of the solution which also gives information about the time for the attainment of the equilibrium through the reactions:

$$UO_2^{2+} + 4\,H^+ + 2\,e = U^{4+} + 2\,H_2O \qquad (3a)$$

$$U^{4+} + 2\,H_2O = UO_2(s) + 4\,H^+ \qquad (3b)$$

The redox potential was measured by using the cell:

$$-Pt/TS/RE^+$$

where RE stands for the reference electrode. In this case we use the internal reference of the CGE. The measuring electrodes were made of platinum. For more details about the experimental methodology the reader is referred to Ref. 11.

We made an attempt to use a $LaF_3(s)$ as a solid state electrode in order to measure the decrease on the La^{3+} concentration with time. Although the electrode works properly for La(III) concentrations over 10^{-4} M, we found that its function was affected by the presence of U(VI) and U(IV).

In Figure 1 we show a plot of of the variation of the CGE and Pt-electrodes with time for one of these experiments. It is possible to see that steady states, in which the Pt electrode potential did not oscillate more than ±0.5mV in one day, were reached in 15-16 days at 25°C. Since we carried out these experiments at very high proton concentration the CGE was insensitive to the U(IV) precipitation and its potential remained constant after the first day.

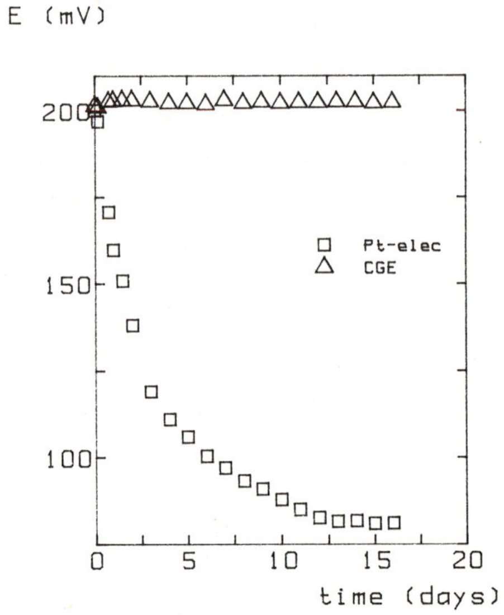

Fig.1

The Dependence of the U(IV)/La(III) Coprecipitation on the pH and the Original $[La(III)]^o/[U(IV)]^o$ Ratio

Once the time dependence of the coprecipitation was established two different types of experiments were performed. In a series of titrations solutions with a relatively high La(III)/U(IV) ratio =5% were investigated.

The experiments were run in the same way as the time dependence study previously described. The coprecipitation was started in the same way as before and once steady redox and CGE potentials were reached samples of the solid and the solution at equilibrium were taken for analysis. The acidity of the test solution was decreased coulometrically and a new equilibrium position was established. The aim of these experiments was to collect the data at well controlled redox conditions since continuous $N_2(g)$ bubbling and monitoring of the redox potential were possible.

In order to collect the data necessary in a reasonable period of time a second series of batch experiments was performed. The conditions were the following:

- we used initial La(III)/U(VI) ratios in the solution ranging between 0.5% to 10%.

- we reduced uranium(VI) to the (IV) state by the procedure we already described.

- we decreased the acidity of the solutions by adding previously degased dilute sodium hydroxide solutions. The pH range 2.0 to 10.5 was studied.

The solutions were then agitated in contact with the precipitated solid for three weeks (the time necessary for the attainment of equilibrium between the solid and the solution). At this point we took samples of the solid and the solution for further analysis. In this way we could investigate a wide range of pH's at different U(IV)/La(III) ratios. The sampling and pH measurement were done under controlled conditions using a continuous nitrogen stream under the CGE measurements and keeping the samples under inert atmosphere prior to the analysis.

The Analytical Procedure

We analyzed the uranium and lanthanum content in the solid and solutions for both continuous and batch experiments by using ICP technique. The analyses were performed in Studsvik using a Perkin Elmer ICP/5500 Spectrophotometer. The solid phase was treated with H_2SO_4 and HNO_3 solutions prior to the analysis. The solution samples were only diluted in order to get the optimal range of concentrations for the analysis. Lanthanum(III) and uranium(VI) in 0.5 $NaClO_4$ solutions in the same conditions as the test samples were used as standards. The following wavelengths 385.96 nm and 408.67 nm were used for uranium and for lanthanum, respectively.

RESULTS

The data measured included pH, in some cases the redox potential, the uranium(IV) and lanthanum(III) concentrations in solution and the uranium and lanthanum fractions in the solid.

Figure 2 is a plot of the logarithm of the total uranium(IV) concentration (log S1) against pH for the whole set of experiments performed and log S2 data for the experiments run with [La(III)]/[U(IV)] = 3% It also includes solubility data obtained at lower pH's from earlier Pu(III/U(IV) coprecipitation experiments[2]. These data agreed well with the present data, a fact which shows that the same UO_2(am) phase was precipitated in both sets of experiments. Data obtained at different initial [La(III)]/[U(IV)] ratios agrees reasonably well at pH's lower than 4. The larger spread of data observed at lower acidities may be due to the formation of two different solid phases as discussed below.

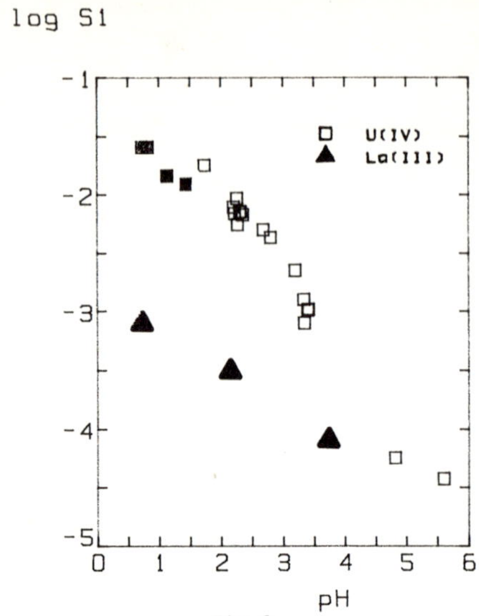

Fig.2

We have already pointed out in the description of the experimental procedure that the experiments run using titrations permitted a better control of the redox conditions. The UO_2 (s) phase was precipitated *in situ* under the optimal conditions of low pH and redox potential. Afterwards, the acidity of the solution previously equilibrated with UO_2(s), was varied stepwise coulometrically. When we tried to obtain a precipitate from neutral solutions a green solid phase was obtained. The colour of this phase remained unchanged even in the presence of H_2(g) and a Pd-catalyst for several weeks. X-ray powder diffraction diagrams of the UO_2(s) and the new phase showed that the former was pure UO_2(s) while the green solid phase showed a diffraction pattern which agreed with the one reported for U_3O_8(s)[12].

In Fig. 3 we have plotted log S1 *vs* pH for the few data obtained for U(IV) and La(III) in alkaline conditions. A striking fact is that the uranium(IV) solubility is constant within the experimental error in the pH range 5.5 pH 10.5 investigated. This indicates that no anionic hydroxo complexes for tetravalent uranium are formed. On the contrary La(III) solubilities depend markedly with pH as is expected from its known hydrolytic properties.

DATA TREATMENT

In a first stage the homogeneous *vs* logarithmic behaviour was tested for the data collected in the pH range between 2 and 5.5 and for the different $[La(III)]^{\circ}/[U(IV)]^{\circ}$ ratios. When the homogeneous distribution law was tested a systematic and large variation in the distribution coefficient D was observed. On the other hand the logarithmic distribution gave a quite constant value for k for the whole set of experiments. No systematic trend with pH or the original ratio [La(III)]/[U(IV)] was observed. Hence, we can conclude that the Doerner-Hoskins[8] logarithmic law is followed by the La(III)/U(IV) hydroxide precipitate.

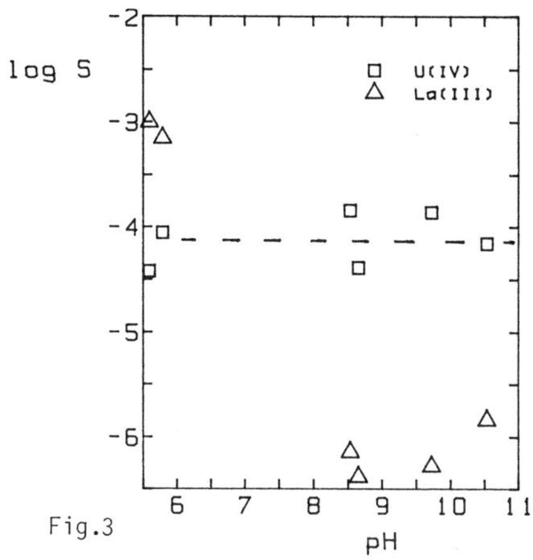

Fig.3

By combining the solubility information obtained for U(IV) and La(III) and using the logarithmic law we can obtain quantitative information on the solubility product of the minor component (La(III)).

The log S1 vs pH plot for the major component U(IV) shows a linear dependence with slope approximately equal to one. This is an indication of the predominance of the solubility equilibrium (See fig 2)

$$UO_2(s) + H^+ + H_2O = U(OH)_3^+ \qquad (3c)$$

with $K_{s1} = [U(OH)_3^+]/[H^+]$(3) and consequently $U(OH)_3^+$ as the predominant U(IV) species in the pH range investigated. Unfortunately. there is no direct determination of the consecutive U(IV) mononuclear hydroxo complexes but the tentative correlation proposed by Baes and Mesmer[10] agrees quite well with our observation ($[U(OH_3^+]$ amounts to more than 05% of total U(IV)).

The different log S2 vs pH plots for La(III) at various [La(III)]/[U(IV)] initial ratios can be fitted with straight lines of slope three (Se fig. 2). This indicates the predominance of the equilibrium:

$$La(OH)_3(s) + 3\ H^+ = La^{3+} + 3\ H_2O$$

with $$K_{s2} = [La^{3+}]/[H^+]^3 \qquad (4)$$

as expected from the hydrolytic behaviour of La(III)[10]. Hence $[U(IV)]=[U(OH)_3^+]$ and $[La(III)]=[La^{3+}]$ at the pH range 2.3 to 5. From equations (3),(4) and (2) we then obtain:

$$k \log([La(III)]^{\circ}/[U(IV)]^{\circ}) = \log\left\{K_{s2}[H^+]^2\, a_{La(OH)_3(s)} \,/\, K_{s1}\, a_{UO_2(s)}\right\}$$

assuming ideal behaviour in the solid phase:

$$a_{La(OH)_3(s)} = X_{La(OH)_3(s)}$$

and

$$a_{UO_2(s)} = X_{UO_2(s)}$$

where X_i stands for the molar fraction of the component i.

By rearrangement we obtain the following expression:

$$\log\left\{[La(III)]^{o}/[U(IV)]^{o}\ X_{UO_2(s)}\ X^{-1}_{La(OH)_3(s)}\right\} = k^{-1}\log(K_{s2}/K_{s1}) + 2k^{-1}\log[H^+]$$

By plotting the experimental function

$$\log Y = \log\left\{[La(III)]^{o}/[U(IV)]^{o}\ X_{UO_2(s)}\ X^{-1}_{La(OH)_3(s)}\right\}$$

against the measured pH we can determine the partition coefficient k and the ratio K_{s2}/K_{s1}.

Figure 4 shows a plot of log Y vs pH for the whole set of data obtained in the pH range between 2 and 5.5 and with initial La(III) to U(IV) ratios between 0.097 and 0.006 (10%,5%,3%,2%,1%). The data shows a linear correlation (r^2=0.90) between log Y and pH, a fact that indicates that a logarithmic distribution is followed. The following results are obtained:

$$2k^{-1} = 0.23 \pm 0.01$$

and $$k^{-1}\log(K_{s2}/K_{s1}) = 1.4 \pm 0.1$$

consequently, $$k = 8.7 \pm 0.2$$

and $$\log(K_{s2}/K_{s1}) = 12.2 \pm 0.2$$

From the solubility data for the major component (U(IV)) we obtain $\log K_{s1}$=0.26±0.2, for the equilibrium (3). We can then calculate the solubility product for the minor component $La(OH)_3(s)$:

$$\log K_{s2} = 12.5 \pm 0.2$$

to be compared to the value tabulated[10] for the equilibrium (4) $\log K_{so}$=20.5.

Fig. 4

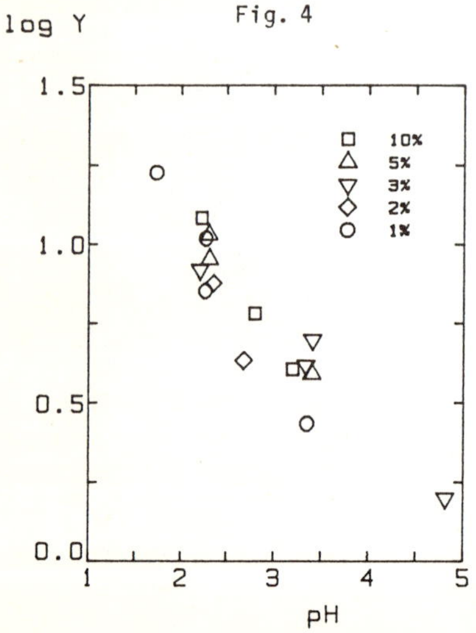

Fig. 5

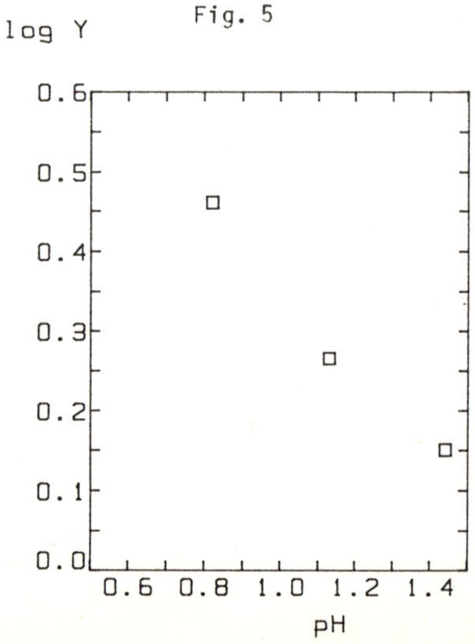

A decrease of eight orders of magnitude referred to a crystalline product is observed in La(III) in the presence of amorphous $UO_2(s)$.

By using the same method on to the very few data previously collected in the system Pu(III)/U(IV) we obtain similar results. In Fig.5 we have plotted log Y vs pH for the data obtained by studying the coprecipitation of $Pu(OH)_3(s)$ with $UO_2(s)$. The work was performed using very near the same conditions as described in the present study. The three experimental points collected fall in a straight line (r^2=0.98). As we have already mentioned the U(IV) solubility vs pH data obtained agreed very well in the log S1-pH plot (Fig.2), indicating that the same kind of $UO_2(s)$ was precipitated.

Since the Pu(III)/U(IV) data was collected at much more acidic pH (0.72 to 1.44), UOH^{3+} is expected to be the predominant species (U^{4+} amounts at most 5% of the total U(IV)). By taking this into account we can assume the following equilibrium:

$$UO_2(s) + 3\ H^+ = UOH^{3+} + H_2O$$

where $$K_{s11} = [UOH^{3+}]/[H^+]$$

The few data available allow us to make a preliminary determination of the stability constant for that process, although it is difficult to establish the magnitude of the experimental error involved. We obtain:

$$\log K_{s11} = 0.72 \pm 0.2$$

By applying eq.7 to the log Y-pH data for the Pu(III)/U(IV) system in a similar way as we did with La(III), we obtain:

$$k = 2 \pm 0.1$$
$$\log K_{s2}(Pu(OH)_3) - \log K_{s11} = 1.7$$

and by using the previous estimate for log K_{s11}

$$\log K_{so}(Pu(OH)_3) = 2.4$$

This would indicate a decrease of 16 orders of magnitude in the solubility product value respect to a crystalline Pu(OH) (s) phase.

DISCUSSION

In the present work we have demonstrated that fresh La(III) $UO_2(s)$ coprecipitate follows a logarithmic rather than a homogeneous distribution. We used the logarithmic distribution and the solubility for the major component to calculate the "effective" solubility product for the minor component, the quantity that should be used when performing equilibrium modelling of the behaviour of radionuclides in ground waters.

The k values obtained are greater than unity. This indicates enrichment of the minor components with respect to the $UO_2(s)$. Hence, in the presence of $UO_2(s)$ the magnitude of the "solubility products" of the minor components are several orders of magnitude lower than the values reported for the corresponding pure phases.

The treatment described has only been tested for the relatively fresh precipitates obtained with microcristaline $UO_2(s)$. Nevertheless, the reproducibility of the phase is demonstrated by the good agreement observed in the U(IV) solubility obtained in various experiments at different precipitation conditions. One question that remains to be answered is the effect of aging on the properties of the solid. This is a critical point when long time scales have to be modelled. It has been established in work related to trace coprecipitation that the overall effect of aging will be a decrease in the degree of the coprecipitation.[3] This is normally related to the fact that the initial logarithmic distributions between solid phases and the solutions shift to more homogeneous ones upon aging, the Berthelot-Nernst distribution law should then be followed. Even in this case the solubilities of the minor components will be lower than those for the pure phases.

Future investigations will tackle this problem. We will continue the work by studying other coprecipitation systems and their behaviour on aging, in order to establish the systematics for its use in the equilibrium modelling of radionuclide migration in ground waters.

ACKNOWLEDGEMENTS

The measurements in the Pu(III)/U(IV) system were performed in the Nuclear Chemistry Department of the Chalmers University of Gothenburg (Sweden). Prof. B. Allard and Dr. B. Torstenfeldt are thanked for a pleasant cooperation. Dr Inga-Kari Björner is gratefully acknowledged for carrying out the ICP measurements of the U(IV)/La(III) system in Studsvik (Sweden).

REFERENCES

1. J.Bruno, R.Forsyth and L.Werme, in Scientific Basis for Nuclear Waste Management VIII, edited by K. Jansen, J. Store and R. Ewing, Mat.Res.Soc.Symp. Proc., 44, 413 (1985).
2. I. Grenthe, I. Puigdomenech and J. Bruno, KBS-Technical Report 83-02 SKB/KBS, Stockholm 1983.
3. L. Gordon, M.L. Salutsky and H.H. Willard, Precipitation from Homogeneous Solutions (John Wiley and Sons, London 1959).
4. M.L. Salutsky, J.G. Stites and A.W. Martin, Anal.Chem., 25, 1677 (1953).
5. J.A. Hermann, US Atomic Energy Commission, AECD-3637, 1954.
6. D.J. Kinsman and H.D. Holland, Geochim. et Cosmochim., Acta, 33, 1 (1969)
7. A. Kroeker, PhD, thesis, Berlin 1892.
8. H.A. Doerner and W.M. Hoskins, J.Am.Chem.Soc., 47, 662 (1925)
9. R.D.Shanon, Acta Crystallogr., A32. 751 (1976).
10. C.F. Baes and R.E. Mesmer, The Hydrolysis of Cations (John Wiley and Sons, New York 1976).
11. J. Bruno and I. Grenthe, Toxicolog. and Environmental Chemistry, 10, 257 (1985)
12. J.Aykan and R.Seight, J.A.Chem.Soc. 53, 427 (1970).

PART X

Solute and Colloid Transport

FORMATION AND TRANSPORT OF AMERICIUM PSEUDOCOLLOIDS IN AQUEOUS SYSTEMS.

U. OLOFSSON*, M. BENGTSSON* AND B. ALLARD**
* Department of Nuclear Chemistry, Chalmers University of Technology, S-412 96 Göteborg, Sweden
** Department of Water in Environment and Society, University of Linköping, S-581 83 Linköping, Sweden

ABSTRACT

The sorption of americium on colloidal quartz and montmorillonite has been studied using a batch technique. Americium appeared to be distributed between all available surfaces in the system (particles and vessel walls). Distribution coefficients, defined as the ratio of the amount of americium per mass of colloidal matter and the concentration in the solution phase, were estimated to be of the same order of magnitude as obtained in measurements on crushed material of much larger particle sizes.

In the presence of sorbents (alumina or granite) the removal of americium from the solution was enhanced, either due to the desorption of the americium from the particle phase and resorption on all available surfaces (sorbent plus vessel wall) or the sorption of the whole colloidal aggregates on the sorbent.

No break-through of americium was observed in column experiments (with alumina or granite packing) after the injection of americium pseudocolloids (quartz or montmorillonite) followed by elution with large volumes of aqueous phase (up to 880 column volumes). However, a minor fraction of the americium passed through the column without significant retention, especially under conditions with high flow rate (short hold-up time in the column).

INTRODUCTION

The transport of radionuclides from a geologic repository for radioactive waste will be delayed due to the interaction of the dissolved trace nuclides with the various solid agents present in the repository and its environments. For cationic hydrolyzed elements like the actinides in their lower oxidation states (III and IV) the sorption on construction materials (e.g. concrete), back-fill materials (e.g. clays) and geologic materials (the host-rock) is very high. However, the formation of low-sorbing complexes (notably with high negative charge) or the sorption of the trace nuclides on negatively charged naturally occurring colloidal matter (forming pseudocolloids) could drastically increase the mobility of the trace radionuclides. It has been observed, that e.g. tetravalent fall-out plutonium from weapons tests tend to be associated with particulates in the soil[1]. It is not known whether the sorption on natural particulates is fully reversible. Neither is the the kinetics known for the desorption of trace elements from particulate matter and the subsequent sorption on exposed macroscopic surfaces. In the present study the sorption of trivalent americium on colloidal quartz and montmorillonite has been studied, especially the effect on the distribution between aqueous phase, particulate phase and solid surfaces when a solid agent is introduced into the system.

EXPERIMENTAL

Radionuclides and chemicals

The radionuclide used in the experiments was Am-241 (half-life 433 y). The americium concentrations in the withdrawn samples were calculated from the gamma activity (NaI well-type scintillation counter).

The colloidal solutions were prepared by stirring crushed quartz and montmorillonite, respectively, with distilled water for two weeks. After settling for about a week, a fraction of the supernatant liquid was separated from the solid material, and the size distribution of the colloidal particles in the liquid was measured by a laser technique [2]. For the quartz system the average particle diameter was about 300 nm and for the montmorillonite about 525 nm.

Crystalline alumina (α-Al_2O_3; particle size of 0.090-0.125 mm or 0.125-0.250 mm) and granite from Stripa (0.090-0.125 mm [3]) were used as sorbent materials in the batch and the column experiments.

Batch experiments

Batch experiments were made in order to investigate the formation of americium pseudocolloids with quartz and montmorillonite, and the sorption of such colloidal particles on solid materials. The solution containing the colloidal quartz or montmorillonite (20 ml) was poured into polypropylene centrifuge tubes (50 ml) and americium was added to a concentration of about 10^{-9} M. Ajustments of pH were made with NaOH or $HClO_4$. The tubes were then centrifuged, sampled, vigorously shaken and sampled again. This sampling procedure was repeated after various storage times. Similar experiments have previously been made in americium solutions initially free from particulate matter [4].

The sorption studies were carried out in a similar way. Solid material (alumina or granite) was added to americium pseudocolloid solutions, and the distrubution between the solid and solution phases was measured as a function of pH and time.

The conditions and experimental procedure can be summarized as follows:

- o pH was ajusted to a value between 6 and 10 at the start of the experiments. No further corrections of pH were made.
- o The initial concentration of colloidal particles was 0.32 mg/ml for the quartz systems and 0.24 mg/ml for the montmorillonite systems.
- o The centrifuging speed was 6000, 10000 or 17000 rpm (centrifuging time 30 min; 17000 rpm corresponds to 27000 g).
- o The amount of sorbent material was 0.2 g, 0.4 g or 0.8 g/tube.
- o The storage time was 6 h, 1 d, 1 w, 6 w and 6 months.

Column experiments

Column experiments were performed in order to study the mobility of americium pseudocolloids in contact with potential sorbents. Quartz or montmorillonite colloids containing americium were injected in columns (ca 11 cm x 4 mm HPLC steel columns) packed with alumina or granite, which was preequilibrated with a synthetic groundwater (ca 120 mg/l of dissolved solids, pH 8.2; [5]). The columns were then eluated with the same synthetic groundwater with variation of the flow-rate (0.05 ml/min or 0.5 ml/min). The column volumes were determined by injection of a pulse of Na-22, which was assumed to pass through the column with little retention.

RESULTS AND DISCUSSION

Removal from the solution phase

The removal of americium from the solution phase by centrifugation at 17000 rpm is illustrated in Figure 1.

In the quartz system, more than 50 % of the added americium was immediately removed from the solution (centrifugation and sampling directly after the addition of the americium stock solution). The removable fraction increased with time and reached a maximum of 90-100 % after 6 w. Further storage (6 months) did not alter the ratio of removable to remaining fractions. Also pH was changing during the first day, giving a pH-range of 6.5-8, which was essentially constant from 1 d to 6 months.

For the montmorillonite system the removable fraction was 80-90 % already at the start, and this fraction did not change significantly with time (up to 6 w). Also in this system the pH range was changing to a final range of 6-7.5 within a day.

The removal of activity from the solution phase was more rapid than in a system free from particles [4]. Also the fact that the removal was essentially independent of pH, which was not the case in the absence of particles, indicate that association to centrifugable particles rather than merely sorption on the vessel walls was occuring in the quartz and montmorillonite systems. However, also in the absence of any initially added particle fraction between 50 and 90 % of the americium would be removed from the solution after 6 months of storage, primarily due to sorption on the vessel walls [4].

The removal of americium from solution was essentially the same at all centrifugation speeds, which indicates that the activity was associated with predominantly large particles (only minor amounts of small particles in the original colloidal solutions, not drastically changing with time during the experiment). The increasing removal of americium with time in the quartz system could either be an indication of an aggregation of centrifugable particles or of slow sorption kinetics (see below).

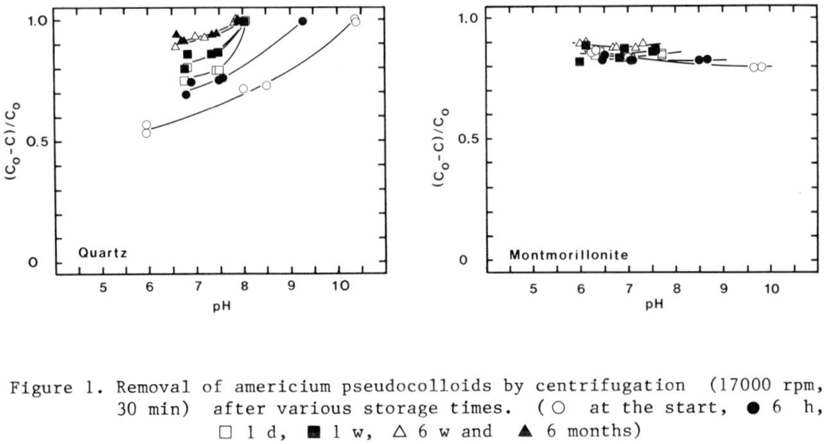

Figure 1. Removal of americium pseudocolloids by centrifugation (17000 rpm, 30 min) after various storage times. (○ at the start, ● 6 h, □ 1 d, ■ 1 w, △ 6 w and ▲ 6 months)

The centrifugable fraction

A comparison of the removable fractions after centrifugation and after vigorous shaking gives some indication of the proportions of the americium that was sorbed on the added particles and on the vessel walls, Figure 2.

After 6 h about 10-40 % of the americium was associated with particles in the quartz system, with the maximum around pH 7.5. Although the total amount of americium that was removed from the solution increased with time (up to 6 w) the particle fraction appeared to be constant. This indicates, that also sorption on the vessel walls (increasing with time) is of importance, just as was observed previously [4].

In the montmorillonite system about 60-80 % of the americium appeared to be associated with the particle fraction. Apparently, the sorption on the particle phase dominates over the sorption on the vessel walls, especially after long storage time (6 w).

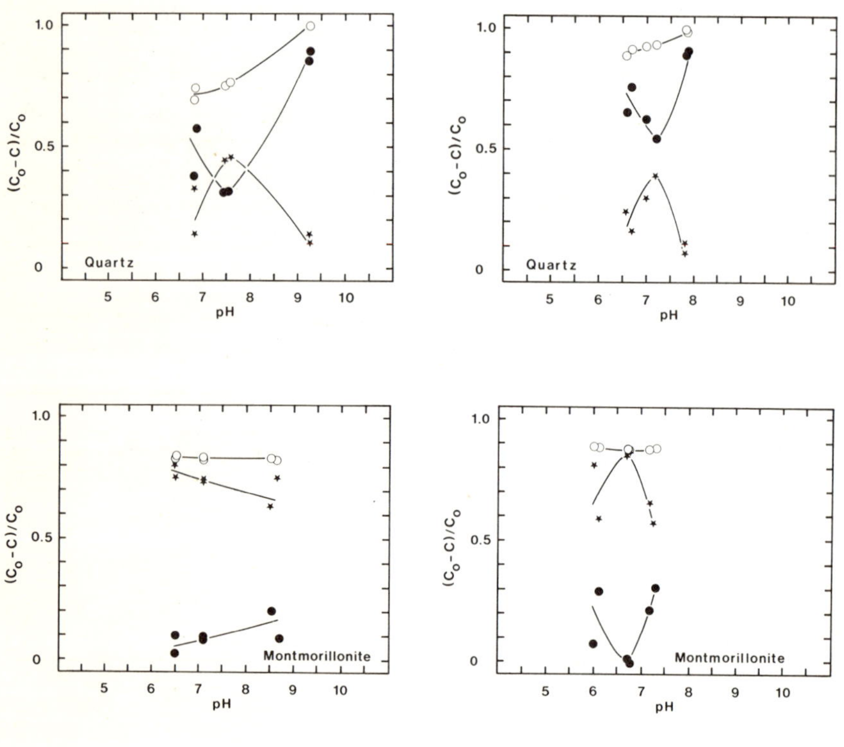

Figure 2. The centrifugable americium pseudocolloid fraction after various storage times (6 h, left, and 6 w, right; ○ after centrifugation, 17000 rpm, ● after shaking, ★ centrifugable fraction)

Both quartz and montmorillonite would have a negative lattice charge in the present pH-range [6-8]. Since americium exists essentially as cationic species in the corresponding solutions [9] the indicated sorption can be expected, considering the charge distribution. An estimation of the total surface provided by the added quartz and montmorillonite aggregates (assuming spherical particles defined by the average diameter) indicates, that the particle surfaces in both systems are of the same order of magnitude (providing that the average size does not change during the experiment). These total surfaces are at least one order of magnitude larger than the surface provided by the vessel wall. Thus, the sorption on the particles can be expected to dominate in the system, although the sorption on the vessel can not be neglected.

From the calculated americium fraction associated with particles (Figure 2) a formal distribution coefficient can be calculated, defined as (concentration in particle phase)/(concentration in solution). For the quartz and montmorillonite systems values in the range 0.2-0.4 m^3/kg and 6-16 m^3/kg respectively, are obtained. These values are in fair agreement with data from batch-wise studies for pure mineral-groundwater systems in the corresponding pH-ranges[6].

Batch-wise sorption studies

The removal of americium from quartz and montmorillonite solutions in contact with alumina is illustrated in Figure 3.

For the quartz system, the removal is much more rapid than in the absence of alumina, and the maximum removal is reaching almost 100 % after 6 w of contact or more. Neither the amount of sorbent (0.2 - 0.8 g) nor the particle size (0.090-0.125 or 0.125-0.250 mm) had any significant influence on the observed removal after long storage time (6 w). Possibly, the removal was more rapid and complete in the low pH-range for the quartz system in the presence of higher amounts of solid sorbent, Figure 4.

The sorption behaviour was very similar in the alumina and granite systems (not illustrated), and no truly significant differences between the two sorbents could be noticed.

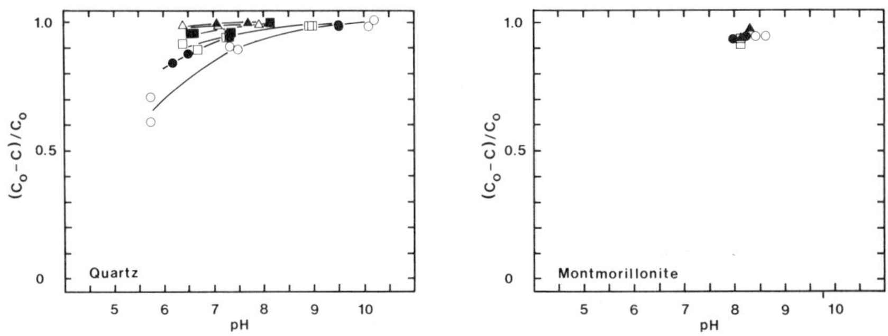

Figure 3. The sorption of americium pseudocolloids on alumina (10 g/1, 0.090- 0.125 mm) after various times (symbols as in Figure 1)

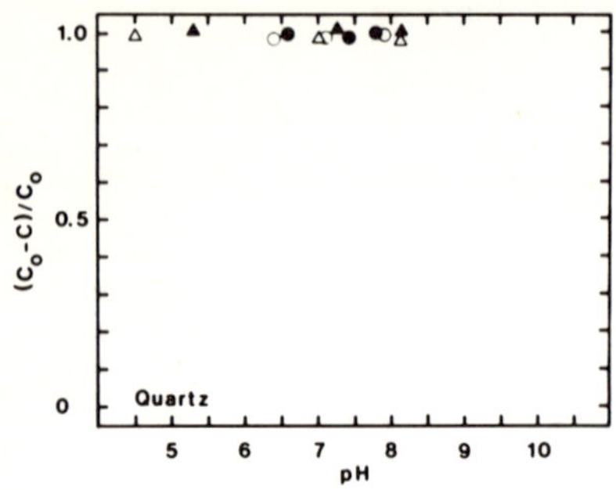

Figure 4. The sorption of americium pseudocolloids on alumina after 1 w (○ 10 g/l, △ 20 g/l, ▲ 40 g/l, ● 10 g/l for 1 w + addition of 10 g/l for 1 w)

For the montmorillonite system the removal was rapid, just as in the system without sorbent. The maximum removal was about 95 %, which was higher than for the system without sorbent (80-90 %) but lower than for quartz.

A desorption af americium from the particle phase (quartz or montmorillonite) and distribution on all available surfaces (particle phase plus sorbent phase) could explain the observed behaviour. The time for such a new distribution appears to be of the order of several weeks. The somewhat lower final value in the alumina-montmorillonite system in comparison with the alumina-quartz system would be expected, considering the much higher sorption on montmorillonite than on alumina and therefore certain preference for montmorillonite in the montmorillonite-alumina mixture.

Evidently, the surfaces provided by 0.2 g of solid sorbent is large enough that no loading effects can be observed. Any effects of particle size or amount of sorbent in the system should therefore have been studied at much lower sorbent/volume ratios.

The observed removal of americium from the solution phase can be due to two different mechanisms:

- o The americium is partly desorbed from the particle phase and redistributed on all available surfaces (particle plus sorbent) in the system.
- o The whole colloidal aggregates (quartz or montmorillonite) are sorbed on the solid.

Possibly both processes are occurring. At least alumina would have a positive surface charge in the pH-range of the experiment [7,8] which would attract the negatively charged colloidal aggregates. Also positively charged americium hydrolysis products would be sorbed on an alumina surface under the present conditions [5,10].

Column experiments

The results from the column experiments are summarized in Table I. In all of the experiments with the high flow rate (0.5 ml/min) and in some of the systems with the low flow rate (0.05 ml/min) a minor fraction of the pseudocolloid passed through the column without significant retention. Evidently the short contact time during especially the high flow rate (a few minutes) is not sufficient for the establishment of a new sorption equilibrium between americium, quartz or montmorillonite particles and column

Table I. Mobility of americium pseudocolloids in columns packed with alumina or granite

Sorbent [a]	Colloidal carrier [b]	Initial pH	Flow rate ml/min	Eluation % [c]	Max. volume [d]
a	Qz	6	0.05	0.2	120
a	Qz	6	0.5	0.3	25
a	Qz	8	0.05	0.5	95 [e]
a	Qz	8	0.5	6.0	200
a	Mo	8	0.05	0	90
a	Mo	8	0.5	6.2	250
b	Qz	8	0.05	0	150
b	Mo	8	0.05	0.2	350
b	Mo	8	0.5	4.4	120
c	Qz	8	0.05	0	140
c	Qz	8	0.5	0.3	220
c	Mo	8	0.05	0.1	380
c	Mo	8	0.5	3.2	210

[a] a = Al_2O_3, 0.090-0.125 mm; b = Al_2O_3, 0.125-0.250 mm; c = Stripa granite, 0.090-0.125 mm

[b] Qz = quartz, Mo = montmorillonite

[c] Fraction passing through the column (average of two runs) with little retention (calibration with Na-22); 0 indicates no significant activity above background

[d] Maximum volume (number of column volumes) that passed through the column (without break-through in any case)

[e] Ca 880 column volumes in one column with the dimensions 23 cm x 4 mm

filling. No break-through was observed in any of the systems (after elution with up to 880 column volumes). No significant differences in performance between the various column filling materials could be observed under the present experimental conditions.

CONCLUSIONS

Americium appeared to be sorbed on quartz and montmorillonite particles of colloidal size. The apparent distribution coefficients related to the distribution beween the particle phase and solution was similar to values obtained with crushed minerals of much larger particle size. Evidently, the americium was distributed over all available surfaces (on particulates as well as vessel walls).

The removal of americium was more rapid and enhanced in the presence of a sorbent (alumina or granite) with macroscopic surfaces. However, the mechanism for the americium distribution (desorption from the particle phase and resorption on all available surfaces or sorption of the whole colloidal quartz or montmorillonite carriers on the sorbent) could not be concluded.

A minor fraction of pseudocolloidal americium passed through a column with alumina or granite with little sorption. Slow kinetics for the desorption-resorption process is a probable reason for the existence of this mobile fraction (very short hold-up time in the column). No break-through was observed after up to 880 column volumes of eluate solution.

A sorption of americium on all exposed surfaces (particles as well as the surfaces and pores of water-carrying fissures etc) would be expected under the conditions in a geologic repository. Formation of a highly mobile, non-sorbing particle fraction does not appear probable.

ACKNOWLEDGEMENT

The help by Dr. S. Wall and M.Sc. L. Björnqvist with the laser measurements is gratefully acknowledged. This work was funded by the Swedish Nuclear Fuel and Waste Management Co (SKB).

REFERENCES

1. B. Allard and J. Rydberg, in Plutonium Chemistry, edited by W.T. Carnall and G.R. Choppin, (ACS Symposium Series Vol. 216, American Chem. Soc., Washington, D.C., 1983), p. 275.
2. L. Björnqvist and M. Larsson, Diploma work, Dep. of Physical Chemistry, Chalmers Univ. of Technology, Göteborg, 1984.
3. B. Torstenfelt, K. Andersson and B. Allard, Chem. Geol. 36, 123 (1982).
4. U. Olofsson, B. Allard, K. Andersson and B. Torstenfelt, in Scientific Basis for Nuclear Waste Management. Vol. 4, edited by S. Topp (Elsevier, New York, 1982), p. 191.
5. B. Allard and G.W. Beall, J. Environ. Sci. Health 6, 507 (1979).
6. D.J. Greenland and M.H.B. Hayes, The Chemistry of Soil Constituents, (John Wiley and Sons, London, 1978).
7. G.A. Parks, in Equilibrium Concepts in Natural Water Systems edited by W. Stumm (Advances in Chem. Series 67, American Chem. Soc., Washington, D.C., 1967), p. 121.
8. W. Stumm and J.J. Morgan, Aquatic Chemistry, (John Wiley and Sons, New York, 1981).
9. B. Allard, U. Olofsson and B. Torstenfelt, Inorg. Chim. Acta 94, 205 (1984).
10. B. Allard, G.W. Beall and T. Krajewski, Nucl. Techn. 49, 474 (1980).

ANALYSIS OF COLLOID TRANSPORT

B. J. TRAVIS * and H. E. NUTTALL **
*Earth and Space Sciences Division, Los Alamos National Laboratory, Los Alamos, New Mexico 87545
**Nuttall and Assoc., Inc., 1445 Honeysuckle, Dr., NE, Albuquerque, NM 87122

ABSTRACT

The population balance methodology is described and applied to the transport and capture of polydispersed colloids in packed columns. The transient model includes particle growth, capture, convective transport, and dispersion. We also follow the dynamic accumulation of captured colloids on the solids. The multidimensional parabolic partial differential equation was solved by a recently enhanced method of characteristics technique. This computational technique minimized numerical dispersion and is computationally very fast. The FORTRAN 77 code ran on a VAX-780 in less than a minute and also runs on an IBM-AT using the Professional FORTRAN compiler. The code was extensively tested against various simplified cases and against analytical models.

The packed column experiments by Saltelli et al. were re-analyzed incorporating the experimentally reported size distribution of the colloid feed material. Colloid capture was modeled using a linear size dependent filtration function. The effects of a colloid size dependent filtration factor and various initial colloid size distributions on colloid migration and capture were investigated. Also, we followed the changing colloid size distribution as a function of position in the column.

Some simple arguments are made to assess the likelihood of colloid migration at a potential NTS Yucca Mountain waste disposal site.

INTRODUCTION

To date both laboratory [1,2] and site studies [3,4] have demonstrated the existence and in some cases the accelerated transport of colloids. Nuttall, Hart, and Travis [5] developed a theory for colloid transport using the population balance concept. Travis and Nuttall [4] applied the population balance theory and the TRACR3D computer code to study the hydrology and transport of radiocolloids at a Los Alamos waste disposal site. Saltelli et al. [1] derived and tested a one-dimensional filtration model and applied it to transient column profile data obtained by injecting ^{241}Am contaminated waste-glass colloids into a 26 mm by 180 mm sand packed column. They measured the size distribution of feed colloids and attempted to model the effects of the polydispersed colloids on the capture profile by assuming a distribution for the filtration factor. This approach did not yield a fundamental understanding of the effect of colloid size, but they did show that one must consider the size distribution in treating the experimental data.

In the following analysis of colloid transport, we used the multidimensional form of the population balance equation to analyze directly the effects of polydispersed colloids. The theory is developed for the transport of polydispersed colloids through a one-dimensional porous region. Conversion from a weight distribution to the corresponding number density function and back to a weight distribution after solution of the population balance equation is treated. We used the model to investigate the effect of two size distributions and to compare the effect of a constant versus size dependent filtration type colloid capture model.

THEORY

Randolph [6] derived the general form of the population balance model in 1962. The population balance theory was derived at about the same time by Hulburt and Katz [7]. The population balance is a number continuity equation which has been applied successfully to crystallization, aerosols, and biological processes. Recently Travis and Nuttall [4] applied this methodology to the problem of colloid transport. In this study, we apply the dynamic microscopic form of the population balance model with convective transport in one spatial dimension and particle growth along a colloid size axis to the problem of radiocolloid transport in nuclear waste disposal.

The colloid population density function, P(t,z,L), is dependent on time, space and the colloid characteristic size which we assume to be the colloid diameter. P(t,z,L) is the number of colloids of a particular size L at a point z in space and time t per unit volume of solution. When P is integrated over the colloid size range we have the total number of colloids per unit volume. In general, colloid charge or any other useful colloidal properties could be modeled by simply adding the appropriate property axes. The population balance also includes a birth/death term f(P,t,z,L) which again accounts for the birth of new colloids as well as the adsorption and release of colloids by the rock matrix. The adsorption and release function treats colloids of arbitrary size. In this study we do not treat the adsorption of radionuclides on the colloids nor are we investigating the growth of colloids. To model colloids of the same size but with differing amounts of radionuclides would require an additional property axis for concentration. Under these assumptions the population balance is:

$$\frac{\partial P}{\partial t} + v\frac{\partial P}{\partial z} - D\frac{\partial^2 P}{\partial z^2} + u\frac{\partial P}{\partial L} = -f \qquad (1)$$

The filtration term is

$$f = \lambda(L)\, v\, P(t,z,L) \quad . \qquad (2)$$

It is sometimes desirable to work with a dimensionless form of the population balance. Carrying out the transformation of Eq. (1) using the dimensionless variables defined in the notation section leads to:

$$\frac{\partial\psi}{\partial\tau} + \phi\frac{\partial\psi}{\partial\beta} - \frac{D}{\bar{z}\,\bar{v}}\frac{\partial^2\psi}{\partial\beta^2} + \upsilon\frac{\partial\psi}{\partial\rho} = \frac{\bar{z}}{P_i\bar{v}}\, f(\psi,\tau,\beta,\rho) \qquad (3)$$

Adsorption of colloids onto the surrounding matrix is assumed to follow the filtration model where the rate of mass accumulation is proportional to the concentration of colloids in the solution. In our model the colloids concentration in solution must be expressed in terms of the population distribution function. The mass concentration c(t,z) of colloids in the fluid is

$$c = \int_0^\infty c(L)\,dL \qquad (4)$$

where

$$c(L) = \rho\, K_v\, P(t,z,L)\, L^3 \quad , \qquad (5)$$

and K_v = volumetric shape factor, $\pi/6$ for a sphere.

The rate of mass build up on the matrix at a point in space for each colloid size is

$$\frac{d\sigma_i}{dt} = \varepsilon\ \lambda(L)\ v\ c(L) \tag{6}$$

The total rate of mass accumulation for all colloid sizes is equal to the integral of Eq. (6) over the colloid size range.

$$\frac{d\sigma}{dt} = \varepsilon\ \rho\ v\ k_v \int_0^{L_m} \lambda(L)\ P(t,z,L)\ L^3\ dL \tag{7}$$

In the general case $\lambda(L)$ will be some function of colloid size. However, if $\lambda(L)$ is size independent then the filtration follows the conventional form showing a straight line profile for the concentration of colloids captured on the sand packing when plotted on semilog paper.

To treat experimental data, it is often necessary to convert between weight distributions and number distributions. In terms of the number density function the distributed weight function is [8]:

$$w(L) = \rho K_v L^3 P(L)/M_T \tag{8}$$

The population balance, Eq. (3), was solved numerically by an improved method of characteristics technique presented by Douglas and Russell [9]. The buildup of colloids on the porous matrix is described by Eq. (7) which was coupled to the population balance and solved as a series of ODEs using the 4th order Runge-Kutta method. Note that σ is a function of column position, z, and time.

In our studies we tested both a gamma and a log-normal distribution to model the initial weight distribution of radiocolloids. The log-normal distribution gave a better fit to the measured feed distribution. The cumulative and distributed forms for the log-normal distribution are respectively: (Randolph and Larson [8] p. 32)

$$W(L) = 0.5 + 0.5\ \mathrm{erf}(z) \tag{9}$$

$$z = [\ln(L/\bar{L}')]/2^{1/2} \ln \sigma'$$

where $\bar{L}'$ = geometric mean size, μm; σ' = width parameter, (geometric STD), μm, and the distributed form of the log-normal function is

$$w(L) = \frac{1}{(2\pi)^{1/2} \ln \sigma'} \exp\left[\frac{-\ln^2(L/\bar{L}')}{2\ln^2 \sigma'}\right] \tag{10}$$

CASE STUDIES

Packed Column Simulations

Saltelli, et at. [1] investigated the migration of polydispersed ^{241}Am colloids through a packed column of glauconitic sand. Colloidal ^{241}Am, leached from a simulated waste glass, was injected into a column. Three experiments of 28, 80, and 170 days duration were run. At the end of each test the column was dismantled, sectioned, and the ^{241}Am profile analyzed. Those authors analyzed their data using the filtration capture model and found that the column profiles could not be matched using the conventional constant parameter filtration model. To fit the profile curves they assumed that the colloid feed exhibited a distribution of filtration coefficients. In this way they hoped to approximate size dependent filtration and by adjusting the filtration distribution parameters they did fit the shape profile curves. Their model also included a Langmuir isotherm model to treat the observed early adsorption of ^{241}Am which occurred throughout the length of the column. This rapid adsorption of ^{241}Am on the porous matrix was at a very low concentration level and the mechanism for this phenomenon has not been directly investigated; therefore, we did not choose to include this mechanism in our model. To match the experimental data we simply assumed an initial adsorbed ^{241}Am concentration of 10^{-10} moles/liter.

The size distribution of the colloid feed was measured by filtration and represented in our analysis by a log-normal distribution. The measured ^{241}Am colloid distribution along with our curve-fitted distributed and cumulative log-normal distributions are illustrated in Fig. 1. The measured range of the feed colloids was from approximately .05 μm to a maximum of 10 μm. Our curve-fit parameters for the log-normal distribution were $\bar{L}' = 0.45$ μm and $\sigma = 7.39$ μm.

The upper colloid size was experimentally controlled by passing the feed solution through a 10 μm filter prior to entering the column. The experimental conditions are give in Table I.

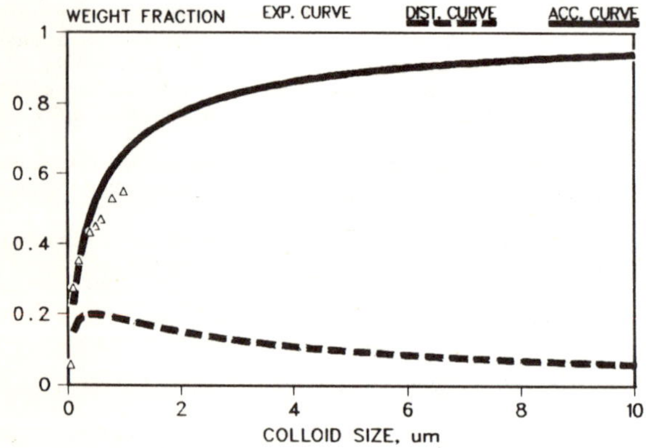

Figure 1. Colloid feed size distribution.

TABLE I - Experimental Parameters

Column Dimensions:	18 cm X 2.6 cm
Porosity:	0.38
Dispersion Coefficient:	62.6 cm^2/d
Darcy Velocity:	20.9 cm/d
Inlet Concentration:	3.8×10^{-10} M
Exit Concentration:	$3.5\text{-}6.5 \times 10^{-13}$ M
Feed Concentration Correction at 170 d:	0.242
Column Packing:	Sand with 20% clay

We solved the population balance model for the above conditions and used the log-normal distribution to represent the colloid feed. We also found that a size dependent λ was required. In our analysis, however, λ represents the actual capture mechanism whereas in Saltelli et al. [1], λ represented capture mechanism and size distribution effects together. The resulting concentration profile for ^{241}Am on the porous matrix is compared with the experimentally measured profile after 170 days of continuous column operation and shown in Fig. 2 . The strong curvature in the calculated profile is created by the linear size dependent filtration function, λ = 2 L. The asymptotic level of concentration at 10^{-10} shown on the far right of the curve was modeled by arbitrarily setting this level of concentration as an initial condition in the column. This assumption did not change the basic shape of our calculated curve. It only affected the plateau portion of the curve at the bottom of the column. The mechanism for this early low level of concentration is still speculative and can be modelled in several ways; however, the true physical mechanism has not been identified.

The numerical solution of the population balance also provides some interesting information on the change in the fluid phase colloid size distribution as a function of position down the vertical column. Fig. 3 illustrates in three dimensions the position varying colloid weight distribution in the liquid. The surface shows both the rapid decrease in the mass of colloids in solution and the shift in the mode of the distribution to small sizes. Because of the size dependent filtration coefficient the larger colloids are preferentially removed early at the top of the column leaving the smaller particles to transport downward through the porous medium. In the case of a constant filtration factor, λ, the nature of the colloid size distribution has no effect on the solid concentration profile since then each particle has the same capture probability independent of size.

The calculated ^{241}Am concentration profile in Fig. 2 follows the trend of the experimental data but does not agree equally well at every point. A major cause of this is the uncertainty in the true form of the filtration function λ(L). An alternative approach to analyzing this experiment is to consider it an inverse problem for λ(L). That is, instead of estimating λ(L) and then calculating colloid transport, we could solve Eqs. (1) and (7) directly for λ(L). That is, we could solve the minimization problem

$$\min \sum_i \sum_j \left(\sigma_o(t_i,z_j) - \sigma_m(t_i,z_j)\right)^2 \tag{11}$$

subject to

$$\frac{d\sigma_m}{dt} = \varepsilon \rho v K_v \int_0^{L_m} \lambda(L)\, P(t,z,L)\, L^3\, dL \tag{12}$$

and

$$\frac{\partial P}{\partial t} + v\frac{\partial P}{\partial z} - D\,\frac{\partial^2 P}{\partial z^2} = -\, v\, \lambda(L)\, P, \tag{13}$$

$$P(t,o,L) = g(L), \quad P(o,z,L) = 0 \tag{14}$$

where $\sigma_o(t_i,z_j)$ is the observed mass accumulation at time t_i and location z_j and σ_m is the calculated value. The solution can be obtained by nonlinear optimization techniques, such as the adjoint method. The solution will be the most likely form for λ(L). We plan in fact to solve the equation set (11) to (14) in the near future. This analysis points out a problem frequently encountered--incomplete data. In this case, a set of experiments, each using a different but uniform colloid size, would allow independent determination of λ(L).

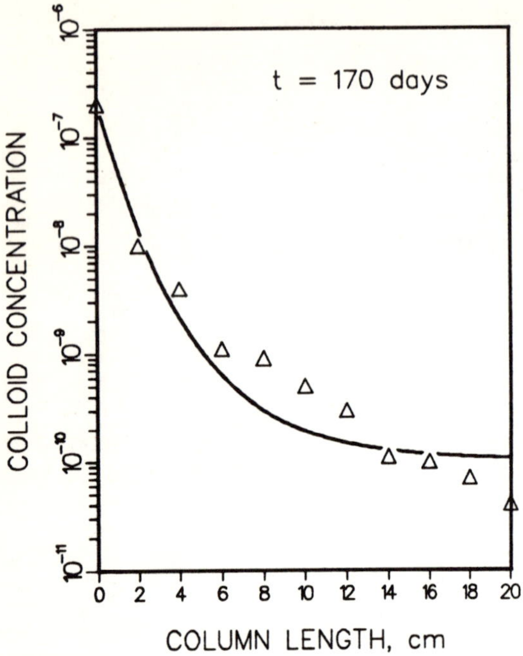

Figure 2. Measured and calculated solid concentration profile for ^{241}Am. (concentration, moles/liter of column.)

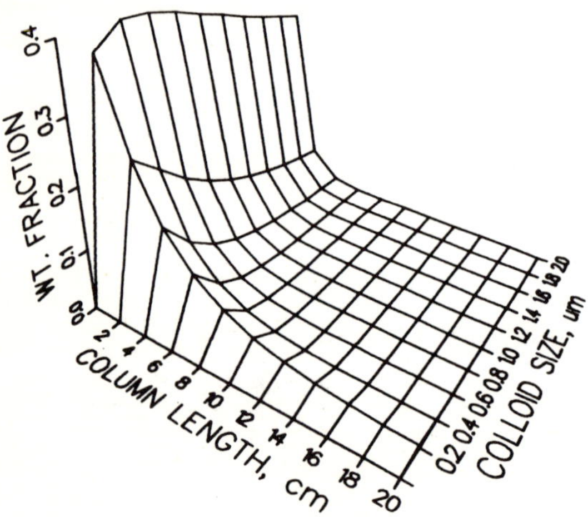

Figure 3. Liquid phase colloid weight distribution versus position along column.

NTS YUCCA MOUNTAIN SITE

Colloids may originate in a waste repository from several sources: by the corrosion of waste glass and canisters, by degradation of the engineered backfills, and by erosion from naturally occurring clays. Hence, there is a reasonable likelihood that colloids will exist; however, the important consideration is whether the geological medium is conducive to colloid transport or whether colloid migration and capture will retard their transport. Colloid transport depends on at least four factors: filtration, pH of the solution, water velocity, and diffusion.

At Yucca Mountain filtration will be an effective barrier in the tuff matrix. Pore size is an important element in filtration. Pseudocolloids will obviously not flow through pores that are smaller than the pseudocolloids themselves. Pore size distributions have been measured for several Yucca Mountain tuff samples. Mercury porosimetry data, measured by Quantachrome Co., on a sample of Topopah Spring tuff showed that most of the pores are less than 1 μm and approximately 50% are smaller than 0.1 μm. Even though a small percentage of pores are larger than 1.0 μm these large pores will not transport colloids because they will be dry. In the Topopah Spring formation at NTS, the saturation is less than 90% and large pores are the last to become saturated. In another NTS site formation, Calico Hills, approximately 50% of the pores are smaller than 0.06 μm. Hence colloids must be on the order of 0.1 μm in diameter or less to pass through the NTS tuff.

Pore water velocities, assuming matrix flow, are estimated at less than 2.0 mm/yr. These low velocities are on the order of molecular diffusion rates thus greatly limiting the migration of colloids within the tuff matrix. Transport of colloids through fractures is possible; however, for the measured fracture apertures of 50 to 250 μm in Topopah Spring tuff the water flow rate would only be about 2 m/yr or still not much greater than molecular diffusion transfer rates. If fracture flow were episodic, corresponding to a few intense rainfalls per year, flow rates in fractures might be much larger but for only very short times.

Colloids greater than 0.1 μm in diameter will likely be removed by gravitational settling [10]. The transporting ability of very small colloids may be diminished by diffusion of radionuclides out of the particles and subsequent adsorption of the nuclides onto the tuff matrix. Diffusivities for some nuclides in some clays have been measured at 10^{-13} to 10^{-16} cm^2/s. Characteristic diffusion times for 0.1 μm particles are consequently only a few days. Migration of small colloids may be possible in a fracture flow regime. Work is continuing to understand the migration mechanisms of the very small colloids.

SUMMARY

The population balance methodology which operates on a number continuity approach is well suited for describing the complex phenomena of colloid transport. In the present study the problem of polydispersed colloid transport in a porous medium was successfully modeled and agreement was found between model results and data. The advantages of this methodology are that it provides a comprehensive and fundamental tool for analyzing the complex mechanisms of colloid migration. The approach requires conversion back and forth between the weight and number distributions; however, this conversion is a straight forward process. In this study, the colloid size distribution had a significant effect on the migration and capture process when a size dependent filtration capture model was used. In the case of size dependent filtration, the larger colloids were rapidly captured while the smaller ones migrate much further through the porous medium. Both a constant and a linear size dependent filtration type capture model were tested. The size dependent model was required to match the measured column profile data. The experimental data was modeled successfully with only one adjustable parameter, s, the slope of the filtration coefficient function. If λ, the filtration factor, is constant then the colloid size distribution has no effect on transport or capture.

A limited study of parameter sensitivity showed that the width of the colloid feed distribution influenced the curvature on a semilog plot of the solid phase concentration profile with narrowing of the distribution converging to the monodispersed model and producing a linear concentration profile on a semilog plot.

In light of our findings, we made a preliminary review of some colloid transport characteristics for a potential NTS nuclear waste disposal site. We found geological factors favorable for retarding colloid migration. However, the potential for very small colloids to migrate in fractures with episodic flow characteristics may be significant and therefore requires further study and analysis.

NOTATION

c	Colloid Concentration in fluid	gm/cm^3
n	Hydrodynamic Dispersion Coefficient	cm^2/d
f	Birth/Death Rate Function	$\#/cm^3 \mu m d$
L	Colloid Size Dimension	μm
M_T	Initial Mass Concentration of Colloids	gm/cm^3
P	Population Density Function	$\#/cm^3 \mu m$
s	Slope of Filtration Function	$cm^{-1} \mu m^{-1}$
t	Time	d
u	Growth Rate Along the Size Axis	$\mu m/d$
v	Darcy Water Velocity	cm/d
w	Weight Distribution Function	$gm/cm^3 \mu m$
W	Cumulative Weight Distribution	gm/cm^3
z	Space Dimension	cm

Greek Symbols

ψ	Population Density Function (P/P_i)	dimensionless
ϕ	Velocity $(v/\bar{v})$	dimensionless
ν	Growth Date $(u\bar{z}/\bar{v}\bar{L})$	dimensionless
β	Space $(z/\bar{z})$	dimensionless
ρ	Colloid Size $(L/\bar{L})$	dimensionless
λ	Filtration Coefficient	cm^{-1}
σ	Mass Concentration of Colloids on the Matrix	gm/cm^3
σ_i	Mass Concentration of Colloids at size L on the Matrix	$gm/cm^3 \mu m$
τ	Time $(t\bar{v}/\bar{z})$	dimensionless
-	Average Properties of Limiting Properties	
ε	Porosity	

ACKNOWLEDGEMENT

"Prepared by Nevada Nuclear Waste Storage Investigations (NNWSI) Project participants as part of the Civilian Radioactive Waste Management Program (CRWM). The NNWSI Project is managed by the Waste Management Program Office (WMPO) of the U. S. Department of Energy, Nevada Operations Office (DOE/NV). NNWSI Project work is sponsored by the Office of Geologic Repositories (OGR) of the DOE Office of Civilian Radioactive Waste Management (OCRWM)."

The authors express their sincere appreciation to Mabel P. Grey for typing this paper and to M. A. Olson for preparing illustrations and graphic presentation.

REFERENCES

1. A. Saltelli, A. Avogadro, and G. Bidoglio, Nuclear Technology, 67, 245-254 (1984).

2. S.M. Fried, et al., Argonne National Laboratory report ANL-76-127, (1976).

3. D.R. Champ, W.F. Merritt, and J.L. Young, Scientific Basis for Nuclear Waste Management, 5, Werner Lutze (Ed), North-Holland, N.Y., 745-754, (1982).

4. B.J. Travis, and H.E. Nuttall, in Scientific Basis for Nuclear Waste Management VIII, 44, edited by C.M. Jantzen, J.A. Stone, R.C. Ewing, (Materials Research Society, Pittsburg, Pennsylvania, 1985), 969-976.

5. H.E. Nuttall, R. Hart, and B.J. Travis, Submitted to Nuclear Technology, (1985).

6. A.D. Randolph, Can. J. Chem. Eng., 42, 280-281, (1962).

7. H.M. Hulburt, and S. Katz, Chemical Engineering Science, 19, 555-574, (1964).

8. A.D. Randolph and M.A. Larson, Theory of Particulate Processes, (Academic Press, New York, 1971).

9. J. Douglas Jr., and T.F. Russell, SIAM Journal of Numerical Analysis, 19 (5), 871-885 (1982).

10. J.A. Apps, et al., prepared for U.S. NRC, NUREG/CR-3062, 1983.

SORPTION AND DIFFUSION OF RADIONUCLIDES IN ROCK MATRIX AND NATURAL FRACTURE SURFACES STUDIED BY AUTORADIOGRAPHY

S. MUURONEN*, E-L. KÄMÄRÄINEN*, T. JAAKKOLA*, S. PINNIOJA* AND A. LINDBERG**
* University of Helsinki, Department of Radiochemistry, Unioninkatu 35, SF-00170 Helsinki, Finland
** Geological Survey of Finland, Kivimiehentie 1, SF-02150 Espoo, Finland

A method based on autoradiography was developed to determine the sorption and diffusion of cesium, strontium, cobalt, nickel, iodine and americium into rock matrix. Samples chosen for this study were filled and unfilled natural fracture surfaces and drill cores having a central drilled hole ("drill core cups"). Rock types were mica gneiss, tonalite and rapakivi granite, which were selected to represent the common rocks and minerals in Finnish bedrock. Distribution coefficients (K_a-values) of cesium and strontium determined for fissure surfaces and drill core cups were of the same order of magnitude. After three months contact time the greatest penetration depth for cesium was 2.5 mm, for a natural fissure surface sample of rapakivi granite. For strontium the penetration depths of 11 mm in three months and 35 mm in twelve months were found for filled natural fissure surface samples of rapakivi granite. The range of D_a-values was $1.5 \times 10^{-15} - 3.2 \times 10^{-14}\ m^2/s$ and $1.4 \times 10^{-14} - 2.1 \times 10^{-13}\ m^2/s$ for cesium and strontium, resp. For cobalt the D_a-values of $5 \times 10^{-16}\ m^2/s$ in tonalite was obtained. In six months the penetration depths of nickel, iodine and americium were too low (< 0.5 mm) to allow calculation of D_a.

INTRODUCTION

According to the concept of the Finnish nuclear waste program the low and medium level waste will be disposed in deep underground repositories. Elements released from their waste package will be able to force their way through barriers and will be transported with groundwater in bedrock along fracture zones and fissures. The migration of radionuclides in groundwater will be delayed by interactions with fracture and fissure surfaces and filling materials. Important processes are sorption and diffusion. Nuclides are mainly sorbed on fissures and porous minerals in rock.

The aim of this study was to determine the sorption and diffusion of radionuclides on rock samples in groundwater conditions. An autoradiographic method was used for determination of penetration depths and diffusion coefficients of Cs-134, Sr-90, Co-60, Ni-63, I-125 and Am-241.

MATERIALS

Samples chosen for the study were sections of drilling cores from mica gneiss, tonalite and rapakivi granite rocks. Mica gneiss and tonalite samples originated from the Nuclear Power plant site in Olkiluoto, Finland, and rapakivi granite samples originated from the plant site at Hästholmen in Loviisa, Finland. Mineralogical composition of the rocks was determined by point counter method (Table I) [1].

Two types of drill core samples were used: "drill core cups" and natural fracture surfaces. The drill core cups were cylinderical, about 10 cm in height and 5 cm in diameter, with a drilled hole (diameter 1.5 cm and depth about 7 cm) in the middle for the tracer solution.

The natural fracture surfaces were both unfilled and filled representing each of the rock types selected for this study. Mineralogical composition of fissure filling materials was determined by X-ray diffraction (Table II) [1].

The water used in the experiments was a synthetic groundwater corresponding to the natural granitic groundwater [2, 3].

The sorption and diffusion of radionuclides were studied using isotopes Cs-134, Sr-90, Co-60, Ni-63 as chlorides, I-125 as sodium iodide and Am-241 in nitrate form. The initial element concentrations varied between 10^{10} and 10^{-6} M.

Table I. Mineralogical composition (Vol-%) of rapakivi granite, tonalite and mica gneiss samples used in matrix diffusion studies. Vol-% was calculated by point counter method.

Minerals	Rapakivi granite	Tonalite	Mica gneiss
K feldspar	38.0		0.5
plagioclase	12.8	48.7	29.3
quartz	36.9	16.6	27.3
biotite	7.5	23.0	27.5
hornblende	2.7	9.6	
opaques	0.7	0.3	0.8
apatite	+	1.4	0.1
chlorite	1.4	0.1	+
fluorite	+		
zircon	+	+	+
sericite	+	0.2	0.3
carbonate		+	
sphene		+	
epidote		0.1	
cordierite			12.9
sillimanite			0.7
tourmaline			+
	100.0	100.0	100.0

+ = optically observed

Table II. Mineralogical composition (Vol-%) of fissure filling materials determined by X-ray diffraction. Compositions are approximate (analyzed by B. Johanson, Geological Survey of Finland).

Samples	Rapakivi granite				Tonalite				Mica gneiss			
Minerals	*)35.75	35.85	35.95	36.05	191.2	191.3	191.8	191.9	93.55	94.3	94.4	94.5
quartz	20	40	50	50	+	30			+	+	+	+
albite	10	10	10		10		10	+				
K feldspar		20	20	30								
biotite		20	10	10				10				
muscovite	+	10	10	10	+		10					
margarite								10				
chlorite					75	60	30	70	+	+	+	+
calcite					10	+	40		90	90	90	95
ankerite	40											
pyrite						10	10					

+ = observed

*) sample number = depth (m) from the ground level

EXPERIMENTAL

The determinations were carried out under atmospheric oxygen conditions and at room temperature (20 ± 2 °C).

The distribution coefficients were determined in equilibrium state. Before the samples were spiked they were kept in contact with groundwater for about two months. During this period the changes in Na, Ca, K and Mg concentrations were followed by atomic adsorption spectrophotometry.

The walls of the fracture surface samples were covered with protective paint to confine the contact of radioactive tracer solution to the natural fracture surface and to prevent sorption and penetration in other surfaces of the samples. The fissure surfaces were then immersed in polypropylene bottles containing the tracer solutions. The radioactivity of the spiking solutions varied from 18.5 to 37 kBq in 20 ml of synthetic groundwater.

In the case of drill core cup samples the tracer solution was injected into the drilled hole of the rock sample. The radioactivities of the tracer solutions varied between 18.5 and 74 kBq in 10 ml synthetic groundwater. Contact times varied from one week to one year. The concentrations of radionuclides in the tracer solutions were adjusted to stay at constant level during the entire contact time [1].

All the rock specimens were rinsed and dried after exposure to the spiked groundwater solution. Radioactivities of tracer and rinsing solutions were measured by liquid scintillation counting (Ni-63, Sr-90) or by γ-spectrometry (Co-60, I-125, Cs-134, Am-241) and the sorption of radionuclides was calculated.

Distribution coefficients K_a were calculated from the equation

$$K_a = \frac{\text{amount of radionuclide sorbed/surface area}}{\text{amount of radionuclide in solution/solution volume}} \; m^3/m^2 \qquad (1)$$

The samples were cut up with a diamond-bladed rock saw (blade 1.8 mm thick), after which they were ground with a lapping and polishing machine. Waterproof silicon carbide paper discs were used to remove thin (0.1 - 0.2 mm) layers from samples. After each removal the surface was studied by autoradiography. All autoradiograms were made with X-ray film X-omat MA. The films were exposed (20 h to 7 days), developed with Kodak DX-80 developer and fixed with Kodak FX-40 fixer.

The autoradiograms were measured by microphotodensitometry. Optical densities were compared with those of standards of known activity.
The concentration of the radionuclides was determined at least at three different depths for calculation of diffusion coefficients of the element (Figure 1).

The penetration depth was defined as the distance from the surface to the greatest depth at which radionuclides can be detected by the autoradiographic method. In our experiments this meant a detection limit about 1/1000 of the activity at the surface of the sample.

Apparent diffusivities (D_a) for cesium and strontium in the rock matrix were calculated with the following equation [4, 5]:

$$c = c_o \operatorname{erfc}(x/2\sqrt{D_a t}) \qquad (2)$$

where c_o is concentration of the diffusion species at the surface at time t = 0 and c is concentration of the diffusing species at distance x from the surface.

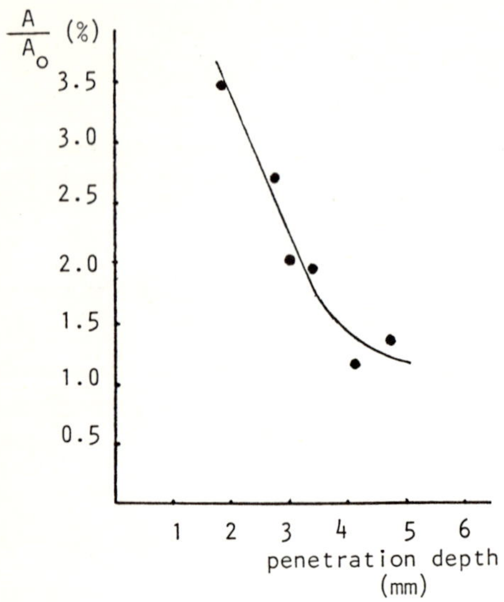

Figure 1. Concentration profile of strontium-90 in the drill core cup sample of tonalite, contact time 12 months.
A = sorbed radioactivity on the surface
A_o = diffused activity at different depths

RESULTS AND DISCUSSION

The distribution coefficients (K_a) of strontium and cesium for natural fracture surfaces and drill core cups are given in Table III. For comparison the K_a-values for different sample types are indicated in Table III. Distribution coefficients for rock samples (fissure surfaces, drill core cups and rock pieces) were of the same order of magnitude. K_a-values for fissure surfaces, drill core cups and rock pieces were one to two orders of magnitude higher than those for thin sections and crushed samples. (The K_a-values for crushed rock samples were calculated by dividing the mass distribution coefficients, K_d (ml/g), by the surface areas, which were determined by BET method based on gas adsorption.) Sorption on natural fissure surfaces was more effective than on fresh core samples, evidently because of higher cation exchange capacity of altered minerals on fissure surfaces. Filling materials on natural fracture surfaces showed incresed sorption compared with pure fracture surfaces, except for calcite, which was filling material on mica gneiss surfaces. K_a-values of strontium and cesium were consistently lower for filled mica gneiss than for unfilled mica gneiss.

To calculate the diffusion coefficients (D_a), the penetration depths were determined after a contact time of three months and one year for natural fissure surfaces and after three, six and twelve months for drill core cups. The penetration depths and some measured apparent diffusion coefficients are given in Table IV.

Table III. K_a-values of cesium-134, strontium-90 and cobalt-60 for different sample types of Finnish bedrock. Sorption time was one week.

Nuclide	Rock sample specimen	K_a (10^{-4} m^3/m^2) Fissure surfaces unfilled	filled	Drill core cups	Rock pieces	Thin sections	Crushed rock (Ø 0.07-2mm)
Cs-134							
	tonalite	1800	2900	1300	1000	190	110
	mica gneiss	6900	1900*	1100	2600	230	-
	rapakivi granite	2200	12000	11100	1200	40	30
Sr-90							
	tonalite	230	3600	100	60	3.6	0.85
	mica gneiss	870	90*	93	180	4.5	-
	rapakivi granite	400	1500	66	40	1.0	0.22
Co-60							
	tonalite	450	20000	1000	300	20	11

*) calcite as filling material (low sorption capacity)

Table IV. Experimental penetration depths and measured apparent diffusion coefficients (D_a) for cesium-134, strontium-90 and cobalt-60. Contact times were, 3, 6 and 12 months.

Sample	Time (months)	Cs-134 Penetration depth (mm)	Cs-134 D_a (m^2/s)	Sr-90 Penetration depth (mm)	Sr-90 D_a (m^2/s)	Co-60 Penetration depth (mm)	Co-60 D_a (m^2/s)
natural fracture surfaces:							
mica gneiss unfilled	3	< 1.0		3.5	9.1×10^{-14}		
mica gneiss filled	3	< 1.0		< 3.7			
tonalite unfilled	3 12	< 1.0		1.9 10	3.5×10^{-14}		
tonalite filled	3 12	3.5	3.2×10^{-14}	2.5 15	5.4×10^{-14}		
rapakivi granite unfilled	3 12	< 1.0		5.7 20	8.6×10^{-14}		
rapakivi granite filled	3 12	2.5	2.7×10^{-14}	7.8 35	2.1×10^{-13}		
drill core cups: tonalite	6 12	0.8 3.0	1.5×10^{-15} 7.2×10^{-15}	4.5 5.5	5.1×10^{-14} 1.4×10^{-14}	0.5 0.9	5.4×10^{-17} 4.8×10^{-16}

In drill core cup samples of tonalite and in natural fissure surface samples of unfilled mica gneiss cesium diffused less than 1 mm in three months and about 2 mm in one year. In the filled tonalite natural fissure surface sample, a mean penetration depth of 3.5 mm was calculated for Cs-134 after contact time of three months, on the basis of autoradiograms taken from different grinding depths. For filled rapakivi granite the penetration depth of 2.5 mm was found.

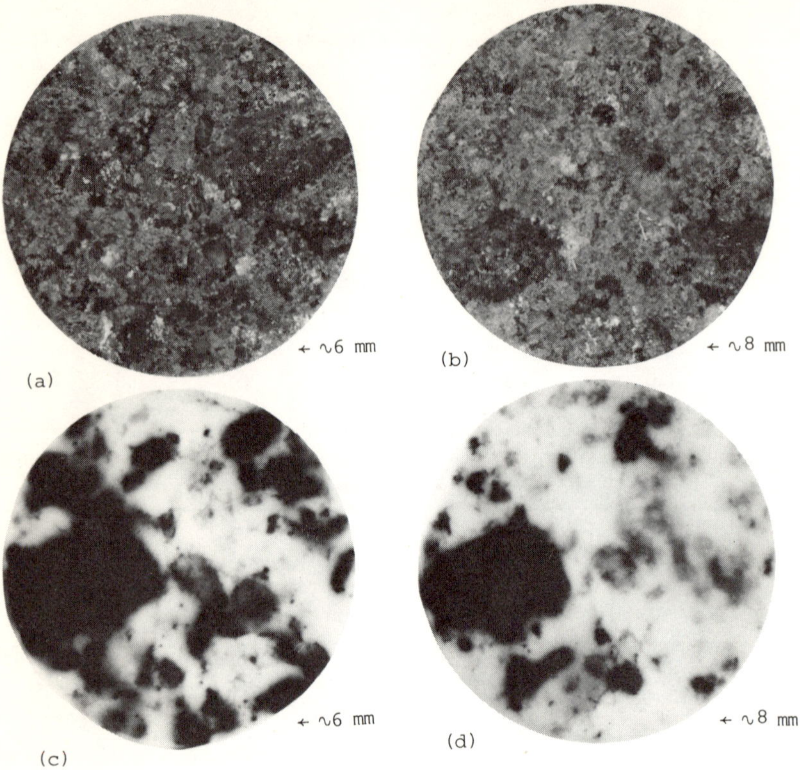

← ∿6 mm

(e)

(f)

← ∿8 mm

(dotted symbol) = altered feldspar (K feldspar and plagioclase)

(dashed symbol) = biotite

Figure 2. The filled rapakivi granite natural fissure surface sample, spiked with Sr-90. Contact time was 3 months. Photographs (a,b) and autoradiograms (c,d) of the sawed surfaces of the sample. The cutting depth was 7 mm. (A diamond blade 1.8 mm in thickness was used in the machine). Porous mineral composition of the cross section surfaces (e,f) is shown.

In all samples strontium diffused deeper than cesium. In one year the penetration depths of strontium for drill core cups of tonalite were as much as 5 mm, depending on the mineral composition, and for unfilled natural fissure surface samples of rapakivi granite as much as 6 mm in three months and as much as 20 mm in one year. In filled natural fissure surface samples of rapakivi granite, the penetration depth was as great as 11 mm (Figure 2) in three months and an exceptionally high 35 mm in one year.

For cobalt a penetration depth of 0.9 mm was found in the drill core cup sample of tonalite after a contact time of on year. For americium, iodine and nickel no diffusion could be observed (depth < 0.5 mm) after a contact time of six months.

Penetration of radionuclides into the rock matrix is a function of time. If the penetration obeys Fick's diffusion theory, penetration is a function of $\sqrt{t}$.

The rate of diffusion was controlled by porous minerals: biotite in tonalite and mica gneiss, altered K-feldspar and biotite in rapakivi granite (Figure 2). Penetration depths obtained in the present study agree with those determined by Torstenfelt et al. [6].

The following effects of filling materials on diffusion of Cs and Sr were found:

- Chlorite and secondary minerals increased the rate of penetration into the mother rock. Radionuclides were sorbed both into the filling material and through fillers into the rock matrix. Radionuclides penetrated deeper in the case of filled samples (Table IV).
- Calcite decreased K_a-values of the samples. Calcite did not sorb significant amounts of Cs and Sr, but they penetrated under the filling material into the mother rock. Penetration depths of Cs and Sr were about the same for filled and unfilled samples.

The measured apparent diffusion coefficients (D_a) were about one order of magnitude lower for cesium than for strontium (Table IV). There was no significant differences between D_a-values obtained for natural fissure surfaces and drill core cups. D_a-values of cobalt measured for drill core cups of tonalite were from one to three orders of magnitude lower than those for cesium and strontium.

The present study will be continued for determination of penetration depths and D_a-values for Co-60, Ni-63, I-125 and Am-241 in each rock type, using contact times up to 1 - 2 years.

The results presented are the initial data obtained from the first of a series of experiments planned to be carried out over next several years. At the present stage of the study, considerable differences is penetration depths in Finnish rocks have been verified. The highest apparent diffusion coefficient was found for Sr-90 in rapakivi granite.

REFERENCES

1. S. Pinnioja, E-L Kämäräinen, T. Jaakkola, M. Siitari, S. Muuronen and A. Lindberg, Sorption and diffusion of cobalt, strontium, cesium and americium in natural fissure surfaces and drill core cups studied by autoradiography, Nuclear Waste Commission of the Finnish Power Companies, YJT-85-16, 1985.

2. B. Allard and J. Beall, Environm. Sci. and Health 6, 507 (1979).

3. J. Hyyppä, The chemical composition of groundwater in Finnish bedrock, Nuclear Waste Commission of the Finnish Power Companies, YJT-84-10, 1984 (in Finnish).

4. R. Barrer, Diffusion in and through solids, (University Press, Cambridge, 1951).

5. V. Mayer, Grundlagen der Kernchemie, (Barth, Leipzig, 1982), pp. 837 - 847.

6. B. Torstenfelt, T. Eliasson, B. Allard, K. Andersson, S. Höglund, T. Ittner and U. Olofsson, in Scientific Basis for Nuclear Waste Management VI, edited by D.G. Brookins (Elsevier Publishers, New York, 1983), pp. 339 - 346.

EXPERIMENTATION AND MODELLING OF U, TH AND LANTHANIDES TRANSPORT IN FISSURED ROCKS: INFLUENCE OF COMPLEXATION

PIERRE TOULHOAT *, **
* Commissariat à l'Energie Atomique. Centre d'Etudes Nucléaires de Saclay. IRDI/DCAEA/SEAIN. 91191 Gif sur Yvette cedex. FRANCE
** Laboratoire de Géochimie Comparée et Systématique, Université Paris VI 4 place Jussieu 75005 Paris. FRANCE

ABSTRACT

Three circulation experiments have been carried out between two drill-holes in a granite after hydraulic fracturation (Le Mayet de Montagne, France). We have injected together uranyl carbonate and Th, La, Sm, Eu, Dy and Yb EDTA complexes. Amounts of tracers were different in each run. Recovery curves are similar for Th and lanthanides, but have a delayed peak and an increased tailing for uranium. For rare earth elements, recovery rates increase with increasing stability of the complex used as tracer. Recovery rates increase also with increasing amount of tracer. Those features are explained by non linear and non reversible sorption. The importance of complexation is demonstrated. Strong organic complexing agents enhance trace elements mobility.

INTRODUCTION

The aim of these experiments was to understand the coupling of mass transfer and solution-solid interactions by using lanthanides and other elements (U and Th) as tracers. Previous studies[1], based only on experimentations, establisned the possibility of using lanthanides chelates. Lanthanides were chosen because they have common chemical properties, all having the same electronic structure (saturated external orbitals). Lanthanides-EDTA complexes are among the most stable chelates in solution[2,3], so their fixation on solids is difficult[4]. In our study the progressive increasing of lanthanides EDTA complexes stability (fig 1) was essential to understand the influence of complexation upon sorption of tracer cations. Uranyl-carbonate complexes are the most stable complexes in natural solutions[5], it was also interesting to compare complexes with different ligands in the same experiment. More information about interaction, in particular linearity and reversibility of sorption was necessary to build accurate models. We have tried to obtain it by injecting different amounts of tracers and carefully studying recovery rates.

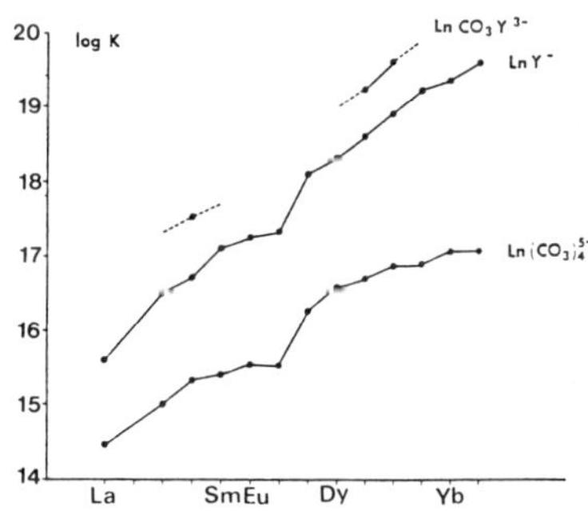

Fig. 1 : Stability constants of Lanthanide complexes[1] Y=EDTA

EXPERIMENTAL PROCEDURE

A circulation between two drill holes has been installed after hydraulic fracturation in a granite (Le Mayet de Montagne, Allier, France), for geothermal studies. The circulation occured approximately at 150 m depth, and was 30 m long. 95% of injected flow was recovered. Flow rates were about 2.9 m^3/h for the first experiment and about 5 m^3/h for the second and the third experiment. Tracers were injected together (about 5 l total volume). Tracer amounts for each run are indicated in table I. In each case, the experiment was monitored by in situ analysis of uranium with a portable laser spectrofluorimeter (approximate accuracy 5%). PH was also monitored for the 1st and 2nd experiments. Lanthanides and thorium were determined in water samples by neutron activation (approximate accuracy 1%) [6]. The total volume for the first run was 238.2 m^3, investigated with 71 samples, 231.7 with 72 samples in the second run, and 165 m^3 with 66 samples in the third run.

EXPERIMENTAL RESULTS

Analysis of recovery curves

Recovery curves give similar results in the three runs. Th and lanthanides have almost identical curves, main peaks are synchronous and tails are identical. On the contrary, U shows a slightly delayed maximum and a more pronounced tailing, especially in run III where injected tracers amounts were low. (fig 2a, 2b, 2c). It already suggests that :

- Th EDTA and lanthanides EDTA complexes have a similar behaviour in such experiments and are likely to interact in the same way with the solid.
- Uranyl carbonate complexes undergo stronger interaction leading to a stronger retardation effect.

Comparison between U and pH recovery curves

Tracers have been injected at a pH of approximately 9.5 to 10, differing considerably from the pH of water (7.5). We have reported on our figures (OH-) concentrations instead of pH in order to compare easily with others that are expressed in concentration or relative concentration. pH and U recovery curves are fairly similar, but pH is somewhat delayed with respect to U, and shows more intense tailing.(fig 4, 5). The most interesting feature is that we can observe the same succession of irregularities in recovery curves (indexed 1 to 4 on figures). It means that these irregularities cannot be attributed to analytical variations, but reflect the existence of several paths. Observed recovery curves actually correspond to superimposed individual recovery curves.

Comparison of lanthanides recovery curves

Lanthanides have almost identical recovery curves when maximum concentration is plotted against volume (fig 3 a, b, c). It means that all five lanthanides have the same apparent retardation factor, that could be interpreted in terms of identical sorption constant Kd (= F/C) We shall see later that, to our mind, this interpretation is in fact erroneous. When we

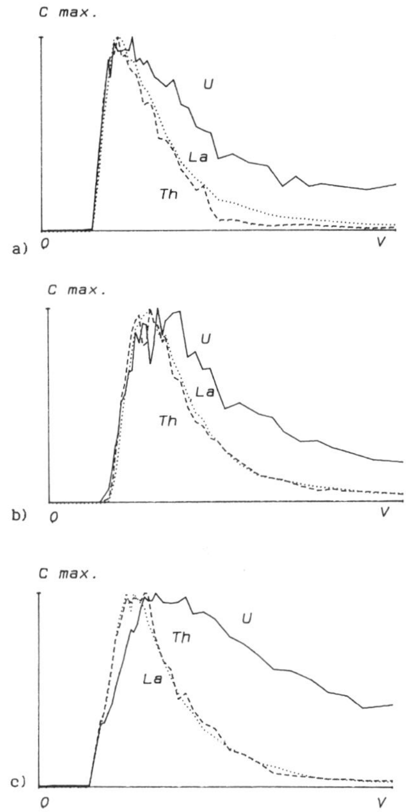

Fig. 2 Comparison of recovery curves for U, Th, and La

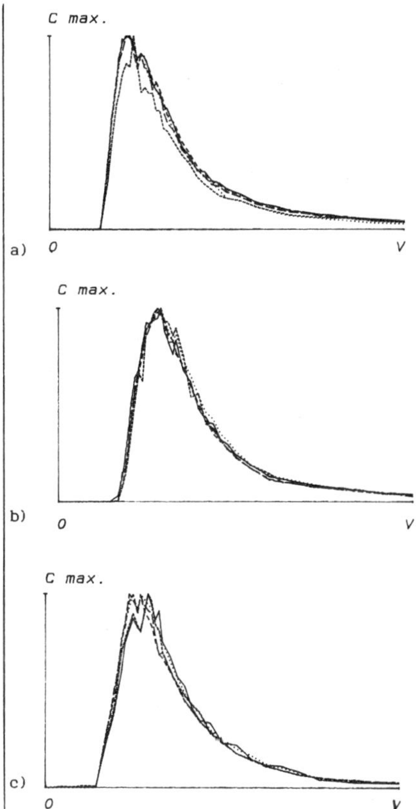

Fig 3 Comparison of recovery curves for lanthanides

a: 1st run b: 2nd run c: 3rd run

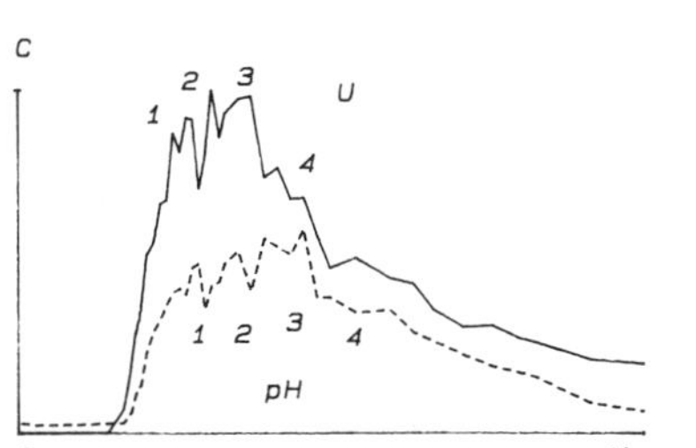

Fig. 4 Comparison of recovery curves for U and pH, 1st run.

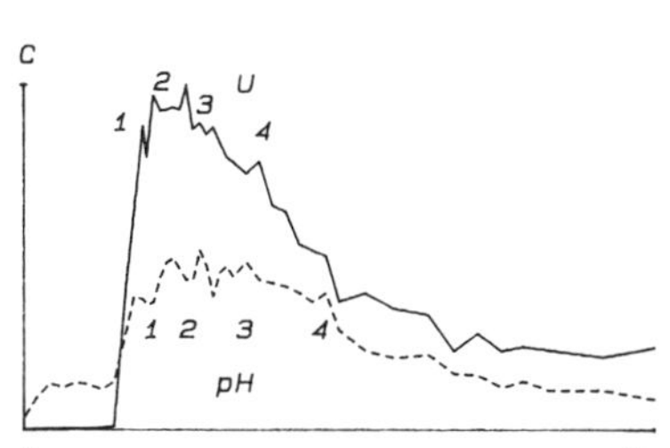

Fig 5 Comparison of recovery curves for U and pH, 2nd run.

N.B. : we have displayed the begining of recovery curves (only 12 m^3)

express concentrations relatively to initial concentrations (or injected amounts)(fig 6 a, b, c) differences appear : recovery is better for heavy rare earths than for light rare earths, and the differences are enhanced when injected amounts are small (run III, fig 6 c). Table I shows the different recovery rates obtained for each run. Same observations were made by Bigot et al.[1]. In first approximation, this is rather easy to explain : heavy rare earths form more stable complexes than light rare earths, so the fixation of heavy rare earths on the substrate is impossible, because they are strongly complexed, on the contrary, some fixation of uncomplexed light rare earths occurs. Fixation would be irreversible or very slowly reversible because recovery is not complete. The kinetics of the decomposition of EDTA complexes could also have some influence on recovery rates.

Evolution of recovery rates with injected tracer amount

Th and Yb have very high and almost constant recovery rates, that means that fixation is very minor. For uranium, recovery is high (about 80%) and do not vary significantly when different quantities are injected. For the other elements, recovery rates increase with increasing injected amounts. It means that the concentration of fixed elements does not depend linearly on the concentration of elements in solution. We apply this conclusion only to EDTA complexed elements. Non linear sorption have been demonstrated with the help of laboratory experiments for some elements : U on clays (Langmuir and Freundlich isotherms)[7] and Cs on granite (Freundlich isotherms)[8] and quartz[9]. In all cases, sorption was found to be highly dependant on pH and chemistry of the solution.

TABLE I : EVOLUTION OF RECOVERY RATES
(Tracer amounts in moles)

TRACER	RUN N°1		RUN N° 2		RUN N° 3	
	amount	recovery	amount	recovery	amount	recovery
U.carb.	0.098	79.8%	0.196	73.3%	0.021	77.6%
Th-EDTA	0.083	91.7%	0.167	94.9%	0.042	93.4%
La-EDTA	1.667	66.5%	3.333	72.2%	0.833	41.4%
Sm-EDTA	0.333	80.9%	0.667	85.5%	0.167	72.5%
Eu-EDTA	0.167	82.6%	0.333	86.2%	0.083	74.4%
Dy-EDTA	0.167	90.5%	0.333	93.9%	0.083	82.6%
Yb-EDTA	0.167	93.0%	0.333	96.1%	0.083	96.7%

MODELLING

We have chosen to work with very simple models, in order to make the evaluation of interactions easier. We have used a conventional dispersion model (D = dispersion coefficient, U = fluid velocity, x = space coordinate, w = porosity, C = concentration in the fluid, F = concentration adsorbed on the solid, μ = fracturation density, t = time. The geometry of the fracture was considered to be plane, vertical and quadrangular; velocities were considered to be uniform within the system.

(Let us recall that we have evidences of several paths, and so models are just indicative). The calculations were done by using finite differences[10]. Hydrodynamical parameters D U have been estimated after modelling based on the hypothesis that Yb was a perfect non-retarded tracer, having the best recovery rate.

$$D(\delta^2C/\delta x^2)-U(\delta C/\delta x) = \omega(\delta C/\delta t)+\mu(\delta F/\delta t) \quad (1)$$

Linear versus non linear sorption

The most common model used in dispersion studies is a linear and reversible model, were surface or volume concentration are linearly dependant on solute concentration. This model is very easy to compute in dispersion equations[11], but has been questioned by several experimental results[7,8,9].

When complexation occurs in solution, simple calculations can account for non linearity. If only the uncomplexed ion (I) is likely to be adsorbed on the solid, we have the following equations (J = ligand concentration)

$$Ks = C / (I . J) \quad (2) \text{ Stability of the complex in solution}$$

$$I = (C / Ks)^{1/2} \quad (3)$$

(if I is approximately equal to J, which is possible, under the present experimental conditions, if only a minor part of the uncomplexed ion is adsorbed). If F is linearly related to I then:

$$F = K . C^{1/2} , \text{ with } K = k / (Ks)^{1/2} \quad (4)$$

K would be a global sorption constant, reflecting sorption and complexation. When complexation occurs in solution, a non linear Freundlich isotherm is thus possible. Such an equation is only an approximation, demonstrating the influence of complexation upon sorption equations. Torstenfelt et al.[8] obtained the following experimental isotherm for Cs on granite:

$$F = K . C^{1/n} \quad (5)$$

with $K = 3.4$ to $4.9 \; 10^{-3}$ and $n = 1.95$ to 2.02

In our case, Lanthanide-EDTA and Th-EDTA complexes are simple complexes LnY-, where Y is EDTA ligand. In our model, we have used a Freundlich equation of type (4). Recovery rates in our experiments were not 100% and their evolution in the case of lanthanides account for irreversible or very slowly reversible adsorption. In fig. 8 we have plotted the evolution of recovery rates with amount of injected tracer both for linear and non linear models for different values of the constant K . These curves have been obtained using a conventional dispersion model (equ. 1). If sorption is linear, recovery rates are constant, with any amount of tracer for a given K. If sorption is non linear (following equation 4), recovery rates increase with increasing amount of tracer. We obtain a good agreement between the model (fig. 8) and experimental results for lanthanides (fig. 7). La-EDTA has the lowest stability constant Ks (fig 1), and so its K (global sorption constant) is the highest, on the contrary, Yb-EDTA has the highest Ks and is expected to have the lowest K. It means that the fraction of uncomplexed La is larger than the fraction of uncomplexed Yb. So much more La is adsorbed

6 a) 1st run

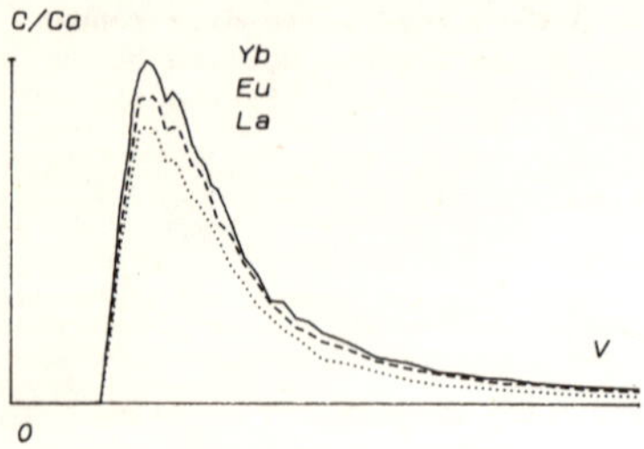

6 b) 2nd run

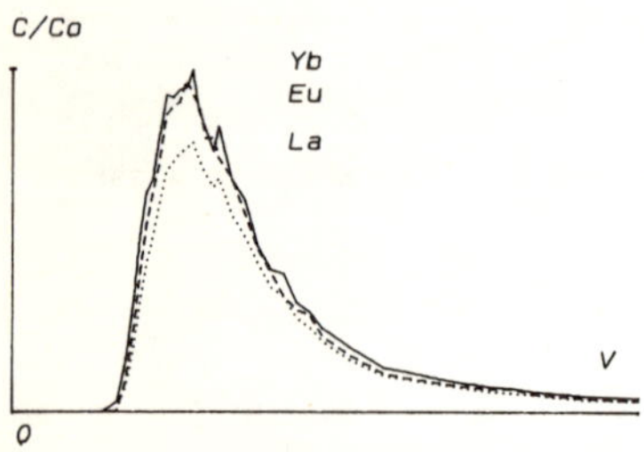

6 c) 3rd run

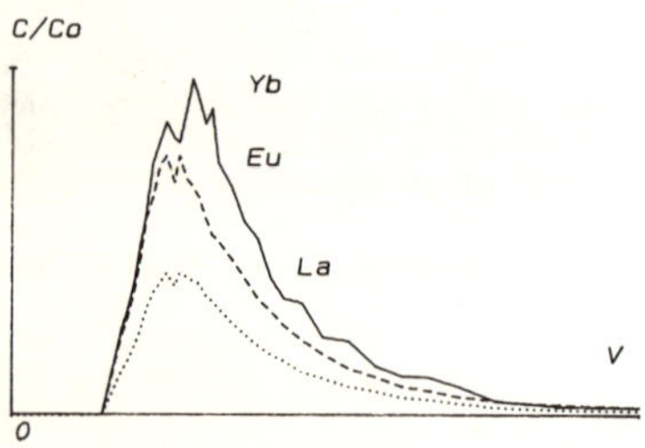

Fig. 6 Comparison of recovery curves for 3 lanthanides, after normalisation to injected amount

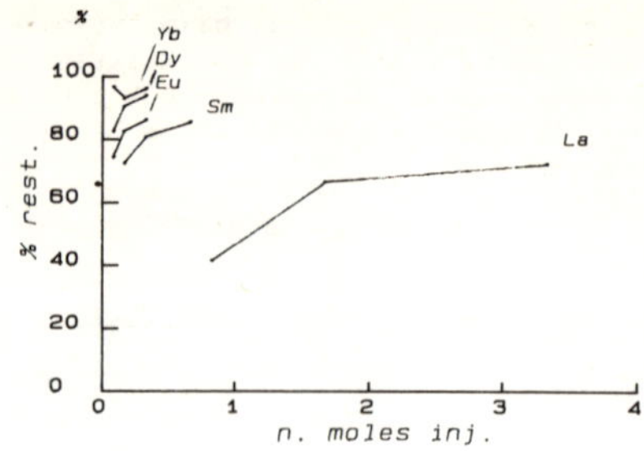

Fig. 7 Evolution of recovery rates with tracer amount experimental results

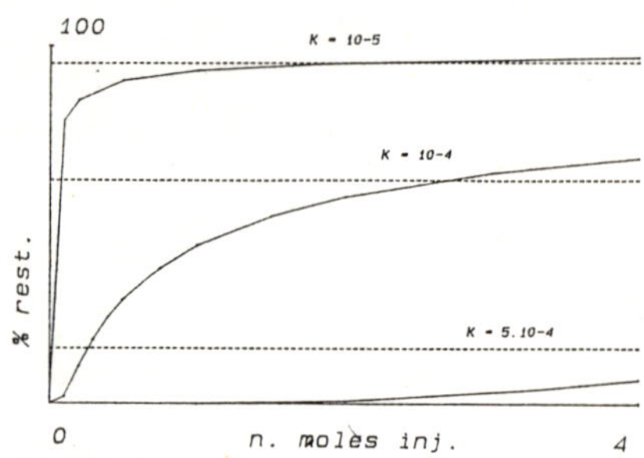

Fig. 8 Evolution of recovery rates with tracer amount theoretical results cont. line : non linear, dashed line : linear.

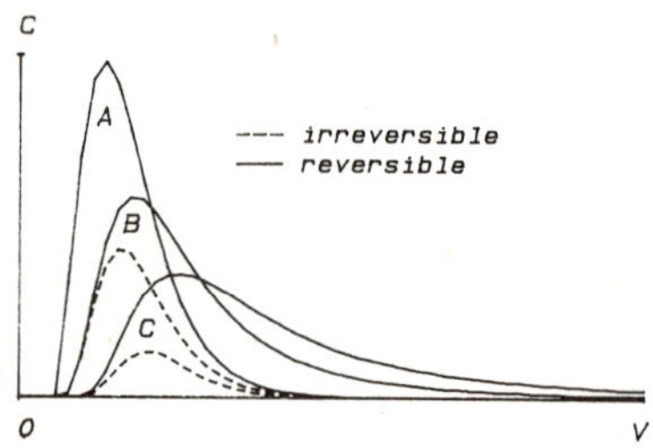

Fig. 9 Comparison of recovery curves with reversible and irreversible sorption

A : $K = 10^{-6}$

B : $K = 10^{-4}$

C : $K = 2 \cdot 10^{-4}$

(arbitrary values, all curves for $K < 10^{-6}$ are the same and correspond to complete recovery) on the solid. If the sorption is not reversible, the recovery rates of La will be smaller than for Yb. This had been already pointed out by Bigot et al.[1]. In the case of tracing experimentations (for repository-sites testing, for instance), an important consequence of the non linearity of sorption model is that if a too small amount of tracer is injected, no or very little response will be observed, not because of analytical problems, but because of the non linearity of the response, especially with tracers that are not very stable.

Influence of irreversibility upon the shape of recovery curves

The fig. 9 allow us to compare recovery curves in the case of non linear sorption for reversible and irreversible sorption. We have used the same parameters in both cases :
- tailing is reduced when sorption is irreversible.
- the delay between peaks corresponding to different values of K is reduced when sorption is irreversible.

The last point could explain the fact that all five lanthanides have identical apparent retardation factors, (fig 3) though they are expected to be different because complexes stabilities are different. Let us recall also the existence of several "channels"that could mask little differences between Kd (The shape of the curve is strongly influenced by "channelling").

Difference between U and lanthanides

The differences between U and other tracers can be explained with a reversible or partially reversible sorption for uranium. Recovery rates are not 100%, but they do not increase with increasing amount of tracer, moreover, they are related to the flow rate in each experiment : the highest recovery rate (79.8%) was obtained in the first experiment (2.8 m^3/h) and the lowest in the second experiment (73.3%), with the highest flow rate (5 m^3/h). It could mean that sorption is reversible, and that some influence of kinetics is visible. Moreover, our model predicts that irreversibility reduces tailing and delay, now the main differences between U and others are increased tailing and delay.

SOLUTION AND SURFACE COMPLEXES

Removal of tracers from the solution by the solid can be described in terms of : ion exchange, precipitation, sorption. Ion exchange models apply easily to alkaline ions on clay minerals[12] but for many other ions, sorption is invoked. Sorption may be explained by either physical (electrostatic interactions) or chemical models. The latter constitutes a powerful tool to understand interactions at the water-solid interface with the help of surface complexes[13,14]. Bourg[14] showed that when Cu-EDTA complexes exist in solution, Cu is adsorbed on silica as Cu free ions, Cu-EDTA complexes being not able to be fixed on solid, but with other complexes such as 2,2' bipyridyl or ethylenediamine, ternary surface complexes, involving the sorption of the ligand, are formed. Such results stress the influence of complexation upon adsorbability of cations. The difference between Ln-EDTA and uranyl-carbonates could be attributed to differences between surface complexes constitution. Uranium could be adsorbed directly as carbonate complexes (carbonate complexes adsorption has been already invoked in natural systems[14]) ; other tracers are fixed only as uncomplexed ions.

CONCLUSION

The main objective of this study was to understand the coupling between hydrodynamics (mass transfer equations) and solid/solution interactions in tracing experiments. Our results stress the importance of complexation, both in solution and on solid surface and its main consequences : when the tracer is complexed in order to be stabilized in solution, non linear sorption laws govern interactions, having important consequences upon the interpretation of tracing experiments. In some cases, the complex can interact itself with the substrate, modifying sorption laws, and possibly the reversibility of sorption reactions. The study of long-lived radionuclides behaviour in geological formations demands in situ experimentations, among which tracing experiments. It is thus very important to study solution and surfaces complexes in order to explain recovery curves, and further, to predict long term sorption properties of radionuclides. The influence of complexing agents upon the mobility of radionuclides and trace elements is of course illustrated by this kind of experimentations, especially in natural systems where organic complexing agents are important.

REFERENCES

1. Bigot S., Treuil M., Dumonceau J., Fromage F. ; J. Hydrol. 70, 133-148 (1984)
2. Wheelwright E.J., Spedding F.H., Schwarzenbach G. ; J. Am. Chem. Soc., 75, 4196-4204 (1953)
3. Dumonceau J., Bigot S., Treuil M., Faucherre J., Fromage F. ; C.R. Acad. Sci., Paris, ser. C., 287, 325-327 (1978)
4. Lacey W.J., De Laguna W. ; Science, 124 402 (1956)
5. Langmuir D. ; Geochim. Cosmochim. Acta, 42, 547-569 (1978)
6. Bigot S., Treuil M. ; J. Radioanal Chem., 59(2), 341-349 (1980)
7. Borovec Z. ; Chem. Geol., 32, 45-58 (1981)
8. Torstenfelt B., Andersson K., Allard B. ; Chem. Geol. 36, 123-137 (1982)
9. Rochon J., Rançon D., Gourmel J.P. in Underground disposal of radioactive wastes, Vol II IAEA Vienna, SM 243/155 (1980) pp 271-314
10. Fried J.J. Groundwater Pollution (Elsevier Scientific publishing company, Amsterdam, 1975), p. 250
11. Neretnieks I. in Environmental migration of long lived radionuclides IAEA Vienna, SM 257/19 (1982) pp 635-657
12. Andersson K., Torstenfelt B., Allard B. in Environmental migration of long lived radionuclides IAEA Vienna, SM 257/20 (1982) pp 111-131
13. Stumm W., Kummert R., Sigg L. ; Croat. Chem. Act. 53, 291-312 (1980)
14. Bourg A.C.M. , in Speciation of trace elements in natural waters edited by Ahrland S. an Nancollas G. IUPAC (1981)

MIGRATION BEHAVIOR OF U(VI) IN TUFF

SEICHI SATO*, HIROTAKA FURUYA*, SADAO ARAYA** and KAZUMORI MATSUO***
*Department of Nuclear Engineering, Faculty of Engineering
Kyushu University Hakozaki, Higashi-ku Fukuoka 812, JAPAN
**Power Reactor and Nuclear Fuel Development Corporation
Akasaka 1-9-13 Minato-ku, Tokyo JAPAN
***Nuclear Safety Research Association 2-2, 1-chome,
Uchisaiwaicho, Chiyoda-ku, Tokyo JAPAN

ABSTRACT

Adsorption coefficient and penetration behavior of U(VI) in Itado tuff were measured. The migration behavior of U(VI) in the tuff was discussed.

Adsorption coefficient of U(VI) was measured by a batch method using a wafer specimen and a powder specimen in pH region from 2 to 7. Adsorption coefficient was 17000 ml/g at the pH value of 5.5.

Penetration behavior of U(VI) in Itado tuff was studied as a function of pH, concentration of CO_3^{2-} and temperature by contacting the rock specimen with U(VI) solution. The concentration profile and fractional amount of U(VI) penetrating into the rock specimen M_t/M were measured. The penetration behavior of U(VI) was seen to be diffusion accompanied by adsorption/desorption process. In the pH region from 3 to 4, it is likely that the penetration process of U(VI) is not sufficiently reversible but is competitive process consisting of diffusion and adsorption. In the pH region above 6, penetration depth was very thin because of both large adsorption coefficient and large adsorption rate constant.

The sectioning method was found to have an advantage to see the penetration behavior directly. But in the pH region from 6 to 7, the method has a shortage that diffusion coefficient, and adsorption and desorption rate constant could not be determined, because of small penetration depth of U(VI). From this point of view, the measurement of the fractional amount of penetration as a function of time should be carried out as well as that of concentration profile.

INTRODUCTION

Radioactive waste forms may be emplaced in a deep geologic formation. Isolation of nuclides from biosphere beyond several thousand years is demanded. When engineered barriers degrade, sorption property and a number of properties relevant to sorption of nuclide finally have an important role to retard migration of nuclides by ground water. Therefore, migration behavior of nuclides in a rock matrix should be clarified in order to assess safety of a disposal system.

In the present work, in an attempt to study the migration behavior of actinides in a monolithic rock, adsorption coefficient of U(VI) and penetration behavior of U(VI) in Itado tuff were studied by making contact between the rock specimen and U(VI) solution. Adsorption coefficients were measured by two kinds of batch method. In one method, powder specimen was exposed to U(VI) solution; in the other, wafer specimen was used. For the study of penetration behavior concentration profile was measured by sectioning the specimen; fractional amount of penetration was also measured by analysis of U(VI) in solution. The penetration behavior of U(VI) was studied as a function of pH-value, concentration of CO_3^{2-} and temperature, so that the migration behavior might be studied in the region of various condition of ground water. The migration behavior of U(VI) was discussed on the basis of the adsorption coefficient, the concentration profile and the fractional amount of penetration.

EXPERIMENTAL

Adsorption coefficient of U(VI) for Itado tuff was measured by two kinds of batch methods. In one of the batch methods, powdered Itado tuff of 32 to 60 mesh was used as a specimen. The specimen of 1.0 g was in contact with 30 ml-uranyl solution in a teflon container for 8.0d. Adsorption coefficient was determined on the basis of the equation

$$K_d = \frac{(C_0 - C)V/W}{C} \qquad (1)$$

, where C_0 is the initial concentration of U(VI), C the concentration of U(VI) after the contact, V solution volume and W weight of the powdered rock. Correction of U(VI) adsorbed on the surface of the teflon container was not carried out , because amount of U(VI) adsorbed on the specimen was found to be much larger than that of the wall.

In the other batch method, a wafer specimen from 0.8 to 1.0 mm in thickness was exposed to U(VI) solution for one week. Sorption equilibrium was confirmed by measuring U(VI) in an aqueous phase as a function of time during the contact.

Penetration of U(VI)

Itado tuff was cut off in the size of 10 x 10 x 10 $(mm)^3$. The rock specimen was washed by distilled water and was dried in a vacuum oven at 100 C for 24h. The specimen was embedded in a resin under a slightly reduced pressure so that the resin may stick to the surface of the specimen. One face of the mechanically polished specimen was in contact with distilled water under a reduced pressure to remove gas existing in a pore in the specimen and to fill the pore with distilled water. A typical

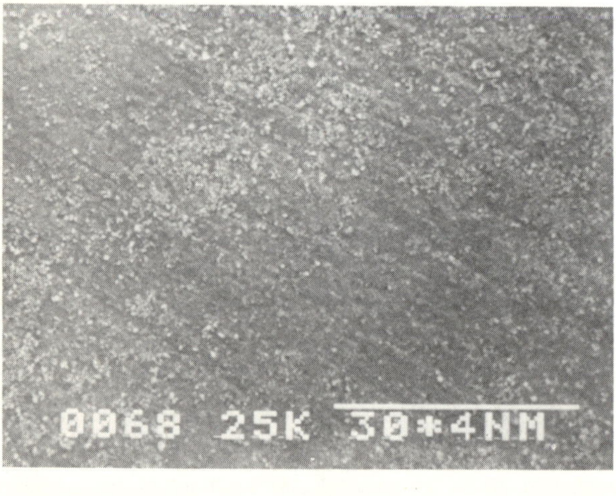

Fig. 1. Scanning electron micrograph of Itado tuff, white line shows 300 μm.

microstructure of the surface of the specimen is shown in Fig. 1. The tuff is a mordenite of high purity whose minerals are quartz, plagioclase, montmorillonite and mica as well as mordenite. The chemical and mineral compositions are available elsewhere(1). A weighed amount of U_3O_8 was dissolved in concentrated nitric acid under a slightly elevated temperature. Then, pH-value of the solution was adjusted using NaOH solution. A prescribed amount of $Na_2CO_3^{2-}$ was added to realize a similar concentration of CO_3^{2-} expected in ground water. Uranyl solution of 4.8 x 10^{-6} M-U(VI) containing 3.0 x 10^{-4} M-$Na_2CO_3^{2-}$ was mainly used in the present experiment.

The rock specimen was exposed to the uranyl solution of 100 ml at a fixed temperature for 8.0 d in a polyethylene container. After the contact, the specimen was mechanically polished every 0.5 mm or 1.0 mm in thickness by abrasive paper. U(VI) was extracted from the powder in 3N-HNO_3.

The acid solution diluted was centrifuged at 4000 rpm for 30 min. The supernatant solution was filtered using Gelman Acro-dish CR filter of 0.2 um pore size. The solution containing U(VI) was evapolated to dryness in a platinum evaporating dish. A prescribed amount of flux consisting of sodium carbonate, potassium carbonate, lithium fluoride and sodium fluoride was put on the dish, melted at 700 C and the melt was quenched. U(VI) in the pellet taken from the platinum dish was determined by Aloka FM-T-4B fluorimeter. Concentration Profile in the rock specimen was determined as a function of penetration depth.

Fractional amount of penetration M_t/M_∞ was measured as a function of pH-value by contacting the specimen of 3.0 mm in thickness with 4.8 x 10^{-6} M-U(VI) solution. Concentration of U(VI) in the solution was determined as a function of time.

RESULTS AND DISCUSSION

Adsorption coefficient of U(VI)

Adsorption coefficients were measured as a function of both pH-value and concentration of CO_3^{2-} by the two kinds of batch method. The results are given in Fig. 2. for pH dependence, and in Table 1 for dependence on CO_3^{2-} concentration.

Adsorption coefficients for both powder specimen with filtration and wafer specimen were a few times as large as that for powder specimen without filtration. This means amount of adsorption of U(VI) on suspended particle can not be disregarded as compared to amount of U(VI) dissolved in solution. Therefore, adsorption coefficient for wafer specimen and for powder specimen with filtration are more reliable than for powder specimen without filtration. Even so, in the measured pH region, adsorption coefficient for wafer specimen is seen in agreement with powder specimen without filtration especially in the pH region from 5 to 7.

Adsorption coefficients are seen to have 17000 ml/g in the solution of pH 5.5, while adsorption coefficient decreased remarkably in acidic solution. When powder specimen was in contact with distilled water, the final pH-value was 4.5. The adsorption coefficient at the pH-value of 4.5 is seen to be 3000 ml/g.

When concentration of CO_3^{2-} ion increases, adsorption coefficient decreased one orders of magnitude from 16000 ml/g at 2.5 x 10^{-5}M to 1800 ml/g at 3.0 x 10^{-3}M. It is regarded to be due to the formation of highly carbonated uranyl complexes which are remarkably stable in solution.

Penetration of U(VI)

The migration behavior in the rock specimen were analyzed by the equation

Table 1. CO_3^{2-} dependence of adsorption coefficient

CO_3^{2-} concentration (mol/l)	pH	Kd (ml/g)
0	5.7	14500
2.5×10^{-5}	5.1	15800
3.0×10^{-4}	5.3	16900
3.0×10^{-3}	6.4	1760

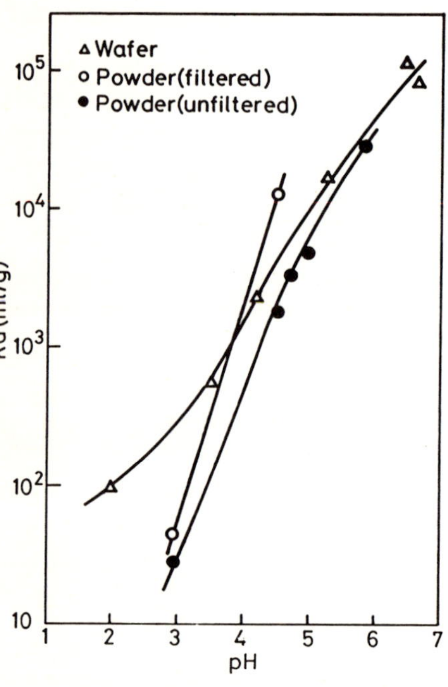

Fig. 2. Adsorption coefficient of Itado tuff as a function of pH value at 30 ℃. The concentration of CO_3^{2-} in the solution in contact with the rock specimen was 3.0×10^{-4} M .

$$\frac{\partial C}{\partial t} = D\frac{\partial^2 C}{\partial x^2} - \frac{\partial S}{\partial t} \qquad (2)$$

with the simultaneous first order reaction of the type

$$\frac{\partial S}{\partial t} = k_1 C - k_2 S$$

,where

$C(mol/cm^3)$ = concentration of the U(VI) in the aqueous phase in a pore in the rock specimen
$S(mol/cm^3)$ = Q_r/ε
$Q_r(mol/cm^3)$ = amount of the U(VI) adsorbed on an inner surface in the unit volume of the rock specimen; Q_{sol},amount of U(VI) in solution phase in the unit volume of the rock
ε = porosity of the rock
$D(cm^2/s)$ = diffusioncoefficient of the U(VI) which includes geometric factor of the specimen pore
$x(cm)$ = position in the specimen from the solution-specimen interface
$t(s)$ = time, and
$k_1(s^{-1})$, $k_2(s^{-1})$ = the rate constants of adsorption and desorption reaction, respectively.

The form of the fundamental equations (2) and (3) are the same as used by Rundberg[2].

The amount of U(VI) in unit volume of the rock specimen $Q_t=Q_r+Q_{sol}$, which is obtained by the sectioning method, can be expressed by the equation

$$\frac{Q_t}{Q_\infty} = 1 - D\left(\frac{a}{\ell} + \frac{1}{1+R}\right)\sum_{n=1}^{\infty}\frac{q_n^2}{p_n}\,\frac{\exp(p_n t)}{1+A_n B_n}\,\frac{\cos q_n x}{\cos k_n a} \qquad (3)$$

$$Q_\infty = \frac{\varepsilon \ell C_0}{a + \ell/(1+R)}$$

$$A_n = 1 + \frac{k_1 k_2}{(p_n + k_2)^2}$$

$$B_n = \frac{1}{2}\left(\frac{a}{\ell} + \frac{p_n}{D q_n^2} + \frac{p_n^2 a}{D^2 q_n^2}\right)$$

,where p_n, k_n are non-zero root of

$$\frac{\ell p_n}{D} = q_n \tan q_n a\,, \qquad k_n^2 = -\frac{p_n + k_1 + k_2}{p_n + k_2}\,\frac{p_n}{D}\,.$$

The initial condition is

$$S = C = 0, \qquad -a < x < a \qquad t = 0,$$

,while boundary condition is

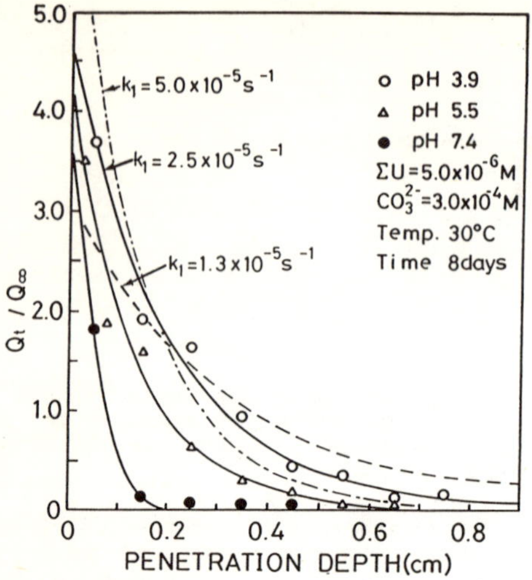

Fig. 3. Relative concentration of U(VI) in the specimen of Itado tuff as a function of penetration depth. Solid line shows the best fitting curve which was obtained using equation (3). Dashed lines for the data of pH 3.9 show sensitivity of the fitting curve as a function of k_1 at the conatant value of K_d = 1500 ml/g.

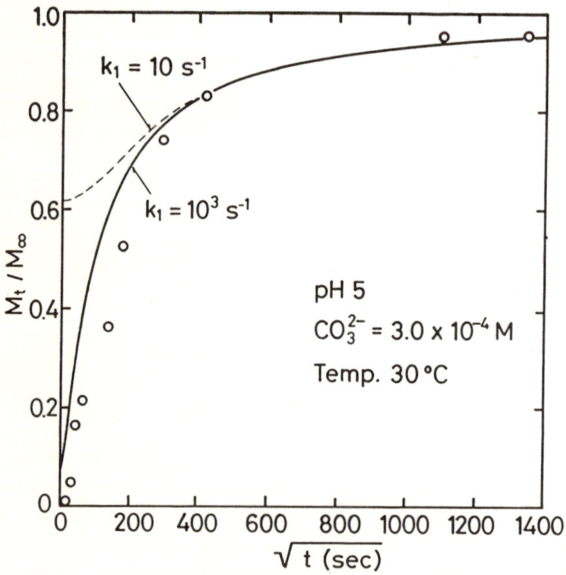

Fig. 4. Fractional amount of penetration of U(VI) in Itado tuff as a function of contact period of time.

$$\ell \frac{\partial C}{\partial t} = \mp D \frac{\partial C}{\partial x}, \qquad x = \pm a, \; t > 0$$

a = specimen thickness

ℓ = effective length of the solution

C_0 = initial concentration

in accordance with the solution given by Crank [3].

Experiemntal results of Q_t/Q_∞ at several pH values by the sectioning method are shown as a function of penetration depth in Fig. 3. ,where fitting curves are also given, using equation (3) numerically. At the pH value of 3.9, penetration rate of U(VI) is seen to be high; at the pH value of 7.4, the rate of U(VI) is seen to be very slow.

At the pH value of 3.9, a fitting curve was also obtained using the diffusion equation with reversible adsorption/desorption process[4]

$$\frac{\partial C}{\partial t} = \frac{\varepsilon D}{K_d \rho} \frac{\partial^2 C}{\partial x^2} \tag{4}$$

,where ρ is density of the rock, ε the porosity. Therefore, the penetration behavior was seen to be diffusion accompanied by adsorption/desorption process. In an attempt to discuss the migration process more in detail, both k_1 and k_2 among the parameters of the fitting curve were varied under the constant K_d value of 1500 ml/g at the pH value, where K_d can be discribed by the equation $K_d = (k_1/k_2)\cdot(\varepsilon/\rho)$. Then the curve was sensitively deviated from the data as can be seen in Fig. 3. This means that penetration behavior is not regarded to be sufficiently reversible but diffusion of U(VI) and adsorption process have a competitive effect on the migration.

In the acidic pH region 3 to 4, it is likely that the main chemical form of U(VI) is UO_2^{2+} [5]. Such K_d values from 100 to 2000 ml/g as is seen in Fig. 2 suggests UO_2^{2+} is the main chemical form concerning in the adsorption/desorption process because UO_2^{2+} exists in an aqueous phase if it does not take part in the adsorption.

On the other hand, in the pH region beyond 5, migration rate of U(VI) is remarkably small as was mentioned above. Adsorption coefficient is very large as is seen in Fig. 2. In this pH region the analysis based on penetration profile is not reliable because of very small penetration depth of U(VI). Therefore fractional amount of penetration was also measured at the pH value of 5.5, which is shown in Fig. 4., though satisfactory fitting curve covering wide time scale was not obtained due to a complex process. It was determined on the basis of the same form of equation used by Rundberg[2], that diffusion coefficient D was 3×10^{-6} cm^2/s, adsorption rate constant k_1 1×10^{3} s^{-1}, and desorption rate constant k_2 4×10^{-3} s^{-1}. The features of the migration of U(VI) in the pH region are very large adsorption coefficient K_d and large adsorption rate constant k_1. Since difffusion coefficient does not change significantly as a function of pH, U(VI) migration was found to be retarded remarkably by adsorption process comparing with U(VI) migration in an acidic solution. In the pH region, UO_2^{2+} and UO_2CO_3 is the main chemical form and UO_2CO_3 remarkably increases with pH rise. If both UO_2^{2+} and UO_2CO_3 does not relate to adsorption, they exists in an aqueous phase. Then the value of K_d will not be such a large of 17000 ml/g at pH 5.5. Therefore pH values from 5 to 6 it is likely that UO_2^{2+} and UO_2CO_3 are closely related with adsorption/desorption process. Since main chemical form appears to be changed remarkably from UO_2^{2+} to UO_2CO_3 with pH rise from 4 to 6, the remarkable increase of both adsorption coefficient and adsorption rate constant may be due to change of chemical form.

To discuss correlation between the chemical form of U(VI) solution and the penetration behavior, penetration profile of U(VI) as a function of concentration of CO_3^{2-} was measured at the pH value of about 6. In the both cases of the solution without CO_3^{2-} and with 2.5 x 10^{-5} M CO_3^{2-}, U(VI) was found to penetrate into the rock remarkably, while U(VI) hardly penetrate in solution of 3.0 x 10^{-3} M CO_3^{2-}. As is seen in Table 1., adsorption coefficient at 3.0 x 10^{-3} M CO_3^{2-} in which $UO_2(CO_3)_3^{4-}$ appears to be a main chemical form, decreases by one order of magnitude comparing with that at 2.5 x 10^{-5} M CO_3^{2-}. The experimental results indicate that kinetic parameters such as k_1 rather than thermodynamic parameter of K_d controls the process at least in the initial stage of the penetration and that the kinetic parameters and the thermodynamic parameter depend on the chemical form sensitively. In such a way, migration behavior was found to be sensitive depending upon chemical form which is a function of pH and CO_3^{2-}.

The temperature dependence of penetration behavior of U(VI) has also been studied from 20 to 60 °C. Remarkable temperature dependence was not observed.

CONCLUSIONS

The penetration behavior of U(VI) was seen to be diffusion accompanied by an adsorption/desorption process. In the pH region from 3 to 4, it is likely that the penetration process of U(VI) is not sufficiently reversible but is competitive process consisting of difusion and adsorption. On the other hand, in the pH region from 6 to 7, penetration depth was found to be very thin that could be explained by both large adsorption coefficient and large adsorption rate constant.

In the migration behavior of U(VI) in the tuff which is in contact with aqueous solution including CO_3^{2-} kinetic parameters such as adsorption rate constant plays an important role as well as adsorption coefficient at least in the initial stage of the contact.

ACKNOWLEDGEMENT

Authors would like to express sincere thanks to Professors M. Kanno and M. Yamawaki at University of Tokyo who explained for us the analytical techinique of uranium using a fluorimeter.

REFERENCES

1. Nuclear Safety and Research Association, Study on Interaction of Geological Media with HLW Radionuclide, 1984 (in Japanese)
2. R. S. Rundberg, "Kinetics of the Adsorption of Radionuclides on Tuff from Yucca mountain" in Scientific Basis for Nuclear Waste Management, edited by G. L. McVay (Elsevier Science Publishers, NewYork 1984) pp. 827
3. J. Crank, The mathematics of Diffusion, 2nd ed. (Oxford University Press, London 1975) pp. 339.
4. I. Neretnieks, J. Geophys. Res. 85, 4373 (1980)
5. D. Langmuir, Geochim. Cosmochim. Acta 42, 547 (1978)

OCEAN FLOOR SEDIMENT AS A REPOSITORY BARRIER: COMPARATIVE DIFFUSION DATA FOR SELECTED RADIONUCLIDES IN SEDIMENTS FROM THE ATLANTIC AND PACIFIC OCEANS*

F. SCHREINER, C. SABAU, A. FRIEDMAN AND S. FRIED
Chemistry Division, Argonne National Laboratory, Argonne, IL 60439, USA

ABSTRACT

Effective diffusion coefficients for selected radionuclides have been measured in ocean floor sediments to provide data for the assessment of barrier effectiveness in subseabed repositories for nuclear waste. The sediments tested include illite-rich and smectite-rich red clays from the mid plate gyre region of the Pacific Ocean, reducing sediment from the continental shelf of the northwest coast of North America, and Atlantic Ocean sediments from the Southern Nares Abyssal Plain and the Great Meteor East region. Results show extremely small effective diffusion coefficients with values less than 10^{-14} m^2s^{-1} for plutonium, americium, curium, thorium, and tin. Radionuclides with high diffusion coefficients of approximately 10^{-10} m^2s^- include the anionic species pertechnetate, TcO_4^-, iodide, I^-, and selenite, SO_3^{-2}. Uranyl(VI) and neptunyl(V) ions, which are stable in solution, have diffusion coefficients around $10^{-12}m^2s^{-1}$. The diffusion behavior of most radionuclides is similar in the oxygenated Pacific sediments and in the anoxic sediments from the Atlantic. An exception is neptunium, which is immobilized by Great Meteor East sediment, but has high mobility in Southern Nares Abyssal Plain sediment. Under stagnant conditions a 30 m thick sediment layer forms an effective geologic barrier isolating radionuclides in a subseabed repository from the biosphere.

INTRODUCTION

The possibility of isolating high-level nuclear waste in the sediment overlying the base rock of the deep oceans has been the subject of research for more than a decade [1,2]. In a subseabed repository the waste, contained in suitable canisters, will be embedded in the sediment at a depth no less than 30 m below the surface. In the water-saturated environment of the subseabed, however, ultimately the canisters are expected to corrode and leaching of the waste form will take place. When this happens, the sediment overlying the repository must be relied upon to isolate the radioactive material from the biosphere. It is therefore important to demonstrate the adequacy of the sediment as a geologic barrier.

In a number of previous reports the results of diffusion tests with selected important radionuclides in several types of sediments for the

*Work performed under the auspices of Sandia National Laboratories contract number 25-5644.

mid-plate gyre region of the Pacific Ocean have been described [3]. In the present paper, new data obtained in two types of sediment from the Atlantic Ocean are added and compared with the earlier results. As will be evident, in aggregate the results point to the sediment column as being a highly effective barrier for the most hazardous transuranium elements plutonium and americium, and an adequate medium for the reduction of radionuclide release even in cases of the more mobile species, such as neptunium, and technetium.

EXPERIMENTAL

The laboratory diffusion tests described here were carried out with five different types of sediments, two from the Atlantic Ocean, and three from the Pacific. The Atlantic Ocean sediments will be referred to as SNAP and GME to indicate their respective origins, The Southern Nares Abyssal Plain, and the Great Meteor East location*. The SNAP sediment was obtained from a core taken at a point 22°54.1' N and 63°29.4' W. It consisted chiefly of clay minerals (70-90%) and sand, with minor components of calcium carbonate (1-5%) and organic carbon (0.3%). The GME sediment came from a point 30°40.9' N and 24°58.9' W. Its composition was 45-50% calcium carbonate, 10-30% clay, 10-30% sand, and 1-1.5% organic carbon.

The three sediments from the Pacific Ocean included two types of red clays, and a reducing sediment. The red clays were collected in the Pacific MPG I study area, which is centered about a point 31°30' N and 158°W [4]. These sediments are not reducing in nature. They are made up of a mixture of clay minerals, hydrous oxides of iron and manganese, and sea water. For the diffusion experiments two sediments were selected, one containing smectite as a major component, the other containing illite. A detailed mineralogical description of the red clays from MPG I can be found in reference 4. In addition, a sediment from the continental shelf off the northwest coast of the United States was tested as a medium providing a reducing chemical environment. In order to preserve the low oxygen fugacity of the sediments from the Atlantic Ocean and from the continental shelf, these materials were kept under an atmosphere of nitrogen in which the oxygen partial pressure never exceeded 50 parts per million (0.00005 atm).

The experimental technique used to determine effective diffusion coefficients for tracer radionuclides in the sediment has been described in a previous publication [5]. It involves the evaluation of concentration-distance curves by measuring the radioactivity of the tracer after allowing appropriate time for diffusion. The individual samples are made up as small right cylinders with a diameter of 14 mm and a height of 12.25 mm, containing the radionuclides to be studied in tracer amounts. The rate of diffusive spreading of the tracer is affected by the tortuosity of the interstitial channels and the potential adsorption on the

*Samples of the Atlantic Ocean sediments were supplied courtesy of Dr. R. E. Cranston, NEA Seabed Working Group, Atlantic Geoscience Centre, Bedford Institute of Oceanography, Dartmouth, Nova Scotia, Canada.

solid matrix of the sediment. Since no effort has been made to separate these effects from the observed overall diffusion rate, the coefficients reported here are referred to as "effective" diffusion coefficients. These numbers, however, are useful for the assessment of barrier effectiveness of the sediment.

Desirable isotopes for the tracer diffusion experiments are those whose decay is associated with the emission of electromagnetic radiation (gamma or X rays), and whose half lives are in the 2 month to 2 year range. This choice eliminates the need for chemical isolation of the radionuclides from the sediment before assay, since the electromagnetic radiation is not significantly attenuated by the sample material.

RESULTS

A summary listing of effective diffusion coefficients for eleven different radionuclides in sediments from the Atlantic and from the Pacific Ocean is given in Table I. These coefficients were obtained by fitting calculated concentration-distance curves to the observed distribution of specific activity in the samples. An example of a concentration profile obtained by the rub-off method is shown in Fig. 1. The relative specific activity in the front 5 mm section of the sample rose from zero to values ranging from 0.2 to 0.3. This level is characteristic for a product value of diffusion coefficient and time D•t, of 1.0 10^{-5} m^2, which in turn implies an effective diffusion coefficient of 5.7 $10^{-11} m^2 s^{-1}$.

The data in Table I show the important transuranium elements plutonium and americium to have extremely low mobilities in all sediments tested. In fact, for these two elements the experimental data do not provide evidence for any diffusive spreading on the time scale of our laboratory tests, i.e. up to 240 days. It follows that their effective diffusion coefficients must be smaller than $8 \cdot 10^{-14}$ $m^2 s^{-1}$.

The highest mobilities are found for anionic species that do not interact significantly with the solid phase, such as pertechnetate, iodide, and selenite. Effective diffusion coefficients for these species are in the range 10^{-9} to $10^{-10} m^2 s^{-1}$. This magnitude is consistent with uninhibited molecular diffusion in the interstitial sea water, with an apparent reduction of the mobility because of tortuosity and channel size.

The large cations NpO_2^+ and UO_2^{+2} also show clear evidence for diffusion, but at a slower rate than that observed for anions. In this case, sorptive interaction with the matrix of the clay and with hydrous oxides reduces the diffusion rates.

An unexpected result of the diffusion measurements in Atlantic sediments is found in the case of neptunium. This element diffuses in the Atlantic SNAP sediment, but is effectively immobilized in the Atlantic GME sediment. It is possible that the relatively high content of organic carbon in the GME sediment provides a sufficiently strong reducing environment in which the pentavalent neptunyl ion is converted to the tetravalent Np^{+4} species, which is hydrolysed to form an insoluble

Table I. Effective Diffusion Coefficients of Selected Radionuclides in Sediments from the Atlantic and Pacific Oceans.

Nuclide	Sediment	No. of Samples	Time (d = days) (h = hours)	Effective Diffusion Coefficient[a] D_{eff}/m^2s^{-1}
^{237}Pu	Atlantic SNAP	3	60.82 d	$< 1.8 \cdot 10^{-14}$
($PuCl_4$,	Atlantic GME	3	60.85 d	$< 1.6 \cdot 10^{-14}$
ox.st. +4)	Pacific smectite	4	53.7 d	$< 1.0 \cdot 10^{-14}$
	Pacific illite	6	119.9 d	$< 1.2 \cdot 10^{-14}$
^{237}Pu (PuO_2Cl_2, ox.st. +6)[d]	Pacific smectite	1	60.0 d	$< 1.9 \cdot 10^{-15}$
^{241}Am	Atlantic SNAP	3	71.95 d	$< 8.0 \cdot 10^{-14}$
($AmCl_3$,	Atlantic GME	3	71.98 d	$< 1.2 \cdot 10^{-14}$
ox.st. +3)	Pacific smectite	1	35.9 d	$< 1.0 \cdot 10^{-14}$
	Pacific illite	2	240.1 d	$< 1.0 \cdot 10^{-14}$
^{235}Np,	Atlantic SNAP	4	53.02 d	$(3.3 \pm 1.7) \cdot 10^{-12}$[b]
^{237}Np	Atlantic GME	4	53.05 d	$< 1.9 \cdot 10^{-14}$
(NpO_2Cl,	Pacific smectite	2	46.2 d	$(2.0 \pm 1.2) \cdot 10^{-12}$
ox.st. +5)	Pacific reducing	2	556.2 d	$(1.4 \pm 1.2) \cdot 10^{-12}$
^{243}Cm	Atlantic SNAP	3	108.0 d	$< 1.0 \cdot 10^{-14}$
($CmCl_3$,	Atlantic GME	3	108.0 d	$< 1.0 \cdot 10^{-14}$
ox.st. +3)	Pacific smectite	2	47.9 d	$< 2.2 \cdot 10^{-15}$
^{95m}Tc	Atlantic SNAP	4	9.3 d	$(3.2 \pm 1.5) \cdot 10^{-11}$
(NH_4TcO_4,	Atlantic GME	4	9.32 d	$(2.3 \pm 1.5) \cdot 10^{-11}$
ox.st. +7)	Pacific smectite	2	17.4 d	$(3.0 \pm 1.5) \cdot 10^{-10}$
	Pacific reducing	2	65.1 d	$(3.0 \pm 1.5) \cdot 10^{-10}$
^{99}Tc	Pacific reducing	2	231.2 d	$(3.0 \pm 1.5) \cdot 10^{-10}$[c]

(Table I) Continued.

Nuclide	Sediment	No. of Samples	Time (d = days) (h = hours)	Effective Diffusion Coefficient[a] D_{eff}/m^2s^{-1}
^{225}Ra	Atlantic SNAP	4	52.2 d	$(5.6\pm2.0)\cdot10^{-12}$
($RaCl_2$,	Atlantic GME	4	52.2 d	$(4.8+2.0)\cdot10^{-12}$
ox.st. +2)	Pacific smectite	2	27.9 d	$< 3.8\cdot10^{-15}$
^{228}Th	Atlantic SNAP	3	99.0 d	$< 1.0\cdot10^{-14}$
($ThCl_4$,	Atlantic GME	3	99.1 d	$< 1.0\cdot10^{-14}$
ox.st. +4)	Pacific smectite	2	55.8 d	$< 1.2\cdot10^{-14}$
^{233}U (UO_2Cl_2, ox.st. +6)	Pacific smectite	2	47.05 d	$(7.4\pm3.5)\cdot10^{-13}$
^{113}Sn ($SnCl_2$, ox.st.+2)	Pacific smectite	2	15.98 d	$< 1.5\cdot10^{-14}$
^{79}Se (Na_2SeO_3, ox.st. +4)	Pacific smectite	2	24.18 h	$(1.7\pm1.0)\cdot10^{-10}$
^{125}I (NaI, ox.st. -1)	Pacific smectite	2	8.0 h	$(2.5\pm1.5)\cdot10^{-9}$

[a] Values preceded by the "<" sign are upper limits for the effective diffusion coefficient.

[b] There is some evidence that the neptunium is immobilized after an initial period of diffusion. The coefficient listed characterizes the mobility during the initial phase of diffusion.

[c] The long term technetium sample indicated complete equilibration of the tracer throughout the sample. The coefficient listed was derived from short-term experiments.

[d] The plutonium had been oxidized to the hexavalent state prior to admixture to the sediment.

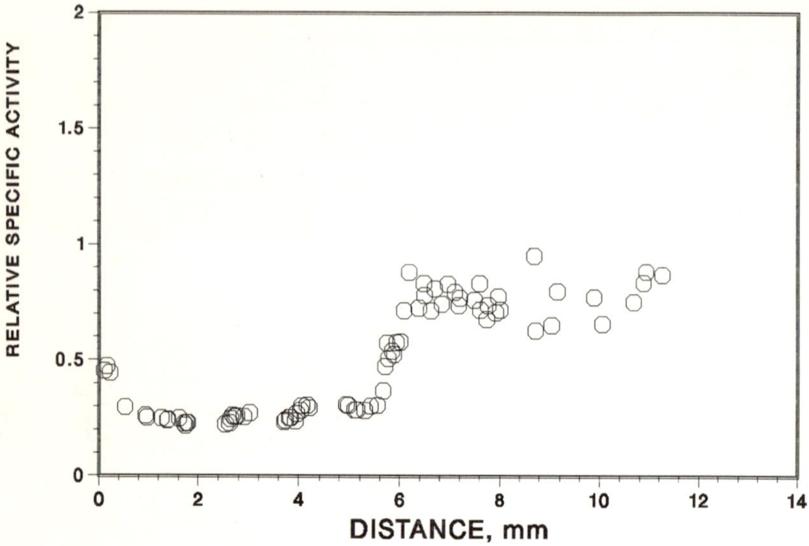

Fig. 1. Concentration curve of ^{99}Tc after 2.05 days of diffusion in Atlantic SNAP sediment.

hydroxide. On the other hand, it seems obvious that heptavalent technetium is not reduced in this milieu, since the pertechnetate ion diffuses in both types of Atlantic Ocean sediment. This example emphasizes the importance of site specific evaluations of repository parameters.

A comparison of the effective diffusion coefficients from Table I with data reported for bentonite clay shows similar mobility characteristics for heavy element ions in both types of substrate. Torstenfelt, et al. [6], list values ranging from $4 \cdot 10^{-15}$ to $6.9 \cdot 10^{-15}$ m^2s^{-1} for plutonium, americium, and thorium, while the numbers for uranium and neptunium are $5.8 \cdot 10^{-13}$ and $2.2 \cdot 10^{-13}$ m^2s^{-1}, respectively.

CONCLUSION

The numbers listed in Table I provide useful guidance in the assessment of the effectiveness of the sediment barrier. Current plans call for an emplacement depth of 30 m for waste canisters in a subseabed repository. It will be assumed that no upward flow of the interstitial water exists in the sediment overlying the repository, and that the only mode of movement for ionic species dissolved in the water is molecular diffusion. The waste canisters will be fabricated to meet the specification set by the U.S. Nuclear Regulatory Commission of retaining their integrity for at least 1000 years. However, in the seawater-saturated environment of the sediment corrosion cannot be avoided, and ultimately leaching of radionuclides will occur. At that point the sediment layer above the repository becomes the principal barrier isolating the radionuclides from the biosphere.

A simple model can serve to visualize the diffusive spreading of radionuclides through the sediment at rates governed by the effective diffusion coefficients listed in Table I. This model assumes the radionuclides initially to be present at a uniform concentration in a sediment layer with a thickness of the expected height of a waste canister, i.e. 3 m, and that this layer is embedded in radionuclide-free sediment at a depth of 30 m. Diffusion outward from the layer is the well-known case of diffusion from a plane sheet of given thickness. Boundary effects arising from the fact that the repository in not infinite in extent will be ignored here. Since the repository is very much larger in lateral extension than in thickness, being a square array of perhaps 10000 canisters spaced 10 m apart from each other, the approximation by a layer of infinite extent involves only a small error. Besides, the error is on the conservative side, since lateral diffusion diminishes the amount of substance diffusing upward.

For land-based repositories, limits for maximum permissible releases of individual radionuclides have been set by the U.S. Environmental Protection Agency [7]. For instance, the cumulative activity permitted to be released over a 10000 year period is 100 curies per 1000 tons of waste for ^{239}Pu and 10000 curies per 1000 tons of waste for ^{99}Tc. These two elements are at opposite ends of the mobility spectrum with effective diffusion coefficients of $2 \cdot 10^{-14}$ m^2s^{-1} for plutonium and $3 \cdot 10^{-10}$ m^2s^{-1} for technetium.

Fuel removed from reactors typically contains 5.9 kg of ^{239}Pu and 0.792 kg of ^{99}Tc per ton of waste [7]. A waste canister ready for emplacement in a repository could contain the waste from as much as 3 tons of spent fuel, or a lesser amount if the fuel is not reprocessed [8]. With a spacing of 10 m between canisters in a square array repository, 1000 tons of fuel require an area of 33300 m^2. The activities of ^{239}Pu and ^{99}Tc are 366800 and 13400 curies, respectively.

Given the numbers of Table I, all but 100 curies of plutonium will be contained within a region bounded by a horizon no more than 0.3 m above the original level of emplacement after a 10000 year period of diffusion. For the more mobile element technetium the corresponding horizon is about 3 m above the original emplacement level, still comfortably within the 30 m thick sediment layer covering the repository. In this case the comparatively high fraction of this nuclide that is considered acceptable to be released, alleviates the problem. Other radionuclides are bracketed by these two extreme cases. If diffusion is the only mode by which radioactivity from the waste can spread, the sea floor sediment provides an adequate geologic barrier for the isolation of nuclear waste.

The simple model calculation does not take into account refinements, such as the inclusion of boundary effects of a finite repository, the canister life, and the radioactive decay during diffusion. However, all of these effects reduce the amount of activity diffusing upward from the repository. The Subseabed Disposal Program has developed the computer codes IONMIG and MARINRAD for the comprehensive assessment of radionuclide releases [9,10]. Conclusions derived from the elaborate calculations with these codes show that in the long term there are only two nuclides of potential concern in a subseabed repository: technetium and neptunium [2].

REFERENCES

1. C. D. Hollister, D. R. Anderson, and G. R. Heath, Science 213, 1321 (1981).
2. "The Subseabed Disposal Program: 1983 Status Report", Sandia Report SAND83-1387, Sandia National Laboratories, Albuquerque, New Mexico, 1983, NTIS(*).
3. a) F. Schreiner, S. Fried, and A. Friedman, Nuclear Technology 59, 429 (1982).
   b) F. Schreiner, S. Fried, and A. Friedman in "Subseabed Annual Report, Chemical Response Studies, October 1981 - September 1982", Sandia Report SAND82-2713, page 239, NTIS(*).
   c) F. Schreiner, S. Fried, and A. Friedman in "Subseabed Annual Report, Chemical Response Studies, October 1982 - September 1983", Sandia Report, in press.
   d) F. Schreiner and C. Sabau in "Subseabed Annual Report, Chemical Response Studies, October 1983 - September 1984", Sandia Report, in press.
4. "Subseabed Disposal Program - A First Year Report, W. Bishop, ed., Sandia Report SAND74-0410, NTIS(*).
5. F. Schreiner, S. Fried, and A. Friedman, Nuclear Technology 59, 429 (1982).
6. B. Torstenfelt, H. Kipatsi, K. Anderson, B. Allard, and U. Olofsson, in "The Scientific Basis for Nuclear Waste Management V", W. Lutze, editor, Elsevier Science Publishing Company, New York, 1982, page 659.
7. "Managing the Nation's Commercial High-Level Radioactive Waste" (Washington, D.C.: U.S. Congress, Office of Technology Assessment, OTA-0-171, March 1985).
8. B. L Cohen, Rev. Mod. Phys. 49, 1 (1977).
9. A. J. Russo, Sandia Report SAND79-1666, Sandia National Laboratories, Albuquerque, New Mexico, 1980, NTIS*.
10. C. L. Koplik, et al., Sandia Report SAND83-7104, Sandia National Laboratories, Albuquerque, New Mexico, 1984, NTIS(*).

*Available from the National Technical Information Service, U. S. Department of Commerce, 5285 Port Royal Road, Springfield, Virginia 22161, USA.

TRANSIENT DIFFUSIONAL RELEASE FROM
WASTE PACKAGES IN A REPOSITORY
IN BASALT

M. S. BENSKY AND D. L. OLIVER
Rockwell Hanford Operations, Richland, Washington 99352

ABSTRACT

Analyses of diffusional release of several typical radionuclides in spent fuel from waste packages emplaced in a repository in basalt were conducted to assess the effects of system characteristics and boundary conditions on computed release rates. Radionuclide releases, including spatial and temporal variations that may be present, represent the source term for transport in the geohydrologic setting and are therefore critical to the assessment of repository acceptability.

Two mathematical approaches were utilized to determine radionuclide release rate versus time characteristics; (1) an analytical solution for one-deimensional diffusion based upon a Dirichlet (constant-concentration) boundary at the waste form surface; and (2) a finite-element numerical solution based upon a Neumann (zero-flux boundary at the waste form surface. The latter method is suitable for radionuclides such as ^{129}I, whose total inventory in spent fuel could be quickly depleted from the waste form and dissolved in the pore spaces of the packing material surrounding the waste form and which, therefore, cannot be adequately represented by a constant concentration at the waste form (i.e., container) surface.

The analysis revealed several system characteristics that are not intuitively obvious. For example, strong sorption in the near-field host rock behaves like a strong mass sink and can yield calculated transient release rates exceeding allowable limits. Similarly, a short half-life effectively removes the radionuclide from the host rock, which induces a steep concentration gradient at the host rock/packing interface and thereby increases the diffusional release rate at that boundary.

Typical results for ^{79}Se and ^{129}I are presented to illustrate these effects. The effects of perturbations to key assumptions are shown to indicate the importance of (1) formulating models that accurately represent the physical system and (2) interpreting analytical results carefully to ensure understanding of the capability of the system.

INTRODUCTION

A geologic repository for isolation of radioactive waste from areas accessible to human activities is comprised of two basic elements; 1) a source of waste material, and 2) pathways connecting the source to the accessible environment. In the United States repository program, regulations have been specified to constrain overall performance of the system by limiting the amount of radionuclides that may enter the accessible environment in the 10,000 year period after repository closure and also to constrain performance of subsystems within each of the two basic elements of the system. The subsystem constraints, stipulated by the U. S. Nuclear Regulatory Commission (NRC), require 1) complete containment within the waste package for a period between 300 and 1,000 years, 2) control of subsequent release rates after loss of containment, and 3) a minimum groundwater travel time of 1,000 years to the accessible environment. The subsystem criteria provide the NRC with additional assurance that the system standard will be met.

This paper describes analyses of release rate, from a single waste package, that have been conducted by the Basalt Waste Isolation Project (BWIP), factors that affect release rates, and difficulties encountered in interpreting and applying the NRC release rate criterion.

DISCUSSION

Radionuclide release from the waste packages in a geologic repository is governed by various characteristics of the host geologic setting, the waste emplacement configuration and the local hydrologic conditions. The ability of the combination of natural and engineered system characteristics to control release rate is a critical facet of the overall safety potential of the repository. The relationship between release rate and safety potential can, however, be misjudged if the analysis fails to address several important aspects of radionuclide behavior after transport into the near-field host rock. The following discussion describes the effects of several key system parameters on waste package fractional release ratus and notes pitfalls in the application of the NRC criterion to certain waste package/host rock conditions.

Release Rate Criterion

The NRC release rate criterion (NRC, 1983) is frequently misunderstood because of the cumbersome wording used in the regulation to describe the allowable release rate for radionuclides having relatively low inventories. The criterion is, in fact, quite straightforward. For high-inventory radionuclides, the allowable annual release rate is a constant fraction, one part in 100,000, of that radionuclide's inventory as measured at 1,000 years after repository closure. For each low-inventory radionuclide (i.e., less than 0,1 percent of the total inventory), the allowable annual release rate is a constant value, equal to one part in 100,000,000 of the total curie inventory of the repository as measured 1,000 years after repository closure.

Figure 1 illustrates the allowable release rates for typical spent fuel. The solid line, comprised of a horizontal segment representing the minimum required release rate (MRRR), which means that the NRC will not require a lower release rate than this for <u>any</u> radionuclide, and a 45 degree segment representing 1E-05 x inventory, depicts the statement from the preceding paragraph. The MRRR is 1E-08 times total curie inventory, or 1E-8 x 1.6E03 = 1.6E-05 Ci/MTU-yr.

For any radionuclide, the allowable fractional release rate is equal to allowable release rate divided by inventory. Along the 45 degree line segment in Figure 1, the value is always 1E-05. At lower inventories, however, the allowable fractional release rate is MRRR divided by inventory. For Se-79, the value is 1.6E-05 divided by 3.5E-01, or 4.6E-05. This is slightly relaxed from the basic "one part in 100,000" constraint. For I-129, MRRR divided by inventory is approximately 5E-04, a fairly significant departure from "one part in 100,000".

Release Rate Analysis

The reference waste emplacement configuration for a repository in basalt is illustrade in Figure 2.

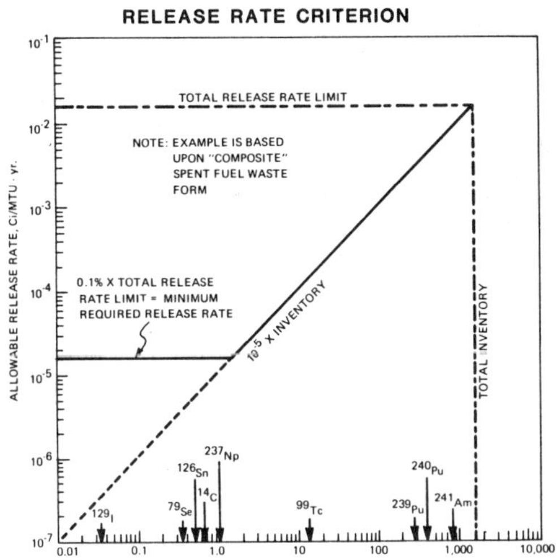

FIGURE 1

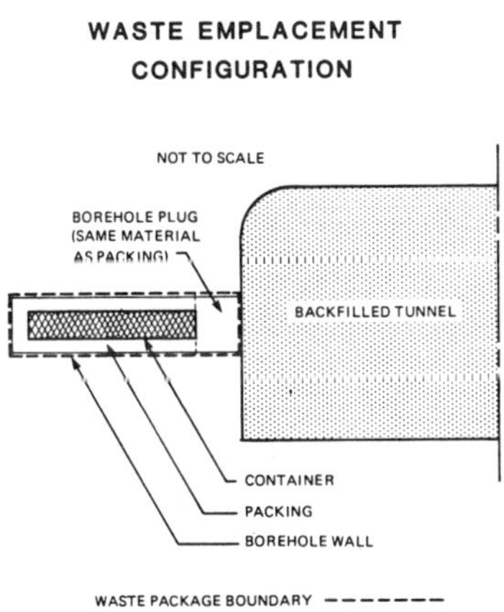

FIGURE 2

The carbon steel container is surrounded by a low-permeability, high-porosity packing material composed of cruched basalt and bentonite clay. Emplacement holes are drilled horizontally into the walls of drifts excavated in a dense entablature layer in the Grande Ronde basalt formation beneath the Hanford site. For the predicted hydrologic conditions controlling groundwater flow in this system, extremely low groundwater velocities occur within the packing and near the waste package, ensuring that molecular diffusion is generally the dominant mode of radionuclide release. Restriction of the release rate analysis to diffusional release, in contrast the combined advective/diffusive release, allows the use of analytical solution methods and substantially simplifies the application of more complex numerical codes.

The system characteristics listed in Table I were utilized as the reference case for preliminary assessment of waste package release rate at the packing/rock interface and as the basis for analysis of perturbations to system variables. The selection of a zero-concentration boundary condition at three meters represents a conservative assumption based upon comparison of test case results between diffusional release cases and combined advection/diffusion release cases in which a zero-flux boundary condition was imposed at the plane of symmetry between adjacent waste packages.

TABLE I
REFERENCE SYSTEM CHARACTERISTICS

Waste form:	Once-through cycle "typical" spent fuel (DOE, 1979)
Waste package contents:	1.8 MTU of spent fuel ^{79}Se 9.04 g ^{129}I 341 g
Container geometry:	D = 50 cm x L = 4 m
Packing thickness:	15 cm
Packing density:	1.8 g/ml
Packing porosity:	0.30
Packing K_D's:	Se = 0.8; I = 0 ml/g
Diffusivity in packing and host rock:	10^{-6} cm^2/sec
Host rock porosity:	0.0002
Host rock K_D (all radionuclides):	0
Selenium element solubility:	10^{-3} molar
^{79}Se isotope effective solubility:	10^{-4} molar
Iodine element solubility = ^{129}I solubility:	10^{0} molar
Zero concentration boundary:	at r = 3 m from package centerline

The effect of utilizing alternative zero-concentration boundaries is illustrated in Figure 3. The nominal value, 300 cm, is equivalent to an assumption of fairly strong advective transport of ^{79}Se away from the system. Relatively steep concentration gradients are thereby maintained and the computed release rates are higher than would actually be expected in a system with very low water velocities. The effect of this boundary location, however, is small and, therefore, contributes less to the uncertainty in the result than do uncertainties in parameters such as solubility, diffusivity and host rock porosity.

Alternative values of packing thickness were considered to assess their impact on the initial transient peak and on the steady state release rate. For ^{79}Se, which can be classified as weakly sorptive, the peak value, as shown in Figure 4, is substantially attenuated by increasing packing thickness. However, the larger emplacement hole required to accommodate the greater amount of packing results in higher steady-state release rates. The buildup of concentration at the packing/rock interface is shown in Figure 5. The higher transient peak for the thinner packing occurs because the packing reaches a concentration close to its solubility limit before appreciable ^{79}Se is transported into the host rock. The resulting concentration gradient is quite steep, and the release rate is high until a significant concentration of ^{79}Se is build up in the near-field host rock.

The dependence of release rate at the packing/rock interface on the buildup of radionuclide in the near-field rock creates several problems relative to using a release rate criterion as the basis for identifying and assessing favorable system characteristics.

EFFECT OF ZERO-CONCENTRATION BOUNDARY LOCATION

^{79}Se

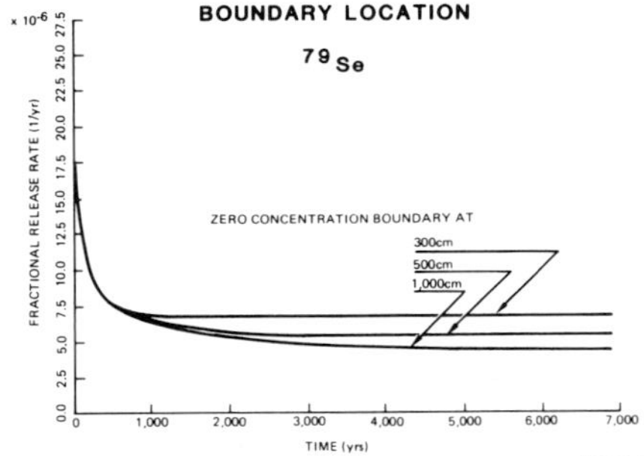

FIGURE 3

EFFECT OF PACKING THICKNESS

^{79}Se

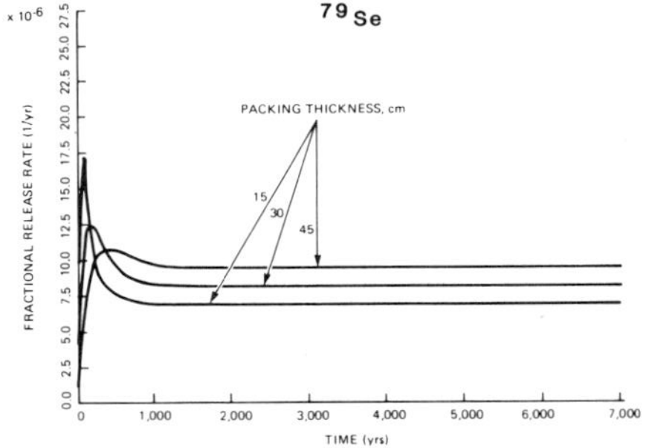

FIGURE 4

CONCENTRATION TRANSIENT AT THE PACKING/ROCK INTERFACE

^{79}Se

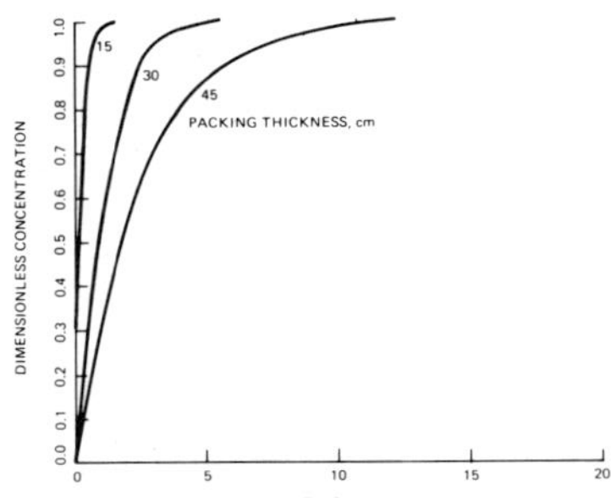

FIGURE

EFFECT OF RADIONUCLIDE HALF-LIFE ON FRACTIONAL RELEASE RATE

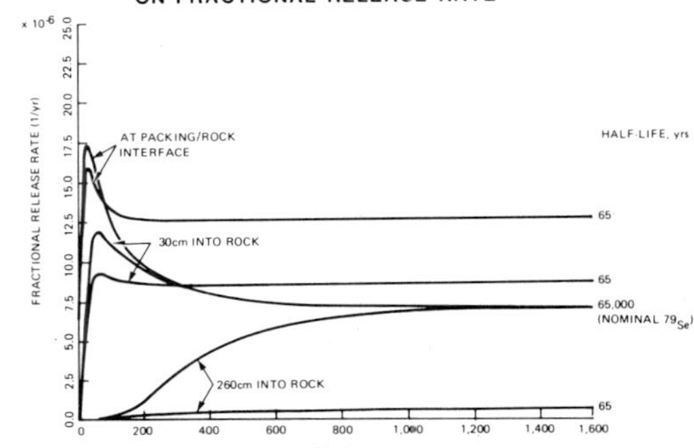

FIGURE 6

The release rates shown in Figure 6 offer an example of how a superior isolation parameter, i.e., short half-life, yields a less favorable result than a long half-life. For the example, the reference ^{79}Se case was re-run with the half-life arbitrarily reduced to 65 years. The reason for the indicated effect is that rapid decay of a radionuclide permanently removes much of that radionuclide from the system, thereby maintaining a relatively steep concentration gradient at the packing/rock interface. Disappearance of the radionuclide is obviously a favorable system characteristic, but its benefit is not evident in a release rate analysis unless the boundary where release rate is monitored is located out in the rock. This mechanism, and sorption effects discussed in the next paragraph, have been described in previous parametric studies (Chambre', 1983).

A similar result is obatined when sorption effects in the host rock are assessed. Sorption is essentially a temporary storage mechanism, and when radionuclide is sorbed onto host rock, it is removed from the system, thereby maintaining a steep concentration gradient for a longer period of time at the packing/rock interface. The effect on transient release rate is illustrated in Figures 7 through 10. Note that, for longlived radionuclides, the steady-state release rate is unaffected by sorption, though in Figure 10, the delay in arrival of ^{79}Se at 260 cm out in the rock prevents its appearance during the 10,000 year period of interest. Antoher interesting aspect of Figure 10 is that the NRC release rate criterion for ^{79}Se is briefly violated at the packing/rock interface. In this instance, strong sorption in the rock, an obviously favorable attribute of the system, creates an apparent violation of the release rate criterion.

The results presented thus far were based upon analysis of radial diffusion from the cylindrical surface of the waste container, neglecting release from the ends. As shown in Figure 2, the diffusional pathway through the borehole plug into the backfilled tunnel is comprised of high-porosity material with low resistance to migration by diffusion. Thus, the end effect at the tunnel end of the container can be quite significant. This is evident in Figure 11, where the release rate across the waste package boundary dominated by release into the tunnel, is shown to be substantially greater than the release rate across the engineered barrier boundary for the reference conditions (Table I). The difference can be significant in assessing compliance if, as has been proposed, the NRC release rate criterion is conservatively applied at the waste package boundary rather than at the engineered barrier boundary.

The assumption of constant, solubility-limited concentration at the waste container boundary is appropriate only so long as the inventory of the radionuclide in the waste form is not depleted during the interval that is analysed. This is particularly important in the case of ^{129}I; the total inventory of 341 grams is less than 1 percent of the amount that can be dissolved in the pore volume of the packing material. As a result, the iodinge concentration in the packing is quickly diluted to less than 1 g/liter (Figure 12), compared to 129 g/liter initially at the waste form surface. As shown in Figure 13, neglecting the effect of inventory limitation yields a steady-state fractional release rate of approximately 3×10^{-3} (exceeding the NRC allowable limit). In contrast, Figure 14 indicates that the inventory limited quasai-steady-state release rate is only about 2×10^{-5}, a factor of 25 below the allowable limit.

FRACTIONAL RELEASE OF ^{79}Se
FOR MINIMAL SORPTION IN PACKING AND ROCK

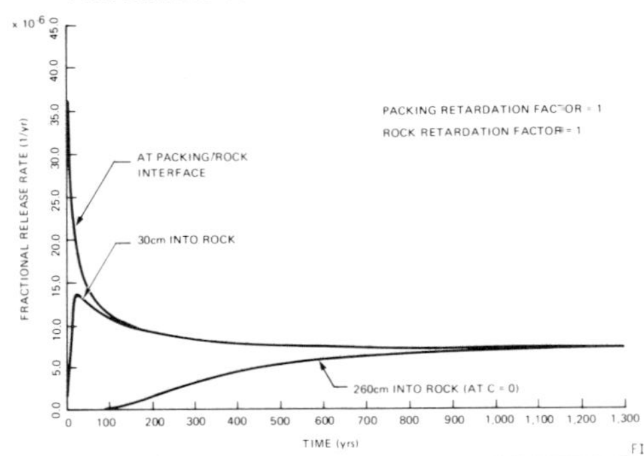

FIGURE 7

FRACTIONAL RELEASE OF ^{79}Se
FOR REFERENCE CONDITIONS

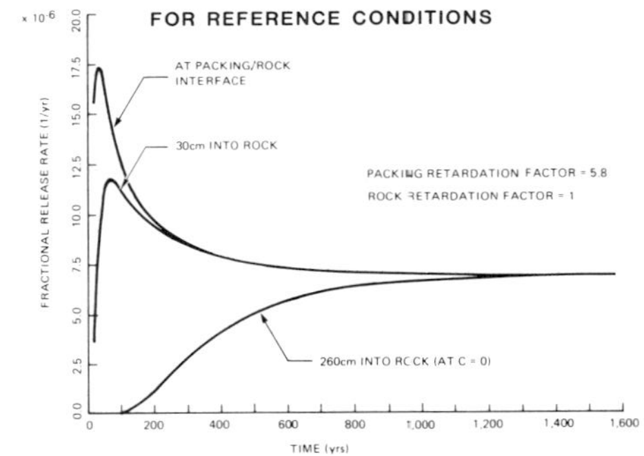

FIGURE 8

FRACTIONAL RELEASE OF ^{79}Se FOR PACKING
SORPTION VALUE IN BOTH PACKING AND ROCK

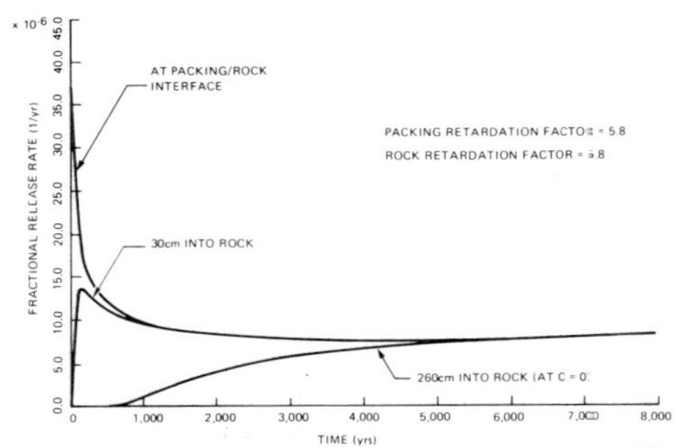

FIGURE 9

FRACTIONAL RELEASE OF ^{79}Se
WITH STRONG SORPTION IN ROCK

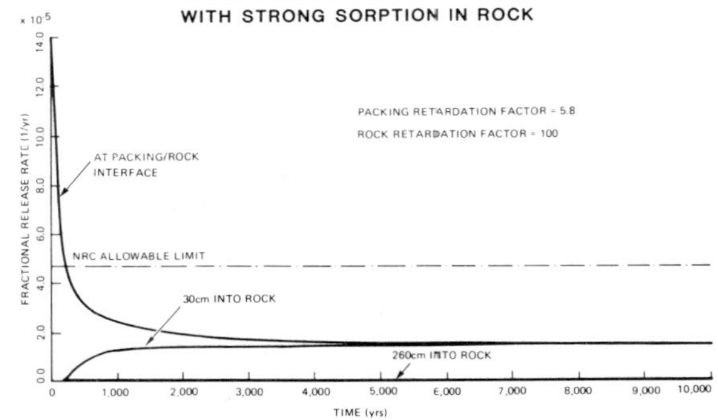

FIGURE 10

FRACTIONAL RELEASE OF ^{79}Se FROM A SINGLE WASTE PACKAGE

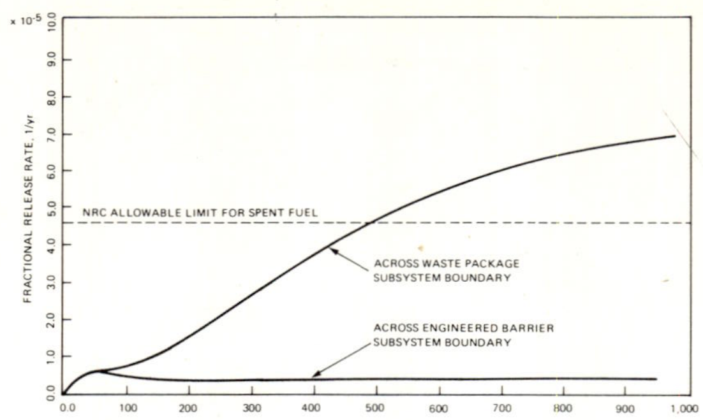

FIGURE 11

^{129}I DIFFUSION

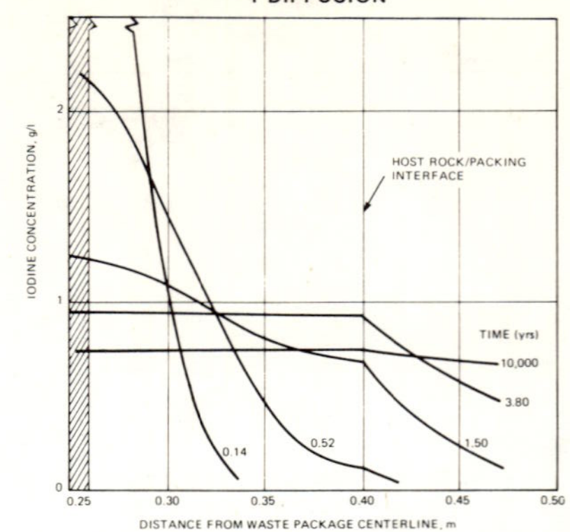

FIGURE 12

FRACTIONAL RELEASE RATE OF ^{129}I FOR UNLIMITED INVENTORY

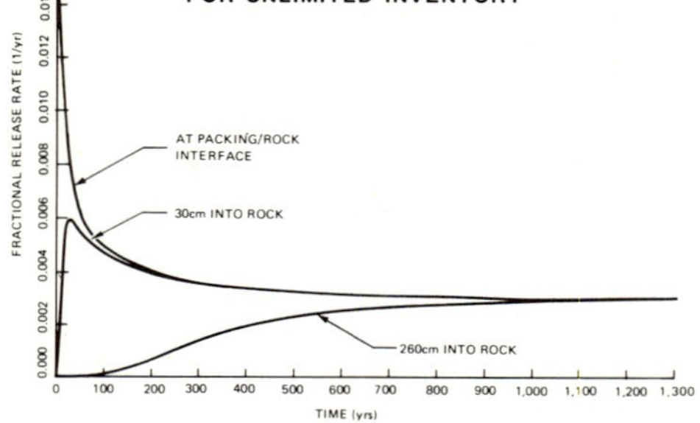

FIGURE 13

INVENTORY-LIMITED FRACTIONAL RELEASE OF ^{129}I

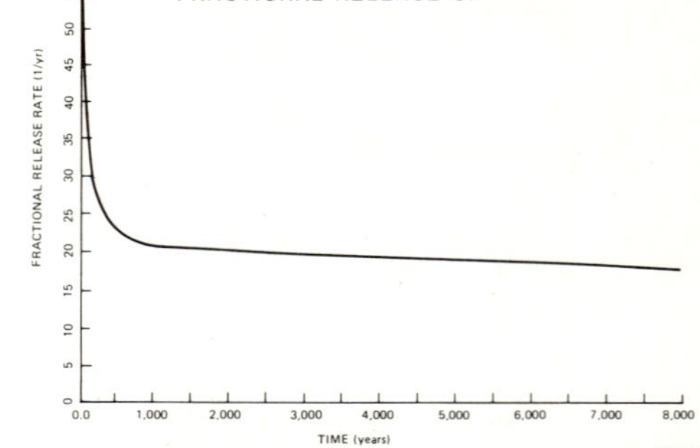

FIGURE 14

CONCLUSIONS

The definition of radionuclide release rates for use as a source term in transport analyses or for judgment of compliance with regulatory criteria must recognize the importance of boundary conditions and system characteristics that affect computed values. The choice of an appropriated boundary at which to measuare release rate is a critical facet of the usefulness of any release rate criterion. Exclusion of some portion of the host rock or of the emplacement tunnels as part of the source can yield incorrect assessment of the isolation capability of the system.

REFERENCES

1. DOE (U. S. Department of Energy), Technology for Commercial Radioactive Waste Management, (Office of Nuclear Waste Management, Washington, D.C., 1979), DOE/ET-0028, Vol. 1.
2. NRC (U. S. Nuclear Regulatory Commission), Disposal of Nuclear Radioactive Waste in Geological Repositories, 1983, Title 10, Code of Federal Regulation, Part 60, Vol. 48, No. 120, Final Rule.
3. P. L. Chambre' and T. H. Pigford, Prediction of Waste Performance in a Geologic Repository, (University of California, Berkeley, California, 1983), UCB-NE-4043, LBL-17272.

PART XI

Geotechnical and Transport Modelling

MECHANICAL BEHAVIOUR OF HOST ROCK CLOSE TO H.L.W. DISPOSAL CAVITIES IN A DEEP GRANITIC FORMATION

J.M. HOORELBEKE* AND M. DOURTHE**
* Commissariat à l'Energie Atomique, IRDI/DERDCA/DIP/SEIP, Centre d'Etudes Nucléaires VALRHO, BP 171, 30205 Bagnols sur Cèze Cedex, France
** Commissariat à l'Energie Atomique, DAM/DArC/DT/STR, C.E.S.T.A., BP 2, Le Barp 33830 Belin-Beliet, France

ABSTRACT

The construction of a H.L.W. repository in a deep granitic formation creates mechanical disturbances in the rock on the scale of the massif and in the nearfield. Amongst all the disturbances noted in the nearfield, this study is concerned with examining the evolution of stresses linked with the excavation of the rock and the rise in temperature in the proximity of the waste packages. Several linear elasticity calculations were made using on the one hand finite element models and on the other simple analytical models. These calculations concern two different storage concepts -"in room" concept and "in floor" concept- whose differences in mechanical behaviour are analysed. A study of sensitivity with regard to the characteristics of the rock and to the initial geostatic stresses is presented. The comparison of the calculated stresses with three-dimensional failure criteria gives a clear indication of the satisfactory behaviour of granite for final storage. However, the need for experimental study and complementary calculation must be emphasised.

PURPOSE OF STUDY

Consecutively to the disposal of H.L.W in a deep granitic formation, mechanical pertubations in nearfield rock are due primarily to the effects of drilling and of stress relaxing around cavities and also to those of the high thermal activity of waste : stress concentrations induced by temperature rise, micro-cracking due to thermal dilatation of minerals.

To estimate the risk of development of high porosity zones close to disposed waste, it is necessary to evaluate stresses and failures in the nearfield rock. The purpose of our study is the calculation of stresses due to rock relaxing after drilling and to thermal loading around disposal cavities in a granitic formation.

Many results of so called homogeneous mechanical calculations have already been presented in France, in which storage is considered as a continuous volume with an homogeneous thermal power [1,2] . Such models give a good evaluation of rock behaviour at a certain distance from disposal, for instance near surface ; but applied stresses on nearfield rock, especially along walls of cavities, cannot be known accurately through these calculations. Consequently, Commissariat à l'Energie Atomique (C.E.A.) has conducted several series of mechanical calculations including the exact geometry of disposal cavities and the nearfield temperature evolution in space and time.

THE CONFIGURATIONS STUDIED

From a mechanical point of view, the main differences between the storage of waste packages in vertical boreholes versus horizontal galleries lie essentially in the initial geostatic stresses in the plane of the linear section of the cavity, and in the interaction with the neighbouring cavities. These differences lead to an examination of the two types of concept. In the study of each of these, by way of hypothesis, and from numerous possibilities, we have chosen a management scenario for vitrified fission products from a reprocessing plant, and the appropriate dimensions of the repository.

The filling materials which assure the thermal continuity between the packages and the rock are assumed to be put in place at the closure of the repository, 100 years after the unloading of the reactor. The temperatures attained are probably higher than those which will be reached in reality.

"In floor" concept

The repository, at a depth of 500 metres comprises 289 vertical boreholes forming a square mesh network every 50 metres (Fig. 1). These boreholes can be filled to a height of 30 metres and measure 3.20 metres in diameter. Each borehole contains six vertical piles of 22 packages of UP3 (La Haque) type glass (reprocessing of 48 000 tU).

"In room" concept

This comprises three main galleries (Fig. 2) in a horizontal plane, at a depth of 500 metres and 80 metres apart. Packages are placed eight abreast in a linear section accross the gallery. The study assumes that these galleries are cut using a tunnel boring machine which supposes a cylindrical profile, eight metres in diameter. The general lay-out is approximately three concentric tori.

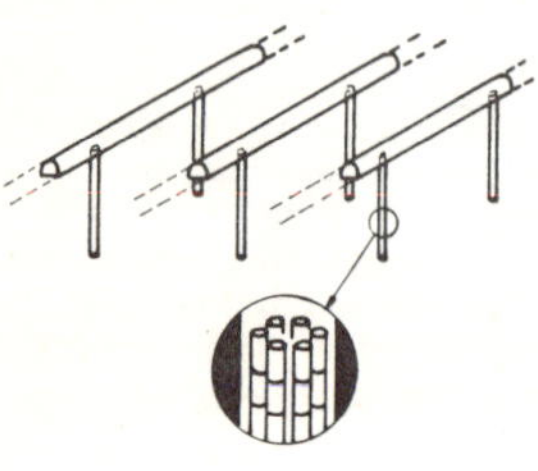

Fig.1 "In Floor" Concept

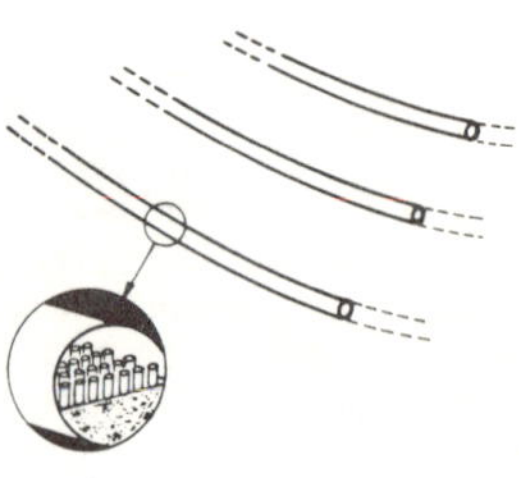

Fig.2 "In Room" Concept

"IN FLOOR" CONCEPT : NUMERIC CALCULATIONS USING FINITE ELEMENT ANALYSIS

The evaluation of the state of stress of the boreholes was made by using an elastic - plastic finite element code for plane deformation developed by the National Civil Engineering Laboratory for Road Construction Works (L.C.P.C.) [3,4]. Given the geometry and the symmetry of the repository, the model was limited to a plane area 25 metres square normal to the benehole axes. The boundary conditions imposed on the calculations consist in a normal zero displacement of the four mesh boundaries which constitute the symetrical planes. The cavity is taken to be mechanically empty, which entails on the one hand the disregard of the efforts assured by the engineered barriers, which are very compressible, on the other hand the swelling pressure eventually developed by the barriers (smectites) in the presence of water.

The mechanical calculations comprise three load increments. The first takes into account the initial geostatic stresses, the drilling of the boreholes and the increase in the stresses linked to the rise in temperature during the first six years following the closure of the repository. The second and third load increments relate to the increase in the stresses of thermal origin 16 and 150 years after closure respectively.

These three dates correspond to temperature maxima in the glass and in the rock, and are associated with the most severe mechanical loading on the nearfield.

Starting from an initial temperature in the rock of 28°C at a depth of 500 metres, Fig. 3 shows the distribution of the temperature around the storage boreholes for the three dates considered. These temperatures were obtained from finite element analysis using the heat capacity of the H.L.W. presented on table I.

Table I : Heat capacity

age of waste (years after unloading)	100	125	150	200	250	300
Heat capacity of the conditioned waste (W/m^3)	915	530	355	200	150	130

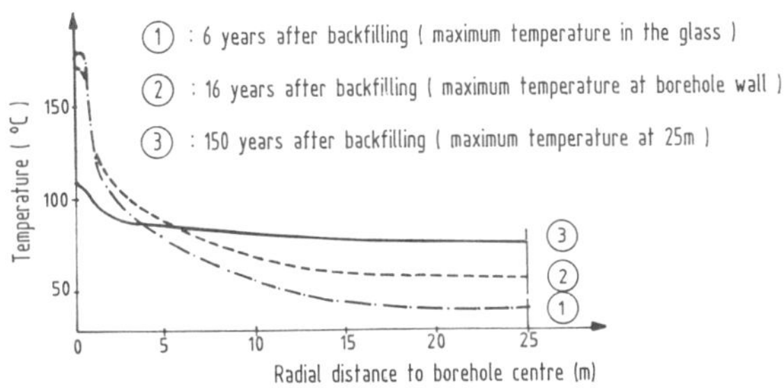

Fig.3 Distribution of temperature around storage boreholes

Calculations of stresses under nominal hypotheses

The assumed nominal hypotheses concerning host granite are as follows :

Characteristics of the rock :

- density (ρ) = 2.64×10^3 kg/m^3
- Young modulus (E) = 50 000 MPa
- Poisson ratio (ν) = 0.27
- Coefficient of linear expansion (α) = 8×10^{-6} °C^{-1}

Initial vertical geostatic stress σ_z^o = 12.9 MPa

Initial horizontal geostatic stress $\sigma_x^o = \sigma_y^o$ = 4.8 MPa

Contrary to ever stable continental pre-cambrian shields (e.g. Baltic shield, emergent for 1.6×10^9 years and very eroded), french granitic terrains have been strongly altered during the tectonic history of the country. Generally the favourable terrains for disposal seem to be most recent granites vis-à-vis the structuring of their host formations [5]. Consequently, the assumption of this study does not consist in horizontal virgin stresses in significant excess of vertical stress. As the nominal hypothesis a pure elastic equilibrium is assumed and the sensitivity study for "in floor" concept takes account of conditions similar to stress distributions measured in french old granitic massifs.

The largest tangential stress is reached at the third load increment after 150 years of storage. The maximum tangential stress in the wall σ_θ is - 77.8 MPa after 6 years, - 99.9 MPa after 16 years and - 104.8 MPa after 150 years.

Figures 4 and 5 show the radial distribution of the horizontal principal stresses σ_1 , σ_2 at the three given dates and in the directions X, Y, X', Y' :

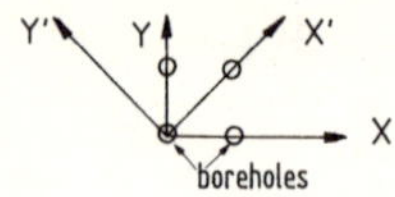

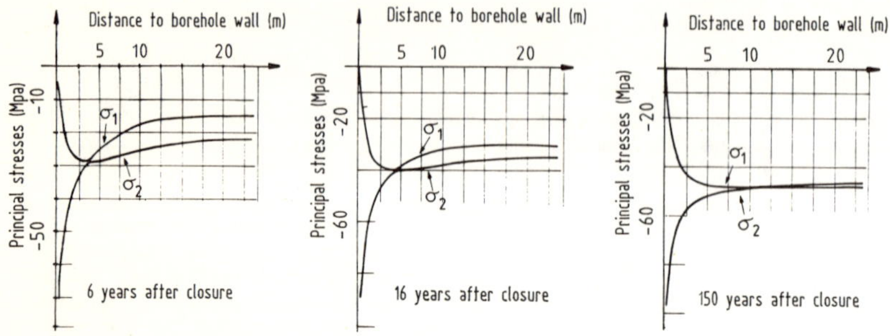

Fig. 4 Radial distribution of the horizontal principal stresses σ_1, σ_2 on both axes X and Y for "in floor" concept

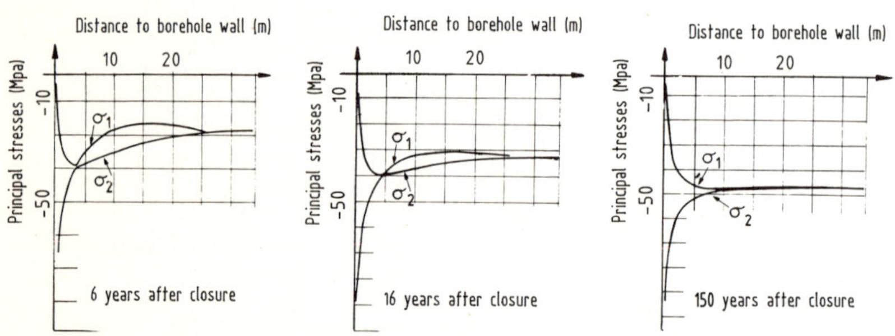

Fig. 5 Radial distribution of the horizontal principal stresses σ_1, σ_2 on both axes X' and Y' for "in floor" concept

The stress values obtained from F.E analysis are systematically introduced in a three dimensional parabolic failure criterion (1) where the vertical principal stress σ_3 normal to the plane (X,Y) is obviously calculated so that the vertical deformation is equal to zero. It shows that no rupture occurs.

$$[(\sigma_1 - \sigma_2)^2 + (\sigma_2 - \sigma_3)^2 + (\sigma_3 - \sigma_1)^2] + 2(|R_c|-|R_t|)(\sigma_1 + \sigma_2 + \sigma_3)$$

$$- 2|R_c||R_t| \geqslant 0 \qquad (1)$$

where $|R_c| = 126$ MPa ; $|R_t| = 6.3$ MPa

R_c is the uni-axial compressive strength and R_t the uni-axial tensile strength.

Sensitivity study

Several successive calculations were made using different hypotheses e.g : 20 % reduction of failure characteristics, allowance for a partial relaxing of the geostatic stresses (σ_x^o/σ_z^o = 0.37 ; σ_y^o/σ_z^o = 0,28), superposition of residual tectonic and geostatic stresses (σ_x^o/σ_z^o = 0.5 ; σ_y^o/σ_z^o = 1 ; this condition is close in value to stress measurements in Massif Armoricain [6,7] and in Massif Central -Auriat site- [8]) and reduction of the modulus of elasticity. No rupture was revealed. In the case defined by σ_z^o = 12.9 MPa, σ_x^o = 4.8 MPa, σ_y^o = 2.4 MPa, E = 50 000MPa, ν = 0.27, the failure only appears if the compressive strength of the rock is equal to 50 MPa and the tensile strength is equal to 2.5 MPa.

"IN ROOM" CONCEPT : NUMERICAL CALCULATION USING FINITE ELEMENT ANALYSIS

Two sets of mechanical calculation were carried out. The first uses a model with plane deformations developed by the University of Montpellier [9], the second uses an axisymetrical model made by the CEA-DAM-DArC [10]. The whole massif is modelled around the storage galleries. In the first calculation the initial geostatic stresses are supposed to be equal to σ_z^o = 12.9 MPa, σ_x^o = σ_y^o = 4.8 MPa whilst the assumption for the second calculation is an hydrostatic distribution σ_z^o = σ_x^o = σ_y^o = 12.9 MPa The thermal and mechanical characteristics of the rock are as follows :
ρ = 2650 kg/m^3, λ = 2.5 W/m/^{o}C, C_p = 940 J/kg, α = 8x10^{-6} ^{o}C^{-1},
E = 68 000 MPa, ν = 0.27, $|R_c|$ = 150 MPa, $|R_t|$ = 7.5 MPa,

The distribution of temperature around the central gallery calculated using a finite element method by CEA-DAM-DArC for several dates is given in Figure 6 [11].

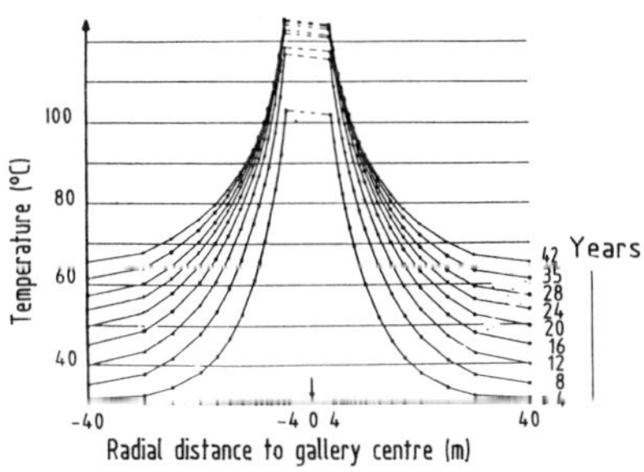

Fig. 6 Temperature distribution around the central gallery as a function of time

Inside the storage volume comprising the three galleries all the stresses are compressive. During the first eight years following the closure of the repository no interaction of stresses between the galleries can be detected. However, between 8 and 14 years after closure the first signs of interaction appear at points equidistant from the galleries. The zone situated around the central gallery is most affected. The evolution within a time-frame of the stresses in the vault and in the sidewall is presented in Figures 7 and 8 corresponding to the first calculation. In the second calculation, for a significantly different distribution of geostatic stresses, the highest stress value is still approximately 80 MPa. No rupture is detected in application of the three-dimensional failure criterion. This result is shown in Figure 9. In the same diagram the failure envelope for granite and the evolution of the maximum loading of the rock within a time frame according to CEA-DAM-DArC are shown. The calculated value of the convergence is very low, which proves a relative mechanical independance between host rock and the compressible backfilling materials put inside the cavities.

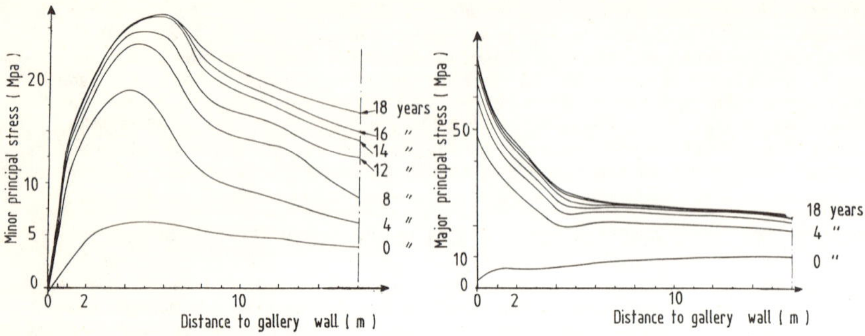

Fig. 7 Radial distribution of the minor and major principal stress in the vault of the central gallery for in "room concept"

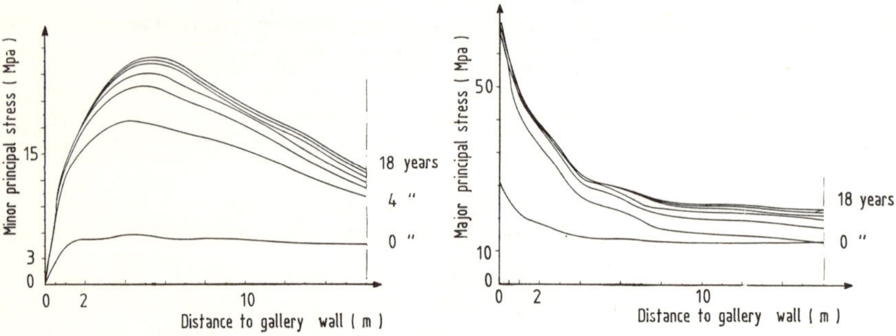

Fig. 8 Radial distribution of the minor and major principal stress in the side walls of the central gallery for "in room" concept

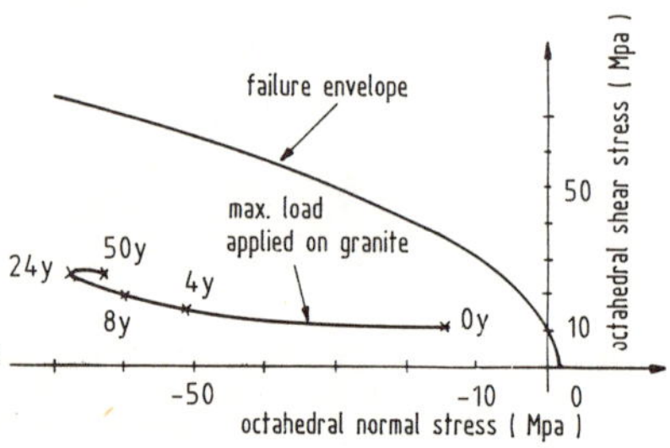

Fig. 9 Loading of the granitic rock for "in room concept". The failure envelope for granite is also shown.

ANALYTICAL SOLUTION

By using the equalities which denote linear elasticity equilibrium it is possible to develop a simple analytical model for integrating the radial decrease in temperature. The storage cavities are assumed to be cylinders with a radius a and of infinite length. At all points of semi-polar coordinates (r, θ, z) the state of stress results from the superposition of geostatic stresses (equations (2) (3) (4)) and of thermal stresses.

Stresses of geostatic origin

The state of stress in the vicinity of a cylindrical opening is :

$$\sigma_{r,g}(r,\theta) = 1/2(\sigma_x^o + \sigma_y^o)(1 - a^2r^{-2}) + 1/2(\sigma_x^o - \sigma_y^o)(1 + 3a^4r^{-4} - 4a^2r^{-2})\cos 2\theta \quad (2)$$

$$\sigma_{\theta,g}(r,\theta) = 1/2(\sigma_x^o + \sigma_y^o)(1 + a^2r^{-2}) - 1/2(\sigma_x^o + \sigma_y^o)(1 + 3a^4r^{-4})\cos 2\theta \quad (3)$$

$$\tau_{r\theta,g}(r,\theta) = -1/2(\sigma_x^o + \sigma_y^o)(1 - 3a^4r^{-4} + 2a^2r^{-2})\sin 2\theta \quad (4)$$

where σ_x^o and σ_y^o are the initial stresses in the plane perpendicular to the cavity.

Stresses of thermal origin

Two cases must be studied : (i) that of an isolated cavity, to which one may adjoin the case of storage in galleries, and (ii) that of a close network of cavities such as storage in vertical boreholes. The first case is obtained by considering as zero the extension of radial stress to infinity whilst the second case supposes a zero radial displacement at a distance b from the centre of the cavities, b being half the distance between the axes of the cavities.

(i) Isolated cavity

According to [4, 12], the radial and tangential stress is governed by :

$$\sigma_{r,t}(r) = \frac{-E\alpha}{1-\nu} r^{-2} \int_a^r \Delta T(\rho)\,\rho\,d\rho \quad (5)$$

$$\sigma_{\theta,T}(r) = \frac{E\alpha}{1-\nu}\left[-\Delta T(r) + r^{-2}\int_a^r \Delta T(\rho)\,\rho\,d\rho\right] \quad (6)$$

and $\tau_{r\theta,T} = 0$

(ii) Close network of cavities [4]

$$\sigma_{r,T}(r) = -\frac{E\alpha}{1-\nu}\left[\frac{(1-a^2r^{-2})}{b^2(1-2\nu+a^2b^{-2})}\int_a^b \Delta T(\rho)\,\rho\,d\rho) + r^{-2}\int_a^r \Delta T(\rho)\,\rho\,d\rho\right] \quad (7)$$

$$\sigma_{\theta,T}(r) = -\frac{E\alpha}{1-\nu}\left[\Delta T(r) + \frac{(1+a^2r^{-2})}{b^2(1-2\nu+a^2\,b^{-2})}\int_a^b \Delta T(\rho)\,\rho\,d\rho - r^{-2}\int_a^r \Delta T(\rho).\rho.d\rho\right] \quad (8)$$

and $\tau_{r\theta,T} = 0$

This simple model makes it possible to ratify the results obtained by means of FE analysis. The coherence of the values obtained from analytical and numerical calculation is excellent : examples are presented on table II.

Table II : Comparison of resultats obtained from FE and analytical calculations

	"In floor concept"					"In room concept"
	Nominal hypotheses			Other hypotheses		
Date (years)	6	16	150	150	150	20
Highest tangential stress in the sidewall obtained from FE analysis (MPa)	- 78	- 100	- 105	- 126	- 73	- 80
Tangential stress in the sidewall obtained from analytical calculation (MPa)	- 80	- 101	- 105	- 125	- 73	- 85

CONCLUSION

The results obtained from different calculations show that nearfield granite considered as a continuous medium does not fail in nearfield sections of a disposal cavity whatever is the design concept.

However, owing to the imposed boundary conditions, the calculations give higher wall stresses for the "in floor" concept than for the "in room" concept. Moreover, the maximum stress appears much later in the "in floor" concept. This may be of some importance for the behaviour of pre-existing fractures or joints. Complementary calculations in three dimensions could complete this study in the case of "in floor" concept in order to verify the boundary conditions and to evaluate the stresses at the intersections between the boreholes and the galleries. At the same time a mechanical study of the nearfield rock will be continued by taking into account the existing of fractures.

REFERENCES

1. M. Dourthe and C. Martinoia, CEA/DAM/DArC/DT Report, 1981.
2. J. Vinsonneau, P. Nogre and C. Duretz, CEA/IRDI/DEDR/DEMT Report, 1982.
3. Groupe de Mécanique des Roches, C.E.T.E. Lyon, 1984.
4. J.M. Hoorelbeke, CEA/IRDI/DERDCA/DIP Report N° 15, 1985.
5. A. Autran and J. Chantraine, in Journée sur le granite (Documents du BRGM, N° 84, 1984) pp 13-34.
6. C. Paquin and C. Froidevaux, Bull. So. geol. Fr. XX, 5, 727-731 (1978).
7. P. Godefroy, BRGM Report N° 80 SGN 032 GEG, 1980.
8. F.H. Cornet, in Journée sur le granite (BRGM, N° 84, 1984) pp 263-277.
9. P. Guichet, T. Guiot and D. Veaux, CEA/DIP - E.N.S.T.I.M.A., 1984.
10. M. Dourthe and Martinoia, CEA/DAM/DArC/DT Report, 1984.
11. J.F. Epherre, CEA/DAM/DArC/DT Report,1984.
12. P. Berest, in Journée sur le granite (BRGM, N° 84,1984), pp 249-262.

HEAT INDUCED FRACTURING OF ROCK IN AN EXISTING UNIAXIAL STRESS FIELD

J.I. MATHIS, O. STEHPANSSON, B. BJARNASON, H. HAKAMI, A. HERDOCIA,
U. MATTILA AND U. SINGH
Dept. of Rock Mechanics, University of Luleå, S-951 87 Luleå, Sweden

ABSTRACT

The thermal fracturing of rock has been the object of several research projects, notably for initial rock breakage in mining [4] as well as crushing [8]. In addition, the process has been studied carefully in regards to the storage of radioactive waste underground where rock fracturing could lead to a loss of radioactivity confinement. The Stripa Project, a project concerning large scale testing of procedures for underground storage of nuclear waste, probably has dealt most thoroughly with this subject by theoretical studies and in-situ heater testing in an attempt to describe the thermal failure process in rock [3].

This project was designed to test the agreement between theoretical and actual rock fracture times of a rock block loaded with a physical as well as a thermal load. Laboratory testing consisted of physically loading center-drilled cubes of rock, 0.3 m on a side, uniaxially from 0 to 25 MPa. These were then thermally loaded with a nominal 3.7 kW (factory rating) cylindrical heater until failure occurred. This time to failure was recorded for comparison with a direct mathematical and a finite element solution. For both cases, calculations were performed at specific time-steps and an estimated failure time calculated from the compiled results.

INTRODUCTION

The basic theory behind the study is quite simple. Suppose a borehole is drilled into a solid rock mass. There exists around the borehole a stress field dependent upon the virgin stresses existing in the rock mass. Now, let us introduce a heater into the borehole. If there was no stress field in the rock mass, a circular uniform stress field would develop around the hole. However, since there is an inherent stress field present, the resulting stress field is a time dependent one composed of the uniform, heater generated component superimposed over the non-uniform in-situ stress field. As this composite stress field changes with time, a stress value will be reached that exceeds the strength of the rock and the rock will fracture. The time necessary before fracturing takes place is dependent upon thermal input, thermal and physical properties of the rock, and the virgin stress field.

THEORY

The problem is best attacked by breaking it into discrete steps. The steps involved are calculating:

1) The transient temperature distribution generated by a cylindrical heater in a cube of rock.

2) The thermal stresses around the cylindrical opening containing the heater due to the temperature field.

3) The stresses around the cylindrical opening containing the heater due to an applied uniaxial load.

Theoretical assumptions

The rock was assumed linearly elastic, physically and thermally isotropic, with constant thermal and physical properties. In addition a perfect contact was assumed between the heater and the rock with a constant power input into the rock from the time of heating.

Temperature distribution

The theoretical temperature distributions in a material with constant thermal properties given a constant power input is roughly parabolic in form, becoming asymptotic at the hole boundary and at an infinite radius (zero temperature).

Assuming a spherical power source of known radius in a thermally linear isotropic media with perfect contact between the source and the heater, the temperature at any point may be expressed as follows [1].

$$T(r) = \frac{Q}{8\pi k r r'} \left[2(kt/\pi)^{1/2} \left(e^{-(r-r')^2/4kt} - e^{-(r+r')^2/4kt}\right) - |r-r'| \operatorname{erfc} \frac{|r-r'|}{2(kt)^{1/2}} + (r+r')\operatorname{erfc} \frac{|r+r'|}{2(kt)^{1/2}} \right] \quad (1)$$

Where:

Q = power input
r' = radius of the power source
r = radius to the point of interest
k = thermal diffusivity
t = time
$T(r)$ = temperature at radius r
$\operatorname{erfc} = 1 - \operatorname{erf}$
$\operatorname{erf} = \frac{2}{\sqrt{\pi}} \int_0^x e^{-\zeta^2} d\zeta$

This, however, is an expression for a spherical power source in an infinite medium. To accurately model our problem the equation was modified. The heater was modelled as a row of spherical sources with a radius equal to the radius of the heater. The temperature distribution at the midplane of the block was therefore the summation of the individual effects from each fictitious 'spherical source' upon the block midplane (Figure 1). The duration of the experiment was considered short enough that the block could be considered infinite.

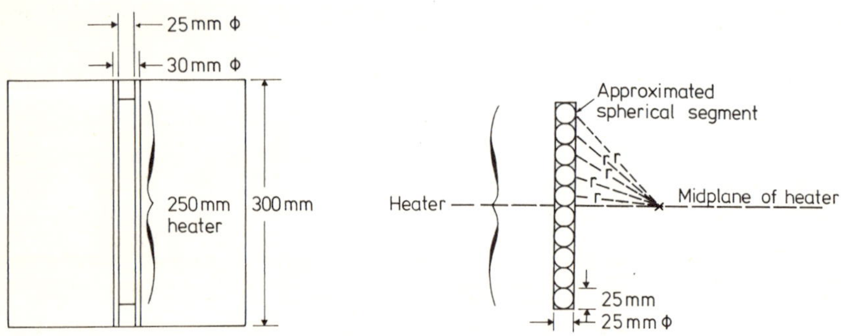

Figure 1. Theoretical heater representation

Thermal stress

Thermal stresses due to temperature loading of a thermally linear isotropic body are dependent upon the temperature field present in the block. Given a thin circular plate of infinite outer radius, the radial (σ_r) and tangential (σ_θ) stresses around a circular hole in the middle of the plane may be calculated as follows [2]:

$$\sigma_r(t) = \frac{E\alpha}{(1-\nu)^2} \int_{r_0}^{r} rT(r)\, dr \quad ; \quad \sigma_\theta(t) = \frac{E\alpha}{(1-\nu)} \left[\frac{1}{r^2} \int_{r_0}^{r} rT(r)\, dr - T(r) \right] \qquad (2)$$

Given: $\sigma_r(t) = 0$ at $r = r_0$; $\sigma_\theta(t) = \sigma_r(t) = 0$ at $r = \infty$

Where:

E	= Young's modulus		r	= radius to point of interest
α	= coefficient of expansion		$T(r)$	= temperature at radius r
ν	= Poisson's ratio		$\sigma_r(t)$	= radial stress at time t
r_0	= hole radius		$\sigma_\theta(t)$	= tangential stress at time t

External stress

The stresses developed around a circular hole in an elastic solid due to some external loading were calculated according to the following [7]:

$$\sigma_r = \frac{S_y}{2} \left[1 - \frac{r_0^2}{r^2} \right] - \frac{S_y}{2} \left[1 + \frac{3r_0^4}{r^4} - \frac{4r_0^2}{r^2} \right] \cos 2\theta \qquad (3)$$

$$\sigma_\theta - \frac{S_y}{2} \left[1 + \frac{r_0^2}{r^2} \right] + \frac{S_y}{2} \left[1 + \frac{3r_0^4}{r^4} \right] \cos 2\theta \qquad (4)$$

Where:

S_y = applied field stress, y direction
r_0 = hole radius
r = radius to point of interest
θ = angle counter-clockwise from +x axis to r
σ_r = radial stress
σ_θ = tangential stress

MATERIAL PROPERTIES

Two different rock types were tested in this study, a fine-grained diorite (hereafter referred to as diabase) and a coarse-grained granite. The physical and thermal properties are given as Table I and the mineral contents as Table II.

EXPERIMENTAL ANALYSIS

Sample preparation

The blocks to be tested were cut to dimensions of 0.30 m per side and polished on two opposite faces. A 0.03 m hole was diamond drilled through the center of the block parallel to the polished surfaces of the cube. Wire type thermocouples were then installed in the block in small holes (2 mm dia., 15 mm length) drilled parallel to the heater hole. To minimize stress problems the blocks were then instrumented with foil type electrical resistance strain gauges (4 per block, axial strain) and painted with a white heat resistant paint as an aid for fracture detection. At this stage, the heater, equipped

Table I: Mechanical and thermal properties of the test rock types

	Granite		Diabase	
	Mean	Standard Deviation	Mean	Standard Deviation
uniaxial compressive strength (MPa)	166.67	7.9	196.16	8.35
tensile strength (MPa)	9.92	0.47	18.04	0.93
Poisson's ratio	0.24	0.04	0.26	0.01
modulus of elasticity (GPa)	58.32	4.24	86.16	1.31
density (kg/m^3)	2635.0	9.7	2932.0	1.6
thermal conductivity (W/m^{0}K)	3.32	0.12	2.84	0.02
thermal expansion (* 10^{-6}) @ 260^{0}C	18.5	7.1	5.5	1.4
specific heat (kJ/kg)	0.721	0.006	0.723	0.009

Table II: Modal composition of the test rock types

Mineral	%
Diabase	
Amphibole	33.5
Plagioclase	26.0
Biotite	17.8
Quartz	15.3
Granite	
Microcline	41.0
Plagioclase	32.0
Quartz	19.0
Biotite-chlorite	6.0

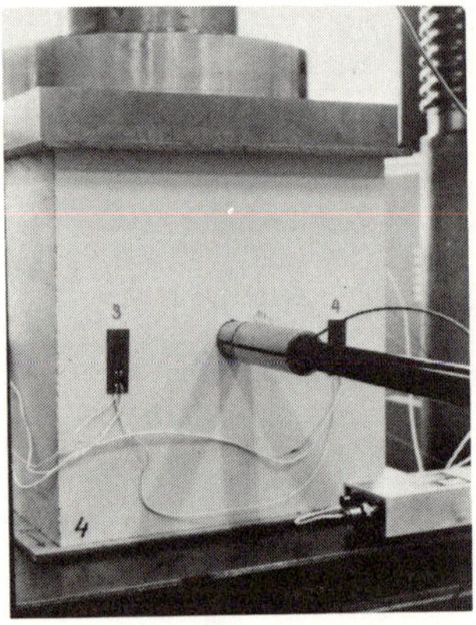

Figure 2. Block prior to testing

with a slotted brass sleeve to fill the difference between hole and heater diameters (5 mm) was inserted in the block. A typical example of a block prior to testing is shown as Figure 2.

Test procedure

Loads applied during testing were 0, 5, 15, and 25 MPa. The block was allowed to stabilize a few minutes once the appropriate load level had been reached. Power was applied to the heater for a fixed time of 45 minutes at a control temperature of 400^0 C. Any fracturing during this time was noted by acoustic emissions and/or visual observations. The block, when cooled, was examined, all visible fractures marked, and photographed.

Acoustic emission

During testing, acoustic emissions from the block were accumulated over a sample interval of 2 seconds and stored on the computer as the number of events per interval. These were later plotted versus test time.

A distinct difference in type of emission was noted between diabase and granite. Diabase, Figure 3a, had very sharp, definite emission peaks of extremely short duration, apparently independent of load. These peaks were often accompanied by audible noise. Granite, Figure 3b, had peaks which were approached by a series of increasing counts culminating in a final large peak. This made it difficult to distinguish an actual time of fracture.

In an attempt to solve the problem of deciding when the block actually fractured, a threshold level of counts based on the maximum peak height was instituted. It appeared from this analysis that the 5% trigger level was likely the best representation of failure in terms of data fit.

Visual observations

Primary fractures occurred on the faces of the block penetrated by the heater hole. These appeared slightly after acoustic emission peaks. The first fracture to occur was a vertical fracture tracing from the hole boundary to the upper or lower block surface. This fracture occurred at slightly different times on the upper and lower half, as well as opposite ends, of the block. Fracture generation was rarely accomplished instantaneously (with the exception of low load levels) but was instead a slow growth process. In the lower load ranges (0-5 MPa) this vertical fracturing was accompanied by audible noise and dilation of the fracture was rapid. With increased load fracture generation became increasingly less violent with the fractures obviously being generated away from the hole boundary about 1 to 2 centimeters. These fractures appeared to propagate to a point on the borehole's horizontal centerline, i.e., tangent to the borehole, after unloading.

Secondary fractures arrived slightly later and appeared to be load dependent. With low uniaxial loads (0-5 MPa) a system of radial fractures developed around the heater hole. For the unloaded blocks, this system was near horizontal, for higher loads, the system shifted to a 45^0 orientation. These fractures did not necessarily originate from, or touch, the borehole surface, but commonly passed through the thermocouple holes.

A third form of fracturing became apparent late during the testing cycle. These fractures initiated on the load faces of the block from the primary fracture (longitudinal along the heater hole) and traveled transversely across the block perpendicular to the heater. Initiation was in the middle section of the load face, with the opposing end of the fracture (those on opposite sides of the primary fracture) always being offset by several centimeters.

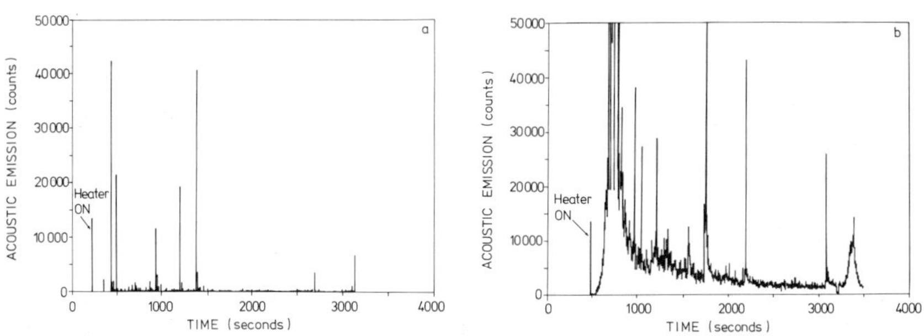

Figure 3. Typical acoustic emission plots a) diabase b) granite

These fractures commonly traveled halfway, and occasionally completely, around the block.

THEORETICAL RESULTS

The analysis was conducted in the order described in the theory section of this article with a constant power input of 0.9 kW. The calculation of erfc and the summation of the individual temperature components on the block midplane were conducted numerically (Figure 4). For these data, a numerical integration of eq. 2 was conducted to yield the thermally induced stress (Figures 5). These were superimposed over the theoretical stress field due to uniaxial loading (Figure 6) to give the resulting stress field around the opening (Figure 7). (Theoretical results are labeled as "discrete" in the accompanying figures.)

Behavior is as would be expected from a theoretical standpoint. The radial stress (σ_r) is always compressive with an increase in magnitude due to heating. The tangential stress (σ_θ) has split tensile and compressive regions. Adjacent to the hole boundary, a very high compressive stress is developed. This changes into a tensile stress field as the distance from the hole increases. The actual magnitude of these stresses is different for granite and diabase based on their different expansion coefficients. Actual location of the stress minima is, of course, time dependent with the tangential stress minima shifting further into the rock with increasing time.

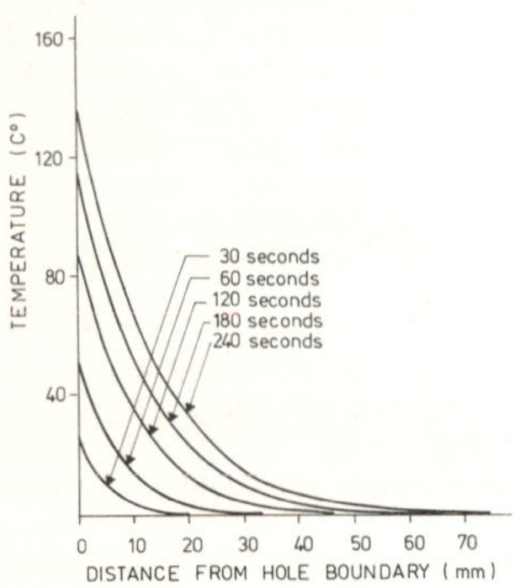

Figure 4. Granite temperature distributions

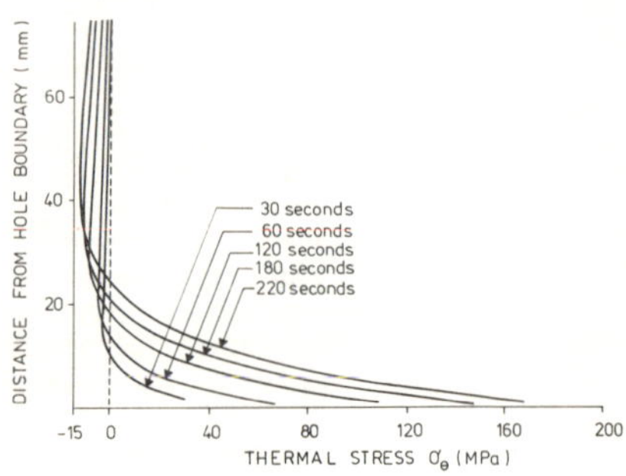

Figure 5. Granite thermal induced tangential stresses

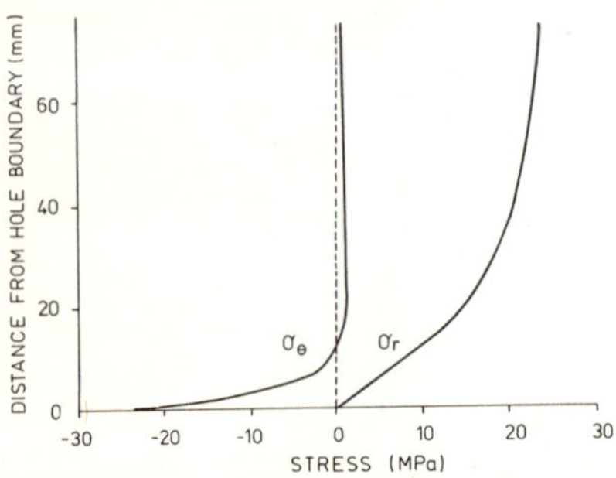

Figure 6. Uniaxial load stresses, 25 MPa

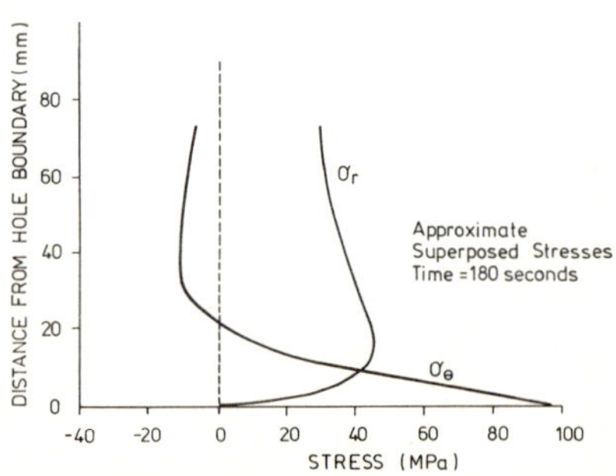

Figure 7. Granite composite stress field

Superposition of the elastic stress field at θ=90⁰ (Figure 7) shows that the developed tangential tensile stresses due to the uniaxial load are effectively cancelled out by the heater developed tangential compressive stresses. Thus, the region of maximum tensile stress is developed away from the hole boundary in the position where the heater developed tangential stresses overlap the weak tangential compressive stress due to uniaxial loading.

FINITE ELEMENT ANALYSIS

Calculations performed in this study were handled by an FEM program called HFEMP [6], a general finite element program for two-dimensional problems developed at the Luleå University of Technology under the years 1980-1983.

Due to the symmetry conditions, only one quarter of the block was actually analyzed. This gave a mesh of 526, 4-node elements.

Physical properties from the laboratory testing, as reported, were utilized. The outer boundary of the block had an initial zero temperature at time zero but was otherwise not restricted. The vertical and horizontal symmetry boundaries were restricted in the X and Y directions respectively. The interior hole boundary was unrestricted as to motion but had a variety of heating assumptions. These were: 1) a constant power input of 0.9 kW for a time equal to the mean laboratory heater ON time, 2) an 11 second forced temperature rise on the hole boundary to 400⁰ C, 3) a constant boundary temperature of 400⁰ C. For an explanation concerning the reasoning behind these boundary conditions, the reader is referred to [5].

Results

Given the linear appearance of the failure time vs. load plots, failure times were evaluated at only two uniaxial load cases, 0 and 25 MPa. Maximum and minimum stresses given by the FEM program were plotted for each time-step given a fixed load. This gave resulting plots such as shown in Figures 8 and 9. The associated strength values were then plotted with these curves and a predicted failure time estimated.

This prediction of failure times was to see if the laboratory times could be predicted by either a compressive or tensile failure criteria. The predicted tensile failure times vs. the actual failure times are shown as Figures 8 and 9.

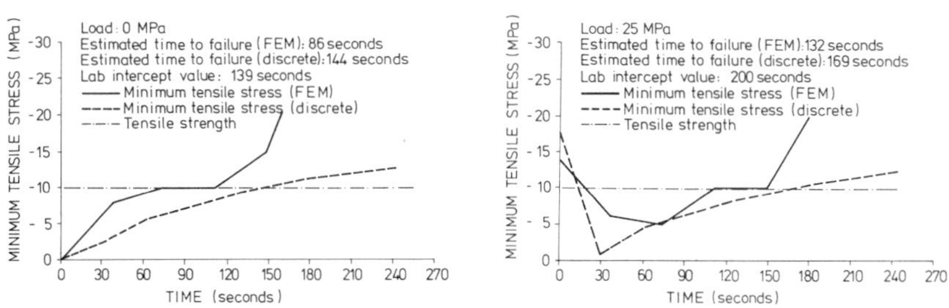

Figure 8. Granite time dependent minimum stress

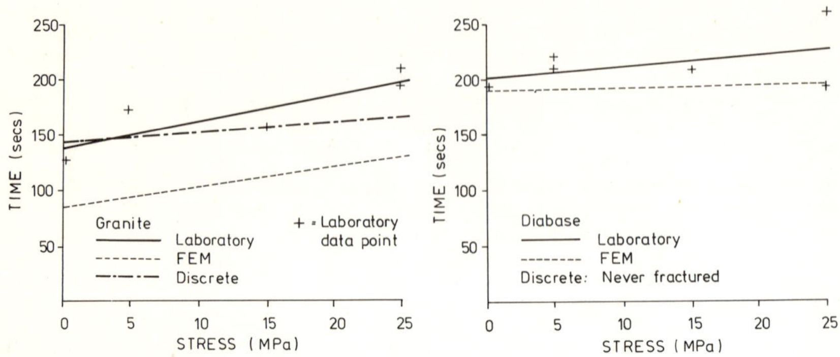

Figure 9. Comparison of laboratory and theoretical results

CONCLUSION

The most important results of this study may be summarized as follows:

1) The mode of failure is most likely tensile, at least in initiation.

2) Failure occurs away from the hole boundary.

3) Failure can be predicted relatively accurately with numerical techniques.

4) Acoustic emissions produced during heating can be used to estimate the initial time of failure.

5) Thermal distributions recorded during the experiments are consistent with the material thermal properties.

6) Tertiary fracturing is of a relatively high energy intensity.

ACKNOWLEDGMENT

The authors gratefully acknowledge the financial support of the Swedish Council for Building Research (grant number 830299-6) which financed this study.

REFERENCES

1. Carlslaw, H.S. & Jaeger, J.C., Conduction of Heat in Solids, 2nd ed. (Oxford University Press, Oxford, 1959).

2. Carlsson, H., A Pilot Heater Test in Stripa Granite. (Lawrence Berkley Laboratory, Technical Report LBL-7086, SAC-06, Berkeley, 1978)

3. Chan,T. & Cook, N.G.W., Calculated Thermally Induced Displacements and Stresses for Heater Experiments at Stripa, Sweden. (Lawrence Berkley Laboratory, Technical Report LBL-7082, SAC-09, Berkeley, 1978)

4. Clark, G.B.& Lehnhoff, T.F., "Thermal-mechanical Fragmentation of Hard Rock for Rapid Excavation". (15th U.S. Symposium on Rock Mechanics, Custer State Park, South Dakota,1973), pp. 501-522.

5. Mathis, J., Stephansson, O., Bjarnasson, B., Hakami, H., Herdocia, A., Mattila, U., Singh, U., Eriksson, L., Heat Induced Fracturing of Rock in an Existing Uniaxial Stress Field, (Swedish Council for Building Research, Stockholm, in press)

6. Nilsson, L. & Oldenburg, M., FEMP - An Interactive, Graphic Finite Element Program for Small and Large Computer Systems (Luleå University of Technology, Technical Report 1983:07 T., Luleå, 1983)

7. Obert, L. & Duvall, W., Rock Mechanics and Design of Structure in Rock. (John Wiley & Sons, Inc., New York, 1967)

8. Thirumali, K., Rock Fragmentation by Creating a Thermal Inclusion with Dialectric Heating (U.S.B.M. R.I. 7424, Washington, D.C., 1970)

NUMERICAL CALCULATION OF THE TIME AND SPACE DEPENDENT TEMPERATURE DISTRIBUTION IN A HLW REPOSITORY OF FINITE EXTENSION AND TEMPORAL STEP-BY-STEP DISPOSAL PROCEDURE

K.HAHNE*, M.SCHLICH* AND E.KORTHAUS**
*RWTH Aachen, Institut für Elektrische Anlagen und Energiewirtschaft, Lehrauftrag Leistungsreaktoren, Schinkelstraße 6, D-5100 Aachen, Federal Republic of Germany
**Kernforschungsanlage Karlsruhe GmbH, Postfach 3640, D-7500 Karlsruhe, Federal Republic of Germany

ABSTRACT

With the program system FAST-STEP the temporal step-by-step disposal procedure of a HLW repository and its finite geometry can be taken into consideration in the calculation of the time and space dependent temperature distribution. This will be obtained through the application of a suitable coupling of calculation models treating both far field and near field phenomena.

After the description of the single steps of the program system the influence of both aspects on the temperature distribution will be demonstrated with the aid of a test calculation. The possible fields of application of FAST-STEP will be pointed out at the end.

INTRODUCTION

In recent years the developements in the field of the disposal of heat generating high-level waste (HLW) in West Germany has been concentrated on the planning and layout of a repository in geological rocksalt formations and on the execution of the necessary thermo-mechanical and safety analysis.

The basis of such a planning is the calculation of the time and space dependent temperature distribution in the disposal area (near field) and in the surrounding rocksalt and rock layers (far field) as a result of the heat generation of the HLW. These temperature calculations provide the input data for the following thermo-mechanical calculations.

The object of the calculation of the temperature distribution is the study of the maximum admissible temperature of those materials which are used for the disposal of the HLW. In the near future additional economical considerations will gain importance to lay out a repository in such a way, that it will be possible to obtain an optimal utilization of the salt volume being available for the disposal.

The physical models usually employed for the calculation of the time and space dependent temperature distribution during the layout of a repository (see Figure 1) overestimate the expected maximum temperature of the materials by simplified assumptions. These underlying simplifications mean tnat

- the disposal area has an infinite horizontal extension (only near field calculation)

- all boreholes are simultaneously filled with HLW

- the entire rocksalt formation is isotropic and completely undisturbed.

The development of the temperature distribution is influenced on the one hand by means of the

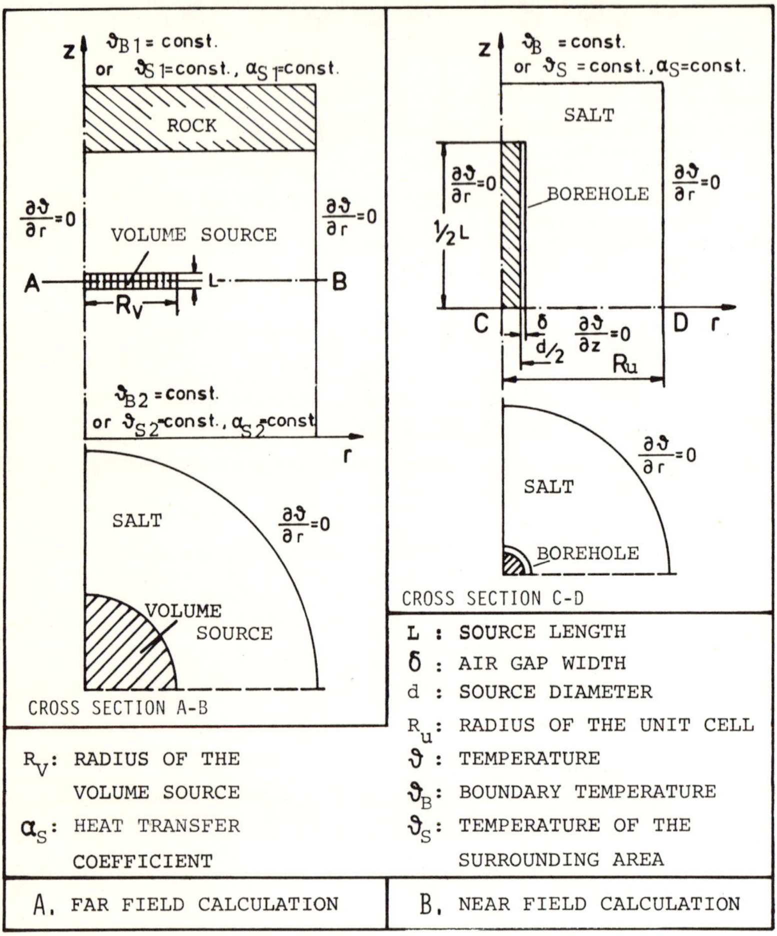

Figure 1. Modelling of heat transfer of HLW
A, Large scale model. B, Unit cell model

- time dependent function of the heat generation
- time between the fuel-element withdrawal and the disposal of the HLW

and on the other hand by means of the geometry

- length and distance of the boreholes
- disposal structure (hexagonal, quadratic).

In the previous analysis a layout of the repository can be found through an appropriate assessment of these effects observing the required boundary conditions.

FINITE EXTENSION OF THE REPOSITORY AND THE TIME DEPENDENT STEP-BY-STEP DISPOSAL PROCEDURE

Description of the physical model

At the time of the final decision about the location of the repository, precise specifications will exist about the geometry and structure of the repository and about the composition and heat generation of the high-level waste. Based on these realistic specifications the calculation of the heat dissipation of a repository has to be improved by a far more detailed modelling, which considers the limited extension of the disposal area as well as the temporal step-by-step load sequence of the boreholes. These aspects cannot be treated with the physical model of the unit cell shown in Figure 1, B.

The only method known at present to realize such a complex computation - using the solution of the instationary, nonlinear differential equation of the heat conduction with temperature dependent material properties (in contrast to the superposition of point or line source calculations) - is based on a procedure, which connects far field with near field calculations in a single calculation sequence (analogous to reactor core design). This concept was realized for the first time in the program system FAST-STEP, developed at the Lehrauftrag Leistungsreaktoren, RWTH Aachen[1]. The develepement of FAST-STEP was done in co-operation with the Kernforschungszentrum Karlsruhe GmbH.

FAST-STEP is based upon a coarse-mesh method which is especially suitable for the three-dimensional calculation of large geometry structures. Peculiarities of this special finite element program system are [2] :

- the consideration of temperature dependent material properties by application of the NEWTON-RAPHSON method
- an implicit time integration with the CRANK-NICOLSON method
- an iterative solution of the linearized equation system by the successive overrelaxation method

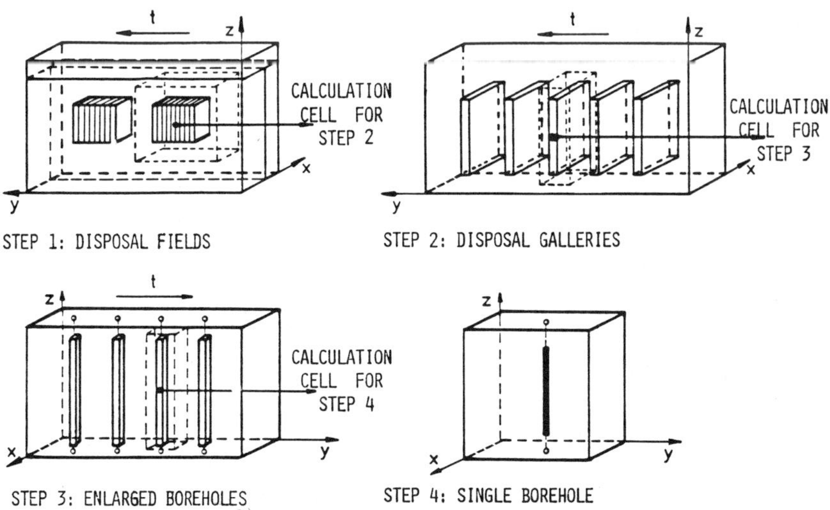

Figure 2. Calculation steps for a repository in the program system FAST-STEP

Figure 2 shows the fundamental idea of such a step-by-step procedure applied to a HLW repository. In the first step - far field calculation - the disposal fields and the entire salt dome, surrounding and overlaying strata are modelled. In each of the following steps 2 - 4 a limited section of the preceding step is considered exclusively. In the single steps of the entire calculation procedure the heat sources are modelled in more and more details (disposal fields, disposal galleries, enlarged boreholes, borehole). The last calculation step yields the temperature distribution around a single borehole, which is now modelled in its true dimensions. The connection of each step with the subsequent one is done by the handing-over of time and space dependent temperatures at the boundaries of the actual calculation cell.

The calculation area of a considered step(i) corresponding to a limited area of the previously considered step(i-1) is fixed through the 'homogenous volume source'. The homogenous volume source means a limited homogenous distribution of the heat generation in the boreholes to more or less extended salt areas (fields, galleries, enlarged boreholes) depending on the respective step. For areas within these volume sources the temperature distribution is - depending on the homogenous heat generation - merely equivalent to an average value of the temperature distribution in the disposal area appearing under real conditions. The influence of the smearing out of the heat production on the temperature distribution is negligible for zones far away from this volume source. In these zones the special structure of the volume source produces nearly the same heat flow as the homogenous source. In the handing over of the temperatures from step(i-1) to step(i) the boundaries of the area picked out in the step(i-1) are therefore placed far away from the homogenous volume source.

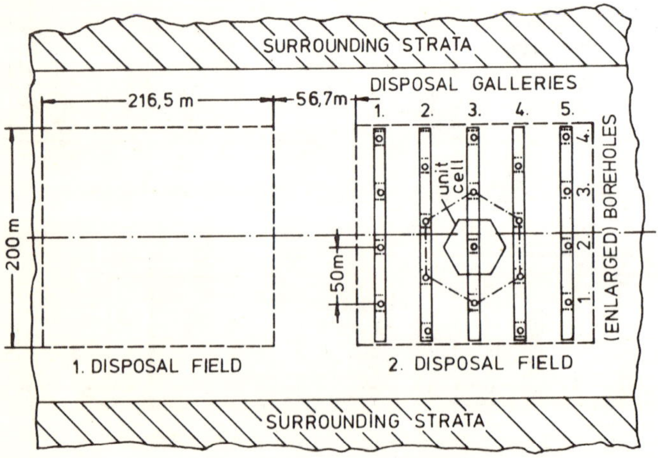

Figure 3. Horizontal section of the repository in the test calculation with FAST-STEP

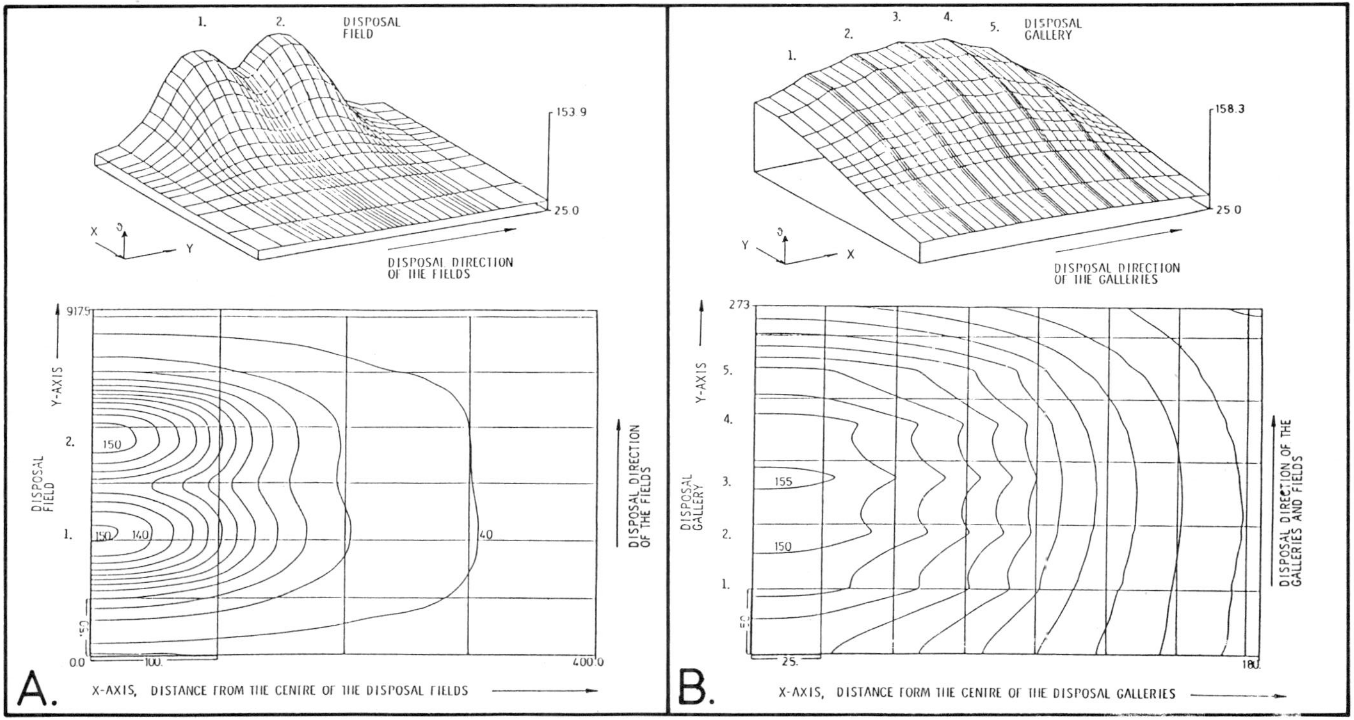

Figure 4. Three-dimensional figures and isotherms for the representation of the temperature distribution obtained from the FAST-STEP 1 and 2 calculations

A, Disposal fields; temperature distribution in the x,y-plane through the centre of the fields

B, Disposal galleries; temperature distribution in the x,y-plane through the centre of the galleries (second field)

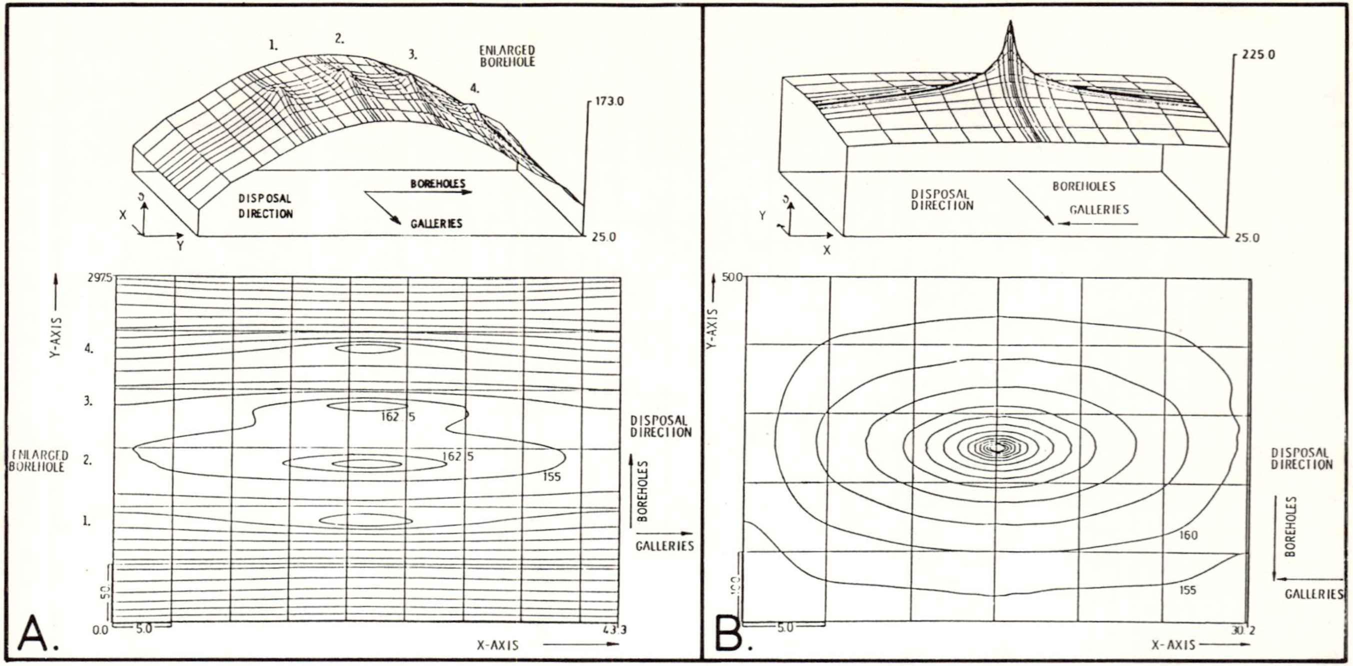

Figure 5. Three-dimensional figures and isotherms for the representation of the temperature distribution obtained from the FAST-STEP 3 and 4 calculations

A, Enlarged boreholes; temperature distribution in the x,y-plane through the centre of the boreholes (third gallery)

B, Single borehole; temperature distribution in the x,y-plane through the centre of the borehole (second enlarged borehole)

Results of a test calculation

Figure 3 shows an outlined horizontal section of a repository consisting of 2 disposal fields, 5 disposal galleries for each field and 4 boreholes for each gallery. This geometry had been calculated with the aim to determine the extreme loaded borehole with applied test values for the geometry, disposal structure and heat generation. The results of such a calculation are presented in Figures 4-6. Figures 4 and 5 show three-dimensional temperature distributions and isotherms of a suitable section of each of the four calculation steps (compare Figure 2). These isotherms highlight the unsymmetric temperature distribution at the moment of the appearence of the maximum temperature in the steps caused by the step-by-step disposal procedure and the finite extension of the disposal area.

Figure 6 presents the maximum temperature at the edge of the extreme loaded borehole vs. time from the FAST-STEP calculation. This temperature curve is compared on the one hand with the result of a calculation for a single source in an infinite extended salt dome. On the other hand it is compared with the results of a unit cell calculation. This calculation is based on the physical model corresponding to Figure 1 and on the simplified assumptions mentioned in the introduction.

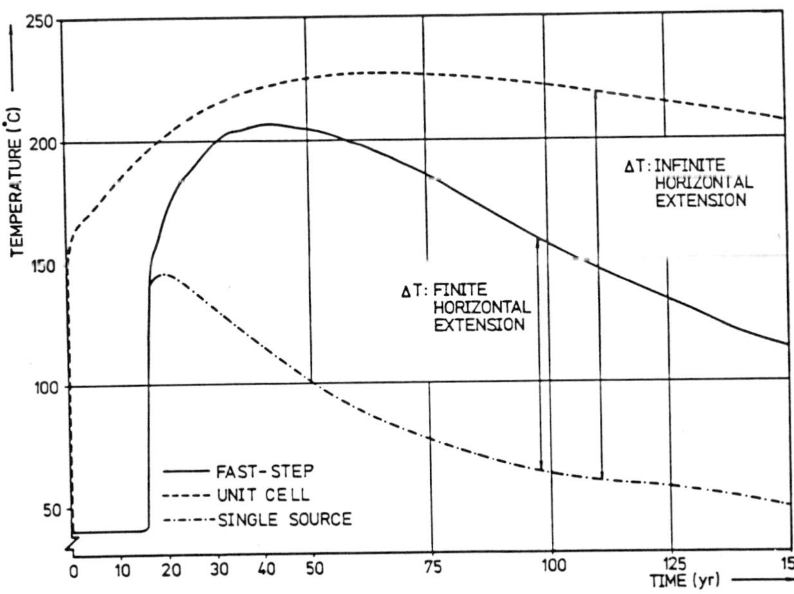

Figure 6. Maximum salt temperature vs. time obtained from different calculations

Compared to the unit cell calculation the FAST-STEP result shows

- a 10% decrease in the maximum temperature
- a temporary earlier appearance of the maximum temperature
- a high decrease in the temperature course following the time of the maximum temperature.

In addition Figure 6 illustrates the difference in the temperature rise calculated for a disposal area with a finite and an infinite geometrical extension respectively.

FUTURE ASPECTS

The important difference between the results of the FAST-STEP calculation and the presently used unit cell calculation gives new fields of application of this new program system.

For the first time FAST-STEP yields the possibility to select disposal strategies to obtain the best solution for the economical utilization of the salt volume needed for the disposal. As it is evident from Figures 5 and 6 of the steps 1-3, the temperature distribution at the boundary regions of the disposal fields remains far below the level of the temperature at the centre of the fields. So, without influencing the maximum temperature in the centre of the disposal fields, the following strategies could be developed to lift up the temperature level in the outer regions of the fields:

- a selection of special disposal sequences
- a reduction of the distances between the boreholes at the edge of the disposal fields
- an increase of the heat generation of the HLW-canister at the boundary region
- a creation of unfilled areas in the centre of the disposal fields
- a mixed disposal of high-level and medium-level waste

The possible reduction of the salt volume leads to an economic improvement in the disposal of HLW.

REFERENCES

1. M. Schlich, et al.; Entwicklung von Rechenmodellen zur Berechnung instationärer Temperatur-, Spannungs- und Dehnungsverteilungen im Endlager nach der Einlagerung hochaktiver Stoffe, Abschlußbericht zum Forschungsauftrag gleichen Namens der Kernforschungszentrum Karlsruhe GmbH, 1. Teilvorhaben: Weiterentwicklung von Rechenmodellen zur Ermittlung von Temperaturfeldern nach der Einlagerung hochaktiver Abfälle, Lehrauftrag Leistungsreaktoren im Institut für Elektrische Anlagen und Energiewirtschaft, RWTH Aachen, Februar 1983
2. P. Ploumen; Numerische Langzeitberechnung dreidimensionaler Temperaturfelder mit Hilfe eines speziellen Finite-Elemente-Verfahrens am Beispiel der Endlagerung hochradioaktiver Abfälle im Salzgestein, Dissertation RWTH Aachen, 1980

DEVELOPMENT OF THE CODE MELODIE FOR LONG TERM RISK ASSESSMENT OF NUCLEAR WASTE REPOSITORIES

P. GOBLET*, P. GUETAT**, J. LEWI***, J-P MANGIN****, G. de MARSILY* and P. RAIMBAULT***
*Centre d'Information Géologique (Ecole Nationale Supérieure des Mines de Paris)
**Commissariat à l'Energie Atomique/Institut de Protection et de Sûreté Nucléaire/Département de Protection Technique (CEA/IPSN/DPT)
***Commissariat à l*Energie Atomique/Institut de Protection et de Sûreté Nucléaire/Département d'Analyse de Sûreté (CEA/IPSN/DAS)
****Commissariat à l'Energie Atomique/Institut de Recherce Technologique et de Développement Industriel/Département de Recherche et Développement Déchets (CEA/IRDI/DRDD)

ABSTRACT

MELODIE is a computer code developed at the CEA/IPSN for risk assessment of nuclear waste repositories in geological formations. The interactive evolution of the source, geosphere and biosphere is studied for periods of time longer than 100 000 years. In its first version, the code can describe a repository in granite rock located at a specific site.

The code is built in a modular form which allows to use different versions of the sub-system models.

The basic model for radionuclide migration and hydrogeology is a subroutine version of the METIS code developed by ENSMP . METIS is a 2D finite element code which solves the advection-dispersion equation for porous media with explicit fracture representation. Linear adsorption kinetics is included as well as matrix diffusion and radionuclide decay chains.

The source model developed at CEA/DRDD**** is derived from CONDIMENT which is a 1D finite difference code describing the behaviour of high level waste packages. Four axisymetric layers are individualized : glass matrix, container, bentonite and granite. The glass leaching is modelled as a dissolution and diffusion process of the individual chemical components.

The biosphere model ABRICOT developed at the CEA/DPT** is based on a detailed description of agricultural activities defined in individual systems.

MELODIE is tested by participation to international exercices such as PAGIS [1], INTRACOIN [2] and HYDROCOIN [3]. Future developments will include introduction of scenarios constructed from geoprospective studies and algorithms for sensitivity studies.

Introduction

In order to perform the evaluation safety of nuclear waste repositories, a global model, called MELODIE, is currently developed, since the middle of the year 1984, in the Institut de Protection et de Sûreté Nucléaire (IPSN) du Commissariat à l'Energie Atomique (CEA), in collaboration with other departments of the CEA, like the Department de Recherche et Développement sur les Déchets (DRDD) and other French organisms, especially the Ecole Nationale Supérieure des Mines de Paris (ENSMP).

The figure 1 shows the various operations contributing to the risk evaluation calculations which will be progressively taken into account in the code MELODIE.

These calculations are based on models representing the different physical and chemical phenomena which are involved in :

- the evolution of each of the different media which constitute the disposal (source, constituted by the waste packages and the other engineered barriers; geosphere; biosphere). Internal stresses (thermal, mechanical, chemical,...) as well as external stresses (corresponding to the effects of scenarii including continuous geological events, random natural phenomena or human intrusions) will be considered in order to determine this evolution.
- the release of the radionuclides from the waste matrixes, and their transfer to man through the engineered barriers, the geosphere and the biosphere.In order to take correctly into account the interactions between the different phenomena involved, the models will be coupled together at each time step.

Two kinds of calculations will be performed :

- best estimate calculations. These ones will be made by complex models, like those which are presented hereafter.
- sensitivity to parameters. An algorithm, based on the latin hypercube method, is currently in development and will be implemented in the code at the end of this year. For this kind of calculations, simplified models will be used by degenerating the complex models used for the best estimate calculations.

In order to obtain the versatility necessary to the different uses of the MELODIE code, a modular structure has been defined : in this structure, a management module calls each computational module independently from the others, with its optimal time step; data are stored in libraries and the relationships between modules are obtained through data exchanges with the libraries.

The individual submodels corresponding to the different phenomena which are involved are developed in specific laboratories and are individually validated. The global model is presently tested through the European project PAGIS (Performance Assessment of Geological Isolation Systems).

We describe hereafter the present state of development of the model and those of the individual submodels which are implemented.

Present status of the code

The actual status of the code, fitted for calculations in a granite formation, is shown in figure 2.This status does not allow to perform the calculation of the modification of the media characteristics or of the external boundary conditions.

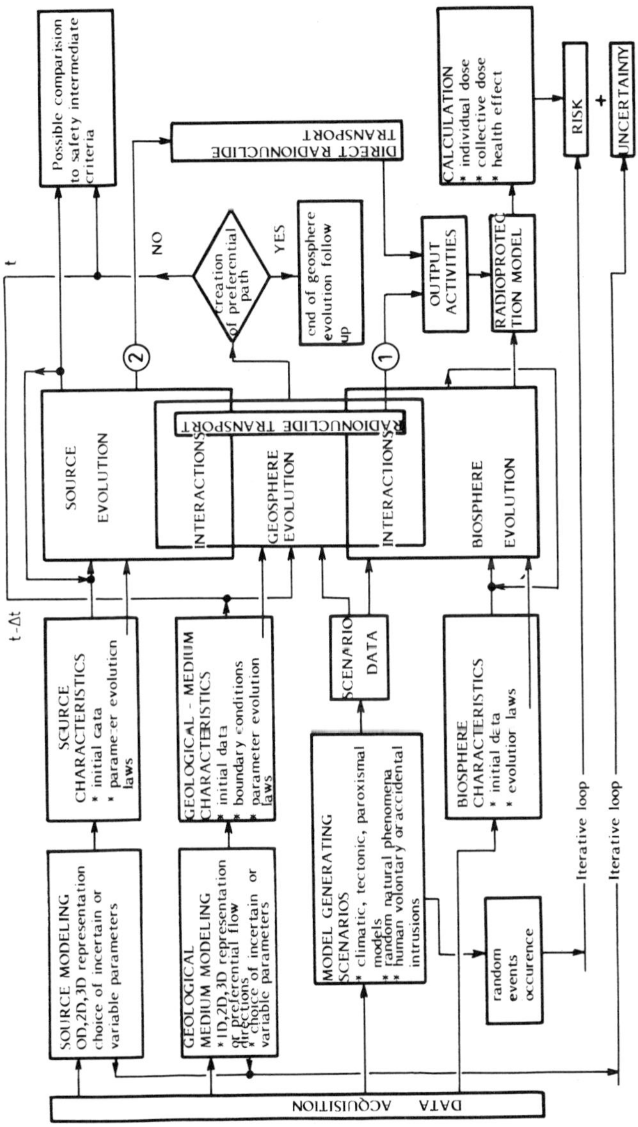

Figure 1. waste repository risk assessment general chart

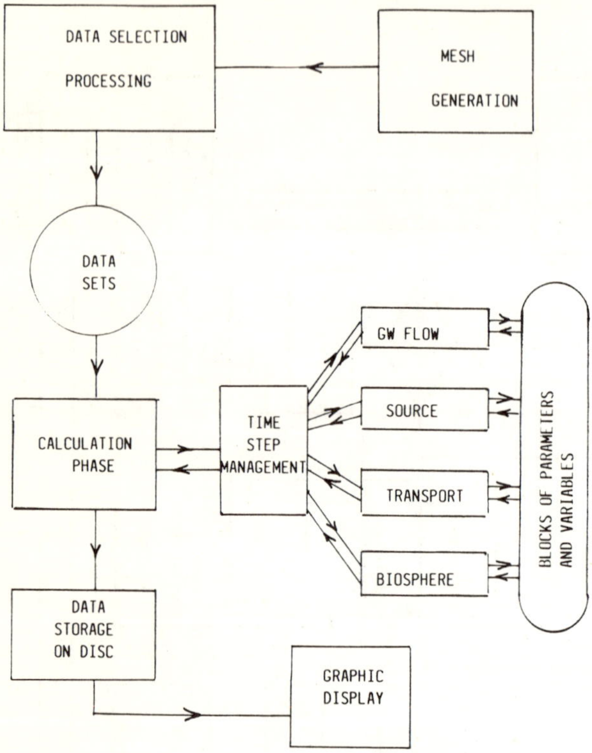

Figure 2. MELODIE Flow chart

The first version of MELODIE is based on the coupling of a groundwater flow model, a source term model, a radionuclide transport model and a biosphere model.

The groundwater flow in the geosphere is calculated as a permanent one; it is determined from the permeabilities of the different zones of the rock formation, the transmittivity of the fractures and the boundary conditions imposed on the geometry of the site. The calculated values may only be modified in the present version of the code through predetermined changes of data included in the input file. The geosphere migration submodel utilizes the water velocity chart, calculated by the groundwater flow model and the radionuclide activity outflow, calculated by the source model, to compute the concentrations in the geosphere as a function of time.

The repository is represented as a series of nodes in a 2D cross-section of the geosphere. At each node, a group of source terms are introduced. These sources are individualized since their outward flow will depend on the local water velocity, and radionuclide concentration.

Since the outward flow from each source is controlled by the local gradient of concentration, the level of interaction between models depends on the ratio of the radionuclide concentration in the source and in the geosphere. Strong coupling occurs when the permeability of the geosphere has a low value, which requires to use in this case special numerical procedures, such as iteration at each time step or predictor-corrector techniques. The problem is magnified if the release from the source is peaked at certain dates and if the characteristic time steps of the source and geosphere submodels are very different.

The exutories of the geosphere are identified from the groundwater flow distribution on the surface. The outward radionuclide activity flow from each of them can thus be calculated as a function of time. The individual or collective doses, are then computed from the concentration in the rivers which originate from these exutories and from the different transfer paths to man.

First calculations are currently performed with the fully coupled submodels for specific fission products such as Cs and Tc.

Source model

The source model, called CONDIMENT, describes the release of radionuclides from the waste matrix (borosilicate glass) and their migration through the backfilling material and a small fraction of granitic rock.

CONDIMENT is a 1D finite difference code in a cylindrical geometry which solves the advection-dispersion equation in the various media.

Geochemical interactions are represented by a linear isotherm model, with retardation factors which depend on the migrating element. After water penetration, the concentration of the element in the glass is set by its solubility limit. The rate of dissolution of the element determines the movement of a variable surface inside the glass which sets the boundary between the region with constant saturation concentration and the diffusion region.

For radionuclides less soluble than silica, the model can still be used with the release from the glass being set by the silica solubility limit.

The input variables are the groundwater velocity and the concentration of radionuclides at the boundary of the granite cylinder which can change with time. Assymetry introduced by the flow of water through the system is taken into account by individualizing two 180° sectors for each radius.

Figures 3 and 4 give examples of calculations performed with the model CONDIMENT. The parameters values are the following :

	Glass	Bentonite	Granite
Permeability	3.10^{-2}	3.10^{-6}	10^{-2} m/yr
Diffusion coef.	3.10^{-5}	10^{-4}	10^{-2} m²/yr
Porosity	10^{-2}	10^{-3}	10^{-2}
Retard. Coef.	10	1	1
Water Velocity			10 m/yr

Concentration in aqueous phase : 10 g/l

Concentration in matrix : 36.2 Kg/m^3

Figure 3 shows the concentration of Cesium as a function of radius in the near field for different times.

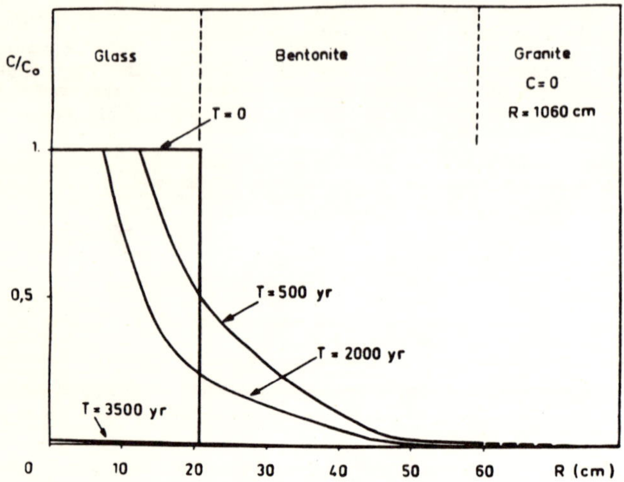

Figure 3. Concentration of Cesium versus radius in the source

Figure 4 indicates the evolution of the outward flow of Cesium from the system as a function of time for zero boundary concentration. We notice the strong peaking of Cesium released from the outward fraction of the glass, then a long time of decreasing flux connected to the small diffusion coefficient in the glass and then the sharp drop when all radionuclides are in solution.

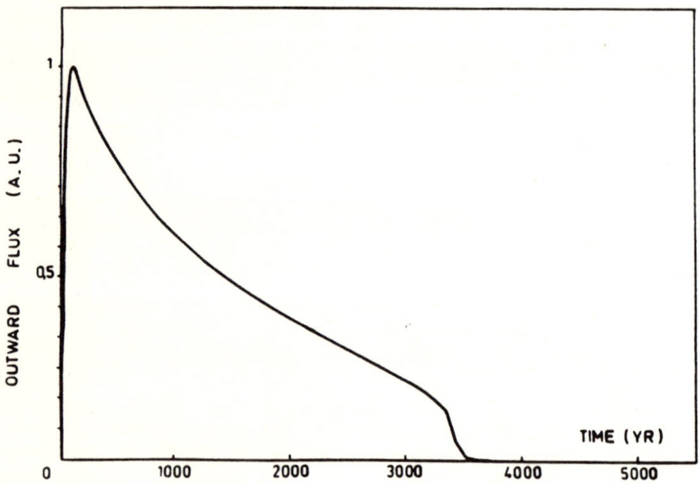

Figure 4. Outward Flux of Cesium from the source versus time

The corrosion of the container cannot presently be described by a realistic model; it is modelled as a failure time, where the glass comes directly in contact with the bentonite. A statistical distribution of container failure times will eventually be introduced.

The future developments of CONDIMENT will allow to take into account chemical interactions such as precipitations reactions related to the chemical properties of the leaching water. Non linear isotherms will be considered as well.

The CONDIMENT model can be used to describe the evolution of temperature in the near field and thus the heat flux to the geosphere. Parameters will be allowed to vary with local temperature.

Groundwater flow and radionuclide transport models

The submodels for groundwater flow and radionuclide transport in MELODIE are derived from the METIS code developed at the Ecole des Mines de Paris. METIS is a 2D finite element code which solves the diffusion equation and the advection-dispersion equation for equivalent porous medium. First order adsorption kinetics as well as matrix diffusion and nuclear decay chains are included.

A recent development of the code was to implement explicit fracture representation for grounwater flow. Fractures are considered as linear elements with specific transmissivity : this simplifies mesh generation and saves computer time. This representation will be extended to radionuclide transport calculations.

METIS is participating to international exercices such as INTRACOIN [2] and HYDROCOIN.[3]

An exemple of 2D transport with filiation, performed in INTRACOIN is given in figures 5,6,7 and 8.

A constant flow of ^{234}U is released from a 10 m long bar for 100 000 years. The ^{234}U migration in the geosphere is calculated as well as those of the ^{230}Th and ^{226}Ra decay products. The water velocity is perpendicular to the bar and equal 1m/year. The dispersivity is 50 m in x, 5 m in y.The retardation factors for Uranium, Thorium and Radium are respectively 60, 500 and 20.

The constant volume activity contour plots are shown in figures 5,6 and 7 for the three radionuclides at time t = 65470 years. The total activity flux at 500 m from the source for Uranium, Thorium and Radium are given in figure 8 as a function of time up to 200 000 years.

The mesh size was 50 m in x and 5 m in y, and the chosen time step t = 333 years is equal to 1/3 of the time necessary for the fastest radionuclide to migrate through one grid element.

The question of time step management is of critical importance for advection-dispersion codes since very large permeability contrasts may exist in the geosphere and the optimum time step depends on the region which is considered. This is also important for coupling other models with different time characteristics. This aspect has yet to be carefully optimized in the METIS code.

The future METIS developments will also consider the introduction of variable water density to solve thermo-convection problems in a coupled or uncoupled way.

SIM.17, U234 . Act.VOL. (1.E5 CI/M3)

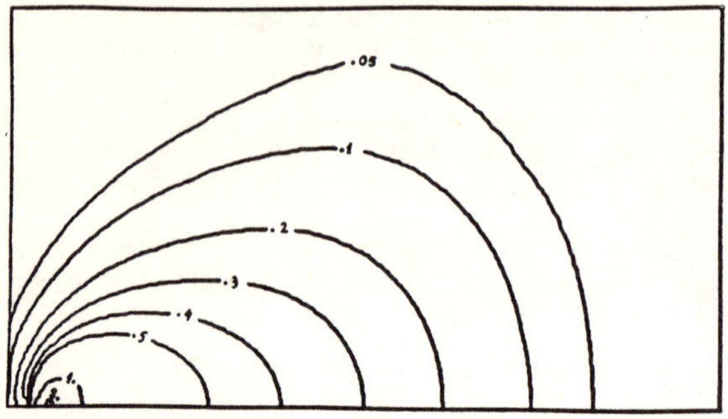

Fig. 5

SIM.17, TH230, ACT.VOL. (1.E5 CI/M3)

Fig. 6

SIM 17, AA226, ACT.Vol. (1.E5 CI/M3)

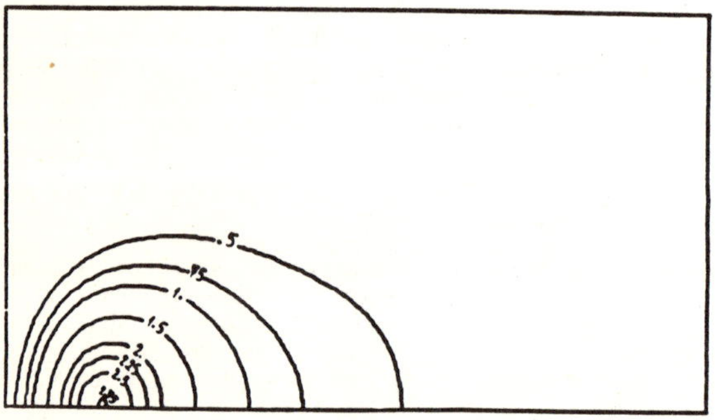

Fig.7

Constant volume activity contour plots at t = 65470 years for ^{234}U, ^{230}Th, ^{226}Ra

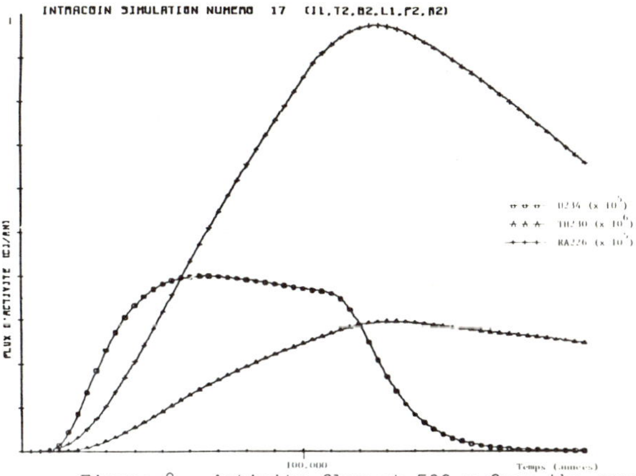

Figure 8 - Activity flux at 500 m from the source

Biosphere model

The biosphere model, called ABRICOT, is based on a water and food consumption description. Individual, regional or collective doses are calculated from the ingestion of radionuclides corresponding to menus the components of which originate from different systems in the pathways of radionuclides.

The radionuclides coming out of the exutories are present in the different rivers which originate from the site with concentrations depending on the water flow. The main paths to man are : drinking water, fishing, transfer through vegetables, cereals, grass and cattle, when irrigation is used.

The accumulation in soils is taken into account by using a distribution coefficient between soil and water.

A global physiological parameter which takes into account the translocation leaves-fruits and leaves-roots is used as well as an aging effect parameter for plants.

Future developments of the model will include calculations of doses from inhalation and direct exposure. A specific biosphere model valid for a coastal site will be developed as well.

Conclusion

In the framework of the methodology for risk assessment connected to high level waste repositories which has been defined in France, a global risk assessment code is under development. The first version of this model is now completed and currently tested through the european PAGIS [1] project. The code couples together the source, the geosphere and the biosphere models, which allows to perform dose calculations to man over very long time periods.

Next developments will be to introduce thermal effects in order to calculate the evolution of the individual model parameters. Mechanical and geochemical effects will be taken into account as well.

Risk calculations will be made after a scenario generator has been implemented. Sensitivity studies will be performed in parallel to these developments. The corresponding algorithm will be implemented in the code at the end of this year.

REFERENCES

1. Pagis. Summary Report of Phase 1 Joint Research Centre - ISPRA EUR 9220 EN. (1984)

2. Intracoin. Final Report level 1, SKI 84 : 3 September 1984

3. Hydrocoin. Progress Report no.1, SKI May 1984 - December 1984

Author Index

Subject Index